Preface

The McGraw-Hill *Dictionary of Chemistry* concentrates on the vocabulary of those disciplines that constitute chemistry and related fields. With more than 8500 terms, it serves as a major compendium of the specialized language that is essential to understanding chemistry. The language of chemistry embraces many unique disciplines which are usually represented in specialized dictionaries and glossaries. Engineers, students, teachers, librarians, writers, and general readers of scientific literature will appreciate the convenience of a single comprehensive reference.

Terms and definitions in the Dictionary represent six fields: analytical chemistry, chemistry, inorganic chemistry, organic chemistry, physical chemistry, and spectroscopy. Each definition is identified by the field in which it is primarily used. When the same definition is used in more than one branch of chemistry, it is identified by the general field label [CHEMISTRY].

The terms selected for this Dictionary are fundamental to understanding chemistry. All definitions were drawn from the *McGraw-Hill Dictionary of Scientific and Technical Terms* (5th ed., 1994). Along with definitions and pronunciations, terms also include synonyms, acronyms, and abbreviations where appropriate. Such synonyms, acronyms, and abbreviations also appear in the alphabetical sequence as cross references to the defining terms.

The *McGraw-Hill Dictionary of Chemistry* is a reference that the editors hope will facilitate the communication of ideas and information, and thus serve the needs of readers with either professional or pedagogical interests in chemistry.

Sybil P. Parker
EDITOR IN CHIEF

Editorial Staff

Sybil P. Parker, Editor in Chief

Arthur Biderman, Senior editor
Jonathan Weil, Editor
Betty Richman, Editor
Patricia W. Albers, Editorial administrator

Dr. Henry F. Beechhold, Pronunciation Editor
Professor of English
Chairman, Linguistics Program
College of New Jersey
Trenton, New Jersey

Joe Faulk, Editing manager
Frank Kotowski, Jr., Senior editing supervisor
Ruth W. Mannino, Senior editing supervisor

Suzanne W. B. Rapcavage, Senior production supervisor

How to Use the Dictionary

ALPHABETIZATION. The terms in the *McGraw-Hill Dictionary of Chemistry* are alphabetized on a letter-by-letter basis; word spacing, hyphen, comma, solidus, and apostrophe in a term are ignored in the sequencing. Also ignored in the sequencing of terms (usually chemical compounds) are italic elements, numbers, small capitals, and Greek letters. For example, the following terms appear within alphabet letter "A":

amino alcohol	***para*-aminophenol**
1-aminoanthraquinone	***n*-amylamine**
γ-aminobutyric acid	**4-AP**

FORMAT. The basic format for a defining entry provides the term in boldface, the field in small capitals, and the single definition in lightface:

term [FIELD] Definition.

A field may be followed by multiple definitions, each introduced by a boldface number:

term [FIELD] **1.** Definition. **2.** Definition. **3.** Definition.

A term may have definitions in two or more fields:

term [PHYSICAL CHEMISTRY] Definition. [SPECTROSCOPY] Definition.

A simple cross-reference entry appears as:

term [FIELD] *See* another term.

A cross reference may also appear in combination with definitions:

term [PHYSICAL CHEMISTRY] Definition. [SPECTROSCOPY] *See* another term.

CROSS REFERENCING. A cross-reference entry directs the user to the defining entry. For example, the user looking up "arachic acid" finds:

arachic acid [ORGANIC CHEMISTRY] *See* eicosanoic acid.

The user then turns to the "E" terms for the definition.
Cross references are also made from variant spellings, acronyms, abbreviations, and symbols.

AES [SPECTROSCOPY] *See* Auger electron spectroscopy.
aluminium [CHEMISTRY] *See* aluminum.
at. wt [CHEMISTRY] *See* atomic weight.
Au [CHEMISTRY] *See* gold.

CHEMICAL FORMULAS. Chemistry definitions may include either an empirical formula (say, for abietic acid, $C_{20}H_{30}O_2$) or a line formula (for acrylonitrile, CH_2CHCN), whichever is appropriate.

ALSO KNOWN AS . . ., etc. A definition may conclude with a mention of a synonym of the term, a variant spelling, an abbreviation for the term, or other such information, introduced by "Also known as . . .," "Also spelled . . .," "Abbreviated . . .," "Symbolized . . .," "Derived from . . .." When a term has more than one definition, the positioning of any of these phrases conveys the extent of applicability. For example:

> **term** [PHYSICAL CHEMISTRY] **1.** Definition. Also known as synonym. **2.** Definition. Symbolized T.

In the above arrangement, "Also known as . . ." applies only to the first definition: "Symbolized . . ." applies only to the second definition.

> **term** [PHYSICAL CHEMISTRY] **1.** Definition. **2.** Definition. [SPECTROSCOPY] Definition. Also known as synonym.

In the above arrangement "Also known as . . ." applies only to the second field.

> **term** [PHYSICAL CHEMISTRY] Also known as synonym. **1.** Definition. **2.** Definition. [SPECTROSCOPY] Definition.

In the above arrangement, "Also known as . . ." applies to both definitions in the first field.

> **term** Also known as synonym. [PHYSICAL CHEMISTRY] **1.** Definition. **2.** Definition. [SPECTROSCOPY] Definition.

In the above arrangement, "Also known as . . ." applies to all definitions in both fields.

Scope of Fields

Analytical Chemistry The science and art of determining composition of materials in terms of elements and compounds which they contain.

Chemistry The scientific study of the properties, composition, and structure of matter, the changes in structure and composition of matter, and accompanying energy changes.

Inorganic Chemistry The branch of chemistry that deals with reactions and properties of all chemical elements and their compounds, excluding hydrocarbons but usually including carbides and other simple carbon compounds (such as CO_2, CO, and HCN).

Organic Chemistry The study of the hydrocarbons and their derivatives, both naturally occurring and synthetic.

Physical Chemistry The description and prediction of chemical behavior by means of physical theory, with extensive use of graphs and mathematical formulas; main subject areas are structure, thermodynamics, and kinetics.

Spectroscopy The branch of physics concerned with the production, measurement, and interpretation of electromagnetic spectra arising from either emission or absorption of radiant energy by various substances.

Pronunciation Key

Vowels

a as in b**a**t, th**a**t
ā as in b**ai**t, cr**a**te
ä as in b**o**ther, f**a**ther
e as in b**e**t, n**e**t
ē as in b**ee**t, tr**ea**t
i as in b**i**t, sk**i**t
ī as in b**i**te, l**igh**t
ō as in b**oa**t, n**o**te
ȯ as in b**ough**t, t**au**t
u̇ as in b**oo**k, p**u**ll
ü as in b**oo**t, p**oo**l
ə as in b**u**t, sof**a**
au̇ as in cr**ow**d, p**ow**er
ȯi as in b**oi**l, sp**oi**l
yə as in form**u**la, spectac**u**lar
yü as in f**ue**l, m**u**le

Semivowels/Semiconsonants

w as in **w**ind, t**w**in
y as in **y**et, on**ion**

Stress (Accent)

ˈ precedes syllable with primary stress
ˌ precedes syllable with secondary stress
¦ precedes syllable with variable or indeterminate primary/secondary stress

Consonants

b as in **b**i**b**, dri**bb**le
ch as in **ch**arge, stre**tch**
d as in **d**og, ba**d**
f as in **f**ix, sa**f**e
g as in **g**ood, si**g**nal
h as in **h**and, be**h**ind
j as in **j**oint, di**g**it
k as in **c**ast, bri**ck**
k̲ as in Ba**ch** (used rarely)
l as in **l**oud, be**ll**
m as in **m**ild, su**mm**er
n as in **n**ew, de**n**t
n̲ indicates nasalization of preceding vowel
ŋ as in ri**ng**, si**ng**le
p as in **p**ier, sli**p**
r as in **r**ed, sca**r**
s as in **s**ign, po**s**t
sh as in **s**ugar, **sh**oe
t as in **t**imid, ca**t**
th as in **th**in, brea**th**
t̲h̲ as in **th**en, brea**th**e
v as in **v**eil, wea**v**e
z as in **z**oo, crui**s**e
zh as in bei**g**e, trea**s**ure

Syllabication

· Indicates syllable boundary when following syllable is unstressed

Contents

A

abalyn [ORGANIC CHEMISTRY] A liquid rosin that is a methyl ester of abietic acid; prepared by treating rosin with methyl alcohol; used as a plasticizer. { 'ab·ə‚lin }

Abegg's rule [CHEMISTRY] An empirical rule, holding for a large number of elements, that the sum of the maximum positive and negative valencies of an element equals eight. { 'ä·begz 'rül }

Abel tester [PHYSICAL CHEMISTRY] A laboratory instrument used in testing the flash point of kerosine and other volatile oils having flash points below 120°F (49°C); the oil is contained in a closed cup which is heated by a fixed flame below and a movable flame above. { ä'bəl 'tes·tər }

abietic acid [ORGANIC CHEMISTRY] $C_{20}H_{30}O_2$ A tricyclic, crystalline acid obtained from rosin; used in making esters for plasticizers. { ‚a·bē'et·ik 'as·əd }

ab initio computation [PHYSICAL CHEMISTRY] Computation of the geometry of a molecule solely from a knowledge of its composition and molecular structure as derived from the solution of the Schrödinger equation for the given molecule. { ‚ab ə'nish·ē·ō ‚käm·pyə'tā·shən }

Abney mounting [SPECTROSCOPY] A modification of the Rowland mounting in which only the slit is moved to observe different parts of the spectrum. { 'ab·nē ‚maùnt·iŋ }

ABS [ORGANIC CHEMISTRY] *See* acrylonitrile butadiene styrene resin.

absolute alcohol [ORGANIC CHEMISTRY] Ethyl alcohol that contains no more than 1% water. Also known as anhydrous alcohol. { 'ab·sə‚lüt 'al·kə·hȯl }

absolute boiling point [CHEMISTRY] The boiling point of a substance expressed in the unit of an absolute temperature scale. { 'ab·sə‚lüt 'bȯil·iŋ ‚pȯint }

absolute configuration [ORGANIC CHEMISTRY] The three-dimensional arrangement of substituents around a chiral center in a molecule. Also known as absolute stereochemistry. { 'ab·sə‚lüt kən‚fig·yə'rā·shən }

absolute density [CHEMISTRY] *See* absolute gravity. { 'ab·sə‚lüt 'dens·ə·dē }

absolute detection limit [ANALYTICAL CHEMISTRY] The smallest amount of an element or compound that is detectable in or on a given sample; expressed in terms of mass units or numbers of atoms or molecules. { 'ab·sə‚lüt di'tek·shən ‚lim·ət }

absolute gravity [CHEMISTRY] Density or specific gravity of a fluid reduced to standard conditions; for example, with gases, to 760 mmHg pressure and 0°C temperature. Also known as absolute density. { 'ab·sə‚lüt 'grav·ə·dē }

absolute method [ANALYTICAL CHEMISTRY] A method of chemical analysis that bases characterization completely on standards defined in terms of physical properties. { 'ab·sə‚lüt 'meth·əd }

absolute reaction rate [PHYSICAL CHEMISTRY] The rate of a chemical reaction as calculated by means of the (statistical-mechanics) theory of absolute reaction rates. { 'ab·sə‚lüt rē'ak·shən ‚rāt }

absolute stereochemistry [ORGANIC CHEMISTRY] *See* absolute configuration. { 'ab·sə‚lüt ‚ster·ē·ō'kem·ə·strē }

absorb [CHEMISTRY] To take up a substance in bulk. { əb'sȯrb }

absorbance [PHYSICAL CHEMISTRY] The common logarithm of the reciprocal of the transmittance of a pure solvent. Also known as absorbancy; extinction. { əb'sȯr·bəns }

absorbancy [PHYSICAL CHEMISTRY] *See* absorbance. { əb'sȯr·bən·sē }

absorbency [CHEMISTRY] Penetration of one substance into another. { əb'sȯr·bən·sē }

absorbency index [ANALYTICAL CHEMISTRY] *See* absorptivity. { əb'sȯr·bən·sē 'in·deks }

absorptiometer [ANALYTICAL CHEMISTRY] **1.** An instrument equipped with a filter system or other simple dispersing system to measure the absorption of nearly monochromatic radiation in the visible range by a gas or a liquid, and so determine the concentration of the absorbing constituents in the gas or liquid. **2.** A device for regulating the thickness of a liquid in spectrophotometry. { əb,sȯrp·tē'ä·məd·ər }

absorptiometric analysis [ANALYTICAL CHEMISTRY] Chemical analysis of a gas or a liquid by measurement of the peak electromagnetic absorption wavelengths that are unique to a specific material or element. { əb,sȯrp·tē·ə'met·rik ə'nal·ə·sis }

absorption [CHEMISTRY] The taking up of matter in bulk by other matter, as in dissolving of a gas by a liquid. { əb'sȯrp·shən }

absorption constant [ANALYTICAL CHEMISTRY] *See* absorptivity. { əb'sȯrp·shən ,käns·tənt }

absorption edge [SPECTROSCOPY] The wavelength corresponding to a discontinuity in the variation of the absorption coefficient of a substance with the wavelength of the radiation. Also known as absorption limit. { əb'sȯrp·shən ,ej }

absorption limit [SPECTROSCOPY] *See* absorption edge. { əb'sȯrp·shən 'lim·ət }

absorption line [SPECTROSCOPY] A minute range of wavelength or frequency in the electromagnetic spectrum within which radiant energy is absorbed by the medium through which it is passing. { əb'sȯrp·shən ,līn }

absorption peak [SPECTROSCOPY] A wavelength of maximum electromagnetic absorption by a chemical sample; used to identify specific elements, radicals, or compounds. { əb'sȯrp·shən ,pēk }

absorption spectrophotometer [SPECTROSCOPY] An instrument used to measure the relative intensity of absorption spectral lines and bands. Also known as difference spectrophotometer. { əb'sȯrp·shən ,spek·trə·fə'täm·ə·dər }

absorption spectroscopy [SPECTROSCOPY] The study of spectra obtained by the passage of radiant energy from a continuous source through a cooler, selectively absorbing medium. { əb'sȯrp·shən ,spek'träs·kə·pē }

absorption spectrum [SPECTROSCOPY] The array of absorption lines and absorption bands which results from the passage of radiant energy from a continuous source through a cooler, selectively absorbing medium. { əb'sȯrp·shən ,spek·trəm }

absorption tube [CHEMISTRY] A tube filled with a solid absorbent and used to absorb gases and vapors. { əb'sȯrp·shən ,tüb }

absorptive power [ANALYTICAL CHEMISTRY] *See* absorptivity. { əb'sȯrp·tiv ,pau̇·ər }

absorptivity [ANALYTICAL CHEMISTRY] The constant a in the Beer's law relation $A = abc$, where A is the absorbance, b the path length, and c the concentration of solution. Also known as absorptive power. Formerly known as absorbency index; absorption constant; extinction coefficient. { ab,sȯrp'tiv·əd·ē }

abstraction reaction [CHEMISTRY] A bimolecular chemical reaction in which an atom that is either neutral or charged is removed from a molecular entity. { ab'strak·shən rē,ak·shən }

Ac [CHEMISTRY] *See* actinium.

acaroid resin [ORGANIC CHEMISTRY] A gum resin from aloelike trees of the genus *Xanthorrhoea* in Australia and Tasmania; used in varnishes and inks. Also known as gum accroides; yacca gum. { 'a·kə,rȯid 'rez·ən }

accelerator mass spectrometer [SPECTROSCOPY] A combination of a mass spectrometer and an accelerator that can be used to measure the natural abundances of very rare radioactive isotopes. { ak¦sel·ə,rād·ər ¦mas spek'träm·əd·ər }

accelofilter [CHEMISTRY] A filtration device that uses a vacuum or pressure to draw or force the liquid through the filter to increase the rate of filtration. { ak'sel·ō,fil·tər }

acceptor [CHEMISTRY] **1.** A chemical whose reaction rate with another chemical increases because the other substance undergoes another reaction. **2.** A species that accepts electrons, protons, electron pairs, or molecules such as dyes. { ak'sep·tər }

acenaphthene [ORGANIC CHEMISTRY] $C_{12}H_{10}$ An unsaturated hydrocarbon whose colorless crystals melt at 92°C; insoluble in water; used as a dye intermediate and as an agent for inducing polyploidy. { ˌas·ə'naf·thēn }

acenaphthequinone [ORGANIC CHEMISTRY] $C_{10}H_6(CO)_2$ A three-ring hydrocarbon in the form of yellow needles melting at 261-263°C; insoluble in water and soluble in alcohol; used in dye synthesis. { ˌas·əˌnaf·thə·kwaˌnōn }

acene [ORGANIC CHEMISTRY] Any condensed polycyclic compound with fused rings in a linear arrangement; for example, anthracene. { ə'sēn }

acenocoumarin [ORGANIC CHEMISTRY] *See* acenocoumarol. { əˌsēn·ə'kü·mə·rən }

acenocoumarol [ORGANIC CHEMISTRY] $C_{19}H_{15}NO_6$ A tasteless, odorless, white, crystalline powder with a melting point of 197°C; slightly soluble in water and organic solvents; used as an anticoagulant. Also known as acenocoumarin. { əˌsēn·ə'kü·mə·rəl }

acephate [ORGANIC CHEMISTRY] $C_4H_{10}NO_3PS$ A white solid with a melting point of 72-80°C; very soluble in water; used as an insecticide for a wide range of aphids and foliage pests. { 'as·ə·fāt }

acephatemet [ORGANIC CHEMISTRY] $CH_3OCH_3SPONH_2$ A white, crystalline solid with a melting point of 39-41°C; limited solubility in water; used as an insecticide to control cutworms and borers on vegetables. { as·ə'fāt·mət }

acetal [ORGANIC CHEMISTRY] **1.** $CH_3CH(OC_2H_5)_2$ A colorless, flammable, volatile liquid used as a solvent and in manufacture of perfumes. Also known as 1,1-diethoxyethane. **2.** Any one of a class of compounds formed by the addition of alcohols to aldehydes. { 'as·əˌtal }

acetaldehyde [ORGANIC CHEMISTRY] C_2H_4O A colorless, flammable liquid used chiefly to manufacture acetic acid. { ˌas·əd'al·dəˌhīd }

acetaldehyde cyanohydrin [ORGANIC CHEMISTRY] *See* lactonitrile. { ˌas·əd'al·dəˌhīd ˌsī·ə·nō'hīd·rən }

acetal resins [ORGANIC CHEMISTRY] Linear, synthetic resins produced by the polymerization of formaldehyde (acetal homopolymers) or of formaldehyde with trioxane (acetal copolymers); hard, tough plastics used as substitutes for metals. Also known as polyacetals. { 'as·əˌtəl 'rez·ənz }

acetamide [ORGANIC CHEMISTRY] CH_3CONH_2 The crystalline, colorless amide of acetic acid, used in organic synthesis and as a solvent. { ə'sed·əˌmīd }

acetamidine hydrochloride [ORGANIC CHEMISTRY] $C_2H_6N_2 \cdot HCl$ Deliquescent crystals that are long prisms with a melting point reported as either 174°C or 164-166°C; soluble in water and alcohol; used in the synthesis of imidazoles, pyrimidines, and triazines. { ə·sed'am·əˌdēn hī·drə'klȯˌrīd }

acetaminophen [ORGANIC CHEMISTRY] $C_8H_9O_2N$ Large monoclinic prisms with a melting point of 169-170°C; soluble in organic solvents such as methanol and ethanol; used in the manufacture of azo dyes and photographic chemicals, and as an analgesic and antipyretic. { əˌsēd·ə'mēn·ə·fən }

acetanilide [ORGANIC CHEMISTRY] An odorless compound in the form of white, shining, crystalline leaflets or a white, crystalline powder with a melting point of 114-116°C; soluble in hot water, alcohol, ether, chloroform, acetone, glycerol, and benzene; used as a rubber accelerator, in the manufacture of dyestuffs and intermediates, as a precursor in penicillin manufacture, and as a painkiller. { ˌa·səd'an·əˌlīd }

acetate [ORGANIC CHEMISTRY] One of two species derived from acetic acid, CH_3COOH; one type is the acetate ion, CH_3COO^-; the second type is a compound whose structure contains the acetate ion, such as ethyl acetate. { 'as·əˌtāt }

acetate dye [CHEMISTRY] **1.** Any of a group of water-insoluble azo or anthroquinone dyes used for dyeing acetate fibers. **2.** Any of a group of water-insoluble amino azo dyes that are treated with formaldehyde and bisulfate to make them water-soluble. { 'as·əˌtāt ˌdī }

acetate of lime [ORGANIC CHEMISTRY] Calcium acetate made from pyroligneous acid and a water suspension of calcium hydroxide. { 'as·əˌtāt əv 'līm }

acetenyl [ORGANIC CHEMISTRY] *See* ethinyl. { ə'sed·əˌnil }

acetic acid [ORGANIC CHEMISTRY] CH_3COOH **1.** A clear, colorless liquid or crystalline mass with a pungent odor, miscible with water or alcohol; crystallizes in deliquescent needles; a component of vinegar. Also known as ethanoic acid. **2.** A mixture of the normal and acetic salts; used as a mordant in the dyeing of wool. { ə'sēd·ik 'as·əd }

acetic anhydride [ORGANIC CHEMISTRY] $(CH_3CO)_2O$ A liquid with a pungent odor that combines with water to form acetic acid; used as an acetylating agent. { ə'sēd·ik an'hīd,rīd }

acetic ether [ORGANIC CHEMISTRY] *See* ethyl acetate. { ə'sēd·ik 'ē·thər }

acetidin [ORGANIC CHEMISTRY] *See* ethyl acetate. { ə'sed·ə·din }

acetin [ORGANIC CHEMISTRY] $C_3H_5(OH)_2OOCCH_3$ A thick, colorless, hygroscopic liquid with a boiling point of 158°C, made by heating glycerol and strong acetic acid; soluble in water and alcohol; used in tanning, as a dye solvent and food additive, and in explosives. Also spelled acetine. { 'as·ə·tin }

acetine [ORGANIC CHEMISTRY] *See* acetin. { 'as·ə,tēn }

acetoacetate [ORGANIC CHEMISTRY] A salt which contains the CH_3COCH_2COO radical; derived from acetoacetic acid. { ¦as·ə,tō·'as·ə,tāt }

acetoacetic acid [ORGANIC CHEMISTRY] CH_3COCH_2COOH A colorless liquid miscible with water; derived from β-hydroxybutyric acid in the body. { ¦as·ə,tō,ə'sēd·ik 'as·əd }

acetoacetic ester [ORGANIC CHEMISTRY] *See* ethyl acetoacetate. { ¦as·ə,tō,ə'sēd·ik 'es·tər }

acetoin [ORGANIC CHEMISTRY] $CH_3COCHOHCH_3$ A slightly yellow liquid, melting point 15°C, used as an aroma carrier in the preparation of flavors and essences; produced by fermentation or from diacetyl by partial reduction with zinc and acid. { ə'sed·ə·wən }

acetol [ORGANIC CHEMISTRY] CH_3COCH_2OH A colorless liquid soluble in water; a reducing agent. { 'as·ə·tōl }

acetolysis [ORGANIC CHEMISTRY] Decomposition of an organic molecule through the action of acetic acid or acetic anhydride. { ,as·ə'täl·ə·səs }

acetone [ORGANIC CHEMISTRY] CH_3COCH_3 A colorless, volatile, extremely flammable liquid, miscible with water; used as a solvent and reagent. Also known as 2-propanone. { 'as·ə,tōn }

acetone cyanohydrin [ORGANIC CHEMISTRY] $(CH_3)_2COHCN$ A colorless liquid obtained from condensation of acetone with hydrocyanic acid; used as an insecticide or as an organic chemical intermediate. { 'as·ə,tōn sī,ə·nō'hīd·rən }

acetone glucose [ORGANIC CHEMISTRY] *See* acetone sugar. { 'as·ə,tōn 'glü·kōs }

acetone number [CHEMISTRY] A ratio used to estimate the degree of polymerization of materials such as drying oils; it is the weight in grams of acetone added to 100 grams of a drying oil to cause an insoluble phase to form. { 'as·ə·tōn 'nəm·bər }

acetone pyrolysis [ORGANIC CHEMISTRY] Thermal decomposition of acetone into ketene. { 'as·ə·tōn pī'räl·ə·səs }

acetone-sodium bisulfite [ORGANIC CHEMISTRY] $(CH_3)_2C(OH)SO_3Na$ Crystals that have a slight sulfur dioxide odor and slightly fatty feel; freely soluble in water, decomposed by acids; used in photography and in textile dyeing and printing. { 'as·ə·tōn 'sōd·ē·əm ,bī'səl,fāt }

acetone sugar [ORGANIC CHEMISTRY] Any reducing sugar that contains acetone; examples are 1,2-monoacetone-D-glucofuranose and 1,2-5,6-diacetone-D-glucofuranose. Also known as acetone glucose. { 'as·ə·tōn 'shu̇g·ər }

acetonitrile [ORGANIC CHEMISTRY] CH_3CN A colorless liquid soluble in water; used in organic synthesis. { ,as·ə·tō'nī,tril }

acetonylacetone [ORGANIC CHEMISTRY] $CH_3COCH_2CH_2COCH_3$ A colorless liquid with a boiling point of 192.2°C; soluble in water; used as a solvent and as an intermediate for pharmaceuticals and photographic chemicals. { ,as·ə,tän·əl'as·ə,tōn }

acetophenone [ORGANIC CHEMISTRY] $C_6H_5COCH_3$ Colorless crystals with a melting point of 19.6°C and a specific gravity of 1.028; used as a chemical intermediate. { ,as·ə,tä·fə'nōn }

acetostearin [ORGANIC CHEMISTRY] A general term for monoglycerides of stearic acid acetylated with acetic anhydride; used as a protective food coating and as plasticizers for waxes and synthetic resins to improve low-temperature characteristics. { ə‚sē·dō′stēr·ən }

acetoxime [ORGANIC CHEMISTRY] $(CH_3)_2CNOH$ Colorless crystals with a chloral-like odor and a melting point of 61°C; soluble in alcohol, ethers, and water; used in organic synthesis and as a solvent for cellulose ethers. { ‚as·ə′täk‚sēm }

aceturic acid [ORGANIC CHEMISTRY] $CH_3CONHCHCH_2COOH$ Long, needlelike crystals with a melting point of 206-208°C; soluble in water and alcohol; forms stable salts with organic bases; used in medicine. { ¦as·ə¦tür·ik ′as·əd }

acetyl [ORGANIC CHEMISTRY] $CH_3CO—$ A two-carbon organic radical containing a methyl group and a carbonyl group. { ə′sēd·əl }

acetylacetone [ORGANIC CHEMISTRY] $CH_3COCH_2OCCH_3$ A colorless liquid with a pleasant odor and a boiling point of 140.5°C; soluble in water; used as a solvent, lubricant additive, paint drier, and pesticide. { ə¦sed·əl′as·ə‚tōn }

acetylating agent [ORGANIC CHEMISTRY] A reagent, such as acetic anhydride, capable of bonding an acetyl group onto an organic molecule. { ə′sed·əl‚āt·iŋ ‚ā·jənt }

acetylation [ORGANIC CHEMISTRY] The process of bonding an acetyl group onto an organic molecule. { ə‚sed·əl′ā·shən }

acetyl benzoyl peroxide [ORGANIC CHEMISTRY] $C_6H_5CO \cdot O_2 \cdot OCCH_3$ White crystals with a melting point of 36.6°C; moderately soluble in ether, chloroform, carbon tetrachloride, and water; used as a germicide and disinfectant. { ə′sed·əl ′ben·zȯil pə′räk‚sīd }

acetyl bromide [ORGANIC CHEMISTRY] CH_3COBr A colorless, fuming liquid with a boiling point of 81°C, soluble in ether, chloroform, and benzene; used in organic synthesis and dye manufacture. { ə′sed·əl ′brō‚mīd }

α-acetylbutyrolactone [ORGANIC CHEMISTRY] $C_6H_8O_3$ A liquid with an esterlike odor; soluble in water; used in the synthesis of 3,4-disubstituted pyridines. { ¦al·fə ə‚sed·əl¦byüd·ə·rō′lak·tōn }

acetyl chloride [ORGANIC CHEMISTRY] CH_3COCl A colorless, fuming liquid with a boiling point of 51-52°C; soluble in ether, acetone, and acetic acid; used in organic synthesis, and in the manufacture of dyestuffs and pharmaceuticals. { ə′sed·əl ′klȯ‚rīd }

acetylene [ORGANIC CHEMISTRY] C_2H_2 A colorless, highly flammable gas that is explosive when compressed; the simplest compound containing a triple bond; used in organic synthesis and as a welding fuel. Also known as ethyne. { ə′sed·əl‚ēn }

acetylene black [ORGANIC CHEMISTRY] A form of carbon with high electrical conductivity; made by decomposing acetylene by heat. { ə′sed·əl‚ēn ′blak }

acetylene series [ORGANIC CHEMISTRY] A series of unsaturated aliphatic hydrocarbons, each containing at least one triple bond and having the general formula C_nH_{2n-2}. { ə′sed·əl‚ēn ′sir·ēz }

acetylene tetrabromide [ORGANIC CHEMISTRY] $CHBr_2CHBr_2$ A yellowish liquid with a boiling point of 239-242°C; soluble in alcohol and ether; used for separating minerals and as a solvent. { ə′sed·əl‚ēn ‚te·trə′brō‚mīd }

acetylenic [ORGANIC CHEMISTRY] Pertaining to acetylene or being like acetylene, such as having a triple bond. { ə‚sed·ə′len·ik }

acetylenyl [ORGANIC CHEMISTRY] *See* ethinyl. { ə‚sed·ə′len·əl }

***N*-acetylethanolamine** [ORGANIC CHEMISTRY] $CH_3CONHC_2H_4OH$ A brown, viscous liquid with a boiling range of 150-152°C; soluble in alcohol, ether, and water; used as a plasticizer, humectant, high-boiling solvent, and textile conditioner. { ¦en ə‚sed·əl‚eth·ə′näl·ə‚mēn }

acetylide [ORGANIC CHEMISTRY] A compound formed from acetylene with the H atoms replaced by metals, as in cuprous acetylide (Cu_2C_2). { ə′sed·əl‚īd }

acetyl iodide [ORGANIC CHEMISTRY] CH_3COI A colorless, transparent, fuming liquid with a boiling point of 105-108°C; soluble in ether and benzene; used in organic synthesis. { ə′sed·əl ′ī·ə‚dīd }

acetylisoeugenol [ORGANIC CHEMISTRY] $C_6H_3(CHCHCH_3)(OCH_3)(OCOCH_3)$ White crystals with a clovelike odor and a congealing point of 77°C; used in perfumery and flavoring. { ə‚sed·əl‚ī·sō′yü·jə‚nȯl }

acetyl number [ANALYTICAL CHEMISTRY] A measure of free hydroxyl groups in fats or oils determined by the amount of potassium hydroxide used to neutralize the acetic acid formed by saponification of acetylated fat or oil. { ə'sed·əl ˌnəm·bər }

acetyl peroxide [ORGANIC CHEMISTRY] $(CH_3CO)_2O_2$ Colorless crystals with a melting point of 30°C; soluble in alcohol and ether; used as an initiator and catalyst for resins. { ə'sed·əl pə'räkˌsīd }

acetyl propionyl [ORGANIC CHEMISTRY] $CH_3COCOCH_2CH_3$ A yellow liquid with a boiling point of 106-110°C; used in butterscotch- and chocolate-type flavors. { ə'sed·əl 'pro·pē·əˌnil }

acetylsalicylic acid [ORGANIC CHEMISTRY] $CH_3COOC_6H_4COOH$ A white, crystalline, weakly acidic substance, with melting point 137°C; slightly soluble in water; used medicinally as an antipyretic. Also known by trade name aspirin. { əˈsed·əlˈsal·əˈsil·ik 'as·əd }

***N*-acetylsulfanilyl chloride** [ORGANIC CHEMISTRY] $C_8H_8ClNO_3S$ Thick, light tan prisms ranging to brown powder or fine crystals with a melting point of 149°C; soluble in benzene, chloroform, and ether; used as an intermediate in the preparation of sulfanilamide and its derivatives. Abbreviated ASC. { ˈen əˈsed·əl·səl'fan·ə·lil 'klȯrˌīd }

acetylurea [ORGANIC CHEMISTRY] $CH_3CONHCONH_2$ Crystals that are colorless and are slightly soluble in water. { əˌsēd·əlˌyü 'rē·ə }

acetyl valeryl [ORGANIC CHEMISTRY] $CH_3COCOC_4H_9$ A yellow liquid used for cheese, butter, and other flavors. Also known as heptadione-2,3. { ə'sēd·əl ˌval·əˌril }

achiral molecules [ORGANIC CHEMISTRY] Molecules which are superposable to their mirror images. { ˈāˌkī·rəl 'mäl·əˌkyülz }

acid [CHEMISTRY] **1.** Any of a class of chemical compounds whose aqueous solutions turn blue litmus paper red, react with and dissolve certain metals to form salts, and react with bases to form salts. **2.** A compound capable of transferring a hydrogen ion in solution. **3.** A substance that ionizes in solution to yield the positive ion of the solvent. **4.** A molecule or ion that combines with another molecule or ion by forming a covalent bond with two electrons from the other species. { 'as·əd }

π-acid [ORGANIC CHEMISTRY] An acid that readily forms stable complexes with aromatic systems. { 'pī 'as·əd }

acid acceptor [ORGANIC CHEMISTRY] A stabilizer compound added to plastic and resin polymers to combine with trace amounts of acids formed by decomposition of the polymers. { 'as·əd ək'sep·tər }

acid alcohol [ORGANIC CHEMISTRY] A compound containing both a carboxyl group (—COOH) and an alcohol group (—CH_2OH, ═CHOH, or ≡COH). { 'as·əd 'al·kə·hȯl }

acid amide [ORGANIC CHEMISTRY] A compound derived from an acid in which the hydroxyl group (—OH) of the carboxyl group (—COOH) has been replaced by an amino group (—NH_2) or a substituted amino group (—NHR or —NHR_2). { 'as·əd 'aˌmīd }

acid anhydride [CHEMISTRY] An acid with one or more molecules of water removed; for example, SO_3 is the acid anhydride of H_2SO_4, sulfuric acid. { 'as·əd ˌan'hīdˌrīd }

acid azide [ORGANIC CHEMISTRY] **1.** A compound in which the hydroxy group of a carboxylic acid is replaced by the azido group (—NH_3). **2.** An acyl or aroyl derivative of hydrazoic acid. Also known as acyl azide. { 'as·əd 'āˌzīd }

acid-base catalysis [CHEMISTRY] The increase in speed of certain chemical reactions due to the presence of acids and bases. { 'as·əd 'bās kə'tal·ə·sis }

acid-base equilibrium [CHEMISTRY] The condition when acidic and basic ions in a solution exactly neutralize each other; that is, the pH is 7. { 'as·əd 'bās ˌik·wə'lib·rē·əm }

acid-base indicator [ANALYTICAL CHEMISTRY] A substance that reveals, through characteristic color changes, the degree of acidity or basicity of solutions. { 'as·əd 'bās 'in·dəˌkād·ər }

acid-base pair [CHEMISTRY] A concept in the Brönsted theory of acids and bases; the pair consists of the source of the proton (acid) and the base generated by the transfer of the proton. { 'as·əd 'bās 'pār }

acid-base titration [ANALYTICAL CHEMISTRY] A titration in which an acid of known concentration is added to a solution of base of unknown concentration, or the converse. { 'as·əd 'bās tī'trā·shən }

acid cell [PHYSICAL CHEMISTRY] An electrolytic cell whose electrolyte is an acid. { 'as·əd ˌsel }

acid chloride [ORGANIC CHEMISTRY] A compound containing the radical —COCl; an example is benzoyl chloride. { 'as·əd 'klȯrˌīd }

acid disproportionation [CHEMISTRY] The self-oxidation of a sample of an oxidized element to the next higher oxidation state and then a corresponding reduction to lower oxidation states. { 'as·əd ˌdis·prəˌpȯr·shə'nā·shən }

acid dye [ORGANIC CHEMISTRY] Any of a group of sodium salts of sulfonic and carboxylic acids used to dye natural and synthetic fibers, leather, and paper. { 'as·əd ˌdī }

acid electrolyte [INORGANIC CHEMISTRY] A compound, such as sulfuric acid, that dissociates into ions when dissolved, forming an acidic solution that conducts an electric current. { 'as·əd ə'lek·trəˌlīt }

acid halide [ORGANIC CHEMISTRY] A compound of the type RCOX, where R is an alkyl or aryl radical and X is a halogen. { 'as·əd 'hāˌlīd }

acid heat test [ANALYTICAL CHEMISTRY] The determination of degree of unsaturation of organic compounds by reacting with sulfuric acid and measuring the heat of reaction. { 'as·əd 'hēt ˌtest }

acidic [CHEMISTRY] **1.** Pertaining to an acid or to its properties. **2.** Forming an acid during a chemical process. { ə'sid·ik }

acidic dye [ORGANIC CHEMISTRY] An organic anion that binds to and stains positively charged macromolecules. { ə¦sid·ik 'dī }

acidic group [ORGANIC CHEMISTRY] The radical COOH, present in organic acids. { ə'sid·ik 'grüp }

acidic oxide [INORGANIC CHEMISTRY] An oxygen compound of a nonmetal, for example, SO_2 or P_2O_5, which yields an oxyacid with water. { ə'sid·ik 'äkˌsīd }

acidic titrant [ANALYTICAL CHEMISTRY] An acid solution of known concentration used to determine the basicity of another solution by titration. { ə'sid·ik 'tī·trənt }

acidification [CHEMISTRY] Addition of an acid to a solution until the pH falls below 7. { əˌsid·ə·fə'kā·shən }

acidimeter [ANALYTICAL CHEMISTRY] An apparatus or a standard solution used to determine the amount of acid in a sample. { ˌas·ə'dim·ə·tər }

acidimetry [ANALYTICAL CHEMISTRY] The titration of an acid with a standard solution of base. { ˌas·ə'dim·ə·trē }

acidity [CHEMISTRY] The state of being acid. { ə'sid·ə·tē }

acidity function [CHEMISTRY] A quantitative scale for measuring the acidity of a solvent system; usually established over a range of compositions. { ə'sid·əd·ē ˌfəŋk·shən }

acid number [CHEMISTRY] *See* acid value. { 'as·əd ¦nəm·bər }

acidolysis [ORGANIC CHEMISTRY] A chemical reaction involving the decomposition of a molecule, with the addition of the elements of an acid to the molecule; the reaction is comparable to hydrolysis or alcoholysis, in which water or alcohol, respectively, is used in place of the acid. Also known as acyl exchange. { ˌas·ə'däl·ə·səs }

acid phosphate [INORGANIC CHEMISTRY] A mono- or dihydric phosphate; for example, M_2HPO_4 or MH_2PO_4, where M represents a metal atom. { 'as·əd 'fasˌfat }

acid potassium phthalate [ORGANIC CHEMISTRY] *See* potassium biphthalate. { 'as·əd pə'tas·ē·əm 'thaˌlāt }

acid potassium sulfate [INORGANIC CHEMISTRY] *See* potassium bisulfate. { 'as·əd pə'tas·ē·əm 'səlˌfāt }

acid reaction [CHEMISTRY] A chemical reaction produced by an acid. { 'as·əd rē'ak·shən }

acid salt [CHEMISTRY] A compound derived from an acid and base in which only a part of the hydrogen is replaced by a basic radical; for example, the acid sulfate $NaHSO_4$. { 'as·əd ˌsȯlt }

acid sodium tartrate [ORGANIC CHEMISTRY] *See* sodium bitartrate. { 'as·əd sōd·ē·əm 'tärˌtrāt }

acid solution [CHEMISTRY] An aqueous solution containing more hydrogen ions than hydroxyl ions. { 'as·əd sə'lü·shən }
acid tartrate [ORGANIC CHEMISTRY] *See* bitartrate. { 'as·əd 'tär‚trāt }
acid value Also known as acid number. [CHEMISTRY] The acidity of a solution expressed in terms of normality. [ORGANIC CHEMISTRY] A number indicating the amount of nonesterified fatty acid present in a sample of fat or fatty oil as determined by alkaline titration. { 'as·əd 'val·yü }
aconitic acid [ORGANIC CHEMISTRY] $C_6H_6O_6$ A white, crystalline organic acid found in sugarcane and sugarbeet; obtained during manufacture of sugar. { ‚ak·ə'nid·ik 'as·əd }
Acree's reaction [ANALYTICAL CHEMISTRY] A test for protein in which a violet ring appears when concentrated sulfuric acid is introduced below a mixture of the unknown solution and a formaldehyde solution containing a trace of ferric chloride. { 'ak·rēz rē'ak·shən }
acridine [ORGANIC CHEMISTRY] $(C_6H_4)_2NCH$ A typical member of a group of organic heterocyclic compounds containing benzene rings fused to the 2,3 and 5,6 positions of pyridine; derivatives include dyes and medicines. { 'ak·rə‚dēn }
acridine dye [ORGANIC CHEMISTRY] Any of a class of basic dyes containing the acridine nucleus that bind to deoxyribonucleic acid. { 'ak·rə‚dēn ¦dī }
acridine orange [ORGANIC CHEMISTRY] A dye with an affinity for nucleic acids; the complexes of nucleic acid and dye fluoresce orange with RNA and green with DNA when observed in the fluorescence microscope. { 'ak·rə‚dēn 'är·inj }
acriflavine [ORGANIC CHEMISTRY] $C_{14}H_{14}N_3Cl$ A yellow acridine dye obtained from proflavine by methylation in the form of red crystals; used as an antiseptic in solution. { ‚ak·rə'flā‚vēn }
acrolein [ORGANIC CHEMISTRY] $CH_2{=}CHCHO$ A colorless to yellow liquid with a pungent odor and a boiling point of 52.7°C; soluble in water, alcohol, and ether; used in organic synthesis, pharmaceuticals manufacture, and as an herbicide and tear gas. { ə'krōl·ē·ən }
acrolein cyanohydrin [ORGANIC CHEMISTRY] $CH_2{:}CHCH(OH)CN$ A liquid soluble in water and boiling at 165°C; copolymerizes with ethylene and acrylonitrile; used to modify synthetic resins. { ə'krōl·ē·ən ‚sī·ə·nō'hī·drən }
acrolein dimer [ORGANIC CHEMISTRY] $C_6H_8O_2$ A flammable, water-soluble liquid used as an intermediate for resins, dyestuffs, and pharmaceuticals. { ə'krōl·ē·ən 'dī·mər }
acrolein test [ANALYTICAL CHEMISTRY] A test for the presence of glycerin or fats; a sample is heated with potassium bisulfate, and acrolein is released if the test is positive. { ə'krōl·ē·ən ‚test }
acrylamide [ORGANIC CHEMISTRY] $CH_2CHCONH_2$ Colorless, odorless crystals with a melting point of 84.5°C; soluble in water, alcohol, and acetone; used in organic synthesis, polymerization, sewage treatment, ore processing, and permanent press fabrics. { ə'kril·ə‚mīd }
acrylamide copolymer [ORGANIC CHEMISTRY] A thermosetting resin formed of acrylamide with other resins, such as the acrylic resins. { ə'kril·ə‚mīd kō'päl·ə·mər }
acrylate [ORGANIC CHEMISTRY] **1.** A salt or ester of acrylic acid. **2.** *See* acrylate resin. { 'ak·rə‚lāt }
acrylate resin [ORGANIC CHEMISTRY] Acrylic acid or ester polymer with a $—CH_2—CH(COOR)—$ structure; used in paints, sizings and finishes for paper and textiles, adhesives, and plastics. Also known as acrylate. { 'ak·rə‚lāt 'rez·ən }
acrylic acid [ORGANIC CHEMISTRY] $CH_2CHCOOH$ An easily polymerized, colorless, corrosive liquid used as a monomer for acrylate resins. { ə'kril·ik 'as·əd }
acrylic ester [ORGANIC CHEMISTRY] An ester of acrylic acid. { ə'kril·ik 'es·tər }
acrylic resin [ORGANIC CHEMISTRY] A thermoplastic synthetic organic polymer made by the polymerization of acrylic derivatives such as acrylic acid, methacrylic acid, ethyl acrylate, and methyl acrylate; used for adhesives, protective coatings, and finishes. { ə'kril·ik 'rez·ən }
acrylic rubber [ORGANIC CHEMISTRY] Synthetic rubber containing acrylonitrile; for example, nitrile rubber. { ə'kril·ik 'rəb·ər }

acrylonitrile [ORGANIC CHEMISTRY] CH_2CHCN A colorless liquid compound used in the manufacture of acrylic rubber and fibers. Also known as vinylcyanide. { ˌak·rəˌlōˈnī·trəl }

acrylonitrile butadiene styrene resin [ORGANIC CHEMISTRY] A polymer made by blending acrylonitrile-styrene copolymer with a butadiene-acrylonitrile rubber or by interpolymerizing polybutadiene with styrene and acrylonitrile; combines the advantages of hardness and strength of the vinyl resin component with the toughness and impact resistance of the rubbery component. Abbreviated ABS. { ˌak·rəˌlōˈnī·trəl ˌbyüd·əˈdīˌēn ˈstī·rēn ˈrez·ən }

acrylonitrile copolymer [ORGANIC CHEMISTRY] Oil-resistant synthetic rubber made by polymerization of acrylonitrile with compounds such as butadiene or acrylic acid. { ˌak·rəˌlōˈnī·trəl kōˈpäl·ə·mər }

actinide series [CHEMISTRY] The group of elements of atomic number 89 through 103. Also known as actinoid elements. { ˈak·təˌnīd ˈsirˌēz }

actinism [CHEMISTRY] The production of chemical changes in a substance upon which electromagnetic radiation is incident. { ¦ak·təˈniz·əm }

actinium [CHEMISTRY] A radioactive element, symbol Ac, atomic number 89; its longest-lived isotope is ^{227}Ac with a half-life of 21.7 years; the element is trivalent; chief use is, in equilibrium with its decay products, as a source of alpha rays. { akˈtin·ē·əm }

actinochemistry [CHEMISTRY] A branch of chemistry concerned with chemical reactions produced by light or other radiation. { ˌak·tə·nōˈkem·ə·strē }

actinoid elements [CHEMISTRY] *See* actinide series. { ˈak·təˌnȯid ˈel·ə·məns }

activated complex [PHYSICAL CHEMISTRY] An energetically excited state which is intermediate between reactants and products in a chemical reaction. Also known as transition state. { ˈak·təˌvād·əd ˈkäm·pleks }

activation [CHEMISTRY] Treatment of a substance by heat, radiation, or activating reagent to produce a more complete or rapid chemical or physical change. { ˌak·təˈvā·shən }

activation energy [PHYSICAL CHEMISTRY] The energy, in excess over the ground state, which must be added to an atomic or molecular system to allow a particular process to take place. { ˌak·təˈvā·shən ˈen·ər·jē }

activator [CHEMISTRY] **1.** A substance that increases the effectiveness of a rubber vulcanization accelerator; for example, zinc oxide or litharge. **2.** A trace quantity of a substance that imparts luminescence to crystals; for example, silver or copper in zinc sulfide or cadmium sulfide pigments. { ˈak·təˌvād·ər }

active center [CHEMISTRY] **1.** Any one of the points on the surface of a catalyst at which the chemical reaction is initiated or takes place. **2.** *See* active site. { ˈak·tiv ˈsen·tər }

active site [CHEMISTRY] The effective site at which a given heterogeneous catalytic reaction can take place. Also known as active center. { ˈak·tiv ˈsīt }

active solid [CHEMISTRY] A porous solid possessing adsorptive properties and used for chromatographic separations. { ˈak·tiv ˈsäl·əd }

activity [PHYSICAL CHEMISTRY] A thermodynamic function that correlates changes in the chemical potential with changes in experimentally measurable quantities, such as concentrations or partial pressures, through relations formally equivalent to those for ideal systems. { ˌakˈtiv·əd·ē }

activity coefficient [PHYSICAL CHEMISTRY] A characteristic of a quantity expressing the deviation of a solution from ideal thermodynamic behavior; often used in connection with electrolytes. { ˌakˈtiv·əd·ē ˌkō·əˈfish·ənt }

activity series [CHEMISTRY] A series of elements that have similar properties—for example, metals—arranged in descending order of chemical activity. { akˈtiv·əd·ē ˌsir·ēz }

actol [ORGANIC CHEMISTRY] *See* silver lactate. { ˈakˌtȯl }

acyclic compound [ORGANIC CHEMISTRY] A chemical compound with an open-chain molecular structure rather than a ring-shaped structure; for example, the alkane series. { āˈsik·lik ˈkämˌpau̇nd }

acyl [ORGANIC CHEMISTRY] A radical formed from an organic acid by removal of a hydroxyl group; the general formula is RCO, where R may be aliphatic, alicyclic, or aromatic. { 'a·səl }

acylation [ORGANIC CHEMISTRY] Any process whereby the acyl group is incorporated into a molecule by substitution. { ˌas·ə'lā·shən }

acyl azide [ORGANIC CHEMISTRY] *See* acid azide. { 'a·səl 'āˌzīd }

acylcarbene [ORGANIC CHEMISTRY] A carbene radical in which at least one of the groups attached to the divalent carbon is an acyl group; for example, acetylcarbene. { ˌa·səl'kärˌbēn }

acyl exchange [ORGANIC CHEMISTRY] *See* acidolysis. { 'a·səl iks'chānj }

acyl halide [ORGANIC CHEMISTRY] One of a large group of organic substances containing the halocarbonyl group; for example, acyl fluoride. { 'a·səl 'halˌīd }

acylnitrene [ORGANIC CHEMISTRY] A nitrene in which the nitrogen is covalently bonded to an acyl group. { ˌa·səl'nīˌtrēn }

acyloin [ORGANIC CHEMISTRY] An organic compound that may be synthesized by condensation of aldehydes; an example is benzoin, $C_6H_5COCHOHC_6H_5$. { ə'sil·ə·wən }

acyloin condensation [ORGANIC CHEMISTRY] The reaction of an aliphatic ester with metallic sodium to form intermediates converted by hydrolysis into aliphatic α-hydroxyketones called acyloins. { ə'sil·ə·wən ˌkänˌden'sā·shən }

adamantane [ORGANIC CHEMISTRY] A $C_{10}H_{16}$ alicyclic hydrocarbon whose structure has the same arrangement of carbon atoms as does the basic unit of the diamond lattice. { ˌad·ə'manˌtān }

adamsite [ORGANIC CHEMISTRY] $C_6H_4 \cdot NH \cdot C_6H_4 \cdot AsCl$ A yellow crystalline arsenical; used in leather tanning and in warfare and riot control to produce skin and eye irritation, chest distress, and nausea;U.S. Army code is DM. Also known as diphenylaminechloroarsine; phenarsazine chloride. { 'a·dəmˌzīt }

adatom [PHYSICAL CHEMISTRY] An atom adsorbed on a surface so that it will migrate over the surface. { 'adˌad·əm }

addition agent [PHYSICAL CHEMISTRY] A substance added to a plating solution to change characteristics of the deposited substances. { ə'di·shən ˌā·jənt }

addition polymer [ORGANIC CHEMISTRY] A polymer formed by the chain addition of unsaturated monomer molecules, such as olefins, with one another without the formation of a by-product, as water; examples are polyethylene, polypropylene, and polystyrene. Also known as addition resin. { ə'di·shən 'päl·ə·mər }

addition reaction [ORGANIC CHEMISTRY] A type of reaction of unsaturated hydrocarbons with hydrogen, halogens, halogen acids, and other reagents, so that no change in valency is observed and the organic compound forms a more complex one. { ə'di·shən rē'ak·shən }

addition resin [ORGANIC CHEMISTRY] *See* addition polymer. { ə'di·shən 'rez·ən }

adduct [CHEMISTRY] **1.** A chemical compound that forms from chemical addition of two species; for example, reaction of butadiene with styrene forms an adduct, 4-phenyl-1-cyclohexene. **2.** The complex compound formed by association of an inclusion complex. { 'aˌdəkt }

adiabatic approximation [PHYSICAL CHEMISTRY] *See* Born-Oppenheimer approximation. { ¦ad·ē·ə¦bad·ik əˌpräk·sə'mā·shən }

adiabatic calorimeter [PHYSICAL CHEMISTRY] An instrument used to study chemical reactions which have a minimum loss of heat. { ¦ad·ē·ə¦bad·ik ˌkal·ə'rim·əd·ər }

adiabatic flame temperature [PHYSICAL CHEMISTRY] The highest possible temperature of combustion obtained under the conditions that the burning occurs in an adiabatic vessel, that it is complete, and that dissociation does not occur. { ¦ad·ē·ə¦bad·ik ¦flām 'tem·prə·chər }

adipate [ORGANIC CHEMISTRY] Salt produced by reaction of adipic acid with a basic compound. { 'ad·əˌpāt }

adipic acid [ORGANIC CHEMISTRY] $HOOC(CH_2)_4COOH$ A colorless crystalline dicarboxylic acid, sparingly soluble in water; used in nylon manufacture. { ə'dip·ik 'as·əd }

adiponitrile [ORGANIC CHEMISTRY] $NC(CH_2)_4CN$ The high-boiling liquid dinitrile of adipic acid; used to make nylon intermediates. { ˌad·ə·pō'nī·trəl }

adjective dye [CHEMISTRY] Any dye that needs a mordant. { ə'jek·tiv ˌdī }

adsorbate [CHEMISTRY] A solid, liquid, or gas which is adsorbed as molecules, atoms, or ions by such substances as charcoal, silica, metals, water, and mercury. { ad'sȯr‚bāt }

adsorbent [CHEMISTRY] A solid or liquid that adsorbs other substances; for example, charcoal, silica, metals, water, and mercury. { ad'sȯr·bənt }

adsorption [CHEMISTRY] The surface retention of solid, liquid, or gas molecules, atoms, or ions by a solid or liquid, as opposed to absorption, the penetration of substances into the bulk of the solid or liquid. { ad'sȯrp·shən }

adsorption catalysis [PHYSICAL CHEMISTRY] A catalytic reaction in which the catalyst is an adsorbent. { ad'sorp·shən kə'tal·ə·səs }

adsorption chromatography [ANALYTICAL CHEMISTRY] Separation of a chemical mixture (gas or liquid) by passing it over an adsorbent bed which adsorbs different compounds at different rates. { ad'sȯrp·shən ‚krō·mə'täg·rə·fē }

adsorption complex [CHEMISTRY] An entity consisting of an adsorbate and that portion of the adsorbent to which it is bound. { ad'sȯrp·shən ‚käm‚pleks }

adsorption indicator [ANALYTICAL CHEMISTRY] An indicator used in solutions to detect slight excess of a substance or ion; precipitate becomes colored when the indicator is adsorbed. An example is fluorescein. { ad'sȯrp·shən ‚in·də‚kād·ər }

adsorption isobar [PHYSICAL CHEMISTRY] A graph showing how adsorption varies with some parameter, such as temperature, while holding pressure constant. { ad'sȯrp·shən 'ī·sō‚bär }

adsorption isotherm [PHYSICAL CHEMISTRY] The relationship between the gas pressure *p* and the amount *w*, in grams, of a gas or vapor taken up per gram of solid at a constant temperature. { ad'sȯrp·shən 'ī·sō‚thərm }

adsorption potential [PHYSICAL CHEMISTRY] A change in the chemical potential that occurs as an ion moves from a gas or solution phase to the surface of an adsorbent. { ad'sȯrp·shən pə‚ten·chəl }

aeration cell [PHYSICAL CHEMISTRY] An electrolytic cell whose electromotive force is due to electrodes of the same material located in different concentrations of dissolved air. Also known as oxygen cell. { e'rā·shən ‚sel }

aerogel [CHEMISTRY] A porous solid formed from a gel by replacing the liquid with a gas with little change in volume so that the solid is highly porous. { 'e·rō‚jel }

aerosol [CHEMISTRY] A gaseous suspension of ultramicroscopic particles of a liquid or a solid. { 'e rə‚sȯl }

AES [SPECTROSCOPY] *See* Auger electron spectroscopy.

affinity [CHEMISTRY] The extent to which a substance or functional group can enter into a chemical reaction with a given agent. Also known as chemical affinity. { ə'fin·əd·ē }

affinity chromatography [ANALYTICAL CHEMISTRY] A chromatographic technique that utilizes the ability of biological molecules to bind to certain ligands specifically and reversibly; used in protein biochemistry. { ə'fin·əd·ē ‚krō·mə'täg·rə·fē }

Ag [CHEMISTRY] *See* silver.

agaric acid [ORGANIC CHEMISTRY] $C_{19}H_{36}(OH)(COOH)_3$ An acid with melting point 141°C; soluble in water, insoluble in benzene; used as an irritant. Also known as agaricin. { ə'gar·ik 'as·əd }

agaricin [ORGANIC CHEMISTRY] *See* agaric acid. { ə'gar·ə·sən }

agavose [ORGANIC CHEMISTRY] $C_{12}H_{22}O_{11}$ A sugar found in the juice of the agave tree; used in medicine as a diuretic and laxative. { 'ag·ə‚vōs }

aggregate [CHEMISTRY] A group of atoms or molecules that are held together in any way, for example, a micelle. { 'ag·rə·gət }

aggregation [CHEMISTRY] A process that results in the formation of aggregates. { ‚ag·rə'gā·shən }

aging [CHEMISTRY] All irreversible structural changes that occur in a precipitate after it has formed. { 'āj iŋ }

air [CHEMISTRY] A predominantly mechanical mixture of a variety of individual gases forming the earth's enveloping atmosphere. { er }

air deficiency [CHEMISTRY] Insufficient air in an air-fuel mixture causing either incomplete fuel oxidation or lack of ignition. { 'er di‚fish·ən·sē }

air-fuel ratio [CHEMISTRY] The ratio of air to fuel by weight or volume which is significant for proper oxidative combustion of the fuel. { 'er 'fyül ˌrā·shō }
air line [SPECTROSCOPY] Lines in a spectrum due to the excitation of air molecules by spark discharges, and not ordinarily present in arc discharges. { 'er ˌlīn }
air-sensitive crystal [CHEMISTRY] A crystal that decomposes when exposed to air. { 'er ¦sen·səd·iv 'krist·əl }
air-slaked [CHEMISTRY] Having the property of a substance, such as lime, that has been at least partially converted to a carbonate by exposure to air. { 'er ˌslākt }
ajmaline [ORGANIC CHEMISTRY] $C_{20}H_{26}N_2O_2$ An amber, crystalline alkaloid obtained from *Rauwolfia* plants, especially *R. serpentina*. { 'aj·məˌlēn }
alanyl [ORGANIC CHEMISTRY] The radical CH_3CHNH_2CO—; occurs in, for example, alanyl alanine, a dipeptide. { 'al·əˌnil }
alchemy [CHEMISTRY] A speculative chemical system having as its central aims the transmutation of base metals to gold and the discovery of the philosopher's stone. { 'al·kə·mē }
alcogel [CHEMISTRY] A gel formed by an alcosol. { 'al·kəˌjel }
alcohol [ORGANIC CHEMISTRY] Any member of a class of organic compounds in which a hydrogen atom of a hydrocarbon has been replaced by a hydroxy (—OH) group. { 'al·kəˌhȯl }
alcoholate [ORGANIC CHEMISTRY] A compound formed by the reaction of an alcohol with an alkali metal. Also known as alkoxide. { ˌal·kə'hȯˌlāt }
alcoholysis [ORGANIC CHEMISTRY] The breaking of a carbon-to-carbon bond by addition of an alcohol. { ˌal·kə'hȯl·ə·səs }
alcosol [CHEMISTRY] Mixture of an alcohol and a colloid. { 'al·kəˌsȯl }
aldehyde [ORGANIC CHEMISTRY] One of a class of organic compounds containing the CHO radical. { 'al·dəˌhīd }
aldehyde ammonia [ORGANIC CHEMISTRY] $CH_3CHOHNH_2$ A white, crystalline solid with a melting point of 97°C; soluble in water and alcohol; used in organic synthesis and as a vulcanization accelerator. { 'al·dəˌhīd ə'mō·nyə }
aldehyde polymer [ORGANIC CHEMISTRY] Any of the plastics based on aldehydes, such as formaldehyde, acetaldehyde, butyraldehyde, or acrylic aldehyde (acrolein). { 'al·dəˌhīd 'päl·ə·mər }
aldicarb [ORGANIC CHEMISTRY] $C_7H_{14}N_2O_2S$ A colorless, crystalline compound with a melting point of 100°C; used as an insecticide, miticide, and nematicide to treat soil for cotton, sugarbeets, potatoes, peanuts, and ornamentals. { 'al·dəˌkärb }
aldohexose [ORGANIC CHEMISTRY] A hexose, such as glucose or mannose, containing the aldehyde group. { ˌal·dō'hekˌsōs }
aldol [ORGANIC CHEMISTRY] $CH_3CH(OH)CH_2CHO$ A colorless, thick liquid with a boiling point of 83°C; used in manufacturing rubber age resistors, accelerators, and vulcanizers. { 'alˌdȯl }
aldol condensation [ORGANIC CHEMISTRY] Formation of a β-hydroxycarbonyl compound by the condensation of an aldehyde or a ketone in the presence of an acid or base catalyst. Also known as aldol reaction. { 'alˌdȯl ˌkän·dən'sā·shən }
aldol reaction [ORGANIC CHEMISTRY] *See* aldol condensation. { 'alˌdȯl rē'ak·shən }
aldose [ORGANIC CHEMISTRY] A class of monosaccharide sugars; the molecule contains an aldehyde group. { 'alˌdōs }
Aldrin [ORGANIC CHEMISTRY] $C_{12}H_8Cl_6$ Trade name for a water-insoluble, white, crystalline compound, consisting mainly of chlorinated dimethanonaphthalene; used as a pesticide. { 'al·drən }
alfin catalyst [ORGANIC CHEMISTRY] A catalyst derived from reaction of an alkali alcoholate with an olefin halide; used to convert olefins (for example, ethylene, propylene, or butylenes) into polyolefin polymers. { 'al·fin 'kad·əˌlist }
algin [ORGANIC CHEMISTRY] *See* sodium alginate. { 'al·jən }
alginic acid [ORGANIC CHEMISTRY] $(C_6H_8O_6)_n$ An insoluble colloidal acid obtained from brown marine algae; it is hard when dry and absorbent when moist. Also known as algin. { al'jin·ik 'as·əd }
alginic acid sodium salt [ORGANIC CHEMISTRY] *See* sodium alginate. { al'jin·ik 'as·əd 'sōd·ē·əm 'sȯlt }

alicyclic [ORGANIC CHEMISTRY] **1.** Having the properties of both aliphatic and cyclic substances. **2.** Referring to a class of saturated hydrocarbon compounds whose structures contain one ring. Also known as cycloaliphatic; cycloalkane. **3.** Any one of the compounds of the alicyclic class. Also known as cyclane. { ¦al·ə¦sī·klik }

aliphatic [ORGANIC CHEMISTRY] Of or pertaining to any organic compound of hydrogen and carbon characterized by a straight chain of the carbon atoms; three subgroups of such compounds are alkanes, alkenes, and alkynes. { ¦al·ə¦fad·ik }

aliphatic acid [ORGANIC CHEMISTRY] Any organic acid derived from aliphatic hydrocarbons. { ¦al·ə¦fad·ik 'as·əd }

aliphatic acid ester [ORGANIC CHEMISTRY] Any organic ester derived from aliphatic acids. { ¦al·ə¦fad·ik 'as·əd 'es·tər }

aliphatic polycyclic hydrocarbon [ORGANIC CHEMISTRY] A hydrocarbon compound in which at least two of the aliphatic structures are cyclic or closed. { ¦al·ə¦fad·ik ˌpä·lə'sī·klik ˌhī·drə'kär·bən }

aliphatic polyene compound [ORGANIC CHEMISTRY] Any unsaturated aliphatic or alicyclic compound with more than four carbons in the chain and with at least two double bonds; for example, hexadiene. { ¦al·ə¦fad·ik 'päl·ēˌēn ˌkämˌpau̇nd }

aliphatic series [ORGANIC CHEMISTRY] A series of open-chained carbon-hydrogen compounds; the two major classes are the series with saturated bonds and the series with the unsaturated bonds. { ¦al·ə¦fad·ik 'sir·ēz }

aliquant [CHEMISTRY] A part of a sample that has been divided into a set of equal parts plus a smaller remainder part. { 'al·əˌkwänt }

aliquot [CHEMISTRY] A part of a sample that has been divided into exactly equal parts with no remainder. { 'al·əˌkwät }

alizarin [ORGANIC CHEMISTRY] $C_{14}H_6O_2(OH)_2$ An orange crystalline compound, insoluble in cold water; made synthetically from anthraquinone, used in the manufacture of dyes and red pigments. { ə'liz·ə·rən }

alizarin dye [ORGANIC CHEMISTRY] Sodium salts of sulfonic acids derived from alizarin. { ə'liz·ə·rən 'dī }

alizarin red [ORGANIC CHEMISTRY] Any of several red dyes derived from anthraquinone. { ə'liz·ə·rən 'red }

alkadiene [ORGANIC CHEMISTRY] *See* diene. { ˌal·kə'dīˌēn }

alkalescence [CHEMISTRY] The property of a substance that is alkaline, that is, having a pH greater than 7. { ˌal·kə'les·əns }

alkali [CHEMISTRY] Any compound having highly basic qualities. { 'al·kəˌlī }

alkali-aggregate reaction [CHEMISTRY] The chemical reaction of an aggregate with the alkali in a cement, resulting in a weakening of the concrete. { 'al·kəˌlī 'ag·rə·gət rē'ak·shən }

alkali alcoholate [ORGANIC CHEMISTRY] A compound formed from an alcohol and an alkali metal base; the alkali metal replaces the hydrogen in the hydroxyl group. { 'al·kəˌlī ˌal·kə'hȯˌlāt }

alkali blue [ORGANIC CHEMISTRY] The sodium salt of triphenylrosanilinesulfonic acid; used as an indicator. { 'al·kəˌlī 'blü }

alkalide [INORGANIC CHEMISTRY] A member of a class of crystalline salts with an alkali metal atom. { 'al·kəˌlīd }

alkali metal [CHEMISTRY] Any of the elements of group I in the periodic table: lithium, sodium, potassium, rubidium, cesium, and francium. { 'al·kəˌlī ˌmed·əl }

alkalimeter [ANALYTICAL CHEMISTRY] **1.** An apparatus for measuring the quantity of alkali in a solid or liquid. **2.** An apparatus for measuring the quantity of carbon dioxide formed in a reaction. { ˌal·kə'lim·əd·ər }

alkalimetry [ANALYTICAL CHEMISTRY] Quantitative measurement of the concentration of bases or the quantity of one free base in a solution; techniques include titration and other analytical methods. { ˌal·kə'lim·ə·trē }

alkaline [CHEMISTRY] **1.** Having properties of an alkali. **2.** Having a pH greater than 7. { 'al·kəˌlīn }

alkaline earth [INORGANIC CHEMISTRY] An oxide of an element of group II in the periodic table, such as barium, calcium, and strontium. Also known as alkaline-earth oxide. { ¦al·kəˌlīn 'ərth }

alkaline-earth metals [CHEMISTRY] The heaviest members of group II in the periodic table; usually calcium, strontium, magnesium, and barium. { ¦al·kə,līn 'ərth 'med·əlz }

alkaline-earth oxide [INORGANIC CHEMISTRY] *See* alkaline earth. { ¦al·kə,līn 'ərth 'äk,sīd }

alkalinity [CHEMISTRY] The property of having excess hydroxide ions in solution. { ,al·kə'lin·ə·dē }

alkaloid [ORGANIC CHEMISTRY] One of a group of nitrogenous bases of plant origin, such as nicotine, cocaine, and morphine. { 'al·kə,lȯid }

alkalometry [ANALYTICAL CHEMISTRY] The measurement of the quantity of alkaloids present in a substance. { ,al·kə'läm·ə·trē }

alkamine [ORGANIC CHEMISTRY] A compound that has both the alcohol and amino groups. Also known as amino alcohol. { 'al·kə,mēn }

alkane [ORGANIC CHEMISTRY] A member of a series of saturated aliphatic hydrocarbons having the empirical formula C_nH_{2n+2}. Also known as paraffin; paraffinic hydrocarbon. { 'al,kān }

alkannin [ORGANIC CHEMISTRY] $C_{16}H_{16}O_5$ A red powder, the coloring ingredient of alkanet; soluble in alcohol, benzene, ether, and oils; used as a coloring agent for fats and oils, wines, and wax. { al'ka·nən }

alkanolamine [ORGANIC CHEMISTRY] One of a group of viscous, water-soluble amino alcohols of the aliphatic series. { ,al·kə'näl·ə,mēn }

alkene [ORGANIC CHEMISTRY] One of a class of unsaturated aliphatic hydrocarbons containing one or more carbon-to-carbon double bonds. { 'al,kēn }

alkoxide [ORGANIC CHEMISTRY] *See* alcoholate. { al'käk,sīd }

alkoxy [ORGANIC CHEMISTRY] An alkyl radical attached to a molecule by oxygen, such as the ethoxy radical. { al'käk·sē }

alkyd resin [ORGANIC CHEMISTRY] A class of adhesive resins made from unsaturated acids and glycerol. { 'al·kəd 'rez·ən }

alkyl [ORGANIC CHEMISTRY] An organic group that results from removal of a hydrogen atom from an acyclic, saturated hydrocarbon; may be represented in a chemical formula by R—. { 'al,kil }

alkylamine [ORGANIC CHEMISTRY] A compound consisting of an alkyl group attached to the nitrogen of an amine; an example is ethylamine, $C_2H_5NH_2$. { ¦al·kəl·ə¦mēn }

alkylaryl sulfonates [ORGANIC CHEMISTRY] General name for alkylbenzene sulfonates. { ¦al·kəl·ə¦rəl 'səl·fə,nāts }

alkylate [ORGANIC CHEMISTRY] A product of the alkylation process in petroleum refining. { 'al·kə,lāt }

alkylation [ORGANIC CHEMISTRY] A chemical process in which an alkyl radical is introduced into an organic compound by substitution or addition. { ,al·kə'lā·shən }

alkylbenzene sulfonates [ORGANIC CHEMISTRY] Widely used nonbiodegradable detergents, commonly dodecylbenzene or tridecylbenzene sulfonates. { ¦al·kəl¦ben,zēn 'səl·fə,nāts }

alkylene [ORGANIC CHEMISTRY] An organic radical formed from an unsaturated aliphatic hydrocarbon; for example, the ethylene radical C_2H_3—. { 'al·kə,lēn }

alkyl halide [ORGANIC CHEMISTRY] A compound consisting of an alkyl group and a halogen; an example is ethylbromide. { 'al·kəl 'hāl,īd }

alkyloxonium ion [ORGANIC CHEMISTRY] $(ROH_2)^+$ An oxonium ion containing one alkyl group. { ¦al·kil,äk¦sō·nē·əm 'ī,än }

alkyne [ORGANIC CHEMISTRY] One of a group of organic compounds containing a carbon-to-carbon triple bond. { 'al,kīn }

allelochemistry [CHEMISTRY] The science of compounds synthesized by one organism that stimulate or inhibit other organisms. { ə¦lē·lō¦kem·ə·strē }

allene [ORGANIC CHEMISTRY] C_3H_4 An unsaturated aliphatic hydrocarbon with two double bonds. Also known as propadiene. { ,a'lēn }

allethrin [ORGANIC CHEMISTRY] An insecticide, a synthetic pyrethroid, more effective than pyrethrin. { 'al·ə·thrən }

allidochlor [ORGANIC CHEMISTRY] $C_8H_{12}NOCl$ An amber liquid having slight solubility in water; used as a preemergence herbicide for vegetable crops, soybeans, sorghum, and ornamentals. { ə'lid·ə‚klór }

allo- [CHEMISTRY] Prefix applied to the stabler form of two isomers. { 'a·lō }

allotriomorphism [CHEMISTRY] *See* allotropy. { ə¦lä·trē·ə¦mor‚fiz·əm }

allotrope [CHEMISTRY] A form of an element showing allotropy. { 'a·lə‚trōp }

allotropism [CHEMISTRY] *See* allotropy. { ‚a·lə'trä‚piz·əm }

allotropy [CHEMISTRY] The assumption by an element of two or more different forms or structures which are most frequently stable in different temperature ranges, such as different crystalline forms of carbon as charcoal, graphite, or diamond. Also known as allotriomorphism; allotropism. { ə'lä·trə·pē }

allulose [ORGANIC CHEMISTRY] $CH_2OHCO(CHOH)_3CH_2OH$ A constituent of cane sugar molasses; it is nonfermentable. { 'al·yə‚lōs }

allyl- [ORGANIC CHEMISTRY] A prefix used in names of compounds whose structure contains an allyl cation. { 'al·əl }

allylacetone [ORGANIC CHEMISTRY] $CH_2CHCH_2CH_2COCH_3$ A colorless liquid, soluble in water and organic solvents; used in pharmaceutical synthesis, perfumes, fungicides, and insecticides. { ‚al·əl'as·ə‚tōn }

allyl alcohol [ORGANIC CHEMISTRY] CH_2CHCH_2OH Colorless, pungent liquid, boiling at 96°C; soluble in water; made from allyl chloride by hydrolysis. { 'al əl 'al kə‚hól }

allylamine [ORGANIC CHEMISTRY] $CH_2CHCH_2NH_2$ A yellow oil that is miscible with water; boils at 58°C; prepared from mustard oil. { ¦al·əl·ə¦mēn }

allyl bromide [ORGANIC CHEMISTRY] C_3H_5Br A colorless to light yellow, irritating toxic liquid with a boiling point of 71.3°C; soluble in organic solvents, used in organic synthesis and for the manufacture of synthetic perfumes. { 'al·əl 'brō‚mīd }

allyl cation [ORGANIC CHEMISTRY] A carbonium cation with a structure usually represented as $CH_2—CH—CH_2^+$; attachment site is the saturated carbon atom. { 'al·əl 'kat‚ī·ən }

allyl chloride [ORGANIC CHEMISTRY] CH_2CHCH_2Cl A volatile, pungent, toxic, flammable, colorless liquid, boiling at 46°C; insoluble in water; made by chlorination of propylene at high temperatures. { 'al·əl 'klór‚īd }

allyl cyanide [ORGANIC CHEMISTRY] C_4H_5N A liquid with an onionlike odor and a boiling point of 119°C; slightly soluble in water; used as a cross-linking agent in polymerization. { 'al·əl 'sī·ə‚nīd }

allylene [ORGANIC CHEMISTRY] $CH_3C{:}CH$ An acetylenic, three-carbon hydrocarbon; a colorless gas boiling at −24°C; soluble in ether. Also known as propyne. { 'al·ə‚lēn }

allylic hydrogen [ORGANIC CHEMISTRY] In an organic molecule, a hydrogen attached to a carbon atom that is adjacent to a double bond. { ə'lil ik 'hī·drə·jən }

allylic rearrangement [ORGANIC CHEMISTRY] In a three carbon molecule, the shifting of a double bond from the 1,2 carbon position to the 2,3 position, with the accompanying migration of an entering substituent or substituent group from the third carbon to the first. { ə'lil·ik rē·ə'rānj·mənt }

allyl isothiocyanate [ORGANIC CHEMISTRY] $CH_2CH{:}CH_2NCS$ A pungent, colorless to pale-yellow liquid; soluble in alcohol, slightly soluble in water; irritating odor; boiling point 152°C; used as a fumigant and as a poison gas. Also known as mustard oil. { 'al·əl ¦ī·sō‚thī·ō'sī·ə‚nat }

allyl mercaptan [ORGANIC CHEMISTRY] CH_2CHCH_2SH A colorless liquid with a boiling point of 67-68°C; soluble in ether and alcohol; used as intermediate in pharmaceutical manufacture. { 'al·əl mər'kap‚tan }

allyl plastic [ORGANIC CHEMISTRY] *See* allyl resin. { 'al·əl 'plas·tik }

allyl resin [ORGANIC CHEMISTRY] Any of a class of thermosetting synthetic resins derived from esters of allyl alcohol or allyl chloride; used in making cast and laminated products. Also known as allyl plastic. { 'al·əl 'rez·ən }

allyl sulfide [ORGANIC CHEMISTRY] $(CH_2CHCH_2)_2S$ A colorless liquid with a garliclike odor and a boiling point of 139°C; used in synthetic oil of garlic. { 'al·əl 'səl‚fīd }

allylthiourea [ORGANIC CHEMISTRY] $C_3H_5NHCSNH_2$ A white, crystalline solid that melts at 78°C; soluble in water; used as a corrosion inhibitor. { ˌal·əlˌthī·ōˌyu̇ˈrē·ə }

allyltrichlorosilane [ORGANIC CHEMISTRY] $CH_2CHCH_2SiCl_3$ A pungent, colorless liquid with a boiling point of 117.5°C; used as an intermediate for silicones. { ˌal·əlˌtrīˌklȯrˈäs·əˌlān }

allylurea [ORGANIC CHEMISTRY] $C_4H_8N_2O$ Crystals with a melting point of 85°C; freely soluble in water and alcohol; used to manufacture allylthiourea and other corrosion inhibitors. { ˌal·əl·yu̇ˈrē·ə }

allyxycarb [ORGANIC CHEMISTRY] $C_{16}H_{22}N_2O_2$ A yellow, crystalline compound used as an insecticide for fruit orchards, vegetable crops, rice, and citrus. { əˈliks·əˌkarb }

alpha cellulose [ORGANIC CHEMISTRY] A highly refined, insoluble cellulose from which sugars, pectin, and other soluble materials have been removed. Also known as chemical cellulose. { ˈal·fə ˈsel·yəˌlōs }

alpha olefin [ORGANIC CHEMISTRY] An olefin where the unsaturation (double bond) is at the alpha position, that is, between the two end carbons of the carbon chain. { ˈal·fə ˈō·ləˌfən }

alpha position [ORGANIC CHEMISTRY] In chemical nomenclature, the position of a substituting group of atoms in the main group of a molecule; for example, in a straight-chain compound such as α-hydroxypropionic acid ($CH_3CHOHCOOH$), the hydroxyl radical is in the alpha position. { ˈal·fə pəˌzish·ən }

alternant hydrocarbon [ORGANIC CHEMISTRY] A member of a class of conjugated molecules whose carbon atoms can be divided into two sets so that members of one set are formally bonded only to members of the other set. { ˈȯl·tər·nənt ˌhī·drəˈkär·bən }

alternating copolymer [ORGANIC CHEMISTRY] A polymer formed of two different monomer molecules that alternate in sequence in the polymer chain. { ˈȯl·tərˌnād·iŋ ˌkōˈpäl·ə·mər }

alternation of multiplicities law [CHEMISTRY] The law that the periodic table arranges the elements in such a sequence that their number of orbital electrons, and hence their multiplicities, alternates between even and odd numbers. { ˌȯl·tərˈnā·shən əv ˌməl·təˈplis·əd·ēz ˌlȯ }

alum [INORGANIC CHEMISTRY] **1.** Any of a group of double sulfates of trivalent metals such as aluminum, chromium, or iron and a univalent metal such as potassium or sodium. **2.** *See* aluminum sulfate; ammonium aluminum sulfate; potassium aluminum sulfate. { ˈal·əm }

alumina [INORGANIC CHEMISTRY] Al_2O_3 The native form of aluminum oxide occurring as corundum or in hydrated forms, as a powder or crystalline substance. { əˈlüm·ə·nə }

aluminate [INORGANIC CHEMISTRY] A negative ion usually assigned the formula AlO_2^- and derived from aluminum hydroxide. { əˈlüm·əˌnāt }

alumina trihydrate [INORGANIC CHEMISTRY] $Al_2O_3 \cdot 3H_2O$, or $Al(OH)_3$ A white powder; insoluble in water, soluble in hydrochloric or sulfuric acid or sodium hydroxide; used in the manufacture of ceramic glasses and in paper coating. Also known as aluminum hydrate; aluminum hydroxide; hydrated alumina; hydrated aluminum oxide. { əˈlüm·ə·nə ˌtrīˈhīˌdrāt }

aluminium [CHEMISTRY] *See* aluminum. { ˌal·yüˈmin·ē·əm }

aluminon [ORGANIC CHEMISTRY] $C_{22}H_{23}N_3O_9$ A yellowish-brown, glassy powder that is freely soluble in water; used for the detection and colorimetric estimation of aluminum in foods, water, and tissues, and as a pharyngeal aerosol spray. { əˈlüm·əˌnän }

aluminosilicate [INORGANIC CHEMISTRY] $3Al_2O_3 \cdot 2SiO_2$ A colorless, crystalline combination of silicate and aluminate in the form of rhombic crystals. { əˈlüm·əˌnōˈsil·əˌkāt }

aluminum [CHEMISTRY] A chemical element, symbol Al, atomic number 13, and atomic weight 26.9815. Also spelled aluminium. { əˈlüm·ə·nəm }

aluminum acetate [ORGANIC CHEMISTRY] $Al(CH_3COO)_3$ A white, amorphous powder that is soluble in water; used in aqueous solution as an antiseptic. { əˈlüm·ə·nəm ˈas·əˌtāt }

aluminum ammonium sulfate [INORGANIC CHEMISTRY] *See* ammonium aluminum sulfate. { ə'lüm·ə·nəm ə'mon·ē·əm 'səl,fāt }

aluminum borohydride [ORGANIC CHEMISTRY] $Al(BH_4)_3$ A volatile liquid with a boiling point of 44.5°C; used in organic synthesis and as a jet fuel additive. { ə'lüm·ə·nəm bòr·ō'hī,drīd }

aluminum chloride [INORGANIC CHEMISTRY] $AlCl_3$ or Al_2Cl_6 A deliquescent compound in the form of white to colorless hexagonal crystals; fumes in air and reacts explosively with water; used as a catalyst. { ə'lüm·ə·nəm 'klòr,īd }

aluminum fluoride [INORGANIC CHEMISTRY] $AlF_3 \cdot 3^1/_2H_2O$ A white, crystalline powder, insoluble in cold water. { ə'lüm·ə·nəm 'flůr,īd }

aluminum fluosilicate [INORGANIC CHEMISTRY] $Al_2(SiF_6)_3$ A white powder that is soluble in hot water; used for artificial gems, enamels, and glass. Also known as aluminum silicofluoride. { ə'lüm·ə·nəm ,flü·ə'sil·i·kət }

aluminum halide [INORGANIC CHEMISTRY] A compound of aluminum with a halogen element, such as aluminum chloride. { ə'lüm·ə·nəm 'ha,līd }

aluminum hydrate [INORGANIC CHEMISTRY] *See* alumina trihydrate.

aluminum hydroxide [INORGANIC CHEMISTRY] *See* alumina trihydrate.

aluminum monostearate [ORGANIC CHEMISTRY] $Al(OH)_2[OOC(CH_2)_{16}CH_3]$ A white to yellowish-white powder with a melting point of 155°C; used in the manufacture of medicine, paint, and ink, in waterproofing, and as a plastics stabilizer. { ə'lüm·ə·nəm ,män·ō'stir,āt }

aluminum nitrate [INORGANIC CHEMISTRY] $Al(NO_3)_3 \cdot 9H_2O$ White, deliquescent crystals with a melting point of 73°C; soluble in alcohol and acetone; used as a mordant for textiles, in leather tanning, and as a catalyst in petroleum refining. { ə'lüm·ə·nəm 'nī,trāt }

aluminum oleate [ORGANIC CHEMISTRY] A soaplike compound of aluminum and oleic acid, used in lubricating oils and greases to improve their viscosity. { ə'lüm·ə·nəm 'ō·lē,āt }

aluminum orthophosphate [INORGANIC CHEMISTRY] $AlPO_4$ White crystals, melting above 1500°C; insoluble in water, soluble in acids and bases; useful in ceramics, paints, pulp, and paper. Also known as aluminum phosphate. { ə'lüm·ə·nəm ,òr·thō'fäs,fāt }

aluminum oxide [INORGANIC CHEMISTRY] Al_2O_3 A compound in the form of a white powder or colorless hexagonal crystals; melts at 2020°C; insoluble in water; used in aluminum production, paper, spark plugs, absorbing gases, light bulbs, artificial gems, and manufacture of abrasives, refractories, ceramics, and electrical insulators. { ə'lüm·ə·nəm 'äk,sīd }

aluminum palmitate [ORGANIC CHEMISTRY] $Al(C_{16}H_{31}O_2) \cdot H_2O$ An aluminum soap used in waterproofing fabrics, paper, and leather and as a drier in paints. { ə'lüm·ə·nəm 'päm·ə,tāt }

aluminum phosphate [INORGANIC CHEMISTRY] *See* aluminum orthophosphate. { ə'lüm·ə·nəm 'fäs,fāt }

aluminum potassium sulfate [INORGANIC CHEMISTRY] *See* potassium aluminum sulfate. { ə'lüm·ə·nəm pə'tas·ē·əm 'səl,fāt }

aluminum silicate [INORGANIC CHEMISTRY] $Al_2(SiO_3)_3$ A white solid that is insoluble in water; used as a refractory in glassmaking. { ə'lüm·ə·nəm 'sil·ə,kāt }

aluminum silicofluoride [INORGANIC CHEMISTRY] *See* aluminum fluosilicate. { ə'lüm·ə·nəm ¦sil·ə·kō¦flůr,īd }

aluminum soap [ORGANIC CHEMISTRY] Any of various salts of higher carboxylic acids and aluminum that are insoluble in water and soluble in oils; used in lubricating greases, paints, varnishes, and waterproofing substances. { ə'lüm·ə·nəm 'sōp }

aluminum sodium sulfate [INORGANIC CHEMISTRY] $AlNa(SO_4)_2 \cdot 12H_2O$ Colorless crystals with an astringent taste and a melting point of 61°C; soluble in water; used as a mordant and for waterproofing textiles, as a food additive, and for matches, tanning, ceramics, engraving, and water purification. Abbreviated SAS. Also known as porous alum; soda alum; sodium aluminum sulfate. { ə'lüm·ə·nəm 'sōd·ē·əm 'səl,fāt }

aluminum stearate [ORGANIC CHEMISTRY] $Al(C_{17}H_{35}COO)_3$ An aluminum soap in the form of a white powder that is insoluble in water and soluble in oils; used for waterproofing fabrics and concrete and as a drier in paints and varnishes. { ə′lüm·ə·nəm ′stir‚āt }

aluminum sulfate [INORGANIC CHEMISTRY] $Al_2(SO_4)_3 \cdot 18H_2O$ A colorless salt in the form of monoclinic crystals that decompose in heat and are soluble in water; used in papermaking, water purification, and tanning, and as a mordant in dyeing. Also known as alum. { ə′lüm·ə·nəm ′səl‚fāt }

aluminum triacetate [ORGANIC CHEMISTRY] $Al(C_2H_3O_2)_3$ A white solid that is very slightly soluble in cold water. { ə′lüm·ə·nəm ‚trī′as·ə‚tāt }

Am [CHEMISTRY] *See* americium; ammonium.

ambident [ORGANIC CHEMISTRY] Pertaining to a chemical species whose molecules possess two reactive sites. { ′am·bə·dənt }

americium [CHEMISTRY] A chemical element, symbol Am, atomic number 95; the mass number of the isotope with the longest half-life is 243. { ‚am·ə′ris·ē·əm }

americyl ion [INORGANIC CHEMISTRY] A dioxo monocation of americium, with the formula $(AmO_2)^-$. { ə′mer·ə·səl ′ī‚än }

Ames test [ANALYTICAL CHEMISTRY] A bioassay that uses a set of histidine auxotrophic mutants of *Salmonella typhimurium* for detecting mutagenic and possibly carcinogenic compounds. { ′āmz ‚test }

amicron [PHYSICAL CHEMISTRY] A particle having a size of 10^{-7} centimeter or less, which is a size in a system of classification of particle sizes in colloid chemistry. { ā′mī‚krän }

amidation [ORGANIC CHEMISTRY] The process of forming an amide; for example, in the laboratory benzyl reacts with methyl amine to form N-methylbenzamide. { ‚am·ə‚dā·shən }

amide [ORGANIC CHEMISTRY] One of a class of organic compounds containing the $CONH_2$ radical. { ′am‚īd }

amide hydrolysis [ORGANIC CHEMISTRY] The cleavage of an amide into its constitutive acid and amine fragments by a net addition of water. { ′am‚īd hī′dräl·ə·səs }

amidine [ORGANIC CHEMISTRY] A compound which contains the radical $CNHNH_2$. { ′am·ə‚dēn }

amido [ORGANIC CHEMISTRY] Indicating the NH_2 radical when it is present in a molecule with the CO radical. { ə′mē‚dō }

amidol [ORGANIC CHEMISTRY] $C_6H_3(NH_2)_2OH \cdot HCl$ A grayish-white crystalline salt; soluble in water, slightly soluble in alcohol; used as a developer in photography and as an analytical reagent. { ′am·i‚dòl }

amination [ORGANIC CHEMISTRY] **1.** The preparation of amines. **2.** A process in which the amino group ($—NH_2$) is introduced into organic molecules. { ‚am·ə′nā·shən }

amine [ORGANIC CHEMISTRY] One of a class of organic compounds which can be considered to be derived from ammonia by replacement of one or more hydrogens by functional groups. { ə′mēn }

aminiform [ORGANIC CHEMISTRY] *See* cystamine. { ə′mē·nō‚fôrm }

amino alcohol [ORGANIC CHEMISTRY] *See* alkamine. { ə′mē‚nō ′al·kə‚hòl }

amino-, amin- [CHEMISTRY] Having the property of a compound in which the group NH_2 is attached to a radical other than an acid radical. { ə′mē‚nō }

1-aminoanthraquinone [ORGANIC CHEMISTRY] $C_{14}H_9NO_2$ Ruby-red crystals with a melting point of 250°C; freely soluble in alcohol, benzene, chloroform, ether, glacial acetic acid, and hydrochloric acid; used in the manufacture of dyes and pharmaceuticals. { ¦wən ə¦mē·nō‚an·thrə·kwē′nōn }

2-amino-1-butanol [ORGANIC CHEMISTRY] $CH_3CH_2CH(NH_2)CH_2OH$ A liquid miscible with water, soluble in alcohols; used in the synthesis of surface-active agents, vulcanizing accelerators, and pharmaceuticals, and as an emulsifying agent for such products as cosmetic creams and lotions. { ¦tü ə¦mē·nō ¦wən ¦byüt·ən‚ól }

γ-aminobutyric acid [ORGANIC CHEMISTRY] $H_2NCH_2CH_2CH_2COOH$ Crystals which are either leaflets or needles, with a melting point of 202°C; thought to be a central nervous system postsynaptic inhibitory transmitter. Abbreviated GABA. { ¦gam·ə ə¦mē·nō‚byü¦tir·ik ′as·əd }

ε-aminocaproic acid [ORGANIC CHEMISTRY] $C_6H_{13}NO_2$ Crystals with a melting point of 204-206°C; freely soluble in water; used as an antifibrinolytic agent and a spacer for affinity chromatography. { ¦ep·sə,lən ə¦mē·nō·kə¦prō·ik 'as·əd }

aminocarb [ORGANIC CHEMISTRY] $C_{11}H_{16}N_2O_2$ A tan, crystalline compound with a melting point of 93-94°C; slightly soluble in water; used as an insecticide for control of forest insects and pests of cotton, tomatoes, tobacco, and fruit crops. { ə'mē·nō,kärb }

aminocide [ORGANIC CHEMISTRY] *See* succinic acid 2,2-dimethylhydrazide. { ə'mē·nō,sīd }

aminodiborane [INORGANIC CHEMISTRY] Any compound derived from diborane (B_2H_6) in which one H of the bridge has been replaced by NH_2. { ə¦mē·nō,dī¦bȯr,ān }

3-amino-2,5-dichlorobenzoic acid [ORGANIC CHEMISTRY] $C_7H_5O_2NCl_2$ A white solid with a melting point of 200-201°C; solubility in water is 700 parts per million at 20°C; used as a preemergence herbicide for soybeans, corn, and sweet potatoes. { ¦thrē ə¦mē·nō ¦tü ¦fīv dī,klȯr·ə,ben¦zō·ik 'as·əd }

aminoethane [ORGANIC CHEMISTRY] *See* ethyl amine. { ə'mē·nō,eth·ən }

amino group [ORGANIC CHEMISTRY] A functional group (—NH_2) formed by the loss of a hydrogen atom from ammonia. { ə'mē·nō ,grüp }

2-amino-2-methyl-1,3-propanediol [ORGANIC CHEMISTRY] $HOCH_2C(CH_3)(NH_2)CH_2OH$ Crystals with a melting point of 109-111°C; soluble in water and alcohol; used in the synthesis of surface-active agents, pharmaceuticals, and vulcanizers, and as an emulsifying agent for cosmetics, leather dressings, polishes, and cleaning compounds. { ¦tü ə'mē·nō ¦tü ¦meth·əl ¦wən ¦thrē ¦prō,pān'dī,ȯl }

3-amino-2-naphthoic acid [ORGANIC CHEMISTRY] $H_2NC_{10}H_6COOH$ Yellow crystals in the shape of scales with a melting point of 214°C; soluble in alcohol and ether; used in the determination of copper, nickel, and cobalt. { ¦thrē ə'mē·nō ¦tü naf'thō·ik 'as·əd }

1-amino-2-naphthol-4-sulfonic acid [ORGANIC CHEMISTRY] $H_2NC_{10}H_5(OH)SO_3H$ White or gray, needlelike crystals; soluble in hot sodium bisulfite solutions; used in the manufacture of azo dyes. { ¦wən ə'mē·nō ¦tü 'naf,thȯl ¦fȯr səl'fän·ik 'as·əd }

2-amino-5-naphthol-7-sulfonic acid [ORGANIC CHEMISTRY] $C_{10}H_5NH_2OHSO_3H$ Gray or white needles that are soluble in hot water; used as a dye intermediate. { ¦tü ə'mē·nō ¦fīv 'naf,thȯl ¦sev·ən səl'fän·ik 'as·əd }

amino nitrogen [CHEMISTRY] Nitrogen combined with hydrogen in the amino group. Also known as ammonia nitrogen. { ə'mē·nō 'nī·trə·jən }

aminophenol [ORGANIC CHEMISTRY] A type of compound containing the NH_2 and OH groups joined to the benzene ring; examples are *para*-aminophenol and *ortho*-hydroxylaniline. { ə,mē·nō'fē,nȯl }

***para*-aminophenol** [ORGANIC CHEMISTRY] $p\text{-}HOC_6H_4NO_2$ A phenol in which an amino (—NH_2) group is located on the benzene ring of carbon atoms para (*p*) to the hydroxyl (—OH) group; used as a photographic developer and as an intermediate in dye manufacture. { ¦par·ə ə,mē·nō'fē·nȯl }

3-aminophthalic hydrazide [ORGANIC CHEMISTRY] *See* luminol. { ¦thrē ə¦mē·nō¦thal·ik 'hī·drə·zīd }

2-aminopropane [ORGANIC CHEMISTRY] *See* isopropylamine. { ¦tü ə,mē·nō'prō pān }

β-aminopyridine [ORGANIC CHEMISTRY] $C_5H_6N_2$ Crystals with a melting point of 64°C; soluble in water, alcohol, and benzene; used in drug and dye manufacture. Also known as 3-aminopyridine. { ¦bād·ə ə,mē·nō'pī·rə,dēn }

3-aminopyridine [ORGANIC CHEMISTRY] *See* β-aminopyridine. { ¦thrē ə,mē·nō'pī·rə,dēn }

4-aminopyridine [ORGANIC CHEMISTRY] $C_5H_6N_2$ White crystals with a melting point of 158.9°C; soluble in water; used as a repellent for birds. Abbreviated 4-AP. { ¦fȯr ə,mē·nō'pī·rə,dēn }

amino resin [ORGANIC CHEMISTRY] A type of resin prepared by condensation polymerization, with an aldehyde, of a compound containing an amino group. { ə'mē·nō 'rez·ən }

2-aminothiazole [ORGANIC CHEMISTRY] $C_3H_4N_2S$ Pale-yellow crystals that melt at 92°C; soluble in cold water, slightly soluble in ethyl alcohol; used as an intermediate in the synthesis of sulfathiazole. { ¦tü ə,mē·nō'thī·ə,zȯl }

aminotriazole [ORGANIC CHEMISTRY] $C_2H_4N_4$ Crystals with a melting point of 159°C; soluble in water, methanol, chloroform, and ethanol; used as an herbicide, cotton plant defoliant, and growth regulator for annual grasses and broadleaf and aquatic weeds. Abbreviated ATA. { ə¦mē·nō′trī·ə‚zȯl }

ammine [INORGANIC CHEMISTRY] One of a group of complex compounds formed by coordination of ammonia molecules with metal ions. { ′a‚mēn }

ammonation [INORGANIC CHEMISTRY] A reaction in which ammonia is added to other molecules or ions by covalent bond formation utilizing the unshared pair of electrons on the nitrogen atom, or through ion-dipole electrostatic interactions. { ‚a·mə′nā·shən }

ammonia [INORGANIC CHEMISTRY] NH_3 A colorless gaseous alkaline compound that is very soluble in water, has a characteristic pungent odor, is lighter than air, and is formed as a result of the decomposition of most nitrogenous organic material; used as a fertilizer and as a chemical intermediate. { ə′mōn·yə }

ammonia alum [INORGANIC CHEMISTRY] *See* ammonium aluminum sulfate. { ə′mōn·yə ′al·əm }

ammoniac [INORGANIC CHEMISTRY] *See* ammoniacal. { ə′mōn·ē‚ak }

ammoniacal [INORGANIC CHEMISTRY] Pertaining to ammonia or its properties. Also known as ammoniac. { ¦a·mə¦nī·ə·kəl }

ammonia dynamite [CHEMISTRY] Dynamite with part of the nitroglycerin replaced by ammonium nitrate. { ə′mōn·yə ′dī·nə‚mīt }

ammoniated mercury [INORGANIC CHEMISTRY] $HgNH_2Cl$ A white powder that darkens on light exposure; insoluble in water and alcohol, soluble in ammonium carbonate solutions and in warm acids; used in pharmaceuticals and as a local anti-infective in medicine. { ə′mōn·ē·ād·əd ′mər·kyə·rē }

ammoniated superphosphate [INORGANIC CHEMISTRY] A fertilizer containing 5 parts of ammonia to 100 parts of superphosphate. { ə′mōn·ē·ād·əd ‚sü·pər′fäs‚fāt }

ammoniation [CHEMISTRY] Treating or combining with ammonia. { ə‚mōn·ē′ā·shən }

ammonia water [CHEMISTRY] A water solution of ammonia; a clear colorless liquid that is basic because of dissociation of NH_4OH to produce hydroxide ions; used as a reagent, solvent, and neutralizing agent. { ə′mōn·yə ‚wȯd·ər }

ammonification [CHEMISTRY] Addition of ammonia or ammonia compounds, especially to the soil. { ə‚män·ə·fə′kā·shən }

ammonium acetate [ORGANIC CHEMISTRY] **1.** CH_3COONH_4 A normal salt formed by the neutralization of acetic acid with ammonium hydroxide; a white, crystalline, deliquescent material used in solution for the standardization of electrodes for hydrogen ions. **2.** $CH_3COONH_4 \cdot CH_3COOH$ An acid salt resulting from the distillation of the neutral salt or from its solution in hot acetic acid; crystallizes in deliquescent needles. **3.** A mixture of the normal and acetic salts; used as a mordant in the dyeing of wool. { ə′mōn·yəm ′as·ə‚tāt }

ammonium alginate [ORGANIC CHEMISTRY] $(C_6H_7O_6 \cdot NH_4)_n$ A high-molecular-weight, hydrophilic colloid; used as a thickening agent/stabilizer in ice cream, cheese, canned fruits, and other food products. { ə′mōn·yəm ′al·jə‚nāt }

ammonium alum [INORGANIC CHEMISTRY] *See* ammonium aluminum sulfate. { ə′mōn·yəm ′al·əm }

ammonium aluminum sulfate [INORGANIC CHEMISTRY] $NH_4Al(SO_4)_2 \cdot 12H_2O$ Colorless, odorless crystals that are soluble in water; used in manufacturing medicines and baking powder, dyeing, papermaking, and tanning. Also known as alum; aluminum ammonium sulfate; ammonia alum; ammonium alum. { ə′mōn·yəm ə¦lü·mə·nəm ′səl‚fāt }

ammonium benzoate [ORGANIC CHEMISTRY] $NH_4C_7H_5O_2$ A salt of benzoic acid prepared as a coarse, white powder; used as a preservative in certain adhesives and rubber latex. { ə′mōn·yəm ′ben·zə‚wāt }

ammonium bicarbonate [INORGANIC CHEMISTRY] NH_4HCO_3 White, crystalline, water-soluble salt; used in baking powders and in fire-extinguishing mixtures. Also known as ammonium hydrogen carbonate. { ə′mōn·yəm bī′kär·bə‚nāt }

ammonium bichromate [INORGANIC CHEMISTRY] *See* ammonium dichromate. { ə'mōn·yəm bī'krō,māt }

ammonium bifluoride [INORGANIC CHEMISTRY] $NH_4F \cdot HF$ A salt that crystallizes in the orthorhombic system and is soluble in water; prepared in the form of white flakes from ammonia treated with hydrogen fluoride; used in solution as a fungicide and wood preservative. Also known as ammonium acid fluoride; ammonium hydrogen fluoride. { ə'mōn·yəm bī'flu̇r,īd }

ammonium bitartrate [ORGANIC CHEMISTRY] $NH_4HC_4H_4O_6$ Colorless crystals that are soluble in water; used to make baking powder and to detect calcium. Also known as monoammonium tartrate. { ə'mōn·yəm bī'tär,trāt }

ammonium borate [INORGANIC CHEMISTRY] NH_4BO_3 A white, crystalline, water-soluble salt which decomposes at 198°C; used as a fire retardant on fabrics. { ə'mōn·yəm 'bȯr,āt }

ammonium bromide [INORGANIC CHEMISTRY] NH_4Br An ammonium halide that crystallizes in the cubic system; made by the reaction of ammonia with hydrobromic acid or bromine; used in photography and for pharmaceutical preparations (sedatives). { ə'mōn·yəm 'brō,mīd }

ammonium carbamate [INORGANIC CHEMISTRY] $NH_4NH_2CO_2$ A salt that forms colorless, rhombic crystals, which are very soluble in cold water; an important, unstable intermediate in the manufacture of urea; found in commercial ammonium carbonate. { ə'mōn·yəm 'kär·bə,māt }

ammonium carbonate [INORGANIC CHEMISTRY] **1.** $(NH_4)_2CO_3$ The normal ammonium salt of carbonic acid, prepared by passing gaseous carbon dioxide into an aqueous solution of ammonia and allowing the vapors (ammonia, carbon dioxide, water) to crystallize. **2.** $NH_4HCO_3 \cdot NH_2COONH_4$ A white, crystalline double salt of ammonium bicarbonate and ammonium carbamate obtained commercially; the principal ingredient of smelling salts. { ə'mōn·yəm 'kär·bə,nāt }

ammonium chloride [INORGANIC CHEMISTRY] NH_4Cl A white crystalline salt that occurs naturally as a sublimation product of volcanic action or is manufactured; used as an electrolyte in dry cells, as a flux for soldering, tinning, and galvanizing, and as an expectorant. { ə'mōn·yəm 'klȯr,īd }

ammonium chromate [INORGANIC CHEMISTRY] $(NH_4)_2CrO_4$ A salt that forms yellow, monoclinic crystals; made from ammonium hydroxide and ammonium dichromate; used in photography as a sensitizer for gelatin coatings. { ə'mōn·yəm 'krō,māt }

ammonium citrate [ORGANIC CHEMISTRY] $(NH_4)_2HC_6H_5O_7$ White, granular material; used as a reagent. { ə'mōn·yəm 'sī,trāt }

ammonium dichromate [INORGANIC CHEMISTRY] $(NH_4)_2Cr_2O_7$ A salt that forms orange, monoclinic crystals; made from ammonium sulfate and sodium dichromate; soluble in water and alcohol; ignites readily; used in photography, lithography, pyrotechnics, and dyeing. Also known as ammonium bichromate. { ə'mōn·yəm dī'krō,māt }

ammonium fluoride [INORGANIC CHEMISTRY] NH_4F A white, unstable, crystalline salt with a strong odor of ammonia; soluble in cold water; used in analytical chemistry, glass etching, and wood preservation, and as a textile mordant. { ə'mon·yəm 'flu̇r,īd }

ammonium fluosilicate [INORGANIC CHEMISTRY] $(NH_4)_2SiF_6$ A toxic, white, crystalline powder; soluble in alcohol and water; used for mothproofing, glass etching, and electroplating. Also known as ammonium silicofluoride. { ə'mōn·yəm ,flü·ə'sil·ə,kāt }

ammonium formate [ORGANIC CHEMISTRY] HCO_2NH_4 Deliquescent crystals or granules with a melting point of 116°C; soluble in water and alcohol; used in analytical chemistry to precipitate base metals from salts of the noble metals. { ə'mōn·yəm 'fȯr·,māt }

ammonium gluconate [ORGANIC CHEMISTRY] $NH_4C_6H_{11}O_7$ A white, crystalline powder made from gluconic acid and ammonia; soluble in water; used as an emulsifier for cheese and salad dressing and as a catalyst in textile printing. { ə'mōn·yəm 'glü·kə,nāt }

ammonium halide [INORGANIC CHEMISTRY] A compound with the ammonium ion bonded to an ion formed from one of the halogen elements. { ə'mōn·yəm 'hal,īd }

ammonium hydrogen carbonate [INORGANIC CHEMISTRY] *See* ammonium bicarbonate. { ə'mōn·yəm 'hi·drə·jən 'kär·bə,nāt }

ammonium hydrogen fluoride [INORGANIC CHEMISTRY] *See* ammonium bifluoride. { ə'mōn·yəm 'hi·drə·jən 'flu̇r,īd }

ammonium hydroxide [INORGANIC CHEMISTRY] NH_4OH A hydrate of ammonia, crystalline below −79°C; it is a weak base known only in solution as ammonia water. Also known as aqua ammonia. { ə'mōn·yəm ,hī'dräk,sīd }

ammonium iodide [INORGANIC CHEMISTRY] NH_4I A salt prepared from ammonia and hydrogen iodide or iodine; it forms colorless, regular crystals which sublime when heated; used in photography and for pharmaceutical preparations. { ə'mōn·yəm 'ī·ə,dīd }

ammonium lactate [ORGANIC CHEMISTRY] $NH_4C_3H_5O_3$ A yellow, syrupy liquid used in finishing leather. { ə'mōn·yəm 'lak,tāt }

ammonium lineolate [ORGANIC CHEMISTRY] $C_{17}H_{31}COONH_4$ A soft, pasty material used as an emulsifying agent in various industrial applications. { ə'mōn·yəm lə'nē·ə,lāt }

ammonium metatungstate [ORGANIC CHEMISTRY] $(NH_4)_6H_2W_{12}O_{40}$ A white powder, soluble in water, used for electroplating. { ə'mōn·yəm ,med·ə'təŋ,stāt }

ammonium molybdate [INORGANIC CHEMISTRY] $(NH_4)_2MoO_4$ White, crystalline salt used as an analytic reagent, as a precipitant of phosphoric acid, and in pigments. { ə'mōn·yəm mə'lib,dāt }

ammonium nickel sulfate [INORGANIC CHEMISTRY] *See* nickel ammonium sulfate. { ə'mōn·yəm ¦nik·əl 'səl,fāt }

ammonium nitrate [INORGANIC CHEMISTRY] NH_4NO_3 A colorless crystalline salt; very insensitive and stable high explosive; also used as a fertilizer. { ə'mōn·yəm 'nī,trāt }

ammonium oxalate [ORGANIC CHEMISTRY] $(NH_4)_2C_2O_4 \cdot H_2O$ A salt in the form of colorless, rhombic crystals. { ə'mōn·yəm 'äk·sə,lāt }

ammonium perchlorate [INORGANIC CHEMISTRY] NH_4ClO_4 A salt that forms colorless or white rhombic and regular crystals, which are soluble in water; it decomposes at 150°C, and the reaction is explosive at higher temperatures. { ə'mōn·yəm pər'klȯr,āt }

ammonium persulfate [INORGANIC CHEMISTRY] $(NH_4)_2S_2O_8$ White crystals which decompose on melting; soluble in water; used as an oxidizing agent and bleaching agent, and in etching, electroplating, food preservation, and aniline dyes. { ə'mōn·yəm pər'səl,fāt }

ammonium phosphate [INORGANIC CHEMISTRY] $(NH_4)_2HPO_4$ A salt of ammonia and phosphoric acid that forms white monoclinic crystals, which are soluble in water; used as a fertilizer and fire retardant. { ə'mōn·yəm 'fäs,fāt }

ammonium picrate [ORGANIC CHEMISTRY] $NH_4C_6H_2O(NO_2)_3$ Compound with stable yellow and metastable red forms of orthorhombic crystals; used as a military explosive for armor-piercing shells. { ə'mōn·yəm 'pik,rāt }

ammonium salt [INORGANIC CHEMISTRY] A product of a reaction between ammonia and various acids; examples are ammonium chloride and ammonium nitrate. { ə'mōn·yəm 'sȯlt }

ammonium silicofluoride [INORGANIC CHEMISTRY] *See* ammonium fluosilicate. { ə'mōn·yəm ¦sil·ə·kō'flu̇r,īd }

ammonium soap [ORGANIC CHEMISTRY] A product from reaction of a fatty acid with ammonium hydroxide; used in toiletry preparations such as soaps and in emulsions. { ə'mōn·yəm 'sōp }

ammonium stearate [ORGANIC CHEMISTRY] $C_{17}H_{35}COONH_4$ A tan, waxlike substance with a melting point of 73-75°C; used in cosmetics and for waterproofing cements, paper, textiles, and other materials. { ə'mōn·yəm 'stir,āt }

ammonium sulfamate [INORGANIC CHEMISTRY] $NH_4OSO_2NH_2$ White crystals with a melting point of 130°C; soluble in water; used for flameproofing textiles, in electroplating, and as an herbicide to control woody plant species. { ə'mōn·yəm 'səl·fə‚māt }

ammonium sulfate [INORGANIC CHEMISTRY] $(NH_4)_2SO_4$ Colorless, rhombic crystals which melt at 140°C and are soluble in water. { ə'mōn·yəm 'səl‚fāt }

ammonium sulfide [INORGANIC CHEMISTRY] $(NH_4)_2S$ Yellow crystals, stable only when dry and below 0°C; decomposes on melting; soluble in water and alcohol; used in photographic developers and for coloring brasses and bronzes. { ə'mōn·yəm 'səl‚fīd }

ammonium tartrate [ORGANIC CHEMISTRY] $C_4H_{12}N_2O_6$ Colorless, monoclinic crystals; used in textiles and in medicine. { ə'mōn·yəm 'tär‚trāt }

ammonium thiocyanate [ORGANIC CHEMISTRY] NH_4SCN Colorless, deliquescent crystals with a melting point of 149.6°C; soluble in water, acetone, alcohol, and ammonia; used in analytical chemistry, freezing solutions, fabric dyeing, electroplating, photography, and steel pickling. { ə'mōn·yəm ‚thī·ō'sī·ə‚nāt }

ammonium vanadate [INORGANIC CHEMISTRY] NH_4VO_3 A white to yellow, water-soluble, crystalline powder; used in inks and as a paint drier and textile mordant. { ə'mōn·yəm 'van·ə‚dāt }

ammonolysis [CHEMISTRY] **1.** A dissociation reaction of the ammonia molecule producing H^+ and NH_2^- species. **2.** Breaking of a bond by addition of ammonia. { ‚a mə'näl·ə·səs }

amount of substance [CHEMISTRY] A measure of the number of elementary entities present in a substance or system; usually measured in moles. { ə'maůnt əv 'səb·stəns }

amperometric titration [PHYSICAL CHEMISTRY] A titration that involves measuring an electric current or changes in current during the course of the titration. { ¦am·pə·rə¦me·trik tī'trā·shən }

amperometry [PHYSICAL CHEMISTRY] Chemical analysis by techniques which involve measuring electric currents. { ‚am·pə'rä·me·trē }

amphipathic molecule [ORGANIC CHEMISTRY] A molecule having both hydrophilic and hydrophobic groups; examples are wetting agents and membrane lipids such as phosphoglycerides. { ‚am·fə'path·ik 'mäl·ə‚kyül }

amphiphile [CHEMISTRY] A molecule which has a polar head attached to a long hydrophobic tail. { 'am·fə‚fīl }

amphiprotic [CHEMISTRY] *See* amphoteric. { ¦am·fə¦präd·ik }

ampholyte [CHEMISTRY] An amphoteric electrolyte. { 'am·fə‚līt }

ampholytic detergent [CHEMISTRY] A detergent that is cationic in acidic solutions and anionic in basic solutions. { ¦am·fə¦lid·ik di'tər·jənt }

amphoteric [CHEMISTRY] Having both acidic and basic characteristics. Also known as amphiprotic. { ¦am·fə¦ter·ik }

amphoterism [CHEMISTRY] The property of being able to react either as an acid or a base. { am'täd·ə‚riz·əm }

amyl [ORGANIC CHEMISTRY] Any of the eight isomeric arrangements of the radical C_5H_{11} or a mixture of them. Also known as pentyl. { 'am·əl }

amyl acetate [ORGANIC CHEMISTRY] $CH_3COO(CH_2)_2CH(CH_3)_2$ A colorless liquid, boiling at 142°C; soluble in alcohol and ether, slightly soluble in water; used in flavors and perfumes. Also known as banana oil; isoamyl acetate. { 'am·əl 'as·ə‚tāt }

amyl alcohol [ORGANIC CHEMISTRY] **1.** A colorless liquid that is a mixture of isomeric alcohols. **2.** An optically active liquid composed of isopentyl alcohol and active amyl alcohol. { 'am·əl 'al·kə‚hōl }

***n*-amylamine** [ORGANIC CHEMISTRY] $C_5H_{11}NH_2$ A colorless liquid with a boiling point of 104.4°C; soluble in water, alcohol, and ether; used in dyestuffs, insecticides, synthetic detergents, corrosion inhibitors, and pharmaceuticals, and as a gasoline additive. { ¦en ə'mil·ə‚mēn }

amyl benzoate [ORGANIC CHEMISTRY] *See* isoamyl benzoate. { 'am·əl 'ben·zə‚wāt }

amylene [ORGANIC CHEMISTRY] C_5H_{10} A highly flammable liquid with a low boiling point, 37.5–38.5°C; often a component of petroleum. Also known as 2-methyl-2-butene. { 'am·ə‚lēn }

amyl ether [ORGANIC CHEMISTRY] **1.** Either of two isomeric compounds, *n*-amyl ether or isoamyl ether; both may be represented by the formula $(C_5H_{11})_2O$. **2.** A mixture mainly of isoamyl ether and *n*-amyl ether formed in preparation of amyl alcohols from amyl chloride; very slightly soluble in water; used mainly as a solvent. { 'am·əl 'ē·thər }

amyl mercaptan [ORGANIC CHEMISTRY] $C_5H_{11}SH$ A colorless to light yellow liquid with a boiling range of 104-130°C; soluble in alcohol; used in odorant for detecting gas line leaks. { 'am·əl mər'kap‚tan }

amyl nitrate [ORGANIC CHEMISTRY] $C_5H_{11}ONO_2$ An ester of amyl alcohol added to diesel fuel to raise the cetane number. { 'am·əl 'nī‚trāt }

amyl nitrite [ORGANIC CHEMISTRY] $(CH_3)_2CH(CH_2)_2NO_2$ A yellow liquid; soluble in alcohol, very slightly soluble in water; fruity odor; it is flammable and the vapor is explosive; used in medicine and perfumes. Also known as isoamyl nitrite. { 'am·əl 'nī‚trīt }

amyl propionate [ORGANIC CHEMISTRY] $CH_3CH_2COOC_5H_{11}$ A colorless liquid with an applelike odor and a distillation range of 135-175°C; used in perfumes, lacquers, and flavors. { 'am·əl 'prō·pē·ə‚nāt }

amyl salicylate [ORGANIC CHEMISTRY] $C_6H_4OHCOOC_5H_{11}$ A clear liquid that occasionally has a yellow tinge; boils at 280°C; soluble in alcohol, insoluble in water; used in soap and perfumes. Also known as isoamyl salicylate. { 'am·əl sə'lis·ə‚lāt }

amyl xanthate [ORGANIC CHEMISTRY] A salt formed by replacing the hydrogen attached to the sulfur in amylxanthic acid by a metal; used as collector agent in the flotation of certain minerals. { 'am·əl 'zan‚thāt }

anabasine [ORGANIC CHEMISTRY] A colorless, liquid alkaloid extracted from the plants *Anabasis aphylla* and *Nicotiana glauca*; boiling point is 105°C; soluble in alcohol and ether; used as an insecticide. { ə'na·bə‚sēn }

analog [CHEMISTRY] A compound whose structure is similar to that of another compound but whose composition differs by one element. { 'an·əl‚äg }

analysis [ANALYTICAL CHEMISTRY] The determination of the composition of a substance. { ə'nal·ə·səs }

analysis line [SPECTROSCOPY] The spectral line used in determining the concentration of an element in spectrographic analysis. { ə'nal·ə·səs ‚līn }

analyte [ANALYTICAL CHEMISTRY] **1.** The sample being analyzed. **2.** The specific component that is being measured in a chemical analysis. { 'an·ə‚līt }

analytical blank [ANALYTICAL CHEMISTRY] *See* blank. { ‚an·əl'id·ə·kəl 'blaŋk }

analytical chemistry [CHEMISTRY] The branch of chemistry dealing with techniques which yield any type of information about chemical systems. { ‚an·əl'id·ə·kəl 'kem·ə·strē }

analytical distillation [ANALYTICAL CHEMISTRY] Precise resolution of a volatile liquid mixture into its components; the mixture is vaporized by heat or vacuum, and the vaporized components are recondensed into liquids at their respective boiling points. { ‚an·əl'id·ə·kəl ‚dis·tə'lā·shən }

analytical extraction [ANALYTICAL CHEMISTRY] Precise transfer of one or more components of a mixture (liquid to liquid, gas to liquid, solid to liquid) by contacting the mixture with a solvent in which the component of interest is preferentially soluble. { ‚an·əl'id·ə·kəl ik'strak·shən }

anaphoresis [PHYSICAL CHEMISTRY] Upon application of an electric field, the movement of positively charged colloidal particles or macromolecules suspended in a liquid toward the anode. { ¦an·ə·fə'rē·səs }

anchimeric assistance [ORGANIC CHEMISTRY] The participation by a neighboring group in the rate-determining step of a reaction; most often encountered in reactions of carbocation intermediates. Also known as neighboring-group participation. { ¦aŋ·kə¦mer·ik ə'sis·təns }

anchored catalyst [CHEMISTRY] *See* immobilized catalyst. { 'aŋ·kərd 'kad·ə‚list }

anethole [ORGANIC CHEMISTRY] $C_{10}H_{12}O$ White crystals that melt at 22.5°C; very slightly soluble in water; affected by light; odor resembles oil of anise; used in perfumes and flavors, and as a sensitizer in color-bleaching processes in color photography. { 'an·ə‚thōl }

angle-resolved photoelectron spectroscopy [SPECTROSCOPY] A type of photoelectron spectroscopy which measures the kinetic energies of photoelectrons emitted from a solid surface and the angles at which they are emitted relative to the surface. Abbreviated ARPES. { 'aŋ·gəl ri'zälvd ¦fōd·ō·ə'lek‚trän ‚spek'träs·kə·pē }

anharmonic oscillator spectrum [SPECTROSCOPY] A molecular spectrum which is significantly affected by anharmonicity of the forces between atoms in the molecule. { ‚an·här¦män·ik ¦äs·ə‚lād·ər ¦spek·trəm }

anhydride [CHEMISTRY] A compound formed from an acid by removal of water. { an'hī‚drīd }

anhydrous [CHEMISTRY] Being without water, especially water of hydration. { an'hī·drəs }

anhydrous alcohol [ORGANIC CHEMISTRY] *See* absolute alcohol. { an'hī·drəs 'al·kə‚hól }

anhydrous ammonia [INORGANIC CHEMISTRY] Liquid ammonia, a colorless liquid boiling at −33.3°C. { an'hī·drəs ə'mōn·yə }

anhydrous ferric chloride [INORGANIC CHEMISTRY] *See* ferric chloride. { an'hī·drəs ‚fer·ik 'klór‚īd }

anhydrous hydrogen chloride [INORGANIC CHEMISTRY] HCl Hazardous, toxic, colorless gas used in polymerization, isomerization, alkylation, nitration, and chlorination reactions; becomes hydrochloric acid in aqueous solutions. { an'hī·drəs ‚hī·drə·jən 'klór‚īd }

anhydrous phosphoric acid [INORGANIC CHEMISTRY] *See* phosphoric anhydride. { an'hī·dres fä'sfórik 'as·əd }

anhydrous plumbic acid [INORGANIC CHEMISTRY] *See* lead dioxide. { an'hī·drəs 'pləmb·ik 'as·əd }

anhydrous sodium carbonate [INORGANIC CHEMISTRY] *See* soda ash. { an'hi·dres ‚sōd·ē·əm 'kärb·ə‚nāt }

anhydrous sodium sulfate [INORGANIC CHEMISTRY] Na_2SO_4 Water-soluble, white crystals with bitter, salty taste; melts at 888°C; used in the manufacture of glass, paper, pharmaceuticals, and textiles, and as an analytical reagent. { an'hī·drəs ‚sōd·ē·əm 'səl‚fāt }

anilazine [ORGANIC CHEMISTRY] $C_9H_5Cl_3N_4$ A tan solid with a melting point of 159–160°C; used for fungal diseases of lawns, turf, and vegetable crops. { ə'nil·ə‚zēn }

anilide [ORGANIC CHEMISTRY] A compound that has the $C_6H_5NH_2$— group; an example is benzanilide, $C_6H_5NHCOC_6H_5$. { 'an·əl‚īd }

aniline [ORGANIC CHEMISTRY] $C_6H_5NH_2$ An aromatic amine compound that is a pale brown liquid at room temperature; used in the dye, pharmaceutical, and rubber industries. { 'an·əl·ən }

aniline black [ORGANIC CHEMISTRY] A black dye produced on certain textiles, such as cotton, by oxidizing aniline or aniline hydrochloride. { 'an·əl·ən 'blak }

aniline dye [ORGANIC CHEMISTRY] A dye derived from aniline. { 'an·əl·ən 'dī }

aniline-formaldehyde resin [CHEMISTRY] A thermoplastic resin made by polymerizing aniline and formaldehyde. { 'an·əl·ən ‚fór'mal·də‚hīd ‚rez·ən }

aniline hydrochloride [ORGANIC CHEMISTRY] $C_6H_5NH_2 \cdot HCl$ White crystals, although sometimes the commercial variety has a greenish tinge, melting point 198°C; soluble in water and ethanol; used in dye manufacture, dyeing, and printing. { 'an·əl·ən ‚hī·drə'klór‚īd }

aniline *N,N*-dimethyl [ORGANIC CHEMISTRY] *See* N,N-dimethylaniline. { 'an·əl·ən ¦en ¦en di'meth·əl }

animal black [CHEMISTRY] Finely divided carbon made by calcination of animal bones or ivory; used for pigments, decolorizers, and purifying agents; varieties include bone black and ivory black. { 'an·ə·məl ‚blak }

animal charcoal [CHEMISTRY] Charcoal obtained by the destructive distillation of animal matter at high temperatures; used to adsorb organic coloring matter. { 'an·ə·məl 'chär‚kōl }

anion [CHEMISTRY] An ion that is negatively charged. { 'a‚nī·ən }

anion exchange [CHEMISTRY] A type of ion exchange in which the immobilized functional groups on the solid resin are positive. { 'a‚nī·ən iks'chānj }

anionic polymerization [ORGANIC CHEMISTRY] A type of polymerization in which Lewis bases, such as alkali metals and metallic alkyls, act as catalysts. { ¦a‚nī¦än·ik pə ‚lim·ə·rə'zā·shən }

anionotropy [CHEMISTRY] The breaking off of an ion such as hydroxyl or bromide from a molecule so that a positive ion remains in a state of dynamic equilibrium. { ¦a‚nī·ə'nä·trə·pē }

anisaldehyde [ORGANIC CHEMISTRY] $C_6H_4(OCH_3)CHO$ A compound with melting point 2.5°C, boiling point 249.5°C; insoluble in water, soluble in alcohol and ether; used in perfumery and flavoring, and as an intermediate in production of antihistamines. { ¦a·nəs'al·də‚hīd }

anisic acid [ORGANIC CHEMISTRY] $CH_3OC_6H_4COOH$ White crystals or powder with a melting point of 184°C; soluble in alcohol and ether; used in medicine and as an insect repellent and ovicide. { ə'nis·ik 'as·əd }

anisic alcohol [ORGANIC CHEMISTRY] $C_8H_{10}O_2$ A colorless liquid that boils in the range 255-265°C; it is obtained by reduction of anisic aldehyde; used in perfumery, and as an intermediate in the manufacture of pharmaceuticals. { ə'nis·ik 'al·kə‚hól }

anisole [ORGANIC CHEMISTRY] $C_6H_5OCH_3$ A colorless liquid that is soluble in ether and alcohol, insoluble in water; boiling point is 155°C; vapors are highly toxic; used as a solvent and in perfumery. { 'an·ə‚sōl }

annular atoms [ORGANIC CHEMISTRY] The atoms in a cyclic compound that are members of the ring. { 'an·yə·lər 'ad·əmz }

annulene [ORGANIC CHEMISTRY] One of a group of monocyclic conjugated hydrocarbons which have the general formula $[—CH{=}CH—]_n$. { 'an·yə‚lēn }

anode [PHYSICAL CHEMISTRY] The positive terminal of an electrolytic cell. { 'a‚nōd }

anode-corrosion efficiency [PHYSICAL CHEMISTRY] The ratio of actual weight loss of an anode due to corrosion to the theoretical loss as calculated by Faraday's law. { 'a‚nōd kə¦rō·zhən i‚fish·ən·sē }

anode effect [PHYSICAL CHEMISTRY] A condition produced by polarization of the anode in the electrolysis of fused salts and characterized by a sudden increase in voltage and a corresponding decrease in amperage. { 'a‚nōd i‚fekt }

anode film [CHEMISTRY] The portion of solution in immediate contact with the anode. { 'a‚nōd ‚film }

anodic polarization [PHYSICAL CHEMISTRY] The change in potential of an anode caused by current flow. { ə'näd·ik pō·lə·rə'zā·shən }

anolyte [CHEMISTRY] The part of the electrolyte at or near the anode that is changed in composition by the reactions at the anode. { 'an·ə‚līt }

anomalous Zeeman effect [SPECTROSCOPY] A type of splitting of spectral lines of a light source in a magnetic field which occurs for any line arising from a combination of terms of multiplicity greater than one; due to a nonclassical magnetic behavior of the electron spin. { ə'näm·ə·ləs 'zā‚män i‚fekt }

anomer [ORGANIC CHEMISTRY] One of a pair of isomers of cyclic carbohydrates; resulting from creation of a new point of symmetry when a rearrangement of the atoms occurs at the aldehyde or ketone position. { 'an·ə·mər }

antacid [CHEMISTRY] Any substance that counteracts or neutralizes acidity. { 'ant 'as·əd }

anthracene [ORGANIC CHEMISTRY] $C_{14}H_{10}$ A crystalline tricyclic aromatic hydrocarbon, colorless when pure, melting at 218°C and boiling at 342°C; obtained in the distillation of coal tar; used as an important source of dyestuffs, and in coating applications. { 'an·thrə‚sēn }

anthracene violet [ORGANIC CHEMISTRY] *See* gallein. { 'an·thrə‚sēn 'vī·lət }

anthraciferous coal [ORGANIC CHEMISTRY] Anthracite-hard coal containing or yielding anthracene. { 'an·thrə‚sif·ə·rəs 'kōl }

anthranilic acid [ORGANIC CHEMISTRY] *o*-$NH_2C_6H_4COOH$ A white or pale yellow, crystalline acid melting at 146°C; used as an intermediate in the manufacture of dyes, pharmaceuticals, and perfumes. { ¦an·thrə¦lin·ik 'as·əd }

anthrapurpurin [ORGANIC CHEMISTRY] $C_6H_3OH(CO)_2C_6H_2(OH)_2$ Orange-yellow, crystalline needles with a melting point of 369°C; soluble in alcohol and alkalies; used in dyeing. { ‚an·thrə'pər·pə‚rin }

anthraquinone [ORGANIC CHEMISTRY] $C_6H_4(CO)_2C_6H_4$ Yellow crystalline diketone that is insoluble in water; used in the manufacture of dyes. { ˌan·thrə·kwi'nōn }

anthrone [ORGANIC CHEMISTRY] $C_{14}H_{10}O$ Colorless needles with a melting point of 156°C; soluble in alcohol, benzene, and hot sodium hydroxide; used as a reagent for carbohydrates. { 'anˌthrōn }

anticatalyst [CHEMISTRY] A material that slows down the action of a catalyst; an example is lead, which inhibits the action of platinum. { ¦an·tē'kad·əl·ist }

antifoaming agent [ORGANIC CHEMISTRY] A substance, such as a silicone, organic phosphate, or alcohol, that inhibits the formation of bubbles in a liquid during its agitation by reducing its surface tension. { ¦an·tē¦fōm·iŋ ˌā·jənt }

antifreeze [CHEMISTRY] A substance added to a liquid to lower its freezing point; the principal automotive antifreeze component is ethylene glycol. { 'an·tēˌfrēz }

antimonate [CHEMISTRY] The radical $[Sb(OH)_6]^-$ in salts derived from antimony pentoxide, Sb_4O_{10}, and bases. { 'an·tə·məˌnāt }

antimonic [CHEMISTRY] Derived from or pertaining to pentavalent antimony. { ¦an·tə¦män·ik }

antimonide [INORGANIC CHEMISTRY] A binary compound of antimony with a more positive compound, for example, H_5Sb. Also known as stibide. { 'an·tə·məˌnīd }

antimonous [CHEMISTRY] Pertaining to antinomy, especially trivalent antimony. { 'an·tə·mə·nəs }

antimony [CHEMISTRY] A chemical element, symbol Sb, atomic number 51, atomic weight 121.75. { 'an·təˌmō·nē }

antimony(III) oxide [INORGANIC CHEMISTRY] Sb_2O_3 Colorless rhombic crystals melting at 656°C; insoluble in water; powerful reducing agent. { 'an·təˌmō·nē ˌthrē 'äkˌsīd }

antimonyl [CHEMISTRY] The inorganic radical SbO^-. { 'an·tə·məˌnil }

antimony pentachloride [INORGANIC CHEMISTRY] $SbCl_5$ A reddish-yellow, oily liquid; hygroscopic, it solidifies after moisture is absorbed and decomposes in excess water; soluble in hydrochloric acid and chloroform; used in analytical testing for cesium and alkaloids, for dyeing, and as an intermediary in synthesis. Also known as antimony perchloride. { 'an·təˌmō·nē ˌpent·ə'klȯrˌīd }

antimony pentafluoride [INORGANIC CHEMISTRY] SbF_5 A corrosive, hygroscopic, moderately viscous fluid; reacts violently with water, forms a clear solution with glacial acetic acid; used in the fluorination of organic compounds. { 'an·təˌmō·nē ˌpent·ə'flu̇rˌīd }

antimony pentasulfide [INORGANIC CHEMISTRY] Sb_2S_5 An orange-yellow powder; soluble in alkali, soluble in concentrated hydrochloric acid, with hydrogen sulfide as a by-product, and insoluble in water; used as a red pigment. Also known as antimony persulfide, antimony red, golden antimony sulfide. { 'an·təˌmō·nē ˌpent·ə'səlˌfīd }

antimony perchloride [INORGANIC CHEMISTRY] *See* antimony pentachloride. { 'an·təˌmō·nē ˌpər'klȯrˌīd }

antimony persulfide [INORGANIC CHEMISTRY] *See* antimony pentasulfide. { 'an·təˌmō·nē ˌpər'səlˌfīd }

antimony potassium tartrate [ORGANIC CHEMISTRY] *See* tartar emetic. { 'an·tə·məˌnil pə'tas·ē·əm 'tärˌtrāt }

antimony red [INORGANIC CHEMISTRY] *See* antimony pentasulfide. { 'an·təˌmō·nē 'red }

antimony sodiate [INORGANIC CHEMISTRY] *See* sodium antimonate. { 'an·təˌmō·nē 'sō·dē·āt }

antimony sulfate [INORGANIC CHEMISTRY] $Sb_2(SO_4)_3$ Antimony(III) sulfate, a white, deliquescent powder; soluble in acids. { 'an·təˌmō·nē 'səlˌfāt }

antimony trichloride [INORGANIC CHEMISTRY] $SbCl_3$ Hygroscopic, colorless, crystalline mass; fumes slightly in air, is soluble in alcohol and acetone, and forms antimony oxychloride in water; used as a mordant, as a chlorinating agent, and in fireproofing textiles. { 'an·təˌmō·nē ˌtrī'klȯrˌīd }

antimony trisulfide [INORGANIC CHEMISTRY] Sb_2S_3 Black and orange-red rhombic crystals; soluble in concentrated hydrochloric acid and sulfide solutions, insoluble in water; melting point 546°C; used as a pigment, and in matches and pyrotechnics. { 'an·tə,mō·nē ,trī'səl,fīd }

antimony yellow [INORGANIC CHEMISTRY] *See* lead antimonite. { 'an·tə,mō·nē 'ye·lō }

antioxidant [CHEMISTRY] An inhibitor, such as ascorbic acid, effective in preventing oxidation by molecular oxygen. { ,an·tē'äk·sə·dənt }

anti-Stokes lines [SPECTROSCOPY] Lines of radiated frequencies which are higher than the frequency of the exciting incident light. { ,an·tē'stōks ,līnz }

4-AP [ORGANIC CHEMISTRY] *See* 4-aminopyridine.

apo- [CHEMISTRY] A prefix that denotes formation from or relationship to another chemical compound. { 'ap·ō *or* 'ap·ə }

apoatropine [ORGANIC CHEMISTRY] $C_{17}H_{21}NO_2$ An alkaloid melting at 61°C with decomposition of the compound; highly toxic; obtained by dehydrating atropine. { ,ap·ō'a·trə,pēn }

apodization [SPECTROSCOPY] A mathematical transformation carried out on data received from an interferometer to alter the instrument's response function before the Fourier transformation is calculated to obtain the spectrum. { ,a·pə·də'zā·shən }

apparent concentration [ANALYTICAL CHEMISTRY] The value of analyte concentration obtained when the interference is not considered. { ə'par·ənt ,kän·sən'trā·shən }

aprotic solvent [CHEMISTRY] A solvent that does not yield or accept a proton. { ā'präd·ik 'säl·vənt }

aqua [CHEMISTRY] Latin for water. { 'äk·wə }

aqua ammonia [INORGANIC CHEMISTRY] *See* ammonium hydroxide. { 'äk·wə ə'mōn·ē·ə }

aqua fortis [INORGANIC CHEMISTRY] *See* nitric acid. { ¦äk·wə'fȯrd·əs }

aquametry [ANALYTICAL CHEMISTRY] Analytical processes to measure the water present in materials; methods include Karl Fischer titration, reactions with acid chlorides and anhydrides, oven drying, distillation, and chromatography. { ə'kwäm·ə·trē }

aqua regia [INORGANIC CHEMISTRY] A fuming, highly corrosive, volatile liquid with a suffocating odor made by mixing 1 part concentrated nitric acid and 3 parts concentrated hydrochloric acid; reacts with all metals, including silver and gold. { ¦äk·wə 'rē·jə }

aquasol [CHEMISTRY] *See* hydrosol. { 'ak·wə,sȯl }

aquation [CHEMISTRY] Formation of a complex that contains water by replacement of other coordinated groups in the complex. { ə'kwā·shən }

aqueous electron [PHYSICAL CHEMISTRY] *See* hydrated electron. { 'āk·wē·əs i'lek,trän }

aqueous solution [CHEMISTRY] A solution with the solvent as water. { 'āk·wē·əs sə'lü·shən }

aquo ion [CHEMISTRY] Any ion containing one or more water molecules. { 'a·kwō 'ī,än }

Ar [CHEMISTRY] *See* argon.

arabine [ORGANIC CHEMISTRY] *See* harman. { 'ar·ə,bēn }

arabite [ORGANIC CHEMISTRY] *See* arabitol. { 'ar·ə,bīt }

arabitol [ORGANIC CHEMISTRY] $CH_2OH(CHOH)_3CH_2OH$ An alcohol that is derived from arabinose; a sweet, colorless crystalline material present in D and L forms; soluble in water; melts at 103°C. Also known as arabite. { ə'rab·ə,tȯl }

arachic acid [ORGANIC CHEMISTRY] *See* eicosanoic acid. { ə'rak·ik 'as·əd }

arachidic acid [ORGANIC CHEMISTRY] *See* eicosanoic acid. { ,a·rə'kid·ik 'as·əd }

aralkyl [ORGANIC CHEMISTRY] A radical in which an aryl group is substituted for an alkyl H atom. Derived from arylated alkyl. { a'ral,kil }

arbutin [ORGANIC CHEMISTRY] $C_{12}H_{16}O_7$ A bitter glycoside from the bearberry and certain other plants; sometimes used as a urinary antiseptic. { är'byüt·ən }

arc spectrum [SPECTROSCOPY] The spectrum of a neutral atom, as opposed to that of a molecule or an ion; it is usually produced by vaporizing the substance in an electric arc; designated by the roman numeral I following the symbol for the element, for example, HeI. { 'ärk ,spek·trəm }

arecoline [ORGANIC CHEMISTRY] $C_8H_{13}O_2N$ An alkaloid from the betel nut; an oily, colorless liquid with a boiling point of 209°C; soluble in water, ethanol, and ether; combustible; used as a medicine. { ə'rek·ə,lēn }

arene [ORGANIC CHEMISTRY] *See* aromatic hydrocarbon. { 'a,rēn }

argentic [CHEMISTRY] Relating to or containing silver. { är'jen·tik }

argentic oxide [INORGANIC CHEMISTRY] *See* silver suboxide. { är'jen·tik 'äk,sīd }

argentocyanides [INORGANIC CHEMISTRY] Complexes formed, for example, in the cyanidation of silver ores and in electroplating, when silver cyanide reacts with solutions of soluble metal cyanides. Also known as dicyanoargentates. { är,jen·tō'sī·ə,nīdz }

argentometry [ANALYTICAL CHEMISTRY] A volumetric analysis that employs precipitation of insoluble silver salts; the salts may be chromates or chlorides. { ,är·jən 'täm·ə·trē }

argentum [CHEMISTRY] Latin for silver. { är'jen·təm }

argon [CHEMISTRY] A chemical element, symbol Ar, atomic number 18, atomic weight 39.998. { 'ar,gän }

aristolochic acid [ORGANIC CHEMISTRY] $C_{17}H_{11}NO_7$ Crystals in the form of shiny brown leaflets that decompose at 281-286°C; soluble in alcohol, chloroform, acetone, ether, acetic acid, and aniline; used as an aromatic bitter. Also known as aristolochine. { ə¦ris·tə¦läk·ik 'as·əd }

aristolochine [ORGANIC CHEMISTRY] *See* aristolochic acid. { a,ris'täl·ə,kēn }

Armstrong's acid [ORGANIC CHEMISTRY] *See* naphthalene-1,5-disulfonic acid. { 'ärm,strȯŋz 'as·əd }

Arndt-Eistert synthesis [ORGANIC CHEMISTRY] A method of increasing the length of an aliphatic acid by one carbon by reacting diazomethane with acid chloride. { ¦ärnt ¦ī·stərt ,sin·thə·səs }

aromatic [ORGANIC CHEMISTRY] **1.** Pertaining to or characterized by the presence of at least one benzene ring. **2.** Describing those compounds having physical and chemical properties resembling those of benzene. { ¦ar·ə¦mad·ik }

aromatic alcohol [ORGANIC CHEMISTRY] Any of the compounds containing the hydroxyl group in a side chain to a benzene ring, such as benzyl alcohol. { ¦ar·ə¦mad·ik 'al·kə,hȯl }

aromatic aldehyde [ORGANIC CHEMISTRY] An aromatic compound containing the CHO radical, such as benzaldehyde. { ¦ar·ə¦mad·ik 'al·də,hīd }

aromatic amine [ORGANIC CHEMISTRY] An organic compound that contains one or more amino groups joined to an aromatic structure. { ¦ar·ə¦mad·ik 'am,ēn }

aromatic hydrocarbon [ORGANIC CHEMISTRY] A member of the class of hydrocarbons, of which benzene is the first member, consisting of assemblages of cyclic conjugated carbon atoms and characterized by large resonance energies. Also known as arene. { ¦ar·ə¦mad·ik ¦hī·drə'kär·bən }

aromatic ketone [ORGANIC CHEMISTRY] An aromatic compound containing the —CO radical, such as acetophenone. { ¦ar·ə¦mad·ik 'kē,tōn }

aromatic nucleus [ORGANIC CHEMISTRY] The six-carbon ring characteristic of benzene and related series, or condensed six-carbon rings of naphthalene, anthracene, and so forth. { ¦ar·ə¦mad·ik 'nü·klē·əs }

aroyl [ORGANIC CHEMISTRY] The radical RCO, where R is an aromatic (benzoyl, napthoyl) group. { 'ar·ə·wəl }

aroylation [ORGANIC CHEMISTRY] A reaction in which the aroyl group is incorporated into a molecule by substitution. { ,ar·ə·wə'lā·shən }

ARPES [SPECTROSCOPY] *See* angle-resolved photoelectron spectroscopy.

Arrhenius equation [PHYSICAL CHEMISTRY] The relationship that the specific reaction rate constant k equals the frequency factor constant s times $\exp(-\Delta H_{act}/RT)$, where ΔH_{act} is the heat of activation, R the gas constant, and T the absolute temperature. { ar'rā·nē·əs i'kwā·zhən }

arsenate [INORGANIC CHEMISTRY] **1.** AsO_4^{3-} A negative ion derived from orthoarsenic acid, $H_3AsO_4 \cdot {}^1/_2H_2O$. **2.** A salt or ester of arsenic acid. { 'ärs·ən,āt }

arsenic [CHEMISTRY] A chemical element, symbol As, atomic number 33, atomic weight 74.9216. { 'ärs·ən·ik }

arsenic acid [INORGANIC CHEMISTRY] $H_3AsO_4 \cdot {}^1/_2H_2O$ White, poisonous crystals, soluble in water and alcohol; used in manufacturing insecticides, glass, and arsenates and as a defoliant. Also known as orthoarsenic acid. { är'sen·ik 'as·əd }

arsenical [CHEMISTRY] **1.** Pertaining to arsenic. **2.** A compound that contains arsenic. { ar'sen·ə·kəl }

arsenic disulfide [INORGANIC CHEMISTRY] As_2S_2 Red, orange, or black monoclinic crystals, insoluble in water; used in fireworks; occurs naturally as realgar. { 'ärs·ən·ik dī'səl‚fīd }

arsenic oxide [INORGANIC CHEMISTRY] **1.** An oxide of arsenic. **2.** *See* arsenic pentoxide. **3.** *See* arsenic trioxide. { 'ärs·ən·ik 'äk‚sīd }

arsenic pentasulfide [INORGANIC CHEMISTRY] As_2S_5 Yellow crystals that are insoluble in water and readily decompose to the trisulfide and sulfur; used as a pigment. { 'ärs·ən·ik ‚pent·ə'səl‚fīd }

arsenic pentoxide [INORGANIC CHEMISTRY] As_2O_5 A white, deliquescent compound that decomposes by heat and is soluble in water. Also known as arsenic oxide. { 'ärs·ən·ik ‚pent'äk‚sīd }

arsenic trichloride [INORGANIC CHEMISTRY] $AsCl_3$ An oily, colorless liquid that dissolves in water; used in ceramics, organic chemical syntheses, and in the preparation of pharmaceuticals. { 'ärs·ən·ik ‚trī'klȯr‚īd }

arsenic trioxide [INORGANIC CHEMISTRY] As_2O_3 A toxic compound, slightly soluble in water; octahedral crystals change to the monoclinic form by heating at 200°C; occurs naturally as arsenolite and claudetite; used in small quantities in some medicinal preparations. Also known as arsenic oxide; arsenious acid. { 'ärs·ən·ik ‚trī'äk‚sīd }

arsenic trisulfide [INORGANIC CHEMISTRY] As_2S_3 An acidic compound in the form of yellow or red monoclinic crystals with a melting point at 300°C; occurs as the mineral orpiment; used as a pigment. { 'ärs·ən·ik ‚trī'səl‚fīd }

arsenide [CHEMISTRY] A binary compound of negative, trivalent arsenic; for example, H_3As or GaAs. { 'ärs·ən‚īd }

arsenin [ORGANIC CHEMISTRY] A heterocyclic organic compound composed of a six-membered ring system in which the carbon atoms are unsaturated and the unique heteroatom is arsenic, with no nitrogen atoms present. { är'sen·ən }

arsenious acid [INORGANIC CHEMISTRY] *See* arsenic trioxide. { är'sēn·ē·əs 'as·əd }

arsenite [INORGANIC CHEMISTRY] **1.** AsO_3^{3-} A negative ion derived from aqueous solutions of As_4O_6. **2.** A salt or ester of arsenious acid. { 'är·sə‚nīt }

arsenobenzene [ORGANIC CHEMISTRY] $C_6H_5As{:}AsC_6H_5$ White needles that melt at 212°C; insoluble in cold water, soluble in benzene; derivatives have some use in medicine. { ‚ärs·ən·ō'ben‚zēn }

arseno compound [ORGANIC CHEMISTRY] A compound containing an As-As bond with the general formula $(RAs)_n$, where R represents a functional group; structures are cyclic or long-chain polymers. { ‚ärs·ən·ō 'käm‚paund }

arsine [INORGANIC CHEMISTRY] H_3As A colorless, highly poisonous gas with an unpleasant odor. { är'sēn }

arsinic acid [INORGANIC CHEMISTRY] An acid of general formula R_2AsO_2H; derived from trivalent arsenic; an example is cacodylic acid, or dimethylarsinic acid, $(CH_3)_2AsO_2H$. { är¦sin·ik 'as·əd }

arsonic acid [INORGANIC CHEMISTRY] An acid derived from orthoarsenic acid, $OAs(OH)_3$; the type formula is generally considered to be $RAsO(OH)_2$; an example is *para*-aminobenzenearsonic acid, $NH_2C_6H_4AsO(OH)_2$. { är¦sän·ik 'as·əd }

arsonium [INORGANIC CHEMISTRY] $—AsH_4$ A radical which may be considered analogous to the ammonium radical in that a compound such as AsH_4OH may form. { är'sōn·ē·əm }

artificial camphor [ORGANIC CHEMISTRY] *See* terpene hydrochloride. { ¦ärd·ə¦fish·əl 'kam·fər }

artificial gold [INORGANIC CHEMISTRY] *See* stannic sulfide. { ¦ärd·ə¦fish·əl 'gōld }

artificial malachite [INORGANIC CHEMISTRY] *See* copper carbonate. { ¦ärd·ə¦fish·əl 'mal·ə‚kīt }

artificial neroli oil [ORGANIC CHEMISTRY] *See* methyl anthranilate. { ¦ärd·ə¦fish·əl nə'rōl·ē ‚ȯil }

artificial scheelite [INORGANIC CHEMISTRY] *See* calcium tungstate. { ¦ärd·ə¦fish·əl ′shā‚līt }

aryl [ORGANIC CHEMISTRY] An organic group derived from an aromatic hydrocarbon by removal of one hydrogen. { ′ar·əl }

aryl acid [ORGANIC CHEMISTRY] An organic acid that has an aryl group. { ′ar·əl ′as·əd }

arylamine [ORGANIC CHEMISTRY] An organic compound formed from an aromatic hydrocarbon that has at least one amine group joined to it, such as aniline. { ′ar·əl·ə‚mēn }

aryl compound [ORGANIC CHEMISTRY] Molecules with the six-carbon aromatic ring structure characteristic of benzene or compounds derived from aromatics. { ′ar·əl ‚käm‚paůnd }

aryl diazo compound [ORGANIC CHEMISTRY] A diazo compound bonded to the ring structure characteristic of benzene or any other aromatic derivative. { ′ar·əl dī′āz·ō ‚käm‚paůnd }

arylene [ORGANIC CHEMISTRY] A radical that is bivalent and formed by removal of hydrogen from two carbon sites on an aromatic nucleus. { ′ar·ə‚lēn }

aryl halide [ORGANIC CHEMISTRY] An aromatic derivative in which a ring hydrogen has been replaced by a halide atom. { ′ar·əl ′hal‚īd }

arylide [ORGANIC CHEMISTRY] A compound formed from a metal and an aryl group, for example, PbR_4, where R is the aryl group. { ′ar·ə‚līd }

aryloxy compound [ORGANIC CHEMISTRY] One of a group of compounds useful as organic weed killers, such as 2,4-dichlorophenoxyacetic acid (2,4-D). { ¦ar·əl¦äk·sē ‚käm‚paůnd }

aryne [ORGANIC CHEMISTRY] An aromatic species in which two adjacent atoms of a ring lack substituents, with two orbitals each missing an electron. Also known as benzyne. { ′a‚rīn }

As [CHEMISTRY] *See* arsenic.

asarone [ORGANIC CHEMISTRY] $C_{12}H_{16}O_3$ A crystalline substance with melting point 67°C; insoluble in water, soluble in alcohol; found in plants of the genus *Asarum*; used as a constituent in essential oils such as calumus oil. { ′as·ə‚rōn }

ASC [ORGANIC CHEMISTRY] *See* N-acetylsulfanilyl chloride.

ascaridole [ORGANIC CHEMISTRY] $C_{10}H_{16}O_2$ A terpene peroxide, explosive when heated; used as an initiator in polymerization. { ə′skar·ə‚dōl }

ascending chromatography [ANALYTICAL CHEMISTRY] A technique for the analysis of mixtures of two or more compounds in which the mobile phase (sample and carrier) rises through the fixed phase. { ə′send·iŋ ‚krō·mə′täg·rə·fē }

ash [CHEMISTRY] The incombustible matter remaining after a substance has been incinerated. { ash }

ashing [ANALYTICAL CHEMISTRY] An analytical process in which the chemical material being analyzed is oven-heated to leave only noncombustible ash. { ′ash·iŋ }

aspartame [ORGANIC CHEMISTRY] $C_{14}H_{18}N_2O_5$ A dipeptide ester about 160 times sweeter than sucrose in aqueous solution; used as a low-calorie sweetener. { a′spär‚tām }

aspirin [ORGANIC CHEMISTRY] *See* acetylsalicylic acid. { ′as·prən }

assay [ANALYTICAL CHEMISTRY] Qualitative or quantitative determination of the components of a material, as an ore or a drug. { ′a‚sā }

association [CHEMISTRY] Combination or correlation of substances or functions. { ə‚sō·sē′ā·shən }

A stage [ORGANIC CHEMISTRY] An early stage in a thermosetting resin reaction characterized by linear structure, solubility, and fusibility of the material. { ′ā ‚stāj }

astatine [CHEMISTRY] A radioactive chemical element, symbol At, atomic number 85, the heaviest of the halogen elements. { ′as·tə‚tēn }

asterism [SPECTROSCOPY] A star-shaped pattern sometimes seen in x-ray spectrophotographs. { ′as·tə‚riz·əm }

astigmatic mounting [SPECTROSCOPY] A mounting designed to minimize the astigmatism of a concave diffraction grating. { ¦a·stig‚mad·ik ′mount·iŋ }

astronomical spectrograph [SPECTROSCOPY] An instrument used to photograph spectra of stars. { ˌas·trə'näm·ə·kəl 'spek·trəˌgraf }

astronomical spectroscopy [SPECTROSCOPY] The use of spectrographs in conjunction with telescopes to obtain observational data on the velocities and physical conditions of astronomical objects. { ˌas·trə'näm·ə·kəl ˌspek'träs·kə·pē }

asymmetric carbon atom [ORGANIC CHEMISTRY] A carbon atom with four different atoms or groups of atoms bonded to it. Also known as chiral carbon atom; stereogenic center. { ¦ā·sə¦me·trik ¦kär·bən 'ad·əm }

asymmetric synthesis [ORGANIC CHEMISTRY] Chemical synthesis of a pure enantiomer, or of an enantiomorphic mixture in which one enantiomer predominates, without the use of resolution. { ¦ā·sə¦me·trik 'sin·thə·səs }

asymmetry [PHYSICAL CHEMISTRY] The geometrical design of a molecule, atom, or ion that cannot be divided into like portions by one or more hypothetical planes. Also known as molecular asymmetry. { ¦ā'sim·ə·trə }

asymmetry effect [PHYSICAL CHEMISTRY] The asymmetrical distribution of the ion cloud around an ion that results from the finite relaxation time for the ion cloud when a voltage is applied; leads to a reduction in ion mobility. { ā'sim·ə·trē iˌfekt }

At [ORGANIC CHEMISTRY] *See* astatine.

ATA [ORGANIC CHEMISTRY] *See* aminotriazole.

atactic [ORGANIC CHEMISTRY] Of the configuration for a polymer, having the opposite steric configurations for the carbon atoms of the polymer chain occur in equal frequency and more or less at random. { ā'tak·tik }

atom [CHEMISTRY] The individual structure which constitutes the basic unit of any chemical element. { 'ad·əm }

atomic connectivity [PHYSICAL CHEMISTRY] The specific pattern of chemical bonds between atoms in a molecule. { ə¦täm·ik kəˌnek'tiv·əd·ē }

atomic emission spectroscopy [SPECTROSCOPY] A form of atomic spectroscopy in which one observes the emission of light at discrete wavelengths by atoms which have been electronically excited by collisions with other atoms and molecules in a hot gas. { ə¦täm·ik ə¦mish·ən spek'träs·kə·pē }

atomic fluorescence spectroscopy [SPECTROSCOPY] A form of atomic spectroscopy in which the sample atoms are first excited by absorbing radiation from an external source containing the element to be detected, and the intensity of radiation emitted at characteristic wavelengths during transitions of these atoms back to the ground state is observed. { ə¦täm·ik flu̇¦res·əns spek'träs·kə·pē }

atomic heat capacity [PHYSICAL CHEMISTRY] The heat capacity of a gram-atomic weight of an element. { ə'täm·ik 'hēt kə'pas·əd·ē }

atomic hydrogen [CHEMISTRY] Gaseous hydrogen whose molecules are dissociated into atoms. { ə'täm·ik 'hī·drə·jən }

atomicity [CHEMISTRY] The number of atoms in a molecule of a compound. { ˌad·ə'mis·əd·ē }

atomic percent [CHEMISTRY] The number of atoms of an element in 100 atoms representative of a substance. { ə'täm·ik pər'sent }

atomic photoelectric effect [PHYSICAL CHEMISTRY] *See* photoionization. { ə'täm·ik ˌfōd·ō·i'lek·trik i'fekt }

atomic polarization [PHYSICAL CHEMISTRY] Polarization of a material arising from the change in dipole moment accompanying the stretching of chemical bonds between unlike atoms in molecules. { ə'täm·ik ˌpōl·ə·rə'zā·shən }

atomic radius [PHYSICAL CHEMISTRY] Also known as covalent radius. **1.** Half the distance between the nuclei of two like atoms that are covalently bonded. **2.** The experimentally determined radius of an atom in a covalently bonded compound. { ə'täm·ik 'rād·ē·əs }

atomic spectroscopy [SPECTROSCOPY] The branch of physics concerned with the production, measurement, and interpretation of spectra arising from either emission or absorption of electromagnetic radiation by atoms. { ə'täm·ik ˌspek'träs·kə·pē }

atomic spectrum [SPECTROSCOPY] The spectrum of radiations due to transitions between energy levels in an atom, either absorption or emission. { ə'täm·ik 'spek·trəm }

atomic theory [CHEMISTRY] The assumption that matter is composed of particles called atoms and that these are the limit to which matter can be subdivided. { ə'täm·ik 'thē·ə·rē }

atomic volume [PHYSICAL CHEMISTRY] The volume occupied by 1 gram-atom of an element in the solid state. { ə'täm·ik 'väl·yəm }

atomic weight [CHEMISTRY] The relative mass of an atom based on a scale in which a specific carbon atom (carbon-12) is assigned a mass value of 12. Abbreviated at. wt. Also known as relative atomic mass. { ə'täm·ik 'wāt }

atomization [ANALYTICAL CHEMISTRY] In flame spectrometry, conversion of a volatilized sample into free atoms. [CHEMISTRY] A process in which the chemical bonds in a molecule are broken to yield separated (free) atoms. { ,ad·ə·mə'za·shən }

atoms-in-molecules method [PHYSICAL CHEMISTRY] The description of the electronic structure of a molecule as a perturbation of the isolated states of its constituent atoms. { ¦ad·əmz in 'mäl·ə,kyülz ,meth·əd }

ATR [SPECTROSCOPY] *See* attenuated total reflectance.

atropisomer [ORGANIC CHEMISTRY] One of two conformations of a molecule whose interconversion is slow enough to allow separation and isolation under predetermined conditions. { ¦a·trō¦pī·ə·mər }

attachment [ORGANIC CHEMISTRY] The conversion of a molecular entity into another molecular structure solely by formation of a single two-center bond with another molecular entity and no other changes in bonding. { ə'tach·mənt }

attenuated total reflectance [SPECTROSCOPY] A method of spectrophotometric analysis based on the reflection of energy at the interface of two media which have different refractive indices and are in optical contact with each other. Abbreviated ATR. Also known as frustrated internal reflectance; internal reflectance spectroscopy. { ə'ten·yə,wād·əd 'tōd·əl ri'flek·təns }

at. wt [CHEMISTRY] *See* atomic weight.

Au [CHEMISTRY] *See* gold.

Aufbau principle [CHEMISTRY] A description of the building up of the elements in which the structure of each in sequence is obtained by simultaneously adding one positive charge (proton) to the nucleus of the atom and one negative charge (electron) to an atomic orbital. { 'auf,bau ¦prin·sə·pəl }

Auger electron spectroscopy [SPECTROSCOPY] The energy analysis of Auger electrons produced when an excited atom relaxes by a radiationless process after ionization by a high-energy electron, ion, or x-ray beam. Abbreviated AES. { ō'zhā i'lek,trän spek'träs·kə·pē }

auramine hydrochloride [ORGANIC CHEMISTRY] $C_{17}H_{22}ClN_3 \cdot H_2O$ A compound melting at 267°C; very soluble in water, soluble in ethanol; used as a dye and an antiseptic. Also known as yellow pyoktanin. { 'ȯr·ə,mēn ,hī·drə'klȯr,īd }

aurantia [ORGANIC CHEMISTRY] $C_{12}H_8N_8O_{12}$ An orange aniline dye, used in stains in biology and in some photographic filters. { ȯ'ranch·ə }

aurantiin [ORGANIC CHEMISTRY] *See* naringin. { ȯ'ran·tē·ən }

auric oxide [INORGANIC CHEMISTRY] *See* gold oxide. { 'ȯr·ik 'äk,sīd }

aurin [ORGANIC CHEMISTRY] $C_{19}H_{14}O_3$ A derivative of triphenylmethane; solid with red-brown color with green luster; melting point about 220°C; insoluble in water; used as a dye intermediate. { 'ȯr·ən }

auroral line [SPECTROSCOPY] A prominent green line in the spectrum of the aurora at a wavelength of 5577 angstroms, resulting from a certain forbidden transition of oxygen. { ə'rȯr·əl ,līn }

autoacceleration [ORGANIC CHEMISTRY] The increase in polymerization rate and molecular weight of certain vinyl monomers during bulk polymerization. { ¦ȯd·ō·ik,sel·ə'rā·shən }

autocatalysis [CHEMISTRY] A catalytic reaction started by the products of a reaction that was itself catalytic. { ¦ōd·ō·kə'tal·ə·səs }

autogenous ignition temperature [CHEMISTRY] *See* ignition temperature. { ȯ'täj·ə·nəs ig'nish·ən ˌtem·prə·chər }
automatic titrator [ANALYTICAL CHEMISTRY] **1.** Titration with quantitative reaction and measured flow of reactant. **2.** Electrically generated reactant with potentiometric, ampherometric, or colorimetric end-point or null-point determination. { ¦ȯd·ə¦mad·ik 'tīˌtrād·ər }
autooxidation [CHEMISTRY] **1.** Oxidation caused by the atmosphere. **2.** An oxidation reaction that is self-catalyzed and spontaneous. **3.** An oxidation reaction begun only by an inductor. { ¦ȯd·ōˌäk·sə'dā·shən }
autopoisoning [CHEMISTRY] *See* self-poisoning. { ˌȯd·ō'pȯiz·ən·iŋ }
autoprotolysis [CHEMISTRY] Transfer of a proton from one molecule to another of the same substance. { ˌȯd·ō·prə'täl·ə·səs }
autoprotolysis constant [CHEMISTRY] A constant denoting the equilibrium condition for the autoprotolysis reaction. { ˌȯd·ō·prə'täl·ə·səs 'kän·stənt }
autoracemization [ORGANIC CHEMISTRY] A racemization process that occurs spontaneously. { ¦ȯd·ōˌrā·sə·mə'zā·shən }
auxiliary electrode [PHYSICAL CHEMISTRY] An electrode in an electrochemical cell used for transfer of electric current to the test electrode. { ȯg'zil·yə·rē i'lekˌtrōd }
auxochrome [CHEMISTRY] Any substituent group such as —NH_2 and —OH which, by affecting the spectral regions of strong absorption in chromophores, enhance the ability of the chromogen to act as a dye. { 'ȯk·səˌkrōm }
available chlorine [CHEMISTRY] The quantity of chlorine released by a bleaching powder when treated with acid. { ə'vāl·ə·bəl 'klȯrˌēn }
average bond dissociation energy [PHYSICAL CHEMISTRY] The average value of the bond dissociation energies associated with the homolytic cleavage of several bonds of a set of equivalent bonds of a molecule. Also known as bond energy. { ¦av·rij ¦bänd di·sō·sē'ā·shən ˌen·ər·jē }
average molecular weight [ORGANIC CHEMISTRY] The calculated number to average the molecular weights of the varying-length polymer chains present in a polymer mixture. { 'av·rij mə'lek·yə·lər 'wāt }
azacrown ether [ORGANIC CHEMISTRY] A crown ether that has nitrogen donor atoms as well as oxygen donor atoms to coordinate to the metal iron. { 'az·əˌkraün 'ē·thər }
9-azafluorene [ORGANIC CHEMISTRY] *See* carbazole. { ¦nīn ə'za·ˌflōr·ēn }
azelaic acid [ORGANIC CHEMISTRY] $HOOC(CH_2)_7CCOH$ Colorless leaflets; melting point 106.5°C; a dicarboxylic acid useful in lacquers, alkyd salts, organic synthesis, and formation of polyamides. { ¦az·ə¦lā·ik 'as·əd }
azelate [ORGANIC CHEMISTRY] A salt of azelaic acid, for example, sodium azelate. { 'az·əlˌāt }
azeotrope [CHEMISTRY] *See* azeotropic mixture. { ā'zē·əˌtrōp }
azeotropic mixture [CHEMISTRY] A solution of two or more liquids, the composition of which does not change upon distillation. Also known as azeotrope. { ¦aˌzē·ə¦träp·ik 'miks·chər }
azide [ORGANIC CHEMISTRY] One of several types of compounds containing the —N_3 group and derived from hydrazoic acid, HN_3. { 'āˌzīd }
azimino [ORGANIC CHEMISTRY] *See* diazoamine. { ā'zim·ē·nō }
azine [ORGANIC CHEMISTRY] A compound of six atoms in a ring; at least one of the atoms is nitrogen, and the ring structure resembles benzene; an example is pyridine. { 'āˌzēn }
azine dyes [ORGANIC CHEMISTRY] Benzene-type dyes derived from phenazine; members of the group, such as nigrosines and safranines, are quite varied in application. { 'āˌzēn ˌdīz }
aziridine [ORGANIC CHEMISTRY] *See* ethyleneimine. { ə'zir·əˌdēn }
azlactone [ORGANIC CHEMISTRY] A compound that is an anhydride of α-acylamino acid; the basic ring structure is the 5-oxazolone type. { az'lakˌtōn }
azo- [ORGANIC CHEMISTRY] A prefix indicating the group —N═N—. { 'a·zō }
azobenzene [ORGANIC CHEMISTRY] $C_6H_5N_2C_6H_5$ A compound existing in cis and trans geometric isomers; the cis form melts at 71°C; the trans form comprises orange-red

leaflets, melting at 68.5°C; used in manufacture of dyes and accelerators for rubbers. { ˌa·zō′benˌzēn }

2,2′-azobisisobutyronitrile [ORGANIC CHEMISTRY] $C_8H_{12}N_4$ Crystals that decompose at 107°C; soluble in methanol and in ethanol; used as an initiator of free radical reactions and as a blowing agent for plastics and elastomers. { ¦tü ¦tüˌprīm ˌa·zō·bīˌī·sōˌbyüd·ə·rə′nī·trəl }

azo compound [ORGANIC CHEMISTRY] A compound having two organic groups separated by an azo group (—N=N—). { ′ā·zō ˌkamˌpaund }

azo dyes [ORGANIC CHEMISTRY] Widely used commercial dyestuffs derived from amino compounds, with the —N— chromophore group; can be made as acid, basic, direct, or mordant dyes. { ′a·zō ˌdīz }

azoic dye [ORGANIC CHEMISTRY] A water-insoluble azo dye that is formed by coupling of the components on a fiber. Also known as ice color; ingrain color. { a′zō·ik ′dī }

azole [ORGANIC CHEMISTRY] One of a class of organic compounds with a five-membered N-heterocycle containing two double bonds; an example is 1,2,4-triazole. { ′āˌzōl }

azotometer [ANALYTICAL CHEMISTRY] *See* nitrometer. { ˌaz·ə′täm·əd·ər }

azoxybenzene [ORGANIC CHEMISTRY] $C_6H_5NO{=}N{-}C_6H_5$ A compound existing in cis and trans forms; the cis form melts at 87°C; the trans form comprises yellow crystals, melting at 36°C, insoluble in water, soluble in ethanol. { ə¦zäk·sē′benˌzēn }

azoxy compound [ORGANIC CHEMISTRY] A compound having an oxygen atom bonded to one of the nitrogen atoms of an azo compound. { ā′zäk·sē ˌkämˌpau̇nd }

azulene [ORGANIC CHEMISTRY] $C_{16}H_{26}O$ The blue coloring matter of wormwood and other essential oils; an oily, blue liquid, boiling at 170°C; insoluble in water; used in cosmetics. { ′azh·əˌlēn }

B

B [CHEMISTRY] *See* boron.

Ba [CHEMISTRY] *See* barium.

Babo's law [PHYSICAL CHEMISTRY] A law stating that the relative lowering of a solvent's vapor pressure by a solute is the same at all temperatures. { 'bä‚bōz ‚lò }

backflash [CHEMISTRY] Rapid combustion of a material occurring in an area that the reaction was not intended for. { 'bak‚flash }

back titration [CHEMISTRY] A titration to return to the end point which was passed. { 'bak tī'trā·shən }

Badger's rule [PHYSICAL CHEMISTRY] An empirical relationship between the stretching force constant for a molecular bond and the bond length. { 'baj·ərz ‚rül }

baeckeol [ORGANIC CHEMISTRY] $C_{13}H_{18}O_4$ A phenolic ketone that is crystalline and pale yellow; found in oils from plants of species of the myrtle family. { 'bāk·ē‚ōl }

Baeyer strain theory [ORGANIC CHEMISTRY] The theory that the relative stability of penta- and hexamethylene ring compounds is caused by a propitious bond angle between carbons and a lack of bond strain. { 'bā·ər 'strān ‚thē·ə·rē }

baking soda [INORGANIC CHEMISTRY] *See* sodium bicarbonate. { 'bāk·iŋ ‚sōd·ə }

balance [CHEMISTRY] To bring a chemical equation into balance so that reaction substances and reaction products obey the laws of conservation of mass and charge. { 'bal·əns }

Balmer continuum [SPECTROSCOPY] A continuous range of wavelengths (or wave numbers or frequencies) in the spectrum of hydrogen at wavelengths less than the Balmer limit, resulting from transitions between states with principal quantum number $n = 2$ and states in which the single electron is freed from the atom. { ¦bal·mər kən'tin·yə·wəm }

Balmer discontinuity [SPECTROSCOPY] *See* Balmer jump. { 'bȯl·mər dis‚känt·ən'ü·əd·ē }

Balmer formula [SPECTROSCOPY] An equation for the wavelengths of the spectral lines of hydrogen, $1/\lambda = R[(1/m^2) - (1/n^2)]$, where λ is the wavelength, R is the Rydberg constant, and m and n are positive integers (with n larger than m) that give the principal quantum numbers of the states between which occur the transition giving rise to the line. { 'bȯl·mər ‚fȯr·myə·lə }

Balmer jump [SPECTROSCOPY] The sudden decrease in the intensity of the continuous spectrum of hydrogen at the Balmer limit. Also known as Balmer discontinuity. { 'bȯl·mər ‚jəmp }

Balmer limit [SPECTROSCOPY] The limiting wavelength toward which the lines of the Balmer series crowd and beyond which they merge into a continuum, at approximately 365 nanometers. { 'bȯl·mər ‚lim·ət }

Balmer lines [SPECTROSCOPY] Lines in the hydrogen spectrum, produced by transitions between $n = 2$ and $n > 2$ levels either in emission or in absorption; here n is the principal quantum number. { 'bȯl·mər ‚līnz }

Balmer series [SPECTROSCOPY] The set of Balmer lines. { 'bȯl·mər ¦sir·ēz }

Bamberger's formula [ORGANIC CHEMISTRY] A structural formula for naphthalene that shows the valencies of the benzene rings pointing toward the centers. { 'bäm‚bər·gərz 'fȯr·myə·lə }

banana oil [ORGANIC CHEMISTRY] **1.** A solution of nitrocellulose in amyl acetate having a bananalike odor. **2.** *See* amyl acetate. { bə'nan·ə ‚ȯil }
band [ANALYTICAL CHEMISTRY] The position and spread of a solute within a series of tubes in a liquid-liquid extraction procedure. Also known as zone. [SPECTROSCOPY] *See* band spectrum. { band }
band head [SPECTROSCOPY] A location on the spectrogram of a molecule at which the lines of a band pile up. { 'band ‚hed }
band spectrum [SPECTROSCOPY] A spectrum consisting of groups or bands of closely spaced lines in emission or absorption, characteristic of molecular gases and chemical compounds. Also known as band. { 'band ‚spek·trəm }
barban [ORGANIC CHEMISTRY] $C_{11}H_9O_2NCl_2$ A white, crystalline compound with a melting point of 75-76°C; used as a postemergence herbicide of wild oats in barley, flax, lentil, mustard, and peas. { 'bär‚ban }
barbital [ORGANIC CHEMISTRY] $C_8H_{12}N_2O_3$ A compound crystallizing in needlelike form from water; has a faintly bitter taste; melting point 188-192°C; used to make sodium barbital, a long-duration hypnotic and sedative. { 'bär·bə‚tȯl }
barbituric acid [ORGANIC CHEMISTRY] $C_4H_4O_3N_2$ 2,4,6-Trioxypyrimidine, the parent compound of the barbiturates; colorless crystals melting at 245°C, slightly soluble in water. { ¦bär·bə¦tu̇r·ik 'as·əd }
Barfoed's test [ANALYTICAL CHEMISTRY] A test for monosaccharides conducted in an acid solution; cupric acetate is reduced to cuprous oxide, a red precipitate. { 'bär·fu̇ts ‚test }
barium [CHEMISTRY] A chemical element, symbol Ba, with atomic number 56 and atomic weight of 137.34. { 'bar·ē·əm }
barium acetate [INORGANIC CHEMISTRY] $Ba(C_2H_3O_2)_2 \cdot H_2O$ A barium salt made by treating barium sulfide or barium carbonate with acetic acids; it forms colorless, triclinic crystals that decompose upon heating; used as a reagent for sulfates and chromates. { 'bar·ē·əm 'as·ə‚tāt }
barium azide [INORGANIC CHEMISTRY] $Ba(N_3)_2$ A crystalline compound soluble in water; used in high explosives. { 'bar·ē·əm 'ā‚zīd }
barium binoxide [INORGANIC CHEMISTRY] *See* barium peroxide. { 'bar·ē·əm bī'näk‚sīd }
barium bromate [INORGANIC CHEMISTRY] $Ba(BrO_3)_2 \cdot H_2O$ A poisonous compound that forms colorless, monoclinic crystals, decomposing at 260°C; used for preparing other bromates. { 'bar·ē·əm 'brō‚māt }
barium bromide [INORGANIC CHEMISTRY] $BaBr_2 \cdot 2H_2O$ Colorless crystals soluble in water and alcohol; used in photographic compounds. { 'bar·ē·əm 'brō‚mīd }
barium carbonate [INORGANIC CHEMISTRY] $BaCO_3$ A white powder with a melting point of 174°C; soluble in acids (except sulfuric acid); used in rodenticides, ceramic flux, optical glass, and television picture tubes. { 'bar·ē·əm 'kär·bə·nət }
barium chlorate [INORGANIC CHEMISTRY] $Ba(ClO_3)_2 \cdot H_2O$ A salt prepared by the reaction of barium chloride and sodium chlorate; it forms colorless, monoclinic crystals, soluble in water; used in pyrotechnics. { 'bar·ē·əm 'klȯr‚āt }
barium chloride [INORGANIC CHEMISTRY] $BaCl_2$ A toxic salt obtained as colorless, water-soluble cubic crystals, melting at 963°C; used as a rat poison, in metal surface treatment, and as a laboratory reagent. { 'bar·ē·əm 'klȯr‚īd }
barium chromate [INORGANIC CHEMISTRY] $BaCrO_4$ A toxic salt that forms yellow, rhombic crystals, insoluble in water; used as a pigment in overglazes. { 'bar·ē·əm 'krō‚māt }
barium citrate [ORGANIC CHEMISTRY] $Ba_3(C_6H_5O_7)_2 \cdot 2H_2O$ A grayish-white, toxic, crystalline powder; used as a stabilizer for latex paints. { 'bar·ē·əm 'sī‚trāt }
barium cyanide [ORGANIC CHEMISTRY] $Ba(CN)_2$ A white, crystalline powder; soluble in water and alcohol; used in metallurgy and electroplating. { 'bar·ē·əm 'sī·ə‚nīd }
barium dioxide [INORGANIC CHEMISTRY] *See* barium peroxide. { 'bar·ē·əm dī'äk‚sīd }
barium fluoride [INORGANIC CHEMISTRY] BaF_2 Colorless, cubic crystals, slightly soluble in water; used in enamels. { 'bar·ē·əm flu̇r‚īd }
barium fluosilicate [INORGANIC CHEMISTRY] $BaSiF_6H$ A white, crystalline powder; in-

soluble in water; used in ceramics and insecticides. Also known as barium silicofluoride. { 'bar·ē·əm ˌflü·ə'sil·əˌkāt }

barium hydroxide [INORGANIC CHEMISTRY] $Ba(OH)_2 \cdot 8H_2O$ Colorless, monoclinic crystals, melting at 78°C; soluble in water, insoluble in acetone; used for fat saponification and fusing of silicates. { 'bar·ē·əm hī'dräkˌsīd }

barium hyposulfite [INORGANIC CHEMISTRY] *See* barium thiosulfate. { 'bar·ē·əm ˌhī·pō'səlˌfīt }

barium manganate [INORGANIC CHEMISTRY] $BaMnO_4$ A toxic, emerald-green powder which is used as a paint pigment. Also known as Cassel green; manganese green. { 'bar·ē·əm 'maŋ·gəˌnāt }

barium mercury iodide [INORGANIC CHEMISTRY] *See* mercuric barium iodide. { 'bar·ē·əm 'mər·kyə·rē 'ī·əˌdīd }

barium molybdate [INORGANIC CHEMISTRY] $BaMoO_4$ A toxic, white powder with a melting point of approximately 1600°C; used in electronic and optical equipment and as a paint pigment. { 'bar·ē·əm mə'libˌdāt }

barium monosulfide [INORGANIC CHEMISTRY] BaS A colorless, cubic crystal that is soluble in water; used in pigments. { 'bar·ē·əm ˌmän·ō'səlˌfīd }

barium monoxide [INORGANIC CHEMISTRY] *See* barium oxide. { 'bar·ē·əm mə'näkˌsīd }

barium nitrate [INORGANIC CHEMISTRY] $Ba(NO_3)_2$ A toxic salt occurring as colorless, cubic crystals, melting at 592°C, and soluble in water; used as a reagent, in explosives, and in pyrotechnics. Also known as nitrobarite. { 'bar·ē·əm 'nīˌtrāt }

barium oxide [INORGANIC CHEMISTRY] BaO A white to yellow powder that melts at 1923°C, it forms the hydroxide with water, may be used as a dehydrating agent. Also known as barium monoxide; barium protoxide. { 'bar·ē·əm 'äkˌsīd }

barium perchlorate [INORGANIC CHEMISTRY] $Ba(ClO_4)_2 \cdot 4H_2O$ Tetrahydrate variety which forms colorless hexagons; used in pyrotechnics. { 'bar·ē·əm pər'klȯrˌāt }

barium permanganate [INORGANIC CHEMISTRY] $Ba(MnO_4)_2$ Brownish-violet, toxic crystals; soluble in water; used as a disinfectant. { 'bar·ē·əm pər'maŋ·gəˌnāt }

barium peroxide [INORGANIC CHEMISTRY] BaO_2 A compound formed as white toxic powder, insoluble in water; used as a bleach and in the glass industry. Also known as barium binoxide; barium dioxide; barium superoxide. { 'bar·ē·əm pər'äkˌsīd }

barium protoxide [INORGANIC CHEMISTRY] *See* barium oxide. { 'bar·ē·əm prō'täkˌsīd }

barium silicide [INORGANIC CHEMISTRY] $BaSi_2$ A compound that has the appearance of metal-gray lumps; melts at white heat; used in metallurgy to deoxidize steel. { 'bar·ē·əm 'sil·əˌsīd }

barium stearate [ORGANIC CHEMISTRY] $Ba(C_{18}H_{35}O_2)_2$ A white, crystalline solid; melting point 160°C; used as a lubricant in manufacturing plastics and rubbers, in greases, and in plastics as a stabilizer against deterioration caused by heat and light. { 'bar·ē·əm 'stirˌāt }

barium sulfate [INORGANIC CHEMISTRY] $BaSO_4$ A salt occurring in the form of white, rhombic crystals, insoluble in water; used as a white pigment, as an opaque contrast medium for roentgenographic processes, and as an antidiarrheal. { 'bar·ē·əm 'səlˌfāt }

barium sulfite [INORGANIC CHEMISTRY] $BaSO_3$ A toxic, white powder; soluble in dilute hydrochloric acid; used in paper manufacturing. { 'bar·ē·əm 'səlˌfīt }

barium superoxide [INORGANIC CHEMISTRY] *See* barium peroxide. { 'bar·ē·əm ˌsü·pər'äkˌsīd }

barium tetrasulfide [INORGANIC CHEMISTRY] $BaS_4 \cdot H_2O$ Red or yellow, rhombic crystals, soluble in water. { 'bar·ē·əm ˌte·trə'səlˌfīd }

barium thiocyanate [INORGANIC CHEMISTRY] $Ba(SCN) \cdot .2H_2O$ White crystals that deliquesce; used in dyeing and in photography. { 'bar·ē·əm ˌthī·ō'sī·əˌnāt }

barium thiosulfate [INORGANIC CHEMISTRY] $BaS_2O_3 \cdot H_2O$ A white powder that decomposes upon heating; used to make explosives and in matches. Also known as barium hyposulfite. { 'bar·ē·əm ˌthī·ō'səlˌfāt }

barium titanate [INORGANIC CHEMISTRY] $BaTiO_3$ A grayish powder that is insoluble in water but soluble in concentrated sulfuric acid; used as a ferroelectric ceramic. { 'bar·ē·əm 'tī·təˌnāt }

barium tungstate [INORGANIC CHEMISTRY] $BaWO_4$ A toxic, white powder used as a pigment and in x-ray photography. Also known as barium white; barium wolframate; tungstate white; wolfram white. { 'bar·ē·əm 'təŋ,stāt }

barium white [INORGANIC CHEMISTRY] *See* barium tungstate. { 'bar·ē·əm 'wīt }

barium wolframate [INORGANIC CHEMISTRY] *See* barium tungstate. { 'bar·ē·əm 'wu̇l·frə,māt }

Barlow's rule [PHYSICAL CHEMISTRY] The rule that the volume occupied by the atoms in a given molecule is proportional to the valences of the atoms, using the lowest valency values. { 'bär,lōz ,rül }

Bart reaction [ORGANIC CHEMISTRY] Formation of an aryl arsonic acid by treating the aryl diazo compound with trivalent arsenic compounds, such as sodium arsenite. { 'bärt rē'ak·shən }

baryta water [CHEMISTRY] A solution of barium hydroxide. { bə'rīd·ə 'wȯd·ər }

base [CHEMISTRY] Any chemical species, ionic or molecular, capable of accepting or receiving a proton (hydrogen ion) from another substance; the other substance acts as an acid in giving of the proton. Also known as Brönsted base. { bās }

base-line technique [ANALYTICAL CHEMISTRY] A method for measurement of absorption peaks for quantitative analysis of chemical compounds in which a base line is drawn tangent to the spectrum background; the distance from the base line to the absorption peak is the absorbence due to the sample under study. { 'bās ,līn tek'nēk }

base metal [CHEMISTRY] Any of the metals on the lower end of the electrochemical series. { 'bās ¦med·əl }

base peak [SPECTROSCOPY] The tallest peak in a mass spectrum; it is assigned a relative intensity value of 100, and lesser peaks are reported as a percentage of it. { 'bās ,pēk }

basic [CHEMISTRY] Of a chemical species that has the properties of a base. { 'bā·sik }

basic group [CHEMISTRY] A chemical group (for example, OH^-) which, when freed by ionization in solution, produces a pH greater than 7. { 'bā·sik ¦grüp }

basic oxide [INORGANIC CHEMISTRY] A metallic oxide that is a base, or that forms a hydroxide when combined with water, such as sodium oxide to sodium hydroxide. { 'bā·sik 'äk,sīd }

basic salt [INORGANIC CHEMISTRY] A compound that is a base and a salt because it contains elements of both, for example, copper carbonate hydroxide, $Cu_2(OH)_2CO_3$. { 'bā·sik 'sȯlt }

basic titrant [CHEMISTRY] A standard solution of a base used for titration. { 'bā·sik 'tī·trənt }

bathochromatic shift [PHYSICAL CHEMISTRY] The shift of the fluorescence of a compound toward the red part of the spectrum due to the presence of a bathochrome radical in the molecule. { ¦bath·ō,krō¦mad·ik 'shift }

battery depolarizer [PHYSICAL CHEMISTRY] *See* depolarizer. { 'bad·ə·rē ,dē'pōl·ə,rīz·ər }

battery electrolyte [PHYSICAL CHEMISTRY] A liquid, paste, or other conducting medium in a battery, in which the flow of electric current takes place by migration of ions. { 'bad·ə·rē i'lek·trə,līt }

battery manganese [INORGANIC CHEMISTRY] *See* manganese dioxide. { 'bad·ə·rē ,maŋ·gə,nēs }

Baumé hydrometer scale [PHYSICAL CHEMISTRY] A calibration scale for liquids that is reducible to specific gravity by the following formulas: for liquids heavier than water, specific gravity = $145 \div (145 - n)$ [at 60°F]; for liquids lighter than water, specific gravity = $140 \div (130 + n)$ [at 60°F]; n is the reading on the Baumé scale, in degrees Baumé; Baumé is abbreviated Bé. { bō'mā hī'dräm·əd·ər ,skāl }

BBC [ORGANIC CHEMISTRY] *See* bromobenzylcyanide.

Be [CHEMISTRY] *See* beryllium.

Bé [PHYSICAL CHEMISTRY] *See* Baumé hydrometer scale.

bead test [ANALYTICAL CHEMISTRY] In mineral identification, a test in which borax is fused to a transparent bead, by heating in a blowpipe flame, in a small loop formed by platinum wire; when suitable minerals are melted in this bead, characteristic glassy colors are produced in an oxidizing or reducing flame and serve to identify elements. { 'bēd ‚test }

beam attenuator [SPECTROSCOPY] An attachment to the spectrophotometer that reduces reference to beam energy to accommodate undersized chemical samples. { 'bēm ə'ten·yə‚wād·ər }

beam-condensing unit [SPECTROSCOPY] An attachment to the spectrophotometer that condenses and remagnifies the beam to provide reduced radiation at the sample. { 'bēm kən'den·siŋ ‚yü·nət }

bebeerine [ORGANIC CHEMISTRY] $C_{36}H_{38}N_2O_6$ An alkaloid derived from the bark of the tropical tree *Nectandra rodiaei*; the dextro form is soluble in acetone, the levo form is soluble in benzene and is an antipyretic; the dextro form is also known as chondrodendrin; the levo, as curine. { bə'bi‚rēn }

Béchamp reduction [ORGANIC CHEMISTRY] Reduction of nitro groups to amino groups by the use of ferrous salts or iron and dilute acid. { bā'shän ri'dək·shən }

Beckmann rearrangement [ORGANIC CHEMISTRY] An intramolecular change of a ketoxime into its isomeric amide when treated with phosphorus pentachloride. { 'bek·män rē·ə'rānj·mənt }

bed [CHEMISTRY] The ion-exchange resin contained in the column in an ion-exchange system. { bed }

Beer-Lambert-Bouguer law [ANALYTICAL CHEMISTRY] *See* Bouguer-Lambert-Beer law. { ¦bā·ər ¦läm·bərt bü'ger ‚lȯ }

Beer's law [PHYSICAL CHEMISTRY] The law which states that the absorption of light by a solution changes exponentially with the concentration, all else remaining the same. { 'bā·ərz ‚lȯ }

behenic acid [ORGANIC CHEMISTRY] *See* docosanoic acid. { bə'hen·ik 'as·əd }

behenyl alcohol [ORGANIC CHEMISTRY] $CH_3(CH_2)_{20}CH_2OH$ A saturated fatty alcohol; colorless, waxy solid with a melting point of 71°C; soluble in ethanol and chloroform; used for synthetic fibers and lubricants. Also known as 1-docosanol. { bə'hen·əl 'al·kə‚hȯl }

bempa [ORGANIC CHEMISTRY] $C_6H_{18}N_3PO$ A white solid soluble in water; used as chemosterilant for insects. Also known as hexamethylphosphorictriamide. { 'bem pə }

Benedict equation of state [PHYSICAL CHEMISTRY] An empirical equation relating pressures, temperatures, and volumes for gases and gas mixtures; superseded by the Benedict-Webb-Rubin equation of state. { 'ben·ə‚dikt i'kwā·zhən əv 'stāt }

Benedict's solution [ANALYTICAL CHEMISTRY] A solution of potassium and sodium tartrates, copper sulfate, and sodium carbonate; used to detect reducing sugars. { 'ben·ə‚diks sə'lü·shən }

benequinox [ORGANIC CHEMISTRY] $C_{13}H_{11}N_3O_2$ A yellow-brown powder that decomposes at 195°C; used as a fungicide for grain seeds and seedlings. { ben'ē·kwə‚näks }

bensulide [ORGANIC CHEMISTRY] $C_{14}H_{24}O_4NPS_3$ An S-(O,O-diisopropyl phosphorodithioate) ester of N-(2-mercaptoethyl)-benzenesulfonamide; an amber liquid slightly soluble in water; melting point is 34.4°C; used as a preemergent herbicide for annual grasses and for broadleaf weeds in lawns and vegetable and cotton crops. { 'ben·sə‚līd }

benthiocarb [ORGANIC CHEMISTRY] $C_{12}H_{16}NOCl$ An amber liquid with a boiling point of 126-129°C; slightly soluble in water; used as an herbicide to control aquatic weeds in rice crops. { ben'thī·ō‚kärb }

benzadox [ORGANIC CHEMISTRY] $C_6H_5CONHOCH_2COOH$ White crystals with a melting point of 140°C; soluble in water; used as an herbicide to control kochia in sugarbeets. { 'ben·zə‚däks }

benzal chloride [ORGANIC CHEMISTRY] $C_6H_5CHCl_2$ A colorless liquid that is refractive and fumes in air; boiling point 207°C; used to make benzaldehyde and cinnamic acid. { ‚benz·əl 'klȯr‚īd }

benzaldehyde [ORGANIC CHEMISTRY] C_6H_5CHO A colorless, liquid aldehyde, boiling at 170°C and possessing the odor of bitter almonds; used as a flavoring agent and an intermediate in chemical syntheses. { benz'al·də›hīd }

benzaldoxime [ORGANIC CHEMISTRY] C_6H_5CHNOH An oxime of benzaldehyde; the antiisomeric form melts at 130°C, the syn form at 34°C; both forms are soluble in ethyl alcohol and ether; used in synthesis of other organic compounds. { ›benz·əl'däk›sēm }

benzalkonium [ORGANIC CHEMISTRY] $C_6H_5CH_2N(CH_3)_2R^+$ An organic radical in which R may range from C_8H_{17} to $C_{18}H_{37}$; found in surfactants, as the chloride salt. { ›benz·əl'kōn·ē·əm }

benzalkonium chloride [ORGANIC CHEMISTRY] $C_6H_5CH_2(CH_3)_2NRCl$ A yellow-white powder soluble in water; used as a fungicide and bactericide; the R is a mixture of alkyls from C_8H_{17} to $C_{18}H_{37}$. { ›benz·əl'kōn·ē·əm 'klȯr›īd }

benzamide [ORGANIC CHEMISTRY] $C_6H_5CONH_2$ A compound with melting point 132.5° to 133.5°C; slightly soluble in water, soluble in ethyl alcohol and carbon tetrachloride; used in chemical synthesis. { ben'za›mīd }

benzanilide [ORGANIC CHEMISTRY] $C_6H_5CONHC_6H_5$ Leaflet crystals with a melting point of 163°C; soluble in alcohol; used to manufacture dyes and perfumes. { benz'an·ə›līd }

benzanthracene [ORGANIC CHEMISTRY] $C_{18}H_{14}$ A weakly carcinogenic material that is isomeric with naphthacene; melting point 162°C; insoluble in water, soluble in benzene. { benz'an·thrə›sēn }

benzanthrone [ORGANIC CHEMISTRY] $C_{17}H_{10}O$ A compound with melting point 170°C; insoluble in water; used in dye manufacture. { benz'an›thrōn }

benzene [ORGANIC CHEMISTRY] C_6H_6 A colorless, liquid, flammable, aromatic hydrocarbon that boils at 80.1°C and freezes at 5.4-5.5°C; used to manufacture styrene and phenol. Also known as benzol. { 'ben›zēn }

benzenediazonium chloride [ORGANIC CHEMISTRY] $C_6H_5N(N)Cl$ An ionic salt soluble in water; used as a dye intermediate. { ›ben›zēn›dī·ə'zōn·ē·əm 'klȯr›īd }

benzenephosphorus dichloride [ORGANIC CHEMISTRY] $C_6H_5PCl_2$ An irritating, colorless liquid with a boiling point of 224.6°C; soluble in inert organic solvents; used in organic synthesis and oil additives. { 'ben›zēn'fäs·fə·rəs ›dī'klȯr›īd }

benzene ring [ORGANIC CHEMISTRY] The six-carbon ring structure found in benzene, C_6H_6, and in organic compounds formed from benzene by replacement of one or more hydrogen atoms by other chemical atoms or radicals. { 'ben›zēn ›riŋ }

benzene series [ORGANIC CHEMISTRY] A series of carbon-hydrogen compounds based on the benzene ring, with the general formula C_nH_{2n-6}, where n is 6 or more; examples are benzene, C_6H_6, toluene, C_7H_8, and xylene, C_8H_{10}. { 'ben›zēn ›sir·ēz }

benzenesulfonate [ORGANIC CHEMISTRY] Any salt or ester of benzenesulfonic acid. { ¦ben›zēn'səl·fə›nāt }

benzenesulfonic acid [ORGANIC CHEMISTRY] $C_6H_5SO_3H$ An organosulfur compound, strongly acidic, water soluble, nonvolatile, and hygroscopic; used in the manufacture of detergents and phenols. { ¦ben›zēn›səl'fän·ik 'as·əd }

1,2,4-benzenetricarboxylic acid [ORGANIC CHEMISTRY] $C_6H_3(COOH)_3$ Crystals with a melting point of 218-220°C; crystallizes from acetic acid or from dilute alcohol; used as an intermediate in the preparation of adhesives, plasticizers, dyes, inks, and resins. { ¦wən ¦tü ¦fȯr ¦ben›zēn·trī¦kär·bäk¦sil·ik 'as·əd }

1,2,4-benzenetriol [ORGANIC CHEMISTRY] $C_6H_3(OH)_3$ Monoprismatic leaflets with a melting point of 141°C; freely soluble in water, ether, alcohol, and ethylacetate; used in gas analysis. { ¦wən ¦tü ¦fȯr ¦ben›zēn'trī›ȯl }

benzenoid [ORGANIC CHEMISTRY] Any substance which has the electronic character of benzene. { 'ben·zə›nȯid }

benzhydrol [ORGANIC CHEMISTRY] $(C_6H_5)_2CHOH$ Colorless needles; melting point 69°C; slightly soluble in water, very soluble in ethanol and ether; used in preparation of other organic compounds including antihistamines. { benz'hī›drȯl }

benzidine [ORGANIC CHEMISTRY] $NH_2C_6H_4C_6H_4NH_2$ An aromatic amine with a melting point of 128°C; used as an intermediate in syntheses of direct dyes for cotton. { 'ben·zə›dēn }

benzil [ORGANIC CHEMISTRY] $C_6H_5COCOC_6H_5$ A yellow powder; melting point 95°C;

insoluble in water, soluble in ethanol, ether, and benzene; used in organic synthesis. { 'ben,zil }

benzilic acid [ORGANIC CHEMISTRY] $(C_6H_5)_2C(OH)CO_2H$ A white, crystalline acid, synthesized by heating benzil with alcohol and potassium hydroxide; used in organic synthesis. { ben'zil·ik 'as·əd }

benzimidazole [ORGANIC CHEMISTRY] $C_7H_6N_2$ Colorless crystals; melting point 170°C; slightly soluble in water, soluble in ethanol; used in organic synthesis. { ,ben·zə'mid·ə,zōl }

benzoate [ORGANIC CHEMISTRY] A salt or ester of benzoic acid, formed by replacing the acidic hydrogen of the carboxyl group with a metal or organic radical. { 'ben·zə,wāt }

benzocaine [ORGANIC CHEMISTRY] *See* ethyl-*para*-aminobenzoate. { 'ben·zə,kān }

benzodihydropyrone [ORGANIC CHEMISTRY] $C_9H_8O_2$ A white to light yellow, oily liquid having a sweet odor; soluble in alcohol, chloroform, and ether; used in perfumery. { 'ben·zō·dī,hī·dra'pī,rōn }

benzoic acid [ORGANIC CHEMISTRY] C_6H_5COOH An aromatic carboxylic acid that melts at 122.4°C, boils at 250°C, and is slightly soluble in water and relatively soluble in alcohol and ether; derivatives are valuable in industry, commerce, and medicine. { ben'zō·ik 'as·əd }

benzoic anhydride [ORGANIC CHEMISTRY] $(C_6H_5CO)_2O$ An acid anhydride that melts at 42°C, boils at 360°C, and crystallizes in colorless prisms; used in synthesis of a variety of organic chemicals, including some dyes. { ben'zō·ik an'hī,drīd }

benzoin [ORGANIC CHEMISTRY] $C_{14}H_{12}O_2I$ An optically active compound; white or yellowish crystals, melting point 137°C; soluble in acetone, slightly soluble in water; used in organic synthesis. { 'ben·zə·wən }

α-benzoin oxime [ORGANIC CHEMISTRY] $C_6H_5CH(OH)C(NOH)C_6H_5$ Prisms crystallized from benzene; melting point is 151-152°C; soluble in alcohol and in aqueous ammonium hydroxide solution; used in the detection and determination of copper, molybdenum, and tungsten. { 'al·fə 'ben·zə·wən 'äk,sēm }

benzol [ORGANIC CHEMISTRY] *See* benzene. { 'ben,zōl }

benzomate [ORGANIC CHEMISTRY] $C_{18}H_{18}O_5N$ A white solid that melts at 71.5-73°C; used as a wettable powder as a miticide. { 'ben·zə,māt }

benzonitrile [ORGANIC CHEMISTRY] C_6H_5CN A colorless liquid with an almond odor; made by heating benzoic acid with lead thiocyanate and used in the synthesis of organic chemicals. Also known as phenyl cyanide. { ¦ben·zō¦nī·trəl }

benzophenone [ORGANIC CHEMISTRY] $C_6H_5COC_6H_5$ A diphenyl ketone, boiling point 305.9°C, occurring in four polymorphic forms (α, β, γ, and δ) each with different melting point; used as a constituent of synthetic perfumes and as a chemical intermediate. Also known as diphenyl ketone; phenyl ketone. { ¦ben·zō·fə'nōn }

benzopyrene [ORGANIC CHEMISTRY] $C_{20}H_{12}$ A five-ring aromatic hydrocarbon found in coal tar, in cigarette smoke, and as a product of incomplete combustion; yellow crystals with a melting point of 179°C; soluble in benzene, toluene, and xylene. { ¦ben·zō¦pī,rēn }

1,2-benzopyrone [ORGANIC CHEMISTRY] *See* coumarin. { ¦wən ¦tü ¦ben·zō¦pī,rōn }

5,6-benzoquinoline [ORGANIC CHEMISTRY] $C_{13}H_9N$ Crystals which are soluble in dilute acids, alcohol, ether, or benzene; melting point is 93°C; used as a reagent for the determination of cadmium. { ¦fīv ¦siks ¦ben·zō'kwin·əl,ēn }

benzoquinone [ORGANIC CHEMISTRY] *See* quinone. { ¦ben·zō,kwō'nōn }

benzoresorcinol [ORGANIC CHEMISTRY] $C_{13}H_{10}O_3$ A compound crystallizing as needles from hot-water solution; used in paints and plastics as an ultraviolet light absorber. Also known as resbenzophenone. { ¦ben·zō·ri'sór·sə,nól }

benzosulfimide [ORGANIC CHEMISTRY] *See* saccharin. { ¦ben·zō'səl·fə,mīd }

benzothiazole [ORGANIC CHEMISTRY] C_6H_4SCHN A thiazole fused to a benzene ring; can be made by ring closure from *o*-amino thiophenols and acid chlorides; derivatives are important industrial products. { ¦ben·zō'thī·ə,zōl }

4-benzothienyl-*N*-methylcarbamate [ORGANIC CHEMISTRY] $C_{10}H_9NO_2S$ A white powder compound with a melting point of 128°C; used as an insecticide for crop insects. { ¦fór ¦ben·zō'thī·ə,nil ¦en¦meth·əl'kär·bə,māt }

benzothiofuran [ORGANIC CHEMISTRY] *See* thianaphthene. { ¦ben‚zō¦thī‚ō'fyu̇‚ran }

1,2,3-benzotriazole [ORGANIC CHEMISTRY] $C_6H_5N_3$ A compound with melting point 98.5°C; soluble in ethanol, insoluble in water; derivatives are ultraviolet absorbers; used as a chemical intermediate. { ¦wən ¦tü ¦thrē ¦ben·zō'trī·ə‚zōl }

benzotrichloride [ORGANIC CHEMISTRY] $C_6H_5CCl_3$ A colorless to yellow liquid that fumes upon exposure to air; has penetrating odor; insoluble in water, soluble in ethanol and ether; used to make dyes. { ¦ben·zō‚trī'klȯr‚īd }

benzotrifluoride [ORGANIC CHEMISTRY] Colorless liquid, boiling point 102.1°C; used for dyes and pharmaceuticals, as solvent and vulcanizing agent, and in insecticides. { ¦ben·zō‚trī'flu̇r‚īd }

benzoyl [ORGANIC CHEMISTRY] The radical $C_6H_5ICO^-$ found, for example, in benzoyl chloride. { 'ben·zə·wəl }

benzoylation [ORGANIC CHEMISTRY] Introduction of the aryl radical (C_6H_5CO) into a molecule. { ‚ben·zō·ə'lā·shən }

benzoyl chloride [ORGANIC CHEMISTRY] C_6H_5COCl Colorless liquid whose vapor induces tears; soluble in ether, decomposes in water; used as an intermediate in chemical synthesis. { 'ben·zə·wəl 'klȯr‚īd }

benzoyl chloride 2,4,6-trichlorophenylhydrazone [ORGANIC CHEMISTRY] C_6H_5CCl-$N_2HC_6H_2Cl_3$ A white to yellow solid with a melting point of 96.5-98°C; insoluble in water; used as an anthelminthic for citrus. { 'ben·zə·wəl 'klȯr‚īd ¦tü ¦fȯr ¦siks ‚trī·klȯr·ə‚fen·əl'hī·drə‚zōn }

benzoyl peroxide [ORGANIC CHEMISTRY] $(C_6H_5CO)_2O_2$ A white, crystalline solid; melting point 103-105°C; explodes when heated above 105°C; slightly soluble in water, soluble in organic solvents; used as a bleaching and drying agent and a polymerization catalyst. { 'ben·zə·wəl pə'räk‚sīd }

benzoylpropethyl [ORGANIC CHEMISTRY] $C_{18}H_{17}Cl_2NO_3$ An off-white, crystalline compound with a melting point of 72°C; used as a preemergence herbicide for control of wild oats. { ¦ben·zə·wəl¦prō·pə·thəl }

3,4-benzpyrene [ORGANIC CHEMISTRY] $C_{20}H_{12}$ A polycyclic hydrocarbon; a chemical carcinogen that will cause skin cancer in many species when applied in low dosage. { ¦thrē ¦fȯr ‚benz'pī‚rēn }

benzthiazuron [ORGANIC CHEMISTRY] $C_9H_9N_3SO$ A white powder that decomposes at 287°C; slightly soluble in water; used as a preemergent herbicide for sugarbeets and fodder beet crops. { ‚benz‚thī'az·yə‚rän }

benzyl [ORGANIC CHEMISTRY] The radical $C_6H_5CH_2^-$ found, for example, in benzyl alcohol, $C_6H_5CH_2OH$. { 'ben·zəl }

benzyl acetate [ORGANIC CHEMISTRY] $C_6H_5CH_2OOCCH_3$ A colorless liquid with a flowery odor; used in perfumes and flavorings and as a solvent for plastics and resins, inks, and polishes. Also known as phenylmethyl acetate. { 'ben·zəl 'as·ə‚tāt }

benzylacetone [ORGANIC CHEMISTRY] $C_6H_5(CH_2)_2COCH_3$ A liquid with a melting point of 233-234°C; used as an attractant to trap melon flies. { ¦ben·zəl'as·ə‚tōn }

benzyl alcohol [ORGANIC CHEMISTRY] $C_6H_5CH_2OH$ An alcohol that melts at 15.3°C, boils at 205.8°C, and is soluble in water and readily soluble in alcohol and ether; valued for the esters it forms with acetic, benzoic, and sebacic acids and used in the soap, perfume, and flavor industries. Also known as phenylmethanol. { 'ben·zəl 'al·kə‚hȯl }

benzylamine [ORGANIC CHEMISTRY] $C_6H_5CH_2NH_2$ A liquid that is soluble in water, ethanol, and ether; boils at 185°C (770 mmHg) and at 84°C (24 mmHg); it is toxic; used as a chemical intermediate in dye production. Also known as aminotoluene. { ¦ben·zəl'am‚ēn }

benzyl benzoate [ORGANIC CHEMISTRY] $C_6H_5COOCH_2C_6H_5$ An oily, colorless liquid ester; used as an antispasmodic drug and as a scabicide. { 'ben·zəl 'ben·zə‚wāt }

benzyl bromide [ORGANIC CHEMISTRY] $C_6H_5CH_2Br$ A toxic, irritating, corrosive clear liquid with a boiling point of 198-199°C; acts as a lacrimator; soluble in alcohol, benzene, and ether; used to make foaming and frothing agents. { 'ben·zəl 'brō‚mīd }

benzyl chloride [ORGANIC CHEMISTRY] $C_6H_5CH_2Cl$ A colorless liquid with a pungent odor produced by the chlorination of toluene. { 'ben·zəl 'klȯr‚īd }

benzyl chloroformate [ORGANIC CHEMISTRY] $C_8H_7ClO_2$ An oily liquid with an acrid odor which causes eyes to tear; boiling point is 103°C (20 mmHg pressure); used to block the amino group in peptide synthesis. { 'ben·zəl ˌklȯr·ə'fȯrˌmāt }

benzyl cinnamate [ORGANIC CHEMISTRY] $C_8H_7COOCH_2C_6H_5$ White crystals; melting point 39°C; insoluble in water, soluble in ethanol; used in perfumery. { 'ben·zəl 'sin·əˌmāt }

benzyl cyanide [ORGANIC CHEMISTRY] $C_6H_5CH_2CN$ A toxic, colorless liquid; insoluble in water, soluble in alcohol and ethanol; boils at 234°C; used in organic synthesis. { 'ben·zəl 'sī·əˌnīd }

benzyl ether [ORGANIC CHEMISTRY] $(C_6H_5CH_2)_2O$ A liquid unstable at room temperature; boiling point 295-298°C; used in perfumes and as a plasticizer for nitrocellulose. Also known as dibenzyl ether. { 'ben·zəl 'ē·thər }

benzyl ethyl ether [ORGANIC CHEMISTRY] $C_6H_5CH_2OC_2H_5$ A colorless, oily, combustible liquid with a boiling point of 185°C; used in organic synthesis and as a flavoring. { 'ben·zəl 'eth·əl 'ē·thər }

benzyl fluoride [ORGANIC CHEMISTRY] $C_6H_5CH_2F$ A toxic, irritating, colorless liquid with a boiling point of 139.8°C at 753 millimeters of mercury; used in organic synthesis. { 'ben·zəl 'flu̇rˌīd }

benzyl formate [ORGANIC CHEMISTRY] $C_6H_5CH_2OOCH$ A colorless liquid with a fruity-spicy odor and a boiling point of 203°C; used in perfumes and as a flavoring. { 'ben·zəl 'fȯrˌmāt }

benzylideneacetone [ORGANIC CHEMISTRY] $C_6H_5CH{=}CHCOCH_3$ A crystalline compound soluble in alcohol, benzene, chloroform, and ether; melting point is 41-45°C; used in perfume manufacture and in organic synthesis. { ben¦zil·əˌdēn'as·əˌtōn }

benzyl isoeugenol [ORGANIC CHEMISTRY] $CH_3CHCHC_6H_3(OCH_3)OCH_2C_6H_5$ A white, crystalline compound with a floral odor; soluble in alcohol and ether; used in perfumery. { 'ben·zəl ¦ī·sō'yü·jəˌnȯl }

benzyl mercaptan [ORGANIC CHEMISTRY] $C_6H_5CH_2SH$ A colorless liquid with a boiling point of 195°C; soluble in alcohol and carbon disulfide; used as an odorant and for flavoring. { 'ben·zəl mər'kap·tan }

benzyl penicillinic acid [ORGANIC CHEMISTRY] $C_{16}H_{18}N_2O_4S$ An amorphous white powder extracted with ether or chloroform from an acidified aqueous solution of benzyl penicillin. { 'ben·zəl ¦pen·ə·sə¦lin·ik 'as·əd }

benzyl propionate [ORGANIC CHEMISTRY] $C_2H_5COOCH_2C_6H_5$ A combustible liquid with a sweet odor and a boiling point of 220°C; used in perfumes and for flavoring. { 'ben·zəl 'prō·pē·əˌnāt }

benzyl salicylate [ORGANIC CHEMISTRY] $C_{14}H_{12}O_3$ A thick liquid with a slight, pleasant odor; used as a fixer in perfumery and in sunburn preparations. { 'ben·zəl sə'lis·əˌlāt }

benzyne [ORGANIC CHEMISTRY] C_6H_4 A chemical species whose structure consists of an aromatic ring in which four carbon atoms are bonded to hydrogen atoms and two adjacent carbon atoms lack substitutents; a member of a class of compounds known as arynes. { 'benˌzīn }

berbamine [ORGANIC CHEMISTRY] $C_{37}H_{40}N_2O_6$ An alkaloid, melting point 170°C; slightly soluble in water, soluble in alcohol and ether. { 'bərˌbəˌmēn }

berberine [ORGANIC CHEMISTRY] $C_{20}H_{19}NO_5$ A toxic compound; melting point 145°C; the anhydrous form is insoluble in water, soluble in alcohol and ether. { 'bər·bəˌrēn }

Berg's diver method [PHYSICAL CHEMISTRY] *See* diver method. { 'bərgz 'dīv·ər ˌmeth·əd }

berkelium [CHEMISTRY] A radioactive element, symbol Bk, atomic number 97, the eighth member of the actinide series; properties resemble those of the rare-earth cerium. { 'bər·klē·əm }

Berthelot equation [PHYSICAL CHEMISTRY] A form of the equation of state which relates the temperature, pressure, and volume of a gas with the gas constant. { 'ber·tə·lō i'kwā·zhən }

Berthelot-Thomsen principle [PHYSICAL CHEMISTRY] The principle that of all chemical reactions possible, the one developing the greatest amount of heat will take place, with certain obvious exceptions such as changes of state. { 'ber·tə·lō ¦täm·sən ˌprin·sə·pəl }

berthollide [CHEMISTRY] A compound whose solid phase exhibits a range of composition. { 'bər·thəˌlīd }

beryllate [INORGANIC CHEMISTRY] **1.** BeO_2^{2-} An ion containing beryllium and oxygen. **2.** A salt produced by the reaction of a strong alkali such as sodium hydroxide with beryllium oxide. { 'ber·əˌlāt }

beryllia [INORGANIC CHEMISTRY] *See* beryllium oxide. { bə'ril·ē·ə }

beryllide [INORGANIC CHEMISTRY] A chemical combination of beryllium with a metal, such as zirconium or tantalum. { bə'ril·əˌdē }

beryllium [CHEMISTRY] A chemical element, symbol Be, atomic number 4, atomic weight 9.0122. { bə'ril·ē·əm }

beryllium fluoride [INORGANIC CHEMISTRY] BeF_2 A hygroscopic, amorphous solid with a melting point of 800°C; soluble in water; used in beryllium metallurgy. { bə'ril·ē·əm ¦flu̇rˌīd }

beryllium nitrate [INORGANIC CHEMISTRY] $Be(NO_3)_2 \cdot 3H_2O$ A compound that forms colorless, deliquescent crystals that are soluble in water; used to introduce beryllium oxide into materials used in incandescent mantles. { bə'ril·ē·əm 'nīˌtrāt }

beryllium nitride [INORGANIC CHEMISTRY] Be_3N_2 Refractory, white crystals with a melting point of 2200±40°C; used in the manufacture of radioactive carbon-14 and in experimental rocket fuels. { bə'ril·ē·əm ¦nīˌtrīd }

beryllium oxide [INORGANIC CHEMISTRY] BeO An amorphous white powder, insoluble in water; used to make beryllium salts and as a refractory. Also known as beryllia. { bə'ril·ē·əm 'äkˌsīd }

betaine [ORGANIC CHEMISTRY] $C_5H_{11}O_2N$ An alkaloid; very soluble in water, soluble in ethyl alcohol and methanol; the hydrochloride is used as a source of hydrogen chloride and in medicine. Also known as lycine; oxyneurine. { 'bēd·əˌēn }

beta-ray spectrometer [SPECTROSCOPY] An instrument used to determine the energy distribution of beta particles and secondary electrons. Also known as beta spectrometer. { 'bād·ə ˌrā spek'träm·əd·ər }

beta spectrometer [SPECTROSCOPY] *See* beta-ray spectrometer. { 'bād·ə spek'träm·əd·ər }

BET equation [PHYSICAL CHEMISTRY] *See* Brunauer-Emmett-Teller equation. { ¦bē¦ē¦tē i'kwā·zhən }

betula oil [ORGANIC CHEMISTRY] *See* methyl salicylate. { 'bech·ə·lə ˌȯil }

betulinic acid [ORGANIC CHEMISTRY] $C_{30}H_{48}O_3$ A dibasic acid, slightly soluble in water, ethyl alcohol, and acetone. { ¦bech·ə¦lin·ik 'as·əd }

BHA [ORGANIC CHEMISTRY] *See* butylated hydroxyanisole.

BHT [ORGANIC CHEMISTRY] *See* butylated hydroxytoluene.

Bi [CHEMISTRY] *See* bismuth.

biacetyl [ORGANIC CHEMISTRY] *See* diacetyl. { ¦bī·ə'sēd·əl }

biamperometry [ANALYTICAL CHEMISTRY] Amperometric titration that uses two polarizing or indicating electrodes to detect the end point of a redox reaction between the substance being titrated and the titrant. { ¦bīˌam·pə'räm·ə·trē }

bias [ANALYTICAL CHEMISTRY] A systematic error occurring in a chemical measurement that is inherent in the method itself or caused by some artifact in the system, such as a temperature effect. { 'bī·əs }

bibenzyl [ORGANIC CHEMISTRY] $C_{14}H_{14}$ A hydrocarbon consisting of two benzene rings attached to ethane. Also known as dibenzyl. { ¦bī'ben·zil }

bicarbonate [INORGANIC CHEMISTRY] A salt obtained by the neutralization of one hydrogen in carbonic acid. { bī'kär·bəˌnət }

bicarbonate of soda [INORGANIC CHEMISTRY] *See* sodium bicarbonate. { bī¦kär·bə·nət əv 'sō·də }

bichloride of mercury [INORGANIC CHEMISTRY] *See* mercuric chloride. { bī'klȯrˌīd əv 'mər·kyə·rē }

bicuculine [ORGANIC CHEMISTRY] $C_{20}H_{17}NO_6$ A convulsant alkaloid found in plants of the family Fumariaceaea. { bī'kü·kyəˌlēn }

bicyclic compound [ORGANIC CHEMISTRY] A compound having two rings which share a pair of bridgehead carbon atoms. { bī'sik·lik 'käm‚paůnd }

bidentate ligand [INORGANIC CHEMISTRY] A chelating agent having two groups capable of attachment to a metal ion. { bī'den‚tāt 'lig·ənd }

Biebrich red [ORGANIC CHEMISTRY] *See* scarlet red. { 'bē‚brik 'red }

bifenox [ORGANIC CHEMISTRY] $C_{14}H_9Cl_2NO_5$ A tan, crystalline compound with a melting point of 84-86°C; insoluble in water; used as a preemergence herbicide for weed control in soybeans, corn, and sorghum, and as a pre- and postemergence herbicide in rice and small greens. { bī'fen‚äks }

bifluoride [INORGANIC CHEMISTRY] An acid fluoride whose formula has the form MHF_2; an example is sodium bifluoride, $NaHF_2$. { bī'flůr‚īd }

bifunctional catalyst [CHEMISTRY] A catalytic substance that possesses two catalytic sites and thus is capable of catalyzing two different types of reactions. Also known as dual-function catalyst. { ¦bī¦fəŋk·shən·əl 'kad·ə‚list }

bifunctional chelating agent [ORGANIC CHEMISTRY] A reagent with a molecular structure that contains a strong metal-chelating group and a chemically reactive functional group. { ¦bī¦fəŋk·shən·əl 'kē‚lād·iŋ ‚ā·jənt }

bilateral slit [SPECTROSCOPY] A slit for spectrometers and spectrographs that is bounded by two metal strips which can be moved symmetrically, allowing the distance between them to be adjusted with great precision. { ¦bī‚lad·ə·rəl 'slit }

bilayer [CHEMISTRY] A layer two molecules thick, such as that formed on the surface of the aqueous phase by phospholipids in aqueous solution. { 'bī‚lā·ər }

bimolecular [CHEMISTRY] Referring to two molecules. { ¦bī·mə'lek·yə·lər }

bimolecular reaction [CHEMISTRY] A chemical transformation or change involving two molecules. { ¦bī·mə'lek·yə·lər rē'ak·shən }

binapacryl [ORGANIC CHEMISTRY] $C_{15}H_{18}O_6N_2$ A light tan solid with a melting point of 68-69°C, insoluble in water, used for powdery mildew and for mites on fruits. { bə'nap·ə‚kril }

binary compound [CHEMISTRY] A compound that has two elements; it may contain two or more atoms; examples are KCl and $AlCl_3$. { 'bīn·ə·rē 'käm‚paůnd }

bioassay [ANALYTICAL CHEMISTRY] A method for quantitatively determining the concentration of a substance by its effect on the growth of a suitable animal, plant, or microorganism under controlled conditions. { ¦bī·ō'as‚ā }

bioautography [ANALYTICAL CHEMISTRY] A bioassay based upon the ability of some compounds (for example, vitamin B_{12}) to enhance the growth of some organisms or compounds and to repress the growth of others; used to assay certain antibiotics. { ¦bī·ō‚ó'täg·rə·fē }

biochemistry [CHEMISTRY] The study of chemical substances occurring in living organisms and the reactions and methods for identifying these substances. { ¦bī·ō'kem·ə·strē }

biologic artifact [ORGANIC CHEMISTRY] An organic compound with a chemical structure that demonstrates the compound's derivation from living matter. { ¦bī·ə¦läj·ik 'ärd·ə‚fakt }

biomimetic catalyst [ORGANIC CHEMISTRY] A synthetic compound that can simulate the mode of action of a natural enzyme by catalyzing a reaction at ambient conditions. { ¦bī·ō·mə'med·ik 'kad·ə‚list }

biosensor [ANALYTICAL CHEMISTRY] An analytical device that converts the concentration of an analyte in an appropriate sample into an electrical signal by means of a biologically derived sensing element intimately connected to, or integrated into, a transducer. { ¦bī·ō¦sen·sər }

biphenyl [ORGANIC CHEMISTRY] $C_{12}H_{10}$ A white or slightly yellow crystalline hydrocarbon, melting point 70.0°C, boiling point 255.9°C, and density 1.9896, which gives plates or monoclinic prismatic crystals; used as a heat-transfer medium and as a raw material for chlorinated diphenyls. Also known as diphenyl, phenylbenzene. { bī'fen·əl }

***para*-biphenylamine** [ORGANIC CHEMISTRY] $C_{12}H_{11}N$ Leaflets with a melting point of 53°C; readily soluble in hot water, alcohol, and chloroform; used in the detection of sulfates and also as a carcinogen in cancer research. { ¦par·ə ‚bī·fə'nil·ə‚mēn }

2,2'-bipyridine [ORGANIC CHEMISTRY] *See* 2,2'-dipyridyl. { ¦tü ¦tü·prīm ¦bī'pir·ə,dēn }

biradical [CHEMISTRY] A chemical species having two independent odd-electron sites. { bī'rad·ə·kəl }

Birge-Sponer extrapolation [SPECTROSCOPY] A method of calculating the dissociation limit of a diatomic molecule when the convergence limit cannot be observed directly, based on the assumption that vibrational energy levels converge to a limit for a finite value of the vibrational quantum number. { ¦bir·gə 'spōn·ər ik,strap·ə'lā·shən }

bis- [CHEMISTRY] A prefix indicating doubled or twice. { bis }

2,2-bis(*para*-chlorophenyl)-1,1-dichloroethane [ORGANIC CHEMISTRY] $C_{14}H_{10}Cl_4$ A colorless, crystalline compound with a melting point of 109-111°C; insoluble in water; used as an insecticide on fruits and vegetables. Also known as DDD; TDE. { ¦tü ¦tü ,bis 'par·ə ,klör·ə'fen·əl ¦wən ¦wən di,klörō'e,thān }

bismuth [CHEMISTRY] A metallic element, symbol Bi, of atomic number 83 and atomic weight 208.980. { 'biz·məth }

bismuthate [INORGANIC CHEMISTRY] A compound of bismuth in which the bismuth has a valence of +5; an example is sodium bismuthate, $NaBiO_3$. { 'biz·mə,thāt }

bismuth chloride [INORGANIC CHEMISTRY] $BiCl_3$ A deliquescent material that melts at 230-232°C and decomposes in water to form the oxychloride; used to make bismuth salts. Also known as bismuth trichloride. { 'biz·məth 'klör,īd }

bismuth chromate [INORGANIC CHEMISTRY] $Bi_2O_3 \cdot Cr_2O_3$ An orange-red powder, soluble in alkalies and acids; used as a pigment. { 'biz·məth 'krō,māt }

bismuth citrate [ORGANIC CHEMISTRY] $BiC_6H_5O_7$ A salt of citric acid that forms white crystals, insoluble in water; used as an astringent. { 'biz·məth 'sī,trāt }

bismuth hydroxide [INORGANIC CHEMISTRY] $Bi(OH)_3$ A water-insoluble, white powder; precipitated by hydroxyl ion from bismuth salt solutions. { 'biz·məth hī'dräk,sīd }

bismuth iodide [INORGANIC CHEMISTRY] BiI_3 A bismuth halide that sublimes in grayish-black hexagonal crystals melting at 408°C, insoluble in water; used in analytical chemistry. { 'biz·məth 'ī·ə,dīd }

bismuth nitrate [INORGANIC CHEMISTRY] $Bi(NO_3)_3 \cdot 5H_2O$ White, triclinic crystals that decompose in water; used as an astringent and antiseptic. { 'biz·məth 'nī,trāt }

bismuth oleate [ORGANIC CHEMISTRY] $Bi(C_{17}H_{33}COO)_3$ A salt of oleic acid obtained as yellow granules; used in medicines to treat skin diseases. { 'biz·məth 'ō·lē,āt }

bismuth oxide [INORGANIC CHEMISTRY] *See* bismuth trioxide. { 'biz·məth 'äk,sīd }

bismuth oxychloride [INORGANIC CHEMISTRY] $BiOCl$ A white powder; insoluble in water, soluble in acid; a toxic material if ingested; used in pigments and cosmetics. { 'biz·məth ,äk·sē'klör,īd }

bismuth phenate [ORGANIC CHEMISTRY] $C_6H_5O \cdot Bi(OH)_2$ An odorless, tasteless, gray-white powder; used in medicine. { 'biz·məth 'fen,āt }

bismuth potassium tartrate [ORGANIC CHEMISTRY] *See* potassium bismuth tartrate. { 'biz·məth pe'tas·ē·əm 'tär,trāt }

bismuth pyrogallate [ORGANIC CHEMISTRY] $Bi(OH)C_6H_3(OH)O_2$ An odorless, tasteless, yellowish-green, amorphous powder; used in medicine as intestinal antiseptic and dusting powder. { 'biz·məth ¦pī·rō'gal,āt }

bismuth subcarbonate [INORGANIC CHEMISTRY] $(BiO)_2CO_3$ or $Bi_2O_3 \cdot CO_2 \cdot {}^1/_2H_2O$ A white powder; dissolves in hydrochloric or nitric acid, insoluble in alcohol and water; used as opacifier in x-ray diagnosis, in ceramic glass, and in enamel fluxes. { 'biz·məth səb'kär·bə,nāt }

bismuth subgallate [ORGANIC CHEMISTRY] $C_6H_2(OH)_3COOBi(OH)_2$ A yellow powder; dissolves in dilute alkali solutions, but is insoluble in water, ether, and alcohol; used in medicine. { 'biz·məth ,səb'gal,āt }

bismuth subnitrate [INORGANIC CHEMISTRY] $4BiNO_3(OH)_2 \cdot BiO(OH)$ A white, hygroscopic powder; used in bismuth salts, perfumes, cosmetics, ceramic enamels, pharmaceuticals, and analytical chemistry. { 'biz·məth ,səb'ni,trāt }

bismuth subsalicylate [INORGANIC CHEMISTRY] $Bi(C_7H_5O)_3Bi_2O_3$ A white powder that is insoluble in ethanol and water; used in medicine and as a fungicide for tobacco crops. { 'biz·məth ,səb·sə'lis·ə,lāt }

bismuth telluride [INORGANIC CHEMISTRY] Bi_2Te_3 Gray, hexagonal platelets with a melting point of 573°C; used for semiconductors, thermoelectric cooling, and power generation applications. { 'biz·məth 'tel·yə‚rīd }

bismuth trichloride [INORGANIC CHEMISTRY] *See* bismuth chloride. { 'biz·məth trī'klȯr‚īd }

bismuth trioxide [INORGANIC CHEMISTRY] Bi_2O_3 A yellow powder; melting point 820°C; insoluble in water, dissolves in acid; used to make enamels and to color ceramics. Also known as bismuth oxide; bismuth yellow. { 'biz·məth trī'äk‚sīd }

bismuth yellow [INORGANIC CHEMISTRY] *See* bismuth trioxide. { 'biz·məth 'yel·ō }

bisphenol A [ORGANIC CHEMISTRY] $(CH_3)_2C(C_6H_5OH)_2$ Brown crystals that are insoluble in water; used in the production of phenolic and epoxy resins. { bī'sfēn·ȯl 'ā }

bistable system [CHEMISTRY] A chemical system with two relatively stable states which permits an oscillation between domination by one of these states to domination by the other. { ¦bī¦stā·bəl 'sis·təm }

bisulfate [INORGANIC CHEMISTRY] A compound that has the HSO_4^- radical; derived from sulfuric acid. { bī'səl‚fāt }

bitartrate [ORGANIC CHEMISTRY] A salt with the radical $HC_4H_4O_6^-$. Also known as acid tartrate. { bī'tär‚trāt }

bithionol [ORGANIC CHEMISTRY] A halogenated form of bisphenol used as an ingredient in germicidal soaps and as a medicine in the treatment of clonorchiases. { bī'thī·ə‚nȯl }

biuret [ORGANIC CHEMISTRY] $NH_2CONHCONH_2$ Colorless needles that are soluble in hot water and decompose at 190°C; a condensation product of urea. { ‚bī·ya'ret }

bivalent [CHEMISTRY] Possessing a valence of two. { bī'vā·lənt }

bixin [ORGANIC CHEMISTRY] $C_{25}H_{30}O_4$ A carotenoid acid occurring in the seeds of *Bixa orellano*; used as a fat and food coloring agent. { 'bik·sən }

Bk [CHEMISTRY] *See* berkelium.

black [CHEMISTRY] Fine particles of impure carbon that are made by the incomplete burning of carbon compounds, such as natural gas, naphthas, acetylene, bones, ivory, and vegetables. { blak }

black cyanide [INORGANIC CHEMISTRY] *See* calcium cyanide. { ¦blak 'sī·ə‚nīd }

black iron oxide [INORGANIC CHEMISTRY] *See* ferrous oxide. { 'blak 'ī·ərn 'äk‚sīd }

Blagden's law [PHYSICAL CHEMISTRY] The law that the lowering of a solution's freezing point is proportional to the amount of dissolved substance. { 'blag·dənz ‚lȯ }

blanc fixe [INORGANIC CHEMISTRY] $BaSO_4$ A commercial name for barium sulfate, with some use in pure form in the paint, paper, and pigment industries as a pigment extender. { ‚bläŋk 'feks }

Blanc rule [ORGANIC CHEMISTRY] The rule that glutaric and succinic acids yield cyclic anhydrides on pyrolysis, while adipic and pimelic acids yield cyclic ketones; there are certain exceptions. { 'bläŋk ‚rül }

blank [ANALYTICAL CHEMISTRY] In a chemical analysis, the measured value that is obtained in the absence of a specified component of a sample and that reflects contamination from sources external to the component; it is deducted from the value obtained when the test is performed with the specified component present. Also known as analytical blank. { blaŋk }

blasticidin-S [ORGANIC CHEMISTRY] A compound with a melting point of 235-236°C; soluble in water; used as a fungicide for rice crops. { ‚blas'tis·ə·dən 'es }

bleaching agent [CHEMISTRY] An oxidizing or reducing chemical such as sodium hypochlorite, sulfur dioxide, sodium acid sulfite, or hydrogen peroxide. { 'blēch·iŋ ‚ā·jənt }

bleed [CHEMISTRY] Diffusion of coloring matter from a substance. { blēd }

blind sample [ANALYTICAL CHEMISTRY] In chemical analysis, a selected sample whose composition is unknown except to the person submitting it; used to test the validity of the measurement process. { ¦blīnd 'samp·əl }

block copolymer [ORGANIC CHEMISTRY] A copolymer in which the like monomer units occur in relatively long alternate sequences on a chain. Also known as block polymer. { 'bläk kō'päl·ə·mər }

blocking [CHEMISTRY] Undesired adhesion of granular particles; often occurs with damp powders or plastic pellets in storage bins or during movement through conduits. { 'bläk·iŋ }

blocking group [ORGANIC CHEMISTRY] In peptide synthesis, a group that is reacted with a free amino or carboxyl group on an amino acid to prevent its taking part in subsequent formation of peptide bonds. { 'bläk·iŋ ˌgrüp }

block polymer [ORGANIC CHEMISTRY] *See* block copolymer. { 'bläk 'päl·ə·mər }

blowpipe reaction analysis [ANALYTICAL CHEMISTRY] A method of analysis in which a blowpipe is used to heat and decompose a compound or mineral; a characteristic color appears in the flame or a colored crust appears on charcoal. { 'blōˌpīp rē'ak·shən ə'nal·ə·səs }

blue tetrazolium [ORGANIC CHEMISTRY] $C_{40}H_{32}Cl_2N_8O_2$ Lemon yellow crystals that decompose at 242-245°C; soluble in chloroform, ethanol, and methanol; used in seed germination research, as a stain for molds and bacteria, and in histochemical studies. { ¦blü te·trə'zōl·ē·əm }

BNOA [ORGANIC CHEMISTRY] *See* β-naphthoxyacetic acid.

boat [CHEMISTRY] A platinum or ceramic vessel for holding a substance for analysis by combustion. { bōt }

boat conformation [ORGANIC CHEMISTRY] A boat-shaped conformation in space which can be assumed by cyclohexane or similar compounds; a relatively unstable form. { 'bōt ˌkän·fər'mā·shən }

Boettger's test [ANALYTICAL CHEMISTRY] A test for the presence of saccharides, utilizing the reduction of bismuth subnitrate to metallic bismuth, a precipitate. { 'betˌgərz ˌtest }

boiler compound [CHEMISTRY] Any chemical used to treat boiler water to prevent corrosion, the fouling of heat-absorbing surfaces, foaming, and the contamination of steam. { 'bȯil·ər ˌkamˌpau̇nd }

boiler scale [CHEMISTRY] Deposits from silica and other contaminants in boiler water that form on the internal surfaces of heat-absorbing components, increase metal temperatures, and result in eventual failure of the pressure parts because of overheating. Also known as scale. { 'bȯil·ər ˌskāl }

boiling [PHYSICAL CHEMISTRY] The transition of a substance from the liquid to the gaseous phase, taking place at a single temperature in pure substances and over a range of temperatures in mixtures. { 'bȯil·iŋ }

boiling point [PHYSICAL CHEMISTRY] Abbreviated bp. **1.** The temperature at which the transition from the liquid to the gaseous phase occurs in a pure substance at fixed pressure. **2.** *See* bubble point. { 'bȯil·iŋ ˌpȯint }

boiling-point elevation [CHEMISTRY] The raising of the normal boiling point of a pure liquid compound by the presence of a dissolved substance, the elevation being in direct relation to the dissolved substance's molecular weight. { 'bȯil·iŋ ˌpȯint el·ə'vā·shən }

boiling range [CHEMISTRY] The temperature range of a laboratory distillation of an oil from start until evaporation is complete. { 'bȯil·iŋ ˌrānj }

boletic acid [ORGANIC CHEMISTRY] *See* fumaric acid. { bə'led·ik 'as·əd }

bond [CHEMISTRY] The strong attractive force that holds together atoms in molecules and crystalline salts. Also known as chemical bond. { bänd }

bond angle [PHYSICAL CHEMISTRY] The angle between bonds sharing a common atom. Also known as valence angle. { 'bänd ˌaŋ·gəl }

bond dissociation energy [PHYSICAL CHEMISTRY] The change in enthalpy that occurs with the homolytic cleavage of a chemical bond under conditions of standard state. { ˌbänd diˌsō·sē¦ā·shən 'en·ər·jē }

bond distance [PHYSICAL CHEMISTRY] The distance separating the two nuclei of two atoms bonded to each other in a molecule. Also known as bond length. { 'bänd ˌdis·təns }

bonded-phase chromatography [ANALYTICAL CHEMISTRY] A type of high-pressure liquid chromatography which employs a stable, chemically bonded stationary phase. { ¦bän·dəd ˌfāz ˌkrō·mə'täg·rə·fē }

bond energy [PHYSICAL CHEMISTRY] **1.** The average value of specific bond dissociation energies that have been measured from different molecules of a given type. **2.** *See* average bond association energy. { 'bänd ˌen·ər·jē }

bond hybridization [CHEMISTRY] The linear combination of two or more simple atomic orbitals. { ¦bänd ˌhī·brəd·ə'zā·shən }

bonding [CHEMISTRY] The joining together of atoms to form molecules or crystalline salts. { 'bän·diŋ }

bonding electron [PHYSICAL CHEMISTRY] An electron whose orbit spans the entire molecule and so assists in holding it together. { 'bän·diŋ i'lekˌträn }

bonding orbital [PHYSICAL CHEMISTRY] A molecular orbital formed by a bonding electron whose energy decreases as the nuclei are brought closer together, resulting in a net attraction and chemical bonding. { 'bän·diŋ 'ȯr·bəd·əl }

bond length [PHYSICAL CHEMISTRY] *See* bond distance. { 'bänd ˌleŋkth }

bond-line formula [ORGANIC CHEMISTRY] A representation of a molecule in which bonds are represented by lines, carbon atoms are represented by line ends and intersections, and atoms other than hydrogen and carbon are represented by their elemental symbols, as is hydrogen when it is bonded to an atom other than hydrogen or carbon. Also known as carbon-skeleton formula; line-segment formula. { 'bänd ˌlīn ˌfȯr·myə·lə }

bond migration [CHEMISTRY] The movement of a bond to a different position within the same molecular entity. { 'bänd mīˌgrā·shən }

bond moment [PHYSICAL CHEMISTRY] The degree of polarity of a chemical bond as calculated from the value of the force of the response of the bond when the bond is subjected to an electric field. { 'bänd ˌmō·mənt }

bond strength [CHEMISTRY] The strength with which a chemical bond holds two atoms together; conventionally measured in terms of the amount of energy, in kilocalories per mole, required to break the bond. { 'bänd ˌstreŋkth }

bone ash [CHEMISTRY] A white ash consisting primarily of tribasic calcium phosphate obtained by burning bones in air; used in cleaning jewelry and in some pottery. { 'bōn ˌash }

boracic acid [INORGANIC CHEMISTRY] *See* boric acid. { bə'ras·ik 'as·əd }

borane [INORGANIC CHEMISTRY] **1.** A class of binary compounds of boron and hydrogen; boranes are used as fuels. Also known as boron hydride. **2.** A substance which may be considered a derivative of a boron-hydrogen compound, such as BCl_3 and $B_{10}H_{12}I_2$. { 'bȯˌrān }

borate [CHEMISTRY] **1.** A generic term referring to salts or esters of boric acid. **2.** Related to boric oxide, B_2O_3, or commonly to only the salts of orthoboric acid, H_3BO_3. { 'bȯˌrāt }

borazole [INORGANIC CHEMISTRY] $B_3N_3H_6$ A colorless liquid boiling at 53°C; with water it hydrolyzes to form boron hydrides; the borazole molecule is the inorganic analog of the benzene molecule. { 'bȯr·əˌzōl }

borazon [INORGANIC CHEMISTRY] A form of boron nitride with a zinc blende structure produced by subjecting the ordinary form to high pressure and temperature. { 'bȯr·əˌzän }

boric acid [INORGANIC CHEMISTRY] H_3BO_3 An acid derived from boric oxide in the form of white, triclinic crystals, melting at 185°C, soluble in water. Also known as boracic acid; orthoboric acid. { ¦bȯr·ik 'as·əd }

boric acid ester [ORGANIC CHEMISTRY] Any compound readily hydrolyzed to yield boric acid and the respective alcohol; for example, trimethyl borate hydrolyzes to boric acid and methyl alcohol. { ¦bȯr·ik 'as·əd 'es·tər }

boric oxide [INORGANIC CHEMISTRY] B_2O_3 A trioxide of boron obtained as rhombic crystals melting at 460°C; used as an intermediate in the production of boron halides and metallic borides and as a thermal neutron absorber in nuclear engineering. Also known as boron oxide. { ¦bȯr·ik 'äkˌsīd }

boride [INORGANIC CHEMISTRY] A binary compound of boron and a metal formed by heating a mixture of the two elements. { 'bȯrˌīd }

borneol [ORGANIC CHEMISTRY] $C_{10}H_{17}OH$ White lumps with camphor odor; insoluble in water, soluble in alcohol; melting point 203°C; used in perfumes, medicine, and chemical synthesis. { 'bȯr·nēˌȯl }

Born equation [PHYSICAL CHEMISTRY] An equation for determining the free energy of solvation of an ion in terms of the Avogadro number, the ionic valency, the ion's electronic charge, the dielectric constant of the electrolytic, and the ionic radius. { 'bórn i'kwā·zhən }

Born-Oppenheimer approximation [PHYSICAL CHEMISTRY] The approximation, used in the Born-Oppenheimer method, that the electronic wave functions and energy levels at any instant depend only on the positions of the nuclei at that instant and not on the motions of the nuclei. Also known as adiabatic approximation. { ¦bórn 'äp·ən,hī·mər ə,präk·sə,mā·shən }

Born-Oppenheimer method [PHYSICAL CHEMISTRY] A method for calculating the force constants between atoms by assuming that the electron motion is so fast compared with the nuclear motions that the electrons follow the motions of the nuclei adiabatically. { ¦bórn 'äp·ən,hīm·ər ,meth·əd }

bornyl acetate [ORGANIC CHEMISTRY] $C_{10}H_{17}OOCCH_3$ A colorless liquid that forms crystals at 10°C; has characteristic piny-camphoraceous odor; used in perfumes and for flavoring. { 'bórn·əl 'as·ə,tāt }

bornyl isovalerate [ORGANIC CHEMISTRY] $C_{10}H_{17}OOC_5H_9$ An aromatic fluid with a boiling point of 255-260°C; soluble in alcohol and ether; used in medicine and as a flavoring. { 'bórn·əl ¦ī·sō'val·ə,rāt }

boron [CHEMISTRY] A chemical element, symbol B, atomic number 5, atomic weight 10.811; it has three valence electrons and is nonmetallic. { 'bó,rän }

boron carbide [ORGANIC CHEMISTRY] Any compound of boron and carbon, especially B_4C (used as an abrasive, alloying agent, and neutron absorber). { 'bó,rän 'kär,bīd }

boron fiber [CHEMISTRY] Fiber produced by vapor-deposition methods; used in various composite materials to impart a balance of strength and stiffness. Also known as boron filament. { 'bó,rän ,fī·bər }

boron filament [CHEMISTRY] *See* boron fiber. { 'bó,rän ,fil·ə·mənt }

boron fluoride [INORGANIC CHEMISTRY] BF_3 A colorless pungent gas in a dry atmosphere; used in industry as an acidic catalyst for polymerizations, esterifications, and alkylations. Also known as boron trifluoride. { 'bó,rän 'flúr,īd }

boron hydride [INORGANIC CHEMISTRY] *See* borane. { 'bó,rän 'hī,drīd }

boron nitride [INORGANIC CHEMISTRY] BN A binary compound of boron and nitrogen, especially a white, fluffy powder with high chemical and thermal stability and high electrical resistance. { 'bó,rän 'nī,trīd }

boron nitride fiber [INORGANIC CHEMISTRY] Inorganic, high-strength fiber, made of boron nitride, that is resistant to chemicals and electricity but susceptible to oxidation above 1600°F (870°C); used in composite structures for yarns, fibers, and woven products. { 'bó,rän 'nī,trīd 'fī·bər }

boron oxide [INORGANIC CHEMISTRY] *See* boric oxide. { 'bó,rän 'äk,sīd }

boron polymer [ORGANIC CHEMISTRY] Macromolecules formed by polymerization of compounds containing, for example, boron-nitrogen, boron-phosphorus, or boron-arsenic bonds. { 'bó,rän 'päl·ə·mər }

boron trichloride [INORGANIC CHEMISTRY] BCl_3 A colorless liquid used as a catalyst and in refining of aluminum, magnesium, zinc, and copper. { 'bó,rän trī'klór,īd }

boron triethoxide [ORGANIC CHEMISTRY] *See* ethyl borate. { 'bó,rän ,trī·ə'thäk,sīd }

boron triethyl [ORGANIC CHEMISTRY] *See* triethylborane. { 'bó,rän trī'eth·əl }

boron trifluoride [INORGANIC CHEMISTRY] *See* boron fluoride. { 'bó,rän trī'flúr,īd }

boron trifluoride etherate [ORGANIC CHEMISTRY] $C_4H_{10}BF_3O$ A fuming liquid hydrolyzed by air immediately; boiling point is 125.7°C; used as a catalyst in reactions involving condensation, dehydration, polymerization, alkylation, and acetylation. { 'bó,rän trī'flúr,īd 'ē·thə,rāt }

bottom steam [CHEMISTRY] Steam piped into the bottom of the still during oil distillation. { 'bäd·əm ,stēm }

boturon [ORGANIC CHEMISTRY] $C_{12}H_{13}N_2OCl$ A white solid with a melting point of 145-146°C; used as pre- and postemergence herbicide in cereals, orchards, and vineyards. Also known as butyron. { 'bäch·ə,rän }

Bouguer-Lambert-Beer law [ANALYTICAL CHEMISTRY] The intensity of a beam of monochromatic radiation in an absorbing medium decreases exponentially with penetration distance. Also known as Beer-Lambert-Bouguer law; Lambert-Beer law. { bü'ger ¦läm·bert ¦ber ˌlȯ }

Bouguer-Lambert law [ANALYTICAL CHEMISTRY] The law that the change in intensity of light transmitted through an absorbing substance is related exponentially to the thickness of the absorbing medium and a constant which depends on the sample and the wavelength of the light. Also known as Lambert's law. { bü'ger ¦läm·bərt ˌlȯ }

boundary line [PHYSICAL CHEMISTRY] On a phase diagram, the line along which any two phase areas adjoin in a binary system, or the line along which any two liquidus surfaces intersect in a ternary system. { 'bau̇n·drē ˌlīn }

boundary value component [PHYSICAL CHEMISTRY] *See* perfectly mobile component. { 'bau̇n·drē ˌval·yü kəmˌpō·nənt }

bound water [CHEMISTRY] Water that is a portion of a system such as tissues or soil and does not form ice crystals until the material's temperature is lowered to about −20°C. { ¦bau̇nd 'wȯd·ər }

Bouvealt-Blanc method [ORGANIC CHEMISTRY] A laboratory method for preparing alcohols by reduction of esters utilizing sodium dissolved in alcohol. { ¦büˌvō ¦blän ˌmeth·əd }

bp [PHYSICAL CHEMISTRY] *See* boiling point.

BPMC [ORGANIC CHEMISTRY] *See* 2-*sec*-butyl phenyl-N-methyl carbamate.

Br [CHEMISTRY] *See* bromine.

Brackett series [SPECTROSCOPY] A series of lines in the infrared spectrum of atomic hydrogen whose wave numbers are given by $R_H[(1/16) - (1/n^2)]$, where R_H is the Rydberg constant for hydrogen and n is any integer greater than 4. { 'brak·ət ˌsir·ēz }

braking effects [PHYSICAL CHEMISTRY] The electrophoretic effect and the asymmetry effect, which together control the speed with which ions drift in a strong electrolyte. { 'brāk·iŋ iˌfeks }

branch [ORGANIC CHEMISTRY] *See* side chain. { branch }

branched chain [ORGANIC CHEMISTRY] *See* side chain. { 'brancht 'chān }

bridge [ORGANIC CHEMISTRY] A connection between two different parts of a molecule consisting of a valence bond, an atom, or an unbranched chain of atoms. { brij }

bridged intermediate [ORGANIC CHEMISTRY] *See* bridged ion. { 'brijd in·tər'mēd·ē·ət }

bridged ion [ORGANIC CHEMISTRY] A reactive intermediate in which an atom from one of the reactants is bonded partially to each of two carbon atoms of a reactant containing a double carbon-to-carbon bond. Also known as bridged intermediate, cyclic ion. { 'brijd 'ī·än }

bridging ligand [ORGANIC CHEMISTRY] A ligand in which an atom or molecular species which is able to exist independently is simultaneously bonded to two or more metal atoms. { 'brij·iŋ ˌlīg·ənd }

bright-line spectrum [SPECTROSCOPY] An emission spectrum made up of bright lines on a dark background. { 'brīt ˌlīn 'spek·trəm }

broadening of spectral lines [SPECTROSCOPY] A widening of spectral lines by collision or pressure broadening, or possibly by Doppler effect. { 'brȯd·ən·iŋ əv ¦spek·trəl 'līnz }

Broenner's acid [ORGANIC CHEMISTRY] *See* Brönner's acid. { 'bren·ərz 'as·əd }

bromacetone [ORGANIC CHEMISTRY] $CH_2BrCOCH_3$ A colorless liquid which is a powerful irritant and lacrimator; used as tear gas and to make other chemicals. { ˌbrōm'as·əˌtōn }

bromacil [ORGANIC CHEMISTRY] 5-Bromo-3-*sec*-butyl-6-methyluracil, a soil sterilant; general at high dosage and selective at low. { 'brom·əˌsil }

bromate [CHEMISTRY] **1.** BrO_3^- A negative ion derived from bromic acid, $HBrO_3$ **2.** A salt of bromic acid. **3.** $C_9H_9ClO_3$ A light brown solid with a melting point of 118–119°C; used as a herbicide to control weeds in crops such as flax, cereals, and legumes. { 'brōˌmāt }

bromcresol green [ORGANIC CHEMISTRY] *See* bromocresol green. { brōm'krē‚sȯl 'grēn }
bromcresol purple [ORGANIC CHEMISTRY] *See* bromocresol purple. { brōm'krē‚sȯl 'pər·pəl }
bromeosin [ORGANIC CHEMISTRY] *See* eosin. { ‚brōm'ē·ə·sən }
bromic acid [INORGANIC CHEMISTRY] $HBrO_3$ A liquid, colorless to slightly yellow; boils with decomposition at 100°C; used in dyes and as a chemical intermediate. { 'brō·mik 'as·əd }
bromide [CHEMISTRY] A compound derived from hydrobromic acid, HBr, with the bromine atom in the 1-oxidation state. { 'brō‚mīd }
brominating agent [CHEMISTRY] A compound capable of introducing bromine into a molecule; examples are phosphorus tribromide, bromine chloride, and aluminum tribromide. { 'brō·mə‚nād·iŋ ‚ā·jənt }
bromination [CHEMISTRY] The process of introducing bromine into a molecule. { brō·mə'nā·shən }
bromine [CHEMISTRY] A chemical element, symbol Br, atomic number 35, atomic weight 79.904; used to make dibromide ethylene and in organic synthesis and plastics. { 'brō‚mēn }
bromine number [ANALYTICAL CHEMISTRY] The amount of bromine absorbed by a fatty oil; indicates the purity of the oil and degree of unsaturation. { 'brō‚mēn ‚nəm·bər }
bromine trifluoride [CHEMISTRY] BrF_3 A liquid with a boiling point of 135°C. { 'brō‚mēn ‚trī'flu̇r‚īd }
bromine water [CHEMISTRY] An aqueous saturated solution of bromine used as a reagent wherever a dilute solution of bromine is needed. { 'brō‚mēn ‚wȯd·ər }
bromo- [CHEMISTRY] A prefix that indicates the presence of bromine in a molecule. { 'brō·mō }
***N*-bromoacetamide** [ORGANIC CHEMISTRY] $CH_3CONHBr$ Needlelike crystals with a melting point of 102-105°C; soluble in warm water and cold ether; used as a brominating agent and in the oxidation of primary and secondary alcohols. { ¦en ¦brō·mō·ə'sed·ə‚mīd }
***para*-bromoacetanilide** [ORGANIC CHEMISTRY] C_8H_8BrNO Crystals with a melting point of 168°C; soluble in benzene, chloroform, and ethyl acetate; insoluble in cold water; used as an analgesic and antipyretic. { ¦par·ə ¦brō·mō‚a·səd'an·əl‚īd }
bromoacetone [ORGANIC CHEMISTRY] $BrCH_2COCH_3$ A colorless liquid used as a lacrimatory agent. { ¦brō·mō'as·ə‚tōn }
bromo acid [ORGANIC CHEMISTRY] *See* eosin. { 'brō·mō 'as·əd }
bromoalkane [ORGANIC CHEMISTRY] An aliphatic hydrocarbon with bromine bonded to it. { ¦brō·mō'al‚kān }
***para*-bromoaniline** [ORGANIC CHEMISTRY] $BrC_6H_4NH_2$ Rhombic crystals with a melting point of 66-66.5°C; soluble in alcohol and in ether; used in the preparation of azo dyes and dihydroquinazolines. { ¦par·ə ¦brō·mō'an·ə·lēn }
***para*-bromoanisole** [ORGANIC CHEMISTRY] C_7H_7BrO Crystals which melt at 9-10°C; used in disinfectants. { ¦par·ə ¦brō·mō'an·ə‚sōl }
bromobenzene [ORGANIC CHEMISTRY] C_6H_5Br A heavy, colorless liquid with a pleasant odor; used as a solvent, in motor fuels and top-cylinder compounds, and to make other chemicals. { ¦brō·mō'ben‚zēn }
***para*-bromobenzyl bromide** [ORGANIC CHEMISTRY] $BrC_6H_4CH_2Br$ Crystals with an aromatic odor and a melting point of 61°C; soluble in cold and hot alcohol, water, and ether; used to identify aromatic carboxylic acids. { ¦par·ə ‚brō·mo'benz·əl 'brō‚mīd }
bromobenzylcyanide [ORGANIC CHEMISTRY] $C_6H_5CHBrCN$ A light yellow oily compound used as a tear gas for training and for riot control. Abbreviated BBC. { ¦brō·mō¦benz·əl'sī·ə‚nīd }
bromochloromethane [ORGANIC CHEMISTRY] $BrCH_2Cl$ A clear, colorless liquid with a boiling point of 67°C; volatile, soluble in organic solvents, with a chloroformlike odor; used in fire extinguishers. { ¦brō·mō¦klȯr·ō'me‚thān }

bromochloroprene [ORGANIC CHEMISTRY] CHCl═CHCH$_2$Br A compound used as a nematicide and soil fumigant. { ˌbrō·mō'klȯr·əˌprēn }

bromocresol green [ORGANIC CHEMISTRY] Tetrabromo-*m*-cresol sulfonphthalein, a gray powder soluble in water or alcohol; used as an indicator between pH 4.5 (yellow) and 5.5 (blue). Also known as bromcresol green. { ¦brō·mō'krēˌsȯl 'grēn }

bromocresol purple [ORGANIC CHEMISTRY] Dibromo-*o*-cresol sulfonphthalein, a yellow powder soluble in water; used as an indicator between pH 5.2 (yellow) and 6.8 (purple). Also known as bromcresol purple. { ¦brō·mō'krēˌsȯl ˌpər·pəl }

bromocriptine [ORGANIC CHEMISTRY] $C_{32}H_{40}BrN_5O_5$ A polypeptide alkaloid that is a derivative of the ergotoxin group of ergot alkaloids and is a dopamine receptor agonist. { ˌbrō·mō'kripˌtēn }

bromocyclen [ORGANIC CHEMISTRY] $C_8H_5BrCl_6$ A compound used as an insecticide for wheat crops. { ˌbrō·mō'sī·klən }

bromofenoxim [ORGANIC CHEMISTRY] $C_{13}H_7N_3O_6Br_2$ A cream-colored powder with melting point 196-197°C; slightly soluble in water; used as herbicide to control weeds in cereal crops. { ¦brō·mō·fə'näk·səm }

bromoform [ORGANIC CHEMISTRY] $CHBr_3$ A colorless liquid, slightly soluble in water; used in the separation of minerals. { ˌbrō·mə'fȯrm }

1-bromonaphthalene [ORGANIC CHEMISTRY] $C_{10}H_7Br$ An oily liquid that is slightly soluble in water and miscible with chloroform, benzene, ether, and alcohol; used in the determination of index of refraction of crystals and for refractometric fat determination. { ¦wən ¦brō·mō'naf·thəˌlēn }

bromonium ion [ORGANIC CHEMISTRY] A halonium ion in which the halogen is bromine; occurs as a bridged structure. { brə'mōn·ē·əm 'īˌän }

1-bromooctane [ORGANIC CHEMISTRY] $CH_3(CH_2)_6CH_2Br$ Colorless liquid that is miscible with ether and alcohol; boiling point is 198-200°C; used in organic synthesis. { ¦wən ¦brō·mō'äkˌtān }

***para*-bromophenacyl bromide** [ORGANIC CHEMISTRY] $C_8H_6Br_2O$ Crystals with a melting point of 109-110°C; soluble in warm alcohol; used in the identification of carboxylic acids and as a protecting reagent for acids and phenols. { ¦par·ə ¦brō·mō·fə'nas·əl 'brōˌmīd }

***para*-bromophenylhydrazine** [ORGANIC CHEMISTRY] $C_6H_7BrN_2$ Needlelike crystals with a melting point of 108-109°C; soluble in benzene, ether, chloroform, and alcohol; used in the preparation of indoleacetic acid derivatives and in the study of transosazonation of sugar phenylosazones. { ¦par·ə ¦brō·mō·fen·əl'hī·drəˌzēn }

bromophos [ORGANIC CHEMISTRY] $C_8H_8SPBrCl_2O_3$ A yellow, crystalline compound with a melting point of 54°C; used as an insecticide and miticide for livestock, household insects, flies, and lice. { 'brō·məˌfäs }

bromopicrin [ORGANIC CHEMISTRY] CBr_3NO_2 Prismatic crystals with a melting point of 103°C; soluble in alcohol, benzene, and ether; used for military poison gas. Also known as nitrobromoform. { ˌbrō·mō'pik·rən }

***N*-bromosuccinimide** [ORGANIC CHEMISTRY] $C_4H_4BrNO_2$ Orthorhombic bisphenoidal crystals with a melting point of 173-175°C; used in the bromination of olefins. { ¦en ¦brō·mōˌsək'sin·əˌmīd }

bromotrifluoroethylene [ORGANIC CHEMISTRY] $BrFC{:}CF_2$ A colorless gas with a freezing point of -168°C and a boiling point of -58°C; soluble in chloroform; used as a refrigerant, in hardening of metals, and as a low-toxicity fire extinguisher. Abbreviated BFE. { ¦brō·mō·trī¦flu̇r·ō'eth·əˌlēn }

bromotrifluoromethane [ORGANIC CHEMISTRY] $CBrF_3$ Fluorine compound that has a molecular weight of 148.93, melting point −180°C, boiling point −59°C; used as a fire-extinguishing agent. { ¦brō·mō·trī¦flu̇r·ō'meˌthān }

α-bromo-*meta*-xylene [ORGANIC CHEMISTRY] $CH_3C_6H_4CH_2Br$ A liquid that is a powerful lacrimator; soluble in alcohol and ether; used in organic synthesis and chemical warfare. { ¦al·fə ¦brō·mō ¦med·ə 'zīˌlēn }

bromoxynil [ORGANIC CHEMISTRY] $C_7H_3OBr_2N$ A colorless solid with a melting point of 194-195°C; slightly soluble in water; used as a herbicide in wheat, barley, oats, rye, and seeded turf. { ˌbrō'mäk·sə·nil }

bromoxynil octanoate [ORGANIC CHEMISTRY] $C_{15}H_{17}Br_2NO_2$ A pale brown liquid, insoluble in water; melting point is 45-46°C; used to control broadleaf weeds. { ˌbrō'mäk·sə·nil ˌäk'tan·əˌwāt }

bromthymol blue [ORGANIC CHEMISTRY] An acid-base indicator in the pH range 6.0 to 7.6; color change is yellow to blue. { ¦brōm'thīˌmȯl 'blü }

Brönner's acid [ORGANIC CHEMISTRY] $C_{10}H_6(NH_2)SO_3H$ A colorless, water-soluble naphthylamine sulfonic acid that forms needle crystals; used in dyes. Also spelled Broenner's acid. { 'bren·ərz 'as·əd }

Brönsted acid [CHEMISTRY] A chemical species which can act as a source of protons. Also known as proton acid; protonic acid. { 'brən·steth *or* 'brenˌsted 'as·əd }

Brönsted base [CHEMISTRY] *See* base. { 'bren·stəth ˌbās }

Brönsted-Lowry theory [CHEMISTRY] A theory that all acid-base reactions consist simply of the transfer of a proton from one base to another. Also known as Brönsted theory. { ¦brən·steth ¦lau̇·rē ˌthē·ə·rē }

Brönsted theory [CHEMISTRY] *See* Brönsted-Lowry theory. { 'brən·steth ˌthē·ə·rē }

brown lead oxide [INORGANIC CHEMISTRY] *See* lead dioxide. { ¦brau̇n ¦led 'äkˌsīd }

brown-ring test [ANALYTICAL CHEMISTRY] A common qualitative test for the nitrate ion; a brown ring forms at the juncture of a dilute ferrous sulfate solution layered on top of concentrated sulfuric acid if the upper layer contains nitrate ion. { 'brau̇nˌriŋ ˌtest }

broxyquinoline [ORGANIC CHEMISTRY] $C_9H_5Br_2NO$ Crystals with a melting point of 196°C; soluble in acetic acid, chloroform, benzene, and alcohol; used as a reagent for copper, iron, and other metals. { ˌbräk·si'kwin·əˌlēn }

brucine [ORGANIC CHEMISTRY] $C_{23}H_{26}N_2O_4$ A poisonous alkaloid from the seeds of plant species such as *Nux vomica*; used in alcohol as a denaturant. { 'brüˌsīn }

Brunauer-Emmett-Teller equation [PHYSICAL CHEMISTRY] An extension of the Langmuir isotherm equation in the study of sorption; used for surface area determinations by computing the monolayer area. Abbreviated BET equation. { ¦brüˌnau̇r ¦em·ət ¦tel·ər i'kwā·zhən }

B stage [ORGANIC CHEMISTRY] An intermediate stage in a thermosetting resin reaction in which the plastic softens but does not fuse when heated, and swells but does not dissolve in contact with certain liquids. { 'bē ˌstāj }

bubble point [PHYSICAL CHEMISTRY] In a solution of two or more components, the temperature at which the first bubbles of gas appear. Also known as boiling point. { 'bəb·əl ˌpȯint }

Bucherer reaction [ORGANIC CHEMISTRY] A method of preparation of polynuclear primary aromatic amines; for example, α-naphthylamine is obtained by heating β-naphthol in an autoclave with a solution of ammonia and ammonium sulfite. { 'bük·ər·ər rē'ak·shən }

buckminsterfullerene [CHEMISTRY] C_{60} A molecule whose 60 carbon atoms are thought to be arranged at the vertices of a truncated icosahedron; the most stable of the fullerenes. { ¦bəkˌmin·stər'fu̇l·əˌrēn }

buffer [CHEMISTRY] A solution selected or prepared to minimize changes in hydrogen ion concentration which would otherwise occur as a result of a chemical reaction. Also known as buffer solution. { 'bəf·ər }

buffer capacity [CHEMISTRY] The relative ability of a buffer solution to resist pH change upon addition of an acid or a base. { 'bəf·ər kə'pas·əd·ē }

buffer solution [CHEMISTRY] *See* buffer. { 'bəf·ər sə'lü·shən }

bufotenine [ORGANIC CHEMISTRY] $C_{12}H_{16}N_2O$ An active pressor agent found in the skin of the common toad; a toxic alkaloid with epinephrinelike biological activity. { ˌbyü·fə'teˌnēn }

bulk sample [ANALYTICAL CHEMISTRY] *See* gross sample. { ¦bəlk ¦sam·pəl }

bulk sampling [ANALYTICAL CHEMISTRY] The taking of samples in arbitrary, irregular units rather than discrete units of uniform size for chemical analysis. { ¦bəlk ¦sam·pliŋ }

bumping [CHEMISTRY] Uneven boiling of a liquid caused by irregular rapid escape of large bubbles of highly volatile components as the liquid mixture is heated. { 'bəm·piŋ }

α-bungarotoxin [ORGANIC CHEMISTRY] A neurotoxin found in snake venom which blocks neuromuscular transmission by binding with acetylcholine receptors on motor end plates. { ¦al·fə ¦bəŋ·gə·rə'täk·sən }

Bunsen-Kirchhoff law [SPECTROSCOPY] The law that every element has a characteristic emission spectrum of bright lines and absorption spectrum of dark lines. { ¦bən·sən 'kir,kȯf ,lȯ }

buret [CHEMISTRY] A graduated glass tube used to deliver variable volumes of liquid; usually equipped with a stopcock to control the liquid flow. { byü'ret }

burning velocity [CHEMISTRY] The normal velocity of the region of combustion reaction (reaction zone) relative to nonturbulent unburned gas, in the combustion of a flammable mixture. { 'bər·niŋ və'läs·əd·ē }

burnt lime [INORGANIC CHEMISTRY] *See* calcium oxide. { ¦bərnt 'līm }

Burstein effect [SPECTROSCOPY] The shift of the absorption edge in the spectrum of a semiconductor to higher energies at high carrier densities in the semiconductor. { 'bər,stīn i,fekt }

1,3-butadiene [ORGANIC CHEMISTRY] C_4H_6 A colorless gas, boiling point −4.41°C, a major product of the petrochemical industry; used in the manufacture of synthetic rubber, latex paints, and nylon. { ¦wən ¦thrē ,byüd·ə'dī·ēn }

butadiene dimer [ORGANIC CHEMISTRY] C_8H_{12} The third ingredient in ethylene-propylene-terpolymer (EPT) synthetic rubbers; isomers include 3-methyl-1,4,6-heptatriene, vinylcyclohexene, and cyclooctadiene. { ,byüd·ə'dī·ēn 'dī·mər }

butadiene rubber [ORGANIC CHEMISTRY] *See* polybutadiene. { ,byüd·ə'dī·ēn 'rəb·ər }

butane [ORGANIC CHEMISTRY] C_4H_{10} An alkane of which there are two isomers, *n* and isobutane, occurs in natural gas and is produced by cracking petroleum. { 'byü,tān }

2,3-butanediol [ORGANIC CHEMISTRY] $CH_3CHOHCHOHCH_3$ A major fermentation product of several species of bacteria. { ¦tü ¦thrē ,byüd·ə·nēd·ē,ȯl }

butanol [ORGANIC CHEMISTRY] Any one of four isomeric alcohols having the formula C_4H_9OH; colorless, toxic liquids soluble in most organic liquids. Also known as butyl alcohol. { 'byüt·ən,ȯl }

butazolidine [ORGANIC CHEMISTRY] *See* phenylbutazone. { ,byüd·ə'zäl·ə,dēn }

butene-1 [ORGANIC CHEMISTRY] $CH_3CH_2CHCH_2$ A colorless, highly flammable gas; insoluble in water, soluble in organic solvents; used to produce polybutenes, butadiene aldehydes, and other organic derivatives. { 'byü,tēn 'wən }

butene-2 [ORGANIC CHEMISTRY] $CH_3CHCHCH_3$ A colorless, highly flammable gas, used to make butadiene and in the synthesis of four- and five-carbon organic molecules; the cis form, boiling point 3.7°C, is insoluble in water, soluble in organic solvents, and is also known as high-boiling butene-2; the trans form, boiling point 0.88°C, is insoluble in water, soluble in most organic solvents, and is also known as low-boiling butene-2. { 'byü,tēn 'tü }

butopyronoxyl [ORGANIC CHEMISTRY] $C_{12}H_{18}O_4$ A yellow to amber liquid with a boiling point of 256–260°C; miscible with ether, glacial acetic acid, alcohol, and chloroform; used as an insect repellent for skin and clothing. { ,byüd·ə,pī·rə'näk·səl }

2-butoxyethanol [ORGANIC CHEMISTRY] $HOCH_2CH_2OC_4H_9$ A liquid with a boiling point of 171–172°C; soluble in most organic solvents and water; used in dry cleaning as a solvent for nitrocellulose, albumin, resins, oil, and grease. { ¦tü ,byü,täk·sē'eth·ə,nȯl }

butyl [ORGANIC CHEMISTRY] Any of the four variations of the hydrocarbon radical C_4H_9: $CH_3CH_2CH_2CH_2—$, $(CH_3)_2CHCH_2—$, $CH_3CH_2CHCH_3—$, and $(CH_3)_3C—$. { 'byüd·əl }

butyl acetate [ORGANIC CHEMISTRY] $CH_3COOC_4H_9$ A colorless liquid slightly soluble in water; used as a solvent. { 'byüd·əl 'as·ə,tāt }

butyl acetoacetate [ORGANIC CHEMISTRY] $C_8H_{14}O_3$ A colorless liquid with a boiling point of 213.9°C; soluble in alcohol and ether; used for synthesis of dyestuffs and pharmaceuticals. { ¦byüd·əl ¦as·ə·tō'as·ə,tāt }

butyl acrylate [ORGANIC CHEMISTRY] $CH_2CHCOOC_4H_9$ A colorless liquid that is nearly insoluble in water and polymerizes readily upon heating; used as an intermediate for organic synthesis, polymers, and copolymers. { ¦byüd·əl 'ak·rə,lāt }

butyl alcohol [ORGANIC CHEMISTRY] *See* butanol. { ¦byüd·əl 'al·kə,hȯl }

***n*-butylamine** [ORGANIC CHEMISTRY] $C_4H_9NH_2$ A colorless, flammable liquid; miscible with water and ethanol; used as an intermediate in organic synthesis and to make insecticides, emulsifying agents, and pharmaceuticals. { ¦en ¦byüd·əl·ə¦mēn }

***sec*-butylamine** [ORGANIC CHEMISTRY] $CH_3CHNH_2C_2H_5$ A flammable, colorless liquid; boils in the range 63-68°C; may be used as an intermediate in organic synthesis. { ¦sek ¦byüd·əl·ə¦mēn }

***tert*-butylamine** [ORGANIC CHEMISTRY] $(CH_3)_3CNH_2$ A flammable liquid; boiling range 63-68°C; may be used in organic synthesis as an intermediate. { ¦tərt ¦byüd·əl·ə¦mēn }

butylate [ORGANIC CHEMISTRY] $C_{11}H_{23}NOS$ A colorless liquid used as an herbicide for preplant control of weeds in corn. { 'byüd·əl,āt }

butylated hydroxyanisole [ORGANIC CHEMISTRY] $(CH_3)_3CC_6H_3OH(OCH_3)$ An antioxidant consisting chiefly of a mixture of 2- and 3-*tert*-butyl-4-hydroxyanisole and used to control rancidity of lard and animal fats in foods. Abbreviated BHA. { 'byüd·əl,ād·əd hī,dräk·sē'an·ə,sōl }

butylated hydroxytoluene [ORGANIC CHEMISTRY] $[(CH_3)_3C]_2C_6H_2(CH_3)OH$ Crystals with a melting point of 72°C; soluble in toluene, methanol, and ethanol; used as an antioxidant in foods, in petroleum products, and for synthetic rubbers. Abbreviated BHT. { 'byüd·əl,ād·əd hī,dräk·sē'täl·yə,wēn }

butylbenzene [ORGANIC CHEMISTRY] $C_6H_5C_4H_9$ A colorless liquid used as a raw material for organic synthesis, especially for insecticides; forms are normal (1-phenylbutane), secondary (2-phenylbutane), and tertiary (2-methyl-2-phenylpropane). { ¦byüd·əl,ben,zēn }

***N-sec*-butyl-4-*tert*-butyl-2,6-dinitroaniline** [ORGANIC CHEMISTRY] $C_{14}H_{21}N_3O_4$ Orange crystals with a melting point of 60-61°C; solubility in water is 1.0 part per million at 24°C; used as a preemergence herbicide. { ¦en ¦sek 'byüd·əl ¦fór ¦tərt 'byüd·əl ¦tü ¦siks ,dī,nī·trō'an·ə,lēn }

butyl carbinol [ORGANIC CHEMISTRY] $(CH_3)_3CCH_2OH$ Colorless crystals that melt at 52°C; slightly soluble in water. { 'byüd·əl 'kär·bə,nól }

butyl chloride [ORGANIC CHEMISTRY] C_4H_9Cl A colorless liquid used as an alkylating agent in organic synthesis, as a solvent, and as an anthelminthic; forms are normal (1-chlorobutane), secondary, and iso or tertiary. { 'byüd·əl 'klór,īd }

***tert*-butyl chloroacetate** [ORGANIC CHEMISTRY] $ClCH_2COOC(CH_3)_3$ A liquid with a boiling point of 155°C; hydrolyzes to *tert*-butyl alcohol and chloroacetic acid; used in glycidic ester condensation. { ¦tərt ¦byüd·əl ,klór·ō'as·ə,tāt }

butyl citrate [ORGANIC CHEMISTRY] $C_3H_5O(COOC_4H_9)_3$ A colorless, odorless, nonvolatile liquid, almost insoluble in water; used as a plasticizer, solvent for cellulose nitrate, and antifoam agent. { 'byüd·əl 'sī,trāt }

butyl diglycol carbonate [ORGANIC CHEMISTRY] $(C_4H_9OCO_2\ CH_2CH_2)_2O$ A colorless, combustible liquid with a boiling range of 164-166°C; used as a plasticizer and solvent and in pharmaceuticals and lubricants manufacture. { 'byüd·əl dī¦glī,kól 'kär·bə·nāt }

butylene [ORGANIC CHEMISTRY] Any of three isomeric alkene hydrocarbons with the formula C_4H_8; all are flammable and easily liquefied gases. { 'byüd·ə,lēn }

1,3-butylene glycol [ORGANIC CHEMISTRY] $HOCH_2CH_2CH(OH)CH_3$ A viscous, colorless, hygroscopic liquid; soluble in water and alcohol; used as a solvent, food additive, and flavoring, and for plasticizers and polyurethanes. { ¦wən ¦thrē 'byüd·ə,lēn 'glī,kól }

1,4-butylene glycol [ORGANIC CHEMISTRY] $HOCH_2CH_2CH_2CH_2OH$ A colorless, combustible, oily liquid with a boiling point of 230°C; soluble in alcohol; used as a solvent and humectant, and in plastics and pharmaceuticals manufacture. { ¦wən ¦fór 'byüd·ə,lēn 'glī,kól }

1,2-butylene oxide [ORGANIC CHEMISTRY] $H_2COCHCH_2CH_3$ A colorless, water-soluble liquid with a boiling point of 63°C; used as an intermediate for various polymers. { ¦wən ¦tü 'byüd·ə,lēn 'äk,sīd }

butyl ether [ORGANIC CHEMISTRY] $C_8H_{18}O$ A colorless liquid, boiling at 142°C, and almost insoluble in water; used as an extracting agent, as a medium for Grignard and other reactions, and for purifying other solvents. { ¦byüd·əl 'ē·thər }

butyl formate [ORGANIC CHEMISTRY] $HCOOC_4H_9$ An ester of formic acid and butyl alcohol. { 'byüd·əl 'fȯr,māt }

***tert*-butylhydroperoxide** [ORGANIC CHEMISTRY] $(CH_3)_3COOH$ A liquid soluble in organic solvents; used as a catalyst in polymerization reactions, to introduce the peroxy group into organic molecules. { ¦tərt ¦byüd·əl,hī·drō·pə'räk,sīd }

butyl lactate [ORGANIC CHEMISTRY] $CH_3CHOHCOOC_4H_9$ A stable liquid, water-white and nontoxic, miscible with many solvents; used as a solvent for resins and gums, in lacquers and varnishes, and as a chemical intermediate. { 'byüd·əl 'lak,tāt }

butyl mercaptan [ORGANIC CHEMISTRY] C_4H_9SH A colorless, odorous liquid, a component of skunk secretion; used commercially as a gas-odorizing agent. { 'byüd·əl mər'kap·tan }

butyl oleate [ORGANIC CHEMISTRY] $C_{22}H_{42}O_2$ A butyl ester of oleic acid; used as a plasticizer. { 'byüd·əl 'ō·lē,āt }

***para-tert*-butylphenol** [ORGANIC CHEMISTRY] $(CH_3)_3CC_6H_4OH$ Needlelike crystals with a melting point of 98°C; soluble in alcohol and ether; used as an intermediate in production of varnish and lacquer resins, an additive in motor oil, and an ingredient in deemulsifiers in oil fields. { ¦par·ə ¦tərt ¦byüd·əl'fē,nȯl }

2-*sec*-butyl phenyl-*N*-methyl carbamate [ORGANIC CHEMISTRY] $C_{12}H_{17}NO_2$ A pale yellow or pale red liquid, insoluble in water; used as an insecticide for pests of rice and cotton. Also known as BPMC. { ¦tü sek ¦byüd·əl ¦fen·əl ¦en ¦meth·əl 'kär·bə,māt }

butyl propionate [ORGANIC CHEMISTRY] $C_2H_5COOC_4H_9$ A colorless aromatic liquid; used in fruit essences. { 'byüd·əl 'prō·pē·ə,nāt }

6-*tert*-butyl-3-propylisoxazolo-(5,4-*d*)pyrimidin-4(5*H*)-one [ORGANIC CHEMISTRY] $C_{12}H_{17}N_3O_2$ A white solid with a melting point of 217–218°C, insoluble in water, used as an herbicide on field corn, sweet corn, and sorghum. { ¦siks ¦tərt ¦byüd·əl ¦thrē ,prōp·əl,ī·säk'saz·ə,lō ¦fīv ¦fȯr ¦dē pə'rim·ə·dən ¦fȯr ¦fīv ¦āch ¦ōn }

butyl stearate [ORGANIC CHEMISTRY] $C_{17}H_{35}COOC_4H_9$ A liquid that solidifies at approximately 19°C; mixes with vegetable oils and is soluble in alcohol and ethers but insoluble in water; used as a lubricant, in polishes, as a plasticizer, and as a dye solvent. { 'byüd·əl 'stir,āt }

butynediol [ORGANIC CHEMISTRY] $HOCH_2C{:}CCH_2OH$ White crystals with a melting point of 58°C, soluble in water, aqueous acids, alcohol, and acetone; used as a corrosion inhibitor, defoliant, electroplating brightener, and polymerization accelerator. { ,byüd·ə'nēd·ē·əl }

butyraldehyde [ORGANIC CHEMISTRY] $CH_3(CH_2)_2CHO$ A colorless liquid boiling at 75.7°C; soluble in ether and alcohol, insoluble in water; derived from the oxo process. { ¦byüd·ər'al·də,hīd }

butyrate [ORGANIC CHEMISTRY] An ester or salt of butyric acid containing the $C_4H_7O_2$ radical. { 'byüd·ə,rāt }

butyric acid [ORGANIC CHEMISTRY] $CH_3CH_2CH_2COOH$ A colorless, combustible liquid with boiling point 163.5°C (757 mmHg); soluble in water, alcohol, and ether; used in synthesis of flavors, in pharmaceuticals, and in emulsifying agents. { byü'tir·ik 'as·əd }

butyric anhydride [ORGANIC CHEMISTRY] $C_8H_{14}O_3$ A colorless liquid that decomposes in water to form butyric acid; exists in two isomeric forms. { byü'tir·ik an'hī,drīd }

butyrolactone [ORGANIC CHEMISTRY] $C_4H_6O_2$ A liquid, the anhydride of butyric acid; used as a solvent in the manufacture of plastics. { ¦byüd·ə·rō'lak,tōn }

butyronitrile [ORGANIC CHEMISTRY] $CH_3(CH_2)_2CN$ A toxic, colorless liquid with a boiling point of 116-117.7°C; soluble in alcohol and ether; used in industrial, chemical, and pharmaceutical products, and in poultry medicines. { 'byüd·ə,rän'ī,tril }

C

C [CHEMISTRY] *See* carbon.
Ca [CHEMISTRY] *See* calcium.
Cabannes' factor [ANALYTICAL CHEMISTRY] An equational factor to correct for the depolarization effect of the horizontal components of scattered light during the determination of molecular weight by optical methods. { kə'bänz ,fak·tər }
cacodyl [ORGANIC CHEMISTRY] $(CH_3)_2As^-$ A radical found in, for example, cacodylic acid, $(CH_3)_2AsOOH$. { 'kak·ə,dil }
cacodylate [ORGANIC CHEMISTRY] Any salt of cacodylic acid. { ,kak·ə'di,lāt }
cacodylic acid [ORGANIC CHEMISTRY] $(CH_3)_2AsOOH$ Colorless crystals that melt at 200°C; soluble in alcohol and water; used as a herbicide. { ¦kak·ə¦dil·ik 'as·əd }
cacotheline [ORGANIC CHEMISTRY] $C_{20}H_{22}N_2O_5(NO_2)_2$ An azoic compound used as a metal indicator in chelometric titrations. { kə'käth·ə,lēn }
cadalene [ORGANIC CHEMISTRY] $C_{15}H_{18}$ A colorless liquid which boils at 291-292°C (720 mmHg; 95,990 pascals) and which is a substituted naphthalene. { 'kad·əl,ēn }
cadinene [ORGANIC CHEMISTRY] $C_{15}H_{24}$ A colorless liquid that boils at 274.5°C, and is a terpene derived from cubeb oil, cade oil, juniper berry oil, and other essential oils. { 'kad·ən,ēn }
cadmium [CHEMISTRY] A chemical element, symbol Cd, atomic number 48, atomic weight 112.40. { 'kad·mē·əm }
cadmium acetate [ORGANIC CHEMISTRY] $Cd(OOCCH_3)_2 \cdot 3H_2O$ A compound that forms colorless monoclinic crystals, soluble in water and in alcohol; used for chemical testing for sulfides, selenides, and tellurides and for producing iridescent effects on porcelain. { 'kad·mē·əm 'as·ə,tāt }
cadmium bromate [INORGANIC CHEMISTRY] $Cd(BrO_3)_2$ Colorless powder, soluble in water; used as an analytical reagent. { 'kad·mē·əm 'brō,māt }
cadmium bromide [INORGANIC CHEMISTRY] $CdBr_2$ A compound produced as a yellow crystalline powder, soluble in water and alcohol; used in photography, process engraving, and lithography. { 'kad·mē·əm 'brō,mīd }
cadmium carbonate [INORGANIC CHEMISTRY] $CdCO_3$ A white crystalline powder, insoluble in water, soluble in acids and potassium cyanide; used as a starting compound for other cadmium salts. { 'kad·mē·əm 'kär·bə,nāt }
cadmium chlorate [INORGANIC CHEMISTRY] $CdClO_3$ White crystals, soluble in water; a highly toxic material. { 'kad·mē·əm 'klȯr,āt }
cadmium chloride [INORGANIC CHEMISTRY] $CdCl_2$ A cadmium halide in the form of colorless crystals, soluble in water, methanol, and ethanol; used in photography, in dyeing and calico printing, and as a solution to precipitate sulfides. { 'kad·mē·əm 'klȯr,īd }
cadmium fluoride [INORGANIC CHEMISTRY] CdF_2 A crystalline compound with a melting point of 1110°C; soluble in water and acids; used for electronic and optical applications and as a starting material for laser crystals. { 'kad·mē·əm 'flu̇r,īd }
cadmium hydroxide [INORGANIC CHEMISTRY] $Cd(OH)_2$ A white powder, soluble in dilute acids; used to prepare negative electrodes for cadmium-nickel storage batteries. { 'kad·mē·əm hī'dräk,sīd }
cadmium iodide [INORGANIC CHEMISTRY] CdI_2 A cadmium halide that forms lustrous,

white, hexagonal scales, consisting of two water-soluble allotropes; used in photography, in process engraving, and formerly as an antiseptic. { 'kad·mē·əm 'ī·ə,dīd }

cadmium nitrate [CHEMISTRY] $Cd(NO_3)_2 \cdot 4H_2O$ White, hygroscopic crystals, soluble in water, alcohol, and liquid ammonia; used to give a reddish-yellow luster to glass and porcelain ware. { 'kad·mē·əm 'nī,trāt }

cadmium oxide [INORGANIC CHEMISTRY] CdO In the cubic form, a brown, amorphous powder, insoluble in water, soluble in acids and ammonia salts; used for cadmium plating baths and in the manufacture of paint pigments. { 'kad·mē·əm 'äk,sīd }

cadmium potassium iodide [INORGANIC CHEMISTRY] *See* potassium tetraiodocadmate. { 'kad·mē·əm pə'tas·ē·əm 'ī·ə,dīd }

cadmium sulfate [INORGANIC CHEMISTRY] $CdSO_4$ A compound that forms colorless, efflorescent crystals, soluble in water; used as an antiseptic and astringent, in the treatment of syphilis, gonorrhea, and rheumatism, and as a detector of hydrogen sulfide and fumaric acid. { 'kad·mē·əm 'səl,fāt }

cadmium sulfide [INORGANIC CHEMISTRY] CdS A compound with two forms: orange, insoluble in water, used as a pigment, and also known as orange cadmium; light yellow, hexagonal crystals, insoluble in water, and also known as cadmium yellow. { 'kad·mē·əm 'səl,fīd }

cadmium telluride [INORGANIC CHEMISTRY] CdTe Brownish-black, cubic crystals with a melting point of 1090°C; soluble, with decomposition, in nitric acid; used for semiconductors. { 'kad·mē·əm 'tel·yə,rīd }

cadmium tungstate [INORGANIC CHEMISTRY] $CdWO_4$ White or yellow crystals or powder; soluble in ammonium hydroxide and alkali cyanides; used in fluorescent paint, x-ray screens, and scintillation counters. { 'kad·mē·əm 'təŋ,stāt }

cadmium yellow [INORGANIC CHEMISTRY] *See* cadmium sulfide. { 'kad·mē·əm 'yel·ō }

caffeic acid [ORGANIC CHEMISTRY] $C_9H_8O_4$ A yellow crystalline acid that melts at 223-225°C with decomposition; soluble in water and alcohol. { ka'fē·ik 'as·əd }

caffeine [ORGANIC CHEMISTRY] $C_8H_{10}O_2N_4 \cdot H_2O$ An alkaloid found in a large number of plants, such as tea, coffee, cola, and mate. { 'kaf,ēn }

cage [PHYSICAL CHEMISTRY] An aggregate of molecules in the condensed phase that surrounds fragments formed by thermal or photochemical dissociation or pairs of molecules in a solution that have collided without reacting. { kāj }

cage compound [CHEMISTRY] *See* clathrate. { ¦kāj ¦käm,pau̇nd }

cage effect [PHYSICAL CHEMISTRY] A phenomenon involving the dissociation of molecules unable to move apart rapidly because of the presence of other molecules, with the result that the dissociation products may recombine. { 'kāj i,fekt }

Cailletet and Mathias law [PHYSICAL CHEMISTRY] The law that describes the relationship between the mean density of a liquid and its saturated vapor at that temperature as being a linear function of the temperature. { kī·ə'tā an mə'thī·əs ,lȯ }

cajeputol [ORGANIC CHEMISTRY] *See* eucalyptol. { 'kaj·ə·pə,tȯl }

calabarine [ORGANIC CHEMISTRY] *See* physostigmine. { kə'lab·ə,rēn }

calcined gypsum [INORGANIC CHEMISTRY] *See* plaster of Paris. { 'kal,sīnd 'jip·səm }

calcined soda [INORGANIC CHEMISTRY] *See* soda ash. { 'kal,sīnd 'sō·də }

calcium [CHEMISTRY] A chemical element, symbol Ca, atomic number 20, atomic weight 40.08; used in metallurgy as an alloying agent for aluminum-bearing metal, as an aid in removing bismuth from lead, and as a deoxidizer in steel manufacture, and also used as a cathode coating in some types of photo tubes. { 'kal·sē·əm }

calcium acetate [ORGANIC CHEMISTRY] $Ca(C_2H_3O_2)_2$ A compound that crystallizes as colorless needles that are soluble in water; formerly used as an important source of acetone and acetic acid; now used as a mordant and as a stabilizer of plastics. { 'kal·sē·əm 'as·ə,tāt }

calcium acrylate [ORGANIC CHEMISTRY] $(CH_2CHCOO)_2Ca$ Free-flowing, water-soluble white powder used for soil stabilization, oil-well sealing, and ion exchange and as a binder for clay products and foundry molds. { 'kal·sē·əm 'ak·rə,lāt }

calcium arsenate [INORGANIC CHEMISTRY] $Ca_3(AsO_4)_2$ An arsenic compound used as an insecticide to control cotton pests. { 'kal·sē·əm 'ärs·ən,āt }

calcium arsenite [INORGANIC CHEMISTRY] $Ca_3(AsO_3)_2$ White granules that are soluble in water; used as an insecticide. { 'kal·sē·əm 'ärs·ən,īt }

calcium bisulfite [INORGANIC CHEMISTRY] $Ca(HSO_3)_2$ A white powder, used as an antiseptic and in the sulfite pulping process. { 'kal·sē·əm bī'səl‚fīt }

calcium bromide [INORGANIC CHEMISTRY] $CaBr_2$ A deliquescent salt in the form of colorless hexagonal crystals that are soluble in water and absolute alcohol. { 'kal·sē·əm 'brō‚mīd }

calcium carbide [INORGANIC CHEMISTRY] CaC_2 An alkaline earth carbide obtained in the pure form as transparent crystals that decompose in water, used to make acetylene gas. { 'kal·sē·əm 'kär‚bīd }

calcium carbonate [INORGANIC CHEMISTRY] $CaCO_3$ White rhombohedrons or a white powder; occurs naturally as calcite; used in paint manufacture, as a dentifrice, as an anticaking medium for table salt, and in manufacture of rubber tires. { 'kal·sē·əm 'kär·bə‚nāt }

calcium chlorate [INORGANIC CHEMISTRY] $Ca(ClO_3)_2 \cdot 2H_2O$ White monoclinic crystals, decomposed by heating. { 'kal·sē·əm 'klȯr‚āt }

calcium chloride [INORGANIC CHEMISTRY] $CaCl_2$ A colorless, deliquescent powder that is soluble in water and ethanol; used as an antifreeze and as an antidust agent. { 'kal·sē·əm 'klȯr‚īd }

calcium chromate [INORGANIC CHEMISTRY] $CaCrO_4 \cdot 2H_2O$ Yellow, monoclinic crystals that are slightly soluble in water; used to make other pigments. { 'kal·sē·əm 'krō‚māt }

calcium cyanamide [INORGANIC CHEMISTRY] $CaCN_2$ In pure form, colorless rhombohedral crystals, the commercial form being a gray material containing 55-70% $CaCN_2$; used as a fertilizer, weed killer, and defoliant. { 'kal·sē·əm sī'an·ə‚mīd }

calcium cyanide [INORGANIC CHEMISTRY] $Ca(CN)_2$ In pure form, a white powder that gives off hydrogen cyanide in air at normal humidity, prepared commercially in impure black or gray flakes; used as an insecticide and rodenticide. Also known as black cyanide. { 'kal·sē·əm 'sī·ə‚nīd }

calcium cyclamate [ORGANIC CHEMISTRY] $C_{12}H_{24}O_6N_2S_2Ca_2 \cdot H_2O$ White crystals with a very sweet taste, soluble in water; has been used as a low-calorie sweetening agent. { 'kal·sē·əm 'sī·klə‚māt }

calcium fluoride [INORGANIC CHEMISTRY] CaF_2 Colorless, cubic crystals that are slightly soluble in water and soluble in ammonium salt solutions; used in etching glass and preparing hydrofluoric acid. { 'kal·sē·əm 'flu̇r‚īd }

calcium gluconate [ORGANIC CHEMISTRY] $Ca(C_6H_{11}O_7)_2 \cdot H_2O$ White powder that loses water at 120°C; soluble in hot water but less soluble in cold water, insoluble in acetic acid and alcohol, used in medicine, as a foaming agent, and as a buffer in foods. { 'kal·sē·əm 'glü·kə‚nāt }

calcium hardness [CHEMISTRY] Presence of calcium ions in water, from dissolved carbonates and bicarbonates, treated in boiler water by introducing sodium phosphate. { 'kal·sē·əm ‚härd·nəs }

calcium hydride [INORGANIC CHEMISTRY] CaH_2 In pure form, white crystals that are insoluble in water, used in the production of chromium, titanium, and zirconium in the Hydromet process. { 'kal·se·əm 'hī‚drīd }

calcium hydrogen phosphate [INORGANIC CHEMISTRY] See calcium phosphate. { ¦kal·sē·əm ¦hī·drə jən 'fäs‚fāt }

calcium hydroxide [INORGANIC CHEMISTRY] $Ca(OH)_2$ White crystals, slightly soluble in water, used in cement, mortar, and manufacture of calcium salts. Also known as hydrated lime. { 'kal·sē·əm hī'dräk‚sīd }

calcium hypochlorite [INORGANIC CHEMISTRY] $Ca(OCl)_2 \cdot 4H_2O$ A white powder, used as a bleaching agent and disinfectant for swimming pools. { 'kal·sē·əm hī·pō'klȯr‚īt }

calcium iodide [INORGANIC CHEMISTRY] CaI_2 A yellow, hygroscopic powder that is very soluble in water; used in photography. { 'kal·sē·əm 'ī·ə‚dīd }

calcium iodobehenate [ORGANIC CHEMISTRY] $Ca(OOCC_{21}H_{42}I)_2$ A yellowish powder that is soluble in warm chloroform; used in feed additives. { 'kal·sē·əm ¦ī·ə·dō ¦bē·ə‚nāt }

calcium lactate [ORGANIC CHEMISTRY] $Ca(C_3H_5O_3)_2 \cdot 5H_2O$ A salt of lactic acid in the form of white crystals that are soluble in water; used in calcium therapy and as a blood coagulant. { 'kal·sē·əm 'lak,tāt }

calcium naphthenate [ORGANIC CHEMISTRY] Calcium derivative of cycloparaffin hydrocarbon (generally cyclopentane or cyclohexane base) that is a light, sticky, water-insoluble mass; used as a hardening agent in plastic compounds, in waterproofing, adhesives, wood fillers, and varnishes. { 'kal·se·əm 'naf·thə,nāt }

calcium nitrate [INORGANIC CHEMISTRY] $Ca(NO_3)_2 \cdot 4H_2O$ Colorless, monoclinic crystals that are soluble in water; the anhydrous salt is very deliquescent; used as a fertilizer and in explosives. Also known as nitrocalcite. { 'kal·se·əm 'nī,trāt }

calcium orthoarsenate [ORGANIC CHEMISTRY] $Ca_3(AsO_4)_2$ A white powder, insoluble in water; used as a preemergence insecticide and herbicide for turf. { 'kal·se·əm ¦ór·thō'ärs·ən,āt }

calcium oxalate [INORGANIC CHEMISTRY] $CaC_2O_4 \cdot H_2O$ A salt of oxalic acid in the form of white crystals that are insoluble in water. { 'kal·se·əm 'äk·sə,lāt }

calcium oxide [INORGANIC CHEMISTRY] CaO A caustic white solid sparingly soluble in water; the commercial form is prepared by roasting calcium carbonate limestone in kilns until all the carbon dioxide is driven off; used as a refractory, in pulp and paper manufacture, and as a flux in manufacture of steel. Also known as burnt lime; calx; caustic lime. { 'kal·se·əm 'äk,sīd }

calcium pantothenate [ORGANIC CHEMISTRY] $(C_9H_{16}NO_5)_2Ca$ White slightly hygroscopic powder; soluble in water, insoluble in chloroform and ether; melts at 170–172°C; found in either the dextro or levo form or in racemic mixtures; used in nutrition and in animal feed. { 'kal·se·əm pan·tə'the,nāt }

calcium peroxide [INORGANIC CHEMISTRY] CaO_2 A cream-colored powder that decomposes in water; used as an antiseptic and a detergent. { 'kal·se·əm pə'räk,sīd }

calcium phosphate [INORGANIC CHEMISTRY] **1.** Any phosphate of calcium. **2.** Any of the following three calcium orthophosphates, all of which are white or colorless in pure form: $Ca(H_2PO_4)_2$ is used as a fertilizer, as a plastics stabilizer, and in baking powder, and is also known as acid calcium phosphate, calcium dihydrogen phosphate, monobasic calcium phosphate, monocalcium phosphate; $CaHPO_4$ is used in pharmaceuticals, animal feeds, and toothpastes, and is also known as calcium hydrogen phosphate, dibasic calcium phosphate, dicalcium orthophosphate, dicalcium phosphate; $Ca_3(PO_4)_2$ is used as a fertilizer, and is also known as tribasic calcium phosphate, tricalcium phosphate. { 'kal·se·əm 'fäs,fāt }

calcium plumbate [INORGANIC CHEMISTRY] $Ca(PbO_3)_2$ Orange crystals that are insoluble in cold water but decompose in hot water; used as an oxidizer in the manufacture of glass and matches. { 'kal·se·əm 'pləm,bāt }

calcium plumbite [INORGANIC CHEMISTRY] $CaPbO_2$ Colorless crystals that are slightly soluble in water. { 'kal·se·əm 'pləm,bīt }

calcium pyrophosphate [INORGANIC CHEMISTRY] $Ca_2P_2O_7$ White, abrasive powder, used in dentifrice polishes, in metal polishes, and as a food supplement. { 'kal·se·əm ,pī·rō'fäs,fāt }

calcium resinate [ORGANIC CHEMISTRY] Yellowish white, amorphous powder that is soluble in acid, insoluble in water; made by boiling rosin with calcium hydroxide and filtering, or by fusion of melted rosin with hydrated lime; used for waterproofing, leather tanning, and the manufacture of paint driers and enamels. Also known as limed rosin. { 'kal·se·əm 'rez·ən,āt }

calcium reversal lines [SPECTROSCOPY] Narrow calcium emission lines that appear as bright lines in the center of broad calcium absorption bands in the spectra of certain stars. { 'kal·se·əm ri'vər·səl ,līnz }

calcium silicate [INORGANIC CHEMISTRY] Any of three silicates of calcium: tricalcium silicate, Ca_3SiO_5; dicalcium silicate, Ca_2SiO_4; calcium metasilicate, $CaSiO_3$. { 'kal·se·əm 'sil·ə,kāt }

calcium stearate [ORGANIC CHEMISTRY] $Ca(C_{18}H_{35}O_2)_2$ A metallic soap produced as a white powder that is insoluble in water but slightly soluble in petroleum, benzene, and toluene. { 'kal·se·əm 'stir,āt }

calcium sulfate [INORGANIC CHEMISTRY] **1.** $CaSO_4$ A white crystalline salt, insoluble in water; used in Keene's cement, in pigments, as a paper filler, and as a drying agent.

2. Either of two hydrated forms of the salt: the dihydrate, $CaSO_4 \cdot 2H_2O$, and the hemihydrate, $CaSO_4 \cdot \frac{1}{2}H_2O$. { 'kal·se·əm 'səl͵fāt }

calcium sulfide [INORGANIC CHEMISTRY] CaS In pure form, white cubic crystals, slightly soluble in water; used as a base for luminescent materials. Also known as hepar calcies; sulfurated lime. { 'kal·se·əm 'səl͵fīd }

calcium sulfite [INORGANIC CHEMISTRY] $CaSO_3 \cdot 2H_2O$ A white powder that is soluble in dilute sulfurous acid; may be dehydrated at 150°C to the anhydrous salt; used in the sulfite process for the manufacture of wood pulp. { 'kal·se·əm 'səl͵fīt }

calcium tungstate [INORGANIC CHEMISTRY] $CaWO_4$ White, tetragonal crystals, slightly soluble in water; used in manufacture of luminous paints. Also known as artificial scheelite; calcium wolframate. { 'kal·se·əm 'təŋ͵stāt }

calcium wolframate [INORGANIC CHEMISTRY] *See* calcium tungstate. { 'kal·se·əm 'wu̇l·frə͵māt }

calculation-based molecular modeling [PHYSICAL CHEMISTRY] The use of computers, together with theoretical chemistry and mathematical expressions, to describe the structure of molecules and predict the most favorable conformation of a molecule or to calculate the energy of interaction between two molecules. { ͵kal·kyə'lā·shən ¦bāst mə'lek·yə·lər 'mäd·əl·iŋ }

calibrant [ANALYTICAL CHEMISTRY] In chemical analysis, a substance used to calibrate the response of a measurement system to the analyte. { 'kal·ə·brənt }

calibration reference [ANALYTICAL CHEMISTRY] Any of the standards of various types that indicate whether an analytical instrument or procedure is working within prescribed limits; examples are test solutions used with pH meters, and solutions with known concentrations (standard solutions) used with spectrophotometers. { 'kal ə͵brā·shən ͵ref·rəns }

californium [CHEMISTRY] A chemical element, symbol Cf, atomic number 98; all isotopes are radioactive. { ͵kal·ə'for·ne·əm }

calixarene [ORGANIC CHEMISTRY] A cyclic structure containing the group $(—Ar—CH_2—)_n$, where Ar represents an aryl group. { kə'lik·sə͵rēn }

calmagite [ORGANIC CHEMISTRY] $C_{17}H_{14}N_2O_5S$ A compound crystallizing from acetone as red crystals that are soluble in water; used as an indicator in the titration of calcium or magnesium with EDTA. { 'kal·mə͵jīt }

calomel electrode [PHYSICAL CHEMISTRY] A reference electrode of known potential consisting of mercury, mercury chloride (calomel), and potassium chloride solution; used to measure pH and electromotive force. Also known as calomel half-cell; calomel reference electrode. { 'kal·ə·məl i'lek͵trōd }

calomel half-cell [PHYSICAL CHEMISTRY] *See* calomel electrode. { 'kal·ə·məl 'haf ͵sel }

calomel reference electrode [PHYSICAL CHEMISTRY] *See* calomel electrode. { 'kal·ə·məl 'ref·rəns i'lek͵trōd }

calorimetric titration [ANALYTICAL CHEMISTRY] *See* thermometric titration. { kə¦lȯr·ə¦me·trik tī'trā·shən }

calx [INORGANIC CHEMISTRY] *See* calcium oxide. { kalks }

camphane [ORGANIC CHEMISTRY] $C_{10}H_{18}$ An alicyclic hydrocarbon; white crystals, soluble in alcohol, with a melting point of 158–159°C. { 'kam͵fān }

camphene [ORGANIC CHEMISTRY] $C_{10}H_{16}$ A bicyclic terpene used as raw material in the synthesis of insecticides such as toxaphene and camphor. { 'kam͵fēn }

camphor [ORGANIC CHEMISTRY] $C_{10}H_{16}O$ A bicyclic saturated terpene ketone that exists in optically active dextro and levo forms and as a racemic mixture of these forms; the dextro form is obtained from the wood and bark of the camphor tree, the levo form is found in some essential oils, and the inactive form is obtained from an Asiatic chrysanthemum or made synthetically from certain terpenes. { 'kam·fər }

***d*-camphorsulfonic acid** [ORGANIC CHEMISTRY] $C_{10}H_{16}O_4S$ A compound crystallizing as prisms from ethyl acetate or glacial acetic acid; slightly soluble in glacial acetic acid and in ethyl acetate; used in the resolution of optically active isomers. Also known as Reychler's acid. { ¦dē ¦kam·fər͵səl'fän·ik 'as·əd }

cane sugar [ORGANIC CHEMISTRY] Sucrose derived from sugarcane. { 'kān ͵shu̇g·ər }

cannabidiol [ORGANIC CHEMISTRY] $C_{21}H_{28}(OH)_2$ A constituent of cannabis which, upon isomerization to a tetrahydrocannabinol, has some of the physiologic activity of marijuana. { ¦kan·ə·bə'dī‚ȯl }

cannabinoid [ORGANIC CHEMISTRY] Any one of the various chemical constituents of cannabis (marijuana), that is, the isomeric tetrahydrocannabinols, cannabinol, and cannabidiol. { kə'nab·ə‚nȯid }

cannabinol [ORGANIC CHEMISTRY] $C_{21}H_{26}O_2$ A physiologically inactive phenol formed by spontaneous dehydrogenation of tetrahydrocannabinol from cannabis. { 'kan·ə·bə‚nȯl }

cannabiscetin [ORGANIC CHEMISTRY] *See* myricetin. { ¦kan·ə'bis·ə‚tēn }

Cannizzaro reaction [ORGANIC CHEMISTRY] The reaction in which aldehydes that do not have a hydrogen attached to the carbon adjacent to the carbonyl group, upon encountering strong alkali, readily form an alcohol and an acid salt. { kän·it'sär·ō rē'ak·shən }

canonical form [ORGANIC CHEMISTRY] **1.** A resonance structure for a cyclic compound in which the bonds do not intersect. **2.** *See* contributing structure. { kə'nän·ə·kəl ‚fȯrm }

canonical structure [ORGANIC CHEMISTRY] *See* contributing structure. { kə'nän·ə·kəl 'strək·chər }

cantharides camphor [ORGANIC CHEMISTRY] *See* cantharidin. { kan'thar·ə‚dēz 'kam·fər }

cantharidin [ORGANIC CHEMISTRY] $C_{10}H_{12}O_4$ Colorless crystals that melt at 218°C; slightly soluble in acetone, chloroform, alcohol, and water; used in veterinary medicine. Also known as cantharides camphor. { kan'thar·ə·dən }

capacity [ANALYTICAL CHEMISTRY] In chromatography, a measurement used in ion-exchange systems to express the adsorption ability of the ion-exchange materials. { kə'pas·əd·ē }

capillary column [ANALYTICAL CHEMISTRY] One of the long, narrow (100 meters by 0.2-0.5 millimeter or 330 feet by 0.008-0.02 inch) columns used for capillary gas chromatography. Also known as open tubular column. { 'kap·ə‚ler·ē ‚käl·əm }

capillary condensation [PHYSICAL CHEMISTRY] Condensation of an adsorbed vapor within the pores of the adsorbate. { 'kap·ə‚ler·ē ‚kän‚den'sā·shən }

capillary gas chromatography [ANALYTICAL CHEMISTRY] A highly efficient type of gas chromatography in which the gaseous sample passes through capillary tubes with internal diameters between 0.2 and 0.5 millimeter and lengths up to 100 meters, and adsorption takes place on a medium that is spread on the inner walls of these tubes. { 'kap·ə‚ler·ē ¦gas krō·mə'täg·rə·fē }

caprate [ORGANIC CHEMISTRY] Any of the salts of capric acid, containing the group $C_9H_{19}COO$—. { 'ka‚prāt }

capric acid [ORGANIC CHEMISTRY] $CH_3(CH_2)_8COOH$ A fatty acid found in oils and animal fats. { 'ka‚prik 'as·əd }

capric anhydride [ORGANIC CHEMISTRY] $(CH_3(CH_2)_8CO)_2O$ White crystals that are insoluble in water; used as a chemical intermediate. { 'ka‚prik an'hī‚drīd }

caproamide [ORGANIC CHEMISTRY] $CH_3(CH_2)_4CONH_2$ An amide, melting point 100-101°C; used as a chemical intermediate. { ¦ka·prō'am‚īd }

caproic acid [ORGANIC CHEMISTRY] $CH_3(CH_2)_4COOH$ A colorless liquid fatty acid found in oils and animal fats; used in synthesizing pharmaceuticals and flavors. { kə'prō·ik 'as·əd }

caproic anhydride [ORGANIC CHEMISTRY] $[CH_3(CH_2)_4COO]_2$ White crystals that are insoluble in water, melting point −40.6°C, boiling point 241-243°C. { kə'prō·ik an'hī ‚drīd }

caprolactam [ORGANIC CHEMISTRY] $(CH_2)_5NH \cdot CO$ White flakes, melting point 68-69°C, made from cyclohexanone; used to make synthetic fiber, particularly nylon-6. { ¦ka·prō¦lak·təm }

ε-caprolactone [ORGANIC CHEMISTRY] $CH_2(CH_2)_4NHCO$ White crystals, used to make synthetic fibers, plastics, films, coatings, and plasticizers; its vapors or fine crystals are respiratory irritants. { ¦ā·də ¦ka·prō¦lak‚tōn }

caprylamide [ORGANIC CHEMISTRY] $CH_3(CH_2)_6CONH_2$ An amine, melting point 105-110°C; decomposes above 200°C; used as a chemical intermediate. { kə'pril·ə‚mīd }

capryl compounds [ORGANIC CHEMISTRY] A misnomer for octyl compounds; that is, the term octyl halide is preferred for caprylic halides, and octanoic acid for caprylic acid. { 'ka,prəl ,käm,paünz }
1-caprylene [ORGANIC CHEMISTRY] *See* 1-octene. { ¦wən 'kap·rə,lēn }
caprylic acid [ORGANIC CHEMISTRY] $C_8H_{16}O_2$ A liquid fatty acid occurring in butter, coconut oil, and other fats and oils. { kə'pril·ik 'as·əd }
caprylic anhydride [ORGANIC CHEMISTRY] $[CH_3(CH_2)_6CO]_2O$ A white solid that melts at −1°C; used as a chemical intermediate. { kə'pril·ik an'hī,drīd }
capsaicin [ORGANIC CHEMISTRY] $C_{18}H_{27}O_3N$ A toxic material extracted from capsicum. { kap'sā·ə·sən }
captan [ORGANIC CHEMISTRY] $C_9H_8O_2NSCl_3$ A buff to white solid with a melting point of 175°C; used as a fungicide for diseases of fruits, vegetables, and flowers. { 'kap,tan }
carbamide [ORGANIC CHEMISTRY] *See* urea. { 'kär·bə,mīd }
carbamoyl [ORGANIC CHEMISTRY] The radical NH_2CO, formed from carbamic acid. { kär'bam·ə,wil }
carbanilide [ORGANIC CHEMISTRY] $(NHC_6H_5)CO(NHC_6H_5)$ Colorless crystals that are very slightly soluble in water, and dissolve in ether and alcohol; used in organic synthesis. { kär·bə'nil,īd }
carbanion [CHEMISTRY] One of the charged fragments which arise on heterolytic cleavage of a covalent bond involving carbon; the fragment carries an unshared pair of electrons and bears a negative charge. { ¦kärb'an,ī·ən }
carbaryl [ORGANIC CHEMISTRY] $C_{12}H_{11}NO_2$ A colorless, crystalline compound with a melting point of 142°C; used as an insecticide for crops, forests, lawns, poultry, and pets. { 'kär·bə,ril }
carbazole [ORGANIC CHEMISTRY] One of a group of organic heterocyclic compounds containing a dibenzopyrrole system. Also known as 9-azafluorene. { 'kär·bə,zōl }
carbene [ORGANIC CHEMISTRY] A compound of carbon which exhibits two valences to a carbon atom; the two valence electrons are distributed in the same valence; an example is CH_2. { 'kär,bēn }
carbenium ion [ORGANIC CHEMISTRY] A cation in which the charged atom is carbon; for example, R_2C^+, where R is an organic group. { kär'bē·nē·əm ,ī·ən }
carbenoid species [ORGANIC CHEMISTRY] A species that is not a free carbene but has the characteristics of a carbene when participating in a chemical reaction. { 'kär·bə,nȯid ,spē·shēz }
carbide [INORGANIC CHEMISTRY] A binary compound of carbon with an element more electropositive than carbon, carbon-hydrogen compounds are excluded. { 'kär,bīd }
carbinol [ORGANIC CHEMISTRY] **1.** A primary alcohol with general formula RCH_2OH. **2.** The radical CH_2OH of primary alcohols. **3.** An alcohol derived from methanol. { 'kär·bə,nȯl }
carbinyl [ORGANIC CHEMISTRY] *See* methyl. { 'kär·bə,nil }
carbocation [ORGANIC CHEMISTRY] A positively charged ion whose charge resides, at least in part, on a carbon atom or group of carbon atoms. { ¦kär·bō'kat,ī·ən }
carbocyclic compound [ORGANIC CHEMISTRY] A compound with a homocyclic ring in which all the ring atoms are carbon, for example, benzene. { ¦kär·bō¦si·klik 'käm,paùnd }
carbodihydrazide [ORGANIC CHEMISTRY] $CO(NHNH_2)_2$ Colorless crystals that melt at 154°C; very soluble in alcohol and water; used in photographic chemicals. { ¦kär·bō,dī'hī·drə,zīd }
carbodiimide [ORGANIC CHEMISTRY] **1.** $HN{=}C{=}NH$ An unstable tautomer of cyanamide. **2.** Any compound with the general formula $RN{=}C{=}NR$ which is a formal derivative of carbodiimide. { ¦kär·bō'dī·ə,mīd }
carbofuran [ORGANIC CHEMISTRY] $C_{12}H_{15}NO_3$ A white solid with a melting point of 150-152°C; soluble in water; used as an insecticide, miticide, and nematicide in many crops. { ,kär·bō'fyůr,än }
carbohydrate gum [ORGANIC CHEMISTRY] A polysaccharide which produces a gel of a viscous solution when it is dispersed in water at low concentrations; examples are

agar, guar gum, xanthan gum, gum arabic, and sodium carboxymethyl cellulose. { ˌkär·bōˈhīˌdrāt ˈgəm }

carbolic acid [ORGANIC CHEMISTRY] *See* phenol. { kärˈbäl·ik ˈas·əd }

carbon [CHEMISTRY] A nonmetallic chemical element, symbol C, atomic number 6, atomic weight 12.01115; occurs freely as diamond, graphite, and coal. { ˈkär·bən }

carbonate [CHEMISTRY] **1.** An ester or salt of carbonic acid. **2.** A compound containing the carbonate (CO_3^{2-}) ion. **3.** Containing carbonates. { ˈkär·bə·nət }

carbonation [CHEMISTRY] Conversion to a carbonate. { ˌkär·bəˈnā·shən }

carbon black [CHEMISTRY] **1.** An amorphous form of carbon produced commercially by thermal or oxidative decomposition of hydrocarbons and used principally in rubber goods, pigments, and printer's ink. **2.** *See* gas black. { ¦kär·bən ¦blak }

carbon dioxide [INORGANIC CHEMISTRY] CO_2 A colorless, odorless, tasteless gas about 1.5 times as dense as air. { ¦kär·bən dīˈäkˌsīd }

carbon dioxide absorption tube [ANALYTICAL CHEMISTRY] An absorbent-packed tube used to capture the carbon dioxide formed during the microdetermination of carbon-hydrogen by the Pragl combustion procedure. { ¦kär·bən dīˈäkˌsīd əbˈsȯrp·shən ˌtüb }

carbon disulfide [ORGANIC CHEMISTRY] CS_2 A sulfide, used as a solvent for oils, fats, and rubbers and in paint removers. { ¦kär·bən dīˈsəlˌfīd }

carbon film [ANALYTICAL CHEMISTRY] Carbon deposited by evaporation onto a specimen to protect and prepare it for electron microscopy. { ¦kär·bən ˈfilm }

carbon-hydrogen analyzer [ANALYTICAL CHEMISTRY] A device used in the quantitative analysis of the carbon and hydrogen content of organic compounds. { ¦kär·bən ¦hī·drə·jən ˈan·əˌlīz·ər }

carbonic acid [INORGANIC CHEMISTRY] H_2CO_3 The acid formed by combination of carbon dioxide and water. { kärˈbän·ik ˈas·əd }

carbonium ion [ORGANIC CHEMISTRY] A carbocation which has a positively charged carbon with a coordination number greater than 3. { kärˈbōn·ē·əm ˌī·ən }

carbonization [CHEMISTRY] The conversion of a carbon-containing substance to carbon or a carbon residue as the destructive distillation of coal by heat in the absence of air, yielding a solid residue with a higher percentage of carbon than the original coal; carried on for the production of coke and of fuel gas. { ˌkär·bə·nəˈzā·shən }

carbon molecular sieve [CHEMISTRY] A molecular sieve that utilizes a special type of activated carbon for the adsorbent. { ¦kär·bən mə¦lek·yə·lər ˈsiv }

carbon monoxide [INORGANIC CHEMISTRY] CO A colorless, odorless gas resulting from the incomplete oxidation of carbon; found, for example, in mines and automobile exhaust; poisonous to animals. { ¦kär·bən məˈnäkˌsīd }

carbon number [ANALYTICAL CHEMISTRY] The number of carbon atoms in a material under analysis; plotted against chromatographic retention volume for compound identification. { ˈkär·bən ˌnəm·bər }

carbon replication [ANALYTICAL CHEMISTRY] A faithful carbon-film, mold of a specimen surface (for example, powders, bones, or crystals) which is thin enough to be studied by electron microscopy. { ˈkär·bən ˌrep·ləˈkā·shən }

carbon-skeleton formula [ORGANIC CHEMISTRY] *See* bond-line formula. { ¦kär·bən ¦skel·ə·tən ˌfȯr·myə·lə }

carbon suboxide [INORGANIC CHEMISTRY] C_3O_2 A colorless lacrimatory gas having an unpleasant odor with a boiling point of −6.8°C. { ˈkär·bən səbˈäkˌsīd }

carbon tetrachloride [ORGANIC CHEMISTRY] CCl_4 Colorless dense liquid, specific gravity 1.595, slightly soluble in water; used as a dry-cleaning agent. { ˈkär·bən te·trəˈklȯrˌīd }

carbon tetrafluoride [ORGANIC CHEMISTRY] CF_4 A colorless gas with a boiling point of −126°C; used as a refrigerant. Also known as tetrafluoromethane. { ˈkär·bən te·trəˈflu̇rˌīd }

carbonyl [ORGANIC CHEMISTRY] A functional group found in organic compounds in which a carbon atom is doubly bonded to an oxygen atom (—CO—). Also known as carbonyl group. { ˈkär·bəˌnil }

carbonylation [CHEMISTRY] Introduction of a carbonyl radical into a molecule. { kärˌbän·əlˈā·shən }

carbonyl bromide [ORGANIC CHEMISTRY] $COBr_2$ A poisonous liquid boiling at 187.83°C; may be used by the military as a toxic suffocant. { 'kär·bə,nil 'brō,mīd }

carbonyl compound [ORGANIC CHEMISTRY] A compound containing the carbonyl group (CO). { 'kär·bə,nil ,käm,paund }

***N,N'*-carbonyldiimidazole** [ORGANIC CHEMISTRY] $C_7H_6N_4O$ Crystals with a melting point of 115.5-116°C; hydrolyzed by water very quickly; used in the synthesis of peptides. { ¦en ¦en,prīm 'kär·bə,nil,dī·i'mid·ə,zōl }

carbonyl fluoride [ORGANIC CHEMISTRY] COF_2 A colorless gas that is soluble in water; used in organic synthesis. { 'kär·bə,nil 'flür,īd }

carbonyl group [ORGANIC CHEMISTRY] *See* carbonyl. { 'kär·bə,nil ,grüp }

carbophenothion [ORGANIC CHEMISTRY] $C_{11}H_{16}ClO_2PS_3$ An amber liquid used to control pests on fruits, nuts, vegetables, and fiber crops. { ¦kär·bō¦fēn·ō'thī,än }

carborane [ORGANIC CHEMISTRY] **1.** Any of a class of compounds containing boron, carbon, and hydrogen. **2.** $B_{10}C_2H_{12}$ A specific member of the class. { 'kär·bə,rān }

carboxin [ORGANIC CHEMISTRY] $C_{12}H_{13}NO_2S$ An off-white solid with a melting point of 91.5-92.5°C; used to treat seeds of barley, oats, wheat, corn, and cotton for fungus diseases. Also known as DCMO. { kär'bäk·sən }

carboxy group [ORGANIC CHEMISTRY] —COOH The functional group of carboxylic acid. Also known as carboxyl group. { kär'bäk·sē ,grüp }

carboxylate anion [ORGANIC CHEMISTRY] An anion with the general formula $(RCO_2)^-$, which is formed when the hydrogen attached to the carboxyl group of a carboxylic acid is removed. { kär'bäk·sə,lāt 'an,ī·ən }

carboxylation [ORGANIC CHEMISTRY] Addition of a carboxyl group into a molecule. { kär,bäk·sə'lā·shən }

carboxyl group [ORGANIC CHEMISTRY] *See* carboxy group. { kär'bäk·səl ,grüp }

carboxylic [CHEMISTRY] Having chemical properties resembling those of carboxylic acid. { ¦kär,bäk¦sil·ik }

carboxylic acid [ORGANIC CHEMISTRY] Any of a family of organic acids characterized by the presence of one or more carboxyl groups. { ¦kär,bäk¦sil·ik 'as·əd }

carboxymethyl cellulose [ORGANIC CHEMISTRY] An acid ether derivative of cellulose used as a sodium salt; a white, odorless, bulky solid used as a stabilizer and emulsifier; negatively charged resin used in ion-exchange chromatography as a cation exchanger. Also known as cellulose gum. { kär,bäk·sē¦meth·əl 'sel·yə,lōs }

carbyne [CHEMISTRY] Elemental carbon in a triply bonded form. { 'kär,bīn }

δ-3-carene [ORGANIC CHEMISTRY] $C_{10}H_{16}$ A clear, colorless, combustible terpene liquid, stable to about 250°C; used as a solvent and in chemical synthesis. { ¦del·tə ¦thrē 'ka,rēn }

Carius method [ANALYTICAL CHEMISTRY] A procedure used to analyze organic compounds for sulfur, halogens, and phosphorus that involves heating the sample with fuming nitric acid in a sealed tube. { 'kär·ē·əs ,meth·əd }

carminic acid [ORGANIC CHEMISTRY] $C_{22}H_{20}O_{13}$ A glucosidal hydroxyanthrapurin that is derived from cochineal; a red crystalline dye used as a stain for biological materials. Also known as cochinilin. { kär'min·ik 'as·əd }

carnaubic acid [ORGANIC CHEMISTRY] $C_{24}H_{48}O_2$ An acid found in carnauba wax and beef kidney. { kär'nȯ·bik 'as·əd }

Carnot's reagent [CHEMISTRY] A solution of sodium bismuth thiosulfate in alcohol used for determining potassium. { kär'noz rē'ā·jənt }

Caro's acid [INORGANIC CHEMISTRY] H_2SO_5 A white solid melting at about 45°C, formed during the acid hydrolysis of peroxydisulfates. { 'kä·roz 'as·əd }

carrageenan [ORGANIC CHEMISTRY] A polysaccharide derived from the red seaweed (Rhodophyceae) and used chiefly as an emulsifying, gelling, and stabilizing agent and as a viscosity builder in foods, cosmetics, and pharmaceuticals. Also spelled carrageenin. { ,kar·ə'gē·nən }

carrageenin [ORGANIC CHEMISTRY] *See* carrageenan. { ,kar·ə'gē·nən }

carrier [CHEMISTRY] A substance that, when associated with a trace of another substance, will carry the trace with it through a chemical or physical process. { 'kar·ē·ər }

carrier gas [ANALYTICAL CHEMISTRY] In gas chromatography, a gas used as an eluant for extracting the sample from the column as the gas passes through. Also known as eluant gas. { 'kar·ē·ər ˌgas }

carvacrol [ORGANIC CHEMISTRY] $(CH_3)_2CHC_6H_3(CH_3)OH$ A colorless liquid, boiling at 237°C; used in perfumes, flavorings, and fungicides. { 'kär·vəˌkrȯl }

carvol [ORGANIC CHEMISTRY] *See* carvone. { 'kärˌvȯl }

carvone [ORGANIC CHEMISTRY] $C_{10}H_{14}O$ A liquid ketone that boils at 231°C; soluble in water and alcohol; it is optically active and occurs naturally in both dextro and levo forms; used in flavorings and perfumery. Also known as carvol. { 'kärˌvōn }

caryophyllene [ORGANIC CHEMISTRY] $C_{15}H_{24}$ A liquid sesquiterpene that is found in some essential oils, particularly clove oil. { ˌkar·ē·ō'fīˌlēn }

caryophyllin [ORGANIC CHEMISTRY] $C_{30}H_{48}O_3$ A ketone, soluble in alcohol, extracted from oil of cloves. { ˌkar·ē·ō'fil·ən }

casein [ORGANIC CHEMISTRY] The protein of milk; a white solid soluble in acids. { 'kaˌsēn }

casein-formaldehyde [ORGANIC CHEMISTRY] A modified natural polymer. { 'kaˌsēn fȯr'mal·dəˌhīd }

Cassel green [INORGANIC CHEMISTRY] *See* barium manganate. { 'kas·əl ˌgrēn }

castor oil acid [ORGANIC CHEMISTRY] *See* ricinoleic acid. { 'kas·tər ¦ȯil 'as·əd }

cata-condensed polycyclic [ORGANIC CHEMISTRY] An aromatic compound in which no more than two rings have a single carbon atom in common. { ˌkad·ə·kən'denst ˌpäl·i'sī·klik }

catalysis [CHEMISTRY] A phenomenon in which a relatively small amount of substance augments the rate of a chemical reaction without itself being consumed. { kə'tal·ə·səs }

catalyst [CHEMISTRY] Substance that alters the velocity of a chemical reaction and may be recovered essentially unaltered in form and amount at the end of the reaction. { 'kad·əl·əst }

catalyst carrier [CHEMISTRY] A neutral material used to support a catalyst, such as activated carbon, diatomaceous earth, or activated alumina. { 'kad·əl·əst ˌkar·ē·ər }

catalyst selectivity [CHEMISTRY] **1.** The relative activity of a catalyst in reference to a particular compound in a mixture. **2.** The relative rate of a single reactant in competing reactions. { 'kad·əl·əst səˌlek'tiv·əd·ē }

cataphoresis [PHYSICAL CHEMISTRY] *See* electrophoresis. { ˌkad·ə·fə'rē·səs }

catechol [ORGANIC CHEMISTRY] One of a group of three isomeric dihydroxy benzenes in which the two hydroxyl groups are ortho to each other. Also known as catechin; pyrocatechol; pyrocatechuic acid. { 'kad·əˌkȯl }

catenane [ORGANIC CHEMISTRY] A compound whose structure has at least two interlocking rings, similar to links in a chain, where the two links remain free of conventional chemical bonds, yet are still fitted together mechanically. { 'kat·ənˌān }

catenation [CHEMISTRY] Formation of a chain structure by the bonding of atoms of the same element, for example, carbon in the hydrocarbons. { ˌkat·ən'ā·shən }

cathode [PHYSICAL CHEMISTRY] The electrode at which reduction takes place in an electrochemical cell, that is, a cell through which electrons are being forced. { 'kathˌōd }

cathodic polarization [PHYSICAL CHEMISTRY] Portion of electric cell polarization occurring at the cathode. { kə'thäd·ik ˌpō·lə·rə'zā·shən }

catholyte [CHEMISTRY] Electrolyte adjacent to the cathode in an electrolytic cell. { 'kath·əˌlīt }

cation [CHEMISTRY] A positively charged atom or group of atoms, or a radical which moves to the negative pole (cathode) during electrolysis. { 'katˌī·ən }

cation analysis [ANALYTICAL CHEMISTRY] Qualitative analysis for cations in aqueous solution. { 'katˌī·ən ə'nal·ə·səs }

cation exchange [CHEMISTRY] A chemical reaction in which hydrated cations of a solid are exchanged, equivalent for equivalent, for cations of like charge in solution. { 'katˌī·ən iks'chānj }

cation exchange resin [ORGANIC CHEMISTRY] A highly polymerized synthetic organic compound consisting of a large, nondiffusible anion and a simple, diffusible cation, which later can be exchanged for a cation in the medium in which the resin is placed. { 'kat‚ī·ən iks'chānj 'rez·ən }

cationic detergent [CHEMISTRY] A member of a group of detergents that have molecules containing a quaternary ammonium salt cation with a group of 12 to 24 carbon atoms attached to the nitrogen atom in the cation; an example is alkyltrimethyl ammonium bromide. { ‚kad·ē'än·ik di'tər·jənt }

cationic hetero atom [CHEMISTRY] A positively charged atom, other than carbon, in an otherwise carbon atomic chain or ring. { ‚kad·ē'än·ik 'hed·ə·rō 'ad·əm }

cationic polymerization [ORGANIC CHEMISTRY] A type of polymerization in which Lewis acids act as catalysts. { ‚kad·ē'än·ik pə‚lim·ə·rə'zā·shən }

cationic reagent [CHEMISTRY] A surface-active agent with active positive ions used for ore beneficiation (flotation via flocculation); an example of a cationic reagent is cetyl trimethyl ammonium bromide. { ‚kad·ē'än·ik rē'ā·jənt }

cationtrophy [CHEMISTRY] The breaking off of an ion, such as a hydrogen ion or metal ion, from a molecule so that a negative ion remains in equilibrium. { ‚kad·ē'än·trə·fē }

caustic [CHEMISTRY] **1.** Burning or corrosive. **2.** A hydroxide of a light metal. { 'kȯ·stik }

caustic alcohol [ORGANIC CHEMISTRY] *See* sodium ethylate. { 'kȯ·stik 'al·kə‚hȯl }

causticity [CHEMISTRY] The property of being caustic. { kȯ'stis·əd·ē }

caustic lime [INORGANIC CHEMISTRY] *See* calcium oxide. { 'kȯ·stik'līm }

caustic potash [INORGANIC CHEMISTRY] *See* potassium hydroxide. { 'kȯ·stik 'päd‚ash }

caustic soda [INORGANIC CHEMISTRY] *See* sodium hydroxide. { 'kȯ·stik 'sōd·ə }

caustic wash [CHEMISTRY] **1.** Treating a product with a solution of caustic soda to remove impurities. **2.** The solution itself. { 'kȯ·stik 'wäsh }

cavitation [CHEMISTRY] Emulsification produced by disruption of a liquid into a liquid-gas two-phase system, when the hydrodynamic pressure of the liquid is reduced to the vapor pressure. { ‚kav·ə'tā·shən }

Cd [CHEMISTRY] *See* cadmium.

Ce [CHEMISTRY] *See* cerium.

cell [PHYSICAL CHEMISTRY] A cup, jar, or vessel containing electrolyte solutions and metal electrodes to produce an electric current (conductiometric or potentiometric) or for electrolysis (electrolytic). { sel }

cell constant [PHYSICAL CHEMISTRY] The ratio of distance between conductance-titration electrodes to the area of the electrodes, measured from the determined resistance of a solution of known specific conductance. { 'sel ‚kän·stənt }

cellobiose [ORGANIC CHEMISTRY] $C_{12}H_{22}O_{11}$ A disaccharide which does not occur freely in nature or as a glucoside; a unit of cellulose and lichenin; crystallizes as minute water-soluble crystals from alcohol. Also known as cellose. { ‚sel·ō'bī‚ōs }

cellose [ORGANIC CHEMISTRY] *See* cellobiose. { 'se‚lōs }

cellosolve [ORGANIC CHEMISTRY] $C_2H_5OCH_2CH_2OH$ An important industrial chemical used in varnish removers, in cleaning solutions, and as a solvent for paints, varnishes, and plastics. Also known as 2-ethoxyethanol. { 'sel·ə‚sälv }

cellulose acetate [ORGANIC CHEMISTRY] An acetic acid ester of cellulose; a tough, flexible, slow-burning, and long-lasting thermoplastic material used as the base for magnetic tape and movie film, in acetate rayon, as a plastic film in food packaging, in lacquers, and for molded receiver cabinets. { 'sel·yə‚lōs 'as·ə‚tāt }

cellulose acetate butyrate [ORGANIC CHEMISTRY] An ester of cellulose formed by the action of a mixture of acetic acid and butyric acid and their anhydrides on purified cellulose; has high impact resistance, clarity, and weatherability; used in making plastic film, lacquer, lenses, and outdoor signs. { 'sel·yə‚lōs 'as·ə‚tāt 'byüd·ə‚rāt }

cellulose diacetate [ORGANIC CHEMISTRY] The ester formed by esterification of two hydroxyl groups of a cellulose molecule with acetic acid. { 'sel·yə‚lōs dī'as·ə‚tāt }

cellulose ester [ORGANIC CHEMISTRY] Cellulose in which the free hydroxyl groups have been replaced wholly or in part by acidic groups. { 'sel·yə‚lōs 'es·tər }

cellulose ether [ORGANIC CHEMISTRY] The product of the partial or complete etherification of the hydroxyl groups in a cellulose molecule. { 'sel·yə‚lōs 'ē·thər }

cellulose fiber [ORGANIC CHEMISTRY] Any fiber based on esters or ethers of cellulose. { 'sel·yə,lōs 'fī·bər }
cellulose gum [ORGANIC CHEMISTRY] *See* carboxymethyl cellulose. { 'sel·yə,lōs 'gəm }
cellulose methyl ether [ORGANIC CHEMISTRY] *See* methylcellulose. { 'sel·yə,lōs 'meth·əl ,ē·thər }
cellulose nitrate [ORGANIC CHEMISTRY] Any of several esters of nitric acid, produced by treating cotton or some other form of cellulose with a mixture of nitric and sulfuric acids; used as explosive and propellant. Also known as nitrocellulose; nitrocotton. { 'sel·yə,lōs 'nī,trāt }
cellulose propionate [ORGANIC CHEMISTRY] An ester of cellulose and propionic acid. { 'sel·yə,lōs 'prō·pē·ə,nāt }
cellulose triacetate [ORGANIC CHEMISTRY] A cellulose resin formed by the complete esterification of the cellulose by acetic acid; used as a base in protective coatings. { 'sel·yə,lōs trī'as·ə,tāt }
cellulose xanthate [ORGANIC CHEMISTRY] A compound formed by reaction of soda cellulose (prepared by treating cellulose with strong sodium hydroxide solution) with carbon disulfide. { 'sel·yə,lōs 'zan,thāt }
cellulosic [ORGANIC CHEMISTRY] Any of the derivatives of cellulose, such as cellulose acetate. { ,sel·yə'lō·sik }
cellulosic resin [ORGANIC CHEMISTRY] Any resin based on cellulose compounds such as esters and ethers. { 'sel·yə,lōs 'rez·ən }
cementation [CHEMISTRY] The setting of a plastic material. { ,sē,men'tā·shən }
21-centimeter line [SPECTROSCOPY] A radio-frequency spectral line of neutral atomic hydrogen at a wavelength of approximately 21 centimeters and a frequency of approximately 1420 megahertz, that results from hyperfine transitions between states in which the spins of the electron and proton are parallel and antiparallel. { ¦twen·tē¦wən 'sen·tə,mēd·ər ,līn }
centrifugation potentials [PHYSICAL CHEMISTRY] Electric potential differences between points at different distances from the axis of rotation of a colloidal solution that is being rapidly rotated in a centrifuge. { sen,trif·ə'gā·shən pə,ten·chəlz }
centrifuge tube [ANALYTICAL CHEMISTRY] Calibrated, tube-shaped glass container used with laboratory centrifuges for volumetric analysis of separable (solid-liquid or immiscible liquid) samples. { 'sen·trə,fyüj ,tüb }
CEPHA [ORGANIC CHEMISTRY] *See* ethephon. { 'sef·ə }
cephaeline [ORGANIC CHEMISTRY] $C_{14}H_{19}O_2N$ An alkaloid, slightly soluble in water, extracted from the root of ipecac; used as an emetic. { sə'fā·ə,lēn }
ceramide [ORGANIC CHEMISTRY] Any of a group of amides formed by linking a fatty acid to sphingosine. { 'ser·ə,mīd }
cerate [ORGANIC CHEMISTRY] A metallic salt or soap made from lard. { 'sir,āt }
ceria [INORGANIC CHEMISTRY] *See* ceric oxide. { 'ser·ē·ə }
ceric oxide [INORGANIC CHEMISTRY] CeO_2 A pale-yellow to white powder; soluble in sulfuric acid, insoluble in dilute acid and water; used in ceramics and as a polish for optical glass. Also known as ceria; cerium dioxide; cerium oxide. { 'sir·ik 'äk,sīd }
ceric sulfate [INORGANIC CHEMISTRY] $Ce(SO_4)_2 \cdot 4H_2O$ Yellow needles forming a basic salt with excess water; used in waterproofing, mildew-proofing, and in dyeing and printing textiles. { 'sir·ik 'səl,fāt }
cerinic acid [ORGANIC CHEMISTRY] *See* cerotic acid. { sə'rēn·ik 'as·əd }
cerium [CHEMISTRY] A chemical element, symbol Ce, atomic number 58, atomic weight 140.12; a rare-earth metal, used as a getter in the metal industry, as an opacifier and polisher in the glass industry, in Welsbach gas mantles, in cored carbon arcs, and as a liquid-liquid extraction agent to remove fission products from spent uranium fuel. { 'sir·ē·əm }
cerium dioxide [INORGANIC CHEMISTRY] *See* ceric oxide. { 'sir·ē·əm dī'äk,sīd }
cerium fluoride [INORGANIC CHEMISTRY] CeF_3 White hexagonal crystals, melting point 1460°C; used in arc carbons to increase the brilliance of carbon-arc lamps. { 'sir·ē·əm 'flür,īd }
cerium oxide [INORGANIC CHEMISTRY] *See* ceric oxide. { 'sir·ē·əm 'äk,sīd }

cerium stearate [ORGANIC CHEMISTRY] $Ce(C_{18}H_{35}O_2)_2$ White, waxy, inert powder, melting point 100-110°C; used in waterproofing compounds. { 'sir·ē·əm 'stir,āt }

cerotic acid [ORGANIC CHEMISTRY] $CH_3(CH_2)_{24}COOH$ A fatty acid derived from carnauba wax or beeswax; melts at 87.7°C. Also known as cerinic acid; hexacosanoic acid. { sə'räd·ik 'as·əd }

certified reference material [ANALYTICAL CHEMISTRY] A reference material, one or more of whose property values are certified by a technically valid procedure, for which a certificate or other documentation has been issued by an appropriate certifying agency. { 'sərd·ə,fīd 'ref·rəns mə'tir·ē·əl }

ceryl alcohol [ORGANIC CHEMISTRY] $C_{26}H_{53}OH$ An alcohol derived from Chinese wax, melting at 79°C and insoluble in water. { 'sir·əl 'al·kə,hòl }

cesium [CHEMISTRY] A chemical element, symbol Cs, atomic number 55, atomic weight 132.905. { 'sē·zē·əm }

cesium bromide [INORGANIC CHEMISTRY] CsBr A colorless, crystalline powder with a melting point of 636°C; soluble in water; used in medicine, for infrared spectroscopy, and in scintillation counters. { 'sē·zē·əm 'brō,mīd }

cesium carbonate [INORGANIC CHEMISTRY] Cs_2CO_3 A white, hygroscopic, crystalline powder; soluble in water; used in specialty glasses. { 'sē·zē·əm 'kär·bə,nāt }

cesium chloride [INORGANIC CHEMISTRY] CsCl Colorless cuboid crystals, melting point 646°C; used in filaments of radio tubes to increase sensitivity, in photoelectric cells, and for photosensitive deposit on cathodes. { 'sē·zē·əm 'klòr,īd }

cesium fluoride [INORGANIC CHEMISTRY] CsF Toxic, irritating, deliquescent crystals with a melting point of 682°C; soluble in water and methanol; used in medicine, mineral water, and brewing. { 'sē·zē·əm 'flùr,īd }

cesium hydroxide [INORGANIC CHEMISTRY] CsOH Colorless or yellow, fused crystalline mass with a melting point of 272.3°C; soluble in water; used as electrolyte in alkaline storage batteries at subzero temperatures. { 'sē·zē·əm ,hī'dräk,sīd }

cesium iodide [INORGANIC CHEMISTRY] CsI A colorless, deliquescent, crystalline powder with a melting point of 621°C; soluble in water and alcohol; crystals used for infrared spectroscopy. { 'sē·zē·əm 'ī·ə,dīd }

cesium perchlorate [INORGANIC CHEMISTRY] $CsClO_4$ A crystalline solid with a melting point of 250°C; soluble in water; used in optics and for specialty glasses. { 'sē·zē·əm pər'klòr,āt }

cesium sulfate [INORGANIC CHEMISTRY] Cs_2SO_4 Colorless crystals with a melting point of 1010°C; soluble in water; used for brewing and in mineral waters. { 'sē·zē·əm 'səl,fāt }

cetane [ORGANIC CHEMISTRY] *See n*-hexadecane. { 'sē,tān }

cetane-number improver [CHEMISTRY] A chemical which has the effect of increasing a diesel fuel's cetane number; examples are nitrates, nitroalkanes, nitrocarbonates, and peroxides. { 'sē,tān ,nəm·bər im'prüv·ər }

cetin [ORGANIC CHEMISTRY] $C_{15}H_{31}COOC_{16}H_{33}$ A white, crystalline, waxy substance with a melting point of 50°C; soluble in alcohol and ether; used as a base for ointments and emulsions and in the manufacture of soaps and candles. { 'sēt·ən }

cetrimonium bromide [ORGANIC CHEMISTRY] $CH_3(CH_2)_{15}N(CH_3)_3Br$ Crystals with a melting point of 237-243°C, soluble in alcohol, water, and sparingly in acetone, used as a cationic detergent, antiseptic, and precipitant for nucleic acids and mucopolysaccharides. { ,se·trə'mōn·ē·əm 'brō,mīd }

cetyl [ORGANIC CHEMISTRY] The radical represented as $C_{16}H_{33}$—. { 'sēd·əl }

cetyl alcohol [ORGANIC CHEMISTRY] $C_{15}H_{33}OH$ A colorless wax, insoluble in water although a solution in kerosine forms an insoluble film on water. { 'sēd·əl 'al·kə,hòl }

cetyl vinyl ether [ORGANIC CHEMISTRY] $C_{16}H_{33}OCO{:}CH_2$ A colorless liquid with a boiling point of 142°C; may be copolymerized with unsaturated monomers to make internally plasticized resins. { 'sēd·əl 'vīn·əl 'ē·thər }

Cf [CHEMISTRY] *See* californium.

CFC [ORGANIC CHEMISTRY] *See* chlorofluorocarbon.

chain [CHEMISTRY] A structure in which similar atoms are linked by bonds. { chān }

chain balance [ANALYTICAL CHEMISTRY] An analytical balance with one end of a fine gold chain suspended from the beam and the other fastened to a device which moves over a graduated vernier scale. { 'chān ˌbal·əns }

chain isomerism [ORGANIC CHEMISTRY] A type of molecular isomerism seen in carbon compounds; as the number of carbon atoms in the molecule increases, the linkage between the atoms may be a straight chain or branched chains producing isomers that differ from each other by possessing different carbon skeletons. { 'chān ˌī'säm·əˌriz·əm }

chain reaction [CHEMISTRY] A chemical reaction in which many molecules undergo chemical reaction after one molecule becomes activated. { ¦chān rē'ak·shən }

chain scission [ORGANIC CHEMISTRY] The cleavage of polymer chains, as in natural rubber as a result of heating. { 'chān ˌsizh·ən }

chair conformation [PHYSICAL CHEMISTRY] *See* chair form. { 'cher ˌkän·fərˌmā·shən }

chair form [PHYSICAL CHEMISTRY] A particular nonplanar conformation of a cyclic molecule with more than five atoms in the ring; for example, in the chair form of cyclohexane, the hydrogens are staggered and directed perpendicularly to the mean plane of the carbons (axial conformation, *a*) or equatorially to the center of the mean plane (equatorial conformation, *e*). Also known as chair conformation. { 'cher ˌfȯrm }

chalcogen [INORGANIC CHEMISTRY] One of the elements that form group VI of the periodic table; included are oxygen, sulfur, selenium, tellurium, and polonium. { 'kal·kə·jən }

chalcogenide [INORGANIC CHEMISTRY] A binary compound containing a chalcogen and a more electropositive element or radical. { 'kal·kə·jəˌnīd }

chalking [CHEMISTRY] **1.** Treating with chalk. **2.** Forming a powder which is easily rubbed off. { 'chȯk·iŋ }

chamber acid [INORGANIC CHEMISTRY] Sulfuric acid made by the obsolete chamber process. { 'chām·bər 'as·əd }

chance cause [ANALYTICAL CHEMISTRY] A cause for variability in a measurement process that occurs randomly and unpredictably and for unknown reasons. { ¦chans 'kȯz }

channel black [CHEMISTRY] *See* gas black. { 'chan·əl ˌblak }

channeling [ANALYTICAL CHEMISTRY] In chromatography, furrows or breaks in an ion-exchange bed which permit a solution to run through without having contact with active groups elsewhere in the bed. { 'chan·əl·iŋ }

characteristic loss spectroscopy [SPECTROSCOPY] A branch of electron spectroscopy in which a solid surface is bombarded with monochromatic electrons, and backscattered particles which have lost an amount of energy equal to the core-level binding energy are detected. Abbreviated CLS. { ˌkar·ik·tə'ris·tik 'lȯs ˌspek'träs·kə·pē }

charge-delocalized ion [ORGANIC CHEMISTRY] A charged species in which the charge is distributed over more than one atom. { 'chärj dē'lōk·əlˌīzd 'ī·ən }

charged species [CHEMISTRY] A chemical entity in which the overall total of electrons is unequal to the overall total of protons. { 'chärjd 'spē·shēz }

charge-localized ion [ORGANIC CHEMISTRY] A charged species in which the charge is centered on a single atom. { 'chärj ˌlō·kəˌlīzd 'ī·ən }

charge population [CHEMISTRY] The net electric charge on a specified atom in a molecule that, while it cannot be observed physically, can be determined by a prescribed definition. { 'chärj ˌpäp·yəˌlā·shən }

charge transfer [PHYSICAL CHEMISTRY] The process in which an ion takes an electron from a neutral atom, with a resultant transfer of charge. { 'chärj ˌtranz·fər }

charge-transfer complexes [CHEMISTRY] Compounds in which electrons move between molecules. { 'chärj ˌtranz·fər 'käm·plek·səs }

chavicol [ORGANIC CHEMISTRY] $C_3H_5C_6H_4OH$ A colorless phenol that is liquid at room temperature; boils at 230°C; soluble in alcohol and water; found in many essential oils. { 'chav·əˌkȯl }

check sample [ANALYTICAL CHEMISTRY] *See* control sample. { 'chek ˌsam·pəl }

check standard [ANALYTICAL CHEMISTRY] In physical calibration, an artifact that is measured at specified intervals. { 'chek ˌstan·dərd }

chelate [ORGANIC CHEMISTRY] A molecular structure in which a heterocyclic ring can be formed by the unshared electrons of neighboring atoms. { 'kēˌlāt }

chelating agent [ORGANIC CHEMISTRY] An organic compound in which atoms form more than one coordinate bond with metals in solution. { 'ke,lād·iŋ ,ā·jənt }

chelating resin [ORGANIC CHEMISTRY] Any of the ion-exchange resins with unusually high selectivity for specific cations; for example, phenol-formaldehyde resin with 8-quinolinol replacing part of the phenol, particularly selective for copper, nickel, cobalt, and iron(III). { 'ke,lād·iŋ 'rez·ən }

chelation [ORGANIC CHEMISTRY] A chemical process involving formation of a heterocyclic ring compound which contains at least one metal cation or hydrogen ion in the ring. { kē'lā·shən }

chelerythrine [ORGANIC CHEMISTRY] $C_{21}H_{17}O_4H$ A poisonous, crystalline alkaloid, slightly soluble in alcohol; it is derived from the seeds of the herb celandine (*Chelidonium majus*) and has narcotic properties. { ,kel·ə'rī,thrēn }

cheletropic reaction [PHYSICAL CHEMISTRY] A chemical reaction involving the elimination of a molecule in which two sigma bonds terminating at a single atom are made or broken. { ,kel·ə'trä·pik rē'ak·shən }

chelidonic acid [ORGANIC CHEMISTRY] $C_7H_4O_6$ A pyran isolated from the perennial herb celandine (*Chelidonium majus*). { ¦kel·ə¦dän·ik 'as·əd }

chelometry [ANALYTICAL CHEMISTRY] Analytical technique involving the formation of 1:1 soluble chelates when a metal ion is titrated with aminopolycarboxylate and polyamine reagents; a form of complexiometric titration. { ke'läm·ə·trē }

chemical [CHEMISTRY] **1.** Related to the science of chemistry. **2.** A substance characterized by definite molecular composition. { 'kem·i·kəl }

chemical affinity [CHEMISTRY] *See* affinity. { 'kem·i·kəl ə'fin·əd·ē }

chemical bond [CHEMISTRY] *See* bond. { 'kem·i·kəl ¦bänd }

chemical cellulose [ORGANIC CHEMISTRY] *See* alpha cellulose. { 'kem·i·kəl 'sel·yə,lōs }

chemical compound [CHEMISTRY] *See* compound. { 'kem·i·kəl 'käm,paủnd }

chemical dating [ANALYTICAL CHEMISTRY] The determination of the relative or absolute age of minerals and of ancient objects and materials by measurement of their chemical compositions. { 'kem·i·kəl 'dād·iŋ }

chemical deposition [CHEMISTRY] Precipitation of a metal from a solution of a salt by introducing another metal. { 'kem·i·kəl ,dep·ə'zish·ən }

chemical dynamics [PHYSICAL CHEMISTRY] A branch of physical chemistry that seeks to explain time-dependent phenomena, such as energy transfer and chemical reactions, in terms of the detailed motion of the nuclei and electrons that constitute the system. { 'kem·ə·kəl dī'nam·iks }

chemical element [CHEMISTRY] *See* element. { 'kem·i·kəl 'el·ə·mənt }

chemical energy [PHYSICAL CHEMISTRY] Energy of a chemical compound which, by the law of conservation of energy, must undergo a change equal and opposite to the change of heat energy in a reaction; the rearrangement of the atoms in reacting compounds to produce new compounds causes a change in chemical energy. { 'kem·i·kəl 'en·ər·jē }

chemical equilibrium [CHEMISTRY] A condition in which a chemical reaction is occurring at equal rates in its forward and reverse directions, so that the concentrations of the reacting substances do not change with time. Also known as equilibrium. { 'kem·i·kəl ,ē·kwə'lib·rē·əm }

chemical exchange process [CHEMISTRY] A method of separating isotopes of the lighter elements by the repetition of a process of chemical change which involves exchange of the isotopes. { 'kem·i·kəl iks'chānj ,präs·əs }

chemical flux [CHEMISTRY] In a chemical reaction, the amount of a given substance per unit volume transformed per unit time. Also known as chemiflux. { 'kem·ə·kəl 'fləks }

chemical formula [CHEMISTRY] A notation utilizing chemical symbols and numbers to indicate the chemical composition of a pure substance; examples are CH_4 for methane and HCl for hydrogen chloride. { 'kem·i·kəl 'fȯr·myə·lə }

chemical indicator [ANALYTICAL CHEMISTRY] **1.** A substance whose physical appearance is altered at or near the end point of a chemical titration. **2.** A substance whose color varies as the concentration of hydrogen ions in the solution to which it is added varies. Also known as indicator. { 'kem·i·kəl 'in·də,kād·ər }

chemical inhibitor [CHEMISTRY] A substance capable of stopping or retarding a chemical reaction. { 'kem·i·kəl in'hib·əd·ər }

chemical kinetics [PHYSICAL CHEMISTRY] That branch of physical chemistry concerned with the mechanisms and rates of chemical reactions. Also known as reaction kinetics. { 'kem·i·kəl kə'ned·iks }

chemically pure [CHEMISTRY] Without impurities detectable by analysis. Abbreviated cp. { 'kem·ik·lē 'pyu̇r }

chemical microscopy [ANALYTICAL CHEMISTRY] Application of the microscope to the solution of chemical problems. { 'kem·i·kəl mī'kräs·kə·pē }

chemical polarity [PHYSICAL CHEMISTRY] Tendency of a molecule, or compound, to be attracted or repelled by electrical charges because of an asymmetrical arrangement of atoms around the nucleus. { 'kem·i·kəl pə'lar·əd·ē }

chemical potential [PHYSICAL CHEMISTRY] In a thermodynamic system of several constituents, the rate of change of the Gibbs function of the system with respect to the change in the number of moles of a particular constituent. { 'kem·i·kəl pə'ten·chəl }

chemical purity [CHEMISTRY] *See* purity. { 'kem·ə·kə 'pyu̇r·ə·dē }

chemical reaction [CHEMISTRY] A change in which a substance (or substances) is changed into one or more new substances; there is only a minute change, Δm, in the mass of the system, given by $\Delta E = \Delta mc^2$, where ΔE is the energy emitted or absorbed and c is the speed of light. { 'kem·i·kəl rē'ak·shən }

chemical reactivity [CHEMISTRY] The tendency of two or more chemicals to react to form one or more products differing from the reactants. { 'kem·i·kəl rē,ak'tiv·əd·ē }

chemical relaxation [CHEMISTRY] The readjustment of a chemical system to a new equilibrium after the equilibrium of a chemical reaction is disturbed by a sudden change, particularly in an external parameter such as pressure or temperature. { 'kem·ə·kəl ,rē,lak'sā·shən }

chemical shift [PHYSICAL CHEMISTRY] Shift in a nuclear magnetic-resonance spectrum resulting from diamagnetic shielding of the nuclei by the surrounding electrons. { 'kem·i·kəl 'shift }

chemical species [CHEMISTRY] *See* species. { 'kem·i·kəl 'spē,shēz }

chemical symbol [CHEMISTRY] A notation for one of the chemical elements, consisting of letters; for example Ne, O, C, and Na represent neon, oxygen, carbon, and sodium. { 'kem·i·kəl 'sim·bəl }

chemical synthesis [CHEMISTRY] The formation of one chemical compound from another. { 'kem·i·kəl 'sin·thə·səs }

chemical thermodynamics [PHYSICAL CHEMISTRY] The application of thermodynamic principles to problems of chemical interest. { 'kem·i·kəl ,thər·mō·də'nam·iks }

chemiclearance [CHEMISTRY] The use of chemical analysis to establish the safe use of a substance. { 'kem·i,klir·əns }

chemiflux [CHEMISTRY] *See* chemical flux. { 'kem·ə,fləks }

chemi-ionization [CHEMISTRY] Ionization that occurs as a result of the collison of a particle with a neutral species, usually excited, such as a metastable atom. { ,kem·ē,ī·ə·nə'zā·shən }

chemiluminescence [PHYSICAL CHEMISTRY] Emission of light as a result of a chemical reaction without an apparent change in temperature. { ,kem·i,lüm·ə'nes·əns }

chemionics [CHEMISTRY] The chemistry of molecular components and devices that operate on photons, electrons, and ions. { ,kem·ē'än·iks }

chemiosmosis [CHEMISTRY] A chemical reaction occurring through an intervening semipermeable membrane. Also known as chemosmosis. { ¦kem·ē,äs¦mō·səs }

chemisorption [PHYSICAL CHEMISTRY] A chemical adsorption process in which weak chemical bonds are formed between gas or liquid molecules and a solid surface. { 'kem·i,sȯrp·shən }

chemist [CHEMISTRY] A scientist specializing in chemistry. { 'kem·əst }

chemosmosis [CHEMISTRY] *See* chemiosmosis. { ,kem,äs'mō·səs }

chinaldine [ORGANIC CHEMISTRY] *See* quinaldine. { ki'näl‚dēn }
Chinese vermilion [INORGANIC CHEMISTRY] *See* mercuric sulfide. { chīn'nēz vər'mil·yən }
Chinese white [CHEMISTRY] A term used in the paint industry for zinc oxide and kaolin used as a white pigment. Also known as zinc white. { chīn'nēz 'wīt }
chinic acid [ORGANIC CHEMISTRY] *See* quinic acid. { 'kin·ik 'as·əd }
chinoidine [ORGANIC CHEMISTRY] *See* quinoidine. { ki'nòi‚dēn }
chinone [ORGANIC CHEMISTRY] *See* quinone. { kin'ōn }
chiral carbon atom [ORGANIC CHEMISTRY] *See* asymmetric carbon atom. { ¦kī·rəl ¦kär·bən 'ad·əm }
chiral center [ORGANIC CHEMISTRY] An atom in a molecule that is attached to four different groups. { 'kī·rəl 'sen·tər }
chirality [CHEMISTRY] The handedness of an asymmetric molecule. { kī'ral·əd·ē }
chiral molecules [CHEMISTRY] Molecules which are not superposable with their mirror images. { 'kī·rəl 'mäl·ə‚kyülz }
chloflurecol methyl ester [ORGANIC CHEMISTRY] $C_{15}H_{11}ClO_3$ A white, crystalline compound with a melting point of 152°C; slight solubility in water; used as a growth regulator for grass and weeds. { klō'flùr·ə‚kòl 'meth·əl 'es·tər }
chloral [ORGANIC CHEMISTRY] CCl_3CHO A colorless, oily liquid soluble in water; used industrially to prepare DDT; a hypnotic. Also known as trichloroacetic aldehyde; trichloroethanol. { 'klòr·əl }
chloralase [ORGANIC CHEMISTRY] $C_8H_{11}Cl_3O_6$ Colorless, water-soluble crystals, melting at 185°C; made by heating chloral with dextrose; used as a hypnotic. { 'klòr·ə‚lās }
chloral hydrate [ORGANIC CHEMISTRY] $CCl_3CH(OH)_2$ Colorless, deliquescent needles with slightly bitter caustic taste, soluble in water, a hypnotic. Also known as crystalline chloral; hydrated chloral. { 'klòr·əl 'hī‚drāt }
chloralkane [ORGANIC CHEMISTRY] Chlorinated aliphatic hydrocarbon of the methane series (C_nH_{2n+2}). { klòr'al‚kān }
chloralosane [ORGANIC CHEMISTRY] *See* chloralose. { ‚klòr·ə'lō‚sān }
chloralose [ORGANIC CHEMISTRY] $C_8H_{11}O_6Cl_3$ A crystalline compound with a melting point of 178°C; used as a repellent for birds. Also known as glucochloralose. { 'klòr·ə‚lōs }
α-chloralose [ORGANIC CHEMISTRY] $C_8H_{11}O_6Cl_3$ Needlelike crystals with a melting point of 87°C; soluble in glacial acetic acid and ether; used on seed grains as a bird repellent and as a hypnotic for animals. Also known as chloralosane; glucochloral; α-D-glucochloralose. { 'al·fə ¦klòr·ə‚lōs }
chloramine T [ORGANIC CHEMISTRY] $CH_3C_6H_4SO_2NClNa \cdot 3H_2O$ A white, crystalline powder that decomposes slowly in air, freeing chlorine; used as an antiseptic, a germicide, and an oxidizing agent and chlorinating agent. { 'klòr·ə‚mēn 'tē }
chloranil [ORGANIC CHEMISTRY] $C_6Cl_4O_2$ Yellow leaflets melting at 290°C; soluble in organic solvents; made from phenol by treatment with potassium chloride and hydrochloric acid; used as an agricultural fungicide and as an oxidizing agent in the manufacture of dyes. { klòr'an·əl }
chloranilic acid [ORGANIC CHEMISTRY] $C_6H_2Cl_2O_4$ A relatively strong dibasic acid whose crystals are red and melt between 283 and 284°C; used in spectrophotometry. { klòr·ə'nil·ik 'as·əd }
chlorate [INORGANIC CHEMISTRY] ClO_3^- **1.** A negative ion derived from chloric acid. **2.** A salt of chloric acid. { 'klòr‚āt }
chlorbenside [ORGANIC CHEMISTRY] $C_{13}H_{10}SCl_2$ White crystals with a melting point of 72°C; used as a miticide for spider mites on fruit trees and ornamentals. { klòr'ben‚sīd }
chlorbromuron [ORGANIC CHEMISTRY] $C_9H_{10}ONBrCl$ A white solid with a melting point of 94-96°C; used as a pre- and postemergence herbicide for annual grass and for broadleaf weeds on crops, soybeans, and Irish potatoes. { ‚klòr·brə'myü·rən }
chlordan [ORGANIC CHEMISTRY] *See* chlordane. { 'klòr‚dan }
chlordane [ORGANIC CHEMISTRY] $C_{10}H_6Cl_8$ A volatile liquid insecticide; a chlorinated hexahydromethanoindene. Also spelled chlordan. { 'klòr‚dān }

chlordimeform [ORGANIC CHEMISTRY] $C_{10}H_{13}ClN_2$ A tan-colored solid, melting point 35°C; used as a miticide and insecticide for fruits, vegetables, and cotton. { ˌklȯr ˈdī·məˌfȯrm }

chlorendic acid [ORGANIC CHEMISTRY] $C_9H_4Cl_6O_4$ White, fine crystals used in fire-resistant polyester resins and as an intermediate for dyes, fungicides, and insecticides. { klȯr′en·dik ′as·əd }

chlorendic anhydride [ORGANIC CHEMISTRY] $C_9H_2Cl_6O_3$ White, fine crystals used in fire-resistant polyester resins, in hardening epoxy resins, and as a chemical intermediate. { klȯr′en·dik an′hīˌdrīd }

chlorfenethol [ORGANIC CHEMISTRY] $C_{14}H_{12}Cl_2O$ A colorless, crystalline compound with a melting point of 69.5-70°C; insoluble in water; used for control of mites in ornamentals and shrub trees. { ˌklȯr′fen·əˌthȯl }

chlorfenpropmethyl [ORGANIC CHEMISTRY] $C_{10}H_{10}OCl_2$ A colorless to brown liquid used as a postemergence herbicide of wild oats, cereals, fodder beets, sugarbeets, and peas. { ¦klȯr·fənˌpräp′meth·əl }

chlorfensulfide [ORGANIC CHEMISTRY] $C_{12}H_6Cl_4N_2S$ A yellow, crystalline compound with a melting point of 123.5-124°C; used as a miticide for citrus. { ˌklȯr·fən′səlˌfīd }

chlorfenvinphos [ORGANIC CHEMISTRY] $C_{12}H_{14}Cl_3O_4P$ An amber liquid with a boiling point of 168-170°C; used as an insecticide for ticks, flies, lice, and mites on cattle. { ˌklȯr·fən′vinˌfäs }

chlorhydrin [ORGANIC CHEMISTRY] *See* chlorohydrin. { klȯr′hī·drən }

chloric acid [INORGANIC CHEMISTRY] $HClO_3$ A compound that exists only in solution and as chlorate salts; breaks down at 40°C. { ′klȯr·ik ′as·əd }

chloride [CHEMISTRY] **1.** A compound which is derived from hydrochloric acid and contains the chlorine atom in the −1 oxidation state. **2.** In general, any binary compound containing chloride. { ′klȯrˌīd }

chloride benzilate [ORGANIC CHEMISTRY] *See* lachesne. { ′klȯrˌīd ′ben·zəˌlāt }

chloridization [CHEMISTRY] *See* chlorination. { ˌklȯr·ə·də′zā·shən }

chlorimide [INORGANIC CHEMISTRY] *See* dichloramine. { ′klȯr·əˌmīd }

chlorinated paraffin [ORGANIC CHEMISTRY] One of a group of chlorine derivatives of paraffin compounds. { ′klȯr·əˌnād·əd ′par·ə·fən }

chlorination [CHEMISTRY] **1.** Introduction of chlorine into a compound. Also known as chloridization. **2.** Water sterilization by chlorine gas. { ˌklȯr·ə′nā·shən }

chlorine [CHEMISTRY] A chemical element, symbol Cl, atomic number 17, atomic weight 35.453; used in manufacture of solvents, insecticides, and many non-chlorine-containing compounds, and to bleach paper and pulp. { ′klȯrˌēn }

chlorine dioxide [INORGANIC CHEMISTRY] ClO_2 A green gas used to bleach cellulose and to treat water. { ′klȯrˌēn dī′äkˌsīd }

chlorine water [CHEMISTRY] A clear, yellowish liquid used as a deodorizer, antiseptic, and disinfectant. { ′klȯrˌēn ˌwȯd·ər }

chlorite [INORGANIC CHEMISTRY] A salt of chlorous acid. { ′klȯrˌīt }

chloritization [CHEMISTRY] The introduction of, production of, replacement by, or conversion into chlorite. { ˌklȯr·əd·ə′za·shən }

chlormephos [ORGANIC CHEMISTRY] $C_5H_{12}O_2S_2ClP$ A liquid used as an insecticide for soil. { ′klȯr·məˌfäs }

chloro- [ORGANIC CHEMISTRY] A prefix describing an organic compound which contains chlorine atoms substituted for hydrogen. { ′klȯr·ō }

chloroacetic acid [ORGANIC CHEMISTRY] $ClCH_2COOH$ White or colorless, deliquescent crystals that are soluble in water, ether, chloroform, benzene, and alcohol; used as an herbicide and in the manufacture of dyes and other organic molecules. { ¦klȯr·ə¦sēd·ik ′as·əd }

chloroacetic anhydride [ORGANIC CHEMISTRY] $C_4H_4Cl_2O_3$ Crystals with a melting point of 46°C; soluble in chloroform and ether; used in the preparation of cellulose chloracetates and in the N-acetylation of amino acids in alkaline solution. { ¦klȯr·ə¦sēd·ik an′hīˌdrīd }

chloroacetone [ORGANIC CHEMISTRY] CH_3COCH_2Cl Pungent, colorless liquid used as military tear gas and in organic synthesis. { klȯr′as·əˌtōn }

chloroacetonitrile [ORGANIC CHEMISTRY] $ClCH_2CN$ A colorless liquid with a pungent odor; soluble in hydrocarbons and alcohols; used as a fumigant. { ¦klȯr·ō‚as·ə·'tän·ə·trəl }

chloroacetophenone [ORGANIC CHEMISTRY] $C_6H_5COCH_2Cl$ Rhombic crystals melting at 59°C; an intermediate in organic synthesis. { ¦klȯr·ō‚as·ə'tä·fə‚nōn }

chloroacrolein [ORGANIC CHEMISTRY] $H_2C{:}ClCHO$ A colorless liquid with a boiling point of 29-31°C; used as a tear gas. { ‚klȯr·ō·ə'krō·lē·ən }

chlorobenzaldehyde [ORGANIC CHEMISTRY] C_6H_4CHOCl A colorless to yellowish liquid (ortho form) or powder (para form) with a boiling range of 209-215°C; soluble in alcohol, ether, and acetone; used in dye manufacture. { ‚klȯr·ō‚ben'zal·də‚hīd }

chlorobenzene [ORGANIC CHEMISTRY] C_6H_5Cl A colorless, mobile, volatile liquid with an almondlike odor; used to produce phenol, DDT, and aniline. { ‚klȯr·ō'ben‚zēn }

chlorobenzilate [ORGANIC CHEMISTRY] $C_{16}H_{14}Cl_2O_3$ A yellow-brown, viscous liquid with a melting point of 35-37°C; used as a miticide in agriculture and horticulture. { ‚klȯr·ō'ben·zə‚lāt }

para-chlorobenzoic acid [ORGANIC CHEMISTRY] ClC_6H_4COOH A white powder with a melting point of 238°C; soluble in methanol, absolute alcohol, and ether; used in the manufacture of dyes, fungicides, and pharamaceuticals. { ¦par·ə ¦klȯr·ō‚ben'zō·ik 'as·əd }

chlorobenzoyl chloride [ORGANIC CHEMISTRY] ClC_6H_4COCl A colorless liquid with a boiling range of 227-239°C; soluble in alcohol, acetone, and water; used in dye and pharmaceuticals manufacture. { ‚klȯr·ō'ben‚zȯil 'klȯr‚īd }

chlorobenzyl chloride [ORGANIC CHEMISTRY] $ClC_6H_4CH_2Cl$ A colorless liquid with a boiling range of 216-222°C, soluble in acetone, alcohol, and ether; used in the manufacture of organic chemicals. { ‚klor·ō'ben·zil 'klȯr‚īd }

chlorobutadiene [ORGANIC CHEMISTRY] *See* chloroprene. { ‚klor·ō‚byüd·ə'dī‚en }

chlorobutanol [ORGANIC CHEMISTRY] $Cl_3CC(CH_3)_2OH$ Colorless to white crystals with a melting point of 78°C; soluble in alcohol, glycerol, ether, and chloroform; used as a plasticizer and a preservative for biological solutions. { ‚klȯr·ō'byüt·ən‚ȯl }

chlorocarbon [ORGANIC CHEMISTRY] A compound of chlorine and carbon only, such as carbon tetrachloride, CCl_4. { ¦klȯr·ō'kär·bən }

chlorochromic anhydride [INORGANIC CHEMISTRY] *See* chromyl chloride. { ¦klȯr·ō ¦krō·mik an'hī‚drīd }

***O*-[2-chloro-1-(2,5-dichlorophenyl)-vinyl]-*O,O*-diethylphosphorothioate** [ORGANIC CHEMISTRY] $C_{12}H_{14}O_3PSCl_2$ A brown liquid with a boiling point of 145°C at 0.005 mmHg (0.667 pascal); used as an insecticide for lawn and turf pests. { ¦ō ¦tü ‚klȯr·ō ¦wən ¦tü ¦fīv dī‚klȯr·ō¦fen·əl 'vīn·əl ¦ō ¦ō dī‚eth·əl fäs‚fȯr·ō'thī·ə‚wāt }

1,1,1-chlorodifluoroethane [ORGANIC CHEMISTRY] CH_3CClF_2 A colorless gas with a boiling point of −130.8°C; used as a refrigerant, solvent, and aerosol propellant. { ¦wən ¦wən ¦wən ‚klȯr·ō‚dī‚flür·ō'eth‚ān }

chlorodifluoromethane [ORGANIC CHEMISTRY] $CHClF_2$ A colorless gas with a boiling point of −40.8°C and freezing point of −160°C; used as an aerosol propellant and refrigerant. { klȯr·ō·dī·flür·ō'meth·ān }

1-chloro-2,4-dinitrobenzene [ORGANIC CHEMISTRY] $C_6H_3ClN_2O_4$ Yellow crystals with a melting point of 52-54°C; soluble in hot alcohol, ether, and benzene; used as a reagent in the determination of pyridine compounds such as nicotinic acid, and nicotinamide. { ¦wən ‚klȯr·ō ¦tü ¦fȯr dī‚nī·trō'ben‚zēn }

chloroethene [ORGANIC CHEMISTRY] *See* vinyl chloride. { ‚klor·ō'eth‚ēn }

chloroethyl alcohol [ORGANIC CHEMISTRY] *See* ethylene chlorohydrin. { ‚klȯr·ō'eth·əl 'al·kə‚hȯl }

chlorofluorocarbon [ORGANIC CHEMISTRY] Any member of a group of substances that are derivatives of methane or ethane with all the hydrogen atoms replaced by combinations of chlorine and fluorine. Abbreviated CFC; FCC. { ¦klȯr·ə¦flür·ə‚kär·bən }

chlorofluoromethane [ORGANIC CHEMISTRY] A chlorofluorocarbon derived from methane, for example, chlorodifluoromethane (ClF_2CH). { ‚klȯr·ə‚flür·ə'meth‚ān }

chloroform [ORGANIC CHEMISTRY] $CHCl_3$ A colorless, sweet-smelling, nonflammable liquid; used at one time as an anesthetic. Also known as trichloromethane. { 'klȯr·ə‚fȯrm }

chlorohydrin [ORGANIC CHEMISTRY] Any of the compounds derived from a group of glycols or polyhydroxy alcohols by chlorine substitution for part of the hydroxyl groups. Also spelled chlorhydrin. { ˌklor·əˈhī·drən }

chlorohydrocarbon [ORGANIC CHEMISTRY] A carbon- and hydrogen-containing compound with chlorine substituted for some hydrogen in the molecule. { ¦klȯr·ōˈhī·drəˌkär·bən }

chlorohydroquinone [ORGANIC CHEMISTRY] $ClC_6H_3(OH)_2$ White to light tan crystals with a melting point of 100°C; soluble in water and alcohol; used as a photographic developer and bactericide and for dyestuffs. { ¦klȯr·ōˌhī·drə·kwinˈōn }

5-chloro-8-hydroxyquinoline [ORGANIC CHEMISTRY] C_9H_6ClNO Crystals with a melting point of 130°C; used as a fungicide and bactericide. { ¦fīv ˌklȯr·ō ¦āt hīˌdräk·sē ˈkwin·əˌlēn }

chloromethane [ORGANIC CHEMISTRY] CH_3Cl A colorless, noncorrosive, liquefiable gas which condenses to a colorless liquid; used as a refrigerant, and as a catalyst carrier in manufacture of butyl rubber. Also known as methyl chloride. { ¦klȯr·ōˈmethˌān }

1-chloronaphthalene [ORGANIC CHEMISTRY] $C_{10}H_7Cl$ An oily liquid used as an immersion medium in the microscopic determination of refractive index of crystals and as a solvent for oils, fats, and DDT. { ¦wən ˌklȯr·ōˈnaf·thəˌlēn }

chloronium ion [ORGANIC CHEMISTRY] A halonium ion in which the halogen is chlorine; sometimes occurs as a bridged form. { kləˈrōn·ē·əm ˈī·ən }

chlorophenol red [ORGANIC CHEMISTRY] $C_{19}H_{12}Cl_2O_5S$ A dye that is used as an acid-base indicator; yellow in acid solution, red in basic solution. Also spelled chlorphenol red. { ¦klȯr·əˈfen·ȯl ˈred }

chloropicrin [INORGANIC CHEMISTRY] CCl_3NO_2 A colorless liquid with a sweet odor whose vapor is very irritating to the lungs and causes vomiting, coughing, and crying; used as a soil fumigant. Also known as nitrochloroform; trichloronitromethane. { ˌklȯr·ōˈpik·rən }

chloroplatinate [INORGANIC CHEMISTRY] **1.** A double salt of platinic chloride and another chloride. **2.** A salt of chloroplatinic acid. Also known as platinochloride. { ˌklȯr·ōˈplat·ənˌāt }

chloroplatinic acid [INORGANIC CHEMISTRY] H_2PtCl_6 An acid obtained as red-brown deliquescent crystals; used in chemical analysis. Also known as platinic chloride. { ¦klȯr·ə·pləˈtin·ik ˈas·əd }

chloroprene [ORGANIC CHEMISTRY] C_4H_5Cl A colorless liquid which polymerizes to chloroprene resin. Also known as chlorobutadiene. { ˈklȯr·əˌprēn }

chloroprene resin [ORGANIC CHEMISTRY] A polymer of chloroprene used to form materials resembling natural rubber. { ˈklȯr·əˌprēn ˈrez·ən }

chloropropane [ORGANIC CHEMISTRY] Propane molecules with chlorine substituted in various amounts for the hydrogen atoms. { ¦klȯr·ōˈprōˌpān }

3-chloro-1,2-propanediol [ORGANIC CHEMISTRY] $ClCH_2CH(OH)CH_2OH$ A sweetish-tasting liquid that has a tendency to turn a straw color; soluble in ether, alcohol, and water; used to manufacture dye intermediates and to lower the freezing point of dynamite. { ¦thrē ˌklȯr·ō ¦wən ¦tü ¦prōˌpānˈdīˌȯl }

chloropropene [CHEMISTRY] Propene molecules with chlorine substituted for some hydrogen atoms. { ¦klȯr·əˈprōˌpēn }

***β*-chloropropionitrile** [ORGANIC CHEMISTRY] $ClCH_2CH_2CN$ A liquid with an acrid odor; miscible with various organic solvents such as ethanol, ether, and acetone; used in polymer synthesis and in the synthesis of pharmaceuticals. { ¦bād·ə ˌklȯr·ō¦prō·pē·ōˈnīˌtrəl }

***N*-chlorosuccinimide** [ORGANIC CHEMISTRY] $C_4H_4ClNO_2$ Orthorhombic crystals with the smell of chlorine; melting point is 150-151°C; soluble in water, benzene, and alcohol; used as a chlorinating agent. { ¦en ˌklȯr·ō·səkˈsin·əˌmīd }

chlorosulfonic acid [INORGANIC CHEMISTRY] $ClSO_2OH$ A fuming liquid that decomposes in water to sulfuric acid and hydrochloric acid; used in pharmaceuticals, pesticides, and dyes, and as a chemical intermediate. { ¦klȯr·ō·səlˈfän·ik ˈas·əd }

chlorothalonil [ORGANIC CHEMISTRY] $C_8Cl_4N_2$ Colorless crystals with a melting point of 250-251°C; used as a fungicide for crops, turf, and ornamental flowers. { ˌklȯr·ə'thal·ə·nəl }

chlorothymol [ORGANIC CHEMISTRY] $CH_3C_6H_2(OH)(C_3H_7)Cl$ White crystals melting at 59-61°C; soluble in benzene alcohol, insoluble in water; used as a bactericide. { ˌklȯr·ə'thīˌmȯl }

***ortho*-chlorotoluene** [ORGANIC CHEMISTRY] $CH_3C_6H_4Cl$ A liquid with a boiling point of 158.97°C; soluble in alcohol, chloroform, benzene, and ether; used in organic synthesis, as a solvent, and as an intermediate in dyestuff manufacture. { ¦ȯr·thō ˌklȯr·ō'täl·yəˌwēn }

4-(4-chloro-*ortho*-tolyl)oxybutyric acid [ORGANIC CHEMISTRY] $C_{10}H_{13}ClO_3$ A white, crystalline compound with a melting point of 99-100°C; used as a postemergence herbicide for peas. { ¦fȯr ¦fȯr ˌklȯr·ō ˌȯr·thō ¦täl·əlˌäk·sē·byü'tir·ik 'as·əd }

chlorotrifluoroethylene polymer [ORGANIC CHEMISTRY] A colorless, noninflammable, heat-resistant resin, soluble in most organic solvents, and with a high impact strength; can be made into transparent filling and thin sheets; used for chemical piping, fittings, and insulation for wire and cables, and in electronic components. Also known as fluorothene; polytrifluorochloroethylene resin. { ¦klȯr·ō·trī¦flu̇r·ō 'eth·əlˌen 'päl·ə·mər }

chlorotrifluoromethane [ORGANIC CHEMISTRY] $CClF_3$ A colorless gas having a boiling point of −81.4°C and a freezing point of −181°C; used as a dielectric and aerospace clinical, refrigerant, and aerosol propellant, and for metals hardening and pharmaceuticals manufacture. { ¦klȯr·ō·trī¦flu̇r·ō'meˌthān }

chloroxine [ORGANIC CHEMISTRY] $C_9H_5Cl_2NO$ Crystals with a melting point of 179-180°C; soluble in benzene and in sodium and potassium hydroxides; used as an analytical reagent. { klə'räk·sən }

4-chloro-3,5-xylenol [ORGANIC CHEMISTRY] $ClC_6H_2(CH_3)_2OH$ Crystals with a melting point of 115.5°C; soluble in water, 95% alcohol, benzene, terpenes, ether, and alkali hydroxides; used as an antiseptic and germicide and to stop mildew; used in humans as a topical and urinary antiseptic and as a topical antiseptic in animals. { ¦fȯr ¦klȯr·ō ¦thrē ¦fīv 'zī·ləˌnȯl }

chlorthiamid [ORGANIC CHEMISTRY] $C_7H_5Cl_2NS$ An off-white, crystalline compound with a melting point of 151-152°C; used as a herbicide for selective weed control in industrial sites. { klȯr'thī·əˌmid }

cholesteric material [PHYSICAL CHEMISTRY] A liquid crystal material in which the elongated molecules are parallel to each other within the plane of a layer, but the direction of orientation is twisted slightly from layer to layer to form a helix through the layers. { kə'les·tə·rik mə'tir·ē·əl }

cholesteric phase [PHYSICAL CHEMISTRY] A form of the nematic phase of a liquid crystal in which the molecules are spiral. { kə'les·tə·rik ˌfāz }

Christiansen effect [ANALYTICAL CHEMISTRY] Monochromatic transparency effect when finely powdered substances, such as glass or quartz, are immersed in a liquid having the same refractive index. { 'kris·chən·sən i'fekt }

chromate [INORGANIC CHEMISTRY] CrO_4^{2-} **1.** An ion derived from the unstable acid H_2CrO_4. **2.** A salt or ester of chromic acid. { 'krōˌmāt }

chromatogram [ANALYTICAL CHEMISTRY] The pattern formed by zones of separated pigments and of colorless substance in chromatographic procedures. { krō'mad·əˌgram }

chromatograph [ANALYTICAL CHEMISTRY] To employ chromatography to separate substances. { krō'mad·əˌgraf }

chromatographic adsorption [ANALYTICAL CHEMISTRY] Preferential adsorption of chemical compounds (gases or liquids) in an ascending molecular-weight sequence onto a solid adsorbent material, such as activated carbon, alumina, or silica gel; used for analysis and separation of chemical mixtures. { krō¦mad·ə¦graf·ik ad'sȯrp·shən }

chromatographic bed [ANALYTICAL CHEMISTRY] Any of the different configurations in which the stationary phase is contained. { krō¦mad·ə¦graf·ik 'bed }

chromatography [ANALYTICAL CHEMISTRY] A method of separating and analyzing mixtures of chemical substances by chromatographic adsorption. { ˌkrō·mə'täg·ˌrə·fē }

chrome alum [INORGANIC CHEMISTRY] $KCr(SO_4)_2 \cdot 12H_2O$ An alum obtained as purple crystals and used as a mordant, in tanning, and in photography in the fixing bath. Also known as potassium chromium sulfate. { 'krōm 'al·əm }

chrome dye [CHEMISTRY] One of a class of acid dyes used on wool with a chromium compound as mordant. { ¦krōm ¦dī }

chrome green [INORGANIC CHEMISTRY] *See* chromic oxide. { ¦krōm ˌgrēn }

chrome red [CHEMISTRY] **1.** A pigment containing basic lead chromate. **2.** Any of several mordant acid dyes. { ¦krōm 'red }

chrome yellow [CHEMISTRY] **1.** A yellow pigment composed of normal lead chromate, $PbCrO_4$, or other lead compounds. **2.** Any of several mordant acid dyes. { ¦krōm 'yel·ō }

chromic acid [INORGANIC CHEMISTRY] H_2CrO_4 The hydrate of CrO_3; exists only as salts or in solution. { ¦krō·mik 'as·əd }

chromic chloride [INORGANIC CHEMISTRY] $CrCl_3$ Crystals that are pinkish violet shimmering plates, almost insoluble in water, but easily soluble in presence of minute traces of chromous chloride; used in calico printing, as a mordant for cotton and silk. { ¦krō·mik 'klȯrˌīd }

chromic fluoride [INORGANIC CHEMISTRY] $CrF_3 \cdot 4H_2O$ Crystals that are green, soluble in water; used in dyeing cottons. { ¦krō·mik 'flu̇rˌīd }

chromic hydroxide [INORGANIC CHEMISTRY] $Cr(OH)_3 \cdot 2H_2O$ Gray-green, gelatinous precipitate formed when a base is added to a chromic salt; the precipitate dries to a bluish, amorphous powder; prepared as an intermediate in the manufacture of other soluble chromium salts. { ¦krō·mik hī'dräkˌsīd }

chromic nitrate [INORGANIC CHEMISTRY] $Cr(NO_3)_3 \cdot 9H_2O$ Purple, rhombic crystals that are soluble in water; used as a mordant in textile dyeing. { ¦krō·mik 'nīˌtrāt }

chromic oxide [INORGANIC CHEMISTRY] Cr_2O_3 A dark green, amorphous powder, forming hexagonal crystals on heating that are insoluble in water or acids; used as a pigment to color glass and ceramic ware and as a catalyst. Also known as chrome green. { ¦krō·mik 'äkˌsīd }

chromium [CHEMISTRY] A metallic chemical element, symbol Cr, atomic number 24, atomic weight 51.996. { 'krō·mē·əm }

chromium carbide [INORGANIC CHEMISTRY] Cr_3C_2 Orthorhombic crystals with a melting point of 1890°C; resistant to oxidation, acids, and alkalies; used for hot-extrusion dies, in spray-coating materials, and as a component for pumps and valves. { 'krō·mē·əm 'kärˌbīd }

chromium chloride [INORGANIC CHEMISTRY] A group of compounds of chromium and chloride; chromium may be in the +2, +3, or +6 oxidation state. { 'krō·mē·əm 'klȯrˌīd }

chromium dioxide [INORGANIC CHEMISTRY] Cr_2O_2 Black, acicular crystals; a semiconducting material with strong magnetic properties used in recording tapes. { 'krō·mē·əm dī'äkˌsīd }

chromium oxide [INORGANIC CHEMISTRY] A compound of chromium and oxygen; chromium may be in the +2, +3, or +6 oxidation state. { 'krō·mē·əm 'äkˌsīd }

chromium oxychloride [INORGANIC CHEMISTRY] *See* chromyl chloride. { 'krō·mē·əm äk·sē'klȯrˌīd }

chromium stearate [ORGANIC CHEMISTRY] $Cr(C_{18}H_{35}O_2)_3$ A dark-green powder, melting at 95–100°C; used in greases, ceramics, and plastics. { 'krō·mē·əm 'stirˌāt }

chromometer [ANALYTICAL CHEMISTRY] *See* colorimeter. { krə'mäm·əd·ər }

chromophore [CHEMISTRY] An arrangement of atoms that gives rise to color in many organic substances. { 'krō·məˌfȯr }

chromotropic acid [ORGANIC CHEMISTRY] $C_{10}H_8O_8S_2$ White, needlelike crystals that are soluble in water; used as an analytical reagent and azo dye intermediate. { ¦krō·mə¦träp·ik 'as·əd }

chromyl chloride [INORGANIC CHEMISTRY] CrO_2Cl_2 A dark-red, toxic, fuming liquid that boils at 116°C; reacts with water to form chromic acid; used to make dyes and chromium complexes. Also known as chlorochromic anhydride; chromium oxychloride. { 'krō·məl 'klȯr‚īd }

chronoamperometry [ANALYTICAL CHEMISTRY] Electroanalysis by measuring at a working electrode the rate of change of current versus time during a titration; the potential is controlled. { ¦krän·ō‚am·pə'räm·ə·trē }

chronopotentiometry [ANALYTICAL CHEMISTRY] Electroanalysis based on the measurement at a working electrode of the rate of change in potential versus time; the current is controlled. { ¦krän·ō·pə‚ten·chē'äm·ə·trē }

chrysazin [ORGANIC CHEMISTRY] *See* 1,8-dihydroxyanthraquinone. { 'krī·sə·sən }

chrysene [ORGANIC CHEMISTRY] $C_{18}H_{12}$ An organic, polynuclear hydrocarbon which when pure gives a bluish fluorescence; a component of short afterglow or luminescent paint. { 'krī‚sēn }

chrysoidine [ORGANIC CHEMISTRY] $C_6H_5NNC_6H_3(NH_2)_2 \cdot HCl$ Large, black crystals or a red-brown powder that melts at 117°C; soluble in water and alcohol; used as an orange dye for silk and cotton. { kri'sō·ə‚dēn }

chrysophanic acid [ORGANIC CHEMISTRY] $C_{15}H_{10}O_4$ Yellow leaves that melt at 196°C; soluble in ether, chloroform, and hot alcohol; extracted from senna leaves and rhubarb root; used in medicine as a mild laxative. { ¦kris·ō'fan·ik 'as·əd }

Chugaev reaction [ORGANIC CHEMISTRY] The thermal decomposition of methyl esters of xanthates to yield olefins without rearrangement. { chü'gā·əv rē‚ak·shən }

cigarette burning [CHEMISTRY] In rocket propellants, black powder, gasless delay elements, and pyrotechnic candles, the type of burning induced in a solid grain by permitting burning on one end only, so that the burning progresses in the direction of the longitudinal axis. { 'sig·ə‚ret ¦bərn·iŋ }

cinchonamine [ORGANIC CHEMISTRY] $C_{19}H_{24}N_2O$ A yellow, crystalline, water-insoluble alkaloid that melts at 184°C; derived from the bark of *Remijia purdieana*, a member of the madder family of shrubs. { siŋ'kän·ə‚mēn }

cinchonine [ORGANIC CHEMISTRY] $C_{19}H_{22}N_2O$ A colorless, crystalline alkaloid that melts at about 245°C; extracted from cinchona bark, it is used as a substitute for quinine and as a spot reagent for bismuth. { 'siŋ·kə‚nēn }

cineol [ORGANIC CHEMISTRY] *See* eucalyptol. { 'sin·ē‚ȯl }

cinnamate [ORGANIC CHEMISTRY] A salt of cinnamic acid, containing the functional group $C_9H_7O_2$—. { 'sin·ə‚māt }

cinnamic acid [ORGANIC CHEMISTRY] $C_6H_5CHCHCOOH$ Colorless, monoclinic acid; forms scales, slightly soluble in water; found in natural balsams. { sə'nam·ik 'as·əd }

cinnamic alcohol [ORGANIC CHEMISTRY] $C_6H_5CH{:}CHCH_2OH$ White needles that congeal upon heating and are soluble in alcohol; used in perfumery. { sə'nam·ik 'al·kə‚hȯl }

cinnamic aldehyde [ORGANIC CHEMISTRY] $C_6H_5CH{:}CHCHO$ A yellow oil with a cinnamon odor, sweet taste, and a boiling point of 248°C; used in flavors and perfumes. { sə'nam·ik 'al·də‚hīd }

cinnamoyl chloride [ORGANIC CHEMISTRY] $C_6H_5CHCHCOCl$ Yellow crystals that melt at 35°C, and decompose in water; used as a chemical intermediate. { 'sin·ə‚mȯil 'klȯr‚īd }

circular chromatography [ANALYTICAL CHEMISTRY] *See* radial chromatography. { 'sər·kyə·lər ‚krō·mə'täg·rə·fē }

circular paper chromatography [ANALYTICAL CHEMISTRY] A paper chromatographic technique in which migration from a spot in the sheet takes place in 360° so that zones separate as a series of concentric rings. { 'sər·kyə·lər 'pā·pər ‚krō·mə'täg·rə·fē }

cis [ORGANIC CHEMISTRY] A descriptive term indicating a form of isomerism in which atoms are located on the same side of an asymmetric molecule. { sis }

cis-trans isomerism [ORGANIC CHEMISTRY] A type of geometrical isomerism found in alkenic systems in which it is possible for each of the doubly bonded carbons to carry two different atoms or groups; two similar atoms or groups may be on the same side (cis) or on opposite sides (trans) of a plane bisecting the alkenic carbons and perpendicular to the plane of the alkenic system. { 'sis¦stranz ī'säm·ə‚riz·əm }

citraconic acid [ORGANIC CHEMISTRY] $C_5H_6O_4$ A dicarboxylic acid; hygroscopic crystals that melt at 91°C; derived from citric acid by heating. { ¦si·trə¦kän·ik 'as·əd }

citral [ORGANIC CHEMISTRY] $C_{10}H_{16}O$ A pale-yellow liquid that in commerce is a mixture of two isomeric forms, alpha and beta; insoluble in water, soluble in glycerin or benzyl benzoate; used in perfumery and as an intermediate to form other compounds. Also known as geranial; geranialdehyde. { 'si,tral }

citronellal hydrate [ORGANIC CHEMISTRY] *See* hydroxycitronellal. { ,si·trə'nel·əl 'hī,drāt }

citronellol [ORGANIC CHEMISTRY] $C_{10}H_{19}OH$ A liquid derived from citronella oil; soluble in alcohol; used in perfumery. { ,si·trə'nel,ȯl }

Cl [CHEMISTRY] *See* chlorine.

Claisen condensation [ORGANIC CHEMISTRY] **1.** Condensation, in the presence of sodium ethoxide, of esters or of esters and ketones to form β-dicarbonyl compounds. **2.** Condensation of arylaldehydes and acylphenones with esters or ketones in the presence of sodium ethoxide to yield unsaturated esters. Also known as Claisen reaction. { 'klās·ən känd·ən'sā·shən }

Claisen flask [CHEMISTRY] A glass flask with a U-shaped neck, used for distillation. { 'klās·ən ,flask }

Claisen reaction [ORGANIC CHEMISTRY] *See* Claisen condensation. { 'klās·ən'äk·shən }

Claisen rearrangement [ORGANIC CHEMISTRY] A thermally induced sigmatrophic shift in which an allyl phenyl ether is rearranged to yield an *ortho*-allylphenol. { 'klā·sən ,rē·ə'rānj·mənt }

Claisen-Schmidt condensation [ORGANIC CHEMISTRY] A reaction employed for preparation of unsaturated aldehydes and ketones by condensation of aromatic aldehydes with aliphatic aldehydes or ketones in the presence of sodium hydroxide. { ¦klās·ən ¦shmit känd·ən'sā·shən }

Clark degree [CHEMISTRY] *See* English degree. { 'klärk də,grē }

clathrate [CHEMISTRY] An inclusion compound in which the guest species is enclosed on all sides by the species forming the crystal lattice. Also known as cage compound; inclusion compound. { 'kla,thrāt }

clathrochelate [INORGANIC CHEMISTRY] A type of coordination compound containing a metal ion both coordinately saturated and encapsulated by a single ligand. { ¦klath·rō'kē,lāt }

Cleveland open-cup tester [ANALYTICAL CHEMISTRY] A laboratory apparatus used to determine flash point and fire point of petroleum products. { 'klev·lənd ,ō·pən 'kəp ,test·ər }

CLS [SPECTROSCOPY] *See* characteristic loss spectroscopy.

Cm [CHEMISTRY] *See* curium.

Co [CHEMISTRY] *See* cobalt.

coacervate [CHEMISTRY] An aggregate of colloidal droplets bound together by the force of electrostatic attraction. { kō'as·ər,vāt }

coacervation [CHEMISTRY] The separation, by addition of a third component, of an aqueous solution of a macromolecule colloid (polymer) into two liquid phases, one of which is colloid-rich (the coacervate) and the other an aqueous solution of the coacervating agent (the equilibrium liquid). { kō,as·ər'vā·shən }

coagulant [CHEMISTRY] An agent that causes coagulation. { kō'ag·yə·lənt }

coagulation [CHEMISTRY] A separation or precipitation from a dispersed state of suspensoid particles resulting from their growth; may result from prolonged heating, addition of an electrolyte, or from a condensation reaction between solute and solvent; an example is the setting of a gel. { kō,ag·yə'lā·shən }

coalescent [CHEMISTRY] Chemical additive used in immiscible liquid-liquid mixtures to cause small droplets of the suspended liquid to unite, preparatory to removal from the carrier liquid. { ,kō·ə'les·ənt }

coal-tar dye [ORGANIC CHEMISTRY] Dye made from a coal-tar hydrocarbon or a derivative such as benzene, toluene, xylene, naphthalene, or aniline. { 'kōl ,tär ,dī }

cobalt [CHEMISTRY] A metallic element, symbol Co, atomic number 27, atomic weight 58.93; used chiefly in alloys. { 'kō,bȯlt }

cobalt blue [CHEMISTRY] A green-blue pigment formed of alumina and cobalt oxide. Also known as cobalt ultramarine; king's blue. { ¦kō‚bȯlt ¦blü }

cobalt bromide [INORGANIC CHEMISTRY] *See* cobaltous bromide. { 'kō‚bȯlt 'brō‚mīd }

cobalt chloride [INORGANIC CHEMISTRY] *See* cobaltous chloride. { 'kō‚bȯlt 'klȯr‚īd }

cobalt nitrate [INORGANIC CHEMISTRY] *See* cobaltous nitrate. { 'kō‚bȯlt 'nī‚trāt }

cobaltous acetate [ORGANIC CHEMISTRY] $Co(C_2H_3O_2)_2 \cdot 4H_2O$ Reddish-violet, deliquescent crystals; soluble in water, alcohol, and acids; used in paint and varnish driers, for anodizing, and as a feed additive mineral supplement. Also known as cobalt acetate. { kō'bȯl·təs 'as·ə‚tāt }

cobaltous bromide [INORGANIC CHEMISTRY] $CoBr_2 \cdot 6H_2O$ Red-violet crystals with a melting point of 47-48°C; soluble in water, alcohol, and ether; used in hygrometers. Also known as cobalt bromide. { kō'bȯl·təs 'brō‚mīd }

cobaltous chloride [INORGANIC CHEMISTRY] $CoCl_2$ or $CoCl_2 \cdot 6H_2O$ A compound whose anhydrous form consists of blue crystals and sublimes when heated, and whose hydrated form consists of red crystals and melts at 86.8°C; both forms are used as an absorbent for ammonia in dyes and as a catalyst. Also known as cobalt chloride. { kō'bȯl·təs 'klȯr‚īd }

cobaltous fluorosilicate [INORGANIC CHEMISTRY] $CoSiF_6 \cdot H_2O$ A water-soluble, orange-red powder, used in toothpastes. { kō'bȯl·təs ¦flu̇r·ō'sil·ə‚kāt }

cobaltous nitrate [INORGANIC CHEMISTRY] $Co(NO_3)_2 \cdot 6H_2O$ A red crystalline compound with a melting point of 56°C; soluble in organic solvents; used in sympathetic inks, as an additive to soils and animal feeds, and for vitamin preparations and hair dyes. Also known as cobalt nitrate. { kō'bȯl·təs 'nī‚trāt }

cobalt oxide [INORGANIC CHEMISTRY] CoO A grayish brown powder that decomposes at 1935°C, insoluble in water; used as a colorant in ceramics and in manufacture of glass. { 'kō‚bȯlt 'äk‚sīd }

cobalt potassium nitrite [INORGANIC CHEMISTRY] $K_3Co(NO_2)_6$ A yellow powder which decomposes at the melting point of 200°C; used in medicine and as a yellow pigment. Also known as cobalt yellow; Fischer's salt; potassium cobaltinitrite. { 'kō‚bȯlt pə'tas·ē·əm 'nī‚trīt }

cobalt sulfate [INORGANIC CHEMISTRY] Any compound of either divalent or trivalent cobalt and the sulfate group; anhydrous cobaltous sulfate, $CoSO_4$, contains divalent cobalt, has a melting point of 96.8°C, is soluble in methanol, and is utilized to prepare pigments and cobalt salts; cobaltic sulfate, $Co_2(SO_4)_3 \cdot 18H_2O$, contains trivalent cobalt, is soluble in sulfuric acid, and functions as an oxidizing agent. { 'kō‚bȯlt 'səl‚fāt }

cobalt trifluoride [INORGANIC CHEMISTRY] CoF_3 A brownish powder that reacts with water to form a precipitate of cobaltic hydroxide; used as a fluorinating agent. Also known as cobaltic fluoride. { 'kō‚bȯlt trī'flu̇r‚īd }

cobalt ultramarine [CHEMISTRY] *See* cobalt blue. { 'kō‚bȯlt ‚əl·trə·mə'rēn }

cobalt yellow [INORGANIC CHEMISTRY] *See* cobalt potassium nitrite. { 'kō‚bȯlt 'yel·ō }

cobaltic fluoride [INORGANIC CHEMISTRY] *See* cobalt trifluoride. { kə'bȯl·tik 'flu̇r‚īd }

cochineal [CHEMISTRY] A red dye made of the dried bodies of the female cochineal insect (*Coccus cacti*), found in Central America and Mexico; used as a biological stain and indicator. { 'käch·ə‚nēl }

cochineal solution [ANALYTICAL CHEMISTRY] An indicator in acid-base titration. { 'käch·ə‚nēl sə'lü·shən }

cochinilin [ORGANIC CHEMISTRY] *See* carminic acid. { kō'chin·ə·lən }

cocodyl oxide [ORGANIC CHEMISTRY] $(CH_3)_2AsOAs(CH_3)_2$ A liquid that has an obnoxious odor; slightly soluble in water, soluble in alcohol and ether; boils at 150°C. Also known as alkarsine; bisdimethyl arsenic oxide; dicacodyl oxide. { 'kō·kə·dəl 'äk‚sīd }

codimer [ORGANIC CHEMISTRY] **1.** A copolymer formed from the polymerization of two dissimilar olefin molecules. **2.** The product of polymerization of isobutylene with one of the two normal butylenes. { ¦kō'dī·mər }

cognac oil [ORGANIC CHEMISTRY] *See* ethyl enanthate. { 'kōn‚yak ‚ȯil }

coherent precipitate [PHYSICAL CHEMISTRY] A precipitate that is a continuation of the lattice structure of the solvent and has no phase or grain boundary. { kō'hir·ənt prə'sip·ə,tāt }

coion [ANALYTICAL CHEMISTRY] Any of the small ions entering a solid ion exchanger and having the same charge as that of the fixed ions. { kō'ī,än }

colchicine [ORGANIC CHEMISTRY] $C_{22}H_{25}O_6N$ An alkaloid extracted from the stem of the autumn crocus; used experimentally to inhibit spindle formation and delay centromere division, and medicinally in the treatment of gout. { 'käl·chə,sēn }

colcothar [INORGANIC CHEMISTRY] Red ferric oxide made by heating ferrous sulfate in the air; used as a pigment and as an abrasive in polishing glass. { 'käl·kə,thär }

collection trap [ANALYTICAL CHEMISTRY] Cooled device to collect gas-chromatographic eluent, holding it for subsequent compound-identification analysis. { kə'lek·shən ,trap }

2,4,6-collidine [ORGANIC CHEMISTRY] $(CH_3)_3C_5H_2N$ A liquid boiling at 170.4°C; slightly soluble in water, soluble in alcohol; used as a chemical intermediate. { ¦tü ¦fȯr ¦siks 'käl·ə,dēn }

colligative properties [PHYSICAL CHEMISTRY] Properties dependent on the number of molecules but not their nature. { kə'lig·ə·div ,präp·ərd·ēz }

collision broadening [SPECTROSCOPY] *See* collision line-broadening. { kə'lizh·ən ,brȯd·ən·iŋ }

collision diameter [PHYSICAL CHEMISTRY] The distance between the centers of two molecules taking part in a collision at the time of their closest approach. { kə'lizh·ən dī,am·əd·ər }

collision line-broadening [SPECTROSCOPY] Spreading of a spectral line due to interruption of the radiation process when the radiator collides with another particle. Also known as collision broadening. { kə'lizh·ən 'līn ,brȯd·ən·iŋ }

collision theory [PHYSICAL CHEMISTRY] Theory of chemical reaction proposing that the rate of product formation is equal to the number of reactant-molecule collisions multiplied by a factor that corrects for low-energy-level collisions. { kə'lizh·ən ,thē·ə·rē }

collodion [ORGANIC CHEMISTRY] Cellulose nitrate deposited from a solution of 60% ether and 40% alcohol, used for making fibers and film and in membranes for dialysis. { kə'lōd·ē·ən }

collodion cotton [ORGANIC CHEMISTRY] *See* pyroxylin. { kə'lōd·ē·ən ,kät·ən }

collodion replication [ANALYTICAL CHEMISTRY] Production of a faithful collodion-film mold of a specimen surface (for example, powders, bones, microorganisms, crystals) which is sufficiently thin to be studied by electron microscopy. { kə'lōd·ē·ən rep·lə'kā·shən }

colloid [CHEMISTRY] The phase of a colloidal system made up of particles having dimensions of 10-10,000 angstroms (1-1000 nanometers) and which is dispersed in a different phase. { 'käl,ȯid }

colloidal crystal [CHEMISTRY] A periodic array of suspended colloidal particles that can arise spontaneously in a monodisperse colloidal system under appropriate conditions. { kə'lȯid·əl 'krist·əl }

colloidal dispersion [CHEMISTRY] *See* colloidal system. { kə'lȯid·əl dis'pər·zhən }

colloidal electrolyte [PHYSICAL CHEMISTRY] An electrolyte that yields at least one type of ion in the colloidal size range. { kə'lȯid·əl i'lek·trə,līt }

colloidal suspension [CHEMISTRY] *See* colloidal system. { kə'lȯid·əl səs'pen·shən }

colloidal system [CHEMISTRY] An intimate mixture of two substances, one of which, called the dispersed phase (or colloid), is uniformly distributed in a finely divided state through the second substance, called the dispersion medium (or dispersing medium); the dispersion medium or dispersed phase may be a gas, liquid, or solid. Also known as colloidal dispersion; colloidal suspension. { kə'lȯid·əl 'sis·təm }

colloid chemistry [PHYSICAL CHEMISTRY] The scientific study of matter whose size is approximately 10 to 10,000 angstroms (1 to 1000 nanometers), and which exists as a suspension in a continuous medium, especially a liquid, solid, or gaseous substance. { ¦käl,ȯid 'kem·ə·strē }

color comparator [ANALYTICAL CHEMISTRY] A photoelectric instrument that compares an unknown color with that of a standard color sample for matching purposes. Also known as photoelectric color comparator. { 'kəl·ər kəm'par·əd·ər }

colorimeter [ANALYTICAL CHEMISTRY] A device for measuring concentration of a known constituent in solution by comparison with colors of a few solutions of known concentration of that constituent. Also known as chromometer. { ˌkəl·ə'rim·əd·ər }

color stability [CHEMISTRY] Resistance of materials to change in color that can be caused by light or aging, as of petroleum or whiskey. { 'kəl·ər stə'bil·əd·ē }

color standard [ANALYTICAL CHEMISTRY] Liquid solution of known chemical composition and concentration, hence of known and standardized color, used for optical analysis of samples of unknown strength. { 'kəl·ər ˌstan·dərd }

color test [ANALYTICAL CHEMISTRY] The quantitative analysis of a substance by comparing the intensity of the color produced in a sample by a reagent with a standard color produced similarly in a solution of known strength. { 'kəl·ər ˌtest }

color throw [ANALYTICAL CHEMISTRY] In an ion-exchange process, discoloration of the liquid passing through the bed. { 'kəl·ər ˌthrō }

column [ANALYTICAL CHEMISTRY] In chromatography, a tube holding the stationary phase through which the mobile phase is passed. { 'käl·əm }

column bleed [ANALYTICAL CHEMISTRY] The loss of carrier liquid during gas chromatography due to evaporation into the gas under analysis. { 'käl·əm ˌblēd }

column chromatography [ANALYTICAL CHEMISTRY] Chromatographic technique of two general types: packed columns usually contain either a granular adsorbent or a granular support material coated with a thin layer of high-boiling solvent (partitioning liquid); open-tubular columns contain a thin film of partitioning liquid on the column walls and have an opening so that gas can pass through the center of the column. { 'käl·əm ˌkrō·mə'täg·rə·fē }

column development chromatography [ANALYTICAL CHEMISTRY] Columnar apparatus for separating or concentrating one or more components from a physical mixture by use of adsorbent packing; as the specimen percolates along the length of the adsorbent, its various components are preferentially held at different rates, effecting a separation. { 'käl·əm də'vel·əp·mənt ˌkrō·mə'täg·rə·fē }

combination principle [SPECTROSCOPY] *See* Ritz's combination principle. { ˌkäm·bə'nā·shən ˌprin·sə·pəl }

combination reaction [CHEMISTRY] A chemical reaction in which two reactions combine to form a single product. { ˌkäm·bə'nā·shən rēˌak·shən }

combination vibration [SPECTROSCOPY] A vibration of a polyatomic molecule involving the simultaneous excitation of two or more normal vibrations. { ˌkäm·bə'nā·shən vī'brā·shən }

combined carbon [CHEMISTRY] Carbon that is chemically combined within a compound, as contrasted with free or uncombined elemental carbon. { kəm'bīnd 'kär·bən }

combined cyanide [ORGANIC CHEMISTRY] The cyanide portion of a complex ion composed of cyanide and a metal. { kəm'bīnd 'sī·əˌnīd }

combining-volumes principle [CHEMISTRY] The principle that when gases take part in chemical reactions the volumes of the reacting gases and those of the products (if gaseous) are in the ratio of small whole numbers, provided that all measurements are made at the same temperature and pressure. Also known as Gay-Lussac's law of volumes. { kəm¦bīn·iŋ ¦väl·yəmz ˌprin·sə·pəl }

combining weight [CHEMISTRY] The weight of an element that chemically combines with 8 grams of oxygen or its equivalent. { kəm'bīn·iŋ ˌwāt }

combustion [CHEMISTRY] The burning of gas, liquid, or solid, in which the fuel is oxidized, evolving heat and often light. { kəm'bəs·chən }

combustion efficiency [CHEMISTRY] The ratio of heat actually developed in a combustion process to the heat that would be released if the combustion were perfect. { kəm'bəs·chən i'fish·ən·sē }

combustion furnace [ANALYTICAL CHEMISTRY] A heating device used in the analysis of organic compounds for elements. { kəm'bəs·chən ˌfər·nəs }

combustion rate [CHEMISTRY] The rate of burning of any substance. { kəm'bəs·chən ˌrāt }

combustion train [ANALYTICAL CHEMISTRY] The arrangement of apparatus for elementary organic analysis. { kəm'bəs·chən ‚trān }
combustion tube [ANALYTICAL CHEMISTRY] A glass, silica, or porcelain tube, resistant to high temperatures, that is a component of a combustion train. { kəm'bəs·chən ‚tüb }
combustion wave [CHEMISTRY] **1.** A zone of burning propagated through a combustible medium. **2.** The zoned, reacting, gaseous material formed when an explosive mixture is ignited. { kəm'bəs·chən ‚wāv }
common cause [ANALYTICAL CHEMISTRY] A cause of variability in a measurement process that is inherent in and common to the process itself. { 'käm·ən 'kȯz }
common-ion effect [CHEMISTRY] The lowering of the degree of ionization of a compound when another ionizable compound is added to a solution; the compound added has a common ion with the other compound. { ¦käm·ən ¦ī‚än i'fekt }
common salt [INORGANIC CHEMISTRY] *See* halite; sodium chloride. { ¦käm·ən 'sȯlt }
comonomer [CHEMISTRY] One of the compounds used to produce a specific polymeric product. { ¦kō'män·ə·mər }
comparator [ANALYTICAL CHEMISTRY] An instrument used to determine the concentration of a solution by comparing the intensity of color with a series of standard colors. { kəm'par·əd·ər }
comparator-densitometer [ANALYTICAL CHEMISTRY] Device that projects a labeled spectrum onto a screen adjacent to an enlarged image of the spectrum to be analyzed, allowing visual comparison. { ¦kəm'par·əd·ər den·sə'täm·əd·ər }
comparison spectrum [SPECTROSCOPY] A line spectrum whose wavelengths are accurately known, and which is matched with another spectrum to determine the wavelengths of the latter. { kəm'par·ə·sən ‚spek·trəm }
compatibilizer [ORGANIC CHEMISTRY] Any polymeric interfacial agent that facilitates formation of uniform blends of normally immiscible polymers with desirable end properties. { kəm'pad·ə·bə‚līz·ər }
competing equilibria condition [CHEMISTRY] The competition for a reactant in a complex chemical system in which several reactions are taking place at the same time. { kəm'pēd·iŋ ‚ē·kwə'lib·rē·ə kən‚dish·ən }
complete combustion [CHEMISTRY] Combustion in which the entire quantity of oxidizable constituents of a fuel is reacted. { kəm'plēt kəm'bəs·chən }
complexation [CHEMISTRY] *See* complexing. { ‚käm‚plek'sā·shən }
complexation analysis [ANALYTICAL CHEMISTRY] The determination of the ligand/metal ratio in a coordination complex. { ‚käm·plek'sā·shən ə‚nal·ə·səs }
complexation indicator [ANALYTICAL CHEMISTRY] *See* metal ion indicator. { ‚käm·plek'sā·shən ‚in·də‚kād·ər }
complexation reaction [CHEMISTRY] A chemical reaction that takes place between a metal ion and a molecular or ionic entity known as a ligand that contains at least one atom with an unshared pair of electrons. { ‚käm·plek'sā·shən rē‚ak·shən }
complex chemical reaction [CHEMISTRY] A chemical system in which a number of chemical reactions take place simultaneously, including reversible reactions, consecutive reactions, and concurrent or side reactions. { 'käm‚pleks ¦kem·i·kəl rē'ak·shən }
complex compound [CHEMISTRY] Any of a group of chemical compounds in which a part of the molecular bonding is of the coordinate type. Also known as coordination complex. { 'käm‚pleks 'käm‚paủnd }
compleximetric titration [ANALYTICAL CHEMISTRY] *See* complexometric titration. { kəm‚plek·sə¦me·trik tī'trā·shən }
complexing [CHEMISTRY] Formation of a complex compound. Also known as complexation. { 'käm‚plek·siŋ }
complexing agent [CHEMISTRY] A substance capable of forming a complex compound with another material in solution. { 'käm‚plek·siŋ ‚ā·jənt }
complex ion [CHEMISTRY] A complex, electrically charged group of atoms or radical, for example, $Cu(NH_3)_2^{+2}$. { 'käm‚pleks 'ī‚än }
complexometric titration [ANALYTICAL CHEMISTRY] A technique of volumetric analysis

in which the formation of a colored complex is used to indicate the end point of a titration. Also known as chelatometry. Also spelled compleximetric titration. { kəm¦plek·sə¦me·trik ˌtīʹtrā·shən }

complex salt [INORGANIC CHEMISTRY] A class of salts in which there are no detectable quantities of each of the metal ions existing in solution; an example is $K_3Fe(CN)_6$, which in solution has K^+ but no Fe^{3+} because Fe is strongly bound in the complex ion, $Fe(CN)_6{}^{3-}$. { ʹkämˌpleks ʹsȯlt }

component [CHEMISTRY] **1.** A part of a mixture. **2.** The smallest number of chemical substances which are able to form all the constituents of a system in whatever proportion they may be present. { kəmʹpō·nənt }

component-substances law [CHEMISTRY] The law that each substance, singly or in mixture, composing a material exhibits specific properties that are independent of the other substances in that material. { kəmʹpō·nənt ʹsub·stən·səs ˌlȯ }

composite sample [ANALYTICAL CHEMISTRY] A sample comprising two or more increments selected to represent the material being analyzed. { kəmʹpäz·ət ʹsam·pəl }

composition [CHEMISTRY] The elements or compounds making up a material or produced from it by analysis. { ˌkäm·pəʹzish·ən }

compound [CHEMISTRY] A substance whose molecules consist of unlike atoms and whose constituents cannot be separated by physical means. Also known as chemical compound. { ʹkämˌpau̇nd }

Compton rule [PHYSICAL CHEMISTRY] An empirical law stating that the heat of fusion of an element times its atomic weight divided by its melting point in degrees Kelvin equals approximately 2. { ʹkäm·tən ˌrül }

computational chemistry [CHEMISTRY] The use of calculations to predict molecular structure, properties, and reactions. { ˌkäm·pyəʹtā·shən·əl ʹkem·ə·strē }

concave grating [SPECTROSCOPY] A reflection grating which both collimates and focuses the light falling upon it, made by spacing straight grooves equally along the chord of a concave spherical or paraboloid mirror surface. Also known as Rowland grating. { ʹkänˌkāv ʹgrād·iŋ }

concentrate [CHEMISTRY] To increase the amount of a dissolved substance by evaporation. { ʹkän·sənˌtrāt }

concentration [CHEMISTRY] In solutions, the mass, volume, or number of moles of solute present in proportion to the amount of solvent or total solution. { ˌkän·sənʹtrā·shən }

concentration cell [PHYSICAL CHEMISTRY] **1.** Electrochemical cell for potentiometric measurement of ionic concentrations where the electrode potential electromotive force produced is determined as the difference in emf between a known cell (concentration) and the unknown cell. **2.** An electrolytic cell in which the electromotive force is due to a difference in electrolyte concentrations at the anode and the cathode. { ˌkän·sənʹtrā·shən ˌsel }

concentration gradient [CHEMISTRY] The graded difference in the concentration of a solute throughout the solvent phase. { ˌkän·sənʹtrā·shən ˌgrād·ē·ənt }

concentration polarization [PHYSICAL CHEMISTRY] That part of the polarization of an electrolytic cell resulting from changes in the electrolyte concentration due to the passage of current through the solution. { ˌkän·sənʹtrā·shən ˌpō·lə·rəʹzā·shən }

concentration potential [CHEMISTRY] Tendency for a univalent electrolyte to concentrate in a specific region of a solution. { ˌkän·sənʹtrā·shən pəʹten·shəl }

concentration scale [CHEMISTRY] Any of several numerical systems defining the quantitative relation of the components of a mixture; for solutions, concentration is expressed as the mass, volume, or number of moles of solute present in proportion to the amount of solvent or total solution. { ˌkän·sənʹtrā·shən ˌskāl }

concerted reaction [ORGANIC CHEMISTRY] A reaction in which there is a simultaneous occurrence of bond making and bond breaking. { kənʹsərd əd rēʹak shən }

concomitant [ANALYTICAL CHEMISTRY] Any species in a material undergoing chemical analysis other than the analyte or the solvent in which the sample is dissolved. { kənʹkäm·ə·tənt }

condensable vapors [CHEMISTRY] Gases or vapors which when subjected to appropriately altered conditions of temperature or pressure become liquids. { kən'den·sə·bəl 'vā·pərz }

condensation [CHEMISTRY] Transformation from a gas to a liquid. { ˌkän·dən'sā·shən }

condensation polymer [ORGANIC CHEMISTRY] A high-molecular-weight compound formed by condensation polymerization. { ˌkän·dən'sā·shən 'päl·ə·mər }

condensation polymerization [ORGANIC CHEMISTRY] The formation of high-molecular-weight polymers from monomers by chemical reactions of the condensation type. { ˌkän·dən'sā·shən pəˌlim·ə·rə'zā·shən }

condensation reaction [CHEMISTRY] One of a class of chemical reactions involving a combination between molecules or between parts of the same molecule. { ˌkän·dən'sā·shən rē'ak·shən }

condensation resin [ORGANIC CHEMISTRY] A resin formed by polycondensation. { ˌkän·dən'sā·shən 'rez·ən }

condensation temperature [ANALYTICAL CHEMISTRY] In boiling-point determination, the temperature established on the bulb of a thermometer on which a thin moving film of liquid coexists with vapor from which the liquid has condensed, the vapor phase being replenished at the moment of measurement from a boiling-liquid phase. { ˌkän·dən'sā·shən 'tem·prə·chər }

condensed phase [PHYSICAL CHEMISTRY] Either the solid or liquid phase of a material. { kən'denst ˌfāz }

condensed structural formula [CHEMISTRY] A structural representation of a compound that includes all of the atoms present in a molecule or other chemical entity but represents only certain bonds as lines in order to emphasize a structural characteristic. { kən¦denst ¦strək·chər·əl 'fȯr·myə·lə }

condensed system [PHYSICAL CHEMISTRY] A chemical system in which the vapor pressure is negligible or in which the pressure maintained on the system is greater than the vapor pressure of any portion. { kən'denst 'sis·təm }

conductance coefficient [PHYSICAL CHEMISTRY] The ratio of the equivalent conductance of an electrolyte, at a given concentration of solute, to the limiting equivalent conductance of the electrolyte as the concentration of the electrolyte approaches 0. { kən¦dək·təns ˌkō·ə'fish·ənt }

conductimetry [CHEMISTRY] The scientific study of conductance measurements of solutions; to avoid electrolytic complications, conductance measurements are usually taken with alternating current. { kän·dək'tim·ə·trē }

conductometric titration [ANALYTICAL CHEMISTRY] A titration in which electrical conductance of a solution is measured during the course of the titration. { kən¦dək·tə¦meˌtrik tī'trā·shən }

configuration [CHEMISTRY] The three-dimensional spatial arrangement of atoms in a stable or isolable molecule. { kənˌfig·yə'rā·shən }

configuration interaction [PHYSICAL CHEMISTRY] Interaction between two different possible arrangements of the electrons in an atom (or molecule); the resulting electron distribution, energy levels, and transitions differ from what would occur in the absence of the interaction. { kənˌfig·yə'rā·shən in·tər'ak·shən }

conformation [ORGANIC CHEMISTRY] In a molecule, a specific orientation of the atoms that varies from other possible orientations by rotation or rotations about single bonds; generally in mobile equilibrium with other conformations of the same structure. Also known as conformational isomer; conformer. { kän·fər'mā·shən }

conformational analysis [PHYSICAL CHEMISTRY] The determination of the arrangement in space of the constituent atoms of a molecule that may rotate about a single bond. { kän·fər'mā·shən·əl ə'nal·ə·səs }

conformational isomer [ORGANIC CHEMISTRY] *See* conformation. { kän·fər'mā·shən·əl 'ī·sə·mər }

conformer [ORGANIC CHEMISTRY] *See* conformation. { kən'fȯr·mər }

congener [CHEMISTRY] A chemical substance that is related to another substance, such as a derivative of a compound or an element belonging to the same family as another element in the periodic table. { 'kän·jə·nər }

conglomerate [ORGANIC CHEMISTRY] *See* racemic mixture. { kən'gläm·ə·rət }

congo red [ORGANIC CHEMISTRY] $C_{32}H_{22}N_6Na_2O_6S_2$ An azo dye, sodium diphenyldiazo-bis-α-naphthylamine sulfonate, used as a biological stain and as an acid-base indicator; it is red in alkaline solution and blue in acid solution. { 'käŋ·gō 'red }

conjugate acid-base pair [CHEMISTRY] An acid and a base related by the ability of the acid to generate the base by loss of a proton. { 'kän·jə·gət ¦as·əd ¦bās 'per }

conjugated diene [ORGANIC CHEMISTRY] An acyclic hydrocarbon with a molecular structure containing two carbon-carbon double bonds separated by a single bond. { 'kän·jə‚gād·əd 'dī‚ēn }

conjugated polyene [ORGANIC CHEMISTRY] An acyclic hydrocarbon with a molecular structure containing alternating carbon-carbon double and single bonds. { 'kän·jə‚gād·əd 'päl·ē‚ēn }

conode [PHYSICAL CHEMISTRY] *See* tie line. { 'kō‚nōd }

conservation of orbital symmetry [ORGANIC CHEMISTRY] *See* Woodward-Hoffmann rule. { ‚kän·sər'vā·shən əv 'ȯr·bəd·əl 'sim·ə·trē }

consolute [CHEMISTRY] Of or pertaining to liquids that are perfectly miscible in all proportions under certain conditions. { 'kan·sə‚lüt }

constant-current electrolysis [CHEMISTRY] Electrolysis in which a constant current flows through the cell; used in electrodeposition analysis. { ¦kän·stənt 'kər·ənt i‚lek'träl·ə·səs }

constant-current titration [ANALYTICAL CHEMISTRY] *See* potentiometric titration. { ¦kän·stənt 'kər·ənt tī'trā·shən }

constant-deviation spectrometer [SPECTROSCOPY] A spectrometer in which the collimator and telescope are held fixed and the observed wavelength is varied by rotating the prism or diffraction grating. { ¦kän·stənt ‚dē·vē¦ā·shən spek'träm·əd·ər }

constant-potential electrolysis [CHEMISTRY] Electrolysis in which a constant voltage is applied to the cell; used in electrodeposition analysis. { ¦kän·stənt pə'ten·chəl i‚lek'träl·ə·səs }

constant series [CHEMISTRY] *See* displacement series. { 'kän·stənt 'sir‚ēz }

constitutional isomers [ORGANIC CHEMISTRY] Isomers which differ in the manner in which their atoms are linked. Also known as structural isomers. { ‚kän·stə'tü·shən·əl 'ī·sə·mərz }

constitutional unit [CHEMISTRY] An atom or group of atoms that is part of a chain in a polymer or oligomer. { ‚kän·stə'tü·shən·əl 'yü·nət }

constitutive property [CHEMISTRY] Any physical or chemical property that depends on the constitution or structure of the molecule. { 'kän·stə‚tüd·iv 'präp·ərd·ē }

contact acid [INORGANIC CHEMISTRY] Sulfuric acid produced by the contact process. { 'kän‚takt 'as·əd }

contemporary carbon [CHEMISTRY] The isotopic carbon content of living matter, based on the assumption of a natural proportion of carbon-14. { kən'tem·pə‚rer·ē 'kär·bən }

continuous phase [CHEMISTRY] The liquid in a disperse system in which solids are suspended or droplets of another liquid are dispersed. Also known as dispersion medium, external phase. { kən¦tin·yə·wəs 'fāz }

continuous spectrum [SPECTROSCOPY] A radiation spectrum which is continuously distributed over a frequency region without being broken up into lines or bands. { kən¦tin·yə·wəs 'spek·trəm }

continuous titrator [ANALYTICAL CHEMISTRY] A titrator so equipped that a reservoir refills the buret. { kən¦tin·yə·wəs 'tī‚trād·ər }

contributing structure [ORGANIC CHEMISTRY] A structural formula that is one of a set of formulas, each contributing to the total wave function of a molecule. Also known as canonical form; canonical structure. { kən¦trib·yəd·ing 'strək·chər }

control sample [ANALYTICAL CHEMISTRY] A material of known composition that is analyzed along with test samples in order to evaluate the accuracy of an analytical procedure. Also known as check sample. { kən'trōl ‚sam·pəl }

convergence limit [SPECTROSCOPY] **1.** The short-wavelength limit of a set of spectral lines that obey a Rydberg series formula; equivalently, the long-wavelength limit of the continuous spectrum corresponding to ionization from or recombination to a given state. **2.** The wavelength at which the difference between successive vibrational bands in a molecular spectrum decreases to 0. { kən'vər·jəns ˌlim·ət }

convergence pressure [PHYSICAL CHEMISTRY] The pressure at which the different constant-temperature K (liquid-vapor equilibrium) factors for each member of a two-component system converge to unity. { kən'vər·jəns ˌpresh·ər }

conversion [CHEMISTRY] Change of a compound from one isomeric form to another. { kən'vər·zhən }

cool flame [CHEMISTRY] A faint, luminous phenomenon observed when, for example, a mixture of ether vapor and oxygen is slowly heated; it proceeds by diffusion of reactive molecules which initiate chemical processes as they go. { ¦kül ¦flām }

coordinate bond [CHEMISTRY] *See* coordinate valence. { kō'ȯrd·ən·ət 'bänd }

coordinated complex [CHEMISTRY] *See* coordination compound. { kō'ȯrd·ənˌād·əd 'kämˌpleks }

coordinate valence [CHEMISTRY] A chemical bond between two atoms in which a shared pair of electrons forms the bond and the pair has been supplied by one of the two atoms. Also known as coordinate bond; dative bond. { kō'ȯrd·ən·ət 'vā·ləns }

coordination chemistry [CHEMISTRY] The chemistry of metal ions in their interactions with other molecules or ions. { kōˌȯrd·ən'ā·shən 'kem·ə·strē }

coordination complex [CHEMISTRY] *See* complex compound. { kōˌȯrd·ən'ā·shən 'kämˌpleks }

coordination compound [CHEMISTRY] A compound with a central atom or ion and a group of ions or molecules surrounding it. Also known as coordinated complex; Werner complex. { kōˌȯrd·ən'ā·shən ˌkämˌpaůnd }

coordination polygon [CHEMISTRY] The symmetrical polygonal chemical structure of simple polyatomic aggregates having coordination numbers of 4 or less. { kōˌȯrd·ən'ā·shən ˌpäl·iˌgän }

coordination polyhedron [CHEMISTRY] The symmetrical polyhedral chemical structure of relatively simple polyatomic aggregates having coordination numbers of 4 to 8. { kōˌȯrd·ən'ā·shən ˌpäl·i·hē·drən }

coordination polymer [ORGANIC CHEMISTRY] Organic addition polymer that is neither free-radical nor simply ionic; prepared by catalysts that combine an organometallic (for example, triethyl aluminum) and a transition metal compound (for example, $TiCl_4$). { kōˌȯrd·ən'ā·shən ˌpäl·ə·mər }

copolymer [ORGANIC CHEMISTRY] A mixed polymer, the product of polymerization of two or more substances at the same time. { kō'päl·i·mər }

copolymerization [CHEMISTRY] A polymerization reaction that forms a copolymer. { ˌkō·pəˌlim·ə·rə'zā·shən }

copper [CHEMISTRY] A chemical element, symbol Cu, atomic number 29, atomic weight 63.546. { 'käp·ər }

copper acetate [ORGANIC CHEMISTRY] *See* cupric acetate. { 'käp·ər 'as·əˌtāt }

copper arsenate [INORGANIC CHEMISTRY] $Cu_3(AsO_4)_2 \cdot 4H_2O$ or $Cu_5H_2(AsO_4)_4 \cdot 2H_2O$ Bluish powder, soluble in ammonium hydroxide and dilute acids, insoluble in water and alcohol; used as a fungicide and insecticide. { 'käp·ər 'ärs·ənˌāt }

copper arsenite [INORGANIC CHEMISTRY] $CuHAsO_3$ A toxic, light green powder which is soluble in acids and decomposes at the melting point; used as a pigment and insecticide. Also known as copper orthoarsenite; cupric arsenite; Scheele's green. { 'käp·ər 'ärs·ənˌīt }

copperas [INORGANIC CHEMISTRY] *See* ferrous sulfate. { 'käp·ə·rəs }

copper blue [INORGANIC CHEMISTRY] *See* mountain blue. { 'käp·ər ¦blü }

copper bromide [INORGANIC CHEMISTRY] *See* cupric bromide; cuprous bromide. { 'käp·ər 'brōˌmīd }

copper carbonate [INORGANIC CHEMISTRY] $Cu_2(OH)_2CO_3$ A toxic, green powder; decomposes at 200°C and is soluble in acids; used in pigments and pyrotechnics and as a fungicide and feed additive. Also known as artificial malachite; cupric carbonate; mineral green. { 'käp·ər 'kär·bəˌnāt }

copper chloride [INORGANIC CHEMISTRY] *See* cupric chloride; cuprous chloride. { 'käp·ər 'klȯrˌīd }

copper chromate [INORGANIC CHEMISTRY] *See* cupric chromate. { 'käp·ər 'krō,māt }

copper cyanide [INORGANIC CHEMISTRY] *See* cupric cyanide. { 'käp·ər sī·ə,nīd }

copper fluoride [INORGANIC CHEMISTRY] *See* cupric fluoride; cuprous fluoride. { 'käp·ər 'flu̇r,īd }

copper gluconate [ORGANIC CHEMISTRY] $[CH_2OH(CHOH)_4COO]_2Cu$ A light blue, crystalline powder; soluble in water; used in medicine and as a dietary supplement. Also known as cupric gluconate. { 'käp·ər 'glü·kə,nāt }

copper hydroxide [INORGANIC CHEMISTRY] *See* cupric hydroxide. { 'käp·ər hī 'dräk,sīd }

copper nitrite [INORGANIC CHEMISTRY] *See* cupric nitrate. { 'käp·ər 'nī,trīt }

copper number [ANALYTICAL CHEMISTRY] The number of milligrams of copper obtained by the reduction of Benedict's or Fehling's solution by 1 gram of carbohydrate. { 'käp·ər ,nəm·bər }

copper oleate [ORGANIC CHEMISTRY] $Cu[OOC(CH_2)_7CH{=}CH(CH_2)_7CH_3]_2$ A green-blue liquid, used as a fungicide for fruits and vegetables. { 'käp·ər 'ō·lē,āt }

copperon [ORGANIC CHEMISTRY] *See* cupferron. { 'käp·ə,rän }

copper orthoarsenite [INORGANIC CHEMISTRY] *See* copper arsenite. { 'käp·ər ¦ór·thō'ärs·ən,īt }

copper oxide [INORGANIC CHEMISTRY] *See* cupric oxide; cuprous oxide. { 'käp·ər 'äk ,sīd }

copper-8-quinolinolate [ORGANIC CHEMISTRY] $C_{18}H_{14}N_2O_2Cu$ A khaki-colored, water-insoluble solid used as a fungicide in fruit-handling equipment. { 'käp·ər ¦āt ¦kwīn·ə¦lin·ə,lāt }

copper resinate [ORGANIC CHEMISTRY] Poisonous green powder, soluble in oils and ether, insoluble in water; made by heating rosin oil with copper sulfate, followed by filtering and drying of the resultant solids; used as a metal-paint preservative and insecticide. { 'käp·ər 'rez·ən,āt }

copper sulfate [INORGANIC CHEMISTRY] *See* cupric sulfate. { 'käp·ər 'səl,fāt }

copper sulfide [INORGANIC CHEMISTRY] CuS Black, monoclinic or hexagonal crystals that break down at 220°C; used in paints on ship bottoms to prevent fouling. { 'käp·ər 'səl,fīd }

coprecipitation [CHEMISTRY] Simultaneous precipitation of more than one substance. { ¦kō·prə,sip·ə'tā·shən }

coriandrol [ORGANIC CHEMISTRY] *See* linalool. { ,kór·ē'an,dról }

Coriolis operator [SPECTROSCOPY] An operator which gives a large contribution to the energy of an axially symmetric molecule arising from the interaction between vibration and rotation when two vibrations have equal or nearly equal frequencies. { kór·ē'ō·ləs ,äp·ə,rād·ər }

Coriolis resonance interactions [SPECTROSCOPY] Perturbation of two vibrations of a polyatomic molecule, having nearly equal frequencies, on each other, due to the energy contribution of the Coriolis operator. { kór·ē'ō·ləs 'rez·ən·əns ,in·tər,ak·shənz }

corresponding states [PHYSICAL CHEMISTRY] The condition when two or more substances are at the same reduced pressures, the same reduced temperatures, and the same reduced volumes. { ,kär·ə'spänd·iŋ 'stāts }

corrosive sublimate [INORGANIC CHEMISTRY] *See* mercuric chloride. { kə'rō·siv 'səb·lə,māt }

cotectic [PHYSICAL CHEMISTRY] Referring to conditions of pressure, temperature, and composition under which two or more solid phases crystallize at the same time, with no resorption, from a single liquid over a finite range of decreasing temperature. { kō'tek·tik }

cotectic crystallization [PHYSICAL CHEMISTRY] Simultaneous crystallization of two or more solid phases from a single liquid over a finite range of falling temperature without resorption. { kō'tek·tik ,krist·əl·ə'zā·shən }

Cotton effect [ANALYTICAL CHEMISTRY] The characteristic wavelength dependence of the optical rotatory dispersion curve or the circular dichroism curve or both in the vicinity of an absorption band. { 'kät·ən i'fekt }

coudé spectrograph [SPECTROSCOPY] A stationary spectrograph that is attached to the tube of a coudé telescope. { kü'dā 'spek·trə͵graf }

coudé spectroscopy [SPECTROSCOPY] The production and investigation of astronomical spectra using a coudé spectrograph. { kü'dā spek'träs·kə·pē }

Coulomb energy [PHYSICAL CHEMISTRY] The energy associated with the electrostatic interaction between two or more electron distributions in terms of which the actual electron distribution of a covalent bond is described. { 'kü͵läm ͵en·ər·jē }

coulometer [PHYSICAL CHEMISTRY] An electrolytic cell for the precise measurement of electrical quantities or current intensity by quantitative determination of chemical substances produced or consumed. Also known as voltameter. { kü'läm·əd·ər }

coulometric analysis [ANALYTICAL CHEMISTRY] A technique in which the amount of a substance is determined quantitatively by measuring the total amount of electricity required to deplete a solution of the substance. { ͵kü·lə'me·trik ə'nal·ə·səs }

coulometric titration [ANALYTICAL CHEMISTRY] The slow electrolytic generation of a soluble species which is capable of reacting quantitatively with the substance sought; some independent property must be observed to establish the equivalence point in the reaction. { ͵kü·lə'me·trik tī'trā·shən }

coulometry [ANALYTICAL CHEMISTRY] A determination of the amount of an electrolyte released during electrolysis by measuring the number of coulombs used. { kə'läm·ə·trē }

coulostatic analysis [PHYSICAL CHEMISTRY] An electrochemical technique involving the application of a very short, large pulse of current to the electrode; the pulse charges the capacitive electrode-solution interface to a new potential, then the circuit is opened, and the return of the working electrode potential to its initial value is monitored; the current necessary to discharge the electrode interface comes from the electrolysis of electroactive species in solution; the change in electrode potential versus time results in a plot, the shape of which is proportional to concentration. { ͵kü·lə'stad·ik ə'nal·ə·səs }

coumachlor [ORGANIC CHEMISTRY] $C_{19}H_{15}ClO_4$ A white, crystalline compound with a melting point of 169-171°C; insoluble in water; used as a rodenticide. { 'kü·mə͵klȯr }

coumarin [ORGANIC CHEMISTRY] $C_9H_6O_2$ The anhydride of *o*-coumaric acid; a toxic, white, crystalline lactone found in many plants and made synthetically; used in making perfume and soap. Also known as 1,2-benzopyrone. { 'kü·mə·rən }

coumarone [ORGANIC CHEMISTRY] C_8H_6O A colorless liquid, boiling point 169°C. { 'kü·mə͵rōn }

coumarone-indene resin [ORGANIC CHEMISTRY] A synthetic resin prepared by polymerization of coumarone and indene. { 'kü·mə͵rōn 'in͵dēn ͵rez·ən }

coumatetralyl [ORGANIC CHEMISTRY] $C_{19}H_{16}O_3$ A yellow-white, crystalline compound with a melting point of 172-176°C; slightly soluble in water; used as a rodenticide. { ¦kü·mə¦te·trə͵lil }

count [CHEMISTRY] An ionizing event. { kau̇nt }

countercurrent cascade [ANALYTICAL CHEMISTRY] An extraction process involving the introduction of a sample, all at once, into a continuously flowing countercurrent system where both phases are moving in opposite directions and are continuously at equilibrium. { 'kau̇nt·ər͵kər·ənt kas'kād }

counterion [PHYSICAL CHEMISTRY] In a solution, an ion with a charge opposite to that of another ion included in the ionic makeup of the solution. { 'kau̇nt·ər͵ī͵än }

couple [CHEMISTRY] Joining of two molecules. { 'kəp·əl }

coupled reaction [CHEMISTRY] A reaction which involves two oxidants with a single reductant, where one reaction taken alone would be thermodynamically unfavorable. { 'kəp·əld rē'ak·shən }

coupling agent [CHEMISTRY] A substance that can react with both reinforcement and matrix components of a composite material to form a binding link at their interface. { 'kəp·liŋ ͵ā·jənt }

covalence [CHEMISTRY] The number of covalent bonds which an atom can form. { kō'vā·ləns }

covalent bond [CHEMISTRY] A bond in which each atom of a bound pair contributes one electron to form a pair of electrons. Also known as electron pair bond. { kō'vā·lənt 'bänd }

covalent hydride [INORGANIC CHEMISTRY] A compound formed from a nonmetal and hydrogen, for example, H_2S and NH_3. { kō'vā·lənt 'hī‚drīd }

covalent radius [PHYSICAL CHEMISTRY] *See* atomic radius. { kō'vā·lənt 'rād·ē·əs }

Cox chart [CHEMISTRY] A straight-line graph of the logarithm of vapor pressure against a special nonuniform temperature scale; vapor pressure-temperature lines for many substances intersect at a common point on the Cox chart. { 'käks ‚chärt }

Cr [CHEMISTRY] *See* chromium.

crack [CHEMISTRY] To break a compound into simpler molecules. { krak }

cream of tartar [ORGANIC CHEMISTRY] *See* potassium bitartrate. { 'krēm əv 'tärd·ər }

creosol [ORGANIC CHEMISTRY] $CH_3O(CH_3)C_6H_3OH$ A combination of isomers, derived from coal tar or petroleum; a yellowish liquid with a phenolic odor; used as a disinfectant, in the manufacture of resins, and in flotation of ore. Also known as hydroxymethylbenzene; methyl phenol. { 'krē·ə‚sōl }

cresol [ORGANIC CHEMISTRY] $CH_3C_6H_4OH$ One of three poisonous, colorless isomeric methyl phenols: *o*-cresol, *m*-cresol, *p*-cresol; used in the production of phenolic resins, tricresyl phosphate, disinfectants, and solvents. { 'krē‚sȯl }

cresol red [ORGANIC CHEMISTRY] $C_{21}H_{18}O_5S$ A compound derived from *o*-cresol and used as an acid-base indicator; color change is yellow to red at pH 0.4 to 1.8, or 7.0 to 8.8, depending on preparation. { 'krē‚sȯl 'red }

cricondenbar [PHYSICAL CHEMISTRY] Maximum pressure at which two phases (for example, liquid and vapor) can coexist. { krə'kän·dən‚bär }

cricondentherm [PHYSICAL CHEMISTRY] Maximum temperature at which two phases (for example, liquid and vapor) can coexist. { krə'kän·dən‚thərm }

critical absorption wavelength [SPECTROSCOPY] The wavelength, characteristic of a given electron energy level in an atom of a specified element, at which an absorption discontinuity occurs. { 'krid·ə·kəl əb'sȯrp·shən 'wāv‚leŋkth }

critical condensation temperature [PHYSICAL CHEMISTRY] The temperature at which the sublimand of a sublimed solid recondenses; used to analyze solid mixtures, analogous to liquid distillation. Also known as true condensing point. { 'krid·ə·kəl ‚kän·dən'sā·shən ‚tem·prə·chər }

critical constant [PHYSICAL CHEMISTRY] A characteristic temperature, pressure, and specific volume of a gas above which it cannot be liquefied. { 'krid·ə·kəl 'kän·stənt }

critical current density [PHYSICAL CHEMISTRY] The amount of current per unit area of electrode at which an abrupt change occurs in a variable of an electrolytic process. { 'krid·ə·kəl 'kər·ənt ‚den·səd·ē }

critical density [CHEMISTRY] The density of a substance exhibited at its critical temperature and critical pressure. { 'krid·ə·kəl 'den·səd·ē }

critical line [PHYSICAL CHEMISTRY] *See* critical locus. { 'krid·ə·kəl 'līn }

critical locus [PHYSICAL CHEMISTRY] The line connecting the critical points of a series of liquid-gas phase-boundary loops for multicomponent mixtures plotted on a pressure versus temperature graph. Also known as critical line. { 'krid·ə·kəl 'lō·kəs }

critical micelle concentration [PHYSICAL CHEMISTRY] The concentration of a micelle (oriented molecular arrangement of an electrically charged colloidal particle or ion) at which the rate of increase of electrical conductance with increase in concentration levels off or proceeds at a much slower rate. { 'krid·ə·kəl mi'sel ‚kän·sən'trā·shən }

critical phenomena [PHYSICAL CHEMISTRY] Physical properties of liquids and gases at the critical point (conditions at which two phases are just about to become one); for example, critical pressure is that needed to condense a gas at the critical temperature, and above the critical temperature the gas cannot be liquefied at any pressure. { 'krid·ə·kəl fə'näm·ə·nə }

critical point [PHYSICAL CHEMISTRY] **1.** The temperature and pressure at which two phases of a substance in equilibrium with each other become identical, forming one phase. **2.** The temperature and pressure at which two ordinarily partially miscible liquids are consolute. { 'krid·ə·kəl 'pȯint }

critical properties [PHYSICAL CHEMISTRY] Physical and thermodynamic properties of materials at conditions of critical temperature, pressure, and volume, that is, at the critical point. { 'krid·ə·kəl 'präp·ərd·ēz }

critical solution temperature [PHYSICAL CHEMISTRY] The temperature at which a mixture of two liquids, immiscible at ordinary temperatures, ceases to separate into two phases. { 'krid·ə·kəl sə'lü·shən ˌtem·prə·chər }

critical state [PHYSICAL CHEMISTRY] Unique condition of pressure, temperature, and composition wherein all properties of coexisting vapor and liquid become identical. { 'krid·ə·kəl 'stāt }

critical temperature [PHYSICAL CHEMISTRY] The temperature of the liquid-vapor critical point, that is, the temperature above which the substance has no liquid-vapor transition. { 'krid·ə·kəl 'tem·prə·chər }

crosscurrent extraction [ANALYTICAL CHEMISTRY] Procedure of batchwise liquid-liquid extraction in a separatory funnel; solvent is added to the sample in the funnel, which is then shaken, and the extract phase is allowed to coalesce, then is drawn off. { 'krȯsˌkər·ənt ik'strak·shən }

cross-linking [ORGANIC CHEMISTRY] The setting up of chemical links between the molecular chains of polymers. { 'krȯs ˌliŋk·iŋ }

crotonaldehyde [ORGANIC CHEMISTRY] C_3H_5CHO A colorless liquid boiling at 104°C, soluble in water; vapors are lacrimatory; used as an intermediate in manufacture of *n*-butyl alcohol and quinaldine. Also known as propylene aldehyde. { ¦krōt·ən'al·dəˌhīd }

crotonic acid [ORGANIC CHEMISTRY] C_3H_5COOH An unsaturated acid, with colorless, monoclinic crystals, soluble in water; used in the preparation of synthetic resins, plasticizers, and pharmaceuticals. { krō'tän·ik 'as·əd }

crown ether [ORGANIC CHEMISTRY] A macrocyclic polyether whose structure exhibits a conformation with a so-called hole capable of trapping cations by coordination with a lone pair of electrons on the oxygen atoms. { 'krau̇n ˌē·thər }

crufomate [ORGANIC CHEMISTRY] $C_{12}H_{19}ClNO_3P$ A white, crystalline compound, with a melting point of 61.8°C, which is insoluble in water; used both internally and externally for cattle parasites. { 'krü·fəˌmāt }

cryohydrate [CHEMISTRY] A salt that contains water of crystallization at low temperatures. Also known as cryosel. { ¦krī·ō'hīˌdrāt }

cryohydric point [PHYSICAL CHEMISTRY] The eutectic point of an aqueous salt solution. { ˌkrī·ō'hīˌdrik ˌpȯint }

cryoscopic constant [ANALYTICAL CHEMISTRY] Equation constant expressed in degrees per mole of pure solvent; used to calculate the freezing-point-depression effects of a solute. { ¦krī·ə¦skäp·ik 'kän·stənt }

cryoscopy [ANALYTICAL CHEMISTRY] A phase-equilibrium technique to determine molecular weight and other properties of a solute by dissolving it in a liquid solvent and then ascertaining the solvent's freezing point. { krī'äs·kəˌpē }

cryosel [CHEMISTRY] *See* cryohydrate. { 'krī·əˌsel }

cryptand [ORGANIC CHEMISTRY] A macropolycyclic polyazo-polyether, where the three-coordinate nitrogen atoms provide the vertices of a three-dimensional structure. { 'kripˌtand }

crystal aerugo [ORGANIC CHEMISTRY] *See* cupric acetate. { 'krist·əl ē'rü·gō }

crystal field theory [PHYSICAL CHEMISTRY] The theory which assumes that the ligands of a coordination compound are the sources of negative charge which perturb the energy levels of the central metal ion and thus subject the metal ion to an electric field analogous to that within an ionic crystalline lattice. { ¦krist·əl 'fēld ˌthē·ə·rē }

crystal grating [SPECTROSCOPY] A diffraction grating for gamma rays or x-rays which uses the equally spaced lattice planes of a crystal. { 'krist·əl ˌgrād·iŋ }

crystalline chloral [ORGANIC CHEMISTRY] *See* chloral hydrate. { 'kris·tə·lən 'klȯr·əl }

crystalline polymer [CHEMISTRY] A polymer whose sections of adjacent chains are packed in a regular array. { 'kris·tə·lən 'päl·i·mər }

crystallinity [ORGANIC CHEMISTRY] The degree to which polymer molecules are oriented into repeating patterns. { ˌkris·təˈlin·əd·ē }

crystal monochromator [SPECTROSCOPY] A spectrometer in which a collimated beam of slow neutrons from a reactor is incident on a single crystal of copper, lead, or other element mounted on a divided circle. { ¦krist·əl ˌmän·əˈkrōˌmād·ər }

crystals of Venus [ORGANIC CHEMISTRY] *See* cupric acetate. { ˈkrist·əlz əv ˈvē·nəs }

crystal violet [ORGANIC CHEMISTRY] *See* methyl violet. { ¦krist·əl ˈvī·lət }

crystogen [ORGANIC CHEMISTRY] *See* cystamine. { ˈkris·tə·jən }

Cs [CHEMISTRY] *See* cesium.

C stage [ORGANIC CHEMISTRY] The final stage in a thermosetting resin reaction in which the material is relatively insoluble and infusible; the resin in a fully cured thermoset molding is in this stage. Also known as resite. { ˈsē ˌstāj }

Cu [CHEMISTRY] *See* copper.

cumene [ORGANIC CHEMISTRY] $C_6H_5CH(CH_3)_2$ A colorless, oily benzenoid hydrocarbon cooling at 152.4°C; used as an additive for high-octane motor fuel. { ˈkyüˌmēn }

cumene hydroperoxide [ORGANIC CHEMISTRY] $C_6H_5C(CH_3)_2OOH$ An isopropyl hydroperoxide of cumene; an oily liquid, used to make phenol and acetone. { ˈkyüˌmēn ˌhī·drō·pəˈräkˌsīd }

cumidine [ORGANIC CHEMISTRY] $C_9H_{13}N$ A colorless, water-insoluble liquid, boiling at 225°C. { ˈkyü·məˌdēn }

cumulated double bonds [CHEMISTRY] Two double bonds on the same carbon atom, as in $>C{=}C{=}C<$. { ˈkyü·myəˌlād·əd ¦dəb·əl ˈbändz }

cumulative double bonds [ORGANIC CHEMISTRY] Double bonds joining at least three contiguous carbon atoms in a single structure, for example, $H_2C{=}C{=}CH_2$ (allene). Also known as twinned double bonds. { ˈkyü·myə·ləd·iv ˈdəb·əl ˈbänz }

cumulene [ORGANIC CHEMISTRY] A compound with a molecular structure which contains two or more double bonds in succession. { ˈkyü·myəˌlēn }

cupferron [ORGANIC CHEMISTRY] $NH_4ONONC_6H_5$ A colorless salt that forms crystals with a melting point of 164°C; its acid solution is a precipitating reagent. Also known as copperon. { ˈkəp·fəˌrän }

cupreine [ORGANIC CHEMISTRY] $C_{19}H_{22}O_2N_2 \cdot H_2O$ Colorless, anhydrous crystals with a melting point of 198°C; soluble in chloroform and ether; used in medicine. Also known as hydroxycinchonine. { ˈkyü·prē·ēn }

cupric [CHEMISTRY] The divalent ion of copper. { ˈkyü·prik }

cupric acetate [ORGANIC CHEMISTRY] $Cu(C_2H_3O_2)_2 \cdot H_2O$ Blue-green crystals, soluble in water; used as a raw material to make paris green. Also known as copper acetate; crystal aerugo; crystals of Venus; verdigris. { ˈkyü·prik ˈas·əˌtāt }

cupric arsenite [INORGANIC CHEMISTRY] *See* copper arsenite. { ˈkyü·prik ˈärs·ənˌīt }

cupric bromide [INORGANIC CHEMISTRY] $CuBr_2$ Black prismatic crystals; used in photography as an intensifier and in organic synthesis as a brominating agent. Also known as copper bromide. { ˈkyü·prik ˈbrōˌmīd }

cupric carbonate [INORGANIC CHEMISTRY] *See* copper carbonate. { ˈkyü·prik ˈkär·bəˌnāt }

cupric chloride [INORGANIC CHEMISTRY] Also known as copper chloride. **1.** $CuCl_2$ Yellowish-brown, deliquescent powder soluble in water, alcohol, and ammonium chloride. **2.** $CuCl_2 \cdot H_2O$ A dihydrate of cupric chloride forming green crystals soluble in water; used as a mordant in dyeing and printing textile fabrics and in the refining of copper, gold, and silver. { ˈkyü·prik ˈklȯrˌīd }

cupric chromate [INORGANIC CHEMISTRY] $CuCrO_4$ A yellow liquid, used as a mordant. Also known as copper chromate. { ˈkyü·prik ˈkrōˌmāt }

cupric cyanide [INORGANIC CHEMISTRY] $Cu(CN)_2$ A green powder, insoluble in water; used in electroplating copper on iron. Also known as copper cyanide. { ˈkyü·prik ˈsī·əˌnīd }

cupric fluoride [INORGANIC CHEMISTRY] CuF_2 White crystalline powder used in ceramics and in the preparation of brazing and soldering fluxes. Also known as copper fluoride. { ˈkyü·prik ˈflu̇rˌīd }

cupric gluconate [ORGANIC CHEMISTRY] *See* copper gluconate. { ˈkyü·prik ˈglü·kəˌnāt }

cupric hydroxide [INORGANIC CHEMISTRY] $Cu(OH)_2$ Blue macro- or microscopic crystals; used as a mordant and pigment, in manufacture of many copper salts, and for staining paper. Also known as copper hydroxide. { 'kyü·prik hī'dräk,sīd }

cupric nitrate [INORGANIC CHEMISTRY] $Cu(NO_3)_2 \cdot 3H_2O$ Green powder or blue crystals soluble in water; used in electroplating copper on iron. Also known as copper nitrate. { 'kyü·prik 'nī,trāt }

cupric oxide [INORGANIC CHEMISTRY] CuO Black, monoclinic crystals, insoluble in water; used in making fibers and ceramics, and in organic and gas analyses. Also known as copper oxide. { 'kyü·prik 'äk,sīd }

cupric sulfate [INORGANIC CHEMISTRY] $CuSO_4$ A water-soluble salt used in copper-plating baths; crystallizes as hydrous copper sulfate, which is blue. Also known as copper sulfate. { 'kyü·prik 'səl,fāt }

cuprous bromide [INORGANIC CHEMISTRY] Cu_2Br_2 White or gray crystals slightly soluble in cold water. Also known as copper bromide. { 'kyü·prəs 'brō,mīd }

cuprous chloride [INORGANIC CHEMISTRY] CuCl or Cu_2Cl_2 Green, tetrahedral crystals, insoluble in water. Also known as copper chloride; resin of copper. { 'kyü·prəs 'klór,īd }

cuprous fluoride [INORGANIC CHEMISTRY] Cu_2F_2 Red, crystalline powder, melting point 908°C. Also known as copper fluoride. { 'kyü·prəs 'flúr,īd }

cuprous oxide [INORGANIC CHEMISTRY] Cu_2O An oxide of copper found in nature as cuprite and formed on copper by heat; used chiefly as a pigment and as a fungicide. Also known as copper oxide. { 'kyü·prəs 'äk,sīd }

cure [CHEMISTRY] To change the properties of a resin material by chemical polycondensation or addition reactions. { kyúr }

curing agent [ORGANIC CHEMISTRY] *See* hardener. { 'kyúr·iŋ ,ā·jənt }

curing temperature [CHEMISTRY] That temperature at which a resin or adhesive is subjected to curing. { 'kyúr·iŋ ,tem·prə·chər }

curing time [CHEMISTRY] The period of time in which a part is subjected to heat or pressure to cure the resin. { 'kyúr·iŋ ,tīm }

curium [CHEMISTRY] An element, symbol Cm, atomic number 96; the isotope of mass 244 is the principal source of this artificially produced element. { 'kyúr·ē·əm }

current efficiency [PHYSICAL CHEMISTRY] The ratio of the amount of electricity, in coulombs, theoretically required to yield a given quantity of material in an electrochemical process, to the amount actually consumed. { 'kər·ənt i,fish·ən·sē }

Curtius reaction [ORGANIC CHEMISTRY] A laboratory method for degrading a carboxylic acid to a primary amine by converting the acid to an acyl azide to give products which can be hydrolyzed to amines. { 'kərd·ē·əs rē,ak·shən }

cyanalcohol [ORGANIC CHEMISTRY] *See* cyanohydrin. { ,sī·ən'al·kə,hól }

cyanamide [INORGANIC CHEMISTRY] NHCNH An acidic compound that forms colorless needles, melting at 46°C, soluble in water. Also known as urea anhydride. { sī'an·ə,mid }

cyanate [INORGANIC CHEMISTRY] A salt or ester of cyanic acid containing the radical CNO. { 'sī·ə,nāt }

cyanazine [ORGANIC CHEMISTRY] $C_9H_{13}N_6Cl$ A white solid with a melting point of 166.5-167°C; used as a pre- and postemergence herbicide for corn, sorghum, soybeans, alfalfa, cotton, and wheat. { sī'an·ə,zēn }

cyanic acid [ORGANIC CHEMISTRY] HCNO A colorless, poisonous liquid, which polymerizes to cyamelide and fulminic acid. { sī'an·ik 'as·əd }

cyanidation [CHEMISTRY] Joining of cyanide to an atom or molecule. { ,si·ə·nə'dā·shən }

cyanide [INORGANIC CHEMISTRY] Any of a group of compounds containing the CN group and derived from hydrogen cyanide, HCN. { 'sī·ə,nīd }

cyanine dye [ORGANIC CHEMISTRY] $C_{29}H_{35}N_2I$ Green metallic crystals, soluble in water; unstable to light, the dye is used in the photography industry as a chemical sensitizer for film. Also known as iodocyanin; quinoline blue. { 'sī·ə·nən ,dī }

cyano- [CHEMISTRY] Combining form indicating the radical CN. { 'sī·ə·nō }

cyanoacetamide [ORGANIC CHEMISTRY] $C_3H_4N_2O$ Needlelike crystals with a melting point of 119.5°C; soluble in water; used in organic synthesis. Also known as malonamide nitrile. { ¦sī·ə·nō·ə'sed·ə·mīd }

cyanoacetic acid [ORGANIC CHEMISTRY] $NCCH_2COOH$ Hygroscopic crystals with a melting point of 66°C; decomposes at 160°C; soluble in ether, water, and alcohol; used in the synthesis of intermediates and in the commercial preparation of barbital. Also known as malonic mononitrile. { ¦sī·ə·nō·ə'sēd·ik 'as·əd }

cyanocarbon [ORGANIC CHEMISTRY] A derivative of hydrocarbon in which all of the hydrogen atoms are replaced by the CN group. { ¦sī·ə·nō'kär·bən }

cyano complex [CHEMISTRY] A coordination compound containing the CN group. { sī'an·ō 'käm‚pleks }

cyanoethylation [ORGANIC CHEMISTRY] A chemical reaction involving the addition of acrylonitrile to compounds with a reactive hydrogen. { ¦sī·ə·nō‚e·thə'lā·shən }

cyanogen [CHEMISTRY] A univalent radical, CN. [INORGANIC CHEMISTRY] C_2N_2 A colorless, highly toxic gas with a pungent odor; a starting material for the production of complex thiocyanates used as insecticides. Also known as dicyanogen. { sī'an·ə·jən }

cyanogen bromide [INORGANIC CHEMISTRY] CNBr White crystals melting at 52°C, vaporizing at 61.3°C, and having toxic fumes that affect nerve centers; used in the synthesis of organic compounds and as a fumigant. { sī'an·ə·jən 'brō‚mīd }

cyanogen chloride [INORGANIC CHEMISTRY] ClCN A poisonous, colorless gas or liquid, soluble in water; used in organic synthesis. { sī'an·ə·jən 'klȯr‚īd }

cyanogen fluoride [INORGANIC CHEMISTRY] CNF A toxic, colorless gas, used as a tear gas. { sī'an·ə·jən 'flu̇r‚īd }

cyanogen iodide [INORGANIC CHEMISTRY] *See* iodine cyanide. { sī'an·ə·jən 'ī·ə‚dīd }

cyanohydrin [ORGANIC CHEMISTRY] A compound containing the radicals CN and OH. Also known as cyanalcohol. { 'sī·ə·nō'hī·drən }

cyanophosphos [ORGANIC CHEMISTRY] $C_{15}H_{14}NO_2PS$ A white, crystalline solid with a melting point of 83°C; used as an insecticide to control larval pests on rice and vegetables. { ‚sī·ə·nō'fäs‚fōs }

cyanoplatinate [INORGANIC CHEMISTRY] *See* platinocyanide. { ¦sī·ə·nō'plat·ən‚āt }

cyanuric acid [ORGANIC CHEMISTRY] $HOC(NCOH)_2N \cdot 2H_2O$ Colorless, monoclinic crystals, slightly soluble in water; formed by polymerization of cyanic acid. Also known as pyrolithic acid. { ¦sī·ə¦nu̇r·ik 'as·əd }

cyclamate [ORGANIC CHEMISTRY] The calcium or sodium salt of cyclohexylsulfamate, an artificial sweetener. { 'sī·klə‚māt }

cyclane [ORGANIC CHEMISTRY] *See* alicyclic. { 'sī‚klān }

cyclethrin [ORGANIC CHEMISTRY] $C_{21}H_{28}O_3$ A viscous, brown liquid, soluble in organic solvents; used as an insecticide. { sī'klē·thrən }

cyclic amide [ORGANIC CHEMISTRY] An amide arranged in a ring of carbon atoms. { 'sīk·lik 'a‚mīd }

cyclic anhydride [ORGANIC CHEMISTRY] A ring compound formed by the removal of water from a compound; an example is phthalic anhydride. { 'sīk·lik an'hī‚drīd }

cyclic chronopotentiometry [ANALYTICAL CHEMISTRY] An analytic electrochemical method in which instantaneous current reversal is imposed at the working electrode, and its potential is monitored with time. { 'sīk·lik ¦krän·ō·pə‚ten·chē'äm·ə·trē }

cyclic coil [PHYSICAL CHEMISTRY] *See* random coil. { 'sīk·lik 'kȯil }

cyclic compound [ORGANIC CHEMISTRY] A compound that contains a ring of atoms. { 'sīk·lik 'käm‚pau̇nd }

cyclic ion [ORGANIC CHEMISTRY] *See* bridged ion. { 'sīk·lik 'ī·ən }

cyclic voltammetry [PHYSICAL CHEMISTRY] An electrochemical technique for studying variable potential at an electrode involving application of a triangular potential sweep, allowing one to sweep back through the potential region just covered. { 'sīk lik vōl'täm·ə·trē }

cyclitol [ORGANIC CHEMISTRY] A cycloalkane that contains one hydroxyl group on each of three or more of the atoms constituting the ring. { 'sī·klə‚tȯl }

cyclization [ORGANIC CHEMISTRY] Changing an open-chain hydrocarbon to a closed ring. { ‚sī·klə'zā·shən }

cycloaddition [ORGANIC CHEMISTRY] A reaction in which unsaturated molecules combine to form a cyclic compound. { ¦sī·klō·ə'dish·ən }

cycloaliphatic [ORGANIC CHEMISTRY] *See* alicyclic. { ¦sī·klō·al·ə'fad·ik }

cycloalkane [ORGANIC CHEMISTRY] *See* alicyclic. { ¦sī·klō'al‚kān }

cycloalkene [ORGANIC CHEMISTRY] An unsaturated, monocyclic hydrocarbon having the formula C_nH_{2n-2}. Also known as cycloolefin. { ¦sī·klō'al‚kēn }

cycloalkylaryl compound [ORGANIC CHEMISTRY] A compound with a multiringed molecular structure containing both aromatic and saturated rings. { ¦sī·klō¦al·kə'lar·əl 'käm‚pau̇nd }

cycloalkyne [ORGANIC CHEMISTRY] A cyclic compound containing one or more triple bonds between carbon atoms. { ¦sī·klō'al‚kīn }

cycloate [ORGANIC CHEMISTRY] $C_{11}H_{21}NOS$ A yellow liquid with limited solubility in water; boiling point is 145-146°C; used as an herbicide to control weeds in sugarbeets, spinach, and table beets. { 'sī·klə‚wāt }

cyclobutadiene [ORGANIC CHEMISTRY] C_4H_4 A cyclic compound containing two alternate double bonds; used in organic synthesis. Also known as butene. { ¦sī·klō ‚byüd·ə'dī‚ēn }

cyclobutane [ORGANIC CHEMISTRY] C_4H_8 An alicyclic hydrocarbon, boiling point 11°C; synthesized as a condensable gas; used in organic synthesis. Also known as tetramethylene. { ¦sī·klō'byü‚tān }

cyclobutene [ORGANIC CHEMISTRY] C_4H_6 An asymmetrical cyclic hydrocarbon occurring in several isomeric forms. Also known as cyclobutylene. { ¦sī·klō'byü‚tēn }

cyclodiolefin [ORGANIC CHEMISTRY] A cycloalkene with two double bonds; sometimes included with alkenes, cycloalkenes, and hydrocarbons containing more than one ethylene bond as olefins in a generic sense. { ¦sī·klō·dī'ō·lə·fən }

cyclododecatriene [ORGANIC CHEMISTRY] $C_{12}H_{18}$ One of two cyclic hydrocarbons with three double bonds; the two forms are stereoisomeric; used to make nylon-6 and nylon-12. { ¦sī·klō‚dō·dek·ə'trī‚ēn }

1,3-cyclohexadiene [ORGANIC CHEMISTRY] C_6H_8 A partly saturated benzene compound with two double bonds; used in organic synthesis. { ¦wən ¦thrē ¦sī·klō‚hek·sə'dī‚ēn }

cyclohexane [ORGANIC CHEMISTRY] C_6H_{12} A colorless liquid that is a cyclic hydrocarbon synthesized by hydrogenation of benzene; used in organic synthesis. Also known as hexamethylene. { ¦sī·klō'hek‚sān }

cyclohexanol [ORGANIC CHEMISTRY] $C_6H_{11}OH$ An oily, colorless, hygroscopic liquid with a camphorlike odor and a boiling point of 160.9°C; used in soapmaking, insecticides, dry cleaning, plasticizers, and germicides. Also known as hexahydrophenol. { ¦sī·klō'hek·sə‚nȯl }

cyclohexanone [ORGANIC CHEMISTRY] $C_6H_{10}O$ An oily liquid with an odor suggesting peppermint and acetone; soluble in alcohol, ether, and other organic solvents; used as an industrial solvent, in the production of adipic acid, and in the preparation of cyclohexanone resins. { ¦sī·klō'hek·sə‚nōn }

cyclohexene [ORGANIC CHEMISTRY] C_6H_{10} A compound that occurs in coal tar; a liquid that is used as an alkylation component; used in the manufacture of hexahydrobenzoic acid, adipic acid, and maleic acid. { ¦sī·klō'hek‚sēn }

cyclohexylamine [ORGANIC CHEMISTRY] $C_6H_{11}NH_2$ A liquid with a strong, fishy, amine odor; miscible with water and common organic solvents; used in organic synthesis and in the manufacture of plasticizers, rubber chemicals, corrosion inhibitors, dyestuffs, dry-cleaning soaps, and emulsifying agents. { ¦sī·klō·hek'sil·ə‚mēn }

cycloidal mass spectrometer [SPECTROSCOPY] Small mass spectrometer of limited mass range fitted with a special-type analyzer that generates a cycloidal-path beam of the sample mass. { sī'klȯid·əl 'mas spek'träm·əd·ər }

cyclonite [ORGANIC CHEMISTRY] $(CH_2)_3N_3(NO_2)_3$ A white, crystalline explosive, consisting of hexahydro-trinitro-triazine, and having high sensitivity and brisance; mixed with other explosives or substances. Abbreviated RDX. Also known as cyclotrimethylenetrinitramine. { 'sī·klə‚nīt }

1,5-cyclooctadiene [ORGANIC CHEMISTRY] C_8H_{12} A cyclic hydrocarbon with two double bonds; prepared from butadiene and used to make cyclooctene and cyclooctane, which are intermediates for the production of plastics, fibers, and so on. { ¦wən ¦fīv ¦sī·klō‚äk·tə'dī‚ēn }

cyclooctane [ORGANIC CHEMISTRY] $(CH_2)_8$ A cyclic alkane melting at 9.5°C; used as an intermediate in production of plastics, fibers, adhesives, and coatings. Also known as octomethylene. { ¦sī·klō'äk‚tān }

cyclooctatetraene [ORGANIC CHEMISTRY] C_8H_8 A cyclic olefin with alternate double bonds; highly reactive; rearranges to styrene. { ¦sī·klō‚äk·tə'te·trə‚ēn }

cycloolefin [ORGANIC CHEMISTRY] *See* cycloalkene. { ¦sī·klō'ō·lə·fən }

1,3-cyclopentadiene [ORGANIC CHEMISTRY] C_5H_6 A colorless liquid boiling at 41.5°C; used to make resins. { ¦wən ¦thrē ¦sī·klō‚pen·tə'dī‚ēn }

cyclopentadienyl anion [ORGANIC CHEMISTRY] C_5H_5— A radical formed from cyclopentadiene. { ¦sī klō‚pen tə‚dī'e nil 'an‚ī ən }

cyclopentane [ORGANIC CHEMISTRY] C_5H_{10} A cyclic hydrocarbon that is a colorless liquid; present in crude petroleum, it is converted during refining to aromatics which improve antiknock and combustion properties of gasoline. { ¦sī·klō'pen‚tān }

cyclopentanoid [ORGANIC CHEMISTRY] A compound whose key structural unit consists of five carbon atoms arranged in a ring. { ¦sī·klō'pen·tə‚nȯid }

cyclopentanol [ORGANIC CHEMISTRY] C_5H_9OH A colorless liquid boiling at 139°C; used as a solvent for perfumes and pharmaceuticals. Also known as cyclopentyl alcohol. { ¦sī·klō'pen·tə‚nȯl }

cyclopentanone [ORGANIC CHEMISTRY] C_5H_8O A saturated monoketone; a colorless liquid boiling at 130°C; used as an intermediate in pharmaceutical preparation. { ¦sī·klō'pen·tə‚nōn }

cyclopentene [ORGANIC CHEMISTRY] $(CH_2)_3CHCH$ A colorless liquid boiling at 45°C; used as a chemical intermediate in petroleum chemistry. { ¦sī·klō'pen‚tēn }

cyclopentyl alcohol [ORGANIC CHEMISTRY] *See* cyclopentanol. { ¦sī·klō'pent·əl 'al·kə‚hȯl }

cyclopropane [ORGANIC CHEMISTRY] C_3H_6 A colorless gas, insoluble in water; used as an anesthetic. { ¦sī·klō'prō‚pān }

cyclopropanoid [ORGANIC CHEMISTRY] A compound whose key structural unit consists of three carbon atoms arranged in a ring. { ¦sī·klō'prō·pə‚nȯid }

α-cyclopropyl-α-(*para*-methoxyphenyl)-5-pyrimidine-methanol [ORGANIC CHEMISTRY] $C_{12}H_{11}N_2O$ White crystals with a melting point of 110–111°C; solubility in water is 650 parts per million; used as a growth regulator for greenhouse plants. { ¦al·fə ¦sī·klō'prō·pəl ¦al·fə ¦par·ə mə‚thäk·sē'fen·əl ¦fīv pī'rim·ə‚dēn'meth·ə‚nȯl }

cyclotrimethylenetrinitramine [ORGANIC CHEMISTRY] *See* cyclonite. { ¦sī·klō‚trī¦meth·əl‚ēn‚trī'nī·trə‚mēn }

cyhexatin [ORGANIC CHEMISTRY] $C_{18}H_{34}OSn$ A whitish solid, insoluble in water; used as a miticide to control plant-feeding mites. { sī'hek sə tən }

cymene [ORGANIC CHEMISTRY] Any of the isomeric hydrocarbons metacymene, paracymene, and orthocymene; paracymene is a liquid that is colorless, has a pleasant odor, and is made from oil of cumin or oil of wild thyme. { 'sī‚mēn }

cystamine [ORGANIC CHEMISTRY] $(CH_2)_6N_4$ A white, crystalline powder, melting at 280°C; used to make synthetic resins. Also known as aminiform; crystogen; cystamine methenamine; hexamethylene tetramine; urotropin. { 'sis·tə‚mēn }

D

2,4-D [ORGANIC CHEMISTRY] *See* 2,4-dichlorophenoxyacetic acid.

dalapon [ORGANIC CHEMISTRY] Generic name for 2,2-dichloropropionic acid, a liquid with a boiling point of 185-190°C at 760 mmHg; soluble in water, alcohol, and ether; used as a herbicide. { 'dal·ə,pän }

Dalton's atomic theory [CHEMISTRY] Theory forming the basis of accepted modern atomic theory, according to which matter is made of particles called atoms, reactions must take place between atoms or groups of atoms, and atoms of the same element are all alike but differ from atoms of another element. { 'dȯl·tənz ə,täm·ik 'thē·ə·rē }

Daniell cell [PHYSICAL CHEMISTRY] A primary cell with a constant electromotive force of 1.1 volts, having a copper electrode in a copper sulfate solution and a zinc electrode in dilute sulfuric acid or zinc sulfate, the solutions separated by a porous partition or by gravity. { 'dan·yəl ,sel }

dansyl chloride [ORGANIC CHEMISTRY] $(CH_3)_2NC_{10}H_6SO_2Cl$ A reagent for fluorescent labeling of amines, amino acids, proteins, and phenols. { 'dans·əl 'klȯr,īd }

DAP [ORGANIC CHEMISTRY] *See* diallyl phthalate.

dark-line spectrum [SPECTROSCOPY] The absorption spectrum that results when white light passes through a substance, consisting of dark lines against a bright background. { 'därk ¦līn 'spek·trəm }

Darzen's procedure [ORGANIC CHEMISTRY] Preparation of alkyl halides by refluxing a molecule of an alcohol with a molecule of thionyl chloride in the presence of a molecule of pyridine. { 'där·zənz prə,sē·jər }

Darzen's reaction [ORGANIC CHEMISTRY] Condensation of aldehydes and ketones with α-haloesters to produce glycidic esters. { 'där·zənz rē,ak·shən }

dative bond [CHEMISTRY] *See* coordinate valence. { ¦dād·iv 'bänd }

dazomet [ORGANIC CHEMISTRY] $C_5H_{10}N_2S_2$ A white, crystalline compound that decomposes at 100°C; used as a herbicide and nematicide for soil fungi and nematodes, weeds, and soil insects. Also known as tetrahydro-3,5-dimethyl-2H-1,3,5-thiadiazine-6-thione. { 'dā· zə·mət }

DBCP [ORGANIC CHEMISTRY] *See* dibromochloropropane.

d-block element [CHEMISTRY] A transition element occupying the first, second, and third long periods of the periodic table. { 'dē ,bläk ,el·ə·mənt }

DCB [INORGANIC CHEMISTRY] *See* 1,4-dichlorobutane.

DCC [ORGANIC CHEMISTRY] *See* dicyclohexylcarbodiimide.

DCCI [ORGANIC CHEMISTRY] *See* dicyclohexylcarbodiimide.

DCNA [ORGANIC CHEMISTRY] *See* 2,6-dichloro-4-nitroaniline.

DCPA [ORGANIC CHEMISTRY] *See* dimethyl-2,3,5,6-tetrachloroterephthalate.

DDD [ORGANIC CHEMISTRY] *See* 2,2-bis(-chlorophenyl)-1,1- dichloroethane.

DDT [ORGANIC CHEMISTRY] Common name for an insecticide; melting point 108.5°C, insoluble in water, very soluble in ethanol and acetone, colorless, and odorless; especially useful against agricultural pests, flies, lice, and mosquitoes. Also known as dichlorodiphenyltrichloroethane.

DDTA [ANALYTICAL CHEMISTRY] *See* derivative differential thermal analysis.

DDVP [ORGANIC CHEMISTRY] *See* dichlorvos.

DEA [ORGANIC CHEMISTRY] *See* diethanolamine.
deacetylation [ORGANIC CHEMISTRY] The removal of an acetyl group from a molecule. { ˌdē·əˌsēd·əl'ā·shən }
deacidification [CHEMISTRY] **1.** Removal of acid. **2.** A process for reducing acidity. { ˌdē·əˌsid·ə·fə'kā·shən }
deactivation [CHEMISTRY] **1.** Rendering inactive, as of a catalyst. **2.** Loss of radioactivity. { dēˌak·tə'vā·shən }
deacylation [ORGANIC CHEMISTRY] Removal of an acyl group from a compound. { dēˌas·ə'lā·shən }
DEAE-cellulose [ORGANIC CHEMISTRY] *See* diethylaminoethyl cellulose. { 'dēˌē'āˌē 'sel·yəˌlōs }
dealkalization [CHEMISTRY] **1.** Removal of alkali. **2.** Reduction of alkalinity, as in the process of neutralization. { dēˌal·kə·lə'zā·shən }
dealkylate [CHEMISTRY] To remove alkyl groups from a compound. { dē'al·kəˌlāt }
dealuminization [CHEMISTRY] Removal of aluminum. { ˌdē·əˌlü·mə·nə'zā·shən }
deamidation [ORGANIC CHEMISTRY] Removal of the amido group from a molecule. { dēˌam·ə'dā·shən }
deamination [ORGANIC CHEMISTRY] Removal of an amino group from a molecule. { dēˌam·ə'nā·shən }
deashing [CHEMISTRY] A form of deionization in which inorganic salts are removed from solution by the adsorption of both the anions and cations by ion-exchange resins. { dē'ash·iŋ }
debenzylation [ORGANIC CHEMISTRY] Removal from a molecule of the benzyl group. { dēˌben·zə'lā·shən }
de Brun-van Eckstein rearrangement [ORGANIC CHEMISTRY] The isomerization of an aldose or ketose when mixed with aqueous calcium hydroxide to form a mixture of various monosaccharides and unfermented ketoses; used to prepare certain ketoses. { də¦brün van'ekˌshtīn ˌrē·əˌrānj·mənt }
Debye-Falkenhagen effect [PHYSICAL CHEMISTRY] The increase in the conductance of an electrolytic solution when the applied voltage has a very high frequency. { də¦bī 'fäl·kənˌhäg·ən iˌfekt }
Debye force [PHYSICAL CHEMISTRY] *See* induction force. { də'bī ˌfórs }
Debye-Hückel theory [PHYSICAL CHEMISTRY] A theory of the behavior of strong electrolytes, according to which each ion is surrounded by an ionic atmosphere of charges of the opposite sign whose behavior retards the movement of ions when a current is passed through the medium. { də¦bī 'hik·əl ˌthē·ə·rē }
Debye relaxation time [PHYSICAL CHEMISTRY] According to the Debye-Hückel theory, the time required for the ionic atmosphere of a charge to reach equilibrium in a current-carrying electrolyte, during which time the motion of the charge is retarded. { də'bī ˌrēˌlak'sā·shən ˌtīm }
decaborane (14) [INORGANIC CHEMISTRY] $B_{10}H_{14}$ A binary compound of boron and hydrogen that is relatively stable at room temperature; melting point 99.5°C, boiling point 213°C. { ¦dek·ə¦bórˌān 'fōr¦tēn }
decahydrate [CHEMISTRY] A compound that has 10 water molecules. { ˌdek·ə'hīˌdrāt }
decahydronaphthalene [ORGANIC CHEMISTRY] $C_{10}H_{18}$ A liquid hydrocarbon, used in some paints and lacquers as a solvent. { ˌdek·əˌhī·drō'naf·thəˌlēn }
decalcification [CHEMISTRY] Loss or removal of calcium or calcium compounds from a calcified material such as bone or soil. { dēˌkal·sə·fə'kā·shən }
decane [ORGANIC CHEMISTRY] $C_{10}H_{22}$ Any of several saturated aliphatic hydrocarbons, especially $CH_3(CH_2)_8CH_3$. { 'dēˌkān }
decanol [ORGANIC CHEMISTRY] *See* decyl alcohol. { 'dek·əˌnól }
decarbonize [CHEMISTRY] To remove carbon by chemical means. { dē'kär·bəˌnīz }
decarboxylate [ORGANIC CHEMISTRY] To remove the carboxyl radical, especially from amino acids and protein. { ˌdē·kär'bäk·səˌlāt }
dechlorination [CHEMISTRY] Removal of chlorine from a substance. { dēˌklór·ə'nā·shən }

decinormal [CHEMISTRY] Pertaining to a chemical solution that is one-tenth normality in reference to a 1 normal solution. { ¦des·ə'nȯr·məl }

decolorizing carbon [CHEMISTRY] Porous or finely divided carbon (activated or bone) with large surface area; used to adsorb colored impurities from liquids, such as lube oils. { dē¦kəl·ə,rīz·iŋ 'kär·bən }

decomposition [CHEMISTRY] The more or less permanent structural breakdown of a molecule into simpler molecules or atoms. { dē,käm·pə'zish·ən }

decomposition potential [PHYSICAL CHEMISTRY] The electrode potential at which the electrolysis current begins to increase appreciably. Also known as decomposition voltage. { dē,käm·pə'zish·ənpə,ten·chəl }

decomposition voltage [PHYSICAL CHEMISTRY] *See* decomposition potential. { dē,käm·pə'zish·ən ,vōl·tij }

decyl [ORGANIC CHEMISTRY] An isomeric grouping of univalent radicals, all with formulas $C_{10}H_{21}$, and derived from the decanes by removing one hydrogen. { 'des·əl }

decyl acetate [ORGANIC CHEMISTRY] $CH_3(CH_2)_9OOCCH_3$ Perfumery liquid with a floral orange-rose aroma. { 'des·əl 'as·ə,tāt }

decyl alcohol [ORGANIC CHEMISTRY] $C_{10}H_{21}OH$ A colorless oil, boiling at 231°C; used in plasticizers, synthetic lubricants, and detergents. Also known as decanol. { 'des·əl 'al·kə,hȯl }

decyl aldehyde [ORGANIC CHEMISTRY] $CH_3(CH_2)_8CHO$ A liquid aldehyde, found in essential oils; used in flavorings and perfumes. { 'des·əl 'al·də,hīd }

decylene [ORGANIC CHEMISTRY] Any of a group of isomeric hydrocarbons with formula $C_{10}H_{20}$; the group is part of the ethylene series. { 'des·ə,lēn }

decyltrichlorosilane [ORGANIC CHEMISTRY] $n\text{-}C_{10}H_{21}SiCl_3$ An organochlorosilane that boils at 183°C at 84 mmHg; used in coupling agents or primers to obtain improved bonding between organic polymers and mineral surfaces. { ¦des·əl·trī¦klȯr·ō'sī,lān }

DEET [ORGANIC CHEMISTRY] *See* diethyltoluamide.

definite-composition law [CHEMISTRY] The law that a given chemical compound always contains the same elements in the same fixed proportions by weight. Also known as definite-proportions law. { ¦def·ə·nət ,käm·pə'zish·ən ,lȯ }

definite-proportions law [CHEMISTRY] *See* definite-composition law. { ¦def·ə·nət prə'pȯr·shənz ,lȯ }

deflagrating spoon [CHEMISTRY] A long-handled spoon used in chemistry to demonstrate deflagration. { 'def·lə,grād·iŋ ,spün }

deflagration [CHEMISTRY] A chemical reaction accompanied by vigorous evolution of heat, flame, sparks, or spattering of burning particles. { ,def·lə'grā·shən }

deflocculant [CHEMISTRY] An agent that causes deflocculation; examples are sodium carbonate and other basic materials used to deflocculate clay slips. { dē'fläk·yə·lənt }

defluorination [CHEMISTRY] Removal of fluorine. { dē,flu̇r·ə'nā·shən }

degasser [PHYSICAL CHEMISTRY] *See* getter. { de'gas·ər }

degradation [ORGANIC CHEMISTRY] Conversion of an organic compound to one containing a smaller number of carbon atoms. { ,deg·rə'dā·shən }

degree [CHEMISTRY] Any one of several units for measuring hardness of water, such as the English or Clark degree, the French degree, and the German degree. { di'grē }

degree of crystallinity [ORGANIC CHEMISTRY] In a fairly large sample of a polymer, the fraction that consists of regions showing long-range three-dimensional order. { di'grē əv ,kris·tə'lin·əd·ē }

degree of freedom [PHYSICAL CHEMISTRY] Any one of the variables, including pressure, temperature, composition, and specific volume, which must be specified to define the state of a system. { di'grē əv 'fre·dəm }

degree of polymerization [ORGANIC CHEMISTRY] The number of structural units in the average polymer molecule in a particular sample. Abbreviated D.P. { di'grē əv pə,lim·ə·rə'zā·shən }

dehydration [CHEMISTRY] Removal of water from any substance. [ORGANIC CHEMIS-

TRY] An elimination reaction in which a molecule loses both a hydroxyl group (OH) and a hydrogen atom (H) that was bonded to an adjacent carbon. { ˌdē·hī'drā·shən }

dehydrator [CHEMISTRY] A substance that removes water from a material; an example is sulfuric acid. { dē'hīˌdrād·ər }

dehydroacetic acid [ORGANIC CHEMISTRY] $C_8H_8O_4$ Crystals that melt at 108.5°C and are insoluble in water, soluble in acetone; used as a fungicide and bactericide. Abbreviated DHA. { dē¦hī·drō·ə¦sēd·ik 'as·əd }

dehydroascorbic acid [ORGANIC CHEMISTRY] $C_6H_6O_6$ A relatively inactive acid resulting from elimination of two hydrogen atoms from ascorbic acid when the latter is oxidized by air or other agents; has potential ascorbic acid activity. { dē¦hī·drō·ə¦skȯr·bik 'as·əd }

dehydrocholic acid [ORGANIC CHEMISTRY] $C_{24}H_{34}O_5$ A white powder melting at 231-240°C, very slightly soluble in water; used as a pharmaceutical intermediate and in medicine. { dē¦hī·drə¦käl·ik 'as·əd }

dehydroepiandrosterone [ORGANIC CHEMISTRY] $C_{19}H_{28}O_2$ Dimorphous crystals with a melting point of 140-141°C, or leaflet crystals with a melting point of 152-153°C; soluble in alcohol, benzene, and ether; used as an androgen. { dē¦hī·drō·eˌpē·ən'dräs·tə ˌrōn }

dehydrogenation [CHEMISTRY] Removal of hydrogen from a compound. { dē¦hī·drə·jə'nā·shən }

dehydrohalogenation [CHEMISTRY] Removal of hydrogen and a halogen from a compound. { dē¦hī·droˌhal·ə·jə'nā·shən }

deionization [CHEMISTRY] An ion-exchange process in which all charged species or ionizable organic and inorganic salts are removed from solution. { dēˌī·ən·ə'zā·shən }

de la Tour method [ANALYTICAL CHEMISTRY] Measurement of critical temperature, involving sealing the sample in a tube and heating it; the temperature at which the meniscus disappears is the critical temperature. { del·ə'tu̇r ˌmeth·əd }

Delepine reaction [ORGANIC CHEMISTRY] Slow ammonolysis of alkyl halides in acid to primary amines in the presence of hexamethylenetetramine. { 'del·əˌpīn rēˌak·shən }

deliquescence [PHYSICAL CHEMISTRY] The absorption of atmospheric water vapor by a crystalline solid until the crystal eventually dissolves into a saturated solution. { del·ə'kwes·əns }

delocalized bond [CHEMISTRY] A type of molecular bonding in which the electron density of delocalized electrons is regarded as being spread over several atoms or over the whole molecule. Also known as nonlocalized bond. { dē'lō·kəˌlīzd 'bänd }

delphidenolon [ORGANIC CHEMISTRY] *See* myricetin. { ¦del·fə¦den·əˌlän }

demal [CHEMISTRY] A unit of concentration, equal to the concentration of a solution in which 1 gram-equivalent of solute is dissolved in 1 cubic decimeter of solvent. { 'dem·əl }

demasking [CHEMISTRY] A process by which a masked substance is made capable of undergoing its usual reactions; can be brought about by a displacement reaction involving addition of, for example, another cation that reacts more strongly with the masking ligand and liberates the masked ion. { dē'mask·iŋ }

demethylation [ORGANIC CHEMISTRY] Removal of the methyl group from a compound. { deˌmeth·ə'lā·shən }

demeton-*S*-methyl [ORGANIC CHEMISTRY] $C_6H_{15}O_3PS_2$ An oily liquid with a 0.3% solubility in water; used as an insecticide and miticide to control aphids. { 'dem·əˌtän ¦es 'meth·əl }

demeton-*S*-methyl sulfoxide [ORGANIC CHEMISTRY] $C_6H_{15}O_4PS_2$ A clear, amber liquid; limited solubility in water; used as an insecticide and miticide for pests of vegetable, fruit, and field crops, ornamental flowers, shrubs, and trees. { 'dem·əˌtän ¦es 'meth·əl səl'fäkˌsīd }

Demjanov rearrangement [ORGANIC CHEMISTRY] A structural rearrangement that accompanies treatment of certain primary aliphatic amines with nitrous acid; the amine will undergo a ring contraction or expansion. { dem'yä·nȯf rē·ə'rānj·mənt }

denaturant [CHEMISTRY] An inert, bad-tasting, or poisonous chemical substance added to a product such as ethyl alcohol to make it unfit for human consumption. { dē'nā·chə·rənt }

denature [CHEMISTRY] **1.** To change a protein by heating it or treating it with alkali or acid so that the original properties such as solubility are changed as a result of the protein's molecular structure being changed in some way. **2.** To add a denaturant, such as methyl alcohol, to grain alcohol to make the grain alcohol poisonous and unfit for human consumption. { dē'nā·chər }

denatured alcohol [CHEMISTRY] Ethyl alcohol containing a poisonous substance, such as methyl alcohol or benzene, which makes it unfit for human consumption. { dē'nā·chərd 'al·kə‚hȯl }

dendritic macromolecule [ORGANIC CHEMISTRY] A macromolecule whose structure is characterized by a high degree of branching that originates from a single focal point (core). { den'drid·ik ‚mak·rō'mäl·ə‚kyül }

denitration [CHEMISTRY] Removal of nitrates or nitrogen. Also known as denitrification. { dē‚nī'trā·shən }

denitrification [CHEMISTRY] *See* denitration. { dē‚nī·trə·fə'kā·shən }

density gradient centrifugation [ANALYTICAL CHEMISTRY] Separation of particles according to density by employing a gradient of varying densities; at equilibrium each particle settles in the gradient at a point equal to its density. { 'den·səd·ē ¦grād·ē·ənt sen‚trif·ə'gā·shən }

deoxidant [CHEMISTRY] *See* deoxidizer. { dē'äk·sə·dənt }

deoxidation [CHEMISTRY] **1.** The condition of a molecule's being deoxidized. **2.** The process of deoxidizing. { dē‚äk·sə'dā·shən }

deoxidize [CHEMISTRY] **1.** To remove oxygen by any of several processes. **2.** To reduce from the state of an oxide. { dē'äk·sə‚dīz }

deoxidizer [CHEMISTRY] Any substance which reduces the amount of oxygen in a substance, especially a metal, or reduces oxide compounds. Also known as deoxidant. { dē'äk·sə‚dīz·ər }

deoxygenation [CHEMISTRY] Removal of oxygen from a substance, such as blood or polluted water. { dē‚äk·sə·jə'nā·shən }

2,4-DEP [ORGANIC CHEMISTRY] *See* tris[2-(2,4-dichlorophenoxy)ethyl]phosphite.

DEPC [ORGANIC CHEMISTRY] *See* diethyl pyrocarbonate.

depolarizer [PHYSICAL CHEMISTRY] A substance added to the electrolyte of a primary cell to prevent excessive buildup of hydrogen bubbles by combining chemically with the hydrogen gas as it forms. Also known as battery depolarizer. { dē'pō·lə‚rīz·ər }

depolymerization [ORGANIC CHEMISTRY] Decomposition of macromolecular compounds into relatively simple compounds. { ‚de·pə‚lim·ə·rə'za·shən }

deposition potential [PHYSICAL CHEMISTRY] The smallest potential which can produce electrolytic deposition when applied to an electrolytic cell. { ‚dep·ə'zish·ən pə'ten·chəl }

deproteinize [ORGANIC CHEMISTRY] To remove protein from a substance. { dē'prō‚tē‚nīz }

depside [ORGANIC CHEMISTRY] One of a class of esters that form from the joining of two or more molecules of phenolic carboxylic acid. { 'dep‚sīd }

depsidone [ORGANIC CHEMISTRY] One of a class of compounds that consists of esters such as depsides, but are also cyclic ethers. { 'dep·sə‚dōn }

derichment [ANALYTICAL CHEMISTRY] In gravimetric analysis by coprecipitation of salts, a system with λ less than unity, when λ is the logarithmic distribution coefficient expressed by the ratio of the logarithms of the ratios of the initial and final solution concentrations of the two salts. { dē'rich·mənt }

derivative [CHEMISTRY] A substance that is made from another substance. { də'riv·əd·iv }

derivative differential thermal analysis [ANALYTICAL CHEMISTRY] A method for precise determination in thermograms of slight temperature changes by taking the first derivative of the differential thermal analysis curve (thermogram) which plots time versus differential temperature as measured by a differential thermocouple. Also known as DDTA. { də'riv·əd·iv dif·ə'ren·chəl 'thər·məl ə'nal·ə·səs }

derivative polarography [ANALYTICAL CHEMISTRY] Polarography technique in which the rate of change of current with respect to applied potential is measured as a function of the applied potential (*di/d*E versus E, where *i* is current and E is applied potential). { də'riv·əd·iv ˌpō·lə'räg·rə·fē }

derivative thermometric titration [ANALYTICAL CHEMISTRY] The use of a special resistance-capacitance network to record first and second derivatives of a thermometric titration curve (temperature versus weight change upon heating) to produce a sharp end-point peak. { də'riv·əd·iv thər·mə'me·trik tī'trā·shən }

descending chromatography [ANALYTICAL CHEMISTRY] A type of paper chromatography in which the sample-carrying solvent mixture is fed to the top of the developing chamber, being separated as it works downward. { di'sen·diŋ krō·mə'täg·rə·fē }

desiccant [CHEMISTRY] *See* drying agent. { 'des·i·kənt }

designated volume [ANALYTICAL CHEMISTRY] The volume of an item of volumetric glassware as calibrated at a given temperature, frequently 20°C (68°F). { ¦dez·igˌnād·əd 'väl·yəm }

desmetryn [ORGANIC CHEMISTRY] $C_9H_{17}N_5S$ A white, crystalline compound with a melting point of 84-86°C; used as a postemergence herbicide for broadleaf and grassy weeds. { dez'me·trən }

desorption [PHYSICAL CHEMISTRY] The process of removing a sorbed substance by the reverse of adsorption or absorption. { dē'sȯrp·shən }

destructive distillation [ORGANIC CHEMISTRY] Decomposition of organic compounds by heat without the presence of air. { di'strək·tiv dis·tə'lā·shən }

desulfonation [ORGANIC CHEMISTRY] Removal of the sulfonate group from an organic molecule. { dēˌsəl·fə'nā·shən }

desyl [ORGANIC CHEMISTRY] The functional group $C_6H_5COCH(C_6H_5—)$; may be formed from desoxybenzoin. { 'des·əl }

DET [ORGANIC CHEMISTRY] *See* diethyltoluamide.

detection limit [ANALYTICAL CHEMISTRY] In chemical analysis, the minimum amount of a particular component that can be determined by a single measurement with a stated confidence level. { di'tek·shən ˌlim·ət }

detergent alkylate [ORGANIC CHEMISTRY] *See* dodecylbenzene. { di'tər·jənt 'al·kəˌlāt }

determination [ANALYTICAL CHEMISTRY] The finding of the value of a chemical or physical property of a compound, such as reaction-rate determination or specific-gravity determination. { dəˌtər·mə'nā·shən }

detonation [CHEMISTRY] An exothermic chemical reaction that propagates with such rapidity that the rate of advance of the reaction zone into the unreacted material exceeds the velocity of sound in the unreacted material; that is, the advancing reaction zone is preceded by a shock wave. { ˌdet·ən'ā·shən }

deuteration [CHEMISTRY] The addition of deuterium to a chemical compound. { ˌdüd·ər'ā·shən }

deuteride [CHEMISTRY] A hydride in which the hydrogen is deuterium. { 'düd·əˌrīd }

deuterium [CHEMISTRY] The isotope of the element hydrogen with one neutron and one proton in the nucleus; atomic weight 2.0144.Designated D, d, H^2, or ^2H. { dü'tir·ē·əm }

deuterium oxide [INORGANIC CHEMISTRY] *See* heavy water. { dü'tir·ē·əm 'äkˌsīd }

developed dye [CHEMISTRY] A direct azo dye that can be further diazotized by a developer after application to the fiber; it couples with the fiber to form colorfast shades. Also known as diazo dye. { də'vel·əpt 'dī }

developer [CHEMISTRY] An organic compound which interacts on a textile fiber to develop a dye. { də'vel·əp·ər }

development [ANALYTICAL CHEMISTRY] In the separation of mixtures by paper chromatography or thin-layer chromatography, the production of colored derivatives of the solutes by spraying the stationary phase with selective reagents in order to establish the location of individual substances. { də'vel·əp·mənt }

devitrification [CHEMISTRY] The process by which the glassy texture of a material is converted into a crystalline texture. { dēˌvi·trə·fə'kā·shən }

devrinol [ORGANIC CHEMISTRY] $C_{17}H_{21}O_2N$ A brown solid with a melting point of 68.5-

70.5°C; slight solubility in water; used as a herbicide for crops. Also known as 2-(α-naphthoxy)-N,N-diethylpropionamide. { 'dev·rə‚nȯl }

Dewar structure [ORGANIC CHEMISTRY] A structural formula for benzene that contains a bond between opposite atoms. { 'dü·ər ‚strək·chər }

dew point [CHEMISTRY] The temperature and pressure at which a gas begins to condense to a liquid. { 'dü ‚pȯint }

dextrinization [ORGANIC CHEMISTRY] Any process that involves dextrinizing. { ‚dek·strə·nə'zā·shən }

dextrinize [ORGANIC CHEMISTRY] To convert a starch into dextrins. { 'dek·strə‚nīz }

dextropimaric acid [ORGANIC CHEMISTRY] $C_{19}H_{29}COOH$ A compound found in particular in oleoresins of pine trees. { ¦dek·strō·pə'mar·ik 'as·əd }

dezincification [CHEMISTRY] Removal of zinc. { dē‚ziŋk·ə·fə'kā·shən }

DHA [ORGANIC CHEMISTRY] *See* dehydroacetic acid; dihydroxyacetone.

Di [CHEMISTRY] *See* didymium.

diacetate [ORGANIC CHEMISTRY] An ester or salt that contains two acetate groups. { dī'as·ə‚tāt }

diacetic ether [ORGANIC CHEMISTRY] *See* ethyl acetoacetate. { dī·ə'sēd‚ik 'ē·thər }

diacetin [ORGANIC CHEMISTRY] $C_3H_5(OH)(CH_3COO)_2$ A colorless, hygroscopic liquid that is soluble in water, alcohol, ether, and benzene; boiling point 259°C; used as a plasticizer and softening agent and as a solvent. Also known as glyceryl diacetate. { dī'as·əd·ən }

diacetone alcohol [ORGANIC CHEMISTRY] $CH_3COCH_2C(CH_3)_2OH$ A colorless liquid used as a solvent for nitrocellulose and resins. { dī'as·ə‚tōn 'al·kə‚hȯl }

diacetyl [ORGANIC CHEMISTRY] **1.** $CH_3COCOCH_3$ A yellowish-green liquid with a boiling point of 88°C; has a strong odor that resembles quinone; occurs naturally in bay oil and butter and is produced from methyl ethyl ketone or by a special fermentation of glucose; used as an aroma carrier in food manufacturing. Also known as biacetyl. **2.** A prefix indicating two acetyl groups. { dī'as·əd·əl }

diacetylurea [ORGANIC CHEMISTRY] $C_5H_8O_3N_2$ An acyl derivative of urea containing two acetyl groups. { dī¦as·əd·əl yu̇'rē·ə }

diacid [CHEMISTRY] An acid that has two acidic hydrogen atoms; an example is oxalic acid. { dī'as·əd }

dialdehyde [ORGANIC CHEMISTRY] A molecule that has two aldehyde groups, such as dialdehyde starch. { dī'al·də‚hīd }

dialifor [ORGANIC CHEMISTRY] $C_{14}H_{17}ClNO_4S_2P$ A white, crystalline compound with a melting point of 67–69°C; insoluble in water; used to control pests in citrus fruits, grapes, and pecans. { dī'al·ə‚fȯr }

dialkyl [ORGANIC CHEMISTRY] A molecule that has two alkyl groups. { dī'al·kəl }

dialkyl amine [ORGANIC CHEMISTRY] An amine that has two alkyl groups bonded to the amino nitrogen. { dī'al·kəl 'a‚mēn }

diallyl phthalate [ORGANIC CHEMISTRY] $C_6H_4(COOCH_2CH{:}CH_2)_2$ A colorless, oily liquid with a boiling range of 158–165°C; used as a plasticizer and for polymerization. Abbreviated DAP. { dī'al·əl 'tha‚lāt }

dialuric acid [ORGANIC CHEMISTRY] $C_4H_4N_2O$ An acid that is derived by oxidation of uric acid or by the reduction of alloxan; may be used in organic synthesis. { ¦dī·ə¦lu̇r·ik 'as·əd }

dialysis [PHYSICAL CHEMISTRY] A process of selective diffusion through a membrane; usually used to separate low-molecular-weight solutes which diffuse through the membrane from the colloidal and high-molecular-weight solutes which do not. { dī'al·ə·səs }

dialyzate [CHEMISTRY] The material that does not diffuse through the membrane during dialysis; alternatively, it may be considered the material that has diffused. { dī'al·ə‚zāt }

diamide [ORGANIC CHEMISTRY] A molecule that has two amide ($—CONH_2$) groups. { 'dī·ə‚mīd }

diamidine [ORGANIC CHEMISTRY] A molecule that has two amidine ($—C{=}NHNH_2$) groups. { dī'am·ə‚dēn }

diamine [ORGANIC CHEMISTRY] Any compound containing two amino groups. { 'dī·ə‚mēn }

diamino [ORGANIC CHEMISTRY] A term used in chemical nomenclature to indicate the presence in a molecule of two amino (-NH_2) groups. { dī'am·ə‚nō }

3,5-diaminobenzoic acid [ORGANIC CHEMISTRY] $C_7H_8N_2O_2$ Monohydrate crystals with a melting point of 228°C; soluble in organic solvents such as alcohol and benzene; used in the detection and determination of nitrites. { ¦thrē ¦fīv dī¦am·ə‚nō‚ben'zō·ik 'as·əd }

2,7-diaminofluorene [ORGANIC CHEMISTRY] $C_{13}H_{12}N_2$ A compound crystallizing as needlelike crystals from water; the melting point is 165°C; soluble in alcohol; used to detect bromide, chloride, nitrate, persulfate, cadmium, zinc, copper, and cobalt. Also known as 2,7-fluorenediamine. { ¦tü ¦sev·ən dī¦am·ə‚nō'flür‚ēn }

diamyl phenol [ORGANIC CHEMISTRY] $(C_5H_{11})_2C_6H_3OH$ A straw-colored liquid with a boiling range of 280-295°C; used in synthetic resins, lubricating oil additives, plasticizers, detergents, and fungicides. { dī'am·əl 'fē‚nȯl }

diamyl sulfide [ORGANIC CHEMISTRY] $(C_5H_{11})_2S$ A combustible, yellow liquid with a distillation range of 170-180°C; used as a flotation agent and an odorant. { dī'am·əl 'səl‚fīd }

diarsine [ORGANIC CHEMISTRY] An arsenic compound containing an As-As bond with the general formula $(R_2As)_2$, where R represents a functional group such as CH_3. { dī'är‚sēn }

diarylamine [ORGANIC CHEMISTRY] A molecule that contains an amine group and two aryl groups joined to the amino nitrogen. { ¦dī·ə'ril·ə‚mēn }

diastereoisomer [ORGANIC CHEMISTRY] One of a pair of optical isomers which are not mirror images of each other. Also known as diastereomer. { ¦dī·ə‚ster·ē·ō'ī·sə·mər }

diastereomer [ORGANIC CHEMISTRY] *See* diastereoisomer. { ¦dī·ə¦ster·ē'ō·mər }

diastereotopic ligand [ORGANIC CHEMISTRY] A ligand whose replacement or addition gives rise to diastereomers. { ¦dī·ə‚ster·ē·ə'täp·ik 'līg·ənd }

diatomic [CHEMISTRY] Consisting of two atoms. { ¦dī·ə'täm·ik }

diazine [ORGANIC CHEMISTRY] **1.** A hydrocarbon consisting of an unsaturated hexatomic ring of two nitrogen atoms and four carbons. **2.** Suffix indicating a ring compound with two nitrogen atoms. { 'dī·ə‚zēn }

diazinon [ORGANIC CHEMISTRY] $C_{12}H_{21}N_2O_3PS$ A light amber to dark brown liquid with a boiling point of 83-84°C; used as an insecticide for soil and household pests, and as an insecticide and nematicide for fruits and vegetables. { dī'a·zə‚nōn }

diazoalkane [ORGANIC CHEMISTRY] A compound with the general formula $R_2C{=}N_2$ in which two hydrogen atoms of an alkane molecule have been replaced by a diazo group. { dī¦a·zō'al‚kān }

diazoamine [ORGANIC CHEMISTRY] The grouping —N=NNH—. Also known as azimino. { dī¦a·zō'a‚mēn }

diazoaminobenzene [ORGANIC CHEMISTRY] $C_6H_5NNNHC_6H_5$ Golden yellow scales with a melting point of 96°C; soluble in alcohol, ether, and benzene; used for dyes and insecticides. { dī¦a·zō¦am·ə·nō'ben‚zēn }

diazoate [ORGANIC CHEMISTRY] A salt with molecular formula of the type $C_6H_5N{=}NOOM$, where M is a nonvalent metal. { dī'a·zə‚wāt }

diazo compound [ORGANIC CHEMISTRY] An organic compound containing the radical —N=N—. { dī'a·zō 'käm‚paùnd }

diazo dye [CHEMISTRY] *See* developed dye. { dī'a·zō ‚dī }

diazo group [ORGANIC CHEMISTRY] A functional group with the formula $=N_2$. { dī'a·zō ‚grüp }

diazoic acid [ORGANIC CHEMISTRY] $C_6H_5N{=}NOOH$ An isomeric form of phenylnitramine. { 'dī·ə‚zō·ik 'as·əd }

diazole [ORGANIC CHEMISTRY] A cyclic hydrocarbon with five atoms in the ring, two of which are nitrogen atoms and three are carbon. { 'dī·ə‚zōl }

diazomethane [ORGANIC CHEMISTRY] CH_2N_2 A poisonous gas used in organic synthesis to methylate compounds. { dī¦a·zō'me‚thān }

diazonium [ORGANIC CHEMISTRY] The grouping $=N{\equiv}N$. { ‚dī·ə'zō·nē·əm }

diazonium salts [ORGANIC CHEMISTRY] Compounds of the type R·X·N:N, where R represents an alkyl or aryl group and X represents an anion such as a halide. { ˌdī·əˈzō·nē·əm ˈsȯls }

diazo oxide [ORGANIC CHEMISTRY] An organic molecule or a grouping of organic molecules that have a diazo group and an oxygen atom joined to ortho positions of an aromatic nucleus. Also known as diazophenol. { dīˈa·zō ˈäkˌsīd }

diazo process [ORGANIC CHEMISTRY] *See* diazotization. { dīˈa·zō ˌpräs·əs }

diazophenol [ORGANIC CHEMISTRY] *See* diazo oxide. { dīǀa·zōˈfeˌnȯl }

diazosulfonate [ORGANIC CHEMISTRY] A salt formed from diazosulfonic acid. { dīǀa·zō ˈsəl·fəˌnāt }

diazosulfonic acid [ORGANIC CHEMISTRY] $C_6H_5N{=}NSO_3H$ Any of a group of aromatic acids containing the diazo group bonded to the sulfonic acid group. { dīǀa·zō·səlˈfän·ik ˈas·əd }

diazotization [ORGANIC CHEMISTRY] Reaction between a primary aromatic amine and nitrous acid to give a diazo compound. Also known as diazo process. { dīˌaz·ət·əˈzā·shən }

dibasic [CHEMISTRY] **1.** Compounds containing two hydrogens that may be replaced by a monovalent metal or radical. **2.** An alcohol that has two hydroxyl groups, for example, ethylene glycol. { dīˈbās·ik }

dibasic acid [CHEMISTRY] An acid having two hydrogen atoms capable of replacement by two basic atoms or radicals. { dīˈbās·ik ˈas·əd }

dibasic calcium phosphate [INORGANIC CHEMISTRY] *See* calcium phosphate. { dīˈbās·ik ˈkal·sē·əm ˈfäsˌfāt }

dibasic magnesium citrate [ORGANIC CHEMISTRY] $MgHC_6H_5O_7 \cdot 5H_2O$ A white or yellowish powder soluble in water; used as a dietary supplement or in medicine. { dīˈbās·ik magˈnē·zē·əm ˈsīˌtrāt }

dibenzyl [ORGANIC CHEMISTRY] *See* bibenzyl. { dīˈben·zil }

dibenzyl disulfide [ORGANIC CHEMISTRY] $C_6H_5CH_2SSCH_2C_6H_5$ A compound crystallizing in leaflets with a melting point of 71-72°C; soluble in hot methanol, benzene, ether, and hot ethanol; used as an antioxidant in compounding of rubber and as an additive to silicone oils. { dīˈben·zil dīˈsəlˌfīd }

dibenzyl ether [ORGANIC CHEMISTRY] *See* benzyl ether. { dīˈben·zil ˈē·thər }

diborane [INORGANIC CHEMISTRY] B_2H_6 A colorless, volatile compound that is soluble in ether; boiling point −92.5°C, melting point −165.5°C; can be used to produce pentaborane and decaborane, proposed for use as rocket fuels; also used to synthesize organic boron compounds. { dīˈbȯrˌan }

dibromide [CHEMISTRY] Indicating the presence of two bromine atoms in a molecule. { dīˈbrōˌmīd }

dibromo- [CHEMISTRY] A prefix indicating two bromine atoms. { dīˈbrō·mō }

dibromochloropropane [ORGANIC CHEMISTRY] $C_3H_5Br_2Cl$ A light yellow liquid with a boiling point of 195°C; used as a nematicide for crops. Abbreviated DBCP. { dīǀbrō·mōˌklȯr·əˈprōˌpān }

dibromodifluoromethane [ORGANIC CHEMISTRY] CF_2Br_2 A colorless, heavy liquid with a boiling point of 24.5°C; soluble in methanol and ether; used in the synthesis of dyes and pharmaceuticals and as a fire-extinguishing agent. { dīǀbrō·mō·dīǀflu̇r·ōˈmeˌthān }

dibromomethane [ORGANIC CHEMISTRY] *See* methylene bromide. { dīǀbrō·mōˈmeˌthān }

2,6-dibromoquinone-4-chlorimide [ORGANIC CHEMISTRY] $C_6H_2Br_2ClNO$ Yellow prisms, soluble in water; used as a reagent for phenol and phosphatases. { ǀtü ǀsiks dīǀbrō·mō·kwəˈnōn ǀfȯr ˈklȯr·əˌmīd }

3,5-dibromosalicylaldehyde [ORGANIC CHEMISTRY] $Br_2C_6H_2(OH)CHO$ Pale yellow crystals with a melting point of 86°C; readily soluble in ether, chloroform, benzene, alcohol, and glacial acetic acid; used as an antibacterial agent. { ǀthrē ǀfīv dīǀbrō·mōǀsal·ə·səlˈal·dəˌhīd }

dibucaine [ORGANIC CHEMISTRY] $C_{20}H_{29}O_2N_3$ A local anesthetic used both as the base and the hydrochloride salt. { ˈdī·byəˌkān }

dibutyl [ORGANIC CHEMISTRY] Indicating the presence of two butyl groupings bonded through a third atom or group in a molecule. { dī'byüd·əl }

dibutyl amine [ORGANIC CHEMISTRY] $C_8H_{19}N$ A colorless, clear liquid with amine aroma; either di-*n*-butylamine, $(C_4H_9)_2NH$, boiling at 160°C, insoluble in water, soluble in hydrocarbon solvents, or di-*sec*-butylamine, $(CH_3CHCH_2CH_3)_2NH$, boiling at 133°C, flammable; used in the manufacture of dyes. { dī'byüd·əl 'a,mēn }

dibutyl maleate [ORGANIC CHEMISTRY] $C_4H_9OOCCHCHCOOC_4H_9$ Oily liquid used for copolymers and plasticizers and as a chemical intermediate. { dī'byüd·əl 'mal·ē,āt }

dibutyl oxalate [ORGANIC CHEMISTRY] $(COOC_4H_9)_2$ High-boiling, water-white liquid with mild odor, used as a solvent and in organic synthesis. { dī'byüd·əl 'äk·sə,lāt }

3,4-di-(*tert*-butyl)-phenyl-*N*-methylcarbamate [ORGANIC CHEMISTRY] $C_{16}H_{25}O_2N$ A white, crystalline compound that melts at 102-103°C; used as an insecticide for sheep blowflies. { ¦thrē ¦fór dī tərt ¦byüd·əl ¦fen·əl ¦en ,meth·əl'kär·bə,māt }

dibutyl phthalate [ORGANIC CHEMISTRY] $C_{16}H_{22}O_4$ A colorless liquid, used as a plasticizer and insect repellent. { dī'byüd·əl 'tha,lāt }

dibutyl succinate [ORGANIC CHEMISTRY] $C_{12}H_{22}O_4$ A colorless liquid, insoluble in water; used as a repellent for cattle flies, cockroaches, and ants around barns. { dī'byüd·əl 'sək·sə,nāt }

dibutyl tartrate [ORGANIC CHEMISTRY] $(COOC_4H_9)_2(CHOH)_2$ Liquid used as a solvent and plasticizer for cellulosics and as a lubricant. { dī'byüd·əl 'tär,trāt }

2,6-di-*tert*-butyl-*para*-tolylmethylcarbamate [ORGANIC CHEMISTRY] $C_{17}H_{27}O_2N$ A colorless solid with a melting point of 200-201°C; solubility in water is 6-7 parts per million; used as a preemergence herbicide on established turf. { ¦tü ¦siks dī tərt ¦byüd·əl ,par·ə ¦tä,lil,meth·əl'kär·bə,māt }

dicalcium [CHEMISTRY] A molecule containing two atoms of calcium. { dī'kal·sē·əm }

dicalcium orthophosphate [INORGANIC CHEMISTRY] *See* calcium phosphate. { dī'kal·sē·əm ,ȯr·thō'fäs,fāt }

dicalcium phosphate [INORGANIC CHEMISTRY] *See* calcium phosphate. { dī'kal·sē·əm 'fäs,fāt }

dicarbocyanine [ORGANIC CHEMISTRY] **1.** A member of a group of dyes termed the cyanine dyes; the structure consists of two heterocyclic rings joined to the five-carbon chain: =CH—CH=CH—CH=CH—. **2.** A particular dicarbocyanine dye containing two quinoline heterocyclic rings. { dī¦kär·bə¦sī·ə,nēn }

dicarboxylic acid [ORGANIC CHEMISTRY] A compound with two carboxyl groups. { dī¦kär·bäk¦sil·ik 'as·əd }

dication [CHEMISTRY] A doubly charged cation with the general formula X^{2+}. { dī'kat,ī·ən }

dichlobenil [ORGANIC CHEMISTRY] $C_7H_3Cl_2N$ A colorless, crystalline compound with a melting point of 139-145°C; used as a herbicide to control weeds in orchards and nurseries. { dī'klō·bə·nəl }

dichlofenthion [ORGANIC CHEMISTRY] $C_{10}H_{13}Cl_2O_3PS$ A white, liquid compound, insoluble in water; used as an insecticide and nematicide for ornamentals, flowers, and lawns. { dī,klō·fən'thī,än }

dichlofluanid [ORGANIC CHEMISTRY] $C_9H_{11}Cl_2FN_2O_2S$ A white powder with a melting point of 105-105.6°C; insoluble in water; used as a fungicide for fruits, garden crops, and ornamental flowers. { ,dī·klō·flü'an·əd }

dichlone [ORGANIC CHEMISTRY] $C_{10}H_4O_2Cl_2$ A yellow, crystalline compound, used as a fungicide for foliage and as an algicide. { 'dī,klōn }

dichloramine [INORGANIC CHEMISTRY] **1.** NH_2Cl_2 An unstable molecule considered to be formed from ammonia by action of chlorine. Also known as chlorimide. **2.** Any chloramine with two chlorine atoms joined to the nitrogen atom. { dī'klȯr·ə,mēn }

dichloride [CHEMISTRY] Any inorganic salt or organic compound that has two chloride atoms in its molecule. { dī'klȯr,īd }

dichloroacetic acid [ORGANIC CHEMISTRY] $CHCl_2COOH$ A strong liquid acid, formed by chlorinating acetic acid; used in organic synthesis. { dī¦klȯr·ō·ə¦sēd·ik 'as·əd }

3,6-dichloro-*ortho*-anisic acid [ORGANIC CHEMISTRY] $C_6H_2Cl_2OCH_3\,COOH$ A light tan, granular solid with a melting point of 114-116°C; used as a herbicide on roadways, crops, and rangelands. { ¦thrē ¦siks dī¦klȯr·ō ¦ȯr·thō ə'nis·ik 'as·əd }

dichlorobenzene [ORGANIC CHEMISTRY] $C_6H_4Cl_2$ Any of a group of substitution products of benzene and two atoms of chlorine; the three forms are *meta*-dichlorobenzene, colorless liquid boiling at 172°C, soluble in alcohol and ether, insoluble in water, or *ortho*-, colorless liquid boiling at 179°C, used as a solvent and chemical intermediate, or *para*-, volatile white crystals, insoluble in water, soluble in organic solvents, used as a germicide, insecticide, and chemical intermediate. { dī¦klȯr·ō'ben‚zēn }

1,4-dichlorobutane [ORGANIC CHEMISTRY] $Cl(CH_2)_4Cl$ A colorless, flammable liquid with a pleasant odor, boiling point 155°C, soluble in organic solvents; used in organic synthesis, including adiponitrile. Abbreviated DCB. { ¦wən ¦fȯr dī¦klȯr·ō'byü‚tān }

dichlorodiethylsulfide [ORGANIC CHEMISTRY] *See* mustard gas. { dī¦klȯr·ō·dī¦eth·əl'səl‚fīd }

dichlorodifluoromethane [ORGANIC CHEMISTRY] CCl_2F_2 A nontoxic, nonflammable, colorless gas made from carbon tetrachloride; boiling point −30°C; used as a refrigerant and as a propellant in aerosols. { dī¦klȯr·ō·dī¦flu̇r·ō'me‚th aman }

***p,p'*-dichlorodiphenylmethyl carbinol** [ORGANIC CHEMISTRY] $(ClC_6H_4)_2C(CH_3)OH$ Crystals with a melting point of 69-69.5°C; soluble in organic solvents; used as an insecticide. Abbreviated DMC. { ¦pē ¦pē‚prīm dī¦klȯr·ō·dī¦fen·əl¦meth·əl 'kär·bə‚nȯl }

dichlorodiphenyltrichloroethane [ORGANIC CHEMISTRY] *See* DDT. { dī¦klȯr·ō·dī¦fen·əl·trī¦klȯr·ō'e‚thān }

***sym*-dichloroethylene** [ORGANIC CHEMISTRY] CHClCHCl Colorless, toxic liquid with pleasant aroma, boiling at 59°C; decomposes in light, air, and moisture; soluble in organic solvents, insoluble in water; exists in cis and trans forms; used as solvent, in medicine, and for chemical synthesis. { ¦sim dī¦klȯr ō'eth·ə‚lēn }

dichloroethyl ether [ORGANIC CHEMISTRY] $ClCH_2CH_2OCH_2CH_2Cl$ A colorless liquid insoluble in water, soluble in organic solvents; used as a solvent in paints, varnishes, lacquers, and as a soil fumigant. { dī‚klȯr·ō'eth·əl'ē·thər }

dichlorofluoromethane [ORGANIC CHEMISTRY] $CHCl_2F$ A colorless, heavy gas with a boiling point of 8.9°C and a freezing point of −135°C; soluble in alcohol and ether; used in fire extinguishers and as a solvent, refrigerant, and aerosol propellant. Also known as fluorocarbon 21; fluorodichloromethane. { dī¦klȯr·ō¦flu̇r·ō'me‚thān }

***α*-dichlorohydrin** [ORGANIC CHEMISTRY] $CH_2ClCHOHCH_2Cl$ Unstable liquid, the commercial product consisting of a mixture of two isomers; used as a solvent and a chemical intermediate. Abbreviated GDCH. { ¦al·fə dī‚klȯr·ō'hī·drən }

2,6-dichloro-4-nitroaniline [ORGANIC CHEMISTRY] $C_6H_4Cl_2N_2O_2$ A yellow, crystalline compound that melts at 192-194°C; used as a fungicide for fruits, vegetables, and ornamental flowers. Abbreviated DCNA. { ¦tü ¦siks dī'klȯr·ō ¦fȯr ‚nī trō'an·ə‚lēn }

dichloropentane [ORGANIC CHEMISTRY] $C_5H_{10}Cl_2$ Mixed dichloro derivatives of normal pentane and isopentane; clear, light-yellow liquid used as solvent, paint and varnish remover, insecticide, and soil fumigant. { dī¦klȯr·ō'pen‚tān }

dichlorophen [ORGANIC CHEMISTRY] $C_{13}H_{10}Cl_2O_2$ A white, crystalline compound with a melting point of 177-178°C; used as an agricultural fungicide, germicide in soaps, and antihelminthic drug in humans. { dī'klȯr·ə·fən }

2,4-dichlorophenoxyacetic acid [ORGANIC CHEMISTRY] $Cl_2C_6H_3OCH_2COOH$ Yellow crystals, melting at 142°C; used as a herbicide and pesticide. Abbreviated 2,4-D. { ¦tü ¦fȯr dī¦klȯr·ō·fə¦näk·sē·ə'sēd·ik 'as·əd }

2,4-dichlorophenyl *para*-nitrophenyl ether [ORGANIC CHEMISTRY] $C_{12}H_7Cl_2NO_3$ A dark brown, crystalline compound with a melting point of 70-71°C; slightly soluble in water; used as a pre- and postemergence herbicide for vegetable crops and paddy rice. { ¦tü ¦fȯr dī¦klȯr·ō'fen·əl ¦par·ə ‚nī·trō'fen·əl 'ē·thər }

1,3-dichloro-2-propanol [ORGANIC CHEMISTRY] $ClCH_2CHOH CH_2Cl$ A liquid soluble in water and miscible with alcohol and ether; used as a solvent for nitrocellulose and hard resins, as a binder for watercolors, in the production of photographic lacquer, and in the determination of vitamin A. { ¦wən ¦thrē dī¦klȯr·ō ¦tü 'prō·pə‚nȯl }

dichlorotoluene [ORGANIC CHEMISTRY] $C_7H_6Cl_2$ A colorless liquid, soluble in organic solvents, insoluble in water; isomers are 2,4-$CH_3C_6H_3Cl_2$, boiling at 200-202°C, and 3,4-($CH_3C_6H_3Cl_2$), boiling at 209°C; used as solvent and chemical intermediate. { dī¦klȯr·ō'täl·yə‚wēn }

dichlorprop [ORGANIC CHEMISTRY] $C_9H_8Cl_2O_3$ A colorless, crystalline solid with a melting point of 117-118°C; used as a herbicide and fumigant for brush control on rangeland and rights-of-way. Abbreviated 2,4-DP. { dī'klȯr‚präp }

dichlorvos [ORGANIC CHEMISTRY] $C_4H_7O_4Cl_2P$ An amber liquid, used as an insecticide and miticide on public health pests, stored products, and flies on cattle. Abbreviated DDVP. { dī'klȯr‚väs }

dichromate [INORGANIC CHEMISTRY] A salt of dichromic acid, usually orange or red. { dī'krō‚māt }

dichromatic dye [CHEMISTRY] Dye or indicator in which different colors are seen, depending upon the thickness of the solution. { dī·krə'mäd·ik 'dī }

dichromic [CHEMISTRY] Pertaining to a molecule with two atoms of chromium. { dī'krō·mik }

dichromic acid [INORGANIC CHEMISTRY] $H_2Cr_2O_7$ An acid known only in solution, especially in the form of dichromates. { dī'krō·mik 'as·əd }

dicovalent carbon [ORGANIC CHEMISTRY] *See* divalent carbon. { ‚dī·kō'vā·lənt 'kär·bən }

dicrotophos [ORGANIC CHEMISTRY] $C_8H_{16}O_2P$ The dimethyl phosphate of 3-hydroxy-N,N-dimethyl-*cis*-crotonamide; a brown liquid with a boiling point of 400°C; miscible with water; used as an insecticide and miticide for cotton, soybeans, seeds, and ornamental flowers. { dī'kräd·ə‚fäs }

dicyandiamide [ORGANIC CHEMISTRY] $NH_2C(NH)(NHCN)$ White crystals with a melting range of 207-209°C; soluble in water and alcohol; used in fertilizers, explosives, oil well drilling muds, pharmaceuticals, and dyestuffs. Also known as cyanoguanidine. { dī‚sī·ən'dī·ə·məd }

dicyanide [CHEMISTRY] A salt that has two cyanide groups. { dī'sī·ə‚nīd }

dicyclohexylamine [ORGANIC CHEMISTRY] $(C_6H_{11})_2NH$ A clear, colorless liquid with a boiling point of 256°C; used for insecticides, corrosion inhibitors, antioxidants, and detergents, and as a plasticizer and catalyst. { dī¦sī·klō‚hek'sil·ə‚mēn }

dicyclohexylcarbodiimide [ORGANIC CHEMISTRY] $C_{13}H_{22}N_2$ Crystals with a melting point of 35-36°C; used in peptide synthesis. Abbreviated DCC; DCCI. { dī¦sī·klō¦hek·səl¦kär·bō'dī·ə·məd }

DIDA [ORGANIC CHEMISTRY] *See* diisodecyl adipate.

didodecyl ether [ORGANIC CHEMISTRY] *See* dilauryl ether. { dī'dō·də·səl 'ē·thər }

DIDP [ORGANIC CHEMISTRY] *See* diisodecyl phthalate.

didymium [CHEMISTRY] A mixture of the rare-earth elements praeseodymium and neodymium. Abbreviated Di. { dī'dim·ē·əm }

dieldrin [ORGANIC CHEMISTRY] $C_{12}H_8Cl_6O$ A white, crystalline contact insecticide obtained by oxidation of aldrin; used in mothproofing carpets and other furnishings. { 'dēl·drən }

dielectric vapor detector [ANALYTICAL CHEMISTRY] Apparatus to measure the change in the dielectric constant of gases or gas mixtures; used as a detector in gas chromatographs to sense changes in carrier gas. { 'dī·ə'lek·trik 'vā·pər di‚tek·tər }

dielectrophoresis [PHYSICAL CHEMISTRY] The ability of an uncharged material to move when subjected to an electric field. { ¦dī·ə‚lek·trō·fə'rē·səs }

Diels-Alder reaction [ORGANIC CHEMISTRY] The 1,4 addition of a conjugated diolefin to a compound, known as a dienophile, containing a double or triple bond; the dienophile may be activated by conjugation with a second double bond or with an electron acceptor. { ¦dēlz ¦äl·dər rē‚ak·shən }

diene [ORGANIC CHEMISTRY] One of a class of organic compounds containing two ethylenic linkages (carbon-to-carbon double bonds) in the molecules. Also known as alkadiene; diolefin. { 'dī‚ēn }

diene resin [ORGANIC CHEMISTRY] Material containing the diene group of double bonds that may polymerize. { 'dī‚ēn 'rez·ən }

diene value [ORGANIC CHEMISTRY] A number that represents the amount of conjugated bonds in a fatty acid or fat. { 'dī‚ēn ‚val·yü }

dienophile [ORGANIC CHEMISTRY] The alkene component of a reaction between an alkene and a diene. { dī'en·ə,fīl }

diester [ORGANIC CHEMISTRY] A compound containing two ester groupings. { ¦dī'es·tər }

diethanolamine [ORGANIC CHEMISTRY] $(HOCH_2CH_2)_2NH$ Colorless, water-soluble, deliquescent crystals, or liquid boiling at 217°C; soluble in alcohol and acetone, insoluble in ether and benzene; used in detergents, as an absorbent of acid gases, and as a chemical intermediate. Also known as DEA. { di·ə·thə'näl·ə,mēn }

diether [ORGANIC CHEMISTRY] A molecule that has two oxygen atoms with ether bonds. { dī'ē·thər }

1,1-diethoxyethane [ORGANIC CHEMISTRY] *See* acetal. { ¦wən ¦wən ,dī·ə¦thäk·sē'e,thān }

diethyl [ORGANIC CHEMISTRY] Pertaining to a molecule with two ethyl groups. { dī'eth·əl }

diethyl adipate [ORGANIC CHEMISTRY] $C_2H_5OCO(CH_2)_4OCOC_2H_5$ Water-insoluble, colorless liquid, boiling at 245°C; used as a plasticizer. { dī'eth·əl 'ad·ə,pāt }

diethylamine [ORGANIC CHEMISTRY] $(C_2H_5)_2NH$ Water-soluble, colorless liquid with ammonia aroma, boiling at 56°C; used in rubber chemicals and pharmaceuticals and as a solvent and flotation agent. { ,dī,eth·əl'a,mēn }

diethylaminoethyl cellulose [ORGANIC CHEMISTRY] A positively charged resin used in ion-exchange chromatography; an anion exchanger. Also known as DEAE-cellulose. { ,dī,eth·əl¦am·ə·nō'eth·əl 'sel·yə,lōs }

diethylbenzene [ORGANIC CHEMISTRY] $C_6H_4(C_2H_5)_2$ Colorless liquid, boiling at 180–185°C; soluble in organic solvents, insoluble in water; usually a mixture of three isomers, which are 1,2- (or *ortho*-diethylbenzene), boiling at 183°C, and 1,3- (or *meta*-), boiling at 181°C, and 1,4- (or *para*-), boiling at 184°C; used as a solvent. { ,dī,eth·əl'ben,zēn }

diethylcarbamazine [ORGANIC CHEMISTRY] $C_{16}H_{29}O_8N_3$ White, water-soluble, hygroscopic crystals, melting at 136°C; used as an anthelminthic. { ,dī,eth·əl,kär'bam·ə,zēn }

diethyl carbinol [ORGANIC CHEMISTRY] $(CH_3CH_2)_2CHOH$ Colorless, alcohol-soluble liquid, boiling at 116°C; slightly soluble in water; used in pharmaceuticals and as a solvent and flotation agent. Also known as sec-*n*-amyl alcohol. { ,dī'eth·əl 'kär·bə,nōl }

diethyl carbonate [ORGANIC CHEMISTRY] $(C_2H_5)_2CO_3$ Stable, colorless liquid with mild aroma, boiling at 126°C; soluble with most organic solvents; used as a solvent and for chemical synthesis. Also known as ethyl carbonate. { ,dī'eth·əl 'kär·bə,nāt }

diethylene glycol [ORGANIC CHEMISTRY] $CH_2OHCH_2OCH_2CH_2OH$ Clear, hygroscopic, water-soluble liquid, boiling at 245°C; soluble in many organic solvents; used as a softener, conditioner, lubricant, and solvent, and in antifreezes and cosmetics. { dī'eth·ə,lēn 'glī,kōl }

diethylene glycol monoethyl ether [ORGANIC CHEMISTRY] $C_6H_{14}O_3$ A hygroscopic liquid used as a solvent for cellulose esters and in lacquers, varnishes, and enamels. { dī'eth·ə,lēn 'glī,kōl ,män·ō'eth·əl 'ē·thər }

diethylenetriamine [ORGANIC CHEMISTRY] $(NH_2C_2H_4)_2NH$ A yellow, hygroscopic liquid with a boiling point of 206.7°C; soluble in water and hydrocarbons; used as a solvent, saponification agent, and fuel component. { dī¦eth·ə,lēn¦trī·ə,mēn }

diethyl ether [ORGANIC CHEMISTRY] $C_4H_{10}O$ A colorless liquid, slightly soluble in water; used as a reagent and solvent. Also known as ethyl ether; ethyl oxide; ethylic ether. { dī'eth·əl'ē·thər }

diethyl maleate [ORGANIC CHEMISTRY] $(HCCOOC_2H_5)_2$ Clear, colorless liquid, boiling at 225°C; slightly soluble in water, soluble in most organic solvents; used as a chemical intermediate. { ,dī'eth·əl 'mal·ē,āt }

diethyl phosphite [ORGANIC CHEMISTRY] $(C_2H_5O)_2HPO$ A colorless liquid with a boiling point of 138°C; soluble in water and common organic solvents; used as a paint solvent, antioxidant, and reducing agent. { dī'eth·əl 'fäs,fīt }

diethyl phthalate [ORGANIC CHEMISTRY] $C_6H_4(CO_2C_2H_5)_2$ Clear, colorless, odorless liquid with bitter taste, boiling at 298°C; soluble in alcohols, ketones, esters, and aromatic hydrocarbons, partly soluble in aliphatic solvents; used as a cellulosic solvent, wetting agent, alcohol denaturant, mosquito repellent, and in perfumes. { dī'eth·əl 'tha,lāt }

diethyl pyrocarbonate [ORGANIC CHEMISTRY] $C_6H_{10}O_5$ A viscous liquid, soluble in alcohols, esters, and ketones; used as a gentle esterifying agent, as a preservative for fruit juices, soft drinks, and wines, and as an inhibitor for ribonuclease. Abbreviated DEPC. { ,dī'eth·əl ,pī·rō'kär·bə,nāt }

diethyl succinate [ORGANIC CHEMISTRY] $(CH_2COOC_2H_5)_2$ Water-white liquid with pleasant aroma, boiling at 216°C; soluble in alcohol and ether, slightly soluble in water; used as a chemical intermediate and plasticizer. { ,dī'eth·əl 'sək·sə,nāt }

diethyl sulfate [ORGANIC CHEMISTRY] $(C_2H_5)_2SO_4$ A colorless oil with a peppermint odor, and boiling at 208°C; used as an intermediate in organic synthesis. Also known as ethyl sulfate. { ,dī'eth·əl 'səl,fāt }

diethyl sulfide [ORGANIC CHEMISTRY] *See* ethyl sulfide. { ,dī'eth·əl 'səl,fīd }

diethyltoluamide [ORGANIC CHEMISTRY] $C_{12}H_{17}ON$ A liquid whose color ranges from off-white to light yellow; used as an insect repellent for people and clothing. Also known as DEET; DET; N,N-diethyl-*meta*-toluamide. { ,dī¦eth·əl,täl·yü'a,mīd }

difference spectrophotometer [SPECTROSCOPY] *See* absorption spectrophotometer. { 'dif·rəns ,spek·trə·fə'täm·əd·ər }

differential aeration cell [PHYSICAL CHEMISTRY] An electrolytic cell whose electromotive force derives from a difference in concentration of atmospheric oxygen at one electrode with reference to another electrode of the same material. Also known as oxygen concentration cell. { ,dif·ə'ren·chəl e'rā·shən ,sel }

differential ebuliometer [ANALYTICAL CHEMISTRY] Apparatus for precise and simultaneous measurement of both the boiling temperature of a liquid and the condensation temperature of the vapors of the boiling liquid. { ,dif·ə'ren·chəl ə,bü·lē 'äm·əd· ər }

differential heat of dilution [PHYSICAL CHEMISTRY] *See* heat of dilution. { ,dif·ə'ren·chəl 'hēt əv də'lü·shən }

differential polarography [ANALYTICAL CHEMISTRY] Technique of polarographic analysis which measures the difference in current flowing between two identical dropping-mercury electrodes at the same potential but in different solutions. { ,dif·ə'ren·chəl ,pō·lə'räg·rə·fē }

differential reaction rate [PHYSICAL CHEMISTRY] The order of a chemical reaction expressed as a differential equation with respect to time; for example, $dx/dt = k(a - x)$ for first order, $dx/dt = k(a - x)(b - x)$ for second order, and so on, where k is the specific rate constant, a is the concentration of reactant A, b is the concentration of reactant B, and dx/dt is the rate of change in concentration for time t. { ,dif·ə¦ren·chəl rē'ak·shən ,rāt }

differential spectrophotometry [SPECTROSCOPY] Spectrophotometric analysis of a sample when a solution of the major component of the sample is placed in the reference cell; the recorded spectrum represents the difference between the sample and the reference cell. { ,dif·ə'ren·chəl ,spek·trō·fə'täm·ə·trē }

differential thermometric titration [ANALYTICAL CHEMISTRY] Thermometric titration in which titrant is added simultaneously to the reaction mixture and to a blank in identically equipped cells. { ,dif·ə'ren·chəl ¦thər·mə¦me·trik tī'trā·shən }

diffraction grating [SPECTROSCOPY] An optical device consisting of an assembly of narrow slits or grooves which produce a large number of beams that can interfere to produce spectra. Also known as grating. { di'frak·shən ,grād·iŋ }

diffraction spectrum [SPECTROSCOPY] Parallel light and dark or colored bands of light produced by diffraction. { di'frak·shən ,spek·trəm }

diffuse series [SPECTROSCOPY] A series occurring in the spectra of many atoms having one, two, or three electrons in the outer shell, in which the total orbital angular momentum quantum number changes from 2 to 1. { də'fyüs 'sir·ēz }

diffuse spectrum [SPECTROSCOPY] Any spectrum having lines which are very broad even when there is no possibility of line broadening by collisions. { də'fyüs 'spek·trəm }

diffusion current [ANALYTICAL CHEMISTRY] In polarography with a dropping-mercury electrode, the flow that is controlled by the rate of diffusion of the active solution species across the concentration gradient produced by the removal of ions or molecules at the electrode surface. { də'fyü·zhən ,kər·ənt }

diffusion flame [CHEMISTRY] A long gas flame that radiates uniformly over its length and precipitates free carbon uniformly. { də'fyü·zhən ,flām }

diffusion potential [PHYSICAL CHEMISTRY] A potential difference across the boundary between electrolytic solutions with different compositions. Also known as liquid junction potential. { də'fyü·zhən pə,ten·chəl }

diffusivity analysis [ANALYTICAL CHEMISTRY] Analysis of difficult-to-separate materials in solution by diffusion effects, using, for example, dialysis, electrodialysis, interferometry, amperometric titration, polarography, or voltammetry. { dif·yü'ziv·əd·ē ə'nal·ə·səs }

difunctional molecule [ORGANIC CHEMISTRY] An organic structure possessing two sites that are highly reactive. { ,dī¦fəŋk·shən·əl 'mäl·ə,kyül }

digallic acid [ORGANIC CHEMISTRY] *See* tannic acid. { dī'gal·ik 'as·əd }

digitoxigenin [ORGANIC CHEMISTRY] $C_{23}H_{34}O_4$ The steroid aglycone formed by removal of three molecules of the sugar digitoxose from digitoxin. { ,dij·ə,täk·sə'jen·ən }

digitoxin [ORGANIC CHEMISTRY] $C_{41}H_{64}O_{13}$ A poisonous steroid glycoside found as the most active principle of digitalis, from the foxglove leaf. { ,dij·ə'täk·sən }

diglycerol [ORGANIC CHEMISTRY] A compound that is a diester of glycerol. { dī'glis·ə,rōl }

diglycine [ORGANIC CHEMISTRY] *See* iminodiacetic acid. { dī'glī,sēn }

diglycolic acid [ORGANIC CHEMISTRY] $O(CH_2COOH)_2$ A white powder that forms a monohydrate; used in the manufacture of plasticizers and in organic synthesis and to break emulsions. { dī·glī'käl·ik 'as·əd }

diglycol laurate [ORGANIC CHEMISTRY] $C_{11}H_{23}COOC_2H_4OC_2H_4OH$ A light, straw-colored, oily liquid; soluble in methanol, ethanol, toluene, and mineral oil; used in emulsions and as an antifoaming agent. { dī'glī,kól 'lór,āt }

diglycol stearate [ORGANIC CHEMISTRY] $(C_{17}H_{35}COOC_2H_4)_2O$ A white, waxy solid with a melting point of 54-55°C; used as an emulsifying agent, suspending medium for powders in the manufacture of polishes, and thickening agent, and in pharmaceuticals. { dī'glī,kól 'stir,āt }

digoxin [ORGANIC CHEMISTRY] $C_{41}H_{64}O_{14}$ A crystalline steroid obtained from a foxglove leaf (*Digitalis lanata*); similar to digitalis in pharmacological effects. { dī'gäk·sən }

dihalide [CHEMISTRY] A molecule containing two atoms of halogen combined with a radical or element. { dī'ha,līd }

dihexy [ORGANIC CHEMISTRY] *See* dodecane. { dī'hek·sē }

dihydrate [CHEMISTRY] A compound with two molecules of water of hydration. { dī'hī,drāt }

dihydrazone [ORGANIC CHEMISTRY] A molecule containing two hydrazone radicals. { dī'hī·drə,zōn }

dihydro- [CHEMISTRY] A prefix indicating combination with two atoms of hydrogen. { dī¦hī drō }

dihydrochloride [CHEMISTRY] A compound containing two molecules of hydrochloric acid. { dī¦hī·drə'klór,īd }

dihydroxy [CHEMISTRY] A molecule containing two hydroxyl groups. { ¦dī,hī¦dräk·sē }

dihydroxyacetone [ORGANIC CHEMISTRY] $(HOCH_2)_2CO$ A colorless, crystalline solid with a melting point of 80°C; soluble in water and alcohol; used in medicine, fungicides, plasticizers, and cosmetics. Abbreviated DHA. { ¦dī,hī¦dräk·sē'as·ə,tōn }

2,4′-dihydroxyacetophenone [ORGANIC CHEMISTRY] $(HO)_2C_6H_3COCH_3$ Needlelike or leafletlike crystals with a melting point of 145-147°C; soluble in pyridine, warm alcohol, and glacial acetic acid; used as a reagent for the determination of iron. { ¦tü ¦fór,prīm ¦dī,hī¦dräk·sē¦as·ə,tä·fə'nōn }

dihydroxy alcohol [ORGANIC CHEMISTRY] *See* glycol. { ¦dī,hī¦dräk·sē 'al·kə,hól }

1,8-dihydroxyanthraquinone [ORGANIC CHEMISTRY] $C_{14}H_8O_4$ Orange, needlelike crystals that dissolve in glacial acetic acid; used as an intermediate in the commercial preparation of indanthrene and alizarin dyestuffs. Also known as chrysazin. { ¦wən ¦āt ¦dī,hī¦dräk·sē,an·thrə·kwə′nōn }

2,2′-dihydroxy-4,4′-dimethoxybenzophenone [ORGANIC CHEMISTRY] $[CH_3OC_6H_3(OH)]_2CO$ Crystals with a melting point of 139-140°C; used in paint and plastics as a light absorber. { ¦tü ¦tü,prīm ¦dī,hī¦dräk·sē ¦fȯr ¦fȯr,prīm ,dī·mə¦thäk·sē¦ben·zō·fə′nōn }

dihydroxymaleic acid [ORGANIC CHEMISTRY] $C_4H_4O_6$ Crystals soluble in alcohol; used in the detection of titanium and fluorides. { ¦dī,hī¦dräk·sē,mə′lā·ik ′as·əd }

3,5-diiodosalicylic acid [ORGANIC CHEMISTRY] $C_7H_4I_2O_3$ Crystals with a sweetish, bitter taste and a melting point of 235-236°C; soluble in most organic solvents; used as a source of iodine in foods and a growth promoter in poultry, hog, and cattle feeds. { ¦thrē ¦fīv dī¦ī·ə,dō,sal·ə′sil·ik ′as·əd }

diiodomethane [ORGANIC CHEMISTRY] *See* methylene iodide. { dī¦ī·ə,dō′me,thān }

diisobutylene [ORGANIC CHEMISTRY] C_8H_{16} Any one of a number of isomers, but most often 2,4,4-trimethylpentene-1 and 2,4,4-trimethylpentene-2; used in alkylation and as a chemical intermediate. { dī¦ī,sō′byüd·əl,ēn }

diisobutyl ketone [ORGANIC CHEMISTRY] $(CH_3)_2CHCH_2COCH_2CH(CH_3)_2$ Stable liquid, boiling at 168°C; soluble in most organic liquids; toxic and flammable; used as a solvent, in lacquers and coatings, and as a chemical intermediate. { dī¦ī,sō′byüd·əl ′kē,tōn }

diisocyanate [ORGANIC CHEMISTRY] A compound that contains two NCO (isocyanate) groups; used to produce polyurethane foams, resins, and rubber. { dī¦ī,sō′sī·ə,nāt }

diisodecyl adipate [ORGANIC CHEMISTRY] $(C_{10}H_{21}OOC)_2(CH_2)_4$ A light-colored, oily liquid with a boiling range of 239-246°C; used as a primary plasticizer for polymers. Abbreviated DIDA. { dī¦ī,sō′de,səl ′ad·ə,pāt }

diisodecyl phthalate [ORGANIC CHEMISTRY] $C_6H_4(COOC_{10}H_{21})_2$ A clear liquid with a boiling point of 250-257°C; used as a plasticizer. Abbreviated DIDP. { dī¦ī,sō′de,səl ′tha,lāt }

diisopropanolamine [ORGANIC CHEMISTRY] $(CH_3CHOHCH_2)_2NH$ A white, crystalline solid with a boiling point of 248.7°C; used as an emulsifying agent for polishes, insecticides, and water paints. Abbreviated DIPA. { dī¦ī,sō,prō·pə′näl·ə,mēn }

diisopropyl [ORGANIC CHEMISTRY] **1.** A molecule containing two isopropyl groups. **2.** *See* 2,3-dimethylbutane. { dī¦ī,sō′prō·pəl }

diisopropyl ether [ORGANIC CHEMISTRY] *See* isopropyl ether. { dī¦ī,sō′prō·pəl ′ē,thər }

diketene [ORGANIC CHEMISTRY] $CH_3COCHCO$ A colorless, readily polymerized liquid with pungent aroma; insoluble in water, soluble in organic solvents; used as a chemical intermediate. { dī′kē,tēn }

diketone [ORGANIC CHEMISTRY] A molecule containing two ketone carbonyl groups. { dī′kē,tōn }

diketopiperazine [ORGANIC CHEMISTRY] **1.** $C_4H_6N_2O_2$ A compound formed by dehydration of two molecules of glycine. **2.** Any of the cyclic molecules formed from α-amino acids other than glycine or by partial hydrolysis of protein. { dī¦kē·dō·pi′per·ə,zēn }

dilactone [ORGANIC CHEMISTRY] A molecule that contains two lactone groups. { dī′lāk·tōn }

dilatancy [CHEMISTRY] The property of a viscous suspension which sets solid under the influence of pressure. { dī′lāt·ən·sē }

dilatant [CHEMISTRY] A material with the ability to increase in volume when its shape is changed. { dī′lāt·ənt }

dilauryl ether [ORGANIC CHEMISTRY] $(C_{12}H_{25})_2NH$ A liquid with a boiling point of 190-195°C; used for electrical insulators, water repellents, and antistatic agents. Also known as didodecyl ether. { dī′lȯr·əl ′ē·thər }

dilauryl thiodipropionate [ORGANIC CHEMISTRY] $(C_{12}H_{25}OOCCH_2CH_2)_2S$ White flakes with a melting point of 40°C; soluble in most organic solvents; used as an antioxidant, plasticizer, and preservative, and in food wraps and edible fats and oils. { dī′lȯr·əl ,thī·ō,dī′prō·pē·ə,nāt }

dilinoleic acid [ORGANIC CHEMISTRY] $C_34H_{62}(COOH)_2$ A light yellow, viscous liquid used as an emulsifying agent and shellac substitute. { dī¦lin·ə¦lā·ik 'as·əd }

diluent [CHEMISTRY] An inert substance added to some other substance or solution so that the volume of the latter substance is increased and its concentration per unit volume is decreased. { 'dil·yə·wənt }

dilute [CHEMISTRY] To make less concentrated. { dī'lüt }

dilution [CHEMISTRY] Increasing the proportion of solvent to solute in any solution and thereby decreasing the concentration of the solute per unit volume. { də'lü·shən }

dimedone [ORGANIC CHEMISTRY] *See* 5,5-dimethyl-1,3-cyclohexanedione. { 'dī·mə ˌdōn }

dimer [CHEMISTRY] A molecule that results from a chemical combination of two entities of the same species, for example, the chlorine molecule (Cl_2) or cyanogen (NCCN). { 'dī·mər }

dimeric water [INORGANIC CHEMISTRY] Water in which pairs of molecules are joined by hydrogen bonds. { dī'mer·ik 'wȯd·ər }

dimerization [CHEMISTRY] A chemical reaction in which two identical molecular entities react to form a single dimer. { ˌdī·mər·ə'zā·shən }

dimetan [ORGANIC CHEMISTRY] The generic name for 5,5-dimethyldehydroresorcinol dimethylcarbamate, a synthetic carbamate insecticide. { 'dī·məˌtan }

dimethachlon [ORGANIC CHEMISTRY] $C_{10}H_7Cl_2NO_2$ A yellowish, crystalline solid with a melting point of 136.5-138°C; insoluble in water; used as a fungicide. { dī·mə'thaˌklän }

dimethoate [ORGANIC CHEMISTRY] $C_5H_{12}NO_3PS_2$ A crystalline compound, soluble in most organic solvents; used as an insecticide. { dī'meth·əˌwāt }

dimethrin [ORGANIC CHEMISTRY] $C_{19}H_{28}O_2$ An amber liquid with a boiling point of 175°C; soluble in petroleum hydrocarbons, alcohols, and methylene chloride; used as an insecticide for mosquitoes, body lice, stable flies, and cattle flies. { dī'me·thrən }

dimethyl [ORGANIC CHEMISTRY] A compound that has two methyl groups. { dī 'meth·əl }

dimethylamine [ORGANIC CHEMISTRY] $(CH_3)_2NH$ Flammable gas with ammonia aroma, boiling at 7°C; soluble in water, ether, and alcohol; used as an acid-gas absorbent, solvent, and flotation agent, in pharmaceuticals and electroplating, and in dehairing hides. { ˌdī¦meth·əl'amˌēn }

***para*-dimethylaminobenzalrhodanine** [ORGANIC CHEMISTRY] $C_{12}H_{12}N_2OS_2$ Deep red, needlelike crystals that decompose at 270°C; soluble in strong acids; used in acetone solution for the detection of ions such as silver, mercury, copper, gold, palladium, and platinum. { ¦par·ə ˌdī¦meth·əl¦am·əˌnōˌben·zəl'rō·dəˌnēn }

2-dimethylaminoethanol [ORGANIC CHEMISTRY] $(CH_3)_2NCH_2CH_2OH$ A colorless liquid with a boiling point of 134.6°C; used for the synthesis of dyestuffs, pharmaceuticals, and corrosion inhibitors, in medicine, and as an emulsifier. { ¦tü ˌdī¦meth·əl¦am·əˌnō'eth·əˌnȯl }

***N,N*-dimethylaniline** [ORGANIC CHEMISTRY] $C_6H_5N(CH_3)_2$ A yellowish liquid slightly soluble in water; used in dyes and solvent and in the manufacture of vanillin. Also known as aniline N,N-dimethyl. { ¦en ¦en ˌdī¦meth·əl'an·əˌlēn }

dimethylbenzene [ORGANIC CHEMISTRY] *See* xylene. { ˌdi¦meth·əl'benˌzēn }

2,3-dimethylbutane [ORGANIC CHEMISTRY] $(CH_3)_2CHCH(CH_3)_2$ A colorless liquid with a boiling point of 57.9°C; used as a high-octane fuel. Also known as diisopropyl. { ¦tü ¦thrē ˌdī¦meth·əl'byüˌtān }

dimethyl carbate [ORGANIC CHEMISTRY] $C_{11}H_{14}O_4$ A colorless liquid with a boiling point of 114-115°C; used as an insect repellent. { ˌdī'meth·əl 'kärˌbāt }

5,5-dimethyl-1,3-cyclohexanedione [ORGANIC CHEMISTRY] $C_8H_{12}O_2$ Crystals that decompose at 148-150°C; soluble in water and inorganic solvents such as methanol and ethanol, used as a reagent for the identification of aldehydes. Also known as dimedone. { ¦fīv ¦fīv ˌdī¦meth·əl ¦wən ¦thrē ˌsī·klō¦hekˌsān 'dīˌōn }

dimethyl diaminophenazine chloride [ORGANIC CHEMISTRY] *See* neutral red. { ˌdī'meth·əl dī¦am·əˌnō'fen·əˌzēn 'klȯrˌīd }

2,2-dimethyl-1,3-dioxolane-4-methanol [ORGANIC CHEMISTRY] $C_6H_{12}O_3$ The acetone ketal of glycerin; a liquid miscible with water and many organic solvents; used as a plasticizer and a solvent. { ¦tü ¦tü ‚dī¦meth·əl ¦wən ¦thrē dī¦äk·sə‚lān ¦fór 'meth·ə‚nól }

dimethyl ether [ORGANIC CHEMISTRY] CH_3OCH_3 A flammable, colorless liquid, boiling at −25°C; soluble in water and alcohol; used as a solvent, extractant, reaction medium, and refrigerant. Also known as methyl ether; wood ether. { ‚dī'meth·əl 'ē·thər }

***N,N*-dimethylformamide** [ORGANIC CHEMISTRY] $HCON(CH_3)_2$ A liquid that boils at 152.8°C; extensively used as a solvent for organic compounds. Abbreviated DMF. { ¦en ¦en ‚dī¦meth·əl'fór·mə‚mīd }

dimethylglyoxime [ORGANIC CHEMISTRY] $(CH_3)_2C_2(NOH)_2$ White, crystalline or powdered solid, used in analytical chemistry as a reagent for nickel. { ‚dī¦meth·əl·glī'äk‚sīm }

***uns*-dimethylhydrazine** [ORGANIC CHEMISTRY] $(CH_3)_2NNH_2$ A flammable, highly toxic, colorless liquid; used as a component of rocket and jet fuels and as a stabilizer for organic peroxide fuel additives. Abbreviated UDMH. { ¦əns dī¦meth·əl'hī·drə‚zēn }

dimethylisopropanolamine [ORGANIC CHEMISTRY] $(CH_3)_2NCN_2CH(OH)CH_3$ A colorless liquid with a boiling point of 125.8°C; soluble in water; used in methadone synthesis. { ‚dī¦meth·əl‚ī·sə‚prō·pə'näl·ə‚mēn }

dimethylolurea [ORGANIC CHEMISTRY] $CO(NHCH_2OH)_2$ Colorless crystals melting at 126°C, soluble in water; used to increase fire resistance and hardness of wood, and in textiles to prevent wrinkles. Also known as 1,3-bis-hydroxymethylurea; DMU. { ‚dī'meth·əl‚ól·yü'rē·ə }

dimethyl phthalate [ORGANIC CHEMISTRY] $C_6H_4(COOCH_3)_2$ Odorless, colorless liquid, boiling at 282°C; soluble in organic solvents, slightly soluble in water; used as a plasticizer, in resins, lacquers, and perfumes, and as an insect repellent. { ‚dī'meth·əl 'tha‚lāt }

dimethyl sebacate [ORGANIC CHEMISTRY] $[(CH_2)_4COOCH_3]_2$ Clear, colorless liquid, boiling at 294°C; used as a vinyl resin, nitrocellulose solvent, or plasticizer. { ‚dī'meth·əl 'seb·ə‚kāt }

dimethyl sulfate [ORGANIC CHEMISTRY] $(CH_3)_2SO_4$ Poisonous, corrosive, colorless liquid, boiling at 188°C; slightly soluble in water, soluble in ether and alcohol; used to methylate amines and phenols. Also known as methyl sulfate. { ‚dī'meth·əl 'səl‚fāt }

dimethyl sulfide [ORGANIC CHEMISTRY] *See* methyl sulfide. { ‚dī'meth·əl 'səl‚fīd }

2,4-dimethylsulfolane [ORGANIC CHEMISTRY] $C_6H_{12}O_2S$ A yellow to colorless liquid miscible with lower aromatic hydrocarbons; used as a solvent in liquid-liquid and vapor-liquid extraction processes. { ¦tü ¦fór ‚dī¦meth·əl'səl·fə‚lān }

dimethyl sulfoxide [ORGANIC CHEMISTRY] $(CH_3)_2SO$ A colorless liquid used as a local analgesic and anti-inflammatory agent, as a solvent in industry, and in laboratories as a medium for carrying out chemical reactions. Abbreviated DMSO. { ‚dī'meth·əl səl'fäk‚sīd }

dimethyl terephthalate [ORGANIC CHEMISTRY] $C_6H_4(COOCH_3)_2$ Colorless crystals, melting at 140°C and subliming above 300°; slightly soluble in water, soluble in hot alcohol and ether; used to make polyester fibers and film. Abbreviated DMT. { ‚dī'meth·əl ‚ter·ə'tha‚lāt }

dimethyl-2,3,5,6-tetrachloroterephthalate [ORGANIC CHEMISTRY] $C_{10}H_6Cl_4O_4$ A colorless, crystalline compound with a melting point of 156°C; used as an herbicide for turf, ornamental flowers, and certain vegetables and berries. Abbreviated DCPA. { ‚dī'meth·əl ¦tü ¦thrē ¦fīv ¦siks ‚te·trə·klór·ō‚ter·ə'tha‚lāt }

dimorphism [CHEMISTRY] Having crystallization in two forms with the same chemical composition. { dī'mór‚fiz·əm }

dineric [PHYSICAL CHEMISTRY] **1.** Having two liquid phases. **2.** Pertaining to the interface between two liquids. { dī'ner·ik }

dinitramine [ORGANIC CHEMISTRY] $C_{11}H_{13}N_3O_4F_3$ A yellow solid with a melting point of 98-99°C; used as a preemergence herbicide for annual grass and broadleaf weeds in cotton and soybeans. { dī'nī·trə‚mēn }

dinitrate [CHEMISTRY] A molecule that contains two nitrate groups. { dī'nī,trāt }
dinitrite [CHEMISTRY] A molecule that has two nitrite groups. { dī'nī,trīt }
2,4-dinitroaniline [ORGANIC CHEMISTRY] $(NO_2)C_6H_3NH_2$ A compound which crystallizes as yellow needles or greenish-yellow plates, melting at 187.5-188°C; soluble in alcohol; used in the manufacture of azo dyes. { ¦tü ¦fȯr dī¦nī·trō'an·ə,lēn }
2,4-dinitrobenzaldehyde [ORGANIC CHEMISTRY] $(NO_2)_2C_6H_3CHO$ Yellow to light brown crystals with a melting point of 72°C; soluble in alcohol, ether, and benzene; used to make Schiff bases. { ¦tü ¦fȯr dī¦nī·tro,ben'zal·də,hīd }
dinitrobenzene [ORGANIC CHEMISTRY] Any one of three isomeric substitution products of benzene having the empirical formula $C_6H_4(NO_2)_2$. { dī,nī·trō'ben,zēn }
2,4-dinitrobenzenesulfenyl chloride [ORGANIC CHEMISTRY] $(NO_2)_2C_6H_3SCl$ Crystals soluble in glacial acetic acid, with a melting point of 96°C; used as a reagent for separation and identification of naturally occurring indoles. { ¦tü ¦fōr dī¦nī·trō,ben,zēn'səl·fə,nil 'klȯr,īd }
3,4-dinitrobenzoic acid [ORGANIC CHEMISTRY] $C_7H_4N_2O_6$ Crystals with a bitter taste and a melting point of 166°C; used in quantitative sugar analysis. { ¦thrē ¦fȯr dī¦nī·trō·ben'zō·ik 'as·əd }
dinitrogen [CHEMISTRY] N_2 The diatomic molecule of nitrogen. { dī'nī·trə·jən }
dinitrogen tetroxide [INORGANIC CHEMISTRY] *See* nitrogen dioxide. { dī'nī·trə·jən te'träk,sīd }
dinitrophenol [ORGANIC CHEMISTRY] Any one of six isomeric substituent products of benzene having the empirical formula $(NO_2)_2C_6H_3OH$. { dī¦nī·trō'fē,nȯl }
2,4-dinitrophenylhydrazine [ORGANIC CHEMISTRY] $(NO_2)_2C_6H_3NHNH_2$ A red, crystalline powder with a melting point of approximately 200°C; soluble in dilute inorganic acids; used as a reagent for determination of ketones and aldehydes. { ¦tü ¦fȯr dī¦nī·trō,fen·əl'hī·drə,zēn }
dinitrotoluene [ORGANIC CHEMISTRY] Any one of six isomeric substitution products of benzene having the empirical formula $CH_3C_6H_3(NO_2)_2$; they are high explosives formed by nitration of toluene. Abbreviated DNT. { dī¦nī·trō'täl·yə,wēn }
dinoseb [ORGANIC CHEMISTRY] $C_{10}H_{12}O_5N_2$ A reddish-brown liquid with a melting point of 32°C; used as an insecticide and herbicide for numerous crops and in fruit and nut orchards. { 'dī·nə,seb }
dinoterb acetate [ORGANIC CHEMISTRY] $C_{12}H_{14}N_2O_6$ A yellow, crystalline compound with a melting point of 133-134°C; used as a preemergence herbicide for sugarbeets, legumes, and cereals, and as a postemergence herbicide for maize, sorghum, and alfalfa. { 'dī·nə,tərb 'as·ə,tāt }
dioctyl [ORGANIC CHEMISTRY] A compound that has two octyl groups. { dī'äkt·əl }
dioctyl phthalate [ORGANIC CHEMISTRY] $(C_8H_{17}OOC)_2C_6H_4$ Pale, viscous liquid, boiling at 384°C; insoluble in water; used as a plasticizer for acrylate, vinyl, and cellulosic resins, and as a miticide in orchards. Abbreviated DOP. { dī¦äkt·əl 'tha,lāt }
dioctyl sebacate [ORGANIC CHEMISTRY] $(CH_2)_8(COOC_8H_{17})_2$ Water-insoluble, straw-colored liquid, boiling at 248°C; used as a plasticizer for vinyl, cellulosic, and styrene resins. { dī'äkt·əl 'seb·ə,kāt }
diodide [CHEMISTRY] A molecule that contains two iodine atoms bonded to an element or radical. { 'dī·ə,dīd }
diolefin [ORGANIC CHEMISTRY] *See* diene. { dī'ō lə,fən }
-dione [ORGANIC CHEMISTRY] Suffix indicating the presence of two keto groups. { 'dī,ōn }
1,4-dioxane [ORGANIC CHEMISTRY] $C_4H_8O_2$ The cyclic ether of ethylene glycol; it is soluble in water in all proportions and is used as a solvent. { ¦wən ¦fȯr dī'äk,sān }
dioxide [CHEMISTRY] A compound containing two atoms of oxygen. { dī'äk,sīd }
dioxin [ORGANIC CHEMISTRY] A member of a family of highly toxic chlorinated aromatic hydrocarbons; found in a number of chemical products as lipophilic contaminants. Also known as polychlorinated dibenzo-*para*-dioxin. { dī'äk·sən }
dioxolane [ORGANIC CHEMISTRY] $C_3H_6O_2$ A cyclic acetal that is a liquid; used as a solvent and extractant. { dī'äk·sə,lān }
dioxopurine [ORGANIC CHEMISTRY] *See* xanthine. { dī¦äk·sō'pyu̇r,ēn }
dioxygen [CHEMISTRY] O_2 Molecular oxygen. { dī'äk·si·jən }

DIPA [ORGANIC CHEMISTRY] *See* diisopropanolamine. { 'dip·ə *or* ¦dē¦ī¦pē¦ā }

dipentene [ORGANIC CHEMISTRY] The racemic mixture of dextro and levo isomers of limonene. { dī'pen,tēn }

dipentene glycol [ORGANIC CHEMISTRY] *See* terpin hydrate. { dī'pen,tēn 'glī,kȯl }

dipentene hydrochloride [ORGANIC CHEMISTRY] *See* terpene hydrochloride. { dī'pen,tēn ,hī·drə'klȯr,īd }

diphacinone [ORGANIC CHEMISTRY] $C_{23}H_{16}O_3$ A yellow powder with a melting point of 145-147°C; used to control rats, mice, and other rodents; acts as an anticoagulant. { də'fas·ə,nōn }

diphenamid [ORGANIC CHEMISTRY] $C_{16}H_{17}ON$ An off-white, crystalline compound with a melting point of 134-135°C; used as a preemergence herbicide for food crops, fruits, and ornamentals. { dī'fen·ə·məd }

diphenatrile [ORGANIC CHEMISTRY] $C_{14}H_{11}N$ A yellow, crystalline compound with a melting point of 73-73.5°C; used as a preemergence herbicide for turf. { dī'fen·ə·trəl }

diphenol [ORGANIC CHEMISTRY] A compound that has two phenol groups, for example, resorcinol. { dī'fē,nȯl }

diphenyl [ORGANIC CHEMISTRY] *See* biphenyl. { dī'fen·əl }

diphenylamine [ORGANIC CHEMISTRY] $(C_6H_5)_2NH$ Colorless leaflets, sparingly soluble in water; melting point 54°C; used as an additive in propellants to increase the storage life by neutralizing the acid products formed upon decomposition of the nitrocellulose. Also known as phenylaniline. { dī¦fen·əl'am,ēn }

diphenylaminechloroarsine [ORGANIC CHEMISTRY] *See* adamsite. { dī¦fen·əl'am,ēn ¦klȯr·ō'är,sēn }

diphenylcarbazide [ORGANIC CHEMISTRY] $CO(NHNHC_6H_5)_2$ White powder, melting point 170°C; used as an indicator, pink for alkalies, colorless for acids. { dī¦fen·əl'kär·bə,zīd }

diphenyl carbonate [ORGANIC CHEMISTRY] $(C_6H_5O)_2CO$ Easily hydrolyzed, white crystals, melting at 78°C; soluble in organic solvents, insoluble in water; used as a solvent, plasticizer, and chemical intermediate. { dī¦fen·əl 'kär·bə,nāt }

diphenylchloroarsine [ORGANIC CHEMISTRY] $(C_6H_5)_2AsCl$ Colorless crystals used during World War I as an antipersonnel device to generate a smoke causing sneezing and vomiting. { dī¦fen·əl,klȯr·ō'är,sēn }

diphenylene oxide [ORGANIC CHEMISTRY] $C_{12}H_8O$ A crystalline solid derived from coal tar; melting point is 87°C; used as an insecticide. { dī¦fen·əl,ēn 'äk,sīd }

diphenyl ether [ORGANIC CHEMISTRY] *See* diphenyl oxide. { dī¦fen·əl 'e,thər }

diphenylethylene [ORGANIC CHEMISTRY] *See* stilbene. { dī¦fen·əl'eth·ə,lēn }

diphenylguanidine [ORGANIC CHEMISTRY] $HNC(NHC_6H_5)_2$ A white powder, melting at 147°C; used as a rubber accelerator. Also known as DPG; melaniline. { dī¦fen·əl'gwän·ə,dēn }

diphenyl ketone [ORGANIC CHEMISTRY] *See* benzophenone. { dī¦fen·əl 'kē,tōn }

diphenylmethane [ORGANIC CHEMISTRY] $(C_6H_5)_2CH_2$ Combustible, colorless crystals melting at 26.5°C; used in perfumery, dyes, and organic synthesis. { dī¦fen·əl'me,thān }

diphenyl oxide [ORGANIC CHEMISTRY] $(C_6H_5)_2O$ A colorless liquid or crystals with a melting point of 27°C and a boiling point of 259°C; soluble in alcohol and ether; used in perfumery, soaps, and resins for laminated electrical insulation. Also known as diphenyl ether; phenyl ether. { dī¦fen·əl 'äk,sīd }

diphenyl phthalate [ORGANIC CHEMISTRY] $C_6H_4(COOC_6H_5)_2$ White powder, melting at 80°C; soluble in chlorinated hydrocarbons, esters, and ketones, insoluble in water; used as a plasticizer for cellulosic and other resins. { dī¦fen·əl 'tha,lāt }

diphosgene [ORGANIC CHEMISTRY] *See* trichloromethyl chloroformate. { dī'fäz,jēn }

diphosphate [CHEMISTRY] A salt that has two phosphate groups. { dī'fäs,fāt }

diphosphoglyceric acid [ORGANIC CHEMISTRY] $C_3H_8O_9P_2$ An ester of glyceric acid, with two molecules of phosphoric acid, characterized by a high-energy phosphate bond. { dī,fäs·fə·glə'ser·ik 'as·əd }

dipicrylamine [ORGANIC CHEMISTRY] $[(NO_2)_3C_6H_2]_2NH$ Yellow, prismlike crystals used in the gravimetric determination of potassium. { dī·pə'kril·ə,mēn }

dipnone [ORGANIC CHEMISTRY] $C_{16}H_{14}O$ A liquid ketone, formed by condensation of two acetophenone molecules; used as a plasticizer. { 'dip,nōn }
dipolar gas [PHYSICAL CHEMISTRY] A gas whose molecules have a permanent electric dipole moment. { dī¦pōl·ər 'gas }
dipolar ion [CHEMISTRY] An ion carrying both a positive and a negative charge. Also known as zwitterion. { 'dī,pō·lər 'ī,än }
dipole-dipole force [PHYSICAL CHEMISTRY] *See* orientation force. { ¦dī,pōl ¦dī,pōl ,fȯrs }
dipole moment [PHYSICAL CHEMISTRY] The vector sum of the bond moments in a molecule, a measure of the polarity of the molecule. { 'dī,pōl ,mō·mənt }
dipping acid [INORGANIC CHEMISTRY] *See* sulfuric acid. { 'dip·iŋ ,as·əd }
dipropyl [ORGANIC CHEMISTRY] A compound containing two propyl groups. { dī'prō·pəl }
dipropylene glycol [ORGANIC CHEMISTRY] $(CH_3CHOHCH_2)_2O$ A colorless, slightly viscous liquid with a boiling point of 233°C; soluble in toluene and in water; used as a solvent and for lacquers and printing inks. { dī'prō·pə,lēn 'glī,kȯl }
diprotic [CHEMISTRY] Pertaining to a chemical structure that has two ionizable hydrogen atoms. { dī'präd·ik }
diprotic acid [CHEMISTRY] An acid that has two ionizable hydrogen atoms in each molecule. { dī'präd·ik 'as·əd }
2,2′-dipyridyl [ORGANIC CHEMISTRY] $C_{10}H_8N_2$ A crystalline substance soluble in organic solvents; melting point is 69.7°C; used as a reagent for the determination of iron. Also known as 2,2′-bipyridine. { ¦tü ¦tü,prīm dī'pir·ə·dəl }
direct effect [PHYSICAL CHEMISTRY] A chemical effect caused by the direct transfer of energy from ionizing radiation to an atom or molecule in a medium. { də'rekt i'fekt }
direct-vision spectroscope [SPECTROSCOPY] A spectroscope that allows the observer to look in the direction of the light source by means of an Amici prism. { də¦rekt ¦vizh·ən 'spek·trə,skōp }
discontinuous phase [CHEMISTRY] *See* disperse phase. { ,dis·kən'tin·yə·wəs 'fāz }
discrete spectrum [SPECTROSCOPY] A spectrum in which the component wavelengths constitute a discrete sequence of values rather than a continuum of values. { di'skrēt 'spek·trəm }
disilane [INORGANIC CHEMISTRY] Si_2H_6 A spontaneously flammable compound of silicon and hydrogen; it exists as a liquid at room temperature. { dī'si,lān }
disilicate [CHEMISTRY] A silicate compound that has two silicon atoms in the molecule. { dī'sil·ə,kāt }
disilicide [CHEMISTRY] A compound that has two silicon atoms joined to a radical or another element. { dī'sil·ə,sīd }
disk colorimeter [ANALYTICAL CHEMISTRY] A device for comparing standard and sample colors by means of rotating color disks. { 'disk kə·lə'rim·əd·ər }
disodium hydrogen phosphate [INORGANIC CHEMISTRY] *See* disodium phosphate. { dī'sōd·ē·əm 'hī·drə·jən 'fäs,fāt }
disodium methylarsonate [ORGANIC CHEMISTRY] $CH_3AsO(ONa)_2$ A colorless, hygroscopic, crystalline solid; soluble in water and methanol; used in pharmaceuticals and as a herbicide. Abbreviated DMA. { dī'sōd·ē·əm ¦meth·əl'ärs·ən,āt }
disodium phosphate [INORGANIC CHEMISTRY] Na_2HPO_4 Transparent crystals, soluble in water; used in the textile processing and other industries to control pH in the range 4-9, as an additive in processed cheese to maintain spreadability, and as a laxative and antacid. Also known as disodium hydrogen phosphate. { dī'sōd·ē·əm 'fäs,fāt }
disodium tartrate [ORGANIC CHEMISTRY] *See* sodium tartrate. { dī'sōd·ē·əm 'tär,trāt }
disperse phase [CHEMISTRY] The phase of a disperse system consisting of particles or droplets of one substance distributed through another system. Also known as discontinuous phase; internal phase. { də'spərs ,fāz }
disperse system [CHEMISTRY] A two-phase system consisting of a dispersion medium and a disperse phase. { də'spərs ,sis·təm }

dispersible inhibitor [CHEMISTRY] An additive that can be dispersed in a liquid with only moderate agitation to retard undesirable chemical action. { di'spər·sə·bəl in'hib·əd·ər }

dispersion [CHEMISTRY] A distribution of finely divided particles in a medium. { də'spər·zhən }

dispersion force [PHYSICAL CHEMISTRY] The force of attraction that exists between molecules that have no permanent dipole. { də'spər·zhən ,fōrs }

dispersion medium [CHEMISTRY] *See* continuous phase. { də'spər·zhən ,mēd·ē·əm }

dispersoid [CHEMISTRY] Matter in a form produced by a disperse system. { də'spər,sȯid }

displacement [CHEMISTRY] A chemical reaction in which an atom, radical, or molecule displaces and sets free an element of a compound. { dis'plās·mənt }

displacement chromatography [ANALYTICAL CHEMISTRY] Variation of column-development or elution chromatography in which the solvent is sorbed more strongly than the sample components; the freed sample migrates down the column, pushed by the solvent. { dis'plās·mənt ,krō·mə'täg·rə·fē }

displacement series [CHEMISTRY] The elements in decreasing order of their negative potentials. Also known as constant series; electromotive series; Volta series. { dis'plās·mənt ,sir·ēz }

disproportionation [CHEMISTRY] The changing of a substance, usually by simultaneous oxidation and reduction, into two or more dissimilar substances. { ,dis·prə,pȯr·shə'nā·shən }

dissociation [PHYSICAL CHEMISTRY] Separation of a molecule into two or more fragments (atoms, ions, radicals) by collision with a second body or by the absorption of electromagnetic radiation. { də,sō·sē'ā·shən }

dissociation constant [PHYSICAL CHEMISTRY] A constant whose numerical value depends on the equilibrium between the undissociated and dissociated forms of a molecule; a higher value indicates greater dissociation. { də,sō·sē'ā·shən ,kän·stənt }

dissociation energy [PHYSICAL CHEMISTRY] The energy required for complete separation of the atoms of a molecule. { də,sō·sē'ā·shən ,en·ər·jē }

dissociation limit [SPECTROSCOPY] The wavelength, in a series of vibrational bands in a molecular spectrum, corresponding to the point at which the molecule dissociates into its constituent atoms; it corresponds to the convergence limit. { də'sō·sē'ā·shən ,lim·ət }

dissociation pressure [PHYSICAL CHEMISTRY] The pressure, for a given temperature, at which a chemical compound dissociates. { də,sō·sē'ā·shən ,presh·ər }

dissociation-voltage effect [PHYSICAL CHEMISTRY] A change in the dissociation of a weak electrolyte produced by a strong electric field. { də,sō·sē!ā·shən 'vōl·tij i,fekt }

dissolution [CHEMISTRY] Dissolving of a material. { ,dis·ə'lü·shən }

dissolve [CHEMISTRY] **1.** To cause to disperse. **2.** To cause to pass into solution. { də'zälv }

dissymmetry coefficient [ANALYTICAL CHEMISTRY] Ratio of the intensities of scattered light at 45 and 135°, used to correct for destructive interference encountered in light-scattering-photometric analyses of liquid samples. { di'sim·ə·trē ,kō·i'fish·ənt }

distillate [CHEMISTRY] The products of distillation formed by condensing vapors. { 'dis·tə,lāt }

distillation [CHEMISTRY] The process of producing a gas or vapor from a liquid by heating the liquid in a vessel and collecting and condensing the vapors into liquids. { ,dis·tə'lā·shən }

distillation column [CHEMISTRY] A still for fractional distillation. { ,dis·tə'lā·shən ,käl·əm }

distillation curve [CHEMISTRY] The graphical plot of temperature versus overhead product (distillate) volume or weight for a distillation operation. { ,dis·tə'lā·shən ,kərv }

distillation loss [CHEMISTRY] In a laboratory distillation, the difference between the volume of liquid introduced into the distilling flask and the sum of the residue and condensate received. { ˌdis·tə'lā·shən ˌlós }

distillation range [CHEMISTRY] The difference between the temperature at the initial boiling point and at the end point of a distillation test. { ˌdis·tə'lā·shən ˌrānj }

distilled mustard gas [ORGANIC CHEMISTRY] A delayed-action casualty gas (mustard gas) that has been distilled, or purified, to greatly reduce the odor and thereby increase its difficulty of detection. { də'stild 'məs·tərd ˌgas }

distilled water [CHEMISTRY] Water that has been freed of dissolved or suspended solids and organisms by distillation. { də'stild 'wód·ər }

distilling flask [CHEMISTRY] A round-bottomed glass flask that is capable of holding a liquid to be distilled. { də'stil·iŋ ˌflask }

distribution coefficient [PHYSICAL CHEMISTRY] The ratio of the amounts of solute dissolved in two immiscible liquids at equilibrium. { ˌdis·trə'byü·shən ˌkō·i'fish·ənt }

distribution law [ANALYTICAL CHEMISTRY] The law stating that if a substance is dissolved in two immiscible liquids, the ratio of its concentration in each is constant. { ˌdis·trə'byü·shən ˌló }

distribution ratio [ANALYTICAL CHEMISTRY] The ratio of the concentrations of a given solute in equal volumes of two immiscible solvents after the mixture has been shaken and equilibrium established. { ˌdi·strə'byü·shən ˌrā·shō }

disubstituted alkene [ORGANIC CHEMISTRY] An alkene with the general formula $R_2C=CH_2$ or $RHC=CHR$, where R is any organic group; a carbon atom is bonded directly to each end of the double bond. { dī'səb·stəˌtüd·əd 'alˌkēn }

disulfate [CHEMISTRY] A compound that has two sulfate radicals. { dī'səlˌfāt }

disulfide [CHEMISTRY] **1.** A compound that has two sulfur atoms bonded to a radical or element. **2.** One of a group of organosulfur compounds RSSR′ that may be symmetrical (R—R′) or unsymmetrical (R and R′, different). { dī'səlˌfīd }

disulfide bond [ORGANIC CHEMISTRY] *See* disulfide bridge. { dī¦səlˌfīd 'bänd }

disulfide bridge [ORGANIC CHEMISTRY] A sulfur-to-sulfur bond linking the sulfur atoms of two polypeptide chains. Also known as disulfide bond. { dī¦səlˌfīd 'brij }

disulfonate [CHEMISTRY] A molecule that has two sulfonate groups. { dī'səl·fəˌnāt }

disulfonic acid [CHEMISTRY] A molecule that has two sulfonic acid groups. { ˌdī·səl'fän·ik 'as·əd }

diterpene [ORGANIC CHEMISTRY] $C_{20}H_{32}$ **1.** A group of terpenes that have twice as many atoms in the molecule as monoterpenes. **2.** Any derivative of diterpene. { dī'tərˌpēn }

dithiocarbamate [ORGANIC CHEMISTRY] **1.** A salt of dithiocarbamic acid. **2.** Any other derivative of dithiocarbamic acid. { ¦dīˌthī·ō'kär·bəˌmāt }

dithiocarbamic acid [ORGANIC CHEMISTRY] NH_2CS_2H A colorless, unstable powder; various metal salts are readily obtained, and used as strong accelerators for rubber. Also known as aminodithioformic acid. { ¦dīˌthī·ōˌkär'bam·ik 'as·əd }

dithioic acid [ORGANIC CHEMISTRY] An organic acid in which sulfur atoms have replaced both oxygen atoms of the carboxy group. { ¦dīˌthī¦ō·ik 'as·əd }

dithionate [CHEMISTRY] Any salt formed from dithionic acid. { dī'thī·əˌnāt }

dithionic acid [INORGANIC CHEMISTRY] $H_2S_2O_6$ A strong acid formed by the oxidation of sulfurous acid, and known only by its salts and in solution. { ¦dīˌthī'än·ik 'as·əd }

dithiooxamide [ORGANIC CHEMISTRY] $NH_2CSCSNH_2$ Red crystals soluble in alcohol; used as a reagent for copper, cobalt, and nickel, and for the determination of osmium. { ¦dīˌthī·ō'äk·səˌmīd }

1,4-dithiothreitol [ORGANIC CHEMISTRY] $C_4H_{10}O_2S_2$ Needlelike crystals soluble in water, ethanol acetone, ethylacetate; used as a protective agent for thiol (SH) groups. { ¦wən ¦fór ¦dīˌthī·ō'thrē·əˌtól }

ditungsten carbide [INORGANIC CHEMISTRY] W_2C A gray powder having hardness approaching that of diamond; forms hexagonal crystals with specific gravity 17.2; melting point 2850°C. { ¦dīˌtəŋ·stən 'kärˌbīd }

divalent carbon [ORGANIC CHEMISTRY] A charged or uncharged carbon atom that has formed only two covalent bonds. Also known as dicovalent carbon. { dī'vā·lənt 'kär·bən }

divalent metal [CHEMISTRY] A metal whose atoms are each capable of chemically combining with two atoms of hydrogen. { dī'vā·lənt 'med·əl }

diver method [PHYSICAL CHEMISTRY] Measure of the size of suspended solid particles; small glass divers of known density sink to the level where the liquid-suspension density is equal to that of the diver, allowing calculation of particle size. Also known as Berg's diver method. { 'dī·vər ,meth·əd }

divinyl [ORGANIC CHEMISTRY] **1.** A molecule that has two vinyl groups. **2.** *See* 1,3-butadiene. { dī'vīn·əl }

divinyl acetylene [ORGANIC CHEMISTRY] C_6H_6 A linear trimer of acetylene, made by passing acetylene into a hydrochloric acid solution that has metallic catalysts; used as an intermediate in neoprene manufacture. { dī'vīn·əl ə'sed·əl,ēn }

divinylbenzene [ORGANIC CHEMISTRY] $C_6H_4(CHCH_2)_2$ Polymerizable, water-white liquid used to make rubbers, drying oils, and ion-exchange resins and other polymers; forms include ortho, meta, and para isomers. Also known as vinylstyrene. { dī¦vīn·əl'ben,zēn }

divinyl ether [ORGANIC CHEMISTRY] *See* vinyl ether. { dī'vīn·əl 'ē·thər }

divinyl oxide [ORGANIC CHEMISTRY] *See* vinyl ether. { dī'vīn·əl 'äk,sīd }

D line [SPECTROSCOPY] The yellow line that is the first line of the major series of the sodium spectrum; the doublet in the Fraunhofer lines whose almost equal components have wavelengths of 5895.93 and 5889.96 angstroms respectively. { 'dē ,līn }

DM [ORGANIC CHEMISTRY] *See* adamsite.

DMA [ORGANIC CHEMISTRY] *See* disodium methylarsonate.

DMC [ORGANIC CHEMISTRY] *See* *p,p'*-dichlorodiphenylmethyl carbinol.

DMDT [ORGANIC CHEMISTRY] *See* methoxychlor.

DMF [ORGANIC CHEMISTRY] *See* N,N-dimethylformamide.

DMSO [ORGANIC CHEMISTRY] *See* dimethyl sulfoxide.

DMT [ORGANIC CHEMISTRY] *See* dimethyl terephthalate.

DMU [ORGANIC CHEMISTRY] *See* dimethylolurea.

Dobbin's reagent [ANALYTICAL CHEMISTRY] A mercuric chloride-potassium iodide reagent used to test for caustic alkalies in soap. { 'däb·ənz rē'ā·jənt }

Dobson spectrophotometer [SPECTROSCOPY] A photoelectric spectrophotometer used in the determination of the ozone content of the atmosphere; compares the solar energy at two wavelengths in the absorption band of ozone by permitting the radiation of each to fall alternately upon a photocell. { 'däb·sən ,spek·trō·fə'täm·əd·ər }

docosane [ORGANIC CHEMISTRY] $C_{22}H_{46}$ A paraffin hydrocarbon, especially the normal isomer $CH_3(CH_2)_{20}CH_3$. { 'däk·ə,sān }

docosanoic acid [ORGANIC CHEMISTRY] $CH_3(CH_2)_{20}CO_2H$ A crystalline fatty acid, melting at 80°C, slightly soluble in water and alcohol, and found in the fats and oils of some seeds such as peanuts. Also known as behenic acid. { ¦dak·ə·sə¦nō·ik 'as·əd }

1-docosanol [ORGANIC CHEMISTRY] *See* behenyl alcohol. { ¦wən də'käs·ə,nól }

docosapentanoic acid [ORGANIC CHEMISTRY] $C_{21}H_{33}CO_2H$ A pale-yellow liquid, boils at 236°C (5 mmHg), insoluble in water, soluble in ether, and found in fish blubber. { ¦däk·ə·sə,pen·tə¦nō·ik 'as·əd }

dodecahydrate [CHEMISTRY] A hydrated compound that has a total of 12 water molecules associated with it. { dō,dek·ə'hī,drāt }

dodecane [ORGANIC CHEMISTRY] $CH_3(CH_2)_{10}CH_3C_{12}H_{26}$ An oily paraffin compound, a colorless liquid, boiling at 214.5°C, insoluble in water; used as a solvent and in jet fuel research. Also known as dihexy; propylene tetramer; tetrapropylene. { 'dō·də,kān }

1-dodecene [ORGANIC CHEMISTRY] $CH_2CH(CH_2)_9CH_3$ A colorless liquid, boiling at 213°C, insoluble in water; used in flavors, dyes, perfumes, and medicines. { ¦wən 'dō·də,sēn }

dodecyl [ORGANIC CHEMISTRY] $C_{12}H_{25}$ A radical derived from dodecane by removing one hydrogen atom; in particular, the normal radical, $CH_3(CH_2)_{10}CH_2$—. { 'dō·də,sil }

dodecylbenzene [ORGANIC CHEMISTRY] Blend of isomeric (mostly monoalkyl) benzenes with saturated side chains averaging 12 carbon atoms; used in the alkyl amyl sulfonate type of detergents. Also known as detergent alkylate. { 'dō·də͵sil 'ben͵zēn }

dodecyl sodium sulfate [ORGANIC CHEMISTRY] *See* sodium lauryl sulfate. { 'dō·də͵sil 'sōd·ē·əm 'səl͵fāt }

dolomol [ORGANIC CHEMISTRY] *See* magnesium stearate. { 'dä·lə͵mòl }

Donnan distribution coefficient [PHYSICAL CHEMISTRY] A coefficient in an expression giving the distribution, on two sides of a boundary between electrolyte solutions in Donnan equilibrium, of ions which can diffuse across the boundary. { ¦dän·ən ͵dis·trə'byü·shən ͵kō·ə͵fish·ənt }

Donnan equilibrium [PHYSICAL CHEMISTRY] The particular equilibrium set up when two coexisting phases are subject to the restriction that one or more of the ionic components cannot pass from one phase into the other; commonly, this restriction is caused by a membrane which is permeable to the solvent and small ions but impermeable to colloidal ions or charged particles of colloidal size. Also known as Gibbs-Donnan equilibrium. { 'dō·nən ē·kwə'lib·rē·əm }

Donnan potential [PHYSICAL CHEMISTRY] The potential difference across a boundary between two electrolytic solutions in Donnan equilibrium. { 'dän·ən pə͵ten·chəl }

DOP [ORGANIC CHEMISTRY] *See* dioctyl phthalate.

Doppler broadening [SPECTROSCOPY] Frequency spreading that occurs in single-frequency radiation when the radiating atoms, molecules, or nuclei do not all have the same velocity and may each give rise to a different Doppler shift. { 'däp·lər ͵bròd·ən·iŋ }

Doppler-free spectroscopy [SPECTROSCOPY] Any of several techniques which make use of the intensity and monochromatic nature of a laser beam to overcome the Doppler broadening of spectral lines and measure their wavelengths with extremely high accuracy. { däp·lər ͵frē spek'träs·kə·pē }

Doppler-free two-photon spectroscopy [SPECTROSCOPY] A version of Doppler-free spectroscopy in which the wavelength of a transition induced by the simultaneous absorption of two photons is measured by placing a sample in the path of a laser beam reflected on itself, so that the Doppler shifts of the incident and reflected beams cancel. { däp·lər ͵frē ¦tü ¦fō͵tän spek'träs·kə·pē }

Dorn effect [PHYSICAL CHEMISTRY] A difference in a potential resulting from the motions of particles through water; the potential exists between the particles and the water. { 'dòrn i͵fekt }

dotriacontane [ORGANIC CHEMISTRY] $C_{32}H_{66}$ A paraffin hydrocarbon, in particular, the normal isomer $CH_3(CH_2)_{30}CH_3$, which is crystalline. { ¦dō͵trī·ə'kän͵tān }

double beam spectrophotometer [SPECTROSCOPY] An instrument that uses a photoelectric circuit to measure the difference in absorption when two closely related wavelengths of light are passed through the same medium. { ¦dəb·əl ¦bēm spek·trō·fə'täm·əd·ər }

double-blind sample [ANALYTICAL CHEMISTRY] In chemical analysis, a sample submitted in such a way that neither its composition nor its identification as a check sample is known to the analyst. { ¦dəb·əl ¦blīnd 'sam·pəl }

double bond [PHYSICAL CHEMISTRY] A type of linkage between atoms in which two pair of electrons are shared equally. { ¦dəb·əl 'bänd }

double-bond isomerism [PHYSICAL CHEMISTRY] Isomerism in which two or more substances possess the same elementary composition but differ in having double bonds in different positions. { ¦dəb·əl ¦bänd ī'säm·ə͵riz·əm }

double-bond shift [ORGANIC CHEMISTRY] In an organic molecular structure, the occurrence when a pair of valence bonds that join a pair of carbons (or other atoms) shifts, via chemical reaction, to a new position, for example, $H_2C{=}C{-}C{-}CH_2$ (butene-1) to $H_2C{-}C{=}C{-}CH_2$ (butene-2). { ¦dəb·əl ¦band 'shift }

double decomposition [CHEMISTRY] The simple exchange of elements of two substances to form two new substances; for example, $CaSO_4 + 2NaCl \rightarrow CaCl_2 + Na_2SO_4$. { ¦dəb·əl dē͵käm·pə'zish·ən }

double layer [PHYSICAL CHEMISTRY] *See* electric double layer. { ¦dəb·əl 'lā·ər }

double nickel salt [INORGANIC CHEMISTRY] *See* nickel ammonium sulfate. { ¦dəb·əl ¦nik·əl 'sȯlt }

double-replacement reaction [CHEMISTRY] A chemical reaction between compounds in which the elements in the reactants recombine to form two different compounds, each of the products having one element from each of the reactants. { ¦dəb·əl ri'plās·mənt rē,ak·shən }

double salt [INORGANIC CHEMISTRY] **1.** A salt that upon hydrolysis forms two different anions and cations. **2.** A salt that is a molecular combination of two other salts. { ¦dəb·əl 'sȯlt }

doublet [PHYSICAL CHEMISTRY] Two electrons which are shared between two atoms and give rise to a nonpolar valence bond. [SPECTROSCOPY] Two closely separated spectral lines arising from a transition between a single state and a pair of states forming a doublet as described in the atomic physics definition. { 'dəb·lət }

downflow [CHEMISTRY] In an ion-exchange system, the direction of the flow of the solution being processed. { 'dau̇n,flō }

2,4-DP [ORGANIC CHEMISTRY] *See* dichlorprop.

D.P. [ORGANIC CHEMISTRY] *See* degree of polymerization.

DPG [ORGANIC CHEMISTRY] *See* diphenylguanidine.

Drew number [PHYSICAL CHEMISTRY] A dimensionless group used in the study of diffusion of a solid material A into a stream of vapor initially composed of substance B, equal to

$$\frac{Z_A(M_A - M_B) + M_B}{(Z_A - Y_{AW})(M_B - M_A)} \cdot \ln \frac{M_V}{M_W}$$

where M_A and M_B are the molecular weights of components A and B, M_V and M_W are the molecular weights of the mixture in the vapor and at the wall, and Y_{AW} and Z_A are the mole fractions of A at the wall and in the diffusing stream, respectively. Symbolized N_D. { 'drü ,nəm·bər }

driving force [CHEMISTRY] In a chemical reaction, the formation of products such as an insoluble compound, a gas, a nonelectrolyte, or a weak electrolyte that enable the reaction to go to completion as a metathesis. { 'drīv·iŋ ,fȯrs }

dropping-mercury electrode [PHYSICAL CHEMISTRY] An electrode consisting of a fine-bore capillary tube above which a constant head of mercury is maintained; the mercury emerges from the tip of the capillary at the rate of a few milligrams per second and forms a spherical drop which falls into the solution at the rate of one every 2-10 seconds. { ¦dräp·iŋ ¦mər·kyə·rē i'lek·trōd }

dropping point [CHEMISTRY] The temperature at which grease changes from a semisolid to a liquid state under standardized conditions. { 'dräp·iŋ ,pȯint }

dry acid [CHEMISTRY] Nonaqueous acetic acid used for oil-well reservoir acidizing treatment. { ¦drī 'as·əd }

dry ashing [ORGANIC CHEMISTRY] The conversion of an organic compound into ash (decomposition) by a burner or in a muffle furnace. { ¦drī 'ash·iŋ }

dry box [CHEMISTRY] A container or chamber filled with argon, or sometimes dry air or air with no carbon dioxide (CO_2), to provide an inert atmosphere in which manipulation of very reactive chemicals is carried out in the laboratory. { 'drī ,bäks }

dry distillation [CHEMISTRY] A process in which a solid is heated in the absence of liquid to release vapors or liquids from the solid, for example, heating a hydrate to produce the anhydrous salt. { ¦drī dis·tə'lā·shən }

dry ice [INORGANIC CHEMISTRY] Carbon dioxide in the solid form, usually made in blocks to be used as a coolant; changes directly to a gas at −78.5°C as heat is absorbed. { ¦drī 'īs }

drying [CHEMISTRY] **1.** An operation in which a liquid, usually water, is removed from a wet solid in equipment termed a dryer. **2.** A process of oxidation whereby a liquid such as linseed oil changes into a solid film. { 'drī·iŋ }

drying agent [CHEMISTRY] Soluble or insoluble chemical substance that has such a great affinity for water that it will abstract water from a great many fluid materials; soluble chemicals are calcium chloride and glycerol, and insoluble chemicals are bauxite and silica gel. Also known as desiccant. { 'drī·iŋ ,ā·jənt }

dry point [ANALYTICAL CHEMISTRY] The temperature at which the last drop of liquid evaporates from the bottom of the flask. { 'drī ˌpȯint }

dual-function catalyst [CHEMISTRY] *See* bifunctional catalyst. { ¦dül ¦fəŋk·shən 'kad·ə·list }

Duhem's equation [PHYSICAL CHEMISTRY] *See* Gibbs-Duhem equation. { dü'emz iˌkwā·zhən }

Dühring's rule [PHYSICAL CHEMISTRY] The rule that a plot of the temperature at which a liquid exerts a particular vapor pressure against the temperature at which a similar reference liquid exerts the same vapor pressure produces a straight or nearly straight line. { 'dir·iŋz ˌrül }

dulcitol [ORGANIC CHEMISTRY] $C_6H_8(OH)_6$ A sugar with a slightly sweet taste; white, crystalline powder with a melting point of 188.5°C; soluble in hot water; used in medicine and bacteriology. { 'dəl·səˌtȯl }

Dumas method [ANALYTICAL CHEMISTRY] A procedure for the determination of nitrogen in organic substances by combustion of the substance. { 'dü·mä ˌmeth·əd }

duplicate measurement [ANALYTICAL CHEMISTRY] An additional measurement made on the same (identical) sample of material to evaluate the variance in the measurement. { ¦dup·lə·kət 'mezh·ər·mənt }

duplicate sample [ANALYTICAL CHEMISTRY] A second sample randomly selected from a material being analyzed in order to evaluate sample variance. { ¦dup·lə·kət 'sam·pəl }

durable-press resin [ORGANIC CHEMISTRY] *See* permanent-press resin. { ¦dür·ə·bəl ¦pres 'rez·ən }

durene [ORGANIC CHEMISTRY] $C_6H_2(CH_3)_4$ Colorless crystals with camphor aroma; boiling point 190°C; soluble in organic solvents, insoluble in water; used as a chemical intermediate. Also known as durol. { 'dùˌrēn }

Dutch liquid [ORGANIC CHEMISTRY] *See* ethylene chloride. { ¦dəch ¦lik·wəd }

Dy [CHEMISTRY] *See* dysprosium.

dye [CHEMISTRY] A colored substance which imparts more or less permanent color to other materials. Also known as dyestuff. { dī }

dyeing assistant [CHEMISTRY] Material such as sodium sulfate added to a dye bath to control or promote the action of a textile dye. { 'dī·iŋ əˌsis·tənt }

dynamic allotropy [CHEMISTRY] A phenomenon in which the allotropes of an element exist in dynamic equilibrium. { dī¦nam·ik ə'lä·trə·pē }

dypnone [ORGANIC CHEMISTRY] $C_6H_5COCHC(CH_3)C_6H_5$ A light-colored liquid with a boiling point of 246°C at 50 mmHg; used as a plasticizer and perfume base and in light-stable coatings. { 'dipˌnōn }

Dyson notation [ORGANIC CHEMISTRY] A notation system for representing organic chemicals developed by G. Malcolm Dyson; the compound is described on a single line, symbols are used for the chemical elements involved as well as for the functional groups and various ring systems; for example, methyl alcohol is C.Q and phenol is B6.Q. { 'dī·sən nōˌtā·shən }

dysprosium [CHEMISTRY] A metallic rare-earth element, symbol Dy, atomic number 66, atomic weight 162.50. { dis'prō·zē·əm }

dystetic mixture [PHYSICAL CHEMISTRY] A mixture of two or more substances that has the highest possible melting point of all mixtures of these substances. { di'sted·ik 'miks·chər }

E

eagle mounting [SPECTROSCOPY] A mounting for a diffraction grating, based on the principle of the Rowland circle, in which the diffracted ray is returned along nearly the same direction as the incident beam. { 'ē·gəl ˌmaủn·tiŋ }

easin [ORGANIC CHEMISTRY] $C_{20}H_6O_5I_4Na_2$ The sodium salt of tetraiodofluorescein; a brown powder, insoluble in water; used as a dye and a pH indicator (hydrogen ion) at pH 2.0. Also known as iodoeasin; sodium tetrafluorescein. { 'ē·ə·zən }

ebulliometer [PHYSICAL CHEMISTRY] The instrument used for ebulliometry. Also known as ebullioscope. { əˌbủ·lē'äm·əd·ər }

ebulliometry [PHYSICAL CHEMISTRY] The precise measurement of the absolute or differential boiling points of solutions. { əˌbủ·lē'äm·ə·trē }

ebullioscope [PHYSICAL CHEMISTRY] *See* ebulliometer. { ə'bủ·lē·əˌskōp }

ebullioscopic constant [PHYSICAL CHEMISTRY] The ratio of the elevation of the boiling point of a solvent caused by dissolving a solute to the molality of the solution, taken at extremely low concentrations. Also known as molal elevation of the boiling point. { e'bü·lē·əˌskop·ik ˌkän·stənt }

ecgonine [ORGANIC CHEMISTRY] $C_9H_{15}NO_3$ An alkaloid obtained in crystalline form by the hydrolysis of cocaine. { 'ek·gəˌnēn }

echelette grating [SPECTROSCOPY] A diffraction grating with coarse groove spacing, designed for the infrared region; has grooves with comparatively flat sides and concentrates most of the radiation by reflection into a small angular coverage. { ¦esh·ə¦let *or* ¦äsh¦let ˌgrād·iŋ }

echelle grating [SPECTROSCOPY] A diffraction grating designed for use in high orders and at angles of illumination greater than 45° to obtain high dispersion and resolving power by the use of high orders of interference. { ā'shel ˌgrād·iŋ }

echelon grating [SPECTROSCOPY] A diffraction grating which consists of about 20 plane-parallel plates about 1 centimeter thick, cut from one sheet, each plate extending beyond the next by about 1 millimeter, and which has a resolving power on the order of 10^6. { 'esh·əˌlän ˌgrād·iŋ }

echinopsine [ORGANIC CHEMISTRY] $C_{10}H_9O$ An alkaloid obtained from *Echinops* species; crystallizes as needles from benzene solution, melts at 152°C; physiological action is similar to that of brucine and strychnine. { ˌek·ə'näpˌsēn }

eclipsed conformation [PHYSICAL CHEMISTRY] A particular arrangement of constituent atoms that may rotate about a single bond in a molecule; for ethane it is such that when viewed along the axis of the carbon-carbon bond the hydrogen atoms of one methyl group are exactly in line with those of the other methyl group. { i'klipst ˌkän·fər'mā·shən }

edge-bridging ligand [ORGANIC CHEMISTRY] A ligand that forms a bridge over one edge of the polyhedron of a metal cluster structure. { 'ej ˌbrij·iŋ 'lī·gənd }

EDTA [ORGANIC CHEMISTRY] *See* ethylenediaminetetraacetic acid.

EDTC [ORGANIC CHEMISTRY] *See* S-ethyl-N,N dipropylthiocarbamate.

EELS [SPECTROSCOPY] *See* electron energy loss spectroscopy.

eff [CHEMISTRY] *See* efficiency.

effective molecular diameter [PHYSICAL CHEMISTRY] The general extent of the electron cloud surrounding a gas molecule as calculated in any of several ways. { ə¦fek·tiv məˌlek·yə·lər dī'am·əd·ər }

effective permeability [PHYSICAL CHEMISTRY] The observed permeability exhibited by a porous medium to one fluid phase when there is physical interaction between this phase and other fluid phases present. { ə¦fek·tiv pər·mē·ə'bil·əd·ē }

effervescence [CHEMISTRY] The bubbling of a solution of an element or chemical compound as the result of the emission of gas without the application of heat; for example, the escape of carbon dioxide from carbonated water. { ˌef·ər'ves·əns }

efficiency Abbreviated [CHEMISTRY] In an ion-exchange system, a measurement of the effectiveness of a system expressed as the amount of regenerant required to remove a given unit of adsorbed material. eff. { ə'fish·ən·sē }

efflorescence [CHEMISTRY] The property of hydrated crystals to lose water of hydration and crumble when exposed to air. { ˌef·lə'res·əns }

effusion [PHYSICAL CHEMISTRY] The movement of a gas through an opening which is small as compared with the average distance which the gas molecules travel between collisions. { e'fyü·zhən }

EGA [ANALYTICAL CHEMISTRY] *See* evolved gas analysis.

EGT [ORGANIC CHEMISTRY] *See* ethylene glycol bis(trichloroacetate).

Ehrlich's reagent [ORGANIC CHEMISTRY] $(CH_3)_2NC_6H_4CHO$ Granular or leafletlike crystals that are soluble in many organic solvents; melting point is 74°C; used in the preparation of dyes, as a reagent for arsphenamine, anthranilic acid, antipyrine, indole, and skatole, and as a differentiating agent between true scarlet fever and serum eruptions. { 'er·liks rēˌā·jənt }

eicosanoic acid [ORGANIC CHEMISTRY] $CH_3(CH_2)_{18}COOH$ A white, crystalline, saturated fatty acid, melting at 75.4°C; a constituent of butter. Also known as arachic acid; arachidic acid. { ¦ī·kə·sə¦nō·ik 'as·əd }

Einchluss thermometer [ANALYTICAL CHEMISTRY] All-glass, liquid-filled thermometer, temperature range −201 to +360°C, used for laboratory test work. { 'īnˌshlüs thərˌmäm·əd·ər }

einsteinium [CHEMISTRY] Synthetic radioactive element, symbol Es, atomic number 99; discovered in debris of 1952 hydrogen bomb explosion; now made in cyclotrons. { īn'stīn·ē·əm }

Einstein photochemical equivalence law [PHYSICAL CHEMISTRY] The law that each molecule taking part in a chemical reaction caused by electromagnetic radiation absorbs one photon of the radiation. Also known as Stark-Einstein law. { 'īnˌstīn ¦fōd·ō¦kem·ə·kəl i'kwiv·ə·ləns ˌlȯ }

Einstein viscosity equation [PHYSICAL CHEMISTRY] An equation which gives the viscosity of a sol in terms of the volume of dissolved particles divided by the total volume. { 'īnˌstīn vis'käs·əd·ē iˌkwā·zhən }

elaidic acid [ORGANIC CHEMISTRY] $CH_3(CH_2)_7CH{:}CH(CH_2)_7COOH$ A transisomer of an unsaturated fatty acid, oleic acid; crystallizes as colorless leaflets, melts at 44°C, boils at 288°C (100 mmHg), insoluble in water, soluble in alcohol and ether; used in chromatography as a reference standard. { ¦el·ə¦id·ik 'as·əd }

elaidinization [ORGANIC CHEMISTRY] The process of changing the geometric cis form of an unsaturated fatty acid or a compound related to it into the trans form, resulting in an acid that is more resistant to oxidation. { ə¦lā·əˌdin·ə'zā·shən }

elaidin reaction [ANALYTICAL CHEMISTRY] A test that differentiates nondrying oils such as olein from semidrying oils and drying oils; nitrous acid converts olein into its solid isomer, while semidrying oils in contact with nitrous acid thicken slowly, and drying oils such as tung oil become hard and resinous. { ə'lā·əd·ən rēˌak·shən }

Elbs reaction [ORGANIC CHEMISTRY] The formation of anthracene derivatives by dehydration and cyclization of diaryl ketone compounds which have a methyl group or methylene group; heating to an elevated temperature is usually required. { 'elbs rēˌak·shən }

electrical calorimeter [ANALYTICAL CHEMISTRY] Device to measure heat evolved (from fusion or vaporization, for example); measured quantities of heat are added electrically to the sample, and the temperature rise is noted. { ə'lek·trə·kəl kal·ə'rim·əd·ər }

electrical equivalent [ANALYTICAL CHEMISTRY] In conductometric analyses of electrolyte solutions, an outside, calibrated current source as compared to (equivalent to) the current passing through the sample under analysis; for example, a Wheatstone-bridge balanced reading. { i'lek·trə·kəl i'kwiv·ə·lənt }

electrically active fluid [PHYSICAL CHEMISTRY] A fluid whose properties are altered by either an electric field (electrorheological fluid) or a magnetic field (ferrofluid). { i'lek·trə·klē ¦ak·tiv 'flü·əd }

electric double layer [PHYSICAL CHEMISTRY] A phenomenon found at a solid-liquid interface; it is made up of ions of one charge type which are fixed to the surface of the solid and an equal number of mobile ions of the opposite charge which are distributed through the neighboring region of the liquid; in such a system the movement of liquid causes a displacement of the mobile counterions with respect to the fixed charges on the solid surface. Also known as double layer. { i'lek·trik ¦dəb·əl 'lā·ər }

electric-field effect [SPECTROSCOPY] *See* Stark effect. { i¦lek·trik ¦fēld i'fekt }

electride [INORGANIC CHEMISTRY] A member of a class of ionic compounds in which the anion is believed to be an electron. { i'lek,trīd }

electrobalance [ANALYTICAL CHEMISTRY] Analytical microbalance utilizing electromagnetic weighing; the sample weight is balanced by the torque produced by current in a coil in a magnetic field, with torque proportional to the current. { i,lek·trō'bal·əns }

electrocatalysis [CHEMISTRY] Any one of the mechanisms which produce a speeding up of half-cell reactions at electrode surfaces. { i,lek·trō·kə'tal·ə·səs }

electrochemical cell [PHYSICAL CHEMISTRY] A combination of two electrodes arranged so that an overall oxidation-reduction reaction produces an electromotive force; includes dry cells, wet cells, standard cells, fuel cells, solid-electrolyte cells, and reserve cells. { i,lek·trō'kem·ə·kəl 'sel }

electrochemical effect [PHYSICAL CHEMISTRY] Conversion of chemical to electric energy, as in electrochemical cells; or the reverse process, used to produce elemental aluminum, magnesium, and bromine from compounds of these elements. { i,lek·trō'kem·ə·kəl i'fekt }

electrochemical emf [PHYSICAL CHEMISTRY] Electrical force generated by means of chemical action, in manufactured cells (such as dry batteries) or by natural means (galvanic reaction). { i,lek·trō'kem·ə·kəl ¦ē¦em'ef }

electrochemical equivalent [PHYSICAL CHEMISTRY] The weight in grams of a substance produced or consumed by electrolysis with 100% current efficiency during the flow of a quantity of electricity equal to 1 faraday (96,487.0±1.6 coulombs). { i,lek·trō'kem·ə·kəl i'kwiv·ə·lənt }

electrochemical potential [PHYSICAL CHEMISTRY] The difference in potential that exists when two dissimilar electrodes are connected through an external conducting circuit and the two electrodes are placed in a conducting solution so that electrochemical reactions occur. { i,lek·trō'kem·ə·kəl pə'ten·chəl }

electrochemical process [PHYSICAL CHEMISTRY] **1.** A chemical change accompanying the passage of an electric current, especially as used in the preparation of commercially important quantities of certain chemical substances. **2.** The reverse change, in which a chemical reaction is used as the source of energy to produce an electric current, as in a battery. { i,lek·tro'kem·ə·kəl 'präs·əs }

electrochemical reduction cell [PHYSICAL CHEMISTRY] The cathode component of an electrochemical cell, at which chemical reduction occurs (while at the anode, chemical oxidation occurs). { i,lek·trō'kem·ə·kəl ri'dək·shən ,sel }

electrochemical series [PHYSICAL CHEMISTRY] A series in which the metals and other substances are listed in the order of their chemical reactivity or electrode potentials, the most reactive at the top and the less reactive at the bottom. Also known as electromotive series. { i,lek·trō'kem·ə·kəl 'sir·ēz }

electrochemical techniques [PHYSICAL CHEMISTRY] The experimental methods developed to study the physical and chemical phenomena associated with electron transfer at the interface of an electrode and solution. { i,lek·trō'kem·ə·kəl tek'nēks }

electrochemiluminescence [PHYSICAL CHEMISTRY] Emission of light produced by an electrochemical reaction. Also known as electrogenerated chemiluminescence. { i,lek·trō,kem·ē·ə,lüm·ə'nes·əns }

electrochemistry [PHYSICAL CHEMISTRY] A branch of chemistry dealing with chemical changes accompanying the passage of an electric current; or with the reverse process, in which a chemical reaction is used to produce an electric current. { i¦lek·trō¦kem·ə·strē }

electrochromatography [ANALYTICAL CHEMISTRY] Type of chromatography that utilizes application of an electric potential to produce an electric differential. Also known as electropherography. { i¦lek·trō,krō·mə'täg·rə·fē }

electrocratic [CHEMISTRY] Referring to the repulsion exhibited by soap films and other colloids in solutions; such repulsion involves a strong osmotic contribution but is largely controlled by electrical forces. { i,lek·trō'krad·ik }

electrocyclic reaction [PHYSICAL CHEMISTRY] The interconversion of a linear π-system containing n π-electrons and a cyclic molecule containing $(n-2)$ π-electrons which is formed by joining the ends of the linear molecule. { i,lek·trō¦sī·klik rē'ak·shən }

electrodecantation [PHYSICAL CHEMISTRY] A modification of electrodialysis in which a cell is divided into three sections by two membranes and electrodes are placed in the end sections; colloidal matter is concentrated at the sides and bottom of the middle section, and the liquid that floats to the top is drawn off. { i¦lek·trō,dē,kan'tā·shən }

electrode efficiency [PHYSICAL CHEMISTRY] The ratio of the amount of metal actually deposited in an electrolytic cell to the amount that could theoretically be deposited as a result of electricity passing through the cell. { i'lek,trōd ə,fish·ən·sē }

electrodeposition analysis [ANALYTICAL CHEMISTRY] An electroanalytical technique in which an element is quantitatively deposited on an electrode. { i¦lek·trō,dep·ə'zish·ən ə'nal·ə·səs }

electrode potential [PHYSICAL CHEMISTRY] The voltage existing between an electrode and the solution or electrolyte in which it is immersed; usually, electrode potentials are referred to a standard electrode, such as the hydrogen electrode. Also known as electrode voltage. { i'lek,trōd pə'ten·chəl }

electrode voltage [PHYSICAL CHEMISTRY] *See* electrode potential. { i'lek,trōd ,vōl·tij }

electrodialysis [PHYSICAL CHEMISTRY] Dialysis that is conducted with the aid of an electromotive force applied to electrodes adjacent to both sides of the membrane. { i¦lek·trō·dī'al·ə·səs }

electrodialyzer [PHYSICAL CHEMISTRY] An instrument used to conduct electrodialysis. { i,lek·trō'dī·ə,līz·ər }

electrofocusing [PHYSICAL CHEMISTRY] *See* isoelectric focusing. { i,lek·trō'fō·kəs·iŋ }

electrogenerated chemiluminescence [PHYSICAL CHEMISTRY] *See* electrochemiluminescence. { i,lek·trō¦jen·ə,rād·əd ,kem·ē·,lüm·ə'nes·əns }

electrogravimetry [ANALYTICAL CHEMISTRY] Electrodeposition analysis in which the quantities of metals deposited may be determined by weighing a suitable electrode before and after deposition. { i,lek·trə·grə'vim·ə·trē }

electrohydraulic effect [PHYSICAL CHEMISTRY] Generation of shock waves and highly reactive species in a liquid as the result of application of very brief but powerful electrical pulses. { i¦lek·trō·hī¦dròl·ik i'fekt }

electrohydrodynamic ionization mass spectroscopy [SPECTROSCOPY] A technique for analysis of nonvolatile molecules in which the nonvolatile material is dissolved in a volatile solvent with a high dielectric constant such as glycerol, and high electric-field gradients at the surface of droplets of the liquid solution induce ion emission. { i¦lek·trō¦hī·drō·dī'nam·ik ,ī·ə·nə'zā·shən ¦mas spek'träs·kə·pē }

electrokinetic phenomena [PHYSICAL CHEMISTRY] The phenomena associated with movement of charged particles through a continuous medium or with the movement of a continuous medium over a charged surface. { i¦lek·trō·kə'ned·ik fə'näm·ə·nə }

electrolysis [PHYSICAL CHEMISTRY] A method by which chemical reactions are carried out by passage of electric current through a solution of an electrolyte or through a molten salt. { i,lek'trä·lə·səs }

electrolyte [PHYSICAL CHEMISTRY] A chemical compound which when molten or dissolved in certain solvents, usually water, will conduct an electric current. { i'lek·trə,līt }

electrolytic analysis [ANALYTICAL CHEMISTRY] Basic electrochemical technique for quantitative analysis of conducting solutions containing oxidizable or reducible material; measurement is based on the weight of material plated out onto the electrode. { i'lek·trə,lid·ik ə'nal·ə·səs }

electrolytic cell [PHYSICAL CHEMISTRY] A cell consisting of electrodes immersed in an electrolyte solution, for carrying out electrolysis. { i'lek·trə,lid·ik 'sel }

electrolytic conductance [PHYSICAL CHEMISTRY] The transport of electric charges, under electric potential differences, by charged particles (called ions) of atomic or larger size. { i'lek·trə,lid·ik kən'dək·təns }

electrolytic conductivity [PHYSICAL CHEMISTRY] The conductivity of a medium in which the transport of electric charges, under electric potential differences, is by particles of atomic or larger size. { i'lek·trə,lid·ik ,kän·dək'tiv·əd·ē }

electrolytic dissociation [CHEMISTRY] The ionization of a compound in a solution. { i'lek·trə,lid·ik di,sō·sē'ā·shən }

electrolytic migration [PHYSICAL CHEMISTRY] The motions of ions in a liquid under the action of an electric field. { i¦lek·trə,lid·ik mī'grā·shən }

electrolytic polarization [PHYSICAL CHEMISTRY] The existence of a minimum potential difference necessary to cause a steady current to flow through an electrolytic cell, resulting from the tendency of the products of electrolysis to recombine. { i¦lek·trə,lid·ik pō·lər·ə'zā·shən }

electrolytic potential [PHYSICAL CHEMISTRY] Difference in potential between an electrode and the immediately adjacent electrolyte, expressed in terms of some standard electrode difference. { i'lek·trə,lid·ik pə'ten·chəl }

electrolytic process [PHYSICAL CHEMISTRY] An electrochemical process involving the principles of electrolysis, especially as relating to the separation and deposition of metals. { i'lek·trə,lid·ik 'präs·əs }

electrolytic separation [PHYSICAL CHEMISTRY] Separation of isotopes by electrolysis, based on differing rates of discharge at the electrode of ions of different isotopes. { i'lek·trə,lid·ik ,sep·ə'rā·shən }

electrolytic solution [PHYSICAL CHEMISTRY] A solution made up of a solvent and an ionically dissociated solute; it will conduct electricity, and ions can be separated from the solution by deposition on an electrically charged electrode. { i'lek·trə,lid·ik sə'lü·shən }

electromigration [ANALYTICAL CHEMISTRY] A process used to separate isotopes or ionic species by the differences in their ionic mobilities in an electric field. { i¦lek·trō·mī'grā·shən }

electromodulation [SPECTROSCOPY] Modulation spectroscopy in which changes in transmission or reflection spectra induced by a perturbing electric field are measured. { i¦lek·trō,mäj·ə'lā·shən }

electromotance [PHYSICAL CHEMISTRY] *See* electromotive force. { i¦lek·trō'mōt·əns }

electromotive force [PHYSICAL CHEMISTRY] **1.** The difference in electric potential that exists between two dissimilar electrodes immersed in the same electrolyte or otherwise connected by ionic conductors. **2.** The resultant of the relative electrode potential of the two dissimilar electrodes at which electrochemical reactions occur. Abbreviated emf. Also known as electromotance. { i¦lek·trə'mōd·iv 'fōrs }

electromotive series [PHYSICAL CHEMISTRY] *See* electrochemical series. { i¦lek·trə'mōd·iv 'sir·ēz }

electron acceptor [PHYSICAL CHEMISTRY] **1.** An atom or part of a molecule joined by a covalent bond to an electron donor. **2.** *See* electrophile. { i'lek,trän ak'sep·tər }

electron-capture detector [ANALYTICAL CHEMISTRY] Extremely sensitive gas chromatography detector that is a modification of the argon ionization detector, with conditions adjusted to favor the formation of negative ions. { i'lek,trän ,kap·chər di'tek·tər }

electron distribution curve [PHYSICAL CHEMISTRY] A curve indicating the electron distribution among the different available energy levels of a solid substance. { i'lek,trän dis·trə'byü·shən ,kərv }

electron donor [PHYSICAL CHEMISTRY] **1.** An atom or part of a molecule which supplies both electrons of a duplet forming a covalent bond. **2.** *See* nucleophile. { i'lek‚trän ‚dō·nər }

electron-dot formula [CHEMISTRY] *See* Lewis structure. { i¦lek‚trän ¦dät ‚fȯr·myə·lə }

electronegative [PHYSICAL CHEMISTRY] Pertaining to an atom or group of atoms that has a relatively great tendency to attract electrons to itself. { i¦lek·trō'neg·əd·iv }

electronegative potential [PHYSICAL CHEMISTRY] Potential of an electrode expressed as negative with respect to the hydrogen electrode. { i¦lek·trō'neg·əd·iv pə'ten·chəl }

electron energy loss spectroscopy [SPECTROSCOPY] A technique for studying atoms, molecules, or solids in which a substance is bombarded with monochromatic electrons, and the energies of scattered electrons are measured to determine the distribution of energy loss. Abbreviated EELS. { i'lek‚trän 'en·ər·jē ‚lȯs spek'träs·kə·pē }

electroneutrality principle [PHYSICAL CHEMISTRY] The principle that in an electrolytic solution the concentrations of all the ionic species are such that the solution as a whole is neutral. { i¦lek·trō·nü'tral·əd·ē ‚prin·sə·pəl }

electron exchanger [ORGANIC CHEMISTRY] *See* redox polymer. { i'lek‚trän iks‚chān·jər }

electronic absorption spectrum [SPECTROSCOPY] Spectrum resulting from absorption of electromagnetic radiation by atoms, ions, and molecules due to excitations of their electrons. { i‚lek'trän·ik əb'sȯrp·shən ‚spek·trəm }

electronic band spectrum [SPECTROSCOPY] Bands of spectral lines associated with a change of electronic state of a molecule; each band corresponds to certain vibrational energies in the initial and final states and consists of numerous rotational lines. { i‚lek'trän·ik 'band ‚spek·trəm }

electronic emission spectrum [SPECTROSCOPY] Spectrum resulting from emission of electromagnetic radiation by atoms, ions, and molecules following excitations of their electrons. { i‚lek'trän·ik i'mish·ən ‚spek·trəm }

electronic energy curve [PHYSICAL CHEMISTRY] A graph of the energy of a diatomic molecule in a given electronic state as a function of the distance between the nuclei of the atoms. { i‚lek'trän·ik 'en·ər·jē ‚kərv }

electronic spectrum [SPECTROSCOPY] Spectrum resulting from emission or absorption of electromagnetic radiation during changes in the electron configuration of atoms, ions, or molecules, as opposed to vibrational, rotational, fine-structure, or hyperfine spectra. { i‚lek'trän·ik 'spek·trəm }

electron nuclear double resonance [SPECTROSCOPY] A type of electron paramagnetic resonance (EPR) spectroscopy permitting greatly enhanced resolution, in which a material is simultaneously irradiated at one of its EPR frequencies and by a second oscillatory field whose frequency is swept over the range of nuclear frequencies. Abbreviated ENDOR. { i'lek‚trän ¦nü·klē·ər ¦dəb·əl 'rez·ən·əns }

electron pair [PHYSICAL CHEMISTRY] A pair of valence electrons which form a nonpolar bond between two neighboring atoms. { i'lek‚trän 'per }

electron pair bond [CHEMISTRY] *See* covalent bond. { i'lek‚trän 'per ‚bänd }

electron probe x-ray microanalysis [ANALYTICAL CHEMISTRY] An analytical technique that uses a narrow electron beam, usually with a diameter less than 1 millimeter, focused on a solid specimen to excite an x-ray spectrum that provides qualitative and quantitative information characteristic of the elements in the sample. Abbreviated EPXMA. { i¦lek‚trän ‚prōb ¦eks‚rā ‚mī·krō·ə'nal·ə·səs }

electron spectroscopy [SPECTROSCOPY] The study of the energy spectra of photoelectrons or Auger electrons emitted from a substance upon bombardment by electromagnetic radiation, electrons, or ions; used to investigate atomic, molecular, or solid-state structure, and in chemical analysis. { i'lek‚trän spek'träs·kə·pē }

electron spectroscopy for chemical analysis [SPECTROSCOPY] *See* x-ray photoelectron spectroscopy. { i'lek‚trän spek'träs·kə·pē fər 'kem·i·kəl ə'nal·ə·səs }

electron spectrum [SPECTROSCOPY] Visual display, photograph, or graphical plot of the intensity of electrons emitted from a substance bombarded by x-rays or other radiation as a function of the kinetic energy of the electrons. { i'lek‚trän 'spek·trəm }

electroosmosis [PHYSICAL CHEMISTRY] The movement in an electric field of liquid with respect to colloidal particles immobilized in a porous diaphragm or a single capillary tube. { i¦lek·trō·äs'mō·səs }

electropherography [ANALYTICAL CHEMISTRY] *See* electrochromatography. { i¦lek·trō·fə'räg·rə·fē }

electrophile [PHYSICAL CHEMISTRY] An electron-deficient ion or molecule that takes part in an electrophilic process. { i'lek·trō,fīl }

electrophilic [PHYSICAL CHEMISTRY] **1.** Pertaining to any chemical process in which electrons are acquired from or shared with other molecules or ions. **2.** Referring to an electron-deficient species. { i¦lek·trō'fil·ik }

electrophilic reagent [PHYSICAL CHEMISTRY] A reactant which accepts an electron pair from a molecule, with which it forms a covalent bond. { i¦lek·trō¦fil·ik rē'a·jənt }

electrophoresis [PHYSICAL CHEMISTRY] An electrochemical process in which colloidal particles or macromolecules with a net electric charge migrate in a solution under the influence of an electric current. Also known as cataphoresis. { i¦lek·trō·fə'rē·səs }

electrophoretic effect [PHYSICAL CHEMISTRY] Retarding effect on the characteristic motion of an ion in an electrolytic solution subjected to a potential gradient, which results from motion in the opposite direction by the ion atmosphere. { i¦lek·trō·fə'red·ik i'fekt }

electropositive [PHYSICAL CHEMISTRY] Pertaining to elements, ions, or radicals that tend to give up or lose electrons. { i¦lek·trə'päz·əd·iv }

electropositive potential [PHYSICAL CHEMISTRY] Potential of an electrode expressed as positive with respect to the hydrogen electrode. { i¦lek·trə¦päz·əd·iv pə'ten·chəl }

electroreflectance [SPECTROSCOPY] Electromodulation in which reflection spectra are studied. Abbreviated ER. { i¦lek·trō·ri'flek·təns }

electrorheological fluid [PHYSICAL CHEMISTRY] A colloidal suspension of finely divided particles in a carrier liquid, usually an insulating oil, whose rheological properties are changed through an increase in resistance when an electric field is applied. { i¦lek·trō,rē·ə¦läj·ə·kəl 'flü·əd }

electrostatic bond [PHYSICAL CHEMISTRY] A valence bond in which two atoms are kept together by electrostatic forces caused by transferring one or more electrons from one atom to the other. { i¦lek·trə'stad·ik 'bänd }

electrostatic valence rule [PHYSICAL CHEMISTRY] The postulate that in a stable ionic structure the valence of each anion, with changed sign, equals the sum of the strengths of its electrostatic bonds to the adjacent cations. { i'lek·trə,stad·ik 'vā·ləns ,rül }

electrosynthesis [CHEMISTRY] A reaction in which synthesis occurs as the result of an electric current. { i¦lek·trō'sin·thə·səs }

electrovalence [PHYSICAL CHEMISTRY] The valence of an atom that has formed an ionic bond. { i¦lek·trō'vā·ləns }

electrovalent bond [PHYSICAL CHEMISTRY] *See* ionic bond. { i¦lek·trō¦vā·lənt 'bänd }

element [CHEMISTRY] A substance made up of atoms with the same atomic number; common examples are hydrogen, gold, and iron. Also known as chemical element. { 'el·ə·mənt }

element 104 [CHEMISTRY] The first element beyond the actinide series, and the twelfth transuranium element; the longest-lived isotope identified has a half-life of 65 seconds and mass number 261. { 'el·ə·mənt ,wən ,ō 'fōr }

element 105 [CHEMISTRY] An artificial element whose isotope of mass number 260 was discovered by bombarding californium-249 with nitrogen-15 ions in a heavy-ion linear accelerator. { 'el·ə·mənt ,wən ,ō 'fīv }

element 106 [CHEMISTRY] An artificial element whose isotope of mass number 263 was discovered by bombarding californium-249 with oxygen-18 ions in a heavy-ion linear accelerator, and whose isotope of mass number 259 was discovered by bombarding lead-207 and lead-208 with chromium-54 ions in a heavy-ion cyclotron. { 'el·ə·mənt ,wən ,ō 'siks }

element 107 [CHEMISTRY] An artificial element whose isotope of mass number 262 was discovered by bombarding bismuth-209 with chromuim-54 ions in a heavy-ion linear accelerator. { 'el·ə·mənt ˌwən ˌō 'sev·ən }

element 108 [CHEMISTRY] An artificial element whose isotope of mass number 265 was discovered by bombarding lead-208 with iron-58 ions in a heavy-ion linear accelerator. { 'el·ə·mənt ˌwən ˌō 'āt }

element 109 [CHEMISTRY] An artificial element; a single atom of its isotope of mass number 266 was produced by bombarding bismuth-266 with iron-58 ions in a heavy-ion linear accelerator. { 'el·ə·mənt ˌwən ˌō 'nīn }

elementary process [PHYSICAL CHEMISTRY] In chemical kinetics, the particular events at the atomic or molecular level which make up an overall reaction. { ˌel·ə'men·trē 'präs·əs }

elementary reaction [ORGANIC CHEMISTRY] A reaction which involves only a single transition state with no intermediates. Also known as step. { ˌel·ə'men·trē rē'ak·shən }

eleostearic acid [ORGANIC CHEMISTRY] $CH_3(CH_2)_7(CH{:}CH)_3(CH_2)_3COOH$ A colorless, water-insoluble, crystalline, unsaturated fatty acid; the glycerol ester is a chief component of tung oil. { ¦el·ē·ō'stir·ik 'as·əd }

elimination reaction [ORGANIC CHEMISTRY] A chemical reaction involving elimination of some portion of a reactant compound, with the production of a second compound. { əˌlim·ə'nā·shən rēˌak·shən }

ellagic acid [ORGANIC CHEMISTRY] $C_{14}H_6O_8$ A compound isolated from tannins as yellow crystals that are minimally soluble in hot water. Also known as gallogen. { e'laj·ik 'as·əd }

eluant [CHEMISTRY] A liquid used to extract one material from another, as in chromatography. { 'el·yə·wənt }

eluant gas [ANALYTICAL CHEMISTRY] *See* carrier gas. { el'yü·ənt ˌgas }

eluate [CHEMISTRY] The solution that results from the elution process. { 'el·yəˌwāt }

elution [CHEMISTRY] The removal of adsorbed species from a porous bed or chromatographic column by means of a stream of liquid or gas. { ē'lü·shən }

emf [PHYSICAL CHEMISTRY] *See* electromotive force.

emission flame photometry [ANALYTICAL CHEMISTRY] A form of flame photometry in which a sample solution to be analyzed is aspirated into a hydrogen-oxygen or acetylene-oxygen flame; the line emission spectrum is formed, and the line or band of interest is isolated with a monochromator and its intensity measured photoelectrically. { i'mish·ən ˌflām fō'täm·ə·trē }

emission lines [SPECTROSCOPY] Spectral lines resulting from emission of electromagnetic radiation by atoms, ions, or molecules during changes from excited states to states of lower energy. { i'mish·ən ˌlīnz }

emission spectrometer [SPECTROSCOPY] A spectrometer that measures percent concentrations of preselected elements in samples of metals and other materials; when the sample is vaporized by an electric spark or arc, the characteristic wavelengths of light emitted by each element are measured with a diffraction grating and an array of photodetectors. { i'mish·ən spek'träm·əd·ər }

emission spectrum [SPECTROSCOPY] Electromagnetic spectrum produced when radiations from any emitting source, excited by any of various forms of energy, are dispersed. { i'mish·ən ˌspek·trəm }

emodin [ORGANIC CHEMISTRY] $C_{14}H_4O_2(OH)_3CH_3$ Orange needles crystallizing from alcohol solution, melting point 256-257°C, practically insoluble in water, soluble in alcohol and aqueous alkali hydroxide solutions, occurs as the rhamnoside in plants such as rhubarb root and alder buckthorn; used as a laxative. { 'em·ə·dən }

empirical formula [CHEMISTRY] A chemical formula that indicates the composition of a compound in terms of the relative numbers and kinds of atoms in the simplest ratio; for example, the empirical formula for fluorobenzene is C_6H_5F. { em'pir·ə·kəl 'fȯr·myə·lə }

emulsification [CHEMISTRY] The process of dispersing one liquid in a second immiscible liquid; the largest group of emulsifying agents are soaps, detergents, and other compounds, whose basic structure is a paraffin chain terminating in a polar group. { əˌməl·sə·fə'kā·shən }

emulsion [CHEMISTRY] A stable dispersion of one liquid in a second immiscible liquid, such as milk (oil dispersed in water). { ə'məl·shən }

emulsion breaking [CHEMISTRY] In an emulsion, the combined sedimentation and coalescence of emulsified drops of the dispersed phase so that they will settle out of the carrier liquid; can be accomplished mechanically (in settlers, cyclones, or centrifuges) with or without the aid of chemical additives to increase the surface tension of the droplets. { ə'məl·shən ˌbrāk·iŋ }

emulsion polymerization [ORGANIC CHEMISTRY] A polymerization reaction that occurs in one phase of an emulsion. { ə'məl·shən pəˌlim·ə·rə'zā·shən }

enamine [ORGANIC CHEMISTRY] An amine in which there is a carbon-to-carbon double bond adjacent to the nitrogen, —C═C—N—; considered to be the nitrogen analog of an enol. { 'en·əˌmēn }

enantiomer [CHEMISTRY] *See* enantiomorph. { əˈnan·teˈo·mər }

enantiomeric excess [ORGANIC CHEMISTRY] In an asymmetric synthesis, a chemical yield that contains more of the desired enantiomer than other products. { əˈnan·tē·ōˈmer·ik ek'ses }

enantiomorph [CHEMISTRY] One of an isomeric pair of either crystalline forms or chemical compounds whose molecules are nonsuperimposable mirror images. Also known as enantiomer; optical antipode; optical isomer. { ə'nan·tē·əˌmȯrf }

enantiomorphism [CHEMISTRY] A phenomenon of mirror-image relationship exhibited by right-handed and left-handed crystals or by the molecular structures of two stereoisomers. { əˈnan·tē·ə'mȯrˌfiz·əm }

enantioselective reaction [ORGANIC CHEMISTRY] *See* stereoselective reaction. { əˈnan·tē·ə·siˈlek·tiv rē'ak·shən }

enantiotopic ligand [ORGANIC CHEMISTRY] A ligand whose replacement or addition gives rise to enantiomers. { əˈnan·tē·əˈtäp·ik 'līg·ənd }

enantiotropy [CHEMISTRY] The relation of crystal forms of the same substance in which one form is stable above the transition-point temperature and the other stable below it, so that the forms can change reversibly one into the other. { əˌnan·te'ä·trə·pē }

encounter [PHYSICAL CHEMISTRY] A group of collisions, each of which consists of two molecules that collide without reacting and do not separate immediately because of the cage of surrounding molecules. { en'kau̇n·tər }

endo- [ORGANIC CHEMISTRY] Prefix that denotes inward-directed valence bonds of a six-membered ring in its boat form. { 'en·dō }

endocyclic double bond [ORGANIC CHEMISTRY] In a molecular structure, a double bond that is part of the ring system. { ˈen·dō'sī·klik ˈdəb·əl 'bänd }

endoergic [PHYSICAL CHEMISTRY] *See* endothermic. { ˈen·dōˈər·jik }

ENDOR [SPECTROSCOPY] *See* electron nuclear double resonance. { 'enˌdȯr }

endosulfan [ORGANIC CHEMISTRY] $C_9H_6Cl_6O_3S$ A tan solid that melts between 70 and 100°C, used as an insecticide and miticide on vegetable and forage crops, on ornamental flowers, and in controlling termites and tsetse flies. { ˈen·dō'səlˌfan }

endotherm [PHYSICAL CHEMISTRY] In differential thermal analysis, a graph of the temperature difference between a sample compound and a thermally inert reference compound (commonly aluminum oxide) as the substances are simultaneously heated to elevated temperatures at a predetermined rate, and the sample compound undergoes endothermal or exothermal processes. { 'en·dəˌthərm }

endothermic [PHYSICAL CHEMISTRY] Pertaining to a chemical reaction which absorbs heat. Also known as endoergic. { ˌen·də'thər·mik }

end point [ANALYTICAL CHEMISTRY] That stage in the titration at which an effect, such as a color change, occurs, indicating that a desired point in the titration has been reached. { 'end ˌpȯint }

end radiation [SPECTROSCOPY] *See* quantum limit. { 'end ˌrād·ēˌā·shən }

endrin [ORGANIC CHEMISTRY] $C_{12}H_8OCl_6$ Poisonous, white crystals that are insoluble in water; it is used as a pesticide and is a stereoisomer of dieldrin, another pesticide. { 'en·drən }

energy profile [PHYSICAL CHEMISTRY] A diagram of the energy changes that take place during a reaction in a chemical system. { 'en·ər·jē ˌprōˌfīl }

English degree [CHEMISTRY] A unit of water hardness, equal to 1 part calcium carbonate to 70,000 parts water; equivalent to 1 grain of calcium carbonate per gallon of water. Also known as Clark degree. { 'iŋ·glish di,grē }

English vermilion [INORGANIC CHEMISTRY] Bright vermilion pigment of precipitated mercury sulfide; in paints, it tends to darken when exposed to light. { 'iŋ·glish vər'mil·yən }

enhanced line [SPECTROSCOPY] *See* enhanced spectral line. { en'hanst 'līn }

enhanced spectral line [SPECTROSCOPY] A spectral line of a very hot source, such as a spark, whose intensity is much greater than that of a line in a flame or arc spectrum. Also known as enhanced line. { en'hanst 'spek·trəl ,līn }

enium ion [ORGANIC CHEMISTRY] A cationic portion of an ionic species in which the valence shell of a positively charged nonmetallic atom has two electrons less than normal, and the charged entity has one covalent bond less than the corresponding uncharged species; used as a suffix with the root name. Also known as ylium ion. { 'en·ē·əm ,ī·ən }

enol [ORGANIC CHEMISTRY] An organic compound with a hydroxide group adjacent to a double bond; varies with a ketone form in the effect known as enol-keto tautomerism; an example is the compound $CH_3COH{=}CHCO_2C_2H_5$. { 'ē,nȯl }

enolate anion [ORGANIC CHEMISTRY] The delocalized anion which is left after the removal of a proton from an enol, or of the carbonyl compound in equilibrium with the enol. { 'ē·nə,lāt 'an,ī·ən }

enol-keto tautomerism [ORGANIC CHEMISTRY] The tautomeric migration of a hydrogen atom from an adjacent carbon atom to a carbonyl group of a keto compound to produce the enol form of the compound; the reverse process of hydrogen atom migration also occurs. { ¦ē·nȯl ¦kēd·ō tȯ'tä·mə,riz·əm }

entering group [ORGANIC CHEMISTRY] An atom or group that becomes bonded to the main portion of the substrate during a chemical reaction. { 'en·tər·iŋ ,grüp }

enthalpimetric analysis [ANALYTICAL CHEMISTRY] Generic designation for a group of modern thermochemical methodologies such as thermometric enthalpy titrations which rely on monitoring the temperature changes produced in adiabatic calorimeters by heats of reaction occurring in solution; in contradistinction, classical methods of thermoanalysis such as thermogravimetry focus primarily on changes occurring in solid samples in response to externally imposed programmed alterations in temperature. { en,thal·pə'me·trik ə'nal·ə·səs }

enthalpy of reaction [PHYSICAL CHEMISTRY] The change in enthalpy accompanying a chemical reaction. { en'thal·pē əv rē'ak·shən }

enthalpy titration [ANALYTICAL CHEMISTRY] *See* thermometric titration. { en'thal·pē tī'trā·shən }

enthalpy of transition [PHYSICAL CHEMISTRY] The change of enthalpy accompanying a phase transition. { en'thal·pē əv tran'zish·ən }

entrance slit [SPECTROSCOPY] Narrow slit through which passes the light entering a spectrometer. { 'en·trəns ,slit }

entropy of activation [PHYSICAL CHEMISTRY] The difference in entropy between the activated complex in a chemical reaction and the reactants. { 'en·trə·pē əv,ak·tə'vā·shən }

entropy of mixing [PHYSICAL CHEMISTRY] After mixing substances, the difference between the entropy of the mixture and the sum of the entropies of the components of the mixture. { 'en·trə·pē əv 'mik·siŋ }

entropy of transition [PHYSICAL CHEMISTRY] The heat absorbed or liberated in a phase change divided by the absolute temperature at which the change occurs. { 'en·trə·pē əv tran'zish·ən }

eosin [ORGANIC CHEMISTRY] $C_{20}H_8O_5Br_4$ **1.** A red fluorescent dye in the form of triclinic crystals that are insoluble in water; used chiefly in cosmetics and as a toner. Also known as bromeosin; bromo acid; eosine; tetrabromofluorescein. **2.** The red to brown crystalline sodium or potassium salt of this dye; used in organic pigments, as a biological stain, and in pharmaceuticals. { 'ē·ə·sən }

ephedrine [ORGANIC CHEMISTRY] $C_{10}H_{15}NO$ A white, crystalline, water-soluble alkaloid present in several *Ephedra* species and also produced synthetically; a sympathomimetic amine, it is used for its action on the bronchi, blood pressure, blood vessels, and central nervous system. { ə'fed·rən }

epi- [ORGANIC CHEMISTRY] A prefix used in naming compounds to indicate the presence of a bridge or intramolecular connection. { 'ep·ē }

epichlorohydrin [ORGANIC CHEMISTRY] C_3H_5OCl A colorless, unstable liquid, insoluble in water; used as a solvent for resins. { ¦ep·ə,klȯr·ə'hī·drən }

epihydrin alcohol [ORGANIC CHEMISTRY] *See* glycidol. ¦ep·ə¦hi·drən 'al·kə,hȯl }

epimer [ORGANIC CHEMISTRY] A type of isomer in which the difference between the two compounds is the relative position of the H (hydrogen) group and OH (hydroxyl) group on the last asymmetric C (carbon) atom of the chain, as in the sugars D-glucose and D-mannose. { 'ep·ə·mər }

epimerization [ORGANIC CHEMISTRY] In an optically active compound that contains two or more asymmetric centers, a process in which only one of these centers is altered by some reaction to form an epimer. { ,e·pim·ə·rə'zā·shən }

EPN [ORGANIC CHEMISTRY] *See* O-ethyl-O-*para*-nitrophenyl phenylphosphonothioate.

epoxidation [ORGANIC CHEMISTRY] Reaction yielding an epoxy compound, such as the conversion of ethylene to ethylene oxide. { e,päk·sə'dā·shən }

epoxide [ORGANIC CHEMISTRY] **1.** A reactive group in which an oxygen atom is joined to each of two carbon atoms which are already bonded. **2.** A three-membered cyclic ether. Also known as oxirane. **3.** *See* ethylene oxide. { e'päk,sīd }

epoxy- [ORGANIC CHEMISTRY] A prefix indicating presence of an epoxide group in a molecule. { ə'päk·sē }

1,2-epoxyethane [ORGANIC CHEMISTRY] *See* ethylene oxide. { ¦wən ¦tü ə¦päk·sē 'e ,thān }

epoxy resin [ORGANIC CHEMISTRY] A polyether resin formed originally by the polymerization of bisphenol A and epichlorohydrin, having high strength, and low shrinkage during curing; used as a coating, adhesive, casting, or foam. { ə'päk·sē 'rez·ən }

EPXMA [ANALYTICAL CHEMISTRY] *See* electron probe x-ray microanalysis.

equation [CHEMISTRY] A symbolic expression that represents in an abbreviated form the laboratory observations of a chemical change; an equation (such as $2H_2 + O_2 \rightarrow 2H_2O$) indicates what reactants are consumed (H_2 and O_2) and what products are formed (H_2O), the correct formula of each reactant and product, and satisfies the law of conservation of atoms in that the symbols for the number of atoms reacting equals the number of atoms in the products. { i'kwā·zhən }

equation of state [PHYSICAL CHEMISTRY] A mathematical expression which defines the physical state of a homogeneous substance (gas, liquid, or solid) by relating volume to pressure and absolute temperature for a given mass of the material. { i'kwā·zhən əv 'stāt }

equidensity technique [ANALYTICAL CHEMISTRY] Interference microscopy technique utilizing the Sabattier effect in photographic emulsions; the equidensities (lines of equal density in a photographic emulsion) are produced by exactly superimposing a positive and a negative of the same interferogram, and making a copy; used to measure photographic film emulsion density. { ¦ē·kwə¦den·səd·ē ,tek,nēk }

equilibrium [CHEMISTRY] *See* chemical equilibrium. { ,ē·kwə'lib·rē·əm }

equilibrium constant [CHEMISTRY] A constant at a given temperature such that when a reversible chemical reaction $cC + bB = gG + hH$ has reached equilibrium, the value of this constant K^0 is equal to

$$\frac{a_G^g a_H^h}{a_C^c a_B^b}$$

where a_G, a_H, a_C, and a_B represent chemical activities of the species G, H, C, and B at equilibrium. { ,ē·kwə'lib·rē·əm ,kän·stənt }

equilibrium diagram [PHYSICAL CHEMISTRY] A phase diagram of the equilibrium relationship between temperature, pressure, and composition in any system. { ,ē·kwə'lib·rē·əm 'dī·ə,gram }

equilibrium dialysis [ANALYTICAL CHEMISTRY] A technique used to determine the degree of ion bonding by protein; the protein solution, placed in a bag impermeable to protein but permeable to small ions, is immersed in a solution containing the diffusible ion whose binding is being studied; after equilibration of the ion across the membrane, the concentration of ion in the protein-free solution is determined; the concentration of ion in the protein solution is determined by subtraction; if binding has occurred, the concentration of ion in the protein solution must be greater. { ˌē·kwə'lib·rē·əm dī'al·ə·səs }

equilibrium film [PHYSICAL CHEMISTRY] A liquid film that is stable or metastable at a certain thickness with respect to small changes in the thickness. { ˌē·kwə'lib·rē·əm ˌfilm }

equilibrium moisture content [PHYSICAL CHEMISTRY] The moisture content in a hydroscopic material that is being dried by contact with air at constant temperature and humidity when a definite, fixed (equilibrium) moisture content in the solid is reached. { ˌē·kwə'lib·rē·əm 'mȯis·chər ˌkän·tent }

equilibrium potential [PHYSICAL CHEMISTRY] A point in which forward and reverse reaction rates are equal in an electrolytic solution, thereby establishing the potential of an electrode. { ˌē·kwə'lib·rē·əm pə'ten·chəl }

equilibrium prism [PHYSICAL CHEMISTRY] Three-dimensional (solid) diagram for multicomponent mixtures to show the effects of composition changes on some key property, such as freezing point. { ˌē·kwə'lib·rē·əm ˌpriz·əm }

equilibrium ratio [PHYSICAL CHEMISTRY] **1.** In any system, relation of the proportions of the various components (gas, liquid) at equilibrium conditions. **2.** *See* equilibrium vaporization ratio. { ˌē·kwə'lib·rē·əm ˌrā·shō }

equilibrium solubility [PHYSICAL CHEMISTRY] The maximum solubility of one material in another (for example, water in hydrocarbons) for specified conditions of temperature and pressure. { ˌē·kwə'lib·rē·əm ˌsäl·yə'bil·əd·ē }

equilibrium still [ANALYTICAL CHEMISTRY] Recirculating distillation apparatus (no product withdrawal) used to determine vapor-liquid equilibria data. { ˌē·kwə'lib·rē·əm ˌstil }

equilibrium vaporization ratio [PHYSICAL CHEMISTRY] In a liquid-vapor equilibrium mixture, the ratio of the mole fraction of a component in the vapor phase (y) to the mole fraction of the same component in the liquid phase (x), or $y/x = K$ (the K factor). Also known as equilibrium ratio. { ˌē·kwə'lib·rē·əm ˌvā·pə·rə'zā·shən ˌrā·shō }

equipartition [CHEMISTRY] **1.** The condition in a gas where under equal pressure the molecules of the gas maintain the same average distance between each other. **2.** The equal distribution of a compound between two solvents. **3.** The distribution of the atoms in an orderly fashion, such as in a crystal. { ¦e·kwə·pär'tish·ən }

equivalence point [CHEMISTRY] The point in a titration where the amounts of titrant and material being titrated are equivalent chemically. { i'kwiv·ə·ləns ˌpȯint }

equivalent conductance [PHYSICAL CHEMISTRY] Property of an electrolyte, equal to the specific conductance divided by the number of gram equivalents of solute per cubic centimeter of solvent. { i'kwiv·ə·lənt kən'dək·təns }

equivalent nuclei [PHYSICAL CHEMISTRY] A set of nuclei in a molecule which are transformed into each other by rotations, reflections, or combinations of these operations, leaving the molecule invariant. { i'kwiv·ə·lənt 'nü·klē·ī }

equivalent weight [CHEMISTRY] The number of parts by weight of an element or compound which will combine with or replace, directly or indirectly, 1.008 parts by weight of hydrogen, 8.00 parts of oxygen, or the equivalent weight of any other element or compound. { i'kwiv·ə·lənt 'wāt }

Er [CHEMISTRY] *See* erbium.

ER [SPECTROSCOPY] *See* electroreflectance.

erbia [INORGANIC CHEMISTRY] *See* erbium oxide. { 'ər·bē·ə }

erbium [CHEMISTRY] A trivalent metallic rare-earth element, symbol Er, of the yttrium subgroup, found in euxenite, gadolinite, fergusonite, and xenotine; atomic number 68, atomic weight 167.26, specific gravity 9.051; insoluble in water, soluble in acids; melts at 1400-1500°C. { 'ər·bē·əm }

erbium halide [INORGANIC CHEMISTRY] A compound of erbium and one of the halide elements. { 'ər·bē·əm'halˌīd }

erbium nitrate [INORGANIC CHEMISTRY] $Er(NO_3)_3 \cdot 5H_2O$ Pink crystals that are soluble in water, alcohol, and acetone; may explode if it is heated or shocked. { 'ər·bē·əm 'nī,trāt }

erbium oxalate [ORGANIC CHEMISTRY] $Er_2(C_2O_4)_3 \cdot 10H_2O$ A red powder that decomposes at 575°C; used to separate erbium from common metals. { 'ər·bē·əm 'äk·sə,lāt }

erbium oxide [INORGANIC CHEMISTRY] Er_2O_3 Pink powder that is insoluble in water; used as an actuator for phosphors and in manufacture of glass that absorbs in the infrared. Also known as erbia. { 'ər·bē·əm 'äk,sīd }

erbium sulfate [INORGANIC CHEMISTRY] $Er_2(SO_4)_3 \cdot 8H_2O$ Red crystals that are soluble in water. { 'ər·bē·əm 'səl,fāt }

erbon [ORGANIC CHEMISTRY] $C_{11}H_9Cl_5O_3$ A white solid with a melting point of 49-50°C; insoluble in water; used as a herbicide for perennial broadleaf weeds. { 'ər,bän }

ergot [ORGANIC CHEMISTRY] Any of the five optically isomeric pairs of alkaloids obtained from this fungus; only the levorotatory isomers are physiologically active. { 'ər·gət }

ergotinine [ORGANIC CHEMISTRY] An alkaloid and an isomer of ergotoxine that is a 1:1:1 mixture of ergocornine, ergocristine, and ergocryptine; crystallizes in long needles from acetone solutions, melting point 229°C, and soluble in chloroform, alcohol, and absolute ether. { ər'gät·ən,ēn }

ergotoxine [ORGANIC CHEMISTRY] An alkaloid and an isomer of ergotinine that is a 1:1:1 mixture of ergocornine, ergocristine, and ergocryptine; crystallizes in orthorhombic crystals, melts at 190°C, and is soluble in methyl alcohol, ethyl alcohol, acetone, and chloroform. { ,ər·gə'täk sēn }

eriodictyol [ORGANIC CHEMISTRY] $C_{15}H_{22}O_6$ A compound isolated from *Eriodictyon californicum* as needlelike crystals from a dilute alcohol solution, sparingly soluble in boiling water, hot alcohol, and glacial acetic acid; used in medicine as an expectorant. { ,er·ē·ō'dik·tē,ȯl }

Erlenmeyer flask [CHEMISTRY] A conical glass laboratory flask, with a broad bottom and a narrow neck. { 'ər·lən,mī·ər 'flask }

Erlenmeyer synthesis [ORGANIC CHEMISTRY] Preparation of cyclic ethers by the condensation of an aldehyde with an α-acylamino acid in the presence of acetic anhydride and sodium acetate. { 'ər·lən,mī·ər 'sin·thə·səs }

erucic acid [ORGANIC CHEMISTRY] $C_{22}H_{42}O_2$ A monoethenoid acid that is the cis isomer of brassidic acid and makes up 40 to 50% of the total fatty acid in rapeseed, wallflower seed, and mustard seed; crystallizes as needles from alcohol solution, insoluble in water, soluble in ethanol and methanol. { ə'rüs·ik 'as·əd }

erythrite [ORGANIC CHEMISTRY] *See* erythritol. { 'er·ə,thrīt }

erythritol [ORGANIC CHEMISTRY] $H(CHOH)_4H$ A tetrahydric alcohol; occurs as tetragonal prisms, melting at 121°C, soluble in water; used in medicine as a vasodilator. Also known as erythrite; erythrol. { ə'rith·rə,tȯl }

erythroidine [ORGANIC CHEMISTRY] $C_{16}H_{19}NO_3$ An alkaloid existing in two forms: α-erythroidine and β-erythroidine, isolated from *Erythrina* species; β-erythroidine has an action similar to that of curare as a skeletal muscle relaxant. { ,er·ə'thrō·ə,dēn }

erythrol [ORGANIC CHEMISTRY] *See* erythritol. { 'er·ə,thrȯl }

erythrophleine [ORGANIC CHEMISTRY] $C_{24}H_{39}NO_5$ An alkaloid isolated from the bark of *Erythrophleum guineense*; used in medicine experimentally for its digitalislike action. { ə,rith·rə'flē·ən }

erythrose [ORGANIC CHEMISTRY] $HOCH_2(CHOH)_2CHO$ A tetrose sugar obtained from erythrol; a syrupy liquid at room temperature. { 'er·ə,thrōs }

erythrosin [ORGANIC CHEMISTRY] $C_{13}H_{18}O_6N_2$ A red compound obtained by reacting tyrosine with nitric acid. { ə'rith·rə·sən }

Es [CHEMISTRY] *See* einsteinium.

escaping tendency [PHYSICAL CHEMISTRY] The tendency of a solute species to escape from solution; related to the chemical potential of the solute. { ə'skap·iŋ ,ten·dən·sē }

Eschka mixture [ANALYTICAL CHEMISTRY] A mixture of two parts magnesium oxide and one part anhydrous sodium carbonate; used as a fusion mixture for determining sulfur in coal. { 'esh·kə ˌmiks·chər }

Eschweiler-Clarke modification [ORGANIC CHEMISTRY] A modification of the Leuckart reaction, involving reductive alkylation of ammonia or amines (except tertiary amines) by formaldehyde and formic acid. { ¦esh·vīl·ər ¦klärk ˌmäd·ə·fə'kā·shən }

eserine [ORGANIC CHEMISTRY] *See* physostigmine. { 'es·əˌrēn }

ester [ORGANIC CHEMISTRY] The compound formed by the elimination of water and the bonding of an alcohol and an organic acid. { 'es·tər }

ester gum [ORGANIC CHEMISTRY] A compound obtained by forming an ester of a natural resin with a polyhydric alcohol; used in varnishes, paints, and cellulosic lacquers. Also known as rosin ester. { 'es·tər ˌgəm }

ester hydrolysis [ORGANIC CHEMISTRY] A reaction in which an ester is converted into its alcohol and acid moieties. Also known as esterolysis. { ¦e·stər hī'dräl·ə·səs }

esterification [ORGANIC CHEMISTRY] A chemical reaction whereby esters are formed. { eˌster·ə·fə'kā·shən }

esterolysis [ORGANIC CHEMISTRY] *See* ester hydrolysis. { ˌe·stər'äl·ə·səs }

estersil [ORGANIC CHEMISTRY] Hydrophobic silica powder, an ester of —SiOH with a monohydric alcohol; used as a filler in silicone rubbers, plastics, and printing inks. { 'es·tərˌsil }

estragole [ORGANIC CHEMISTRY] $C_6H_4(C_3H_5)(OCH_3)$ A colorless liquid with the odor of anise, found in basil oil, estragon oil, and anise bark oil; used in perfumes and flavorings. { 'es·trəˌgōl }

Etard reaction [ORGANIC CHEMISTRY] Direct oxidation of an aromatic or heterocyclic bound methyl group to an aldehyde by utilizing chromyl chloride or certain metallic oxides. { ā'tär rēˌak·shən }

ethamine [ORGANIC CHEMISTRY] *See* ethyl amine. { 'eth·ə·mēn }

ethane [ORGANIC CHEMISTRY] CH_3CH_3 A colorless, odorless gas belonging to the alkane series of hydrocarbons, with freezing point of −183.3°C and boiling point of −88.6°C; used as a fuel and refrigerant and for organic synthesis. { 'ethˌān }

1,2-ethanedithiol [ORGANIC CHEMISTRY] $HSCH_2CH_2SH$ A liquid, freely soluble in alcohol and in alkalies; used as a metal complexing agent. { ¦wən ¦tü ¦ethˌān'dī·əˌmēn }

ethanoic acid [ORGANIC CHEMISTRY] *See* acetic acid. { ¦eth·ə¦nō·ikˌas·əd }

ethanol [ORGANIC CHEMISTRY] C_2H_5OH A colorless liquid, miscible with water, boiling point 78.32°C; used as a reagent and solvent. Also known as ethyl alcohol; grain alcohol. { 'eth·əˌnȯl }

ethanolamine [ORGANIC CHEMISTRY] $NH_2(CH_2)_2OH$ A colorless liquid, miscible in water; used in scrubbing hydrogen sulfide (H_2S) and carbon dioxide (CO_2) from petroleum gas streams, for dry cleaning, in paints, and in pharmaceuticals. { ˌeth·ə'näl·əˌmēn }

ethanolurea [ORGANIC CHEMISTRY] $NH_2CONHCH_2CH_2OH$ A white solid; its formaldehyde condensation products are thermoplastic and water-soluble. { ¦eth·əˌnȯl·yu̇'rē·ə }

ethene [ORGANIC CHEMISTRY] *See* ethylene. { 'eˌthēn }

ethenol [ORGANIC CHEMISTRY] *See* vinyl alcohol. { 'eth·əˌnȯl }

ethephon [ORGANIC CHEMISTRY] $C_2H_6ClO_3P$ A white solid with a melting point of 74.75°C; very soluble in water; used as a growth regulator for tomatoes, apples, cherries, and walnuts. Also known as CEPHA. { 'eth·əˌfän }

ether [ORGANIC CHEMISTRY] **1.** One of a class of organic compounds characterized by the structural feature of an oxygen linking two hydrocarbon groups (such as R—O—R). **2.** $(C_2H_5)_2O$ A colorless liquid, slightly soluble in water; used as a reagent, intermediate, anesthetic, and solvent. Also known as ethyl ether. { 'e·thər }

etherification [ORGANIC CHEMISTRY] The process of making an ether from an alcohol. { ēˌthir·ə·fə'kā·shən }

ethidine [ORGANIC CHEMISTRY] *See* ethylidine. { 'eth·əˌdēn }

ethidium bromide [ORGANIC CHEMISTRY] $C_{21}H_{20}BrN_3$ Dark red crystals with a melting point of 238-240°C; used in treating trypanosomiasis in animals and as an inhibitor of deoxyribonucleic and ribonucleic acid synthesis. Also known as homidium bromide. { e'thid·ē·əm 'brō‚mīd }

ethinyl [ORGANIC CHEMISTRY] The CH_3:C-radical from acetylene. Also known as acetenyl; acetylenyl; ethynyl. { e'thīn·əl }

ethiolate [ORGANIC CHEMISTRY] $C_7H_{15}ONS$ A yellow liquid with a boiling point of 206°C; used as a preemergence herbicide for corn. { ə'thī·ə‚lāt }

ethionic acid [ORGANIC CHEMISTRY] $HO \cdot SO_2 \cdot CH_2 \cdot CH_2 \cdot SO_2OH$ An unstable diacid, known only in solution. Also known as ethylene sulfonic acid. { 'eth·ē‚än·ik 'as·əd }

ethohexadiol [ORGANIC CHEMISTRY] $C_8H_{18}O_2$ A slightly oily liquid, used as an insect repellent. { ¦eth·ō‚hek·sə'dī·ȯl }

ethoprop [ORGANIC CHEMISTRY] $C_8H_{19}O_2PS_2$ A pale yellow liquid compound, insoluble in water; used as an insecticide for soil insects and as a nematicide for plant parasitic nematodes. { 'ē·thō‚präp }

ethoxide [ORGANIC CHEMISTRY] A compound formed from ethanol by replacing the hydrogen of the hydroxy group by a monovalent metal. Also known as ethylate. { e'thäk‚sīd }

ethoxy [ORGANIC CHEMISTRY] The C_2H_5O— radical from ethyl alcohol. Also known as ethyoxyl. { e'thäk·sē }

2-ethoxyethanol [ORGANIC CHEMISTRY] *See* cellosolve. { ¦tü e¦thäk·sē'eth·ə‚nȯl }

ethoxyquin [ORGANIC CHEMISTRY] $C_{14}H_{19}NO$ A dark liquid, used as a growth regulator to protect apples and pears in storage.

ethyl [ORGANIC CHEMISTRY] **1.** The hydrocarbon radical C_2H_5. **2.** Trade name for the tetraethyllead antiknock compound in gasoline. { 'eth·əl }

ethyl acetate [ORGANIC CHEMISTRY] $CH_3COOC_2H_5$ A colorless liquid, slightly soluble in water; boils at 77°C; a medicine, reagent, and solvent. Also known as acetic ester; acetic ether; acetidin. { 'eth·əl 'as·ə‚tāt }

ethyl acetoacetate [ORGANIC CHEMISTRY] $CH_3COCH_2COOC_2H_5$ A colorless liquid, boiling at 181°C; used as a reagent, intermediate, and solvent. Also known as acetoacetic ester; diacetic ether. { 'eth·əl¦as·ə·tō¦as·ə‚tāt }

ethyl acetylene [ORGANIC CHEMISTRY] Compound with boiling point 8.1°C; insoluble in water, soluble in alcohol; used in organic synthesis. { ə'sed·əl‚en }

ethyl acrylate [ORGANIC CHEMISTRY] $C_5H_8O_2$ A colorless liquid, boiling at 99°C; used to manufacture chemicals and resins. { 'eth·əl 'ak·rə‚lāt }

ethyl alcohol [ORGANIC CHEMISTRY] *See* ethanol. { 'eth·əl 'al·kə‚hȯl }

ethyl amine [ORGANIC CHEMISTRY] A colorless liquid, boiling at 15°C, water-soluble; used as a solvent, as a dye intermediate, and in organic synthesis. Also known as aminoethane; ethamine. { 'eth·əl 'am‚ēn }

ethyl *para* aminobenzoate [ORGANIC CHEMISTRY] $C_6H_4NH_2CO_2C_2H_5$ A white powder, melting point 88–92°C, slightly soluble in ethanol and ether, very slightly soluble in water; used as a local anesthetic. Also known as benzocaine. { 'eth·əl ¦par·ə ¦am·ə·nō'ben·zə‚wāt }

ethyl amyl ketone [ORGANIC CHEMISTRY] $C_8H_{16}O$ A colorless liquid, almost insoluble in water; used in perfumery. { ¦eth·əl ¦am·əl 'ke‚tōn }

ethylate [ORGANIC CHEMISTRY] *See* ethoxide. { 'eth·ə‚lāt }

ethylation [ORGANIC CHEMISTRY] Formation of a new compound by introducing the ethyl functional group (C_2H_5). { ‚eth·ə'lā·shən }

ethyl benzene [ORGANIC CHEMISTRY] $C_6H_5C_2H_5$ A colorless liquid that boils at 136°C, insoluble in water; used in organic synthesis, as a solvent, and in making styrene. { 'eth·əl 'ben‚zēn }

ethyl benzoate [ORGANIC CHEMISTRY] $C_6H_5COOCH_2CH_3$ Colorless, aromatic liquid, boiling at 213°C, insoluble in water; used as a solvent, in flavoring extracts, and in perfumery. { 'eth·əl 'ben·zə‚wāt }

ethyl borate [ORGANIC CHEMISTRY] $B(OC_2H_5)_3$ A salt of ethanol and boric acid; colorless, flammable liquid; used in antiseptics, disinfectants, and fireproofing. Also known as boron triethoxide; triethylic borate. { 'eth·əl 'bȯr‚āt }

ethyl bromide [ORGANIC CHEMISTRY] C_2H_5Br A colorless liquid, boiling at 39°C; used as a refrigerant and in organic synthesis. { 'eth·əl 'brō‚mīd }

2-ethylbutene [ORGANIC CHEMISTRY] $CH_3CH_2(C_2H_5)CCH_2$ Colorless liquid, soluble in alcohol and organic solvents, insoluble in water; used in organic synthesis. { ¦tü ˌeth·əl'byüˌtān }
2-ethylbutyl acetate [ORGANIC CHEMISTRY] $C_2H_5CH(C_2H_5)CH_2O_2CCH_3$ Colorless liquid with mild odor; used as a solvent for resins, lacquers, and nitrocellulose. { ¦tü ¦eth·əl¦byüd·əl 'as·əˌtāt }
2-ethylbutyl alcohol [ORGANIC CHEMISTRY] $(C_2H_5)_2CHCH_2OH$ A stable, colorless liquid, miscible in most organic solvents, slightly water-soluble; used as a solvent for resins, waxes, and dyes, and in the synthesis of perfumes, drugs, and flavorings. { ¦tü ¦eth·əl¦byüd·əl 'al·kəˌhȯl }
ethyl butyl ketone [ORGANIC CHEMISTRY] $C_2H_5COC_4H_9$ A colorless liquid, boiling at 147°C; used in solvent mixtures. Also known as 3-heptanone. { ¦eth·əl ¦byüd·əl 'kēˌtōn }
ethyl butyrate [ORGANIC CHEMISTRY] $C_3H_7COOC_2H_5$ A colorless liquid, boiling at 121°C; used in flavoring extracts and perfumery. { ¦eth·əl 'byüd·əˌrāt }
ethyl caprate [ORGANIC CHEMISTRY] $CH_3(CH_2)_8COOC_2H_5$ A colorless liquid, used in the manufacture of wine bouquets and cognac essence. { ¦eth·əl 'kaˌprāt }
ethyl caproate [ORGANIC CHEMISTRY] $C_5H_{11}COOC_2H_5$ A colorless to yellow liquid, boiling at 167°C, soluble in ether and alcohol, and having a pleasant odor; used as a chemical intermediate and in the food industry as an artificial fruit essence. Also known as ethyl hexanoate; ethyl hexoate. { ¦eth·əl kə'prōˌāt }
ethyl caprylate [ORGANIC CHEMISTRY] $CH_3(CH_2)_6COOC_2H_5$ A clear, colorless liquid with a pineapple odor; used to make fruit ethers. Also known as ethyl octanoate. { ¦eth·əl 'kap·rəˌlāt }
ethyl carbamate [ORGANIC CHEMISTRY] *See* urethane. { ¦eth·əl 'kär·bəˌmāt }
ethyl carbinol [ORGANIC CHEMISTRY] *See* propyl alcohol. { ¦eth·əl'kär·bəˌnȯl }
ethyl carbonate [ORGANIC CHEMISTRY] *See* diethyl carbonate. { ¦eth·əl 'kär·bəˌnāt }
ethyl cellulose [ORGANIC CHEMISTRY] The ethyl ester of cellulose; it has film-forming properties and is inert to alkalies and dilute acids; used in adhesives, lacquers, and coatings. { ¦eth·əl'sel·yəˌlōs }
ethyl chloride [ORGANIC CHEMISTRY] C_2H_5Cl A colorless gas, liquefying at 12.2°C, slightly soluble in water; used as a solvent, in medicine, and as an intermediate. Also known as chloroethane. { ¦eth·əl 'klȯrˌīd }
ethyl chloroacetate [ORGANIC CHEMISTRY] $CH_2ClCOOC_2H_5$ A colorless liquid, boiling at 145°C; used as a poison gas, solvent, and chemical intermediate. { ¦eth·əl ˌklȯr·ō'as·əˌtāt }
ethyl cinnamate [ORGANIC CHEMISTRY] $C_6H_5CH{=}CHCOOC_2H_5$ An oily liquid with a faint cinnamon odor; used as a fixative for perfumes. Also known as ethyl phenylacrylate. { ¦eth·əl 'sin·əˌmāt }
ethyl crotonate [ORGANIC CHEMISTRY] $CH_3CHCHCO_2C_2H_5$ A compound with a pungent aroma; boiling point of 143-147°C, soluble in water, soluble in ether; one of two isomeric forms used as an organic intermediate, a solvent for cellulose esters, and as a plasticizer for acrylic resins. { ¦eth·əl 'krōt·ənˌāt }
ethyl crotonic acid [ORGANIC CHEMISTRY] $CH_3CHCC_2H_5COOH$ Colorless monoclinic crystals, subliming at 40°C; used as a peppermint flavoring. { ¦eth·əl krə'tän·ik 'as·əd }
ethyl cyanide [ORGANIC CHEMISTRY] C_2H_5CN A colorless liquid that boils at 97.1°C; poisonous. { ¦eth·əl 'sī·əˌnīd }
***S*-ethyl-*N,N*-dipropylthiocarbamate** [ORGANIC CHEMISTRY] $C_9H_{19}NOS$ An amber liquid soluble in water at 370 parts per million; used as a pre- and postemergence herbicide on vegetable crops. Abbreviated EDTC. { ¦es ¦eth·əl ¦en ¦en dī¦prō·pəlˌthī·ō'kär·bəˌmāt }
ethyl enanthate [ORGANIC CHEMISTRY] $CH_3(CH_2)_5COOC_2H_5$ A clear, colorless oil with a boiling point of 187°C; soluble in alcohol, chloroform, and ether; taste and odor are fruity; used as a flavor for liqueurs and soft drinks. Also known as cognac oil; ethyl heptanoate; ethyl oenanthate. { ¦eth·əl ə'nanˌthāt }
ethylene [ORGANIC CHEMISTRY] C_2H_4 A colorless, flammable gas, boiling at −102.7°C;

used as an agricultural chemical, in medicine, and for the manufacture of organic chemicals and polyethylene. Also known as ethene; olefiant gas. { 'eth·ə,lēn }

ethylene bromide [ORGANIC CHEMISTRY] *See* ethylene dibromide. { 'eth·ə,lēn 'brō,mīd }

ethylene carbonate [ORGANIC CHEMISTRY] $(CH_2O)_2CO$ Odorless, colorless solid with low melting point; soluble in water and organic solvents; used as a polymer and resin solvent, in solvent extraction, and in organic syntheses. { 'eth·ə,lēn 'kär·bə,nāt }

ethylene chloride [ORGANIC CHEMISTRY] $ClCH_2CH_2Cl$ A colorless, oily liquid, boiling at 83.7°C; used as a solvent and fumigant, for organic synthesis, and for ore flotation. Also known as Dutch liquid, ethylene dichloride. { 'eth·ə,lēn 'klȯr,īd }

ethylene chlorobromide [ORGANIC CHEMISTRY] CH_2BrCH_2Cl Volatile, colorless liquid with chloroformlike odor; soluble in ether and alcohol but not in water; general-purpose solvent for cellulosics; used in organic synthesis. { 'eth·ə,lēn klȯr·ə'brō,mīd }

ethylene chlorohydrin [ORGANIC CHEMISTRY] $ClCH_2CH_2OH$ A colorless, poisonous liquid, boiling at 129°C; used as a solvent and in organic synthesis. Also known as chloroethyl alcohol. { 'eth·ə,lēn klȯr·ə'hī·drən }

ethylene cyanide [ORGANIC CHEMISTRY] $C_2H_4(CN)_2$ Colorless crystals, melting at 57°C; used in organic synthesis. Also known as succinonitrile. { 'eth·ə,lēn 'sī·ə,nīd }

ethylene cyanohydrin [ORGANIC CHEMISTRY] C_3H_5ON A colorless liquid that is miscible with water and boils at 221°C. { 'eth·ə,lēn ,sī·ə·nō'hī·drən }

ethylene diacetate [ORGANIC CHEMISTRY] *See* ethylene glycol diacetate. { 'eth·ə,lēn dī'as·ə,tāt }

ethylenediamine [ORGANIC CHEMISTRY] $NH_2CH_2CH_2NH_2$ Colorless liquid, melting at 8.5°C, soluble in water; used as a solvent, corrosion inhibitor, and resin and in adhesive manufacture. { ¦eth·ə·,lēn'dī·ə,mēn }

ethylenediaminetetraacetic acid [ORGANIC CHEMISTRY] $(HOOCCH_2)_2NCH_2CH_2N(CH_2COOH)$ White crystals, slightly soluble in water and decomposing above 160°C; the sodium salt is a strong chelating agent, reacting with many metallic ions to form soluble nonionic chelate. Abbreviated EDTA. { ¦eth·ə·lēn¦dī·ə,mēn,te·trə·ə'sēd·ik 'as·əd }

ethylene dibromide [ORGANIC CHEMISTRY] $BrCH_2CH_2Br$ A colorless, poisonous liquid, boiling at 131°C; insoluble in water; used in medicine, as a solvent in organic synthesis, and in antiknock gasoline. Also known as ethylene bromide. { 'eth·ə·lēn dī'brō,mīd }

ethylene dichloride [ORGANIC CHEMISTRY] *See* ethylene chloride. { 'eth·ə·lēn dī'klȯr,īd }

ethylene glycol bis(trichloroacetate) [ORGANIC CHEMISTRY] $C_4H_4Cl_6O_4$ A white solid with a melting point of 40.3°C; used as a herbicide for cotton and soybeans. Abbreviated EGT. { 'eth·ə·lēn 'glī,kȯl ,bis·trī,klȯr·ō'as·ə,tāt }

ethylene glycol diacetate [ORGANIC CHEMISTRY] $CH_3COOCH_2CH_2OOCCH_3$ A liquid used as a solvent for oils, cellulose esters, and explosives. Also known as ethylene diacetate; glycol diacetate. { 'eth·ə·lēn 'glī,kȯl dī'as·ə,tāt }

ethyleneimine [ORGANIC CHEMISTRY] C_2H_4NH Highly corrosive liquid, colorless and clear; miscible with organic solvents and water; used as an intermediate in fuel oil production, refining lubricants, textiles, and pharmaceuticals. Also known as aziridine. { ,eth·ə'lēn·ə,mīn }

ethylene nitrate [ORGANIC CHEMISTRY] $(CH_2NO_3)_2$ An explosive yellow liquid, insoluble in water. Also known as glycol dinitrate. { 'eth·ə,lēn 'nī,trāt }

ethylene oxide [ORGANIC CHEMISTRY] **1.** $(CH_2)_2O$ A colorless gas, soluble in organic solvents and miscible in water, boiling point 11°C; used in organic synthesis, for sterilizing, and for fumigating. Also known as 1,2-epoxyethane. **2.** *See* epoxide. **3.** *See* oxirane. { 'eth·ə,lēn 'äk,sīd }

ethylene resin [ORGANIC CHEMISTRY] A thermoplastic material composed of polymers of ethylene; the resin is synthesized by polymerization of ethylene at elevated temperatures and pressures in the presence of catalysts. Also known as polyethylene; polyethylene resin. { 'eth·ə,lēn 'rez·ən }

ethylene sulfonic acid [ORGANIC CHEMISTRY] *See* ethionic acid. { 'eth·ə‚lēn səl¦fän·ik 'as·əd }

ethylethanolamine [ORGANIC CHEMISTRY] $C_2H_5NHCH_2CH_2OH$ Water-white liquid with amine odor; soluble in alcohol, ether, and water; used in dyes, insecticides, fungicides, and surface-active agents. { ¦eth·əl‚eth·ə'näl·ə‚mēn }

ethyl ether [ORGANIC CHEMISTRY] *See* ether. { ¦eth·əl 'ē·thər }

ethyl formate [ORGANIC CHEMISTRY] $HCOOC_2H_5$ A colorless liquid, boiling at 54.4°C; used as a solvent, fumigant, and larvicide and in flavors, resins, and medicines. { ¦eth·əl 'fór‚māt }

ethyl hexanoate [ORGANIC CHEMISTRY] *See* ethyl caproate. { ¦eth·əl hek'san·ə‚wāt }

ethyl hexoate [ORGANIC CHEMISTRY] *See* ethyl caproate. { ¦eth·əl 'hek·sə‚wāt }

2-ethyl hexoic acid [ORGANIC CHEMISTRY] $C_4H_9CH(C_2H_5)COOH$ A liquid that is slightly soluble in water, boils at 226.9°C, and has a mild odor; used as an intermediate to make metallic salts for paint and varnish driers, esters for plasticizers, and light metal salts for conversion of some oils to grease. { ¦tü ¦eth·əl hek'sō·ik 'as·əd }

2-ethylhexyl acetate [ORGANIC CHEMISTRY] $CH_3COOCH_2CHC_2H_5C_4H_9$ Water-white, stable liquid; used as a solvent for nitrocellulose, resins, and lacquers. Also known as octyl acetate. { ¦tü ¦eth·əl ¦hek·səl'as·ə‚tāt }

2-ethylhexyl acrylate [ORGANIC CHEMISTRY] $CH_2CHCOOCH_2CH(C_2H_5)C_4H_9$ Pleasant-smelling liquid; used as monomer for plastics, protective coatings, and paper finishes. { ¦tü ¦eth·əl 'ak·rə‚lāt }

2-ethylhexyl alcohol [ORGANIC CHEMISTRY] $C_4H_9CH(C_2H_5)CH_2OH$ Colorless, slightly viscous liquid; used as a defoaming or wetting agent, as a solvent for protective coatings, waxes, and oils, and as a raw material for plasticizers. Also known as octyl alcohol. { ¦tü ¦eth·əl 'al·kə‚hól }

2-ethylhexylamine [ORGANIC CHEMISTRY] $C_4H_9CH(C_2H_5)CH_2NH_2$ Water-white liquid with slight ammonia odor; slightly water-soluble; used to synthesize detergents, rubber chemicals, and oil additives. { ¦tü¦eth·əl·hek'sil·ə‚mēn }

2-ethylhexyl bromide [ORGANIC CHEMISTRY] $C_4H_9CH(C_2H_5)CH_2Br$ Water-white, water-insoluble liquid; used to prepare pharmaceuticals and disinfectants. { ¦tü ¦eth·əl¦hek·səl 'brō‚mīd }

2-ethylhexyl chloride [ORGANIC CHEMISTRY] $C_4H_9CH(C_2H_5)CH_2Cl$ Colorless liquid; used to synthesize cellulose derivatives, pharmaceuticals, resins, insecticides, and dyestuffs. { ¦tü ¦eth·əl¦hek·səl 'klór‚īd }

ethyl-*para*-hydroxybenzoate [ORGANIC CHEMISTRY] $HOC_6H_4COOC_2H_5$ Crystals with a melting point of 116°C that are soluble in water, alcohol, and ether; used as a preservative for pharmaceuticals. Also known as ethylparaben. { ¦eth·əl ¦par·ə hī‚dräk·sē'ben·zə‚wāt }

ethyl-2-hydroxypropionate [ORGANIC CHEMISTRY] *See* ethyl lactate. { ¦eth·əl ¦tü hī‚dräk·sē'prō·pē·ə‚nāt }

ethylic compound [ORGANIC CHEMISTRY] Generic term for ethyl compounds. { e'thil·ik 'käm‚paůnd }

ethylic ether [ORGANIC CHEMISTRY] *See* diethyl ether. { e'thil·ik 'ē·thər }

ethylidine [ORGANIC CHEMISTRY] The $CH_3 \cdot CH=$ radical from ethane, C_2H_5. Also known as ethidine. { e'thil·ə‚dēn }

ethyl iodide [ORGANIC CHEMISTRY] C_2H_5I A colorless liquid, boiling at 72.3°C; used in medicine and in organic synthesis. Also known as hydroiodic ether; iodoethane. { ¦eth·əl 'ī·ə‚dīd }

ethyl isobutylmethane [ORGANIC CHEMISTRY] *See* 2-methylhexane. { ¦eth·əl ‚ī·sō‚byüd·əl'me‚thān }

ethyl isovalerate [ORGANIC CHEMISTRY] $(CH_3)_2CHCH_2COOC_2H_5$ A colorless, oily liquid with an apple odor, soluble in water and miscible with alcohol, benzene, and ether; used for flavoring beverages and confectioneries. { ¦eth·əl ī·sō'val·ə‚rāt }

ethyl lactate [ORGANIC CHEMISTRY] $CH_3CHOHCOOC_2H_5$ A colorless liquid that boils at 154°C, has a mild odor, and is miscible with water and organic solvents such as alcohols, ketones, esters, and hydrocarbons; used as a flavoring and as a solvent for cellulose compounds such as nitrocellulose, cellulose acetate, and cellulose ethers. Also known as ethyl-2-hydroxypropionate. { ¦eth·əl 'lak‚tāt }

ethyl malonate [ORGANIC CHEMISTRY] $CH_2(COOC_2H_5)_2$ A colorless liquid, boiling at 198°C; used as an intermediate and a plasticizer. Also known as malonic ester. { ¦eth·əl 'mal·ə‚nāt }

ethyl mercaptan [ORGANIC CHEMISTRY] C_2H_5SH A colorless liquid, boiling at 36°C. Also known as ethyl sulfhydrate, thioethyl alcohol. { ¦eth·əl mər'kap·tan }

ethyl methacrylate [ORGANIC CHEMISTRY] $CH_2CCH_3COOC_2H_5$ Colorless, easily polymerized liquid, water-insoluble; used to produce polymers and chemical intermediates. { ¦eth·əl me'thak·rə‚lāt }

ethyl methyl ketone [ORGANIC CHEMISTRY] *See* methyl ethyl ketone. { ¦eth·əl ¦meth·əl 'kē‚tōn }

ethyl nitrate [ORGANIC CHEMISTRY] $C_2H_5NO_3$ A colorless, flammable liquid, boiling at 87.6°C; used in perfumes, drugs, and dyes and in organic synthesis. { ¦eth·əl 'nī‚trāt }

ethyl nitrite [ORGANIC CHEMISTRY] $C_2H_5NO_2$ A colorless liquid, boiling at 16.4°C; used in medicine and in organic synthesis. Also known as sweet spirits of niter. { ¦eth·əl 'nī‚trīt }

***O*-ethyl-*O-para*-nitrophenyl phenylphosphonothioate** [ORGANIC CHEMISTRY] $C_2H_5O_4$-NPS A yellow, crystalline compound with a melting point of 36°C; used as an insecticide and miticide on fruit crops. Abbreviated EPN. { ¦ō ¦eth·əl ¦ō ¦par·ə ‚nī·trō'fen·əl‚fen·əl·fäs¦fä·nō 'thī·ə‚wāt }

ethyl octanoate [ORGANIC CHEMISTRY] *See* ethyl caprylate. { ¦eth·əl äk'tan·ə‚wāt }

ethyl oenanthate [ORGANIC CHEMISTRY] *See* ethyl enanthate. { ¦eth·əl ē'nan‚thāt }

ethyl oleate [ORGANIC CHEMISTRY] $C_{20}H_{38}O_2$ A yellow oil, insoluble in water; used as a solvent, plasticizer, and lubricant. { ¦eth·əl 'ō·lē‚āt }

ethyl orthosilicate [ORGANIC CHEMISTRY] *See* ethyl silicate. { ¦eth·əl ‚ȯr·thō'sil·ə‚kāt }

ethyl oxalate [ORGANIC CHEMISTRY] $(COOC_2H_5)_2$ Oily, unstable, colorless liquid that is combustible; miscible with organic solvents, very slightly soluble in water; used as a solvent for cellulosics and resins, and as an intermediate for dyes and pharmaceuticals. { ¦eth·əl 'äk·sə‚lāt }

ethyl oxide [ORGANIC CHEMISTRY] *See* diethyl ether. { ¦eth·əl 'äk‚sīd }

ethylparaben [ORGANIC CHEMISTRY] *See* ethyl-*para*-hydroxybenzoate. { ¦eth·əl'par·ə·bən }

ethyl phenylacrylate [ORGANIC CHEMISTRY] *See* ethyl cinnamate. { ¦eth·əl ‚fen·əl'ak·rə‚lāt }

***N*-ethyl-5-phenylisoxazolium-3′-sulfonate** [ORGANIC CHEMISTRY] $C_{11}H_{11}NO_4S$ Crystals that decompose at 207-208°C; used to form peptide bonds. Also known as Woodward's Reagent K. { ¦en ¦eth·əl ¦fīv ¦fen·əl‚ī·säk·sə'zō·lē·əm ¦thrē‚prīm 'səl·fə‚nāt }

ethyl propionate [ORGANIC CHEMISTRY] $C_2H_5COOC_2H_5$ A colorless liquid, slightly soluble in water, boiling at 99°C; used as solvent and pyroxylin cutting agent. Also known as propionic ether. { ¦eth·əl 'prō·pē·ə‚nāt }

ethyl salicylate [ORGANIC CHEMISTRY] $(HO)C_6H_4COOC_2H_5$ A clear liquid with a pleasant odor; used in commercial preparation of artificial perfumes. Also known as sal ethyl; salicylic acid ethyl ether; salicylic ether. { ¦eth·əl sə'lis·əl‚āt }

ethyl silicate [ORGANIC CHEMISTRY] $(C_2H_5)_4SiO_4$ A colorless, flammable liquid, hydrolyzed by water; used as a preservative for stone, brick, and masonry, in lacquers, and as a bonding agent. Also known as ethyl orthosilicate. { ¦eth·əl 'sil·ə‚kāt }

ethyl sulfate [ORGANIC CHEMISTRY] *See* diethyl sulfate. { ¦eth·əl 'səl‚fāt }

ethyl sulfhydrate [ORGANIC CHEMISTRY] *See* ethyl mercaptan. { ¦eth·əl ‚səlf'hī‚drāt }

ethyl sulfide [ORGANIC CHEMISTRY] $(C_2H_5)_2S$ A colorless, oily liquid, boiling at 92°C; used as a solvent and in organic synthesis. Also known as diethyl sulfide; ethylthioethane. { ¦eth·əl 'səl‚fīd }

ethylthioethane [ORGANIC CHEMISTRY] *See* ethyl sulfide. { ¦eth·əl‚thī·ō'e‚thān }

***ortho*-ethyl(*O*-2,4,5-trichlorophenyl)ethylphosphonothioate** [ORGANIC CHEMISTRY] $C_{10}H_{12}OPSCl_2$ An amber liquid with a boiling point of 108°C at 0.01 mmHg; solubility in water is 50 parts per million; used as an insecticide for vegetable crops and soil pests on meadows. Also known as trichloronate. { ¦ȯr·thō ¦eth·əl ¦ō ¦tü ¦fȯr ¦fīv ¦trī‚klȯr·ō¦fen·əl¦eth·əl‚fäs¦fan·ō'thī·ə‚wāt }

ethyl urethane [ORGANIC CHEMISTRY] *See* urethane. { ¦eth·əl 'yu̇r·ə‚thān }

ethyl vanillin [ORGANIC CHEMISTRY] $C_2H_5O(OH)C_6H_3CHO$ A compound, crystallizing in fine white crystals that melt at 76.5°C, has a strong vanilla odor and four times the flavor of vanilla, soluble in organic solvents such as alcohol, chloroform, and ether; used in the food industry as a flavoring agent to replace or fortify vanilla. { ¦eth·əl və'nil·ən }

ethyne [ORGANIC CHEMISTRY] *See* acetylene. { 'e‚thēn }

ethynyl [ORGANIC CHEMISTRY] *See* ethinyl. { 'eth·ə‚nil }

ethynylation [ORGANIC CHEMISTRY] Production of an acetylenic derivative by the condensation of acetylene with a compound such as an aldehyde; for example, production of butynediol from the union of formaldehyde withacetylene. { ‚eth·ən·əl'ā·shən }

ethyoxyl [ORGANIC CHEMISTRY] *See* ethoxy. { ‚eth·əl'äk·səl }

etioporphyrin [ORGANIC CHEMISTRY] $C_{31}H_{34}N_4$ A synthetic porphyrin that has four ethyl and four methyl groups in a red-pigmented compound whose crystals melt at 280°C. { ‚ēd·ē·ō'pȯr·fə·rən }

Eu [CHEMISTRY] *See* europium.

eucalyptol [ORGANIC CHEMISTRY] $C_{10}H_{18}O$ A colorless oil with a camphorlike odor; boiling point is 174-177°C; used in pharmaceuticals, perfumery, and flavoring. Also known as cajeputol; cineol. { ‚yü·kə'lip‚tȯl }

eugenol [ORGANIC CHEMISTRY] $CH_2CHCH_2C_6H_3(OCH_3)OH$ A colorless or yellowish aromatic liquid with spicy odor and taste, soluble in organic solvents, and extracted from clove oil; used in flavors, perfumes, medicines, and the manufacture of vanilla. { 'yü·jə‚nȯl }

europium [CHEMISTRY] A member of the rare-earth elements in the cerium subgroup, symbol Eu, atomic number 63, atomic weight 151.96, steel gray and malleable, melting at 1100-1200°C. { yu̇'rō·pē·əm }

europium halide [INORGANIC CHEMISTRY] Any of the compounds of the element europium and the halogen elements; for example, europium chloride, $EuCl_3 \cdot xH_2O$. { yu̇'rō·pē·əm 'ha‚līd }

europium oxide [INORGANIC CHEMISTRY] Eu_2O_3 A white powder, insoluble in water; used in red- and infrared-sensitive phosphors. { yu̇'rō·pē·əm 'äk‚sīd }

eutectic [PHYSICAL CHEMISTRY] An alloy or solution that has the lowest possible constant melting point. { yü'tek·tik }

eutectic mixture [PHYSICAL CHEMISTRY] *See* eutectic system. { yü¦tek·tik 'miks·chər }

eutectic point [PHYSICAL CHEMISTRY] The point in the constitutional diagram indicating the composition and temperature of the lowest melting point of a eutectic. { yü'tek·tik 'pȯint }

eutectic system [PHYSICAL CHEMISTRY] The particular composition and temperature of materials at the eutectic point. Also known as eutectic mixture. { yü'tek·tik 'sis·təm }

eutectic temperature [PHYSICAL CHEMISTRY] The temperature at the lowest melting point of a eutectic. { yü'tek·tik 'tem·prə·chər }

eutectogenic system [PHYSICAL CHEMISTRY] A multicomponent liquid-solid mixture in which pure solid phases of each component are in equilibrium with the remaining liquid mixture at a specific (usually minimum) temperature for a given composition, that is, the eutectic point. { yü¦tek·tə¦jen·ik 'sis·təm }

eutectoid [PHYSICAL CHEMISTRY] The point in an equilibrium diagram for a solid solution at which the solution on cooling is converted to a mixture of solids. { yü'tek‚tȯid }

evolved gas analysis [ANALYTICAL CHEMISTRY] An analytical technique in which the characteristics or the amount of volatile products released by a substance and its reaction products are determined as a function of temperature while the sample is subjected to a series of controlled temperature changes. Abbreviated EGA. { ē¦välvd 'gas ə‚nal·ə·səs }

exchange broadening [SPECTROSCOPY] The broadening of a spectral line by some type of chemical or spin exchange process which limits the lifetime of the absorbing or emitting species and produces the broadening via the Heisenberg uncertainty principle. { iks'chānj 'bród·ən·iŋ }

exchange narrowing [SPECTROSCOPY] The phenomenon in which, when a spectral line is split and thereby broadened by some variable perturbation, the broadening may be narrowed by a dynamic process that exchanges different values of the perturbation. { iks'chānj 'nar·ə·wiŋ }

exchange reaction [CHEMISTRY] Reaction in which two atoms or ions exchange places either in two different molecules or in the same molecule. { iks'chānj rē,ak·shən }

exchange velocity [CHEMISTRY] In an ion-exchange process, the speed with which one ion is displaced from an exchanger in favor of another ion. { iks'chānj və'läs·əd·ē }

excimer [CHEMISTRY] An excited diatomic molecule where both atoms are of the same species and are dissociated in the ground state. { 'ek·sə·mər }

exciplex [CHEMISTRY] An excited electron donor-acceptor complex which is dissociated in the ground state. { 'ek·sə,pleks }

excitation index [SPECTROSCOPY] In emission spectroscopy, the ratio of intensities of a pair of extremely nonhomologous spectra lines; used to provide a sensitive indication of variation in excitation conditions. { ,ek,sī'tā·shən ,in,deks }

excitation purity [ANALYTICAL CHEMISTRY] The ratio of the departure of the chromaticity of a specified color to that of the reference source, measured on a chromaticity diagram; used as a guide of the wavelength of spectrum color needed to be mixed with a reference color to give the specified color. { ,ek,sī'tā·shən ,pyůr·əd·ē }

excitation spectrum [SPECTROSCOPY] The graph of luminous efficiency per unit energy of the exciting light absorbed by a photoluminescent body versus the frequency of the exciting light. { ,ek,sī'tā·shən ,spek·trəm }

exciting line [SPECTROSCOPY] The frequency of electromagnetic radiation, that is, the spectral line from a noncontinuous source, which is absorbed by a system in connection with some particular process. { ek'sīd·iŋ ,līn }

exhaustion point [CHEMISTRY] In an ion-exchange process, the state of an adsorbent at which it no longer can produce a useful ion exchange. { ig'zȯs·chən ,pȯint }

exo- [ORGANIC CHEMISTRY] A conformation of carbon bonds in a six-membered ring such that the molecule is boat-shaped with one or more substituents directed outward from the ring. { 'ek·sō }

exocyclic double bond [ORGANIC CHEMISTRY] A double bond that is connected to and external to a ring structure. { ¦ek·sō¦sī·klik ¦dəb·əl 'bänd }

explosion [CHEMISTRY] A chemical reaction or change of state which is effected in an exceedingly short space of time with the generation of a high temperature and generally a large quantity of gas. { ik'splō·zhən }

extender [CHEMISTRY] A material used to dilute or extend or change the properties of resins, ceramics, paints, rubber, and so on. { ik'sten·dər }

extensive property [PHYSICAL CHEMISTRY] A noninherent property of a system, such as volume or internal energy, that changes with the quantity of material in the system; the quantitative value equals the sum of the values of the property for the individual constituents. { ik'sten·siv 'präp·ərd·ē }

external circuit [PHYSICAL CHEMISTRY] All connecting wires, devices, and current sources which achieve desired conditions within an electrolytic cell. { ek'stərn·əl 'sər·kət }

external phase [CHEMISTRY] *See* continuous phase. { ek'stərn·əl 'fāz }

extinction [PHYSICAL CHEMISTRY] *See* absorbance. { ek'stiŋk·shən }

extinction coefficient [ANALYTICAL CHEMISTRY] *See* absorptivity. { ek'stiŋk·shən ,kō·i,fish·ənt }

extract [CHEMISTRY] Material separated from liquid or solid mixture by a solvent. { 'ek,strakt (noun) *or* ik'strakt (verb) }

extractant [CHEMISTRY] The liquid used to remove a solute from another liquid. { ik'strak·tənt }

extracting agent [CHEMISTRY] In a liquid-liquid distribution, the reagent forming a complex or other adduct that has different solubilities in the two immiscible liquids of the extraction system. { ik'strak·tiŋ ,ā·jənt }

extraction [CHEMISTRY] A method of separation in which a solid or solution is contacted with a liquid solvent (the two being essential mutually insoluble) to transfer one or more components into the solvent. { ik'strak·shən }

extreme narrowing approximation [SPECTROSCOPY] A mathematical approximation in the theory of spectral-line shapes to the effect that the exchange narrowing of a perturbation is complete. { ek'strēm 'nar·ə·wiŋ ə,präk·sə'mā·shən }

extrinsic sol [PHYSICAL CHEMISTRY] A colloid whose stability is attributed to electric charge on the surface of the colloidal particles. { ek¦strinz·ik 'säl }

Eyring equation [PHYSICAL CHEMISTRY] An equation, based on statistical mechanics, which gives the specific reaction rate for a chemical reaction in terms of the heat of activation, entropy of activation, the temperature, and various constants. { 'ī·riŋ i,kwā·zhən }

F

F [CHEMISTRY] *See* fluorine.

face-bridging ligand [ORGANIC CHEMISTRY] A ligand that forms a bridge over one triangular face of the polyhedron of a metal cluster structure. { 'fās ˌbrij·iŋ 'līg·ənd }

false body [PHYSICAL CHEMISTRY] The property of certain colloidal substances, such as paints and printing inks, of solidifying when left standing. { ¦fȯls 'bäd·ē }

family [CHEMISTRY] A group of elements whose chemical properties, such as valence, solubility of salts, and behavior toward reagents, are similar. { 'fam·lē }

famphur [ORGANIC CHEMISTRY] $C_{10}H_{16}NO_5PS_2$ A crystalline compound with a melting point of 55°C; slightly soluble in water; used as an insecticide for lice and grubs of reindeer and cattle. { 'fam·fər }

faradaic current [CHEMISTRY] *See* faradic current. { ˌfar·ə¦dā·ik ¦kər·ənt }

Faraday's laws of electrolysis [PHYSICAL CHEMISTRY] **1.** The amount of any substance dissolved or deposited in electrolysis is proportional to the total electric charge passed. **2.** The amounts of different substances dissolved or desposited by the passage of the same electric charge are proportional to their equivalent weights. { 'far·əˌdāz ¦lȯz əv iˌlek'träl·ə·səs }

faradic current [CHEMISTRY] An electric current that corresponds to the reduction or oxidation of a chemical species. Also spelled faradaic current. { fə'rad·ik ˌkə·rənt }

fast chemical reaction [PHYSICAL CHEMISTRY] A reaction with a half-life of milliseconds or less; such reactions occur so rapidly that special experimental techniques are required to observe their rate. { 'fast ¦kem·ə·kəl rē'ak·shən }

fatty acid [ORGANIC CHEMISTRY] An organic monobasic acid of the general formula $C_nH_{2n+1}COOH$ derived from the saturated series of aliphatic hydrocarbons, examples are palmitic acid, stearic acid, and oleic acid; used as a lubricant in cosmetics and nutrition, and for soaps and detergents. { ¦fad·ē 'as·əd }

fatty alcohol [ORGANIC CHEMISTRY] A high-molecular-weight, straight-chain primary alcohol derived from natural fats and oils; includes lauryl, stearyl, oleyl, and linoleyl alcohols; used in pharmaceuticals, cosmetics, detergents, plastics, and lube oils and in textile manufacture. { 'fad·ē 'al·kəˌhȯl }

fatty amine [ORGANIC CHEMISTRY] RCH_2NH_2 A normal aliphatic amine from oils and fats; used as a plasticizer, in medicine, as a chemical intermediate, and in rubber manufacture. { 'fad·ē 'amˌēn }

fatty ester [ORGANIC CHEMISTRY] $RCOOR'$ A fatty acid in which the alkyl group (R′) of a monohydric alcohol replaces the active hydrogen; for example, $RCOOCH_3$ from reaction of RCOOH with methane. { 'fad·ē 'es·tər }

fatty nitrile [ORGANIC CHEMISTRY] RCN An ester of hydrogen cyanide derived from fatty acid; used in lube oil additives and plasticizers, and as a chemical intermediate. { 'fad·ē 'nīˌtrəl }

Favorskii rearrangement [ORGANIC CHEMISTRY] A reaction in which α-halogenated ketones undergo rearrangement in the presence of bases, with loss of the halogen and formation of carboxylic acids or their derivatives with the same number of carbon atoms. { fa'vȯr·skē ˌrē·ə'rānj·mənt }

FCC [ORGANIC CHEMISTRY] *See* chlorofluorocarbon.

Fe [CHEMISTRY] *See* iron.

feedback [CHEMISTRY] In a stepwise reaction, the formation of a substance in one step that affects the rate of a previous step. { 'fēd‚bak }
Fehling's reagent [ANALYTICAL CHEMISTRY] A solution of cupric sulfate, sodium potassium tartrate, and sodium hydroxide, used to test for the presence of reducing compounds such as sugars. { 'fāl·iŋz rē‚ā·jənt }
fenaminosulf [ORGANIC CHEMISTRY] $C_8H_{10}N_3SO_3Na$ A yellow-brown powder, decomposing at 200°C; used as a fungicide for seeds and seedlings in crops. { ‚fen'am·ə·nō‚səlf }
fenazaflor [ORGANIC CHEMISTRY] $C_{15}H_7Cl_2F_3N_2O_2$ A greenish-yellow, crystalline compound with a melting point of 103°C; used as an insecticide and miticide for spider mites and eggs. { fə'naz·ə‚flōr }
fenbutatin oxide [ORGANIC CHEMISTRY] $C_{60}H_{78}OSn_2$ A white, crystalline compound, insoluble in water; used to control mites in deciduous and citrus fruits. { fen'byüd·əd·ən 'äk‚sīd }
fenchol [ORGANIC CHEMISTRY] *See* fenchyl alcohol. { 'fen·chȯl }
fenchone [ORGANIC CHEMISTRY] $C_{10}H_{16}O$ An isomer of camphor; a colorless oil that boils at 193°C and is soluble in ether; a constituent of fennel oil; used as a flavoring. { 'fen‚chōn }
fenchyl alcohol [ORGANIC CHEMISTRY] $C_{10}H_{18}O$ A colorless solid or oily liquid, boiling at 198-204°C, isolated from pine oil and turpentine and also made synthetically; used as a solvent, an intermediate in organic synthesis, and as a flavoring. Also known as fenchol. { 'fen·chəl 'al·kə‚hȯl }
fenitrothion [ORGANIC CHEMISTRY] $C_9H_{12}NO_5PS$ A yellow-brown liquid, insoluble in water; used as a miticide and insecticide for rice, orchards, vegetables, cereals, and cotton, and for fly and mosquito control. { ‚fen·ə·trō'thī‚än }
fensulfothion [ORGANIC CHEMISTRY] $C_{11}H_{17}S_2O_2P$ A brown liquid with a boiling point of 138-141°C; used as an insecticide and nematicide in soils. { ‚fen‚səl·fō'thī‚än }
fentinacetate [ORGANIC CHEMISTRY] $C_{20}H_{18}O_2Sn$ A yellow to brown, crystalline solid that melts at 124-125°C; used as a fungicide, molluscicide, and algicide for early and late blight on potatoes, sugarbeets, peanuts, and coffee. Also known as triphenyltinacetate. { ‚fent·ən'as·ə‚tāt }
fenuron [ORGANIC CHEMISTRY] $C_9H_{12}N_2O$ A white, crystalline compound with a melting point of 133-134°C; soluble in water; used as a herbicide to kill weeds and bushes. { ‚fen'yü‚rän }
fenuron-TCA [ORGANIC CHEMISTRY] $C_{11}H_{13}Cl_3N_2O_3$ A white, crystalline compound with a melting point of 65-68°C; moderately soluble in water; used as a herbicide for noncrop areas. { ‚fen'yü‚rän ¦tē¦sē¦ā }
FEP resin [ORGANIC CHEMISTRY] *See* fluorinated ethylene propylene resin. { ¦ef¦ē¦pē 'rez·ən }
Fermi resonance [PHYSICAL CHEMISTRY] In a polyatomic molecule, the relationship of two vibrational levels that have in zero approximation nearly the same energy; they repel each other, and the eigenfunctions of the two states mix. { 'fer·mē ‚rez·ən·əns }
fermium [CHEMISTRY] A synthetic radioactive element, symbol Fm, with atomic number 100; discovered in debris of the 1952 hydrogen bomb explosion, and now made in nuclear reactors. { 'fer·mē·əm }
ferrate [INORGANIC CHEMISTRY] A multiple iron oxide with another oxide, for example, Na_2FeO_4. { 'fe‚rāt }
ferric [INORGANIC CHEMISTRY] The term for a compound of trivalent iron, for example, ferric bromide, $FeBr_3$. { 'fer·ik }
ferric acetate [ORGANIC CHEMISTRY] $Fe_2(C_2H_3O_2)_3$ A brown compound, soluble in water; used as a tonic and dye mordant. { 'fer·ik 'as·ə‚tāt }
ferric ammonium alum [INORGANIC CHEMISTRY] *See* ferric ammonium sulfate. { 'fer·ik ə'mōn·ē·əm 'al·əm }
ferric ammonium citrate [ORGANIC CHEMISTRY] $Fe(NH_4)_3(C_6H_5O_7)_2$ Red, deliquescent scales or granules; odorless, water soluble, and affected by light; used in medicine and blueprint photography. { 'fer·ik ə'mōn·ē·əm 'sī‚trāt }
ferric ammonium oxalate [ORGANIC CHEMISTRY] $(NH_4)_3Fe(C_2O_4)_3 \cdot 3H_2O$ Green, crys-

talline material, soluble in water and alcohol, sensitive to light; used in blueprint photography. { 'fer·ik ə'mōn·ē·əm 'äk·sə‚lāt }

ferric ammonium sulfate [INORGANIC CHEMISTRY] $FeNH_4(SO_4)_2 \cdot 12H_2O$ Efflorescent, water-soluble crystals; used in medicine, in analytical chemistry, and as a mordant in textile dyeing. Also known as ferric ammonium alum; iron ammonium sulfate. { 'fer·ik ə'mōn·ē·əm 'səl‚fāt }

ferric arsenate [INORGANIC CHEMISTRY] $FeAsO_4 \cdot 2H_2O$ A green or brown powder, insoluble in water, soluble in dilute mineral acids; used as an insecticide. { 'fer·ik 'ärs·ən‚āt }

ferric bromide [INORGANIC CHEMISTRY] $FeBr_3$ Red, deliquescent crystals that decompose upon heating; soluble in water, ether, and alcohol; used in medicine and analytical chemistry. Also known as ferric sesquibromide; ferric tribromide; iron bromide. { 'fer·ik 'brō‚mīd }

ferric chloride [INORGANIC CHEMISTRY] $FeCl_3$ Brown crystals, melting at 300°C, that are soluble in water, alcohol, and glycerol; used as a coagulant for sewage and industrial wastes, as an oxidizing and chlorinating agent, as a disinfectant, in copper etching, and as a mordant. Also known as anhydrous ferric chloride; ferric trichloride; flores martis; iron chloride. { 'fer·ik 'klȯr‚īd }

ferric citrate [ORGANIC CHEMISTRY] $FeC_6H_5O_7 \cdot 3H_2O$ Red scales that react to light; soluble in water, insoluble in alcohol; used as a medicine for certain blood disorders, and for blueprint paper. Also known as iron citrate. { 'fer·ik 'sī‚trāt }

ferric dichromate [INORGANIC CHEMISTRY] $Fe_2(CrO_4)_3$ A red-brown, granular powder, miscible in water; used as a mordant. { 'fer·ik dī'krō‚māt }

ferric ferrocyanide [INORGANIC CHEMISTRY] $Fe_4[Fe(CN)_6]_3$ Dark-blue crystals, used as a pigment, and with oxalic acid in blue ink. Also known as iron ferrocyanide. { 'fer·ik ‚fer·ə'sī·ə‚nīd }

ferric fluoride [INORGANIC CHEMISTRY] FeF_3 Green, rhombohedral crystals, soluble in water and acids; used in porcelain and pottery manufacture. Also known as iron fluoride. { 'fer·ik 'flu̇r‚īd }

ferric hydrate [INORGANIC CHEMISTRY] *See* ferric hydroxide. { 'fer·ik 'hī‚drāt }

ferric hydroxide [INORGANIC CHEMISTRY] $Fe(OH)_3$ A brown powder, insoluble in water; used as arsenic poisoning antidote, in pigments, and in pharmaceutical preparations. Also known as ferric hydrate; iron hydroxide. { 'fer·ik hī'dräk‚sīd }

ferric nitrate [INORGANIC CHEMISTRY] $Fe(NO_3)_3 \cdot 9H_2O$ Colorless crystals, soluble in water and decomposed by heat; used as a dyeing mordant, in tanning, and in analytical chemistry. Also known as iron nitrate. { 'fer·ik 'nī‚trāt }

ferric oxalate [ORGANIC CHEMISTRY] $Fe_2(COO)_3$ Yellow scales, soluble in water, decomposing when heated at about 100°C; used as a catalyst and in photographic printing papers. { 'fer·ik 'äk·sə‚lāt }

ferric oxide [INORGANIC CHEMISTRY] Fe_2O_3 Red, hexagonal crystals or powder, insoluble in water and soluble in acids, melting at 1565°C; used as a catalyst and pigment for metal polishing, in metallurgy, and in medicine. Also known as ferric oxide red; jeweler's rouge; red ocher. { 'fer·ik 'äk‚sīd }

ferric oxide red [INORGANIC CHEMISTRY] *See* ferric oxide. { 'fer·ik 'äk‚sīd 'red }

ferric phosphate [INORGANIC CHEMISTRY] $FePO_4 \cdot 2H_2O$ Yellow, rhombohedral crystals, insoluble in water, soluble in acids; used in medicines and fertilizers. Also known as iron phosphate. { 'fer·ik 'fäs‚fāt }

ferric resinate [ORGANIC CHEMISTRY] Reddish-brown, water-insoluble powder; used as a drier for paints and varnishes. Also known as iron resinate. { 'fer·ik 'rez·ən‚āt }

ferric sesquibromide [INORGANIC CHEMISTRY] *See* ferric bromide. { 'fer·ik ‚ses·kwə'brō‚mīd }

ferric stearate [ORGANIC CHEMISTRY] $Fe(C_{18}H_{35}O_2)_3$ A light-brown, water-insoluble powder; used as a varnish drier. Also known as iron stearate. { 'fer·ik 'stir‚āt }

ferric sulfate [INORGANIC CHEMISTRY] $Fe_2(SO_4)_3 \cdot 9H_2O$ Yellow, water-soluble, rhombohedral crystals, decomposing when heated; used as a chemical intermediate, disinfectant, soil conditioner, pigment, and analytical reagent, and in medicine. Also known as iron sulfate. { 'fer·ik 'səl‚fāt }

ferric tribromide [INORGANIC CHEMISTRY] *See* ferric bromide. { 'fer·ik trī'brō‚mīd }

ferric trichloride [INORGANIC CHEMISTRY] *See* ferric chloride. { 'fer·ik trī'klȯr,īd }

ferric vanadate [INORGANIC CHEMISTRY] $Fe(VO_3)_3$ Grayish-brown powder, insoluble in water and alcohol; used in metallurgy. Also known as iron metavanadate. { 'fer·ik 'van·ə,dāt }

ferricyanic acid [INORGANIC CHEMISTRY] $H_3Fe(CN)_6$ A red-brown unstable solid. { ,fer·i·sī'an·ik 'as·əd }

ferricyanide [INORGANIC CHEMISTRY] A salt containing the radical $Fe(CN)_6{}^{3-}$. { fer·i'sī·ə,nīd }

ferrisulphas [INORGANIC CHEMISTRY] *See* ferrous sulfate. { ¦fe·ri'səl·fəs }

ferrite [INORGANIC CHEMISTRY] An unstable compound of a strong base and ferric oxide which exists in alkaline solution, such as $NaFeO_2$. { 'fe,rīt }

ferrocene [ORGANIC CHEMISTRY] $(CH_2)_5Fe(CH_2)_5$ Orange crystals that are soluble in ether, melting point 174°C; used as a combustion control additive in fuels, and for heat stabilization in greases and plastics. { 'fer·ə,sēn }

ferrocyanic acid [INORGANIC CHEMISTRY] $H_4Fe(CN)_6$ A white solid obtained by treating ferrocyanides with acid. { ¦fe·rō·sī'an·ik 'as·əd }

ferrocyanide [INORGANIC CHEMISTRY] A salt containing the radical $Fe(CN)_6{}^{4-}$. { ¦fe·rō'sī·ə,nīd }

ferrofluid [PHYSICAL CHEMISTRY] A colloidal suspension that becomes magnetized in a magnetic field because of a disperse phase consisting of ferromagnetic or ferrimagnetic particles. { 'fe·rō,flü·əd }

ferrous [CHEMISTRY] The term or prefix used to denote compounds of iron in which iron is in the divalent (2+) state. { 'fer·əs }

ferrous acetate [ORGANIC CHEMISTRY] $Fe(CH_3COO)_2 \cdot 4H_2O$ Soluble green crystals, soluble in water and alcohol, that are combustible and that oxidize to basic ferric acetate in air; used as textile dyeing mordant, as wood preservative, and in medicine. Also known as iron acetate. { 'fer·əs 'as·ə,tāt }

ferrous ammonium sulfate [INORGANIC CHEMISTRY] $Fe(SO_4) \cdot (NH)_2SO_4 \cdot 6H_2O$ Light-green, water-soluble crystals; used in medicine, analytical chemistry, and metallurgy. Also known as iron ammonium sulfate; Mohr's salt. { 'fer·əs ə'mōn·ē·əm 'səl,fāt }

ferrous arsenate [INORGANIC CHEMISTRY] $Fe_3(AsO_4)_2 \cdot 6H_2O$ Water-insoluble, toxic green amorphous powder, soluble in acids; used in medicine and as an insecticide. Also known as iron arsenate. { 'fer·əs 'ärs·ən,āt }

ferrous carbonate [INORGANIC CHEMISTRY] $FeCO_3$ Green rhombohedral crystals that are soluble in carbonated water and decompose when heated; used in medicine. { 'fer·əs 'kär·bə,nāt }

ferrous chloride [INORGANIC CHEMISTRY] $FeCl_2 \cdot 4H_2O$ Green, monoclinic crystals, soluble in water; used as a mordant in dyeing, for sewage treatment, in metallurgy, and in pharmaceutical preparations. Also known as iron chloride; iron dichloride. { 'fer·əs 'klȯr,īd }

ferrous hydroxide [INORGANIC CHEMISTRY] $Fe(OH)_2$ A white, water-insoluble, gelatinous solid that turns reddish-brown as it oxidizes to ferric hydroxide. { 'fer·əs hī'dräk,sīd }

ferrous oxalate [ORGANIC CHEMISTRY] $Fe(COO)_2$ A water-soluble, yellow powder; used in photography and medicine. Also known as iron oxalate. { 'fer·əs 'äk·sə,lāt }

ferrous oxide [INORGANIC CHEMISTRY] FeO A black powder, soluble in water, melting at 1419°C. Also known as black iron oxide; iron monoxide. { 'fer·əs 'äk,sīd }

ferrous sulfate [INORGANIC CHEMISTRY] $FeSO_4 \cdot 7H_2O$ Blue-green, water-soluble, monoclinic crystals; used as a mordant in dyeing wool, in the manufacture of ink, and as a disinfectant. Also known as copperas; ferrisulphas; green copperas; green vitriol; iron sulfate. { 'fer·əs 'səl,fāt }

ferrous sulfide [INORGANIC CHEMISTRY] FeS Black crystals, insoluble in water, soluble in acids, melting point 1195°C; used to generate hydrogen sulfide in ceramics manufacture. Also known as iron sulfide. { 'fer·əs 'səl,fīd }

ferrum [CHEMISTRY] Latin term for iron; derivation of the symbol Fe. { 'fer·əm }

ferulic acid [ORGANIC CHEMISTRY] $C_{10}H_{10}O_4$ A compound widely distributed in small amounts in plants, having two isomers: the cis form is a yellow oil, and the trans form is obtained from water solutions as orthorhombic crystals. { fə'rül·ik 'as·əd }

Féry spectrograph [SPECTROSCOPY] A spectrograph whose only optical element consists of a back-reflecting prism with cylindrically curved faces. { ¦fār·ē 'spek·trə,graf }

Feulgen reaction [ANALYTICAL CHEMISTRY] An aldehyde specific reaction based on the formation of a purple-colored compound when aldehydes react with fuchsin-sulfuric acid; deoxyribonucleic acid gives this reaction after removal of its purine bases by acid hydrolysis; used as a nuclear stain. { 'fȯil·gən rē,ak·shən }

ficin [ORGANIC CHEMISTRY] A proteolytic enzyme obtained from fig latex or sap; hydrolyzes casein, meat, fibrin, and other proteinlike materials; used in the food industry and as a diagnostic aid in medicine. { 'fī·sən }

field-desorption mass spectroscopy [SPECTROSCOPY] A technique for analysis of nonvolatile molecules in which a sample is deposited on a thin tungsten wire containing sharp microneedles of carbon on the surface; a voltage is applied to the wire, thus producing high electric-field gradients at the points of the needles, and moderate heating then causes desorption from the surface or molecular ions, which are focused into a mass spectrometer. { ¦fēld dē'sȯrp·shən ¦mas spek'trä·skə·pē }

figure of merit [ANALYTICAL CHEMISTRY] A performance characteristic of an analytical chemical method that influences its choice for a specific type of determination, such as selectivity, sensitivity, detection limit, precision, and bias. { 'fig·yər əv 'mer·ət }

film-development chromatography [ANALYTICAL CHEMISTRY] Liquid-analysis chromatographic technique in which the stationary phase (adsorbent) is a strip or layer, as in paper or thin-layer chromatography. { film di'vel·əp·mənt ,krō·mə'täg·rə·fē }

film tension [PHYSICAL CHEMISTRY] The contractile force per unit length that is exerted by an equilibrium film in contact with a supporting substrate. { 'film ,ten·chən }

filter flask [CHEMISTRY] A flask with a side arm to which a vacuum can be applied; usually filter flasks have heavy side walls to withstand high vacuum. { 'fil·tər ,flask }

filter photometry [ANALYTICAL CHEMISTRY] **1.** Colorimetric analysis of solution colors with a filter applied to the eyepiece of a conventional colorimeter. **2.** Inspection of a pair of Nessler tubes through a filter. { 'fil·tər fə'täm·ə·trē }

filter-press cell [PHYSICAL CHEMISTRY] An electrolytic cell consisting of several units in series, as in a filter press, in which each electrode, except the two end ones, acts as an anode on one side and a cathode on the other, and the space between electrodes is divided by porous asbestos diaphragms. { 'fil·tər ,pres ,sel }

filter spectrophotometer [SPECTROSCOPY] Spectrophotographic analyzer of spectral radiations in which a filter is used to isolate narrow portions of the spectrum. { 'fil·tər spek·trə·tə'täm·əd·ər }

fingerprint [ANALYTICAL CHEMISTRY] Evidence for the presence or the identity of a substance that is obtained by techniques such as spectroscopy, chromatography, or electrophoresis. { 'fiŋ·gər,print }

fire [CHEMISTRY] The manifestation of rapid combustion, or combination of materials with oxygen. { fīr }

fire point [CHEMISTRY] The lowest temperature at which a volatile combustible substance vaporizes rapidly enough to form above its surface an air-vapor mixture which burns continuously when ignited by a small flame. { 'fīr ,pȯint }

first-order reaction [PHYSICAL CHEMISTRY] A chemical reaction in which the rate of decrease of concentration of component A with time is proportional to the concentration of A. { ¦fərst ,ȯrd·ər rē'ak·shən }

first-order spectrum [SPECTROSCOPY] A spectrum, produced by a diffraction grating, in which the difference in path length of light from adjacent slits is one wavelength. { ¦fərst ,ȯrd·ər 'spek·trəm }

Fischer-Hepp rearrangement [ORGANIC CHEMISTRY] The rearrangement of a nitroso derivative of a secondary aromatic amine to a *p*-nitrosoarylamine; the reaction is brought about by an alcoholic solution of hydrogen chloride. { ¦fish ər ¦hep rē·ə'rānj·mənt }

Fischer indole synthesis [ORGANIC CHEMISTRY] A reaction to form indole derivatives by means of a ring closure of aromatic hydrazones. { ˌfish·ər ˈinˌdōl ˌsin·thə·səs }

Fischer polypeptide synthesis [ORGANIC CHEMISTRY] A synthesis of peptides in which α-amino acids or those peptides with a free amino group react with acid halides of α-haloacids, followed by amination with ammonia. { ˈfish·ər ¦päl·ēˈpepˌtīd ˌsin·thə·səs }

Fischer projection [ORGANIC CHEMISTRY] A method for representing the spatial arrangement of groups around chiral carbon atoms; the four bonds to the chiral carbon are represented by a cross, with the assumption that the horizontal bonds project toward the viewer and the vertical bonds away from the viewer. { ˈfish·ər prəˌjek·shən }

Fischer's salt [INORGANIC CHEMISTRY] *See* cobalt potassium nitrite. { ¦fish·ərz ˈsȯlt }

fissiochemistry [CHEMISTRY] The process of producing chemical change by means of nuclear energy. { ¦fish·ō¦kem·ə·strē }

Fittig's synthesis [ORGANIC CHEMISTRY] The synthesis of aromatic hydrocarbons by the condensation of aryl halides with alkyl halides, using sodium as a catalyst. { ˈfid·iks ˌsin·thə·səs }

fixed carbon [CHEMISTRY] Solid, combustible residue remaining after removal of moisture, ash, and volatile materials from coal, coke, and bituminous materials; expressed as a percentage. { ¦fikst ˈkär·bən }

fixed ion [ANALYTICAL CHEMISTRY] An ion in the lattice of a solid ion exchanger. [PHYSICAL CHEMISTRY] One of a group of nonexchangeable ions in an ion exchanger that have a charge opposite to that of the counterions. { ¦fikst ˈīˌän }

flame [CHEMISTRY] A hot, luminous reaction front (or wave) in a gaseous medium into which the reactants flow and out of which the products flow. { flām }

flame emission spectroscopy [SPECTROSCOPY] A flame photometry technique in which the solution containing the sample to be analyzed is optically excited in an oxyhydrogen or oxyacetylene flame. { ˈflām iˌmish·ən spekˈträs·kə·pē }

flame excitation [SPECTROSCOPY] Use of a high-temperature flame (such as oxyacetylene) to excite spectra emission lines from alkali and alkaline-earth elements and metals. { ¦flām ˌek·sīˈtā·shən }

flame ionization detector [ANALYTICAL CHEMISTRY] A device in which the measured change in conductivity of a standard flame (usually hydrogen) due to the insertion of another gas or vapor is used to detect the gas or vapor. { ¦flām ˌī·ə·nəˈzā·shən diˌtek·tər }

flame photometer [SPECTROSCOPY] One of several types of instruments used in flame photometry, such as the emission flame photometer and the atomic absorption spectrophotometer, in each of which a solution of the chemical being analyzed is vaporized; the spectral lines resulting from the light source going through the vapors enters a monochromator that selects the band or bands of interest. { ˈflām ˌfəˈtäm·əd·ər }

flame photometry [SPECTROSCOPY] A branch of spectrochemical analysis in which samples in solution are excited to produce line emission spectra by introduction into a flame. { ˈflām fəˈtäm·ə·tre }

flame propagation [CHEMISTRY] The spread of a flame in a combustible environment outward from the point at which the combustion started. { ¦flām ˌpräp·əˈgā·shən }

flame spectrometry [SPECTROSCOPY] A procedure used to measure the spectra or to determine wavelengths emitted by flame-excited substances. { ¦flām spekˈträm·ə·trē }

flame spectrophotometry [SPECTROSCOPY] A method used to determine the intensity of radiations of various wavelengths in a spectrum emitted by a chemical inserted into a flame. { ¦flām ¦spek·trə·fəˈtäm·ə·trē }

flame spectrum [SPECTROSCOPY] An emission spectrum obtained by evaporating substances in a nonluminous flame. { ˈflām ˌspek·trəm }

flame speed [CHEMISTRY] The rate at which combustion moves through an explosive mixture. { ˈflām ˌspēd }

flammability [CHEMISTRY] A measure of the extent to which a material will support combustion. Also known as inflammability. { ˌflam·əˈbil·əd·ē }

flammability limits [CHEMISTRY] The stoichiometric composition limits (maximum and minimum) of an ignited oxidizer-fuel mixture what will burn indefinitely at given conditions of temperature and pressure without further ignition. { ˌflam·ə'bil·əd·ē ˌlim·əts }

flash photolysis [PHYSICAL CHEMISTRY] A method of studying fast photochemical reactions in gas molecules; a powerful lamp is discharged in microsecond flashes near a reaction vessel holding the gas, and the products formed by the flash are observed spectroscopically. { 'flash fəˌtäl·ə·səs }

flash point [CHEMISTRY] The lowest temperature at which vapors from a volatile liquid will ignite momentarily upon the application of a small flame under specified conditions; test conditions can be either open- or closed-cup. { 'flash ˌpȯint }

flash spectroscopy [SPECTROSCOPY] The study of the electronic states of molecules after they absorb energy from an intense, brief light flash. { ¦flash spek'träs·kə·pē }

flask [CHEMISTRY] A long-necked vessel, frequently of glass, used for holding liquids. { flask }

F line [SPECTROSCOPY] A green-blue line in the spectrum of hydrogen, at a wavelength of 486.133 nanometers. { 'ef ˌlīn }

floc [CHEMISTRY] Small masses formed in a fluid through coagulation, agglomeration, or biochemical reaction of fine suspended particles. { fläk }

flocculant [CHEMISTRY] *See* flocculating agent. { 'fläk·yə·lənt }

flocculate [CHEMISTRY] To cause to aggregate or coalesce into a flocculent mass. { 'fläk·yəˌlət (adjective) *or* 'fläk·yəˌlāt (verb) }

flocculating agent [CHEMISTRY] A reagent added to a dispersion of solids in a liquid to bring together the fine particles to form flocs. Also known as flocculant. { 'fläk·yəˌlād·iŋ ˌā·jənt }

flocculent [CHEMISTRY] Pertaining to a material that is cloudlike and noncrystalline. { 'fläk·yə·lənt }

floc point [ANALYTICAL CHEMISTRY] The temperature at which wax or solids separate from kerosine and other illuminating oils as a definite floc. { 'fläk ˌpȯint }

floc test [ANALYTICAL CHEMISTRY] A quantitative test applied to kerosine and other illuminating oils to detect substances rendered insoluble by heat. { 'fläk ˌtest }

Flood's equation [PHYSICAL CHEMISTRY] A relation used to determine the liquidus temperature in a binary fused salt system. { 'flədz iˌkwā·zhən }

flores [CHEMISTRY] A form of a chemical compound made by the process of sublimation. { 'flȯr·ēz }

flores martis [INORGANIC CHEMISTRY] *See* ferric chloride. { 'flȯr·ēz 'märd·əs }

flotation agent [CHEMISTRY] A chemical which alters the surface tension of water or which makes it froth easily. { flō'tā·shən ˌā·jənt }

flow birefringence [PHYSICAL CHEMISTRY] Orientation of long, thin asymmetric molecules in the direction of flow of a solution forced to flow through a capillary tube. { ¦flō ˌbī·rə'frin·jəns }

flowers of tin [INORGANIC CHEMISTRY] *See* stannic oxide. { 'flaů·ərz əv 'tin }

flow-programmed chromatography [ANALYTICAL CHEMISTRY] A chromatographic procedure in which the rate of flow of the mobile phase is periodically changed. { ¦flō ˌprōˌgramd ˌkrō·mə'täg·rə·fē }

fluoborate [INORGANIC CHEMISTRY] *See* fluoroborate. { ˌflü·ə'bȯrˌāt }

fluometuron [ORGANIC CHEMISTRY] $C_{10}H_{11}F_3N_2O$ A white, crystalline solid with a melting point of 163-164.5°C; used as a herbicide for cotton and sugarcane. Also known as 1,1-dimethyl-3-(α,α,α-trifluoro-*meta*-tolyl)urea. { ¦flü·ō'me·chə ˌrän }

fluoranthene [ORGANIC CHEMISTRY] $C_{10}H_{10}$ A tetracyclic hydrocarbon found in coal tar fractions and petroleum, forming needlelike crystals, boiling point 250°C, and soluble in organic solvents such as ether and benzene. { flü'ranˌthēn }

fluorene [ORGANIC CHEMISTRY] $C_{13}H_{10}$ A hydrocarbon chemical present in the middle oil fraction of coal tar, insoluble in water, soluble in ether and acetone, melting point 116-117°C; used as the basis for a group of dyes. Also known as 2,3-benzindene; diphenylenemethane. { 'flůˌrēn }

fluorescein [ORGANIC CHEMISTRY] $C_{20}H_{12}O_5$ A yellowish to red powder, melts and decomposes at 290°C, insoluble in water, benzene, and chloroform, soluble in glacial acetic acid, boiling alcohol, ether, dilute acids, and dilute alkali; used in medicine, in oceanography as a marker in seawater, and in textiles to dye silk and wool. { ˌflu̇'re·sē·ən }

fluorescence analysis [ANALYTICAL CHEMISTRY] *See* fluorometric analysis. { flu̇'res·əns əˌnal·ə·səs }

fluorescence spectra [SPECTROSCOPY] Emission spectra of fluorescence in which an atom or molecule is excited by absorbing light and then emits light of characteristic frequencies. { flu̇'res·əns ˌspek·trə }

fluorescent dye [CHEMISTRY] A highly reflective dye that serves to intensify color and add to the brilliance of a fabric. { flu̇¦res·ənt 'dī }

fluorescent pigment [CHEMISTRY] A pigment capable of absorbing both visible and nonvisible electromagnetic radiations and releasing them quickly as energy of desired wavelength; examples are zinc sulfide or cadmium sulfide. { flu̇¦res·ənt 'pig·mənt }

fluoride [INORGANIC CHEMISTRY] A salt of hydrofluoric acid, HF, in which the fluorine atom is in the −1 oxidation state. { 'flu̇rˌīd }

fluorinated ethylene propylene resin [ORGANIC CHEMISTRY] Copolymers of tetrafluoroethylene and hexafluoropropylene. Abbreviated FEP resin. { 'flu̇r·əˌnād·əd 'eth·əˌlēn 'prō·pəˌlēn 'rez·ən }

fluorination [CHEMISTRY] A chemical reaction in which fluorine is introduced into a chemical compound. { ˌflu̇r·ə'nā·shən }

fluorine [CHEMISTRY] A gaseous or liquid chemical element, symbol F, atomic number 9, atomic weight 18.998; a member of the halide family, it is the most electronegative element and the most chemically energetic of the nonmetallic elements; highly toxic, corrosive, and flammable; used in rocket fuels and as a chemical intermediate. { 'flu̇rˌēn }

fluoroacetate [ORGANIC CHEMISTRY] Acetate in which carbon-connected hydrogen atoms are replaced by fluorine atoms. { ¦flu̇r·ō'as·əˌtāt }

fluoroacetic acid [ORGANIC CHEMISTRY] CH_2FCOOH A poisonous, crystalline compound obtained from plants, such as those of the Dichapetalaceae family, South Africa, soluble in water and alcohol, and burns with a green flame; the sodium salt is used as a water-soluble rodent poison. Also known as gifblaar poison. { ¦flu̇r·ō·ə'sēd·ik 'as·əd }

fluoroalkane [ORGANIC CHEMISTRY] Straight-chain, saturated hydrocarbon compound (or analog thereof) in which some of the hydrogen atoms are replaced by fluorine atoms. { ¦flu̇r·ō'alˌkān }

para-fluoroaniline [ORGANIC CHEMISTRY] $FC_6H_4NH_2$ A liquid that is an intermediate in the manufacture of herbicides and plant growth regulators. { ¦par·ə ˌflu̇r·ō'an·əˌlēn }

fluorobenzene [ORGANIC CHEMISTRY] C_6H_5F A colorless liquid with a boiling point of 84.9°C; used as an insecticide intermediate. Also known as phenyl fluoride. { ¦flu̇r·ō¦benˌzēn }

fluoroborate [INORGANIC CHEMISTRY] **1.** Any of a group of compounds related to the borates in which one or more oxygens have been replaced by fluorine atoms. **2.** The BF_4^- ion, which is derived from fluoroboric acid, HBF_4. Also known as fluoborate. { ˌflu̇r·ə'bȯrˌāt }

fluoroboric acid [INORGANIC CHEMISTRY] HBF_4 Colorless, clear, water-miscible acid; used for electrolytic brightening of aluminum and for forming stabilized diazo salts. { ¦flu̇r·əˌbȯr·ik 'as·əd }

fluorocarbon [ORGANIC CHEMISTRY] A hydrocarbon in which part or all hydrogen atoms have been replaced by fluorine atoms; can be liquid or gas and is nonflammable and heat-stable; used as refrigerant, aerosol propellant, and solvent. Also known as fluorohydrocarbon. { ¦flu̇r·ō'kär·bən }

fluorocarbon-11 [ORGANIC CHEMISTRY] *See* trichlorofluoromethane. { ¦flu̇r·ō'kär·bən ə'lev·ən }

fluorocarbon-21 [ORGANIC CHEMISTRY] *See* dichlorofluoromethane. { ¦flu̇r·ō'kär·bən ˌtwen·tē'wən }

fluorocarbon fiber [ORGANIC CHEMISTRY] Fiber made from a fluorocarbon resin, such as poly(tetrafluoroethylene). { ¦flu̇r·ō'kär·bən 'fī·bər }

fluorocarbon resin [ORGANIC CHEMISTRY] Polymeric material made up of carbon and fluorine with or without other halogens (such as chlorine) or hydrogen; the resin is extremely inert and more dense than corresponding fluorocarbons such as poly(tetrafluoroethylene). { ¦flůr·ō'kär·bən 'rez·ən }

fluorochemical [CHEMISTRY] Any chemical compound containing fluorine; usually refers to the fluorocarbons. { ¦flůr·ō'kem·ə·kəl }

fluorodichloromethane [ORGANIC CHEMISTRY] *See* dichlorofluoromethane. { ¦flůr·ō·dī‚klȯr·ō'meth‚ān }

fluorodifen [ORGANIC CHEMISTRY] $C_{13}H_7F_3N_2O_4$ A yellow, crystalline compound with a melting point of 93°C; used as a pre- and postemergence herbicide for food crops. { flü'räd·ə·fen }

1-fluoro-2,4-dinitrobenzene [ORGANIC CHEMISTRY] $(NO_2)_2C_6H_3F$ Crystals that are soluble in benzene, propylene glycol, and ether; used as a reagent for labeling terminal amino acid groups and in the detection of phenols. Also known as Sanger's reagent. { ¦wən ¦flůr·ō ¦tü ¦fȯr dī‚nī·trō'ben‚zēn }

fluoroform [ORGANIC CHEMISTRY] CHF_3 A colorless, nonflammable gas, boiling point 84°C at 1 atmosphere (101,325 pascals), freezing point 160°C at 1 atmosphere; used in refrigeration and as an intermediate in organic synthesis. Also known as propellant 23; refrigerant 23; trifluoromethane. { 'flůr·ə‚fȯrm }

fluorogenic substrate [CHEMISTRY] A nonfluorescent material that is acted upon by an enzyme to produce a fluorescent compound. { 'flůr·ə‚jen·ik 'səb‚strāt }

fluorohydrocarbon [ORGANIC CHEMISTRY] *See* fluorocarbon. { ¦flůr·ō‚hī·drə'kär·bən }

fluorometric analysis [ANALYTICAL CHEMISTRY] A method of chemical analysis in which a sample, exposed to radiation of one wavelength, absorbs this radiation and reemits radiation of the same or longer wavelength in about 10^{-9} second; the intensity of reemited radiation is almost directly proportional to the concentration of the fluorescing material. Also known as fluorescence analysis; fluorometry. { ¦flůr·ə¦me·trik ə'nal·ə·səs }

fluorometry [ANALYTICAL CHEMISTRY] *See* fluorometric analysis. { flü'räm·ə·trē }

***para*-fluorophenylacetic acid** [ORGANIC CHEMISTRY] $FC_6H_4CH_2COOH$ Crystals with a melting point of 86°C; used as an intermediate in the manufacture of fluorinated anesthetics. { ¦par·ə ¦flü·rə‚fen·əl·ə'sēd·ik 'as·əd }

fluorophosphoric acid [INORGANIC CHEMISTRY] H_2PO_3F A colorless, viscous liquid that is miscible with water, used in metal cleaners and as a catalyst. { ¦flůr·ō‚fäs'fȯr·ik 'as·əd }

fluorothene [ORGANIC CHEMISTRY] *See* chlorotrifluoroethylene polymer. { 'flůr·ə‚thēn }

fluorotrichloromethane [ORGANIC CHEMISTRY] *See* trichlorofluoromethane. { ¦flůr·ō·trī¦klȯr·ō'meth‚ān }

fluosilicate [INORGANIC CHEMISTRY] A salt derived from fluosilicic acid, H_2SiF_6, and containing the SiF_6^{-2} ion. { ¦flü·ə'sil·ə‚kāt }

fluosilicic acid [INORGANIC CHEMISTRY] H_2SiF_6 A colorless acid, soluble in water, which attacks glass and stoneware; highly corrosive and toxic; used in water fluoridation and electroplating. Also known as hydrofluorosilicic acid; hydrofluosilicic acid. { ¦flü·ə·sə'lis·ik 'as·əd }

fluosulfonic acid [INORGANIC CHEMISTRY] HSO_3F Colorless, corrosive, fuming liquid; soluble in water with partial decomposition; used as organic synthesis catalyst and in electroplating. { ¦flü·ə·səl'fän·ik 'as·əd }

flurenol [ORGANIC CHEMISTRY] $C_{18}H_{18}O_3$ A solid, crystalline compound with a melting point of 70-71°C; used as an herbicide for vegetables, cereals, and ornamental flowers. { 'flůr·ə‚nȯl }

fluxional compound [ORGANIC CHEMISTRY] **1.** Any of a group of molecules which undergo rapid intramolecular rearrangements in which the component atoms are interchanged among equivalent structures. **2.** Molecules in which bonds are broken and reformed in the rearrangement process. { 'flək·shən·əl ‚käm‚paůnd }

Fm [CHEMISTRY] *See* fermium.

foam [CHEMISTRY] An emulsionlike two-phase system where the dispersed phase is gas or air. { fōm }

folic acid sodium salt [ORGANIC CHEMISTRY] *See* sodium folate. { ¦fō·lik ¦as·əd ¦sōd·ē·əm 'sȯlt }

folimat [ORGANIC CHEMISTRY] $C_5H_{12}NO_4PS$ An oily liquid that decomposes at 135°C; soluble in water; used as an insecticide and miticide on fruit and vegetable crops and on ornamental flowers. Also known as omethioate. { 'fä·lə,mat }

Folin solution [ANALYTICAL CHEMISTRY] An aqueous solution of 500 grams of ammonium sulfate, 5 grams of uranium acetate, and 6 grams of acetic acid in a volume of 1 liter; used to test for uric acid. { 'fō·lən sə,lü·shən }

folpet [ORGANIC CHEMISTRY] $C_9H_4Cl_3NO_2S$ A buff or white, crystalline compound with a melting point of 177-178°C; insoluble in water; used as a fungicide on fruits, vegetables, and ornamental flowers. { 'fäl·pet }

foot's oil [CHEMISTRY] The oil sweated out of slack wax; it takes its name from the fact that it goes to the bottom, or foot, of the pan when sweated. { 'fu̇ts ,ȯil }

force constant [PHYSICAL CHEMISTRY] An expression for the force acting to restrain the relative displacement of the nuclei in a molecule. { 'fȯrs ,kän·stənt }

force field method [PHYSICAL CHEMISTRY] *See* molecular mechanics. { 'fȯrs ¦fēld ,meth·əd }

forensic chemistry [CHEMISTRY] The application of chemistry to the study of materials or problems in cases where the findings may be presented as technical evidence in a court of law. { fə'ren·sik 'kem·ə·strē }

formal charge [PHYSICAL CHEMISTRY] The apparent charge of an element in a compound; for example, magnesium has a formal charge of +2 in MgO and oxygen has a charge of −2. { ¦fȯr·məl ¦chärj }

formaldehyde [ORGANIC CHEMISTRY] HCHO The simplest aldehyde; a gas at room temperature, and a poisonous, clear, colorless liquid solution with pungent odor; used to make synthetic resins by reaction with phenols, urea, and melamine, as a chemical intermediate, as an embalming fluid, and as a disinfectant. Also known as formol; methanal; methylene oxide. { fȯr'mal·də,hīd }

formaldehyde sodium bisulfite [ORGANIC CHEMISTRY] CH_3NaO_4S A compound used as a fixing agent for fibers containing keratin, in metallurgy for flotation of lead-zinc ores, and in photography. { fȯr'mal·də,hīd ¦sōd·ē·əm bī'səl,fīt }

formality [CHEMISTRY] A concentration scale that gives the number of formula weights of solute per liter of solution; designated by F preceded by a number to show solute concentration. { fȯr'mal·əd·ē }

formamide [ORGANIC CHEMISTRY] **1.** A compound containing the radical HCONH. **2.** $HCONH_2$ A clear, colorless hygroscopic liquid, boiling at 200-212°C; soluble in water and alcohol; used as a solvent, softener, and chemical intermediate. Also known as formylamine; methanamide. { fȯrm'am·əd }

formamidinesulfinic acid [ORGANIC CHEMISTRY] $H_2NC(NH)SO_2H$ A reagent for the reduction of ketones to secondary alcohols. { fȯr¦mam·ə,dēn·səl'fin·ik 'as·əd }

formate [ORGANIC CHEMISTRY] A compound containing the HCOO— functional group. { 'fȯr,māt }

formic acid [ORGANIC CHEMISTRY] HCOOH A colorless, pungent, toxic, corrosive liquid melting at 8.4°C; soluble in water, ether, and alcohol; used as a chemical intermediate and solvent, in dyeing and electroplating processes, and in fumigants. Also known as methanoic acid. { ¦fȯr·mik 'as·əd }

formol [ORGANIC CHEMISTRY] *See* formaldehyde. { 'fȯr,mȯl }

formonitrile [INORGANIC CHEMISTRY] *See* hydrocyanic acid. { ¦fȯr·mō¦nī·trəl }

formula [CHEMISTRY] **1.** A combination of chemical symbols that expresses a molecule's composition. **2.** A reaction formula showing the interrelationship between reactants and products. { 'fȯr·myə·lə }

formulation [CHEMISTRY] The particular mixture of base chemicals and additives required for a product. { ,fȯr·myə'lā·shən }

formula weight [CHEMISTRY] **1.** The gram-molecular weight of a substance. **2.** In the case of a substance of uncertain molecular weight such as certain proteins, the molecular weight calculated from the composition, assuming that the element present in the smallest proportion is represented by only one atom. { 'fȯr·myə·lə ,wāt }

formyl [ORGANIC CHEMISTRY] The formic acid radical, HCO—; it is characteristic of aldehydes. { 'fȯr,mil }

formylamine [ORGANIC CHEMISTRY] *See* formamide. { ,fȯr·məl'am,ēn }

Fortrat parabola [SPECTROSCOPY] Graph of wave numbers of lines in a molecular spectral band versus the serial number of the successive lines. { ,fȯrträ pə'rab·ə·lə }

Foulger's test [ANALYTICAL CHEMISTRY] A test for fructose in which urea, sulfuric acid, and stannous chloride are added to the solution to be tested, the solution is boiled, and in the presence of fructose a blue coloration forms. { fül,jäz ,test }

four-degree calorie [CHEMISTRY] The heat needed to change the temperature of 1 gram of water from 3.5 to 4.5°C. { ¦fȯr di¦grē 'kal·ə·rē }

Fourier transform spectroscopy [SPECTROSCOPY] A spectroscopic technique in which all pertinent wavelengths simultaneously irradiate the sample for a short period of time, and the absorption spectrum is found by mathematical manipulation of the Fourier transform so obtained. { ¦fu̇r·ē¦ā 'tranz,fȯrm spek'träs·kə·pē }

fp [PHYSICAL CHEMISTRY] *See* freezing point.

Fr [CHEMISTRY] *See* francium.

fraction [CHEMISTRY] One of the portions of a volatile liquid within certain boiling point ranges, such as petroleum naphtha fractions or gas-oil fractions. { 'frak·shən }

fractional condensation [CHEMISTRY] Separation of components of vaporized liquid mixtures by condensing the vapors in stages (partial condensation); highest-boiling-point components condense in the first condenser stage, allowing the remainder of the vapor to pass on to subsequent condenser stages. { ¦frak·shən·əl ,kän·den'sā·shən }

fractional distillation [CHEMISTRY] A method to separate a mixture of several volatile components of different boiling points; the mixture is distilled at the lowest boiling point, and the distillate is collected as one fraction until the temperature of the vapor rises, showing that the next higher boiling component of the mixture is beginning to distill; this component is then collected as a separate fraction. { ¦frak·shən·əl dis·tə'lā·shən }

fractional precipitation [ANALYTICAL CHEMISTRY] Method for separating elements or compounds with similar solubilities by a series of analytical precipitations, each one improving the purity of the desired element. { ¦frak·shən·əl prə,sip·ə'tā·shən }

fractionating column [CHEMISTRY] An apparatus used widely for separation of fluid (gaseous or liquid) components by vapor-liquid fractionation or liquid-liquid extraction or liquid-solid adsorption. { 'frak·shə,nād·iŋ ,käl·əm }

fractionation [CHEMISTRY] Separation of a mixture in successive stages, each stage removing from the mixture some proportion of one of the substances, as by differential solubility in water-solvent mixtures. { ,frak·shə'nā·shən }

francium [CHEMISTRY] A radioactive alkali metal element, symbol Fr, atomic number 87, atomic weight distinguished by nuclear instability; exists in short-lived radioactive forms, the chief isotope being francium-223. { 'fran·sē·əm }

Franck-Condon principle [PHYSICAL CHEMISTRY] The principle that in any molecular system the transition from one energy state to another is so rapid that the nuclei of the atoms involved can be considered to be stationary during the transition. { ¦fräŋk 'kän·dən ,prin·sə·pəl }

Franck-Rabinowitch hypothesis [PHYSICAL CHEMISTRY] The hypothesis that the decreased quantum efficiencies of certain photochemical reactions observed in the dissolved or liquid state are due to the formation of a cage of solvent molecules around the molecule which has been excited by absorption of a photon. { ¦fräŋk rə'bin·ə,wich hī,päth·ə·səs }

Frankland's method [ORGANIC CHEMISTRY] Reaction of dialkyl zinc compounds with alkyl halides to form hydrocarbons; may be used to form paraffins containing a quaternary carbon atom. { 'fraŋk·lənz ,meth·əd }

Fraude's reagent [INORGANIC CHEMISTRY] *See* perchloric acid. { 'frōdz rē,ā·jənt }

fraunhofer [SPECTROSCOPY] A unit for measurement of the reduced width of a spectrum line such that a spectrum line's reduced width in fraunhofers equals 10^6 times its equivalent width divided by its wavelength. { 'frau̇n,hōf·ər }

Fraunhofer lines [SPECTROSCOPY] The dark lines constituting the Fraunhofer spectrum. { 'fraůn,hōf·ər ,līnz }
Fraunhofer spectrum [SPECTROSCOPY] The absorption lines in sunlight, due to the cooler outer layers of the sun's atmosphere. { 'fraůn,hōf·ər ,spek·trəm }
freeboard [ANALYTICAL CHEMISTRY] The space provided above the resin bed in an ion-exchange column to allow for expansion of the bed during backwashing. { ¦frē,bȯrd }
free cyanide [CHEMISTRY] Cyanide not combined as part of an ionic complex. { ¦frē 'sī·ə,nīd }
free ion [PHYSICAL CHEMISTRY] An ion, such as found in an ionized gas, whose properties, such as spectrum and magnetic moment, are not significantly affected by other atoms, ions, or molecules nearby. { ¦frē 'ī,än }
free molecule [PHYSICAL CHEMISTRY] A molecule, as in a gas, whose properties, such as spectrum and magnetic moment, are not affected by other atoms, ions, and molecules nearby. { ¦frē 'mäl·ə,kyül }
free radical [CHEMISTRY] An atom or a diatomic or polyatomic molecule which possesses one unpaired electron. Also known as a radical. { ¦frē 'rad·ə·kəl }
free-radical reaction [ORGANIC CHEMISTRY] *See* homolytic cleavage. { ¦frē ¦rad·ə·kəl rē'ak·shən }
free water [CHEMISTRY] The volume of water that is not contained in suspension in a vessel containing both water and a suspension of water and another liquid. { 'frē 'wȯd·ər }
freeze [PHYSICAL CHEMISTRY] To solidify a liquid by removal of heat. { frēz }
freezing mixture [PHYSICAL CHEMISTRY] A mixture of substances whose freezing point is lower than that of its constituents. { 'frēz·iŋ ,miks·chər }
freezing point [PHYSICAL CHEMISTRY] The temperature at which a liquid and a solid may be in equilibrium. Abbreviated fp. { 'frēz·iŋ ,pȯint }
freezing-point depression [PHYSICAL CHEMISTRY] The lowering of the freezing point of a solution compared to the pure solvent; the depression is proportional to the active mass of the solute in a given amount of solvent. { 'frēz·iŋ ,pȯint di,presh·ən }
frequency factor [PHYSICAL CHEMISTRY] The constant A (or ν) in the Arrhenius equation, which is the relation between reaction rate and absolute temperature T; the equation is $k = Ae - (\Delta H_{act}/RT)$, where k is the specific rate constant, ΔH_{act} is the heat of activation, and R is the gas constant. { 'frē·kwən·sē ,fak·tər }
Freund method [ORGANIC CHEMISTRY] A method for preparation of cycloparaffins in which dihalo derivatives of the paraffins are treated with zinc to produce the cycloparaffin. { 'frȯind ,meth·əd }
Friedel-Crafts reaction [ORGANIC CHEMISTRY] A substitution reaction, catalyzed by aluminum chloride in which an alkyl (R—) or an acyl (RCO—) group replaces a hydrogen atom of an aromatic nucleus to produce hydrocarbon or a ketone. { frē¦del 'krafs rē,ak·shən }
Friedlander synthesis [ORGANIC CHEMISTRY] A synthesis of quinolines; the method is usually catalyzed by bases and consists of condensation of an aromatic *o*-aminocarbonyl derivative with a compound containing a methylene group in the alpha position to the carbonyl. { 'frēd,lan·dər ,sin·thə·səs }
Fries rearrangement [ORGANIC CHEMISTRY] The conversion of a phenolic ester into the corresponding *o*- and *p*-hydroxyketone by treatment with catalysts of the type of aluminum chloride. { 'frēz rē·ə'rānj·mənt }
Fries' rule [ORGANIC CHEMISTRY] The rule that the most stable form of the bonds of a polynuclear compound is that arrangement which has the maximum number of rings in the benzenoid form, that is, three double bonds in each ring. { 'frēz ,rül }
frontier orbitals [PHYSICAL CHEMISTRY] Orbitals of two molecules that are spatially arranged so that a significant amount of overlap occurs between them. { frən¦tir 'ȯr·bə·təlz }
frother [CHEMISTRY] Substance used in flotation processes to make air bubbles sufficiently permanent, principally by reducing surface tension. { 'frȯ·thər }
froth promoter [CHEMISTRY] A chemical compound used with a frothing agent. { 'frȯth prə,mōd·ər }

frustrated internal reflectance [SPECTROSCOPY] *See* attenuated total reflectance. { 'frəs‚trād·əd in‚tərn·əl ri'flek·təns }

fuchsin [ORGANIC CHEMISTRY] $C_{20}H_{19}N_3$ Brownish-red crystals, used as a dye or in the commercial preparation of other dyes, and as an antifungal drug. Also known as magenta; rosaniline. { 'fyük·sən }

fuel-cell catalyst [CHEMISTRY] A substance, such as platinum, silver, or nickel, from which the electrodes of a fuel cell are made, and which speeds the reaction of the cell; it is especially important in a fuel cell which does not operate at high temperatures. { 'fyül ‚sel 'kad·ə‚list }

fuel-cell electrolyte [CHEMISTRY] The substance which conducts electricity between the electrodes of a fuel cell. { 'fyül ‚sel i'lek·trə‚līt }

fuel-cell fuel [CHEMISTRY] A substance, such as hydrogen, carbon monoxide, sodium, alcohol, or a hydrocarbon, which reacts with oxygen to generate energy in a fuel cell. { 'fyül ‚sel 'fyül }

fugitive dye [CHEMISTRY] A dye that is unstable, that is, not fast; used in the textile processing for purposes of identity. { ¦fyü·jəd·iv 'dī }

Fulcher bands [SPECTROSCOPY] A group of bands in the spectrum of molecular hydrogen that are preferentially excited by a low-voltage discharge. { 'fəl·chər ‚banz }

fullerene [CHEMISTRY] A large molecule composed entirely of carbon, with the chemical formula C_n, where n is any even number from 32 to over 100; believed to have the structure of a hollow spheroidal cage with a surface network of carbon atoms connected in hexagonal and pentagonal rings. { 'fúl·ə‚rēn }

fulminate [ORGANIC CHEMISTRY] **1.** A salt of fulminic acid. **2.** $HgC_2N_2O_2$ An explosive mercury compound derived from the fulminic acid; used for the caps or exploders by means of which charges of gunpowder, dynamite, and other explosives are fired. Also known as mercury fulminate. { 'fúl·mə‚nāt }

fulminic acid [ORGANIC CHEMISTRY] CNOH An unstable isomer of cyanic acid, whose salts are known for their explosive characteristics. { fúl'min·ik 'as·əd }

fulminuric acid [ORGANIC CHEMISTRY] $CN \cdot CH(NO_2) \cdot CONH_2$ A trimer of cyanuric acid; a water-soluble compound, crystallizing in colorless needles, melting at 138°C, and exploding at 145°C. Also known as isocyanuric acid. { ¦fúl·mə¦núr·ik 'as·əd }

fulvene [ORGANIC CHEMISTRY] C_6H_6 A yellow oil, an isomer of benzene. { 'fúl·vēn }

fumaric acid [ORGANIC CHEMISTRY] $C_4H_4O_4$ A dicarboxylic organic acid produced commercially by synthesis and fermentation; the trans isomer of maleic acid, colorless crystals, melting point 287°C; used to make resins, paints, varnishes, and inks, in foods, as a mordant, and as a chemical intermediate. Also known as boletic acid. { fyü'mar·ik 'as·əd }

fume hood [CHEMISTRY] A fume-collection device over an enclosed shelf or table, so that experiments involving poisonous or unpleasant fumes or gases may be conducted away from the experimental area. { 'fyüm ‚húd }

fumes [CHEMISTRY] Particulate matter consisting of the solid particles generated by condensation from the gaseous state, generally after volatilization from melted substances, and often accompanied by a chemical reaction, such as oxidation. { fyümz }

fumigant [CHEMISTRY] A chemical compound which acts in the gaseous state to destroy insects and their larvae and other pests; examples are dichlorethyl ether, *p*-dichlorobenzene, and ethylene oxide. { 'fyü·mə·gənt }

fuming nitric acid [INORGANIC CHEMISTRY] Concentrated nitric acid containing dissolved nitrogen dioxide; may be prepared by adding formaldehyde to concentrated nitric acid. { ¦fyüm·iŋ 'nī‚trik ‚as·əd }

fuming sulfuric acid [INORGANIC CHEMISTRY] Concentrated sulfuric acid containing dissolved sulfur trioxide. Also known as oleum. { ¦fyüm·iŋ səl'fyúr·ik ‚as·əd }

functional group [ORGANIC CHEMISTRY] An atom or group of atoms, acting as a unit, that has replaced a hydrogen atom in a hydrocarbon molecule and whose presence imparts characteristic properties to this molecule; frequently represented as R—. Also known as functionality. { ¦fəŋk·shən·əl 'grüp }

functionality [ORGANIC CHEMISTRY] *See* functional group. { ‚fəŋk·shə'nal·əd·ē }

fundamental series [SPECTROSCOPY] A series occurring in the line spectra of many atoms and ions having one, two, or three electrons in the outer shell, in which the total orbital angular momentum quantum number changes from 3 to 2. { ¦fən·də¦ment·əl 'sir·ēz }

funicular distribution [CHEMISTRY] The distribution of a two-phase, immiscible liquid mixture (such as oil and water, one a wetting phase, the other nonwetting) in a porous system when the wetting phase is continuous over the surface of the solids. { fə'nik·yə·lər dis·trə'byü·shən }

2-furaldehyde [ORGANIC CHEMISTRY] *See* furfural. { ¦tü fə'ral·də,hīd }

furan [ORGANIC CHEMISTRY] **1.** One of a group of organic heterocyclic compounds containing a diunsaturated ring of four carbon atoms and one oxygen atom. **2.** $C_4H_4O_4$ The simplest furan type of molecule; a colorless, mildly toxic liquid, boiling at 32°C, insoluble in water, soluble in alcohol and ether; used as a chemical intermediate. Also known as furfuran; tetrol. { 'fyůr,an }

furancarboxylic acid [ORGANIC CHEMISTRY] *See* furoic acid. { ¦fyůr·ən,kar,bäk'sil·ik 'as·əd }

2,5-furandione [ORGANIC CHEMISTRY] *See* maleic anhydride. { ¦tü ¦fīv ¦fyůr·ən'dī,ōn }

furanoside [ORGANIC CHEMISTRY] A glycoside whose cyclic sugar component resembles that of furan. { fyə'ran·ə,sīd }

furan resin [ORGANIC CHEMISTRY] A liquid, thermosetting resin in which the furan ring is an integral part of the polymer chain, made by the condensation of furfuryl alcohol; used as a cement and adhesive, casting resin, coating, and impregnant. { 'fyůr,an ,rez·ən }

furfural [ORGANIC CHEMISTRY] C_4H_3OCHO When pure, a colorless liquid, soluble in organic solvents, slightly soluble in water; used as a lube oil-refining solvent, in cellulosic formulations, in making resins, as a weed killer, as a fungicide, and as a chemical intermediate. Also known as 2-furaldehyde; furfuraldehyde; furfurol; furol. { 'fər·fə,ral }

furfuraldehyde [ORGANIC CHEMISTRY] *See* furfural. { 'fər·fə'ral·də,hīd }

furfuran [ORGANIC CHEMISTRY] *See* furan. { 'fər·fə,ran }

furfurol [ORGANIC CHEMISTRY] *See* furfural. { 'fər·fə,rōl }

furfuryl [ORGANIC CHEMISTRY] The functional group C_5H_6O— from furfural. { 'fər·fə,ril }

furfuryl alcohol [ORGANIC CHEMISTRY] $C_5H_6O_2$ A liquid with a faint burning odor and bitter taste, soluble in alcohol and ether, usually prepared from furfural; used as a solvent in the manufacturing of wetting agents and resins. { 'fər·fə,ril 'al·kə,hól }

furnace black [CHEMISTRY] A carbon black formed by partial combustion of liquid and gaseous hydrocarbons in a closed furnace with a deficiency of oxygen; used as a reinforcing filler for synthetic rubber. { 'fər·nəs ,blak }

furoic acid [ORGANIC CHEMISTRY] $C_5H_4O_3$ Long monoclinic prisms crystallized from the water solution, soluble in ether and alcohol; used as a preservative and bactericide. Also known as furancarboxylic acid; pyromucic acid. { fyů'rō·ik 'as·əd }

furol [ORGANIC CHEMISTRY] *See* furfural. { 'fyů,rōl }

fused aromatic ring [ORGANIC CHEMISTRY] A molecular structure in which two or more aromatic rings have two carbon atoms in common. { ¦fyüzd ar·ə¦mad·ik 'riŋ }

fused potassium sulfide [INORGANIC CHEMISTRY] *See* potassium sulfide. { 'fyüzd pə'tas·ē·əm 'səl,fīd }

fused-salt electrolysis [PHYSICAL CHEMISTRY] Electrolysis with use of purified fused salts as raw material and as an electrolyte. { ¦fyüzd ¦sólt i,lek'trä·lə·səs }

fusion [PHYSICAL CHEMISTRY] A change of the state of a substance from the solid phase to the liquid phase. Also known as melting. { 'fyü·zhən }

fusion tube [ANALYTICAL CHEMISTRY] Device used for the analysis of the elements in a compound by fusing them with another compound; for example, analysis of nitrogen in organic compounds by fusing the compound with sodium and analyzing for sodium cyanide. { 'fyü·zhən ,tüb }

G

Ga [CHEMISTRY] *See* gallium.

GABA [ORGANIC CHEMISTRY] *See* γ-aminobutyric acid. { ′ga·bə *or* ˌjēˌāˌbē′ā }

Gabriel's synthesis [ORGANIC CHEMISTRY] A synthesis of primary amines by the hydrolysis of N-alkylphthalimides; the latter are obtained from potassium phthalimide and alkyl halides. { ′gā·brē·əlz ˌsin·thə·səs }

adoleic acid [ORGANIC CHEMISTRY] $C_{20}H_{38}O_2$ A fatty acid derived from cod liver oil, and melting at 20°C. { ¦gad·ə¦lē·ik ′as·əd }

gadolinium [CHEMISTRY] A rare-earth element, symbol Gd, atomic number 64, atomic weight 157.25; highly magnetic, especially at low temperatures. { ˌgad·əl′in·ē·əm }

galipol [ORGANIC CHEMISTRY] $C_{15}H_{26}O$ A terpene alcohol derived from the oil of the angostura bark; colorless crystals that melt at 89°C. { ′gal·əˌpȯl }

gallacetophenone [ORGANIC CHEMISTRY] $C_8H_8O_4$ A white to brownish-gray, crystalline powder, melting at 173°C, soluble in water, alcohol, and ether; used as an antiseptic. { ¦gȯl¦as·ə·tä·fə′nōn }

gallein [ORGANIC CHEMISTRY] $C_{20}H_{10}O_7$ A brown powder or green scales, broken down by heat; used as a pH indicator in the analysis of phosphates in urine and as an intermediate in the manufacture of dyes. Also known as anthracene violet; gallin; pyrogallolphthalein. { ′gal·ē·ən }

gallic acid [ORGANIC CHEMISTRY] $C_7H_6O_5$ A crystalline compound that forms needles from solutions of absolute methanol or chloroform, dissolves in water, alcohol, ether, and glycerol; obtained from nutgall tannins or from *Penicillium notatum* fermentation; used to make antioxidants and ink dyes and in photography { ′gal·ik ′as·əd }

gallin [ORGANIC CHEMISTRY] *See* gallein. { ′gal·ən }

gallium [CHEMISTRY] A chemical element, symbol Ga, atomic number 31, atomic weight 69.72. { ′gal·ē·əm }

gallium arsenide [INORGANIC CHEMISTRY] GaAs A crystalline material, melting point 1238°C; frequently alloys of this material are formed with gallium phosphide or indium arsenide. { ′gal·ē·əm ′ärs·ənˌīd }

gallium halide [INORGANIC CHEMISTRY] A compound formed by bonding of gallium to either chlorine, bromine, iodine, fluorine, or astatine. { ′gal·ē·əm ′haˌlīd }

gallium phosphide [INORGANIC CHEMISTRY] GaP Transparent crystals made by reacting phosphorus and gallium suboxide at low temperature { ′gal·ē·əm ′fäsˌfīd }

gallocyanine [ORGANIC CHEMISTRY] $C_{15}H_{13}ClN_2O_5$ Green crystals soluble in alcohol, glacial acetic acid, alkali carbonates, and concentrated hydrochloric acid; used as a dye and as a reagent for the determination of lead. { ˌga·lō′sī·əˌnēn }

gallotannic acid [ORGANIC CHEMISTRY] *See* tannic acid. { ¦ga·lō¦tan·ik ′as·əd }

gallotannin [ORGANIC CHEMISTRY] *See* tannic acid. { ¦ga·lō¦tan·ən }

galvanic series [CHEMISTRY] The relative hierarchy of metals arranged in order from magnesium (least noble) at the anodic, corroded end through platinum (most noble) at the cathodic, protected end. { gal′van·ik ′sir·ēz }

gamma [CHEMISTRY] The gamma position (the third carbon atom in an aliphatic carbon chain) on a chemical compound. { ′gam·ə }

gamma acid [ORGANIC CHEMISTRY] $C_{10}H_5NH_2OHSO_3H$ White crystals, slightly soluble in water; an intermediate in dyestuff manufacture. Also known as 2-amino-8-naphthol-6-sulfonic acid; 7-amino-1-naphthol-3-sulfonic acid; 2,5-naphthylamine sulfonic acid; 3-sulfonic acid; 6-sulfonic acid. { 'gam·ə 'as·əd }

gamma-ray spectrum [SPECTROSCOPY] The set of wavelengths or energies of gamma rays emitted by a given source. { 'gam·ə ˌrā ˌspek·trəm }

gamma transition [PHYSICAL CHEMISTRY] *See* glass transition. { 'gam·ə tran'zish·ən }

gammil [CHEMISTRY] A unit of concentration, equal to a concentration of 1 milligram of solute in 1 liter of solvent. Also known as micril; microgammil. { 'gam·əl }

gas adsorption [PHYSICAL CHEMISTRY] The concentration of a gas upon the surface of a solid substance by attractive forces between the surface and the gas molecules. { ¦gas ad'sorp·shən }

gas analysis [ANALYTICAL CHEMISTRY] Analysis of the constituents or properties of a gas (either pure or mixed); composition can be measured by chemical adsorption, combustion, electrochemical cells, indicator papers, chromatography, mass spectroscopy, and so on; properties analyzed for include heating value, molecular weight, density, and viscosity. { 'gas əˌnal·ə·səs }

gas black [CHEMISTRY] Fine particles of carbon formed by partial combustion or thermal decomposition of natural gas; used to reinforce rubber products such as tires. Also known as carbon black; channel black. { 'gas ˌblak }

gas chromatograph [ANALYTICAL CHEMISTRY] The instrument used in gas chromatography to detect volatile compounds present; also used to determine certain physical properties such as distribution or partition coefficients and adsorption isotherms, and as a preparative technique for isolating pure components or certain fractions from complex mixtures. { ¦gas krō'mad·əˌgraf }

gas chromatography [ANALYTICAL CHEMISTRY] A separation technique involving passage of a gaseous moving phase through a column containing a fixed adsorbent phase; it is used principally as a quantitative analytical technique for volatile compounds. { ¦gas ˌkrō·mə'täg·rə·fē }

gas-condensate liquid [ORGANIC CHEMISTRY] A hydrocarbon, such as propane, butane, and pentane, obtained as condensate when wet natural gas is compressed or refrigerated. { ¦gas 'känd·ənˌsāt ˌlik·wəd }

gas generator [CHEMISTRY] A device used to generate gases in the laboratory. { 'gas ˌjen·əˌrād·ər }

gas-liquid chromatography [ANALYTICAL CHEMISTRY] A form of gas chromatography in which the fixed phase (column packing) is a liquid solvent distributed on an inert solid support. Abbreviated GLC. Also known as gas-liquid partition chromatography. { 'gas ˌlik·wəd ˌkrō·mə'täg·rə·fē }

gas-liquid partition chromatography [ANALYTICAL CHEMISTRY] *See* gas-liquid chromatography. { 'gas ˌlik·wəd pär'tish·ən ˌkrō·mə'täg·rə·fē }

gasometric method [ANALYTICAL CHEMISTRY] An analytical technique for gases; the gas may be measured by instrumental methods or through chemical reactions with specific reagents. { ˌgas·ə¦me·trik 'meth·əd }

gas-solid chromatography [ANALYTICAL CHEMISTRY] A form of gas chromatography in which the moving phase is a gas and the stationary phase is a surface-active sorbent (charcoal, silica gel, or activated alumina). Abbreviated GSC. { ¦gas ¦säl·əd ˌkrō·mə'täg·rə·fē }

gas solubility [PHYSICAL CHEMISTRY] The extent that a gas dissolves in a liquid to produce a homogeneous system. { ¦gas ˌsäl·yə'bil·əd·ē }

Gatterman-Koch synthesis [ORGANIC CHEMISTRY] A synthesis of aldehydes; aldehydes form when an aromatic hydrocarbon is heated in the presence of hydrogen chloride, certain metallic chloride catalysts, and either carbon monoxide or hydrogen cyanide. { ¦gäd·ər·män 'ko̅k ˌsin·thə·səs }

Gatterman reaction [ORGANIC CHEMISTRY] **1.** Reaction of a phenol or phenol ester, and hydrogen chloride or hydrogen cyanide, in the presence of a metallic chloride such as aluminum chloride to form, after hydrolysis, an aldehyde. **2.** Reaction of an aqueous ethanolic solution of diazonium salts with precipitated copper powder or other reducing agent to form diaryl compounds. { 'gäd·ər·män rēˌak·shən }

gaultheria oil [ORGANIC CHEMISTRY] *See* methyl salicylate. { gȯl'thir·ē·ə ˌȯil }

Gay-Lussac's law of volumes [CHEMISTRY] *See* combining volumes principle. { ˌgā·lüˌsäks ¦lȯ əv ′väl·yəmz }

Gd [CHEMISTRY] *See* gadolinium.

GDCH [ORGANIC CHEMISTRY] *See* α-dichlorohydrin.

Ge [CHEMISTRY] *See* germanium.

gel [CHEMISTRY] A two-phase colloidal system consisting of a solid and a liquid in more solid form than a sol. { jel }

gelatin [ORGANIC CHEMISTRY] A protein derived from the skin, white connective tissue, and bones of animals; used as a food and in photography, the plastics industry, metallurgy, and pharmaceuticals. { ′jel·ət·ən }

gelation [CHEMISTRY] **1.** The act or process of freezing. **2.** Formation of a gel from a sol. { jə′lā·shən }

gel electrophoresis [CHEMISTRY] Electrophoresis performed in silica gel, which is a porous, inert medium. { ¦jel iˌlek·trō·fə′rē·səs }

gel filtration [ANALYTICAL CHEMISTRY] A type of column chromatography which separates molecules on the basis of size; higher-molecular-weight substances pass through the column first. Also known as molecular exclusion chromatography; molecular sieve chromatography. { ¦jel fil′trā·shən }

gel permeation chromatography [ANALYTICAL CHEMISTRY] Analysis by chromatography in which the stationary phase consists of beads of porous polymeric material such as a cross-linked dextran carbohydrate derivative sold under the trade name Sephadex; the moving phase is a liquid. { ¦jel ˌpər·mē′ā·shən ˌkrō·mə′täg·rə·fē }

gel point [PHYSICAL CHEMISTRY] Stage at which a liquid begins to exhibit elastic properties and increased viscosity. { ′jel ˌpȯint }

geminal [ORGANIC CHEMISTRY] Referring to like atoms or groups attached to the same atom in a molecule. { ′jem·ə·nəl }

general formula [CHEMISTRY] A formula that can apply not only to one specific compound but to a series of related compounds; for example, the general formula for an aldehyde RCHO, where R is hydrogen in formaldehyde (the simplest aldehyde) and is a hydrocarbon radical for other aldehydes in the series such as CH_3 for acetaldehyde and C_2H_5 for proprionaldehyde. { ¦jen·rəl ′fȯr·myə·lə }

Geneva system [ORGANIC CHEMISTRY] An international system of nomenclature for organic compounds based on hydrocarbon derivatives; names correspond to the longest straight carbon chain in the molecule. { jə¦nē·və ′sis·təm }

genicide [ORGANIC CHEMISTRY] $C_{13}H_8O_2$ A compound with needlelike crystals and a melting point of 174°C; insoluble in water; used as an insecticide, miticide, and ovicide. Also known as oxoxanthone; 9-xanthenone; xanthone. { ′jen·əˌsīd }

genistin [ORGANIC CHEMISTRY] $C_{21}H_{20}O_{10}$ A pale-yellow glucoside derived from soybean meal, crystallizes from 80% methanol solution, melting point 256°C, soluble in hot 80% ethanol, hot 80% methanol, and hot acetone. Also known as 7-D-glucoside. { jə′nis·tən }

gentian violet [ORGANIC CHEMISTRY] *See* methyl violet. { ′jen·chən ′vī·lət }

gentisic acid [ORGANIC CHEMISTRY] $C_7H_6O_4$ A crystalline compound that forms monoclinic prisms from a water solution, sublimes at 200°C, melts at 250°C, and is soluble in water, alcohol, ether, sodium, and salt; used in medicine. Also known as gentianic acid. { jen′tis·ik ′as·əd }

geometrical isomerism [PHYSICAL CHEMISTRY] The phenomenon in which isomers contain atoms attached to each other in the same order and with the same bonds but with different spatial, or geometrical, relationships; the explicit geometry imposed upon a molecule by, say, a double bond between carbon atoms makes possible the existence of these isomers. { ¦jē·ə¦me·trə·kəl ī′sä·məˌriz·əm }

geranial [ORGANIC CHEMISTRY] *See* citral. { jə′rā·nē·əl }

geranialdehyde [ORGANIC CHEMISTRY] *See* citral. { jəˌrā·nē′al·dəˌhīd }

geraniol [ORGANIC CHEMISTRY] $(CH_3)_2CCH(CH_2)_2C(CH_3)CHCH_2OH$ A colorless to pale-yellow liquid, an alcohol and a terpene, boiling point 230°C; soluble in alcohol and ether, insoluble in water; used in perfumery and flavoring. { jə′rā·nēˌȯl }

geranyl [ORGANIC CHEMISTRY] $C_{10}H_{17}$ The functional group from geraniol, $(CH_3)_2$: $CHCH_2CH_2 \cdot CHCH_3$:$CH \cdot CH_2OH$. { jə′ran·əl }

Gerard reagent [CHEMISTRY] The quaternary ammonium compounds, acethydrazide-pyridinium chloride and trimethylacethydrazide ammonium chloride; used to separate aldehydes and ketones from oily or fatty natural materials and to extract sex hormones from urine. { jə'rärd rē,ā·jənt }

germane [INORGANIC CHEMISTRY] **1.** A hydride of germanium whose general formula is Ge_nH_{2n+2}. **2.** The compound GeH_4, a hydride of germanium, a colorless gas that is combustible in air and burns with a blue flame. { ¦jər¦mān }

germanide [INORGANIC CHEMISTRY] A compound of an alkaline earth or alkali metal with germanium; an example is magnesium germanide, Mg_2Ge; the germanides are reactive with water. { 'jər·mə,nīd }

germanium [CHEMISTRY] A brittle, water-insoluble, silvery-gray metallic element in the carbon family, symbol Ge, atomic number 32, atomic weight 72.59, melting at 959°C. { jər'mān·ē·əm }

germanium halide [INORGANIC CHEMISTRY] A dihalide or tetrahalide of fluorine, chlorine, bromine, or iodine with germanium. { jər'mān·ē·əm 'ha,līd }

germanium oxide [INORGANIC CHEMISTRY] The monoxide GeO or dioxide GeO_2; a study of GeO indicates it exists in polymeric form; GeO_2 is a white powder, soluble in alkalies; used in special glass and in medicine. { jər'mān·ē·əm 'äk,sīd }

getter [CHEMISTRY] *See* scavenger. [PHYSICAL CHEMISTRY] **1.** A substance, such as thallium, that binds gases on its surface and is used to maintain a high vacuum in a vacuum tube. **2.** A special metal alloy that is placed in a vacuum tube during manufacture and vaporized after the tube has been evacuated; when the vaporized metal condenses, it absorbs residual gases. Also known as degasser. { 'ged·ər }

ghost image [SPECTROSCOPY] A false image of a spectral line produced by irregularities in the ruling of a diffraction grating. { 'gōst ,im·ij }

Gibbs adsorption equation [PHYSICAL CHEMISTRY] A formula for a system involving a solvent and a solute, according to which there is an excess surface concentration of solute if the solute decreases the surface tension, and a deficient surface concentration of solute if the solute increases the surface tension. { 'gibz ad'sorp·shən i,kwā·zhən }

Gibbs adsorption isotherm [PHYSICAL CHEMISTRY] An equation for the surface pressure of surface monolayers,

$$\phi = RT \int_0^p \Gamma d(\ln p)$$

where ϕ is surface pressure, T is absolute temperature, R is the gas constant, Γ is the number of molecules adsorbed per gram per unit surface area, and p is the pressure of the gas. { 'gibz ad'sorp·shən 'ī·sō,thərm }

Gibbs-Donnan equilibrium [PHYSICAL CHEMISTRY] *See* Donnan equilibrium. { ¦gibz 'dän·ən ē·kwə'lib·rē·əm }

Gibbs-Duhem equation [PHYSICAL CHEMISTRY] A relation that imposes a condition on the composition variation of the set of chemical potentials of a system of two or more components,

$$S dT - V dP + \sum_{i=1}^{r} n_i d\mu_i = 0$$

where S is entropy, T absolute temperature, P pressure, n_i the number of moles of the ith component, and μ_i is the chemical potential of the ith component. Also known as Duhem's equation. { ¦gibz 'dü·əm i,kwā·zhən }

Gibbs-Helmholtz equation [PHYSICAL CHEMISTRY] An expression for the influence of temperature upon the equilibrium constant of a chemical reaction, $(d \ln K°/dT)_P = \Delta H°/RT^2$, where K° is the equilibrium constant, ΔH° the standard heat of the reaction at the absolute temperature T, and R the gas constant. { 'gibz 'helm,hōlts i,kwā·zhən }

Gibbs phase rule [PHYSICAL CHEMISTRY] A relationship used to determine the number of state variables F, usually chosen from among temperature, pressure, and species composition in each phase, which must be specified to fix the thermodynamic state of a system in equilibrium: $F = C - P - M + 2$, where C is the number of chemical species presented at equilibrium, P is the number of phases, and M is the number of independent chemical reactions. Also known as Gibbs rule; phase rule. { 'gibz 'fāz ˌrül }

Gibbs-Poynting equation [PHYSICAL CHEMISTRY] An expression relating the effect of the total applied pressure P upon the vapor pressure p of a liquid, $(dp/dP)_T = V_l/V_g$, where V_l and V_g are molar volumes of the liquid and vapor. { ¦gibz 'pȯint·iŋ iˌkwā·zhən }

Gibbs rule [PHYSICAL CHEMISTRY] *See* Gibbs phase rule. { 'gibz ˌrül }

Giemsa stain [CHEMISTRY] A stain for hemopoietic tissue and hemoprotozoa consisting of a stock glycerol methanol solution of eosinates of Azure B and methylene blue with some excess of the basic dyes. { 'gēm·sə ˌstān }

gifblaar poison [ORGANIC CHEMISTRY] *See* fluoroacetic acid. { 'gifˌblär ˌpȯiz·ən }

Gillespie equilibrium still [ANALYTICAL CHEMISTRY] A recirculating equilibrium distillation apparatus used to establish azeotropic properties of liquid mixtures. { gə 'les pē ˌē·kwə'librē·əm ˌstil }

gitonin [ORGANIC CHEMISTRY] The gitogenin tetraglycoside in *Digitalis purpurea* seed; resembles digitonin. { jə'tōn·ən }

glacial acetic acid [ORGANIC CHEMISTRY] CH_3COOH Pure acetic acid (containing less than 1% water); a clear, colorless, caustic hygroscopic liquid, boiling at 118°C, soluble in water, alcohol, and ether, and crystallizing readily; used as a solvent for oils and resins. { ¦glā·shəl ə¦sēd·ik 'as·əd }

glass electrode [PHYSICAL CHEMISTRY] An electrode or half cell in which potential measurements are made through a glass membrane, which acts as a cation-exchange membrane; thus, the potential arises from phase-boundary and diffusion potentials which, depending on the composition of the glass, are logarithmic functions of the activity of the cations such as H^+, Na^+, or K^+ of the solutions in which the electrode is immersed. { ¦glas i'lekˌtrōd }

glass transition [PHYSICAL CHEMISTRY] The change in an amorphous region of a partially crystalline polymer from a viscous or rubbery condition to a hard and relatively brittle one; usually brought about by changing the temperature. Also known as gamma transition; glassy transition. { 'glas tranˌzish·ən }

glass transition temperature [PHYSICAL CHEMISTRY] The temperature at which a liquid changes to an amorphous or glassy solid. { ¦glas ˌtran'zish·ən ˌtem·prə·chər }

glassy transition [PHYSICAL CHEMISTRY] *See* glass transition. { ¦glas·ē tran'zish·ən }

Glauber's salt [INORGANIC CHEMISTRY] $Na_2SO_4 \cdot 10H_2O$ Crystalline hydrated sodium sulfate, loses water when exposed to air; water soluble, alcohol insoluble; used in textile dyeing and medicine. { 'glaů·bərz ˌsȯlt }

glaze stain [INORGANIC CHEMISTRY] Colorant for ceramic glazes; made of a finely ground calcined oxide such as of cobalt, copper, manganese, or iron. { 'glāz ˌstān }

GLC [ANALYTICAL CHEMISTRY] *See* gas-liquid chromatography.

glucinium [CHEMISTRY] The former name for the element beryllium, coined because the salts of beryllium are sweet-tasting. { glü'sin·ē·əm }

glucochloral [ORGANIC CHEMISTRY] *See* chloralose. { ¦glü·kō¦klȯr·əl }

glucochloralose [ORGANIC CHEMISTRY] *See* chloralose. { ¦glü·kō¦klȯr·əˌlōs }

α-D-glucochloralose [ORGANIC CHEMISTRY] *See* chloralose. { ¦al·fə ¦dē ¦glü·kō¦klȯr·əˌlōs }

gluconate [ORGANIC CHEMISTRY] A salt of gluconic acid. { glü·kəˌnāt }

gluconic acid [ORGANIC CHEMISTRY] $C_6H_{12}O_7$ A crystalline acid obtained from glucose by oxidation; used in cleaning metals. { glü'kän·ik 'as·əd }

gluconic acid sodium salt [ORGANIC CHEMISTRY] *See* sodium gluconate. { glü'kän·ik 'as·əd 'sōd·ē·əm 'sȯlt }

gluside [ORGANIC CHEMISTRY] *See* saccharin. { 'glüˌsīd }

glutaraldehyde [ORGANIC CHEMISTRY] $OHC(CH_2)_3CHO$ A liquid with a boiling point of 188°C; soluble in water and alcohol; used as a biological solution (50) and for leather tanning. { ˌglüd·ə'ral·dəˌhīd }

glycerin [ORGANIC CHEMISTRY] *See* glycerol. { 'glis·ə·rən }

glycerol [ORGANIC CHEMISTRY] $CH_2OHCHOHCH_2OH$ The simplest trihedric alcohol; when pure, it is a colorless, odorless, viscous liquid with a sweet taste; it is completely soluble in water and alcohol but only partially soluble in common solvents such as ether and ethyl acetate; used in manufacture of alkyd resins, explosives, antifreezes, medicines, inks, perfumes, cosmetics, soaps, and finishes. Also known as glycerin; glycyl alcohol. { 'glis·ə‚röl }

glyceryl [ORGANIC CHEMISTRY] $OCH_2OCHOCH_2\equiv$ The functional group from glycerol, $(CH_2OH)_2CHOH$. { 'glis·ə·rəl }

glyceryl diacetate [ORGANIC CHEMISTRY] *See* diacetin. { 'glis·ə·rəl dī'as·ə‚tāt }

glyceryl tristearate [ORGANIC CHEMISTRY] *See* stearin. { 'glis·ə·rəl trī'stir‚āt }

glycidic acid [ORGANIC CHEMISTRY] $C_2H_3O \cdot CO_2H$ A volatile liquid. Also known as epoxy-propionic acid. { glə'sid·ik 'as·əd }

glycidol [ORGANIC CHEMISTRY] $C_3H_6O_2$ A colorless, liquid epoxide that boils at 162°C and is miscible with water; used in organic synthesis. Also known as epihydrin alcohol. { 'glis·ə‚dȯl }

glycin [ORGANIC CHEMISTRY] $C_8H_9NO_3$ A crystalline compound that forms shiny leaflets from water solution, melts at 245-247°C, and is soluble in alkalies and mineral acids; used as a photographic developer and in the analytical determination of iron, phosphorus, and silicon. Also known as photoglycine. { 'glī·sən }

glyco- [ORGANIC CHEMISTRY] Chemical prefix indicating sweetness, or relating to sugar or glycine. { 'glī·kō }

glycol [ORGANIC CHEMISTRY] **1.** $C_nH_{2n}(OH)_2$ An organic chemical with two hydroxyl groups on an essentially aliphatic carbon chain. Also known as dihydroxy alcohol. **2.** $HOCH_2CH_2OH$ A colorless dihydroxy alcohol used as an antifreeze, in hydraulic fluids, and in the manufacture of dynamites and resins. Also known as ethlene glycol. { 'glī‚kȯl }

glycol diacetate [ORGANIC CHEMISTRY] *See* ethylene glycol diacetate. { 'glī‚kȯl dī'as·ə‚tāt }

glycol dinitrate [ORGANIC CHEMISTRY] *See* ethylene nitrate. { 'glī‚kȯl dī'nī‚trāt }

glycol ester [ORGANIC CHEMISTRY] Chemical compound composed of the reaction products of a glycol, $C_nH_{2n}(OH)_2$, and an organic acid; an example is ethylene glycol diacetate, the product of ethylene glycol and acetic acid. { 'glī‚kȯl 'es·tər }

glycol ether [ORGANIC CHEMISTRY] A colorless liquid used as a solvent, in detergents, and as a diluent; a typical example is ethylene glycol diethyl ether, $C_2H_5OCH_2CH_2OC_2H_5$. { 'glī‚kȯl 'ē·thər }

glycolic acid [ORGANIC CHEMISTRY] $CH_2OHCOOH$ Colorless, deliquescent leaflets, decomposing about 78°C; soluble in water, alcohol, and ether; used as a chemical intermediate in fabric dyeing. Also known as hydroxyacetic acid. { glī'käl·ik 'as·əd }

glycolythiourea [ORGANIC CHEMISTRY] *See* 2-thiohydantoin. { ¦gli·kȯl¦thī·ō·yu̇'rē·ə }

glycolyurea [ORGANIC CHEMISTRY] *See* hydantoin. { ¦glī‚kōl·yu̇'rē·ə }

glycyl [ORGANIC CHEMISTRY] $NH_2CH_2COO—$ or $NHCH_2COO\equiv$ The radical from glycine, NH_2CH_2COOH; found in peptides. { 'glī·səl }

glycyl alcohol [ORGANIC CHEMISTRY] *See* glycerol. { 'glī·səl 'al·kə‚hȯl }

glyoxal [ORGANIC CHEMISTRY] $(CHO)_2$ Colorless, deliquescent powder or liquid with mild odor, melting point 15°C, boiling point 51°C; used to insolubilize starches, cellulosic materials, and proteins, in embalming fluids, for leather tanning, and for rayon shrinkproofing. { glī'äk‚sal }

glyoxalic acid [ORGANIC CHEMISTRY] CHOCOOH Colorless crystals that are soluble in water, forming glyoxylic acid. { ¦glī‚äk¦sal·ik 'as·əd }

glyphosate [ORGANIC CHEMISTRY] $C_3H_8NO_5P$ A white solid with a melting point of 200°C; slight solubility in water; used as an herbicide in postharvest treatment of crops. { 'glif·ə‚sāt }

glyphosine [ORGANIC CHEMISTRY] $C_4H_{11}NO_8P_2$ A white solid with a melting point of 203°C; quite soluble in water; used as a growth regulator in sugarcane. { 'glif·ə‚sēn }

glyptal resin [ORGANIC CHEMISTRY] A phthalic anhytxride glycerol made from an emulsion of an alkyd resin; used in lacquers and insulation. { 'glipt·əl 'rez·ən }

gold [CHEMISTRY] A chemical element, symbol Au, atomic number 79, atomic weight 196.967; soluble in aqua regia; melts at 1065°C. { gōld }

H

H [CHEMISTRY] *See* hydrogen.

^{2}H [CHEMISTRY] *See* deuterium.

H^2 [CHEMISTRY] *See* deuterium.

Ha [CHEMISTRY] *See* hahnium.

H acid [ORGANIC CHEMISTRY] $H_2NC_{10}H_4(OH)(SO_3H)_2$ A gray powder or crystalline substance that is soluble in water, ether, and alcohol; used as a dye intermediate. { ¦āch 'as·əd }

hafnium [CHEMISTRY] A metallic element, symbol Hf, atomic number 72, atomic weight 178.49; melting point 2000°C, boiling point above 5400°C. { 'haf·nē·əm }

hafnium carbide [INORGANIC CHEMISTRY] HfC Gray powder, melting at 3887°C; used in the control rods of nuclear reactors. { 'haf·nē·əm 'kär,bīd }

Haggenmacher equation [CHEMISTRY] Equation to calculate latent heats of vaporizations of pure compounds by using critical conditions with Antoine constants. { ¦häg·ən¦mäk·ər i,kwā·zhən }

hahnium [CHEMISTRY] The name suggested by workers in the United States for element 105. Symbolized Ha. { 'hän·ē·əm }

halazone [ORGANIC CHEMISTRY] $COOHC_6H_4SO_2NCl_2$ White crystals, with strong chlorine aroma; slightly soluble in water and chloroform; used as water disinfectant. { 'hal·ə,zōn }

half-cell [PHYSICAL CHEMISTRY] A single electrode immersed in an electrolyte. { 'haf ¦sel }

half-cell potential [PHYSICAL CHEMISTRY] In electrochemical cells, the electrical potential developed by the overall cell reaction; can be considered, for calculation purposes, as the sum of the potential developed at the anode and the potential developed at the cathode, each being a half-cell. { 'haf ¦sel pə'ten·chəl }

half-life [CHEMISTRY] The time required for one-half of a given material to undergo chemical reactions. { 'haf ,līf }

halide [CHEMISTRY] A compound of the type MX, where X is fluorine, chlorine, iodine, bromine, or astatine, and M is another element or organic radical. { 'ha,līd }

haloalkane [ORGANIC CHEMISTRY] Halogenated aliphatic hydrocarbon. { ¦ha·lō¦al,kān }

halocarbon [ORGANIC CHEMISTRY] A compound of carbon and a halogen, sometimes with hydrogen. { ¦ha·lō¦kär·bən }

halocarbon plastic [ORGANIC CHEMISTRY] Plastic made from halocarbon resins. { ¦ha·lō¦kär·bən 'plas·tik }

halocarbon resin [ORGANIC CHEMISTRY] Resin produced by the polymerization of monomers made of halogenated hydrocarbons, such as tetrafluoroethylene, C_2F_4, and trifluorochloroethylene, C_2F_3Cl. { ¦ha·lō¦kär·bən 'rez·ən }

haloform [ORGANIC CHEMISTRY] CHX_3 A compound made by reaction of acetaldehyde or methyl ketones with NaOX, where X is a halogen; an example is iodoform, HCI_3, or bromoform, $HCBr_3$ or chloroform, $HCCl_3$. { 'hal·ə,fórm }

haloform reaction [ORGANIC CHEMISTRY] Halogenation of acetaldehyde or a methyl ketone in aqueous basic solution; the reaction is characteristic of compounds containing a CH_3CO group linked to a hydrogen or to another carbon. { 'hal·ə,fórm rē,ak·shən }

halogen [CHEMISTRY] Any of the elements of the halogen family, consisting of fluorine, chlorine, bromine, iodine, and astatine. { 'hal·ə·jən }

halogen acid [INORGANIC CHEMISTRY] A compound composed of hydrogen bonded to a halogen element, for example, hydrochloric acid. { 'hal·ə·jən ˌas·əd }

halogenated hydrocarbon [ORGANIC CHEMISTRY] One of a group of halogen derivatives of organic hydrogen- and carbon-containing compounds; the group includes monohalogen compounds(alkyl or aryl halides) and polyhalogen compounds that contain the same or different halogen atoms. { 'hal·ə·jəˌnād·əd ˌhī·drə'kär·bən }

halogenation [ORGANIC CHEMISTRY] A chemical process or reaction in which a halogen element is introduced into a substance, generally by the use of the element itself. { ˌhal·ə·jə'nā·shən }

halohydrin [ORGANIC CHEMISTRY] A compound with the general formula X—R—OH where X is a halide such as Cl^-; an example is chlorohydrin. { ˌhal·ə'hī·drən }

halon [ORGANIC CHEMISTRY] A fluorocarbon that has one or more bromine atoms in its molecule. { 'haˌlän }

Hammett acidity function [CHEMISTRY] An expression for the acidity of a medium, defined as $h_0 = K_{BH^+}[BH^+]/[B]$, where K_{BH^+} is the dissociation constant of the acid form of the indicator, and $[BH^+]$ and $[B]$ are the concentrations of the protonated base and the unprotonated base respectively. { 'ham·ət ə'sid·əd·ē ˌfəŋk·shən }

hand sugar refractometer [ANALYTICAL CHEMISTRY] Portable device to read refractive indices of sugar solutions. Also known as proteinometer. { 'hand 'shu̇g·ər ˌrēˌfrak'täm·əd·ər }

Hansa yellow [ORGANIC CHEMISTRY] Group of organic azo pigments with strong tinting power, but poor opacity in paints; used where nontoxicity is important. { 'hän·sə 'yel·ō }

Hantzsch synthesis [ORGANIC CHEMISTRY] The reaction whereby a pyrrole compound is formed when a β-ketoester, chloroacetone, and a primary amine condense. { 'hänsh ˌsin·thə·səs }

Hanus solution [ANALYTICAL CHEMISTRY] Iodine monobromide in glacial acetic acid; used to determine iodine values in oils containing unsaturated organic compounds. { 'han·əs səˌlü·shən }

hard acid [CHEMISTRY] A Lewis acid of low polarizability, small size, and high positive oxidation state; it does not have easily excitable outer electrons; some examples are H^+, Li^+, and Al^+. { 'härd 'as·əd }

hard base [CHEMISTRY] A Lewis base (electron donor) that has high polarizability and low electronegativity, is easily oxidized, or possesses lowlying empty orbitals; some examples are H_2O, HO^-, OCH_3^-, and F^-. { 'härd ¦bās }

hard detergent [CHEMISTRY] A nonbiodegradable detergent. { 'härd di'tər·jənt }

hardener [ORGANIC CHEMISTRY] Compound reacted with a resin polymer to harden it, such as the amines or anhydrides that react with epoxides to cure or harden them into plastic materials. Also known as curing agent. { 'härd·ən·ər }

hardness [CHEMISTRY] The amount of calcium carbonate dissolved in water, usually expressed as parts of calcium carbonate per million parts of water. { 'härd·nəs }

hardness test [ANALYTICAL CHEMISTRY] A test to determine the calcium and magnesium content of water. { 'härd·nəs ˌtest }

hard-sphere collision theory [PHYSICAL CHEMISTRY] A theory for calculating reaction rate constants for biomolecular gas-phase reactions in which the molecules are considered to be colliding, hard spheres. { 'härd ˌsfir kə'lizh·ən ˌthē·ə·rē }

hard water [CHEMISTRY] Water that contains certain salts, such as those of calcium or magnesium, which form insoluble deposits in boilers and form precipitates with soap. { 'härd ¦wo̍d·ər }

Hardy-Schulz rule [PHYSICAL CHEMISTRY] An increase in the charge of ions results in a large increase in their flocculating power. { 'härd·ē 'shu̇lts ˌrül }

Haring cell [PHYSICAL CHEMISTRY] An electrolytic cell with four electrodes used to measure electrolyte resistance and polarization of electrodes. { 'her·iŋ ˌsel }

harman [ORGANIC CHEMISTRY] $C_{12}H_{10}N_2$ Crystals that melt at 237-238°C; inhibits growth of molds and certain bacteria. Also known as arabine; loturine; passiflorin. { 'här·mən }

harmonic vibration-rotation band [SPECTROSCOPY] A vibration-rotation band of a molecule in which the harmonic oscillator approximation holds for the vibrational levels, so that the vibrational levels are equally spaced. { här'män·ik vī'brā·shən rō'tā·shən ˌband }

Hartmann diaphragm [ANALYTICAL CHEMISTRY] Comparison device for positive-element-identification readings from emission spectra. { 'härt·män ˌdī·əˌfram }

Hartmann test [SPECTROSCOPY] A test for spectrometers in which light is passed through different parts of the entrance slit; any resulting changes of the spectrum indicate a fault in the instrument. { 'härt·män ˌtest }

Hartman's solution [ANALYTICAL CHEMISTRY] Solution of thymol, ethyl alcohol, and sulfuric ether; used for selective dentin analysis. { 'härt·mənz səˌlü·shən }

HCB [ORGANIC CHEMISTRY] *See* hexachlorobenzene.

HDPE [ORGANIC CHEMISTRY] *See* high density polyethylene. { ¦āch¦dē¦pē'dē }

He [CHEMISTRY] *See* helium.

heat of activation [PHYSICAL CHEMISTRY] The increase in enthalpy when a substance is transformed from a less active to a more reactive form at constant pressure. { 'hēt əv ˌak·tə'vā·shən }

heat of association [PHYSICAL CHEMISTRY] Increase in enthalpy accompanying the formation of 1 mole of a coordination compound from its constituent molecules or other particles at constant pressure. { 'hēt əv əˌsō·sē'ā·shən }

heat of atomization [PHYSICAL CHEMISTRY] The change in enthalpy accompanying the conversion of 1 mole of an element or a compound at 298 K (77°F) and 1 atmosphere (10^5 pascals) into free atoms. { ¦hēt əv ˌad·ə·mə'zā·shən }

heat of combustion [PHYSICAL CHEMISTRY] The amount of heat released in the oxidation of 1 mole of a substance at constant pressure, or constant volume. Also known as heat value; heating value. { 'hēt əv kəm'bəs·chən }

heat of decomposition [PHYSICAL CHEMISTRY] The change in enthalpy accompanying the decomposition of 1 mole of a compound into its elements at constant pressure. { 'hēt əv dēˌkäm·pə'zish·ən }

heat of dilution [PHYSICAL CHEMISTRY] **1.** The increase in enthalpy accompanying the addition of a specified amount of solvent to a solution of constant pressure. Also known as integral heat of dilution; total heat of dilution. **2.** The increase in enthalpy when an infinitesimal amount of solvent is added to a solution at constant pressure. Also known as differential heat of dilution. { 'hēt əv də'lü·shən }

heat of dissociation [PHYSICAL CHEMISTRY] The increase in enthalpy at constant pressure, when molecules break apart or valence linkages rupture. { 'hēt əv diˌsō·sē'ā·shən }

heat of formation [PHYSICAL CHEMISTRY] The increase in enthalpy resulting from the formation of 1 mole of a substance from its elements at constant pressure. { 'hēt əv fȯr'mā·shən }

heat of hydration [PHYSICAL CHEMISTRY] The increase in enthalpy accompanying the formation of 1 mole of a hydrate from the anhydrous form of the compound and from water at constant pressure. { 'hēt əv hī'drā·shən }

heating value [PHYSICAL CHEMISTRY] *See* heat of combustion. { 'hēd·iŋ ˌval·yü }

heat of ionization [PHYSICAL CHEMISTRY] The increase in enthalpy when 1 mole of a substance is completely ionized at constant pressure. { 'hēt əv ˌī·ən·ə'zā·shən }

heat of linkage [PHYSICAL CHEMISTRY] The bond energy of a particular type of valence linkage between atoms in a molecule, as determined by the energy required to dissociate all bonds of the type in 1 mole of the compound divided by the number of such bonds in a compound. { 'hēt əv 'liŋk·ij }

heat of reaction [PHYSICAL CHEMISTRY] **1.** The negative of the change in enthalpy accompanying a chemical reaction at constant pressure. **2.** The negative of the change in internal energy accompanying a chemical reaction at constant volume. { 'hēt əv rē'ak·shən }

heat of solution [PHYSICAL CHEMISTRY] The enthalpy of a solution minus the sum of the enthalpies of its components. Also known as integral heat of solution; total heat of solution. { 'hēt əv sə'lü·shən }

heat value [PHYSICAL CHEMISTRY] *See* heat of combustion. { 'hēt ˌval·yü }

heavy acid [INORGANIC CHEMISTRY] *See* phosphotungstic acid. { ¦hev·ē 'as·əd }

heavy water [INORGANIC CHEMISTRY] A compound of hydrogen and oxygen containing a higher proportion of the hydrogen isotope deuterium than does naturally occurring water. Also known as deuterium oxide. { 'hev·ē 'wȯd·ər }

Hefner lamp [CHEMISTRY] A flame lamp that burns amyl acetate. { 'hef·nər ˌlamp }

Hehner number [ANALYTICAL CHEMISTRY] Weight percent of water-insoluble fatty acids in fats and oils. { 'hān·ər ˌnəm·bər }

Heitler-London covalence theory [PHYSICAL CHEMISTRY] A calculation of the binding energy and the distance between the atoms of a diatomic hydrogen molecule, which assumes that the two electrons are in atomic orbitals about each of the nuclei, and then combines these orbitals into a symmetric or antisymmetric function. { 'hīt·lər 'lən·dən kō'vā·ləns ˌthē·ə·rē }

helicate [ORGANIC CHEMISTRY] Any member of a group of synthetic, helical arrays of molecules formed by the chemical recognition and organization of metals and organic bases. { 'hel·iˌkāt }

helicin [ORGANIC CHEMISTRY] *See* salicylaldehyde. { 'hel·ə·sən }

heliotropin [ORGANIC CHEMISTRY] *See* piperonal. { ¦hē·lē·ə¦trō·pən }

helium [CHEMISTRY] A gaseous chemical element, symbol He, atomic number 2, and atomic weight 4.0026; one of the noble gases in group 0 of the periodic table. { 'hē·lē·əm }

helium spectrometer [SPECTROSCOPY] A small mass spectrometer used to detect the presence of helium in a vacuum system; for leak detection, a jet of helium is applied to suspected leaks in the outer surface of the system. { 'hē·lē·əm spek'träm·əd·ər }

Hell-Volhard-Zelinsky reaction [ORGANIC CHEMISTRY] Preparation of an ester or α-halo substituted acid (chloro or bromo) by reacting the halogen on the acid in the presence of phosphorus or phosphorus halide, and then followed by hydrolysis or alcoholysis of the haloacyl halide resulting. { ¦hel ¦fōlˌhärt zə'lins·kē rēˌak·shən }

Helmholtz equation [PHYSICAL CHEMISTRY] The relationship stating that the emf (electromotive force) of a reversible electrolytic cell equals the work equivalent of the chemical reaction when charge passes through the cell plus the product of the temperature and the derivative of the emf with respect to temperature. { 'helmˌhōlts iˌkwā·zhən }

hematin [ORGANIC CHEMISTRY] $C_{34}H_{33}O_5N_4Fe$ The hydroxide of ferriheme derived from oxidized heme. { 'hē·məd·ən }

hematoxylin [ORGANIC CHEMISTRY] $C_{16}H_{14}O_6$ A colorless, crystalline compound occurring in hematoxylon; upon oxidation, it is converted to hematein which forms deeply colored lakes with various metals; used as a stain in microscopy. { ˌhē·mə'täk·sə·lən }

hemiacetal [ORGANIC CHEMISTRY] A class of compounds that have the grouping $>C(OH)—(OR)$ and that result from the reaction of an aldehyde and alcohol. { ¦he·mē'as·əˌtal }

hemiketal [ORGANIC CHEMISTRY] A carbonyl compound that results from the addition of an alcohol to the carbonyl group of a ketone, with the general formula (R)(R')C(OH)(OR). { ¦he·mē'ked·əl }

hemimellitic acid [ORGANIC CHEMISTRY] $C_6H_3(COOH)_3$ A compound crystallizing in colorless needles; melting point 196°C; slightly soluble in water. { ¦he·mē·mə¦lid·ik 'as·əd }

hendecanal [ORGANIC CHEMISTRY] *See* undecanal. { hen'dek·ə·nəl }

hendecane [ORGANIC CHEMISTRY] *See* undecane. { 'hen·dəˌkān *or* hen'deˌkān }

hendecyl [ORGANIC CHEMISTRY] *See* undecyl. { hen'des·əl }

Henderson equation for pH [PHYSICAL CHEMISTRY] An equation for the pH of an acid during its neutralization: $pH = pK_a + \log [salt]/[acid]$, where pK_a is the logarithm to base 10 of the reciprocal of the dissociation constant of the acid; the equation is found to be useful for the pH range 4-10, providing the solutions are not too dilute. { 'hen·dər·sən i¦kwā·zhən fər ¦pē'āch }

heneicosane [ORGANIC CHEMISTRY] $C_{21}H_{44}$ Saturated hydrocarbon of the methane series; the crystals melt at 40°C and boil at 215°C (at 15 mmHg). { hen'ī·kə,sān }

Henry's law [PHYSICAL CHEMISTRY] The law that at sufficiently high dilution in a liquid solution, the fugacity of a nondissociating solute becomes proportional to its concentration. { 'hen·rēz ,lȯ }

hentriacontane [ORGANIC CHEMISTRY] $C_{31}H_{64}$ A hydrocarbon; a crystalline material melting at 68°C and boiling at 302°C (at 15 mmHg); derived from roots of *Oenanthe crocata* and found in beeswax. { ,hen,trī·ə'kän,tān }

hepar calcies [INORGANIC CHEMISTRY] *See* calcium sulfide. { 'hē,pär 'kal,sēz }

hepar sulfuris [INORGANIC CHEMISTRY] *See* potassium sulfide. { 'hē,pär səl'fyu̇r əs }

heptachlor [ORGANIC CHEMISTRY] $C_{10}H_7Cl_7$ An insecticide; a white to tan, waxy solid; insoluble in water, soluble in alcohol and xylene; melts at 95-96°C. { 'hep·tə,klȯr }

heptacosane [ORGANIC CHEMISTRY] $C_{27}H_{56}$ A hydrocarbon; water-insoluble crystals melting at 60°C and boiling at 270°C (at 15 mmHg); soluble in alcohol; found in beeswax. { hep'täk·ə,sān *or* ,hep·tə'kō,sān }

heptadecane [ORGANIC CHEMISTRY] $C_{17}H_{36}$ A hydrocarbon; water-insoluble, alcohol-soluble solid melting at 23°C and boiling at 303°C; used as a chemical intermediate. { ,hep·tə'de,kān }

***n*-heptadecanoic acid** [ORGANIC CHEMISTRY] $CH_3(CH_2)_{15}COOH$ A fatty acid that is saturated; soluble in ether and alcohol, insoluble in water; colorless crystals melt at 61°C. Also known as margaric acid. { ¦en ¦hep·tə¦dek·ə¦nō·ik 'as·əd }

heptadecanol [ORGANIC CHEMISTRY] $C_{17}H_{35}OH$ An alcohol; colorless liquid boiling at 309°C; slightly soluble in water; used as a chemical intermediate, as a perfume fixative, in cosmetics and soaps, and to manufacture surfactants. { ¦hep·tə'dek·ə,nȯl }

heptaldehyde [ORGANIC CHEMISTRY] $C_6H_{13}CHO$ An aldehyde; ether-soluble, colorless oil with fruity aroma; slightly soluble in water; boils at 153°C; used as a chemical intermediate and for perfumes and pharmaceuticals. Also known as heptanal. { ,hep·'tal·də,hīd }

heptadione-2,3 [ORGANIC CHEMISTRY] *See* acetyl valeryl. { ,hep·tə'dī,ōn ¦tü ¦thrē }

heptanal [ORGANIC CHEMISTRY] *See* heptaldehyde. { 'hep·tə,nal }

heptane [ORGANIC CHEMISTRY] $CH_3(CH_2)_5CH_3$ A hydrocarbon; water-insoluble, flammable, colorless liquid boiling at 98°C; soluble in alcohol, chloroform, and ether; used as an anesthetic, solvent, and chemical intermediate, and in standard octane-rating tests. { 'hep,tān }

heptanoic acid [ORGANIC CHEMISTRY] $CH_3(CH_2)_5COOH$ Clear oil boiling at 223°C, soluble in alcohol and ether, insoluble in water; used as a chemical intermediate. { ¦hep·tə¦nō·ik 'as·əd }

1-heptanol [ORGANIC CHEMISTRY] $C_7H_{15}OH$ An alcohol; a fragrant, colorless liquid boiling at 174°C; soluble in water, ether, or alcohol; used as a chemical intermediate, as a solvent, and in cosmetics. Also known as heptyl alcohol. { ¦wən 'hep·tə,nȯl }

3-heptanol [ORGANIC CHEMISTRY] $CH_3CH_2CH(OH)C_4H_9$ An alcohol; a liquid boiling at 156°C; used as a coating, solvent, and diluent, as a chemical intermediate, and as a flotation frother. { ¦thrē 'hep·tə,nȯl }

2-heptanone [ORGANIC CHEMISTRY] *See* methyl *n*-amyl ketone. { ¦tü 'hep·tə,nōn }

3-heptanone [ORGANIC CHEMISTRY] *See* ethyl butyl ketone. { ¦thrē 'hep·tə,nōn }

4-heptanone [ORGANIC CHEMISTRY] $(CH_3CH_2CH_2)_2CO$ A colorless liquid that is stable and has a pleasant odor; boils at approximately 98°C; used to put nitrocellulose and raw and blown oils into solution, and used in lacquers and as a flavoring in foods. { ¦fȯr 'hep·tə,nōn }

heptene [ORGANIC CHEMISTRY] $C_{17}H_{14}$ A liquid that is a mixture of isomers; boils at 189.5°C; used as an additive in lubricants, as a catalyst, and as a surface active agent. Also known as heptylene. { 'hep,tēn }

heptoxide [CHEMISTRY] An oxide whose molecule contains seven atoms of oxygen. { hep'täk,sīd }

heptyl [ORGANIC CHEMISTRY] $CH_3(CH_2)_6$— The functional group from heptane, $CH_3(CH_2)_5CH_3$. { 'hep·təl }

heptyl alcohol [ORGANIC CHEMISTRY] *See* 1-heptanol. { 'hep·təl 'al·kə,hȯl }

heptylene [ORGANIC CHEMISTRY] *See* heptene. { 'hep·tə,lēn }

Hercules trap [ANALYTICAL CHEMISTRY] Water-measuring liquid trap used in aquametry when the material collected is heavier than water. { 'hər·kyə,lēz ,trap }

Hess's law [PHYSICAL CHEMISTRY] The law that the evolved or absorbed heat in a chemical reaction is the same whether the reaction takes one step or several steps. Also known as the law of constant heat summation. { 'hes·əz ,lȯ }

hetero- [CHEMISTRY] Prefix meaning different; for example, a heterocyclic compound is one in which the ring is made of more than one kind of atom. { 'hed·ə·rō }

heteroatom [ORGANIC CHEMISTRY] In an organic compound, any atom other than carbon or hydrogen. { 'hed·ə·rō,ad·əm }

heteroazeotrope [CHEMISTRY] Liquid mixture that is not completely miscible in all proportions in the liquid phase, yet does not form an azeotrope. Also known as heterogeneous zeotrope. { ¦hed·ə·rō·ā'zē·ə,trōp }

heterocyclic compound [ORGANIC CHEMISTRY] Compound in which the ring structure is a combination of more than one kind of atom; for example, pyridine, C_5H_5N. { ,hed·ə·rō'sī·klik ¦käm,pau̇nd }

heterogeneous [CHEMISTRY] Pertaining to a mixture of phases such as liquid-vapor, or liquid-vapor-solid. { ,hed·ə'räj·ə·nəs }

heterogeneous catalysis [CHEMISTRY] Catalysis occurring at a phase boundary, usually a solid-fluid interface. { ,hed·ə·rə'jē·nē·əs kə'tal·ə·səs }

heterogeneous chemical reaction [CHEMISTRY] Chemical reaction system in which the reactants are of different phases; for example, gas with liquid, liquid with solid, or a solid catalyst with liquid or gaseous reactants. { ,hed·ə·rə'jē·nē·əs ¦kem·ə·kəl rē'ak·shən }

heterolysis [ORGANIC CHEMISTRY] *See* heterolytic cleavage. { ,hed·ə'räl·ə·səs }

heterolytic bond dissociation energy [PHYSICAL CHEMISTRY] The change in enthalpy that occurs when a chemical bond undergoes heterolytic cleavage. { ,hed·ə·rō¦lid·ik ,bänd ,dis·ə,sō·sē'ā·shən ,en·ər·jē }

heterolytic cleavage [ORGANIC CHEMISTRY] The breaking of a single (two-electron) chemical bond in which both electrons remain on one of the atoms. Also known as heterolysis. { ,hed·ə·rō¦lid·ik 'klēv·ij }

heteronuclear molecule [CHEMISTRY] A diatomic molecule having atoms of different elements. { ¦hed·ə·rə,nü·klē·ər 'mäl·ə,kyül }

heteropolar bond [PHYSICAL CHEMISTRY] A covalent bond whose total dipole moment is not 0. { ,hed·ə·rə¦pō·lər 'bänd }

heteropoly acid [INORGANIC CHEMISTRY] Complex acids of metals, whose specific gravity is greater than 4, with phosphoric acid; an example is phosphomolybdic acid. { ,hed·ə'räp·ə·lē 'as·əd }

heteropoly compound [INORGANIC CHEMISTRY] Polymeric compounds of molybdates with anhydrides of other elements such as phosphorus; the yellow precipitate $(NH_4)_3P(Mo_3O_{10})_4$ is such a compound. { ,hed·ə'räp·ə·lē 'käm,pau̇nd }

heterotopic faces [ORGANIC CHEMISTRY] On molecules, faces of double bonds where addition gives rise to isomeric structures. { ¦hed·ə·rō¦täp·ik 'fās·əz }

heterotopic ligands [CHEMISTRY] Constitutionally identical ligands whose separate replacement by a different ligand gives rise to isomeric structures. { ¦hed·ə·rō¦täp·ik 'lī g·ənz }

hexachlorobenzene [ORGANIC CHEMISTRY] C_6Cl_6 Colorless, needlelike crystals with a melting point of 231°C; used in organic synthesis and as a fungicide. Abbreviated HCB. { ¦hek·sə¦klȯr·ō'ben,zēn }

hexachlorobutadiene [ORGANIC CHEMISTRY] $Cl_2C{:}CClCCl{:}CCl_2$ A colorless liquid with mild aroma, boiling at 210-220°C; soluble in alcohol and ether, insoluble in water; used as solvent, heat-transfer liquid, and hydraulic fluid. { ¦hek·sə¦klȯr·ō,byüd·ə'dī,ēn }

1,2,3,4,5,6-hexachlorocyclohexane [ORGANIC CHEMISTRY] $C_6H_6Cl_6$ A white or yellow powder or flakes with a musty odor; a systemic insecticide toxic to flies, cockroaches, aphids, and boll weevils. Abbreviated TBH. { ¦wən ¦tü ¦thrē ¦fȯr ¦fīv ¦siks ¦hek·sə¦klȯr·ō,sī·klō'hek,sān }

hexachloroethane [ORGANIC CHEMISTRY] Cl_3CCCl_3 Colorless crystals with a camphorlike odor, melting point 185°C, toxic; used in organic synthesis, as a retarding agent in fermentation, and as a rubber accelerator. { ¦hek·sə¦klȯr·ō'e,thān }

hexachlorophene [ORGANIC CHEMISTRY] $(C_6HCl_3OH)_2CH_2$ A white powder melting at 161°C; soluble in alcohol, ether, acetone, and chloroform, insoluble in water; bacteriostat used in antiseptic soaps, cosmetics, and dermatologicals. { ,hek·sə'klȯr·ə,fēn }

hexachloropropylene [ORGANIC CHEMISTRY] $CCl_3CCl{\cdot}CCl_2$ Water-white liquid boiling at 210°C, soluble in alcohol, ether, and chlorinated solvents, insoluble in water; used as a solvent, plasticizer, and hydraulic fluid. { ¦hek·sə¦klȯr·ō'prō·pə,lēn }

hexacontane [ORGANIC CHEMISTRY] $C_{60}H_{122}$ Solid, saturated hydrocarbon of the methane series; melts at 101°C. { ,hek·sə'kän,tān }

hexacosane [ORGANIC CHEMISTRY] $C_{26}H_{54}$ Saturated hydrocarbon of the methane series; colorless crystals melting at 57°C. { ,hek·sə'kō,sān }

hexacosanoic acid [ORGANIC CHEMISTRY] *See* cerotic acid. { ¦hek·sə,kō·sə'nō·ik 'as·əd }

***n*-hexadecane** [ORGANIC CHEMISTRY] $C_{16}H_{34}$ A colorless, solid hydrocarbon, melting point 20°C; a standard reference fuel in determining the ignition quality (cetane number) of diesel fuels. Also known as cetane. { ¦en¦hek·sə'de,kān }

1-hexadecene [ORGANIC CHEMISTRY] $CH_3(CH_2)_{13}CH{:}CH_2$ A colorless liquid made by treating cetyl alcohol with phosphorus pentoxide; boils at 274°C; soluble in organic solvents such as alcohol, ether, and petroleum; used as an intermediate in organic synthesis. { ¦wən ,hek·sə'de,sēn }

hexadentate ligand [INORGANIC CHEMISTRY] A chelating agent having six groups capable of attachment to a metal ion. Also known as sexadentate ligand. { ,hek·sə'den,tāt 'līg·ənd }

hexadiene [ORGANIC CHEMISTRY] C_6H_{10} A group of unsaturated hydrocarbons with two double bonds; some members of the group are 1,4-hexadiene, 1,5-hexadiene, and 2,4-hexadiene. { ,hek·sə'dī,ēn }

hexahydric alcohol [ORGANIC CHEMISTRY] A member of the mannitol-sorbitol-dulcitol sugar group; isomer of $C_6H_8(OH)_6$. { ¦hek·sə¦hī·drik 'al·kə,hȯl }

hexahydrotoluene [ORGANIC CHEMISTRY] *See* methyl cyclohexane. { ¦hek·sə¦hī·drō'täl·yə,wēn }

***n*-hexaldehyde** [ORGANIC CHEMISTRY] $CH_3(CH_2)_4CHO$ Colorless liquid with sharp aroma, boiling at 128.6°C; used as an intermediate for plasticizers, dyes, insecticides, resins, and rubber chemicals. { ¦en ,heks'al·də,hīd }

hexametapol [ORGANIC CHEMISTRY] $C_6H_{18}N_3OP$ A liquid used as a solvent in organic synthesis, as a deicing additive for jet engine fuel, and as an insect pest chemosterilant and chemical mutagen. { ,hek·sə'med·ə,pȯl }

hexamethylene [ORGANIC CHEMISTRY] *See* cyclohexane. { ,hek·sə'meth·ə,lēn }

hexamethylenediamine [ORGANIC CHEMISTRY] $H_2N(CH_2)_6NH_2$ Colorless solid boiling at 205°C; slightly soluble in water, alcohol, and ether; used to make nylon and other high polymers. { ,hek·sə'meth·ə·lēn'dī·ə,mēn }

hexamethylene tetramine [ORGANIC CHEMISTRY] *See* cystamine. { ,hek·sə'meth·ə,lēn 'te·trə,mēn }

hexamethylphosphoric triamide [ORGANIC CHEMISTRY] *See* hempa. { ,hek·sə,meth·əl'fäs'fȯr·ik trī'am·əd }

hexane [ORGANIC CHEMISTRY] C_6H_{14} Water-insoluble, toxic, flammable, colorless liquid with faint aroma; forms include: *n*-hexane, a straight-chain compound boiling at 68.7°C and used as a solvent, paint diluent, alcohol denaturant, and polymerization-reaction medium; isohexane, a mixture of hexane isomers boiling at 54-61°C and used as a solvent and freezing-point depressant; and neohexane. { 'hek,sān }

1,6-hexanediol [ORGANIC CHEMISTRY] $HO(CH_2)_6OH$ A crystalline substance, soluble in water and alcohol; used in gasoline refining, as an intermediate in nylon manufacturing, and in making polyesters and polyurethanes. { ¦wən ¦siks ,hek,sān'dī,ȯl }

hexanitrodiphenyl amine [ORGANIC CHEMISTRY] $(NO_2)_3C_6H_2NHC_6H_2(NO_2)_3$ Explosive, yellow solid melting at 238-244°C; insoluble in water, ether, alcohol, or benzene; soluble in alkalies and acetic and nitric acids; used as an explosive and in potassium analysis. { ¦hek·sə¦nī·trō·dī'fen·əl 'am,ēn }

hexaphenylethane [ORGANIC CHEMISTRY] $(C_6H_5)_3CC(C_6H_5)_3$ The dimer of triphenylmethyl radical. { ¦hek·sə¦fen·əl'eth‚ān }

1-hexene [ORGANIC CHEMISTRY] $CH_3(CH_2)_3HC{:}CH_2$ Colorless, olefinic hydrocarbon boiling at 64°C; soluble in alcohol, acetone, ether, and hydrocarbons, insoluble in water; used as a chemical intermediate and for resins, drugs, and insecticides. Also known as hexylene. { ¦wən 'hek‚sēn }

hexone [ORGANIC CHEMISTRY] *See* methyl isobutyl ketone. { 'hek‚sōn }

***n*-hexyl acetate** [ORGANIC CHEMISTRY] $CH_3COOC_6H_{13}$ Colorless liquid boiling at 169°C; soluble in alcohol and ether, insoluble in water; used as a solvent for resins and cellulosic esters. { ¦en 'hek·səl 'as·ə‚tāt }

hexyl alcohol [ORGANIC CHEMISTRY] $CH_3(CH_2)_4CH_2OH$ Colorless liquid boiling at 156°C; soluble in alcohol and ether, slightly soluble in water; used as a chemical intermediate for pharmaceuticals, perfume esters, and antiseptics. { 'hek·səl 'al·kə‚hȯl }

hexylamine [ORGANIC CHEMISTRY] $CH_3(CH_2)_5NH_2$ Poisonous, water-white liquid with amine aroma; boils at 129°C; a ptomaine base from the autolysis of protoplasm. { hek'sil·ə‚mēn }.

hexylene [ORGANIC CHEMISTRY] *See* 1-hexene. { 'hek·sə‚lēn }

hexylene glycol [ORGANIC CHEMISTRY] $C_6H_{14}O_2$ Water-miscible, colorless liquid boiling at 198°C; used in hydraulic brake fluids, in printing inks, and in textile processing. { 'hek·sə‚lēn 'glī‚kȯl }

***n*-hexyl ether** [ORGANIC CHEMISTRY] $C_6H_{13}OC_6H_{13}$ Faintly colored liquid with a characteristic odor, only slightly water-soluble; used in solvent extraction and in the manufacture of collodion and various cellulosic products. { ¦en 'hek·səl 'ē·thər }

hexylresorcinol [ORGANIC CHEMISTRY] $C_6H_{13}C_6H_3(OH)_2$ Sharp-tasting, white to yellowish crystals melting at 64°C; slightly soluble in water, soluble in glycerin, vegetable oils, and organic solvents; used in medicine. { ¦hek·səl·ri'sȯrs·ən‚ȯl }

1-hexyne [ORGANIC CHEMISTRY] C_4H_9CCH A colorless, water-white liquid, either *n*-butylacetylene, boiling at 71.5°C, or methylpropylacetylene, boiling at 84°C. { ¦wən 'hek‚sēn }

Hf [CHEMISTRY] *See* hafnium.

hfs [SPECTROSCOPY] *See* hyperfine structure.

Hg [CHEMISTRY] *See* mercury.

high-density polyethylene [ORGANIC CHEMISTRY] A thermoplastic polyolefin with a density of 0.941-0.960 gram per cubic centimeter (0.543-0.555 ounce per cubic inch). Abbreviated HDPE. { ‚hī ¦den·səd·ē ‚päl·ē'eth·ə‚lēn }

high-energy bond [PHYSICAL CHEMISTRY] Any chemical bond yielding a decrease in free energy of at least 5 kilocalories per mole. { 'hī ‚en·ər·jē 'bänd }

high-frequency titration [ANALYTICAL CHEMISTRY] A conductimetric titration in which two electrodes are mounted on the outside of the beaker or vessel containing the solution to be analyzed and an alternating current source in the megahertz range is used to measure the course of a titration. { 'hī ¦frē·kwən·sē tī'trā·shən }

high-performance liquid chromatography [ANALYTICAL CHEMISTRY] A type of column chromatography in which the solvent is conveyed through the column under pressure. Abbreviated HPLC. { ¦hī pər'fȯrm·əns ‚lik·wəd ‚krō·mə'täg·rə·fē }

high polymer [ORGANIC CHEMISTRY] A large molecule (of molecular weight greater than 10,000) usually composed of repeat units of low-molecular-weight species; for example, ethylene or propylene. { 'hī 'päl·ə·mər }

high-pressure chemistry [PHYSICAL CHEMISTRY] The study of chemical reactions and phenomena that occur at pressures exceeding 10,000 bars (a bar is nearly equivalent to a kilogram per square centimeter), mainly concerned with the properties of the solid state. { 'hī ¦presh·ər 'kem·ə·strē }

high-resolution electron energy loss spectroscopy [SPECTROSCOPY] A type of electron energy loss spectroscopy in which electron scattering is performed by using a monoenergetic beam and electron energy analyzers to achieve a resolution of 5 to 10 millielectronvolts. Abbreviated HREELS. { 'hī ‚rez·ə'lü·shən i'lek‚trän 'en·ər·jē ‚lȯs spek'träs·kə·pē }

high-temperature chemistry [PHYSICAL CHEMISTRY] The study of chemical phenomena occurring above about 500 K. { 'hī ,tem·prə·chər 'kem·ə·strē }

Hill reaction [ORGANIC CHEMISTRY] Production of substituted phenylacetic acids by the oxidation of the corresponding alkylbenzene by potassium permanganate in the presence of acetic acid. { 'hil rē,ak·shən }

Hinsberg test [ANALYTICAL CHEMISTRY] A test to distinguish between primary and secondary amines; it involves reaction of an amine with benzene disulforyl chloride in alkaline solution; primary amines give sulfonamides that are soluble in basic solution; secondary amines give insoluble derivatives; tertiary amines do not react with the reagent. { 'hinz·bərg ,test }

hippuric acid [ORGANIC CHEMISTRY] $C_6H_5CONHCH_2 \cdot COOH$ Colorless crystals melting at 188°C; soluble in hot water, alcohol, and ether; used in medicine and as a chemical intermediate. { hi'pyu̇r·ik 'as·əd }

Hittorf method [PHYSICAL CHEMISTRY] A procedure for determining transference numbers in which one measures changes in the composition of the solution near the cathode and near the anode of an electrolytic cell, due to passage of a known amount of electricity. { 'hi·dȯrf ,meth·əd }

Ho [CHEMISTRY] *See* holmium.

Hofmann amine separation [ORGANIC CHEMISTRY] A technique to separate a mixture of primary, secondary, and tertiary amines; they are heated with ethyl oxalate; there is no reaction with tertiary amines, primary amines form a diamide, and the secondary amines form a monoamide; when the reaction mixture is distilled, the mixture is separated into components. { 'häf·mən 'am,ēn ,sep·ə,rā·shən }

Hofmann degradation [ORGANIC CHEMISTRY] The action of bromine and an alkali on an amide so that it is converted into a primary amine with one less carbon atom. { 'häf·mən deg·rə'dā·shən }

Hofmann exhaustive methylation reaction [ORGANIC CHEMISTRY] The thermal decomposition of quaternary ammonium hydroxide compounds to yield an olefin and water; an exception is tetramethylammonium hydroxide, which decomposes to give an alcohol. { 'häf·mən ig'zȯs·tiv ,meth·ə'lā·shən rē,ak·shən }

Hofmann mustard-oil reaction [ORGANIC CHEMISTRY] Preparation of alkylisothiocyanates by heating together a primary amine, mercuric chloride, and carbon disulfide. { 'häf·mən 'məs·tərd ,ȯil rē,ak·shən }

Hofmann reaction [ORGANIC CHEMISTRY] A reaction in which amides are degraded by treatment with bromine and alkali (caustic soda) to amines containing one less carbon, used commercially in the production of nylon. { 'häf·mən rē,ak·shən }

Hofmann rearrangement [ORGANIC CHEMISTRY] A chemical rearrangement of the hydrohalides of N-alkylanilines upon heating to give aminoalkyl benzenes. { 'häf·mən ,rē·ə'rānj·mənt }

Hofmeister series [CHEMISTRY] An arrangement of anions or cations in order of decreasing ability to produce coagulation when their salts are added to lyophilic sols. Also known as lyotropic series. { 'hōf,mīs·tər ,sir·ēz }

hole-burning spectroscopy [SPECTROSCOPY] A method of observing extremely narrow line widths in certain ions and molecules embedded in crystalline solids, in which broadening produced by crystal-site-dependent statistical field variations is overcome by having a monochromatic laser temporarily remove ions or molecules at selected crystal sites from their absorption levels, and observing the resulting dip in the absorption profile with a second laser beam. { 'hōl ,bərn·iŋ spek'träs·kə·pē }

holmium [CHEMISTRY] A rare-earth element belonging to the yttrium subgroup, symbol Ho, atomic number 67, atomic weight 164.93, melting point 1400-1525°C. { 'hōl·mē·əm }

homatropine [ORGANIC CHEMISTRY] $C_{16}H_{21}O_3N$ An alkaloid that causes pupil dilation and paralysis of accommodation. { hōm'a·trə,pēn }

homidium bromide [ORGANIC CHEMISTRY] *See* ethidium bromide. { hə'mid·ē·əm 'brō,mīd }

homo- [ORGANIC CHEMISTRY] **1.** Indicating the homolog of a compound differing in formula from the latter by an increase of one CH_2 group. **2.** Indicating a homopolymer made up of a single type of monomer, such as polyethylene from ethylene. **3.** Indicating that a skeletal atom has been added to a well-known structure. { 'hō·mō }

homocyclic compound [ORGANIC CHEMISTRY] A ring compound that has one type of atom in its structure; an example is benzene. { ¦hä·mə'sī·klik 'käm‚paünd }

homogeneous [CHEMISTRY] Pertaining to a substance having uniform composition or structure. { ‚hä·mə'jē·nē·əs }

homogeneous catalysis [CHEMISTRY] Catalysis occurring within a single phase, usually a gas or liquid. { ‚hä·mə'jē·nē·əs kə'tal·ə·səs }

homogeneous chemical reaction [CHEMISTRY] Chemical reaction system in which all constitutents (reactants and catalyst) are of the same phase. { ‚hä·mə'jē·nē·əs ¦kem·i·kəl rē'ak·shən }

homologation [ORGANIC CHEMISTRY] A type of hydroformylation in which carbon monoxide reacts with certain saturated alcohols to yield either aldehydes or alcohols (or a mixture of both) containing one more carbon atom than the parent. { hə‚mäl·ə'gā·shən }

homology [CHEMISTRY] The relation among elements of the same group, or family, in the periodic table. [ORGANIC CHEMISTRY] That state, in a series of organic compounds that differ from each other by a CH_2 such as the methane series C_nH_{2n+2}, in which there is a similarity between the compounds in the series and a graded change of their properties. { hə'mäl·ə·jē }

homolysis [ORGANIC CHEMISTRY] *See* homolytic cleavage.

homolytic cleavage [ORGANIC CHEMISTRY] The breaking of a single (two-electron) bond in which one electron remains on each of the atoms. Also known as free-radical reaction; homolysis. { ¦häm·ə‚lid·ik 'klēv·ij }

homomorphs [CHEMISTRY] Chemical molecules that are similar in size and shape, but not necessarily having any other characteristics in common. { 'hä·mə‚mȯrfs }

homonuclear molecule [CHEMISTRY] A diatomic molecule, both of whose atoms are of the same element. { ‚hō·mō¦nü·klē·ər 'mäl·ə‚kyül }

homopolar bond [PHYSICAL CHEMISTRY] A covalent bond whose total dipole moment is zero. { ¦hä·mə'pō·lər 'bänd }

homopolymer [ORGANIC CHEMISTRY] A polymer formed from a single monomer; an example is polyethylene, formed by polymerization of ethylene. { ‚hä·mō'päl·ə·mər }

homozeotrope [CHEMISTRY] Mixture in which the liquid components are miscible in all proportions in the liquid phase, and may be separated by ordinary distillation. { ¦hä·mə'zē·ə‚trōp }

Hopkins-Cole reaction [ANALYTICAL CHEMISTRY] The appearance of a violet ring when concentrated sulfuric acid is added to a mixture that includes a protein and glyoxylic acid; however, gelatin and zein do not show the reaction. { 'häp·kənz'kōl rē‚ak·shən }

horizontal chromatography [ANALYTICAL CHEMISTRY] Paper chromatography in which the chromatogram is horizontal instead of vertical. { ‚här·ə'zänt·əl ‚krō·mə'täg·rə·fē }

Hortvet sublimator [ANALYTICAL CHEMISTRY] Device for the determination of the condensation temperature (sublimation point) of sublimed solids. { ¦hȯrt¦vet 'səb·lə‚mād·ər }

host-guest complexation chemistry [ORGANIC CHEMISTRY] The design, synthesis, and study of highly structured organic molecular complexes that mimic biological complexes. { 'hōst 'gest ‚käm·plek'sā·shən ‚kem·ə·strē }

host structure [CHEMISTRY] The crystal structure that forms the cage in which the guest molecule is trapped in a clathrate compound. Also known as host substance. { ¦hōst 'strək·chər }

host substance [CHEMISTRY] *See* host structure. { ¦hōst 'səb·stəns }

Houben-Hoesch synthesis [ORGANIC CHEMISTRY] Condensation of cyanides with polyhydric phenols in the presence of hydrogen chloride and zinc chloride to yield phenolic ketones. { 'hü·bən 'hərsh 'sin·thə·səs }

HPLC [ANALYTICAL CHEMISTRY] *See* high-performance liquid chromatography.

HREELS [SPECTROSCOPY] *See* high-resolution electron energy loss spectroscopy. { 'āch ˌrēlz }

Huber's reagent [ANALYTICAL CHEMISTRY] Aqueous solution of ammonium molybdate and potassium ferrocyanide used as a reagent to detect free mineral acid. { 'hyü·bərz rēˌā·jənt }

Hubl's reagent [ANALYTICAL CHEMISTRY] Solution of iodine and mercuric chloride in alcohol; used to determine the iodine content of oils and fats. { 'həb·əlz rēˌā·jənt }

Hull cell [PHYSICAL CHEMISTRY] An electrodeposition cell that operates within a simultaneous range of known current densities. { 'həl ˌsel }

humectant [CHEMISTRY] A substance which absorbs or retains moisture; examples are glycerol, propylene glycol, and sorbitol; used in preparing confectioneries and dried fruit. { hyü'mek·tənt }

humic acid [ORGANIC CHEMISTRY] Any of various complex organic acids obtained from humus; insoluble in acids and organic solvents. { 'hyü·mik 'as·əd }

humidity indicator [INORGANIC CHEMISTRY] Cobalt salt (for example, cobaltous chloride) that changes color as the surrounding humidity changes; changes from pink when hydrated, to greenish-blue when anhydrous. { hyü'mid·əd·ē ˌin·dəˌkād·ər }

humin [ORGANIC CHEMISTRY] An insoluble pigment formed in the acid hydrolysis of a protein that contains tryptophan. { 'hyü·mən }

Humphreys series [SPECTROSCOPY] A series of lines in the infrared spectrum of atomic hydrogen whose wave numbers are given by $R_H(1/36) - (1/n^2)$, where R_H is the Rydberg constant for hydrogen, and n is any number greater than 6. { 'həm·frēz ˌsir·ēz }

Hundsdieke reaction [ORGANIC CHEMISTRY] Production of an alkyl halide by boiling a silver carboxylate with an equivalent weight of bromine in carbon tetrachloride. { 'hənzˌdēk·ə rēˌak·shən }

hybridization [PHYSICAL CHEMISTRY] The mixing together on the same atom of two or more orbitals that have similar energies, forming a hybrid orbital. { ˌhī·brəd·ə'zā·shən }

hybridized orbital [PHYSICAL CHEMISTRY] A molecular orbital which is a linear combination of two or more orbitals of comparable energy (such as 2*s* and 2*p* orbitals), is concentrated along a certain direction in space, and participates in formation of a directed valence bond. { 'hī·brədˌīzd 'ȯr·bəd·əl }

hybrid orbital [PHYSICAL CHEMISTRY] An orbital formed by the combination of two or more atomic orbitals on a single atom. { ¦hī·brəd 'ȯr·bəd·əl }

hydantoin [ORGANIC CHEMISTRY] $C_3N_2O_2H$ A white, crystalline compound, melting point 220°C; used as an intermediate in certain pharmaceutical manufacturing and as a textile softener and lubricant. Also known as glycolyurea. { hī'dant·ə·wən }

hydnocarpic acid [ORGANIC CHEMISTRY] $C_{16}H_{28}O_2$ A nonedible fat and oil isolated from chaulmoogra oil, forming white crystals that melt at 60°C; used to treat Hansen's disease. { ¦hīd·nə¦kär·pik 'as·əd }

hydracrylic acid [ORGANIC CHEMISTRY] $CH_2OH \cdot CH_2COOH$ An oily liquid that is an isomer of lactic acid and that breaks down on heating to acrylic acid. { ¦hī·drə¦kril·ik 'as·əd }

hydrastine [ORGANIC CHEMISTRY] $C_{21}H_{21}NO_6$ An alkaloid isolated from species of the family Ranunculaceae and from *Hydrastis canadensis*; orthorhombic prisms crystallize from alcohol solution, melting point 132°C; highly soluble in acetone and benzene, soluble in chloroform, less soluble in ether and alcohol. { 'hī·draˌstēn }

hydrastinine [ORGANIC CHEMISTRY] $C_{11}H_{13}O_3N$ A compound formed by the decomposition of hydrastine; crystallizes as needles from petroleum-ether solution, soluble in organic solvents such as alcohol, chloroform, and ether; used in medicine as a stimulant in coronary disease and as a hemostatic in uterine hemorrhage. { hī'dras·təˌnēn }

hydrate [CHEMISTRY] **1.** A form of a solid compound which has water in the form of H_2O molecules associated with it; for example, anhydrous copper sulfate is a white solid with the formula $CuSO_4$, but when crystallized from water a blue crystalline solid with formula $CuSO_4 \cdot 5H_2O$ results, and the water molecules are an integral part of the crystal. **2.** A crystalline compound resulting from the combination of water and a gas; frequently a constituent of natural gas that is under pressure. { 'hī,drāt }

hydrated alumina [INORGANIC CHEMISTRY] *See* alumina trihydrate. { 'hī,drād·əd ə'lü·mə·nə }

hydrated aluminum oxide [INORGANIC CHEMISTRY] *See* alumina trihydrate. { 'hī,drāt ə'lü·mə·nəm 'äk,sīd }

hydrated chloral [ORGANIC CHEMISTRY] *See* chloral hydrate. { 'hī,drād·əd 'klȯr·əl }

hydrated electron [PHYSICAL CHEMISTRY] An electron released during ionization of a water molecule by water and surrounded by water molecules oriented so that the electron cannot escape. Also known as aqueous electron. { 'hī,drād·əd i'lek,trän }

hydrated lime [INORGANIC CHEMISTRY] *See* calcium hydroxide. { 'hī,drād·əd 'līm }

hydrated manganic hydroxide [INORGANIC CHEMISTRY] *See* manganic hydroxide. { 'hī,drād·əd maŋ'gan·ik hī'dräk,sīd }

hydrated mercurous nitrate [INORGANIC CHEMISTRY] $Hg_2(NO_3)_2 \cdot 2H_2O$ Poisonous, light-sensitive crystals, soluble in warm water, decomposes at 70°C; used as an analytical reagent and in cosmetics and medicine. { 'hī,drād·əd mər'kyu̇r·əs 'nī,trāt }

hydrated silica [INORGANIC CHEMISTRY] *See* silicic acid. { 'hī,drād·əd 'sil·ə·kə }

hydrate inhibitor [CHEMISTRY] A material (such as alcohol or glycol) added to a gas stream to prevent the formation and freezing of gas hydrates in low-temperature systems. { 'hī,drāt in,hib·əd·ər }

hydration [CHEMISTRY] The incorporation of molecular water into a complex molecule with the molecules or units of another species; the complex may be held together by relatively weak forces or may exist as a definite compound. { hī'drā·shən }

hydrazide [INORGANIC CHEMISTRY] An acyl hydrazine; a compound of the formula

$$R-\overset{\displaystyle O}{\overset{\|}{C}}-NH-NH_2$$

where R may be an alkyl group. { 'hī·drə,zīd }

hydrazine [INORGANIC CHEMISTRY] H_2NNH_2 A colorless, hygroscopic liquid, boiling point 114°C, with an ammonialike odor; it is reducing, decomposable, basic, and bifunctional; used as a rocket fuel, in corrosion inhibition in boilers, and in the synthesis of biologically active materials, explosives, antioxidants, and photographic chemicals. { 'hī·drə,zēn }

hydrazine hydrate [ORGANIC CHEMISTRY] $H_2NNH_2OH_2O$ A colorless, fuming liquid that boils at 119.4°C; used as a component in jet fuels and as an intermediate in organic synthesis. { 'hī·drə,zēn 'hī,drāt }

hydrazinobenzene [ORGANIC CHEMISTRY] *See* phenylhydrazine. { hi¦draz·ə·nō'ben,zēn }

2-hydrazinoethanol [ORGANIC CHEMISTRY] *See* 2-hydroxyethylhydrazine. { ¦tü hī¦draz·ə·nō'e·thə,nȯl }

hydrazobenzene [ORGANIC CHEMISTRY] $C_{12}H_{12}N_2$ A colorless, crystalline compound, melts at 132°C, slightly soluble in water, soluble in alcohol; used as an intermediate in the synthesis of benzidine. { ¦hī·drə·zō'ben,zēn }

hydrazoic acid [INORGANIC CHEMISTRY] NHN:N Explosive liquid, a strong protoplasmic poison boiling at 37°C. { ¦hī·drə¦zō·ik 'as·əd }

hydrazone [ORGANIC CHEMISTRY] A compound containing the grouping —NH · N:C—, and obtained from a condensation reaction involving hydrazines with aldehydes or ketones; has been used as an exotic fuel. { 'hī·drə,zōn }

hydride [INORGANIC CHEMISTRY] A compound containing hydrogen and another element; examples are H_2S, which is a hydride although it may be properly called hydrogen sulfide, and lithium hydride, LiH. { 'hī,drīd }

hydrindantin [ORGANIC CHEMISTRY] $C_{18}H_{10}O_6$ A compound used as a reagent for the photometric determination of amino acids. { ,hī·drən'dant·ən }

hydriodic acid [INORGANIC CHEMISTRY] A yellow liquid that is a water solution of the gas hydrogen iodide; a solution of 59% hydrogen iodide produces a liquid that is constant-boiling; it is a strong acid used in organic synthesis and as a reagent in analytical chemistry. { 'hī·drē,äd·ik 'as·əd }

hydriodic acid gas [INORGANIC CHEMISTRY] *See* hydrogen iodide. { 'hī·drē‚äd·ik ¦as·əd 'gas }

hydrobenzoin [ORGANIC CHEMISTRY] $C_{14}H_{14}O_2$ A colorless, crystalline compound formed by action of sodium amalgam on benzaldehyde, melts at 136°C, and is slightly soluble in water. { ¦hī·drō'ben·zə·wən }

hydroboration [ORGANIC CHEMISTRY] The process of producing organoboranes by the addition of a compound with a B-H bond to an unsaturated hydrocarbon; for example, the reaction of diborane ion with a carbonyl compound. { ¦hī·drō·bə'rā·shən }

hydrobromic acid [INORGANIC CHEMISTRY] HBr A solution of hydrogen bromide in water, usually 40%; a clear, colorless liquid; used in medicine, analytical chemistry, and synthesis of organic compounds. { ¦hī·drə'brō·mik 'as·əd }

hydrocarbon [ORGANIC CHEMISTRY] One of a very large group of chemical compounds composed only of carbon and hydrogen; the largest source of hydrocarbons is from petroleum crude oil. { ¦hī·drə'kär·bən }

hydrocarbon resins [ORGANIC CHEMISTRY] Brittle or gummy materials prepared by the polymerization of several unsaturated constituents of coal tar, rosin, or petroleum; they are inexpensive and find uses in rubber and asphalt formulations and in coating and caulking compositions. { ¦hī·drə'kär·bən 'rez·ənz }

hydrochinone [ORGANIC CHEMISTRY] *See* hydroquinone. { ¦hī·drə·kə'nōn }

hydrochloric acid [INORGANIC CHEMISTRY] HCl A solution of hydrogen chloride gas in water; a poisonous, pungent liquid forming a constant-boiling mixture at 20% concentration in water; widely used as a reagent, in organic synthesis, in acidizing oil wells, ore reduction, food processing, and metal cleaning and pickling. Also known as muriatic acid. { ¦hī·drə'klȯr·ik 'as·əd }

hydrocinnamic acid [ORGANIC CHEMISTRY] $C_6H_5CH_2CH_2COOH$ A compound whose crystals have a floral odor (hyacinth-rose) and melt at 46°C; used in perfumes and flavoring. { ¦hī·drō·si'nam·ik 'as·əd }

hydrocinnamic alcohol [ORGANIC CHEMISTRY] *See* phenylpropyl alcohol. { ¦hī·drō·si'nam·ik 'al·kə‚hȯl }

hydrocinnamic aldehyde [ORGANIC CHEMISTRY] *See* phenylpropyl aldehyde. { ¦hī·drō·si'nam·ik 'al·də‚hīd }

hydrocrackate [ORGANIC CHEMISTRY] The product of a hydrocracker. { ¦hī·drō'kra‚kāt }

hydrocyanic acid [INORGANIC CHEMISTRY] HCN A highly toxic liquid that has the odor of bitter almonds and boils at 25.6°C; used to manufacture cyanide salts, acrylonitrile, and dyes, and as a fumigant in agriculture. Also known as formonitrile; hydrogen cyanide, prussic acid. { ¦hī·drō‚sī'an‚ik 'as·əd }

hydrofluoric acid [INORGANIC CHEMISTRY] An aqueous solution of hydrogen fluoride, HF; colorless, fuming, poisonous liquid; extremely corrosive, it is a weak acid as compared to hydrochloric acid, but will attack glass and other silica materials; used to polish, frost, and etch glass, to pickle copper, brass, and alloy steels, to clean stone and brick, to acidize oil wells, and to dissolve ores. { ¦hī·drə'flu̇r·ik 'as·əd }

hydrofluorosilicic acid [INORGANIC CHEMISTRY] *See* fluosilicic acid. { ¦hī·drō¦flu̇r·ō·sə'lis·ik 'as·əd }

hydrogel [CHEMISTRY] The formation of a colloid in which the disperse phase (colloid) has combined with the continuous phase (water) to produce a viscous jellylike product; for example, coagulated silicic acid. { 'hī·drə‚jel }

hydrogen [CHEMISTRY] The first chemical element, symbol H, in the periodic table, atomic number 1, atomic weight 1.00797; under ordinary conditions it is a colorless, odorless, tasteless gas composed of diatomic molecules, H_2; used in manufacture of ammonia and methanol, for hydrofining, for desulfurization of petroleum products, and to reduce metallic oxide ores. { 'hī·drə·jən }

hydrogenated oil [ORGANIC CHEMISTRY] Unsaturated liquid vegetable oil that has had hydrogen catalytically added so as to convert the oil to a hydrogen-saturated solid. { 'hī·drə·jə‚nād·əd ¦ȯil }

hydrogenation [ORGANIC CHEMISTRY] Catalytic reaction of hydrogen with other compounds, usually unsaturated; for example, unsaturated cottonseed oil is hydrogenated to form solid fats. { hī‚dräj·ə'nā·shən }

hydrogen bond [PHYSICAL CHEMISTRY] A type of bond formed when a hydrogen atom bonded to atom A in one molecule makes an additional bond to atom B either in the same or another molecule; the strongest hydrogen bonds are formed when A and B are highly electronegative atoms, such as fluorine, oxygen, or nitrogen. { 'hī·drə·jən 'bänd }

hydrogen bromide [INORGANIC CHEMISTRY] HBr A hazardous, toxic gas used as a chemical intermediate and as an alkylation catalyst; forms hydrobromic acid in aqueous solution. { 'hī·drə·jən 'brō‚mīd }

hydrogen chloride [INORGANIC CHEMISTRY] HCl A fuming, highly toxic, colorless gas soluble in water, alcohol, and ether; used in the production of vinyl chloride and alkyl chloride, and in polymerization, isomerization, and other reactions. { 'hī·drə·jən 'klȯr‚īd }

hydrogen cyanide [INORGANIC CHEMISTRY] *See* hydrocyanic acid. { 'hī·drə·jən 'sī·ə‚nīd }

hydrogen cycle [CHEMISTRY] The complete process of a cation-exchange operation in which the adsorbent is used in the hydrogen or free acid form. { 'hī·drə·jən ‚sī·kəl }

hydrogen disulfide [INORGANIC CHEMISTRY] *See* hydrogen sulfide. { 'hī·drə·jən dī'səl‚fīd }

hydrogen electrode [PHYSICAL CHEMISTRY] A noble metal (such as platinum) of large surface area covered with hydrogen gas in a solution of hydrogen ion saturated with hydrogen gas; metal is used in a foil form and is welded to a wire sealed in the bottom of a hollow glass tube, which is partially filled with mercury; used as a standard electrode with a potential of zero to measure hydrogen ion activity. { 'hī·drə·jən i'lek‚trōd }

hydrogen equivalent [CHEMISTRY] The number of replaceable hydrogen atoms or hydroxyl groups in a molecule of an acid or a base. { 'hī·drə·jən i'kwiv·ə·lənt }

hydrogen fluoride [INORGANIC CHEMISTRY] HF The hydride of fluoride; anhydrous HF is a mobile, colorless, liquid that fumes in air, melts at −83°C, boils at 19.8°C; used to make fluorine-containing refrigerants (such as Freon) and organic fluorocarbon compounds, as a catalyst in alkylate gasoline manufacture, as a fluorinating agent, and in preparation of hydrofluoric acid. { 'hī·drə·jən 'flu̇r‚īd }

hydrogen iodide [INORGANIC CHEMISTRY] HI A water-soluble, colorless gas that may be used in organic synthesis and as a reagent. Also known as hydriodic acid gas. { 'hī·drə·jən 'ī·ə‚dīd }

hydrogen ion [INORGANIC CHEMISTRY] *See* hydronium ion. { 'hī·drə·jən 'ī‚än }

hydrogen ion concentration [CHEMISTRY] The normality of a solution with respect to hydrogen ions, H^+; it is related to acidity measurements in most cases by pH = log 1/2 [1/(H^+)], where (H^+) is the hydrogen ion concentration in gram equivalents per liter of solution. { 'hī·drə·jən 'ī‚än ‚käns·ən‚trā·shən }

hydrogen ion exponent [CHEMISTRY] A way of expressing pH; namely, pH = $-\log c_H$, where c_H = hydrogen ion concentration. { 'hī·drə·jən 'ī‚än ik'spō·nənt }

hydrogen line [SPECTROSCOPY] A spectral line emitted by neutral hydrogen having a frequency of 1420 megahertz and a wavelength of 21 centimeters; radiation from this line is used in radio astronomy to study the amount and velocity of hydrogen in the Galaxy. { 'hī·drə·jən ‚līn }

hydrogenolysis [CHEMISTRY] A reaction in which hydrogen gas causes a chemical change that is similar to the role of water in hydrolysis. { 'hī·drə·jə'näl·ə·səs }

hydrogenous [CHEMISTRY] Of, pertaining to, or containing hydrogen. { hī'dräj·ə·nəs }

hydrogen peroxide [INORGANIC CHEMISTRY] H_2O_2 Unstable, colorless, heavy liquid boiling at 158°C; soluble in water and alcohol; used as a bleach, chemical intermediate, rocket fuel, and antiseptic. Also known as peroxide. { 'hī·drə·jən pə'räk‚sīd }

hydrogen selenide [INORGANIC CHEMISTRY] H_2Se A toxic, colorless gas, soluble in water, carbon disulfide, and phosgene; used to make metallic selenides and organoselenium compounds and in the preparation of semiconductor materials. { 'hī·drə·jən 'sel·ə‚nīd }

hydrogen sulfide [INORGANIC CHEMISTRY] H_2S Flammable, toxic, colorless gas with offensive odor, boiling at −60°C; soluble in water and alcohol; used as an analytical reagent, as a sulfur source, and for purification of hydrochloric and sulfuric acids. Also known as hydrogen disulfide. { 'hī·drə·jən 'səl‚fīd }

hydrogen tellurate [INORGANIC CHEMISTRY] *See* telluric acid. { 'hī·drə·jən 'tel·yə‚rāt }

hydroiodic ether [ORGANIC CHEMISTRY] *See* ethyl iodide. { ‚hī·drȯi'äd·ik 'ē·thər }

hydrolysis [CHEMISTRY] **1.** Decomposition or alteration of a chemical substance by water. **2.** In aqueous solutions of electrolytes, the reactions of cations with water to produce a weak base or of anions to produce a weak acid. { hī'dräl·ə·səs }

hydrolytic process [CHEMISTRY] A reaction of both organic and inorganic chemistry wherein water effects a double decomposition with another compound, hydrogen going to one compound and hydroxyl to another. { ¦hī·drə¦lid·ik 'prä·səs }

hydronium ion [INORGANIC CHEMISTRY] H_3O^+ An oxonium ion consisting of a proton combined with a molecule of water; found in pure water and in all aqueous solutions. Also known as hydrogen ion. { hī'drō·nē·əm ¦ī‚än }

hydrophile-lipophile balance [ORGANIC CHEMISTRY] The relative simultaneous attraction of an emulsifier for two phases of an emulsion system; for example, water and oil. { 'hī·drə‚fīl 'lip·ə‚fīl ‚bal·əns }

hydrophilic [CHEMISTRY] Having an affinity for, attracting, adsorbing, or absorbing water. { ‚hī·drə'fil·ik }

hydrophobic [CHEMISTRY] Lacking an affinity for, repelling, or failing to adsorb or absorb water. { ¦hī·drə'fō·bik }

hydroquinol [ORGANIC CHEMISTRY] *See* hydroquinone. { ¦hī·drō'kwi‚nȯl }

hydroquinone [ORGANIC CHEMISTRY] $C_6H_4(OH)_2$ White crystals melting at 170°C and boiling at 285°C, soluble in alcohol, ether, and water, used in photographic dye chemicals, in medicine, as an antioxidant and inhibitor, and in paints, varnishes, and motor fuels and oils. Also known as hydrochinone; hydroquinol; quinol. { ¦hī·drə·kwə'nōn }

hydroquinone dimethyl ether [ORGANIC CHEMISTRY] $C_6H_4(OCH_3)_2$ White flakes with a melting point of 56°C; used as a weathering agent in paint, as a flavoring, and in dyes and cosmetics. { ¦hī·drə·kwə'nōn dī¦meth·əl 'ē·thər }

hydroquinone monomethyl ether [ORGANIC CHEMISTRY] $CH_3OC_6H_4OH$ A white, waxy solid with a melting point of 52.5°C; soluble in benzene, acetone, and alcohol; used for antioxidants, pharmaceuticals, and dyestuffs. { ¦hī·drə·kwə'nōn ‚män·ō¦meth·əl 'ē·thər }

hydrosilylation [ORGANIC CHEMISTRY] The addition of a Si-H bond to a C-C double bond of an olefin. { ¦hī·drō‚sil·ē'ā·shən }

hydrosol [CHEMISTRY] A colloidal system in which the dispersion medium is water, and the dispersed phase may be a solid, a gas, or another liquid. Also known as aquasol. { 'hī·drə‚sȯl }

hydrosulfide [CHEMISTRY] A compound that has the SH— radical; for example, sulfhydrates, sulfhydryls, thioalcohols, thiols, sulfur alcohols, and mercaptans. { ¦hī·drə'səl‚fīd }

hydrotrope [CHEMISTRY] Compound with the ability to increase the solubilities of certain slightly soluble organic compounds. { 'hī·drə‚trōp }

hydrous [CHEMISTRY] Indicating the presence of an indefinite amount of water. { 'hī·drəs }

hydroxamic acid [ORGANIC CHEMISTRY] An organic compound that contains the group —C(=O)NHOH. { ¦hī‚dräk¦sam·ik 'as·əd }

hydroxide [CHEMISTRY] Compound containing the OH^- group; the hydroxides of metals are usually bases and those of nonmetals are usually acids; a hydroxide can be organic or inorganic. { hī'dräk‚sīd }

hydroximino [CHEMISTRY] *See* nitroso. { ¦hī‚dräk'sim·ə·nō }

hydroxisoxazole [ORGANIC CHEMISTRY] $C_4H_5NO_2$ A colorless, crystalline compound with a melting point of 86–87°C; used as a fungicide in soil and as a growth regulator for seeds. Also known as 3-hydroxy-5-methylisoxazole; hymexazol. { hī¦dräk·sə¦säk·sə‚zōl }

hydroxy- [ORGANIC CHEMISTRY] Chemical prefix indicating the OH^- group in an organic compound, such as hydroxybenzene for phenol, C_6H_5OH; the use of just oxy- for the prefix is incorrect. Also spelled hydroxyl-. { hī'dräk·sē }

hydroxyacetic acid [ORGANIC CHEMISTRY] *See* glycolic acid. { hī¦dräk·sə·ə'sēd·ik 'as·əd }

hydroxy acid [ORGANIC CHEMISTRY] Any organic acid, with an OH^- group, such as hydroxyacetic acid. { hī'dräk·sē 'as·əd }

hydroxybenzoic acid [ORGANIC CHEMISTRY] $C_7H_6O_3$ Any one of three crystalline derivatives of benzoic acid: ortho, meta, and para forms; the ester of the para compound is used as a bacteriostatic agent. { hī¦dräk·sē·ben'zō·ik 'as·əd }

***para*-hydroxybenzoic acid** [ORGANIC CHEMISTRY] $C_6H_4(OH)COOH \cdot 2H_2O$ Colorless crystals melting at 210°C; soluble in alcohol, water, and ether; used as a chemical intermediate and for synthetic drugs. { ¦par·ə hī¦dräk·sē·ben'zō·ik 'as·əd }

2-hydroxybiphenyl [ORGANIC CHEMISTRY] *See* phenylphenol. { ¦tü hī¦dräk·sē·bī'fen·əl }

hydroxycarbonyl compound [ORGANIC CHEMISTRY] A compound possessing one or more hydroxy (—OH) groups and one or more carbonyl (═C═O) groups. { hī¦dräk·sē'kär·bə·nəl 'käm‚paủnd }

hydroxycholine [ORGANIC CHEMISTRY] *See* muscarine. { hī¦dräk·sē'kō‚lēn }

hydroxycinchonine [ORGANIC CHEMISTRY] *See* cupreine. { hī¦dräk·sē'siŋ·kə‚nēn }

hydroxycitronellal [ORGANIC CHEMISTRY] $C_{10}H_{20}O_2$ A colorless or light yellow, viscous liquid with a boiling range of 94-96°C; soluble in 50% alcohol and fixed oils; used in perfumery and flavoring. Also known as citronellal hydrate. { hī¦dräk·sē‚sī·trə'nel·əl }

2-(hydroxydiphenyl)methane [ORGANIC CHEMISTRY] $C_6H_5CH_2C_6H_4OH$ A crystalline substance with a melting point of 20.2-20.9°C, or a liquid; used as a germicide, preservative, and antiseptic. { ¦tü hī¦dräk·sē·dī'fen·əl 'meth‚ān }

2-hydroxyethylhydrazine [ORGANIC CHEMISTRY] $HOCH_2CH_2NHNH_2$ A colorless, slightly viscous liquid with a melting point of −70°C; soluble in lower alcohols; used as an abscission agent in fruit. Also known as 2-hydrazinoethanol. { ¦tü hī¦dräk·sē¦eth·əl'hī·drə‚zēn }

hydroxyl- [ORGANIC CHEMISTRY] *See* hydroxy-. { hī'dräk·səl }

hydroxylamine [INORGANIC CHEMISTRY] NH_2OH A colorless, crystalline compound produced commercially by acid hydrolysis of nitroparaffins, decomposes on heating, melts at 33°C; used in organic synthesis and as a reducing agent. { ‚hī‚dräk'sil·ə‚mēn }

hydroxylamine hydrochloride [ORGANIC CHEMISTRY] $(NH_2OH)Cl$ A crystalline substance with a melting point of 151°C; soluble in glycerol and propylene glycol; used as a reducing agent in photography and in synthetic and analytic chemistry, as an antioxidant in fatty acids and soaps, and as a reagent for enzyme reactivation. { ‚hī‚dräk'sil·ə‚mēn ‚hī·drə'klȯr‚īd }

***ortho*-hydroxylaniline** [ORGANIC CHEMISTRY] $C_6H_4NH_2OH$ White crystals that turn brownish upon standing for some time; melts at 172-173°C, and will sublime upon more heating; soluble in cold water and benzene; used as a dye for hair and furs, and as a dye intermediate. Also known as *ortho*-aminophenol; oxammonium. { ¦ȯr·thō ‚hī¦dräk·səl'an·əl·ən }

hydroxylation reaction [ORGANIC CHEMISTRY] One of several types of reactions used to introduce one or more hydroxyl groups into organic compounds; an oxidation reaction as opposed to hydrolysis. { hī‚dräk·sə'lā·shən rē‚ak·shən }

β-hydroxynaphthoic acid [ORGANIC CHEMISTRY] $C_{10}H_6OHCOOH$ A yellow solid that is soluble in ether and alcohol and melts at about 218°C; used as a dye and a pigment. { ¦bād·ə hī¦dräk·sē·naf'thō·ik 'as·əd }

4-hydroxy-3-nitrobenzenearsonic acid [ORGANIC CHEMISTRY] $HOC_6H_3(NO_2)AsO(OH)_2$ Crystals used as a reagent for zirconium; also used to control enteric infections and to improve growth and feed efficiency in animals. Also known as roxarsone. { ¦fȯr hī¦dräk·sē ¦thrē ‚nī·trō¦ben‚zēn·är'sän·ik 'as·əd }

8-hydroxyquinoline [ORGANIC CHEMISTRY] C_9H_6NOH White crystals or powder that darken on exposure to light, slightly soluble in water, soluble in benzene, melting at

73-75°C; used in preparing fungicides and in the separation of metals by acting as a precipitating agent. Also known as oxine; oxyquinoline; 8-quinolinol. { ¦āt hī¦dräk·sē'kwin·ə·lən }

3-hydroxytyramine hydrobromide [ORGANIC CHEMISTRY] $(HO)_2\ C_6H_3CH_2CH_2$-$NH_2 \cdot HBr$ A source of dopamine for the synthesis of catecholamine analogs. { ¦thrē hī¦dräk·sē'tī·rə‚mēn ‚hī·drə'brō‚mīd }

hygroscopic [CHEMISTRY] **1.** Possessing a marked ability to accelerate the condensation of water vapor; applied to condensation nuclei composed of salts which yield aqueous solutions of a very low equilibrium vapor pressure compared with that of pure water at the same temperature. **2.** Pertaining to a substance whose physical characteristics are appreciably altered by effects of water vapor. **3.** Pertaining to water absorbed by dry soil minerals from the atmosphere; the amounts depend on the physicochemical character of the surfaces, and increase with rising relative humidity. { ¦hī·grə¦skäp·ik }

hygroscopic depression [CHEMISTRY] The measure of a desiccant's capacity to take on water. { ¦hī·grə¦skäp·ik di'presh·ən }

hymecromone [ORGANIC CHEMISTRY] $C_{10}H_8O_3$ A crystalline substance with a melting point of 194-195°C; soluble in methanol and glacial acetic acid; used as choleretic and antispasmodic drugs and as a standard for the fluorometric determination of enzyme activity. { hī'mek·rə‚mōn }

hymexazol [ORGANIC CHEMISTRY] *See* hydroxisoxazole. { hī'mek·sə‚zȯl }

hyoscyamine [ORGANIC CHEMISTRY] $C_{17}H_{23}O_3N$ A white, crystalline alkaloid isolated from henbane, belladonna, and other plants of the family Solanaceae, which is freely soluble in alcohol and dilute acids, used in medicine as an anticholinergic. { ‚hī·ə'sī·ə‚mēn }

hyperchromicity [PHYSICAL CHEMISTRY] An increase in the absorption of ultraviolet light by polynucleotide solutions due to a loss of the ordered secondary structure. { ‚hī·pər·krō'mis·əd·ē }

hyperconjugation [PHYSICAL CHEMISTRY] An arrangement of bonds in a molecule that is similar to conjugation in its formulation and manifestations, but the effects are weaker; it occurs when a CH_2 or CH_3 group (or in general, an AR_2 or AR_3 group where A may be any polyvalent atom and R any atom or radical) is adjacent to a multiple bond or to a group containing an atom with a lone π-electron, π-electron pair or quartet, or π-electron vacancy; it can be sacrificial (relatively weak) or isovalent (stronger). { ‚hī·pər‚kän·jə'gā·shən }

hyperfine structure [SPECTROSCOPY] A splitting of spectral lines due to the spin of the atomic nucleus or to the occurrence of a mixture of isotopes in the element. Abbreviated hfs. { 'hī·pər‚fīn 'strək·chər }

hypergolic [CHEMISTRY] Capable of igniting spontaneously upon contact. { ¦hī·pər¦gal·ik }

hypervalent atom [CHEMISTRY] A central atom in a single-bonded structure that imparts more than eight valence electrons in forming covalent bonds. { ‚hī·pər'vā·lənt 'ad·əm }

hypo [INORGANIC CHEMISTRY] *See* sodium thiosulfate. { 'hī·pō }

hypochlorite [INORGANIC CHEMISTRY] ClO_3^- A negative ion derived from hypochlorous acid, HClO; the ion is an oxidizing agent and a constituent of bleaching agents. { ‚hī·pə'klȯr‚īt }

hypochlorous acid [INORGANIC CHEMISTRY] HOCl Weak, unstable acid existing in solution only; its salts (such as calcium hypochlorite) are used as bleaching agents. { ¦hī·pə'klȯr·əs 'as·əd }

hypochromicity [PHYSICAL CHEMISTRY] A decrease in the absorption of ultraviolet light by polynucleotide solutions due to the formation of an ordered secondary structure. { ‚hī·pə·krə'mis·əd·ē }

hypoiodous acid [INORGANIC CHEMISTRY] HIO A very weak unstable acid that occurs as the result of the weak hydrolysis of iodine in water. { ¦hī·pō‚ī'ōd·əs 'as·əd }

I

I [CHEMISTRY] *See* iodine.

IBA [ORGANIC CHEMISTRY] *See* indolebutyric acid.

IBIB [ORGANIC CHEMISTRY] *See* isobutyl isobutyrate.

ibogaine [ORGANIC CHEMISTRY] $C_{26}H_{32}O_2N_2$ An alkaloid isolated from the stems and leaves of the shrub *Tabernanthe iboga*, crystallizing from absolute ethanol as prismatic needles, melting at 152-153°C, soluble in ethanol, ether, and chloroform; used in medicine. { ə'bo·gə,en }

ice [PHYSICAL CHEMISTRY] **1.** The dense substance formed by the freezing of water to the solid state; has a melting point of 32°F (0°C) and commonly occurs in the form of hexagonal crystals. **2.** A layer or mass of frozen water. { īs }

ice color [ORGANIC CHEMISTRY] *See* azoic dye. { 'īs ,kəl·ər }

ice crystal [PHYSICAL CHEMISTRY] Any one of a number of macroscopic crystalline forms in which ice appears, including hexagonal columns, hexagonal platelets, dendritic crystals, ice needles, and combinations of these forms; although the crystal lattice of ice is hexagonal in its symmetry, varying conditions of temperature and vapor pressure can lead to growth of crystalline forms in which the simple hexagonal pattern is almost undiscernible. { 'īs ,krist·əl }

ice needle [PHYSICAL CHEMISTRY] A long, thin ice crystal whose cross section perpendicular to its long dimension is typically hexagonal. Also called ice spicule. { 'īs ,nēd·əl }

ice point [PHYSICAL CHEMISTRY] The true freezing point of water; the temperature at which a mixture of air-saturated pure water and pure ice may exist in equilibrium at a pressure of 1 standard atmosphere (101,325 pascals). { 'īs ,pȯint }

ice spicule [PHYSICAL CHEMISTRY] *See* ice needle. { 'īs ¦spik·yəl }

ice splinters [PHYSICAL CHEMISTRY] Minute, electrically charged fragments of ice which have been observed under laboratory conditions to be torn away from dendritic crystals or spatial aggregates exposed to moving air. { 'īs ¦splin·tərz }

ICP-AES [SPECTROSCOPY] *See* inductively coupled plasma atomic emission spectroscopy.

IDA [ORGANIC CHEMISTRY] *See* iminodiacetic acid.

ideal solution [CHEMISTRY] A solution that conforms to Raoult's law over all ranges of temperature and concentration and shows no internal energy change on mixing and no attractive force between components. { ī'dēl sə'lü·shən }

ignite [CHEMISTRY] To start a fuel burning. { ig'nīt }

ignition [CHEMISTRY] The process of starting a fuel mixture burning, or the means for such a process. { ig'nish·ən }

ignition point [CHEMISTRY] *See* ignition temperature. { ig'nish·ən ,pȯint }

ignition temperature [CHEMISTRY] The lowest temperature at which combustion begins and continues in a substance when it is heated in air. Also known as autogenous ignition temperature; ignition point. { ig'nish·ən ,tem·prə·chər }

Ilkovic equation [ANALYTICAL CHEMISTRY] Mathematical relationship between diffusion current, diffusion coefficient, and active-substance concentration; used for polarographic analysis calculations. { 'il·kə,vich i,kwā·zhən }

imbibition [PHYSICAL CHEMISTRY] Absorption of liquid by a solid or a semisolid material. { ,im·bə'bish·ən }

imidazole [ORGANIC CHEMISTRY] $C_3H_4N_2$ One of a group of organic heterocyclic compounds containing a five-membered diunsaturated ring with two nonadjacent nitrogen atoms as part of the ring; the particular compound imidazole is a member of the group. { ˌim·əˈdaˌzōl }

imidazolyl [ORGANIC CHEMISTRY] $C_3H_3N_2\cdot$ A free radical derived from imidazole. { ˌim·əˈda·zəˌlil }

imide [ORGANIC CHEMISTRY] **1.** A compound derived from acid anhydrides by replacing the oxygen (O) with the =NH group. **2.** A compound that has either the =NH group or a secondary amine in which R is an acyl functional group, as R_2NH. { ˈiˌmīd }

imine [ORGANIC CHEMISTRY] A class of compounds that are the product of condensation reactions of aldehydes or ketones with ammonia or amines; they have the NH radical attached to the carbon with the double bond, as R—HC=NH; an example is benzaldimine. { ˈiˌmēn }

imino acid [ORGANIC CHEMISTRY] Organic acid in which the =NH group is attached to one or two carbons; such as acetic acid, $NH(CH_2COOH)_2$. { ˈim·əˌnō ˈas·əd }

imino compound [ORGANIC CHEMISTRY] A compound that has the =NH radical attached to one or two carbon atoms. { ˈim·əˌnō ˌkämˌpau̇nd }

iminodiacetic acid [ORGANIC CHEMISTRY] $C_4H_7NO_4$ A crystalline substance used as an intermediate in the manufacture of chelating agents, surface-active agents, and complex salts. Abbreviated IDA. Also known as diglycine; iminodiethanoic acid. { ˌim·ə·nō·dī·əˈsēd·ik ˈas·əd }

imino nitrogen [ORGANIC CHEMISTRY] Nitrogen combined with hydrogen in the imino group. { ˈim·əˌnō ˈnī·trə·jən }

immersion sampling [ANALYTICAL CHEMISTRY] Collection of a liquid sample for laboratory or other analysis by immersing a container in the liquid and filling it. { əˈmər·zhən ˌsam·pliŋ }

immiscible [CHEMISTRY] Pertaining to liquids that will not mix with each other. { iˈmis·ə·bəl }

immobilized catalyst [CHEMISTRY] A molecular catalyst that is bound without substantial change in its structure to an insoluble solid to prevent solution of the catalyst in the contacting liquid. Also known as anchored catalyst. { iˈmō·bəˌlīzd ˈkad·əˌlist }

imperial red [INORGANIC CHEMISTRY] Any of the red varieties of ferric oxide used as pigment. { imˈpir·ē·əl ˈred }

implosion [CHEMISTRY] The sudden reduction of pressure by chemical reaction or change of state which causes an inrushing of the surrounding medium. { imˈplō·zhən }

In [CHEMISTRY] *See* indium.

inactive tartaric acid [ORGANIC CHEMISTRY] *See* racemic acid. { inˈak·tiv tärˈtär·ik ˈas·əd }

incineration [CHEMISTRY] The process of burning a material so that only ashes remain. { inˌsin·əˈrā·shən }

inclusion complex [CHEMISTRY] An unbonded association in which the molecules of one component are contained wholly or partially within the crystal lattice of the other component. { inˈklü·zhən ˈkämˌpleks }

inclusion compound [CHEMISTRY] *See* clathrate. { inˈklü·zhən ˈkämˌpau̇nd }

incomplete combustion [CHEMISTRY] Combustion in which oxidation of the fuel is incomplete. { ˌin·kəmˈplēt kəmˈbəs·shən }

increment [ANALYTICAL CHEMISTRY] An individual portion of material of a group of samples collected by a single operation of a sampling device from parts of a lot that are separated in time or space. { ˈiŋ·krə·mənt }

incubation [CHEMISTRY] Maintenance of chemical mixtures at specified temperatures for varying time periods to study chemical reactions, such as enzyme activity. { ˌiŋ·kyəˈbā·shən }

indamine [ORGANIC CHEMISTRY] $HN{:}C_6H_4{:}N\cdot C_6H_4NH_2$ An unstable dye obtained by the reaction of *para*-phenylenediamine and aniline. Also known as phenylene blue. { ˈin·dəˌmēn }

indan [ORGANIC CHEMISTRY] $C_6H_4(CH_2)_3$ Colorless liquid boiling at 177°C; soluble in alcohol and ether, insoluble in water; derived from coal tar. { 'in͵dan }

indanthrone [ORGANIC CHEMISTRY] $C_{28}H_{14}N_2O_4$ A blue pigment or vat dye soluble in dilute base solutions; used in cotton dyeing and as a pigment in paints and enamels. { in'dan͵thrōn }

indene [ORGANIC CHEMISTRY] C_9H_8 A colorless, liquid, polynuclear hydrocarbon; boils at 181°C and freezes at −2°C; derived from coal tar distillates; copolymers with benzofuran have been manufactured on a small scale for use in coatings and floor coverings. { 'in͵dēn }

independent migration law [ANALYTICAL CHEMISTRY] The law that each ion in a conductiometric titration contributes a definite amount to the total conductance, irrespective of the nature of the other ions in the electrolyte. { ͵in·də'pen·dənt mī'grā·shən ͵lȯ }

index of unsaturation [ORGANIC CHEMISTRY] A numerical value that represents the number of rings or double bonds in a molecule; a triple bond is considered to have the numerical value of 2. { 'in͵deks əv ¦ən͵sach·ə'ra·shən }

indican [ORGANIC CHEMISTRY] $C_{14}H_{17}O_6N$ A glucoside of indoxyl occurring in the indigo plant; on hydrolysis indican gives rise to indoxyl, which is oxidized to indigo by air. { 'in·də͵kan }

indicator [CHEMISTRY] *See* chemical indicator. { 'in·də͵kād·ər }

indigo [ORGANIC CHEMISTRY] **1.** A blue dye extracted from species of the *Indigofera* bush. **2.** *See* indigo blue. { 'in·də·gō }

indigo blue [ORGANIC CHEMISTRY] $C_{16}H_{10}O_2N_2$ A component of the dye indigo, crystallizing as dark-blue rhomboids that break down at 300°C, that are soluble in hot aniline and hot chloroform, and that are also made synthetically; used as a reagent and a dye. Also known as indigo. { 'in·də·gō 'blü }

indigo carmine [ORGANIC CHEMISTRY] $C_{16}H_8N_2Na_2O_8S_2$ A dark blue powder with coppery luster; used as a dye in testing kidney function and as a reagent in detecting chlorate and nitrate. Also known as soluble indigo blue. { 'in·də·gō 'kär·mən }

indigoid dye [ORGANIC CHEMISTRY] Any of the vat dyes with $C_{16}H_{10}O_2N_2$ (indigo) or $C_{16}H_8S_2O_2$ (thioindigo) groupings; used to dye cotton and rayon, sometimes silk. { 'in·də͵gȯid ͵dī }

indigo red [ORGANIC CHEMISTRY] $C_{16}H_{10}O_2N_2$ A red isomer of indigo obtained in the manufacture of indigo. Also known as indirubin. { 'in·də·gō 'red }

indirect effect [PHYSICAL CHEMISTRY] A chemical effect of ionizing radiation on a dilute solution caused by the interaction of solute molecules with highly reactive transient molecules or ions formed by reaction of the radiation with the solvent. { ͵in·də'rekt i'fekt }

indirubin [ORGANIC CHEMISTRY] *See* indigo red. { ͵in·də'rü·bən }

indium [CHEMISTRY] A metallic element, symbol In, atomic number 49, atomic weight 114.82; soluble in acids; melts at 156°C, boils at 1450°C. { 'in·dē·əm }

indium antimonide [INORGANIC CHEMISTRY] InSb Crystals that melt at 535°C; an intermetallic compound having semiconductor properties and the highest room-temperature electron mobility of any known material; used in Hall-effect and magnetoresistive devices and as an infrared detector. { 'in·dē·əm ͵an'tim·ə͵nīd }

indium arsenide [INORGANIC CHEMISTRY] InAs Metallic crystals that melt at 943°C; an intermetallic compound having semiconductor properties; used in Hall-effect devices. { 'in·dē·əm 'ärs·ən͵īd }

indium chloride [INORGANIC CHEMISTRY] $InCl_3$ Hygroscopic white powder, soluble in water and alcohol. { 'in·dē·əm 'klȯr͵īd }

indium phosphide [INORGANIC CHEMISTRY] InP A metallic mass that is brittle and melts at 1070°C; an intermetallic compound having semiconductor properties. { 'in·dē·əm 'fäs͵fīd }

indium sulfate [INORGANIC CHEMISTRY] $In_2(SO_4)_3$ Deliquescent, water-soluble, grayish powder; decomposes when heated. { 'in·dē·əm 'səl͵fāt }

indogen [ORGANIC CHEMISTRY] The functional group $C_6H_4(NH)COC{=}$; it occurs, for example, in the molecule indigo. { 'in·də·jən }

indogenide [ORGANIC CHEMISTRY] A compound containing the function group $C_6H_4(NH) \cdot CO \cdot C=$ from indogen. { 'in·də·jə,nīd }

indole Also known as 2,3-benzopyrrole. [ORGANIC CHEMISTRY] Carcinogenic, white to yellowish scales with unpleasant aroma; soluble in alcohol, ether, hot water, and fixed oils; melt at 52°C; used as a chemical reagent and in perfumery and medicine. { 'in,dōl }

indolebutyric acid [ORGANIC CHEMISTRY] $C_{12}H_{13}O_2N$ A crystalline acid similar to indoleacetic acid in auxin activity. Abbreviated IBA. { ¦in,dōl·byü'tir·ik 'as·əd }

indoxyl [ORGANIC CHEMISTRY] $(C_8H_6N)OH$ A yellow crystalline glycoside, used as an intermediate in the manufacture of indigo. { in'däk·səl }

induction force [PHYSICAL CHEMISTRY] A type of van der Waals force resulting from the interaction of the dipole moment of a polar molecule and the induced dipole moment of a nonpolar molecule. Also known as Debye force. { in'dək·shən ,fórs }

induction period [PHYSICAL CHEMISTRY] A time of acceleration of a chemical reaction from zero to a maximum rate. { in'dək·shən ,pir·ē·əd }

inductive effect [PHYSICAL CHEMISTRY] In a molecule, a shift of electron density due to the polarization of a bond by a nearby electronegative or electropositive atom. { in'dək·tiv ə'fekt }

inductively coupled plasma-atomic emission spectroscopy [SPECTROSCOPY] A type of atomic spectroscopy in which the light emitted by atoms and ions in an inductively coupled plasma is observed. Abbreviated ICP-AES. { in'dək·tiv·lē ¦kəp·əld ¦plaz·mə ə¦täm·ik i¦mish·ən spek'träs·kə·pē }

industrial alcohol [ORGANIC CHEMISTRY] Ethyl alcohol that has been denatured by acetates, ketones, gasoline, or other additives to make it unfit for beverage purposes. { in'dəs·trē·əl 'al·kə,hól }

inert gas [CHEMISTRY] *See* noble gas. { i'nərt 'gas }

inflammability [CHEMISTRY] *See* flammability. { in,flam·ə'bil·əd·ē }

infrared spectrometer [SPECTROSCOPY] Device used to identify and measure the concentrations of heteroatomic compounds in gases, in many nonaqueous liquids, and in some solids by arc or spark excitation and subsequent measurement of the electromagnetic emissions in the wavelength range of 0.78 to 300 micrometers. { ¦in·frə¦red spek'träm·əd·ər }

infrared spectrophotometry [SPECTROSCOPY] Spectrophotometry in the infrared region, usually for the purpose of chemical analysis through measurement of absorption spectra associated with rotational and vibrational energy levels of molecules. { ¦in·frə¦red ¦spek·trə·fə'täm·ə·trē }

infrared spectroscopy [SPECTROSCOPY] The study of the properties of material systems by means of their interaction with infrared radiation; ordinarily the radiation is dispersed into a spectrum after passing through the material. { ¦in·frə¦red spek'träs·kə·pē }

infusion [CHEMISTRY] The aqueous solution of a soluble constituent of a substance as the result of the substance's steeping in the solvent for a period of time. { in 'fyü·zhən }

ingrain color [ORGANIC CHEMISTRY] *See* azoic dye. { 'in,grān ,kəl·ər }

inhibitor [CHEMISTRY] A substance which is capable of stopping or retarding a chemical reaction; to be technically useful, it must be effective in low concentration. { in 'hib·əd·ər }

initiation step [CHEMISTRY] The reaction that causes a chain reaction to begin but is not itself the principal source of products. { i,nish·ē'ā·shən ,step }

initiator [CHEMISTRY] The substance or molecule (other than reactant) that initiates a chain reaction, as in polymerization; an example is acetyl peroxide. { i'nish·ē ,ād·ər }

inorganic [INORGANIC CHEMISTRY] Pertaining to or composed of chemical compounds that do not contain carbon as the principal element (excepting carbonates, cyanides, and cyanates), that is, matter other than plant or animal. { ¦in·ór¦gan·ik }

inorganic acid [INORGANIC CHEMISTRY] A compound composed of hydrogen and a nonmetal element or radical; examples are hydrochloric acid, HCl, sulfuric acid, H_2SO_4, and carbonic acid, H_2CO_3. { ¦in·ór¦gan·ik 'as·əd }

inorganic chemistry [CHEMISTRY] The study of chemical reactions and properties of all the elements and their compounds, with the exception of hydrocarbons, and usually including carbides, oxides of carbon, metallic carbonates, carbon-sulfur compounds, and carbon-nitrogen compounds. { ¦in·ȯr¦gan·ik ′kem·ə·strē }

inorganic peroxide [INORGANIC CHEMISTRY] An inorganic compound containing an element at its highest state of oxidation (such as perchloric acid, $HClO_4$), or having the peroxy group, —O—O— (such as perchromic acid, $H_3CrO_8 \cdot 2H_2O$). { ¦in·ȯr¦gan·ik pə′räk‚sīd }

inorganic pigment [INORGANIC CHEMISTRY] A natural or synthetic metal oxide, sulfide, or other salt used as a coloring agent for paints, plastics, and inks. { ¦in·ȯr¦gan·ik ′pig·mənt }

inorganic polymer [INORGANIC CHEMISTRY] Large molecules, usually linear or branched chains with atoms other than carbon in their backbone; an example is glass, an inorganic polymer made up of rings and chains of repeating silicate units. { ¦in·ȯr¦gan·ik ′päl·ə·mər }

inositol [ORGANIC CHEMISTRY] $C_6H_6(OH)_6 \cdot 2H_2O$ A water-soluble alcohol often grouped with the vitamins; there are nine stereoisomers of hexahydroxycyclohexane, and the only one of biological importance is optically inactive *meso*-inositol, comprising white crystals, widely distributed in animals and plants; it serves as a growth factor for animals and microorganisms. { i′näs·ə‚tȯl }

insol [CHEMISTRY] *See* insoluble. { ′in‚säl }

insoluble [CHEMISTRY] Incapable of being dissolved in another material; usually refers to solid-liquid or liquid-liquid systems. Abbreviated insol. { in′säl·yə·bəl }

insoluble anode [CHEMISTRY] An anode that resists dissolution during electrolysis. { in′säl·yə·bəl ′an‚ōd }

inspissation [CHEMISTRY] The process of thickening a liquid by evaporation. { ‚in·spi′sā·shən }

integral heat of dilution [PHYSICAL CHEMISTRY] *See* heat of dilution. { ′int·ə·grəl ¦hēt əv də′lü·shən }

integral heat of solution [PHYSICAL CHEMISTRY] *See* heat of solution. { ′int·ə·grəl ¦hed əv sə′lü·shən }

integral procedure decomposition temperature [PHYSICAL CHEMISTRY] Decomposition temperatures derived from graphical integration of the thermogravimetric analysis of a polymer. { ′int·ə·grəl prə¦sē·jər dē‚käm·pə′zish·ən ‚tem·prə·chər }

intensive properties [CHEMISTRY] Properties independent of the quantity or shape of the substance under consideration; for example, temperature, pressure, or composition. { in′ten·siv ′präp·ərd·ēz }

intercalibration [ANALYTICAL CHEMISTRY] A state achieved by a group of laboratories engaged in a monitoring program in which they produce and maintain compatible data outputs. { ‚in·tər‚kal·ə′brā·shən }

interdiffusion [PHYSICAL CHEMISTRY] The self-mixing of two fluids, initially separated by a diaphragm. { ¦in·tər·də′fyü·zhən }

interface [PHYSICAL CHEMISTRY] The boundary between any two phases; among the three phases (gas, liquid, and solid), there are five types of interfaces: gas-liquid, gas-solid, liquid-liquid, liquid-solid, and solid-solid. { ′in·tər‚fās }

interface mixing [PHYSICAL CHEMISTRY] The mixing of two immiscible or partially miscible liquids at the plane of contact (interface). { ′in·tər‚fās ¦mik·siŋ }

interference [ANALYTICAL CHEMISTRY] A systematic error in measurement that occurs when concomitants are present in the sample being analyzed. { ‚in·tər′fir·əns }

interference spectrum [SPECTROSCOPY] A spectrum that results from interference of light, as in a very thin film. { ‚in·tər′fir·əns ¦spek·trəm }

interferogram [SPECTROSCOPY] A graph of the variation of the output signal from an interferometer as the condition for interference within the interferometer is varied. { ‚in·tə′fir·ə‚gram }

interhalogen [INORGANIC CHEMISTRY] Any of the compounds formed from the elements of the halogen family that react with each other to form a series of binary compounds; for example, iodine monofluoride. { ¦in·tər′hal·ə·jən }

interionic attraction [PHYSICAL CHEMISTRY] The coulomb attraction between ions of opposite sign in a solution. { ˌin·tir·ē'än·ik ə¦trak·shən }

intermediate [CHEMISTRY] A precursor to a desired product; ethylene is an intermediate for polyethylene, and ethane is an intermediate for ethylene. { ˌin·tər'mēd·ē·ət }

intermolecular force [PHYSICAL CHEMISTRY] The force between two molecules; it is that negative gradient of the potential energy between the interacting molecules, if energy is a function of the distance between the centers of the molecules. { ˌin·tər·mə'lek·yə·lər 'fōrs }

internal phase [CHEMISTRY] *See* disperse phase. { in'tərn·əl ¦fāz }

internal reflectance spectroscopy [SPECTROSCOPY] *See* attenuated total reflectance. { in'tərn·əl ri¦flek·təns spek'träs·kə·pē }

internal standard [SPECTROSCOPY] The principal line in spectrum analysis by the logarithmic sector method, a quantitative spectroscopy procedure. { in'tərn·əl 'stan·dərd }

internuclear distance [PHYSICAL CHEMISTRY] The distance between two nuclei in a molecule. { ¦in·tər¦nü·klē·ər ˌdis·təns }

interphase [CHEMISTRY] A region between the two phases of a newly created interface that contains particles of both phases. { 'in·tərˌfāz }

interpolymer [ORGANIC CHEMISTRY] A mixed polymer made from two or more starting materials. { ¦in·tər'päl·ə·mər }

interstitial compound [CHEMISTRY] A compound of a transition metal and hydrogen, boron, carbon, or nitrogen whose crystals have a close-packed structure of the metal ions, with the nonmetal atoms being located in the interstices. { ¦in·tər¦stish·əl 'kämˌpau̇nd }

intimate ion pair [ORGANIC CHEMISTRY] *See* tight ion pair. { 'in·tə·mət 'īˌän ˌper }

intracavity absorption spectroscopy [SPECTROSCOPY] A highly sensitive technique in which an absorbing sample is placed inside the resonator of a broad-band dye laser, and absorption lines are detected as dips in the laser emission spectrum. { ¦in·trə'kav·əd·ē əb¦sȯrp·shən spek'träs·kə·pē }

intrinsic viscosity [PHYSICAL CHEMISTRY] The ratio of a solution's specific viscosity to the concentration of the solute, extrapolated to zero concentration. Also known as limiting viscosity number. { in'trin·sik vi'skäs·əd·ē }

introfaction [CHEMISTRY] Change in fluidity and specific wetting properties (for impregnation acceleration) of an impregnating compound, caused by an introfier (impregnation accelerator). { ¦in·trə¦fak·shən }

inverse micelle [PHYSICAL CHEMISTRY] *See* inverted micelle. { 'inˌvərs mī'sel }

inverse Stark effect [SPECTROSCOPY] The Stark effect as observed with absorption lines, in contrast to emission lines. { 'inˌvərs 'stärk iˌfekt }

inverse Zeeman effect [SPECTROSCOPY] A splitting of the absorption lines of atoms or molecules in a static magnetic field; it is the Zeeman effect observed with absorption lines. { 'inˌvərs 'zē·mən iˌfekt }

inversion [CHEMISTRY] Change of a compound into an isomeric form. { in'vər·zhən }

inversion spectrum [SPECTROSCOPY] Lines in the microwave spectra of certain molecules (such as ammonia) which result from the quantum-mechanical analog of an oscillation of the molecule between two configurations which are mirror images of each other. { in'vər·zhən ˌspek·trəm }

inverted micelle [PHYSICAL CHEMISTRY] An aggregate of colloidal dimension in which the polar groups are concentrated in the interior and the lipophilic groups extend outward into the solvent. Also known as inverse micelle. { in¦vərd·əd mī'sel }

iodate [INORGANIC CHEMISTRY] A salt of iodic acid containing the IO_3^- radical; sodium and potassium iodates are the most important salts and are used in medicine. { 'ī·əˌdāt }

iodic acid [INORGANIC CHEMISTRY] HIO_3 Water-soluble, moderately strong acid; colorless or white powder or crystals; decomposes at 110°C; used in analytical chemistry and medicine. { ī'äd·ik 'as·əd }

iodic acid anhydride [INORGANIC CHEMISTRY] *See* iodine pentoxide. { ī'äd·ik 'as·əd an'hīˌdrīd }

iodide [CHEMISTRY] **1.** A compound which contains the iodine atom in the −1 oxidation state and which may be considered to be derived from hydriodic acid (HI); examples are KI and NaI. **2.** A compound of iodine, such as CH_3CH_2I, in which the iodine has combined with a more electropositive group. { ī·ə,dīd }

iodine [CHEMISTRY] A nonmetallic halogen element, symbol I, atomic number 53, atomic weight 126.9044; melts at 114°C, boils at 184°C; the poisonous, corrosive, dark plates or granules are readily sublimed; insoluble in water, soluble in common solvents; used as germicide and antiseptic, in dyes, tinctures, and pharmaceuticals, in engraving lithography, and as a catalyst and analytical reagent. { 'ī·ə,dīn }

iodine bisulfide [INORGANIC CHEMISTRY] *See* sulfur iodine. { 'ī·ə,dīn bī'səl,fīd }

iodine cyanide [INORGANIC CHEMISTRY] ICN Poisonous, colorless needles with pungent aroma and acrid taste; melts at 147°C; soluble in water, alcohol, and ether; used in taxidermy as a preservative. Also known as cyanogen iodide. { 'ī·ə,dīn 'sī·ə,nīd }

iodine disulfide [INORGANIC CHEMISTRY] *See* sulfur iodine. { 'ī·ə,dīn dī'səl,fīd }

iodine number [ANALYTICAL CHEMISTRY] A measure of the iodine absorbed in a given time by a chemically unsaturated material, such as a vegetable oil or a rubber; used to measure the unsaturation of a compound or mixture. Also known as iodine value. { 'ī·ə,dīn ,nəm·bər }

iodine pentoxide [INORGANIC CHEMISTRY] I_2O_5 White crystals, decomposing at 275°C, very soluble in water, insoluble in absolute alcohol, ether, and chloroform; used as an oxidizing agent to oxidize carbon monoxide to dioxide at ordinary temperatures, and in organic synthesis. Also known as iodic acid anhydride. { 'ī·ə,dīn pen'täk,sīd }

iodine test [ANALYTICAL CHEMISTRY] Placing a few drops of potassium iodide solution on a sample to detect the presence of starch; test is positive if sample turns blue. { 'ī·ə,dīn ,test }

iodine value [ANALYTICAL CHEMISTRY] *See* iodine number. { 'ī·ə,dīn ,val·yü }

iodoacetic acid [ORGANIC CHEMISTRY] CH_2ICOOH White or colorless crystals that are soluble in water and alcohol, and melt at 82-83°C; used in biological research for its inhibitive effect on enzymes. { ī¦ō·dō·ə¦sēd·ik 'as·əd }

iodoalkane [ORGANIC CHEMISTRY] An alkane hydrocarbon in which an iodine atom replaces one or more hydrogen atoms in the molecule; an example is iodomethane, CH_3I, better known as methyl iodide. { ī¦ō·dō·al'kān }

iodcyanin [ORGANIC CHEMISTRY] *See* cyanine dye. { ¦ī·əd¦sī·ə·nən }

iodoeasin [ORGANIC CHEMISTRY] *See* easin. { ī¦ō·dō'ē·ə·sən }

iodoethane [ORGANIC CHEMISTRY] *See* ethyl iodide. { ī¦ō·dō'eth,ān }

iodoethylene [ORGANIC CHEMISTRY] *See* tetraiodoethylene. { ī¦ō·dō'eth·ə,lēn }

iodoform [ORGANIC CHEMISTRY] CHI_3 A yellow, hexagonal solid; melting point 119°C; soluble in chloroform, ether, and water, has weak bactericidal qualities and is used in ointments for minor skin diseases. Also known as triiodomethane. { ī'ō·də,fōrm }

iodohydrocarbon [ORGANIC CHEMISTRY] A hydrocarbon in which an iodine atom replaces one or more hydrogen atoms in the molecule, as in an alkane, aromatic, or olefin. { ī¦ō·də,hī·drə'kär·bən }

iodomethane [ORGANIC CHEMISTRY] *See* methyl iodide. { ī¦ō·də'meth,ān }

iodometry [ANALYTICAL CHEMISTRY] An application of iodine chemistry to oxidation-reduction titrations for the quantitative analysis in certain chemical compounds, in which iodine is used as a reductant and the iodine freed in the associated reaction is titrated, usually in neutral or slightly acid mediums with a standard solution of a reductant such as sodium thiosulfate or sodium arsenite; examples of chemicals analyzed are copper(III), gold(VI), arsenic(V), antimony(V), chlorine, and bromine. { ,ī·ə'däm·ə·trē }

iodophor [CHEMISTRY] Any compound that is a carrier of iodine. { ī'äd·ə,fōr }

iodosobenzene [ORGANIC CHEMISTRY] C_6H_5IO A yellowish-white amorphous solid that explodes at 200°C, soluble in hot water and alcohol; a strong oxidizing agent. { ,ī·ə'dō·sō'ben,zēn }

iodoxybenzene [ORGANIC CHEMISTRY] $C_6H_5IO_2$ Clear white crystals that explode at

227-228°C, slightly soluble in water, insoluble in chloroform, acetone, and benzene; a strong oxidizing agent. { ¦ī·ə¦däk·sē′ben‚zēn }

ion [CHEMISTRY] An isolated electron or positron or an atom or molecule which by loss or gain of one or more electrons has acquired a net electric charge. { ′ī‚än }

ion atmosphere [PHYSICAL CHEMISTRY] *See* ion cloud. { ′ī‚än ′at·mə‚sfir }

ion cloud [PHYSICAL CHEMISTRY] A slight preponderance of negative ions around a positive ion in an electrolyte, and vice versa, according to the Debye-Hückel theory. Also known as ion atmosphere. { ′ī‚än ‚klau̇d }

ion-cyclotron-resonance mass spectrometer [SPECTROSCOPY] A device for detecting and measuring the mass distribution of ions orbiting in an applied magnetic field, either by applying a constant radio-frequency signal and varying the magnetic field to bring ion frequencies equal to the applied radio frequency sequentially into resonance, or by rapidly varying the radio frequency and applying Fourier transform techniques. { ′ī‚än ′sī·klə‚trän ′rez·ən·əns ′mas spek′träm·əd·ər }

ion detector [ANALYTICAL CHEMISTRY] Device for detection of presence or concentration of liquid solution ions, such as with a pH meter or by conductimetric techniques. { ′ī‚än di‚tek·tər }

ion exchange [PHYSICAL CHEMISTRY] A chemical reaction in which mobile hydrated ions of a solid are exchanged, equivalent for equivalent, for ions of like charge in solution; the solid has an open, fishnetlike structure, and the mobile ions neutralize the charged, or potentially charged, groups attached to the solid matrix; the solid matrix is termed the ion exchanger. { ′ī‚än iks‚chānj }

ion-exchange chromatography [ANALYTICAL CHEMISTRY] A chromatographic procedure in which the stationary phase consists of ion-exchange resins which may be acidic or basic. { ′ī‚än iks‚chānj ‚krō·mə′täg·rə·fē }

ion exchanger [PHYSICAL CHEMISTRY] A solid or liquid material containing ions that are exchangeable with other ions with a like charge that are present in a solution in which the material is insoluble. { ′ī‚än iks‚chānj·ər }

ion exclusion [CHEMISTRY] Ion-exchange resin system in which the mobile ions in the resin-gel phase electrically neutralize the immobilized charged functional groups attached to the resin, thus preventing penetration of solvent electrolyte into the resin-gel phase; used in separations where electrolyte is to be excluded from the resin, but not nonpolar materials, as the separation of salt from nonpolar glycerin. { ′ī‚än iks‚klü·zhən }

ion-exclusion chromatography [ANALYTICAL CHEMISTRY] Chromatography in which the adsorbent material is saturated with the same mobile ions (cationic or anionic) as are present in the sample-carrying eluent (solvent), thus repelling the similar sample ions. { ′ī‚än iks‚klü·zhən ‚krō·mə′täg·rə·fē }

ionic bond [PHYSICAL CHEMISTRY] A type of chemical bonding in which one or more electrons are transferred completely from one atom to another, thus converting the neutral atoms into electrically charged ions; these ions are approximately spherical and attract one another because of their opposite charge. Also known as electrovalent bond. { ī′än·ik ′bänd }

ionic conductance [PHYSICAL CHEMISTRY] The contribution of a given type of ion to the total equivalent conductance in the limit of infinite dilution. { ī′än·ik kən′dək·təns }

ionic dissociation [PHYSICAL CHEMISTRY] Dissociation that results in the production of ions. { i′än·ik di‚sō·sē′ā·shən }

ionic equilibrium [PHYSICAL CHEMISTRY] The condition in which the rate of dissociation of nonionized molecules is equal to the rate of combination of the ions. { ī′än·ik ‚ē·kwə′lib·rē·əm }

ionic equivalent conductance [PHYSICAL CHEMISTRY] The contribution made by each ion species of a salt toward an electrolyte's equiviconductance. { ī′än·ik i¦kwiv·ə·lənt kən′dək·təns }

ionic gel [CHEMISTRY] A gel with ionic groups attached to the structure of the gel; the groups cannot diffuse out into the surrounding solution. { ī′än·ik ′jel }

ionicity [CHEMISTRY] The ionic character of a solid. { ‚ī·ə′nis·əd·ē }

ionic polymerization [ORGANIC CHEMISTRY] Polymerization that proceeds via ionic intermediates (carbonium ions or carbanions) rather than through neutral species (olefins or acetylenes). { ī′än·ik pə′lim·ə·rə′zā·shən }

ionic radii [PHYSICAL CHEMISTRY] Radii which can be assigned to ions because the rapid variation of their repulsive interaction with distance makes them repel like hard spheres; these radii determine the dimensions of ionic crystals. { ī′än·ik ′rād·ē‚ī }

ionic strength [PHYSICAL CHEMISTRY] A measure of the average electrostatic interactions among ions in an electrolyte; it is equal to one-half the sum of the terms obtained by multiplying the molality of each ion by its valence squared. { ī′än·ik ′streŋkth }

ionization [CHEMISTRY] A process by which a neutral atom or molecule loses or gains electrons, thereby acquiring a net charge and becoming an ion; occurs as the result of the dissociation of the atoms of a molecule in solution ($NaCl \rightarrow Na^+ + Cl^-$) or of a gas in an electric field ($H_2 \rightarrow 2H^+$). { ‚ī·ə·nə′zā·shən }

ionization constant [PHYSICAL CHEMISTRY] Analog of the dissociation constant, where $k = [H^+][A^-]/[HA]$; used for the application of the law of mass action to ionization; in the equation HA represents the acid, such as acetic acid. { ‚ī·ə·nə′zā·shən ¦kän·stənt }

ionization degree [PHYSICAL CHEMISTRY] The proportion of potential ionization that has taken place for an ionizable material in a solution or reaction mixture. { ‚ī·ə·nə′zā·shən di‚grē }

ionization isomer [CHEMISTRY] One of two or more compounds that have identical molecular formulas but different ionic forms. { ‚ī·ə·nə′zā·shən ′ī·sə·mər }

ionized atom [CHEMISTRY] An atom with an excess or deficiency of electrons, so that it has a net charge. { ′ī·ə‚nīzd ′ad·əm }

ion kinetic energy spectrometry [SPECTROSCOPY] A spectrometric technique that uses a beam of ions of high kinetic energy passing through a field-free reaction chamber from which ionic products are collected and energy analyzed; it is a generalization of metastable ion studies in which both unimolecular and bimolecular reactions are considered. { ′ī‚än ki¦ned·ik ′en·ər·jē spek′träm·ə·trē }

ion mean life [PHYSICAL CHEMISTRY] The average time between the ionization of an atom or molecule and its recombination with one or more electrons, or its loss of excess electrons. { ′ī‚än ¦mēn ′līf }

ionogenic group [PHYSICAL CHEMISTRY] A fixed group of atoms in an ion exchanger that is either ionized or capable of dissociation into fixed ions and mobile counterions. { ‚ī·ə·nə′jen·ik ‚grup }

ionography [ANALYTICAL CHEMISTRY] A type of electrochromatography involving migration of ions. { ‚ī·ə′näg·rə·fē }

ionomer [ORGANIC CHEMISTRY] Polymer with covalent bonds between the elements of the chain, and ionic bonds between the chains. { ī′än·ə·mər }

ionomer resin [ORGANIC CHEMISTRY] A polymer which has ethylene as the major component, but which contains both covalent and ionic bonds. { ī′än·ə·mər ′rez·ən }

ionone [ORGANIC CHEMISTRY] $C_{13}H_{20}O$ A colorless to light yellow liquid with a boiling point of 126-128°C at 12 mmHg (1600 pascals); soluble in alcohol, ether, and mineral oil; used in perfumery, flavoring, and vitamin A production. Also known as irisone. { ′ī·ə‚nōn }

ion scattering spectroscopy [SPECTROSCOPY] A spectroscopic technique in which a low-energy (about 1000 electronvolts) beam of inert-gas ions in directed at a surface, and the energies and scattering angles of the scattered ions are used to identify surface atoms. Abbreviated ISS. { ¦ī‚än ¦skad·ə·riŋ spek′träs·kə·pē }

ioxynil [ORGANIC CHEMISTRY] $C_7H_3I_2NO$ A colorless solid with a melting point of 212-213°C; used for postemergence control of seedling weeds in cereals and sports turf. { ī′äk·sə‚nil }

ioxynil octanoate [ORGANIC CHEMISTRY] $C_{15}H_{17}I_2NO_2$ A waxy solid with a melting point of 59-60°C; insoluble in water; used as an insecticide for cereals and sugarcane. { ī′äk·sə‚nil ‚äk·tə ′nō·ət }

IPC [ORGANIC CHEMISTRY] *See* propham.

Ir [CHEMISTRY] *See* iridium.

iridescent layer [PHYSICAL CHEMISTRY] *See* schiller layer. { ‚i·ri′des·ənt ‚lā·ər }

iridic chloride [INORGANIC CHEMISTRY] $IrCl_4$ A hygroscopic brownish-black mass, soluble in water and alcohol; used to analyze for nitric acid, HNO_3, and in analytical microscopic work. Also known as iridium chloride; iridium tetrachloride. { i'rid·ik 'klȯr,īd }

iridium [CHEMISTRY] A metallic element, symbol Ir, atomic number 77, atomic weight 192.2, in the platinum group; insoluble in acids, melting at 2454°C. { i'rid·ē·əm }

iridium chloride [INORGANIC CHEMISTRY] *See* iridic chloride. { i'rid·ē·əm 'klȯr,īd }

iridium tetrachloride [INORGANIC CHEMISTRY] *See* iridic chloride. { i'rid·ē·əm ,te·trə'klȯr,īd }

irisone [ORGANIC CHEMISTRY] *See* ionone. { 'ī·rə,sōn }

iron [CHEMISTRY] A silvery-white metallic element, symbol Fe, atomic number 26, atomic weight 55.847, melting at 1530°C. { 'ī·ərn }

iron acetate [ORGANIC CHEMISTRY] *See* ferrous acetate. { 'ī·ərn 'as·ə,tāt }

iron ammonium sulfate [INORGANIC CHEMISTRY] *See* ferric ammonium sulfate; ferrous ammonium sulfate. { 'ī·ərn ə'mō·nē·əm 'səl,fāt }

iron arsenate [INORGANIC CHEMISTRY] *See* ferrous arsenate. { 'ī·ərn 'ärs·ən,āt }

iron black [CHEMISTRY] Fine black antimony powder used to give a polished-steel look to papier-maché and plaster of paris; made by reaction of zinc with acid solution of an antimony salt and precipitation of black antimony powder. { 'ī·ərn 'blak }

iron blue [INORGANIC CHEMISTRY] Ferric ferrocyanide used as blue pigment by the paint industry for permanent body and trim paints; also used in blue ink, in paper dyeing, and as a fertilizer ingredient. { 'ī·ərn 'blü }

iron bromide [INORGANIC CHEMISTRY] *See* ferric bromide. { 'ī·ərn 'brō,mīd }

iron carbonyl [INORGANIC CHEMISTRY] *See* iron pentacarbonyl. { 'ī·ərn 'kär·bə,nil }

iron chloride [INORGANIC CHEMISTRY] *See* ferric chloride; ferrous chloride. { 'ī·ərn 'klȯr,īd }

iron citrate [ORGANIC CHEMISTRY] *See* ferric citrate. { 'ī·ərn 'sī,trāt }

iron dichloride [INORGANIC CHEMISTRY] *See* ferrous chloride. { 'ī·ərn dī'klȯr,īd }

irone [ORGANIC CHEMISTRY] $C_{14}H_{22}O$ A colorless liquid terpene; a component of essential oil from the orrisroot; used in perfumes. { 'ī,rōn }

iron ferrocyanide [INORGANIC CHEMISTRY] *See* ferric ferrocyanide. { 'ī·ərn ,fer·ə'sī·ə,nīd }

iron fluoride [INORGANIC CHEMISTRY] *See* ferric fluoride. { 'ī·ərn 'flu̇r,īd }

iron hydroxide [INORGANIC CHEMISTRY] *See* ferric hydroxide. { 'ī·ərn hī'dräk,sīd }

iron metavanadate [INORGANIC CHEMISTRY] *See* ferric vanadate. { 'ī·ərn ,med·ə'van·ə,dāt }

iron monoxide [INORGANIC CHEMISTRY] *See* ferrous oxide. { 'ī·ərn mə'näk,sīd }

iron nitrate [INORGANIC CHEMISTRY] *See* ferric nitrate. { 'ī·ərn 'nī,trāt }

iron nonacarbonyl [INORGANIC CHEMISTRY] $Fe_2(CO)_9$ Orange-yellow crystals that break down at 100°C to yield tetracarbonyl, slightly soluble in alcohol and acetone, almost insoluble in water, ether, and benzene. { 'ī·ərn ,nō·nə'kär·bə,nil }

iron oxalate [ORGANIC CHEMISTRY] *See* ferrous oxalate. { 'ī·ərn 'äk·sə,lāt }

iron oxide [INORGANIC CHEMISTRY] Any of the hydrated, synthetic, or natural oxides of iron: ferrous oxide, ferric oxide, ferriferous oxide. { 'ī·ərn 'äk,sīd }

iron pentacarbonyl [INORGANIC CHEMISTRY] $Fe(CO)_5$ An oily liquid that decomposes upon exposure to light, soluble in most organic solvents; used as a source of a pure iron catalyst and for magnet cores. Also known as iron carbonyl. { 'ī·ərn ,pen·tə'kär·bə,nil }

iron phosphate [INORGANIC CHEMISTRY] *See* ferric phosphate. { 'ī·ərn 'fäs,fāt }

iron resinate [ORGANIC CHEMISTRY] *See* ferric resinate. { 'ī·ərn 'rez·ən,āt }

iron stearate [ORGANIC CHEMISTRY] *See* ferric stearate. { 'ī·ərn 'stir,āt }

iron sulfate [INORGANIC CHEMISTRY] *See* ferric sulfate; ferrous sulfate. { 'ī·ərn 'səl ,fāt }

iron sulfide [INORGANIC CHEMISTRY] *See* ferrous sulfide. { 'ī·ərn 'səl,fīd }

iron tetracarbonyl [INORGANIC CHEMISTRY] $Fe_3(CO)_{12}$ Dark-green lustrous crystals that break down at 140-150°C; soluble in organic solvents. Also known as tri-iron dodecacarbonyl. { 'ī·ərn ,te·trə'kär·bə,nil }

irregular polymer [CHEMISTRY] A polymer whose molecular structure does not consist of only one species of constitutional unit in a single sequential arrangement. { i¦reg·yə·lər 'päl·i·mər }

isatin [ORGANIC CHEMISTRY] $C_6H_5NO_2$ An indole substituted with oxygen at carbon position 2 and 3; crystallizes as red needles that are soluble in hot water; used in dye manufacture. { 'ī·sə·tən }

isethionic acid [ORGANIC CHEMISTRY] $CH_2OH \cdot CH \cdot SO_2OH$ A water-soluble liquid, boiling at 100°C; used in the manufacture of detergents. { ¦īs·ə·thī¦än·ik 'as·əd }

iso- [CHEMISTRY] A prefix indicating an isomer of an element in which there is a difference in the nucleus when compared to the most prevalent form of the element. [ORGANIC CHEMISTRY] A prefix indicating a single branching at the end of the carbon chain. { 'ī·sō }

isoactyl thioglycolate [ORGANIC CHEMISTRY] $HSCH_2COOCH_2C_7H_{15}$ A colorless liquid with a slight fruity odor and a boiling point of 125°C; used in antioxidants, insecticides, oil additives, and plasticizers. { ¦ī·sō'akt·əl ˌthī·ə'glī·kəˌlāt }

isoalkane [ORGANIC CHEMISTRY] An alkane with a branched chain whose next-to-last carbon atom is bonded to a single methyl group. { ¦ī·sō'alˌkān }

isoalkyl group [ORGANIC CHEMISTRY] A group of atoms resulting from the removal of a hydrogen atom from a methyl group situated at the end of the straight-chain segment of an isoalkane. { ¦ī·sō'al·kəl ˌgrüp }

isoamyl acetate [ORGANIC CHEMISTRY] *See* amyl acetate. { ¦ī·sō¦am·əl 'as·əˌtāt }

isoamyl alcohol [ORGANIC CHEMISTRY] *See* isobutyl carbinol. { ¦ī·sō¦am·əl 'al·kə ˌhȯl }

isoamyl benzoate [ORGANIC CHEMISTRY] $C_6H_5COOC_5H_{11}$ Colorless liquid with fruity aroma, boils at 260°C, soluble in alcohol, insoluble in water, used in flavors and perfumes. Also known as amyl benzoate. { ¦ī·sō¦am·əl 'ben·zəˌwāt }

isoamyl bromide [ORGANIC CHEMISTRY] $(CH_3)_2CHCH_2CH_2Br$ A colorless liquid with a boiling point of 120-121°C; miscible with alcohol and with ether; used in organic synthesis. { ¦ī·sō¦am·əl 'brōˌmīd }

isoamyl butyrate [ORGANIC CHEMISTRY] $C_5H_{11}COOC_3H_7$ A water-white liquid boiling at 150-180°C; soluble in alcohol and ether; used as a solvent and plasticizer for cellulose acetate and in flavor extracts. { ¦ī·sō¦am·əl 'byüd·əˌrāt }

isoamyl chloride [ORGANIC CHEMISTRY] $C_5H_{11}Cl$ Water-insoluble, colorless liquid boiling at 100°C; it can be any one of several compounds, such as 1-chloro-3-methylbutane, $(CH_3)_2CH(CH_2)_2Cl$, or mixtures thereof; used as a solvent, in inks, for soil fumigation, and as a chemical intermediate. { ¦ī·sō¦am·əl 'klȯrˌīd }

isoamyl nitrite [ORGANIC CHEMISTRY] *See* amyl nitrite. { ¦ī·sō¦am·əl 'nīˌtrīt }

isoamyl salicylate [ORGANIC CHEMISTRY] *See* amyl salicylate. { ¦ī·sō¦am·əl sə'lis·əˌlāt }

isoamyl valerate [ORGANIC CHEMISTRY] $C_4H_9CO_2C_5H_{11}$ Clear liquid with apple aroma; boils at 204°C; soluble in alcohol and ether, insoluble in water; used in medicine and fruit flavors. { ¦ī·sō¦am·əl 'val·əˌrāt }

isobornyl acetate [ORGANIC CHEMISTRY] $C_{10}H_{17}OOCCH_3$ A colorless liquid with an odor of pine needles and a boiling point of 220-224°C; soluble in fixed oils and mineral oil; used in toiletries and soaps and antiseptics, and as a flavoring agent. { ¦ī·sə'bȯrn·əl 'as·əˌtāt }

isobornyl thiocyanoacetate [ORGANIC CHEMISTRY] $C_{10}H_{17}OOCCH_2SCN$ An oily, yellow liquid; soluble in alcohol, benzene, chloroform, and ether; used in medicine and as an insecticide. { ¦ī·sə'bȯrn·əl ¦thī·ə¦sī·ə·nō'as·əˌtāt }

isobutane [ORGANIC CHEMISTRY] $(CH_3)_2CHCH_3$ A colorless, stable gas, noncorrosive to metals, nonreactive with water; boils at −11.7°C; used as a chemical intermediate, refrigerant, and fuel. { ¦ī·sō'byüˌtān }

isobutanol [ORGANIC CHEMISTRY] *See* isobutyl alcohol. { ¦ī·sō'byüt·ənˌȯl }

isobutene [ORGANIC CHEMISTRY] *See* isobutylene. { ¦ī·sō'byüˌtēn }

isobutyl [ORGANIC CHEMISTRY] The radical $(CH_3)_2CHCH_2$—, occurring, for example, in isobutanol (isobutyl alcohol), $(CH_3)_2CHCH_2OH$. { ¦ī·sō'byüd·əl }

isobutyl acetate [ORGANIC CHEMISTRY] $C_4H_9OOCCH_3$ Colorless liquid with fruitlike aroma; soluble in alcohols, ether, and hydrocarbons, insoluble in water; boils at 116°C; used as a solvent for lacquer and nitrocellulose. { ¦ī·sō'byüd·əl 'as·əˌtāt }

isobutyl alcohol [ORGANIC CHEMISTRY] $(CH_3)_2CHCH_2OH$ A colorless liquid that is a by-product of the synthetic production of methanol, boils at 107°C; soluble in water, ether, and alcohol; used as a solvent in paints and lacquers, in organic synthesis, and in resin coatings. Also known as isobutanol; isopropylcarbinol; 2-methyl-l-propanol. { ¦ī·sō'byüd·əl 'al·kə‚hȯl }

isobutyl aldehyde [ORGANIC CHEMISTRY] $(CH_3)_2CHCHO$ Colorless, transparent liquid with pungent aroma; soluble in alcohol, insoluble in water; boils at 64°C; used as a chemical intermediate. Also known as isobutyraldehyde. { ¦ī·sō'byüd·əl 'al·də‚hīd }

isobutyl carbinol [ORGANIC CHEMISTRY] $(CH_3)_2CH(CH_2)_2OH$ Colorless liquid with pungent taste and disagreeable aroma; soluble in alcohol and ether, slightly soluble in water; boils at 132°C; used as a chemical intermediate and solvent, and in pharmaceutical products and medicines. Also known as isoamyl alcohol. { ¦ī·sō'byüd·əl 'kär·bə‚nȯl }

isobutylene [ORGANIC CHEMISTRY] $(CH_3)_2CCH_2$ Flammable, colorless, volatile liquid boiling at −7°C; easily polymerized; used in gasolines, as a chemical intermediate, and to make butyl rubber. Also known as isobutene. { ¦ī·sō'byüd·əl‚ēn }

isobutyl isobutyrate [ORGANIC CHEMISTRY] $(CH_3)_2CHCOOCH_2CH(CH_3)_2$ A colorless liquid with a fruity odor and a boiling point of 148.7°C; soluble in alcohol and ether; used for flavoring and as an insect repellent. Abbreviated IBIB. { ¦ī·sō'byüd·əl ¦ī·sō'byüd·ə‚rāt }

isobutyraldehyde [ORGANIC CHEMISTRY] *See* isobutyl aldehyde. { ¦ī·sō‚byüd·ə'ral·də‚hīd }

isobutyric acid [ORGANIC CHEMISTRY] $(CH_3)_2CHCOOH$ Colorless liquid boiling at 154°C; soluble in water, alcohol, and ether; used as a chemical intermediate and disinfectant, in flavor and perfume bases, and for leather treating. { ¦ī·sō·byü'tir·ik 'as·əd }

isobutyryl [ORGANIC CHEMISTRY] $(CH_3)_2C \cdot CHO$ The radical group from isobutyric acid, $(CH_3)_2CHCOOH$. { ¦ī·sō'byüd·ə·rəl }

isocetyl laurate [ORGANIC CHEMISTRY] $C_{11}H_{23}COOC_{16}H_{33}$ An oily, combustible liquid, soluble in most organic solvents; used in cosmetics and pharmaceuticals and as a plasticizer and textile softener. { ¦ī·sə¦sēd·əl 'lȯ‚rāt }

isocyanate [ORGANIC CHEMISTRY] **1.** One of a group of neutral derivatives of primary amines; its formula is R—N=C=O, where R may be an alkyl or aryl group; an example is 2,4-toluene diisocyanate. **2.** Any compound containing the isocyanato functional group. { ¦ī·sō'sī·ə‚nāt }

isocyanate resin [ORGANIC CHEMISTRY] A linear alkyd resin lengthened by reaction with isocyanates, then treated with a glycol or diamine to cross-link the molecular chain; the product has good abrasion resistance. { ¦ī·sō'sī·ə‚nāt 'rez·ən }

isocyanato group [ORGANIC CHEMISTRY] A functional group (—N=C—O) which forms isocyanates by replacing the hydrogen atom of a hydrocarbon. { ¦ī·sō·sī'an·ə‚tō ‚grüp }

isocyanic acid [ORGANIC CHEMISTRY] HN=C=O One of two forms of cyanic acid; a gas used as an intermediate in the preparation of polyurethane and other resins. { ¦ī·sō·sī'an·ik 'as·əd }

isocyanide [ORGANIC CHEMISTRY] A compound with the general formula RN≡C in which the hydrogen of a hydrocarbon has been replaced by the —N≡C group. { ¦ī·sō'sī·ə‚nīd }

isocyanine [ORGANIC CHEMISTRY] Any one of a series of dyes whose structure has two heterocyclic or quinoline rings connected by an odd number chain of carbon atoms containing conjugated double bonds; for example, cyanine blue. { ¦ī·sō'sī·ə‚nēn }

isocyanuric acid [ORGANIC CHEMISTRY] *See* fulminuric acid. { ¦ī·sō‚sī·ə¦nůr·ik 'as·əd }

isocyclic compound [ORGANIC CHEMISTRY] A compound in which the ring structure is made up of one kind of atom. { ¦ī·sō'sī·klik 'käm‚paůnd }

isodecyl chloride [ORGANIC CHEMISTRY] $C_{10}H_{21}Cl$ A colorless liquid with a boiling point of 210.6°C; used as a solvent and in extractants, cleaning compounds, pharmaceuticals, insecticides, and plasticizers. { ‚ī·sə'des·əl 'klȯr‚īd }

isodisperse [CHEMISTRY] **1.** Having dispersed particles, of colloidal dimensions, that are all of the same size. **2.** Dispersible in solutions with the same pH value. { ˌīs·ə·di'spərs }

isoelectric focusing [PHYSICAL CHEMISTRY] Protein separation technique in which a mixture of protein molecules is resolved into its components by subjecting the mixture to an electric field in a supporting gel having a previously established pH gradient. Also known as electrofocusing. { ¦ī·sō·i'lek·trik 'fō·kəs·iŋ }

isoelectric point [PHYSICAL CHEMISTRY] The pH value of the dispersion medium of a colloidal suspension at which the colloidal particles do not move in an electric field. { ¦ī·sō·i'lek·trik 'pȯint }

isoelectric precipitation [CHEMISTRY] Precipitation of materials at the isoelectric point (the pH at which the net charge on a molecule in solution is zero); proteins coagulate best at this point. { ¦ī·sō·i'lek·trik prəˌsip·ə'tā·shən }

isoelectronic principle [CHEMISTRY] The concept that molecules having the same number of electrons and the same number of atoms whose atomic masses are greater than that of hydrogen (heavy atoms) tend to have similar electronic structures, similar chemical properties, and heavy-atom geometries. { ¦ī·sō·iˌlek'trän·ik 'prin·sə·pəl }

isoelectronic sequence [SPECTROSCOPY] A set of spectra produced by different chemical elements ionized so that their atoms or ions contain the same number of electrons. { ¦ī·sō·iˌlek'trän·ik 'sē·kwəns }

isoeugenol [ORGANIC CHEMISTRY] $C_{10}H_{12}O_2$ An oily liquid prepared from eugenol by heating, slightly soluble in water; used in the manufacture of vanillin. { ¦ī·sō'yü·jəˌnȯl }

isohexane [ORGANIC CHEMISTRY] C_6H_{14} A liquid mixture of isomeric hydrocarbons, flammable and explosive, insoluble in water, soluble in most organic solvents, boils at 54-61°C; used as a solvent, freezing-point depressant, and chemical intermediate. { ¦ī·sō'hekˌsān }

isohydric [CHEMISTRY] Referring to a set of solutions with the same hydrogen ion concentration and not affecting the conductivity of each of the various solutions on mixing. { ¦ī·sə¦hī·drik }

isokinetic relationship [PHYSICAL CHEMISTRY] A linear relationship that exists between the enthalpies and entropies of activation of a series of related reactions. { ˌi·sə·ki¦ned·ik ri'lā·shənˌship }

isokinetic temperature [PHYSICAL CHEMISTRY] The actual or virtual temperature at which rates of all members of a series of related reactions are equal. { ˌī·sə·ki¦ned·ik 'tem·prə·chər }

isolation [CHEMISTRY] Separation of a pure chemical substance from a compound or mixture; as in distillation, precipitation, or absorption. { ˌī·sə'lā·shən }

isomer [CHEMISTRY] One of two or more chemical substances having the same elementary percentage composition and molecular weight but differing in structure, and therefore in properties; there are many ways in which such structural differences occur; one example is provided by the compounds *n*-butane, $CH_3(CH_2)_2CH_3$, and isobutane, $CH_3CH(CH_3)_2$. { 'ī·sə·mər }

isomeric shift [PHYSICAL CHEMISTRY] Shift in the Mössbauer resonance caused by the effect of the valence of the atom on the interaction of the electron density at the nucleus with the nuclear charge. Also known as chemical shift. { ¦ī·sə¦mer·ik 'shift }

isomerism [CHEMISTRY] The phenomenon whereby certain chemical compounds have structures that are different although the compounds possess the same elemental composition. { ī'säm·əˌriz·əm }

isomerization [CHEMISTRY] A process whereby a compound is changed into an isomer; for example, conversion of butane into isobutane. { īˌsäm·ə·rə'zā·shən }

isomolecule [ORGANIC CHEMISTRY] *See* nonlinear molecule. { ¦ī·sō'mäl·əˌkyül }

isomorphism [PHYSICAL CHEMISTRY] A condition present when an ion at high dilution is incorporated by mixed crystal formation into a precipitate, even though such formation would not be predicted on the basis of crystallographic and ionic radii; an example is coprecipitation of lead with potassium chloride. { ¦ī·sə¦mȯrˌfiz·əm }

isonicotinic acid [ORGANIC CHEMISTRY] $C_6H_5NO_2$ White platelets or powder, slightly soluble in water, sublimes at 260°C; used in the manufacture of isonicotinic acid hydrazide, an antitubercular agent. { ¦ī·sə,nik·ə'tin·ik 'as·əd }

isonitrosoacetophenone [ORGANIC CHEMISTRY] $C_8H_7NO_2$ Platelike crystals with a melting point of 126-128°C; soluble in alkalies and alkali carbonates; used to detect ferrous ions and palladium. { ¦ī·sə¦nī·trə·sō,as·ə'täf·ə,nōn }

isooctane [ORGANIC CHEMISTRY] $(CH_3)_2CHCH_2C(CH_3)_3$ Flammable, colorless liquid boiling at 99°C; slightly soluble in alcohol and ether, insoluble in water; used in motor fuels and as a chemical intermediate. { ¦ī·sō'äk,tān }

isooctyl alcohol [ORGANIC CHEMISTRY] $C_7H_{15}CH_2OH$ Mixture of isomers from oxo-process synthesis; boils at 182-195°C; used as a chemical intermediate, resin solvent, emulsifier, and antifoaming agent. { ¦ī·sō'äkt·əl 'al·kə,hȯl }

isoparaffin [ORGANIC CHEMISTRY] A branched-chain version of a straight-chain (normal) saturated hydrocarbon; for example, isooctane, or 2,2,4-trimethyl pentane, $(CH_3)_3C_5H_9$, is the branched-chain version of *n*-octane, $CH_3(CH_2)_6CH_3$. { ¦ī·sō'par·ə·fən }

isopentane [ORGANIC CHEMISTRY] $CH_3CHCH_3CH_2CH_3$ Flammable, colorless liquid with pleasant aroma; boils at 28°C; soluble in oils, ether, and hydrocarbons, insoluble in water; used as a solvent and chemical intermediate. Also known as 2-methylbutane. { ¦ī·sō'pen,tān }

isopentanoic acid [ORGANIC CHEMISTRY] C_4H_9COOH A colorless, combustible liquid with a boiling point of 183.2°C; used for manufacture of plasticizers, pharmaceuticals, and synthetic lubricants. { ¦ī·sə,pen·tə'nō·ik 'as·əd }

isopentyl unit [ORGANIC CHEMISTRY] *See* isoprene unit. { ,ī·sə'pent·əl ,yü·nət }

isophorone [ORGANIC CHEMISTRY] $COCHC(CH_3)CH_2C(CH_3)_2CH_2$ A water-white liquid boiling at 215°C; used as a solvent for lacquers and polyvinyl and nitrocellulose resins. { ¦ī·sə'fȯ,rōn }

isophthalic acid [ORGANIC CHEMISTRY] $C_6H_4(COOH)_2$ Colorless crystals subliming at 345°C; slightly soluble in water, soluble in alcohol and acetic acid, and insoluble in benzene; used as an intermediate for polyester and polyurethane resins, and as a plasticizer. Also known as *meta*-phthalic acid. { ¦ī·sō¦thal·ik 'as·əd }

isopolymolybdate [INORGANIC CHEMISTRY] A class of compounds formed by the acidification of a molybdate solution, or in some cases by heating normal molybdates. { ¦ī·sō,päl·i·mə'lib,dāt }

isopolytungstate [INORGANIC CHEMISTRY] A compound formed by the condensation of tungstate compounds, usually classified into metatungstates, such as $Na_6W_{12}O_{40} \cdot xH_2O$, and paratungstates, such as $Na_{10}W_{12}O_{41} \cdot xH_2O$. { ¦ī·sō,päl·i'təŋ ,stāt }

isoprene [ORGANIC CHEMISTRY] C_5H_8 A conjugated diolefin; a mobile, colorless liquid having a boiling point of 34.1°C; insoluble in water, soluble in alcohol and ether; polymerizes readily to form dimers and high-molecular-weight elastomer resins. { 'ī·sə,prēn }

isoprene unit [ORGANIC CHEMISTRY] The five-carbon structural unit characteristic of terpenes. Also known as isopentyl unit. { ¦ī·sə,prēn ,yü·nət }

isoprenoid [ORGANIC CHEMISTRY] *See* terpene. { ,ī·sə·'prē,nȯid }

isopropaline [ORGANIC CHEMISTRY] $C_{15}H_{23}N_3O_4$ An orange liquid with limited solubility in water; used as a preemergence herbicide for control of grass and broadleaf weeds on tobacco. { ¦ī·sə'prō·pə,lēn }

isopropanol [ORGANIC CHEMISTRY] *See* isopropyl alcohol. { ¦ī·sə'prō·pə,nȯl }

isopropanolamine [ORGANIC CHEMISTRY] $CH_3CH(OH)CH_2NH_2$ A combustible liquid with a faint ammonia odor and a boiling point of 159.9°C; soluble in water; used as an emulsifying agent and for dry-cleaning soaps, wax removers, cosmetics, plasticizers, and insecticides. { ¦ī·sə,prō·pə'nal·ə,mēn }

isopropenyl acetate [ORGANIC CHEMISTRY] $CH_3CO_2C(CH_3){=}CH_2$ A liquid with a boiling point of 97°C; used for acylation of potential enols. { ,ī·sə'prō·pə·nəl 'as·ə,tāt }

2-isopropoxyphenyl *N*-methylcarbamate [ORGANIC CHEMISTRY] $C_{11}H_{15}O_3N$ A colorless solid with a melting point of 91°C; used as an insecticide for cockroaches, flies, mosquitoes, and lawn insects. { ¦tü ¦ī·sō·prə¦päk·sē'fen·əl ¦en ˌmeth·əl'kär·bəˌmāt }

isopropyl [ORGANIC CHEMISTRY] The radical $(CH_3)_2CH$, from isopropane; an example of its occurrence is in isopropyl alcohol, $(CH_3)_2CHOH$. { ¦ī·sə'prō·pəl }

isopropyl acetate [ORGANIC CHEMISTRY] $CH_3COOCH(CH_3)_2$ A colorless, aromatic liquid with a boiling point of 89.4°C, used as a solvent and for paints and printing inks. { ¦ī·sə'prō·pəl 'as·əˌtāt }

isopropyl alcohol [ORGANIC CHEMISTRY] $(CH_3)_2CHOH$ A colorless liquid that boils at 82.4°C; soluble in water, ether, and ethanol; used in manufacturing of acetone and its derivatives, of glycerol, and as a solvent. Also known as isopropanol; 2-propanol; *sec*-propyl alcohol. { ¦ī·sə'prō·pəl 'al·kəˌhòl }

isopropylamine [ORGANIC CHEMISTRY] $(CH_3)_2CHNH_2$ A volatile, colorless liquid with a boiling point of 32.4°C; used as a solvent and in the manufacture of pharmaceuticals, dyes, insecticides, and bactericides. Also known as 2-aminopropane. { ¦ī·sə·prō'pil·əˌmēn }

isopropyl-2-(*N*-benzoyl-3-chloro-4-fluoroanilino)propionate [ORGANIC CHEMISTRY] $C_{19}H_{19}O_3NClF$ Off-white crystals with a melting point of 56-57°C; used as a postemergence herbicide for wild oats and barley. { ¦ī·sə¦prō·pəl 'tü ¦en ¦ben·zə·wəl ¦thrē ¦klòr·ō ¦fòr ˌflu̇r·ō¦an·ə·lō 'prō·pē·əˌnat }

isopropyl 4,4'-dibromobenzilate [ORGANIC CHEMISTRY] $C_{17}H_{16}O_3Br_2$ A brownish solid with a melting point of 77°C; solubility in water is less than 0.5 part per million at 20°C; used as a miticide for deciduous fruit and citrus. { ¦ī·sə¦prō·pəl ¦fòr ¦fòrˌprīm dīˌbrō·mō'ben·zəˌlāt }

isopropyl 4,4'-dichlorobenzilate [ORGANIC CHEMISTRY] $C_{17}H_{16}O_3Cl_2$ A white powder with a melting point of 70-72°C; solubility in water is less than 10 parts per million at 20°C; used as a miticide for spider mites on apple and pear trees. { ¦ī·sə¦prō·pəl ¦fòr ¦fòrˌprīm dīˌklòr·ō'ben·zəˌlāt }

isopropyl ether [ORGANIC CHEMISTRY] $(CH_3)_2CHOCH(CH_3)_2$ Water-soluble, flammable, colorless liquid with etherlike aroma; boils at 68°C; used as a solvent and extractant, in paint and varnish removers, and in spotting formulas. Also known as diisopropyl ether. { ¦ī·sə¦prō·pəl 'ē·thər }

***N*-4-isopropylphenyl-*N'*,*N'*-dimethylurea** [ORGANIC CHEMISTRY] $(CH_3)_2CHC_6H_4NHCON(CH_3)_2$ A crystalline solid with a melting point of 151-153°C; solubility in water is 170 parts per million; used as an herbicide for wheat, barley, and rye. { ¦en ¦fòr ¦ī·sə¦prō·pəl¦fen·əl ¦enˌprīm ¦enˌprīm dīˌmeth·əl·yu̇'rē·ə }

***ortho*-isopropylphenyl-methylcarbamate** [ORGANIC CHEMISTRY] $C_{11}H_{15}O_2N$ A white, crystalline compound with a melting point of 88-89°C; used as an insecticide for rice and cacao crops. Also known as MIPC. { ¦òr·thō ¦ī·sə¦prō·pəl¦fen·əl ˌmeth·əl'kär·bəˌmāt }

isopulegol [ORGANIC CHEMISTRY] $C_{10}H_{17}OH$ An alcohol derived from terpene as a water-white liquid that has a mintlike odor, used in making perfumes. { ¦ī·sō'pyü·ləˌgòl }

isoquinoline [ORGANIC CHEMISTRY] $C_6H_4CHNCHCH$ Colorless liquid boiling at 243°C; soluble in most organic solvents and dilute mineral acids, insoluble in water, derived from coal tar or made synthetically; used to make dyes, insecticides, pharmaceuticals, and rubber accelerators, and as a chemical intermediate. { ¦ī·sə'kwin·əˌlēn }

isosafrole [ORGANIC CHEMISTRY] $C_{10}H_{10}O_2$ A liquid with the odor of anise that is obtained from safrole, and that boils at 253°C; used to make perfumes and flavors. { ¦ī·sō'saˌfrōl }

isosteric [CHEMISTRY] Referring to similar electronic arrangements in chemical compounds. { ¦ī·sə¦ster·ik }

isosterism [PHYSICAL CHEMISTRY] A similarity in the physical properties of ions, compounds, or elements, as a result of electron arrangements that are identical or similar. { ī'säs·təˌriz·əm }

isosynthesis [ORGANIC CHEMISTRY] A process in which mixtures of hydrogen and carbon monoxide are reacted over a thorium oxide catalyst (sometimes mixed with additional substances) to produce branched hydrocarbons. { ¦ī·sō'sin·thə·səs }

isotachophoresis [PHYSICAL CHEMISTRY] A variant of electrophoresis in which ionic species move with equal velocity in the presence of an electric field. { ¦ī·sə,tak·ə·fə'rē·səs }

isotactic [ORGANIC CHEMISTRY] Designating crystalline polymers in which substituents in the asymmetric carbon atoms have the same (rather than random) configuration in relation to the main chain. { ¦ī·sə¦tak·tik }

isothiocyanate [ORGANIC CHEMISTRY] A compound of the type R—N=C=S, where R may be an alkyl or aryl group; an example is mustard oil. Also known as sulfocarbimide. { ,ī·sə,thī·ō'sī·ə,nāt }

isotope-dilution analysis [ANALYTICAL CHEMISTRY] Variation on paper-chromatography analysis; a labeled radioisotope of the same type as the one being quantitated is added to the solution, then quantitatively analyzed afterward via radioactivity measurement. { 'ī·sə,tōp də¦lü·shən ə,nal·ə·səs }

isotope effect [PHYSICAL CHEMISTRY] The effect of difference of mass between isotopes of the same element on nonnuclear physical and chemical properties, such as the rate of reaction or position of equilibrium, of chemical reactions involving the isotopes. { 'ī·sə,tōp i,fekt }

isotope-exchange reaction [CHEMISTRY] A chemical reaction in which interchange of the atoms of a given element between two or more chemical forms of the element occurs, the atoms in one form being isotopically labeled so as to distinguish them from atoms in the other form. { 'ī·sə,tōp iks¦chānj rē,ak·shən }

isotope shift [SPECTROSCOPY] A displacement in the spectral lines due to the different isotopes of an element. { 'ī·sə,tōp ,shift }

isotopic carrier [CHEMISTRY] A carrier that differs from the trace it is carrying only in isotopic composition. { ¦ī·sə¦täp·ik 'kar·ē·ər }

isotopic exchange [PHYSICAL CHEMISTRY] A process in which two atoms belonging to different isotopes of the same element exchange valency states or locations in the same molecule or different molecules. { ¦ī·sə¦täp·ik iks'chānj }

isotopic indicator [CHEMISTRY] *See* isotopic tracer. { ¦ī·sə¦täp·ik 'in·də,kād·ər }

isotopic label [CHEMISTRY] *See* isotopic tracer. { ¦ī·sə¦täp·ik 'lā·bəl }

isotopic tracer [CHEMISTRY] An isotope of an element, either radioactive or stable, a small amount of which may be incorporated into a sample material (the carrier) in order to follow the course of that element through a chemical, biological, or physical process, and also follow the larger sample. Also known as isotopic indicator; isotopic label; tag. { ¦ī·sə¦täp·ik 'trā·sər }

isovalent conjugation [PHYSICAL CHEMISTRY] An arrangement of bonds in a conjugated molecule such that alternative structures with an equal number of bonds can be written; an example occurs in benzene. { ¦ī·sə¦vā·lənt kən'jəŋk·shən }

isovalent hyperconjugation [PHYSICAL CHEMISTRY] An arrangement of bonds in a hyperconjugated molecule such that the number of bonds is the same in the two resonance structures but the second structure is energetically less favorable than the first structure; examples are $H_3{\equiv}C{-}C^{+}H_2$ and $H_3{\equiv}C{-}CH_2$. { ¦ī·sə¦vā·lənt ,hī·pər,kän·jə'gā·shən }

isovaleraldehyde [ORGANIC CHEMISTRY] $(CH_3)_2CHCH_2CHO$ A colorless liquid with an applelike odor and a boiling point of 92°C; soluble in alcohol and ether; used in perfumes and pharmaceuticals and for flavoring. { ¦ī·sō¦val·ər'al·də,hīd }

isovaleric acid [ORGANIC CHEMISTRY] $(CH_3)_2CHCH_2COOH$ Color-less liquid with disagreeable taste and aroma; boils at 176°C; soluble in alcohol and ether; found in valeriana, hop, tobacco, and other plants; used in flavors, perfumes, and medicines. { ¦ī·sō·və'ler·ik 'as·əd }

2-isovaleryl-1,3-indandione [ORGANIC CHEMISTRY] $C_{14}H_{14}O_3$ A yellow, crystalline compound with a melting point of 67-68°C; insoluble in water; used as a rodenticide. { ¦tü ¦ī·sō'val·ə,ril ¦wən ¦thrē ,in·dən'dī,ōn }

ISS [SPECTROSCOPY] *See* ion scattering spectroscopy.

itaconic acid [ORGANIC CHEMISTRY] $CH_2{:}C(COOH)CH_2COOH$ A colorless crystalline compound that decomposes at 165°C, prepared by fermentation with *Aspergillus terreus*; used as an intermediate in organic synthesis and in resins and plasticizers. { ¦id·ə¦kän·ik 'as·əd }

itatartaric acid [ORGANIC CHEMISTRY] $C_5H_8O_6$ A compound produced experimentally by fermentation; formed as a minor product, 5.8% of total acidity produced, of an itaconic-acid producing strain of *Aspergillus niger*. { ¦id·ə¦tär·də·rik ′as·əd }

ium ion [ORGANIC CHEMISTRY] A positively charged group of atoms in which a charged nonmetallic ion other than carbon or silicon possesses a closed-shell electron configuration; often joined to a root word, as in carbonium ion. { ′ī·əm ′ī‚än }

Ivanov reagent [ORGANIC CHEMISTRY] A reagent that is similar to a Grignard reagent, and that is formed by reacting an arylacetic acid or its sodium salt with isopropyl magnesium halide. { ē·və·nȯf rē‚ā·jənt }

J

Jacquemart's reagent [ANALYTICAL CHEMISTRY] Analytical reagent used to test for ethyl alcohol; consists of an aqueous solution of mercuric nitrate and nitric acid. { zhak'märz rē,ā·jənt }

Jahn-Teller effect [PHYSICAL CHEMISTRY] The effect whereby, except for linear molecules, degenerate orbital states in molecules are unstable. { 'yän 'tel·ər i,fekt }

jasmone [ORGANIC CHEMISTRY] $C_{11}H_{16}O$ A liquid ketone found in jasmine oil and other essential oils from plants. { 'jaz,mōn }

jeweler's rouge [INORGANIC CHEMISTRY] *See* ferric oxide. { 'jü·lərz 'rüzh }

jodfenphos [ORGANIC CHEMISTRY] $C_8H_8O_3Cl_2IPS$ A crystalline compound with a melting point of 76°C; slight solubility in water; used as an insecticide in homes, farm buildings, and industrial sites. { 'yōd fen,fäs }

Jones reductor [CHEMISTRY] A device used to chemically reduce solutions, such as ferric salt solutions, consisting of a vertical tube containing granular zinc into which the solution is poured. { 'jōnz ri,dək·tər }

juniperic acid [ORGANIC CHEMISTRY] $C_{16}H_{32}O_3$ A crystalline hydroxy acid that melts at 95°C, obtained from waxy exudations from conifers. { ¦jü·nə¦per·ik 'as·əd }

K

K [CHEMISTRY] *See* cathode; potassium.

K acid [ORGANIC CHEMISTRY] $C_{10}H_4NH_2OH(SO_3H)_2$ An acid derived from naphthylamine trisulfonic acid; used in dye manufacture. { 'kā ˌas·əd }

kalium [CHEMISTRY] *See* potassium. { 'kāl·ē·əm }

karbutilate [ORGANIC CHEMISTRY] $C_{14}H_{21}N_3O_3$ An off-white solid with a melting point of 176-177°C; used as a herbicide on noncroplands, railroad rights-of-way, and plant sites. { kär'byüd·əlˌāt }

Karl Fischer reagent [ANALYTICAL CHEMISTRY] A solution of 8 moles pyridine to 2 moles sulfur dioxide, with the addition of about 15 moles methanol and then 1 mole iodine; used to determine trace quantities of water by titration. { 'kärl 'fish·ər rē'ā·jənt }

Karl Fischer technique [ANALYTICAL CHEMISTRY] A method of determining trace quantities of water by titration; the Karl Fischer reagent is added in small increments to a glass flask containing the sample until the color changes from yellow to brown or a change in potential is observed at the end point. { 'kärl 'fish·ər tek'nēk }

kauri-butanol value [ANALYTICAL CHEMISTRY] The measure of milliliters of paint or varnish petroleum thinner needed to cause cloudiness in a solution of kauri gum in butyl alcohol. { 'kaù·rē 'byüt·ənˌól ˌval·yü }

kayser [SPECTROSCOPY] A unit of reciprocal length, especially wave number, equal to the reciprocal of 1 centimeter. Also known as rydberg. { 'kī·zər }

Keesom force [PHYSICAL CHEMISTRY] *See* orientation force. { 'kā·səm ˌfórs }

Keesom relationship [PHYSICAL CHEMISTRY] An equation for the potential energy associated with the interaction of the dipole moments of two polar molecules. { 'kā·səm ri'lā·shənˌship }

Kekulé structure [ORGANIC CHEMISTRY] A molecular structure of a cyclic conjugated system that is depicted with alternating single and double bonds. { 'kā·kəˌlā ˌstrək·chər }

ketal [ORGANIC CHEMISTRY] **1.** Former term for the =CO group, as in dimethyl ketal (acetone). **2.** Any of the ketone acetates from condensation of alkyl orthoformates with ketones in the presence of alcohols. { 'kēˌtal }

ketene [ORGANIC CHEMISTRY] C_2H_2O A colorless, toxic, highly reactive gas, with disagreeable taste; boils at −56°C, soluble in ether and acetone, and decomposes in water and alcohol; used as an acetylating agent in organic synthesis. { 'kēˌtēn }

ketimide [ORGANIC CHEMISTRY] A compound that is represented by R_2:C:NX, where X is an acyl radical. { 'ked·əˌmīd }

ketimine [ORGANIC CHEMISTRY] An organic compound that contains the divalent group >C=NH; a Schiff base is an example. { 'ked·əˌmēn }

keto- [ORGANIC CHEMISTRY] Organic chemical prefix for the keto or carbonyl group, C:O, as in a ketone. { 'kēd·ō }

keto acid [ORGANIC CHEMISTRY] A compound that is both an acid and a ketone; an example is β-acetoacetic acid. { 'kēd·ō ˌas·əd }

ketoglutarate [ORGANIC CHEMISTRY] A salt or ester of ketoglutaric acid. { ˌkēd·ə 'glüd·əˌrāt }

ketone [ORGANIC CHEMISTRY] One of a class of chemical compounds of the general formula RR'CO, where R and R' are alkyl, aryl, or heterocyclic radicals; the groups R and R' may be the same or different, or incorporated into a ring; the ketones, acetone, and methyl ethyl ketone are used as solvents, and ketones in general are important intermediates in the synthesis of organic compounds. { 'kē‚tōn }

Kiliani reaction [ORGANIC CHEMISTRY] A method of synthesizing a higher aldose from a lower aldose; monosaccharides, such as aldehydes and ketones, react with hydrogen cyanide to form cyanohydrins, which are hydrolyzed to hydroxy acids, converted to lactones, and reduced to aldoses with sodium amalgams. { ‚kil·ē'an·ē rē‚ak·shən }

kilogram-equivalent weight [CHEMISTRY] A unit of mass 1000 times the gram-equivalent weight. { 'kil·ə‚gram i'kwiv·ə·lənt 'wāt }

king's blue [CHEMISTRY] *See* cobalt blue. { 'kiŋz 'blü }

kinic acid [ORGANIC CHEMISTRY] *See* quinic acid. { 'kin·ik 'as·əd }

Kistiakowsky-Fishtine equation [PHYSICAL CHEMISTRY] An equation to calculate latent heats of vaporization of pure compounds; useful when vapor pressure and critical data are not available. { ‚kis·tē·ə'kȯf·skē fə'shtīn i‚kwā·zhən }

kitol [ORGANIC CHEMISTRY] $C_{40}H_{60}O_2$ One of the provitamins of vitamin A derived from whale liver oil; crystallizes from methanol solution. { 'kē‚tȯl }

Kjeldahl method [ANALYTICAL CHEMISTRY] Quantitative analysis of organic compounds to determine nitrogen content by interaction with concentrated sulfuric acid; ammonia is distilled from the NH_4SO_4 formed. { 'kel‚däl ‚meth·əd }

Klein-Rydberg method [PHYSICAL CHEMISTRY] A method for determining the potential energy function of the distance between the nuclei of a diatomic molecule from the molecule's vibrational and rotational levels. { 'klīn 'rid‚berg ‚meth·əd }

Klein's reagent [CHEMISTRY] Saturated solution of borotungstate; used to separate minerals by specific gravity. { 'klīnz rē‚ā·jənt }

Knoevenagel reaction [ORGANIC CHEMISTRY] The condensation of aldehydes with compounds containing an activated methylene ($=CH_2$) group. { kə'nē·və‚näg·əl rē‚ak·shən }

Knorr synthesis [ORGANIC CHEMISTRY] A condensation reaction carried out in either glacial acetic acid or an aqueous alkali in which an α-aminoketone combines with an α-carbonyl compound to form a pyrrole; possibly the most versatile pyrrole synthesis. { 'nȯr ‚sin·thə·səs }

knot [ORGANIC CHEMISTRY] A chiral structure in which rings containing 50 or more members have a knotlike configuration. { nät }

Knudsen cell [PHYSICAL CHEMISTRY] A vessel used to measure very low vapor pressures by measuring the mass of vapor which escapes when the vessel contains a liquid in equilibrium with its vapor. { kə'nüd·sən ‚cel }

Kohlrausch law [PHYSICAL CHEMISTRY] **1.** The law that every ion contributes a definite amount to the equivalent conductance of an electrolyte in the limit of infinite dilution, regardless of the presence of other ions. **2.** The law that the equivalent conductance of a very dilute solution of a strong electrolyte is a linear function of the concentration. { 'kōl‚rau̇sh ‚lȯ }

Kohlrausch method [PHYSICAL CHEMISTRY] A method of measuring the electrolytic conductance of a solution using a Wheatstone bridge. { 'kōl‚rau̇sh ‚meth·əd }

Kojic acid [ORGANIC CHEMISTRY] $C_6H_6O_4$ A crystalline antibiotic with a melting point of 152-154°C; soluble in water, acetone, and alcohol; used in insecticides and as an antifungal and antimicrobial agent. { 'kō·jik ‚as·əd }

Kolbe hydrocarbon synthesis [ORGANIC CHEMISTRY] The production of an alkane by the electrolysis of a water-soluble salt of a carboxylic acid. { 'kōl·bə ‚hī·drə'kär·bən ‚sin·thə·səs }

Kolbe-Schmitt synthesis [ORGANIC CHEMISTRY] The reaction of carbon dioxide with sodium phenoxide at 125°C to give salicyclic acid. { 'kōl·bə 'shmit ‚sin·thə·səs }

Konowaloff rule [PHYSICAL CHEMISTRY] An empirical rule which states that in the vapor over a liquid mixture there is a higher proportion of that component which, when added to the liquid, raises its vapor pressure, than of other components. { ‚kȯ·nə'vä·lȯf ‚rül }

Kopp's law [PHYSICAL CHEMISTRY] The law that for solids the molal heat capacity of a compound at room temperature and pressure approximately equals the sum of heat capacities of the elements in the compound. { 'käps ˌlö }

Korner's method [ORGANIC CHEMISTRY] A method for determining the absolute position of substituents for positional isomers in benzene by the experimental production of positional isomers from a given disubstituted benzene. { 'kȯr·nərz ˌmeth·əd }

Korshun method [ANALYTICAL CHEMISTRY] Microdetermination of carbon and hydrogen in organic compounds; the sample is prepyrolyzed (cracked) in a shortage of oxygen, then oxidized in an excess of oxygen. { 'kȯr·shən ˌmeth·əd }

Kossel-Sommerfeld law [SPECTROSCOPY] The law that the arc spectra of the atom and ions belonging to an isoelectronic sequence resemble each other, especially in their multiplet structure. { 'käs·əl 'zȯm·ərˌfelt ˌlö }

Kovat's retention indexes [ANALYTICAL CHEMISTRY] Procedure to identify compounds in gas chromatography; the behavior of a compound is indicated by its position on a scale of normal alkane values (for example, methane = 100, ethane = 200). { 'kō·vats ri'ten·chən ˌinˌdek·səs }

krypton [CHEMISTRY] A colorless, inert gaseous element, symbol Kr, atomic number 36, atomic weight 83.80; it is odorless and tasteless; used to fill luminescent electric tubes. { 'krip·tän }

Kundt rule [SPECTROSCOPY] The rule that the optical absorption bands of a solution are displaced toward the red when its refractive index increases because of changes in composition or other causes. { 'ku̇nt ˌrül }

kurchatovium [CHEMISTRY] The name suggested by workers in the Soviet Union for element 104. Symbolized Ku. { ˌkər·chə'tō·vē·əm }

L

La [CHEMISTRY] *See* lanthanum.

laboratory sample [ANALYTICAL CHEMISTRY] A sample of a material to be tested or analyzed that is prepared from a gross sample and retains the latter's composition. { ¦lab·rə‚tór·ē ¦sam·pəl }

lachesne [ORGANIC CHEMISTRY] $C_{20}H_{26}ClNO_3$ A compound that crystallizes from a solution of ethanol and acetone, and whose melting point is 213°C; used in ophthalmology. Also known as chloride benzilate. { lə'shēn }

lactam [ORGANIC CHEMISTRY] An internal (cyclic) amide formed by heating gamma (γ) and delta (δ) amino acids; thus γ-aminobutyric acid readily forms γ-butyrolactam lactam (pyrrolidone); many lactams have physiological activity. { 'lak‚tam }

lactate [ORGANIC CHEMISTRY] A salt or ester of lactic acid in which the acidic hydrogen of the carboxyl group has been replaced by a metal or an organic radical. { 'lak‚tāt }

lactide [ORGANIC CHEMISTRY] A cyclic, intermolecular, double ester formed from α-hydroxy acids; most lactides are relatively low melting solids and are easily hydrolyzed by base to form salts of the parent acid, such as sodium lactate. { 'lak‚tīd }

lactim [ORGANIC CHEMISTRY] A tautomeric enol form of a lactam with which it forms an equilibrium whenever the lactam nitrogen carries a free hydrogen. { 'lak·təm }

lactone [ORGANIC CHEMISTRY] An internal cyclic mono ester formed by gamma (γ) or delta (δ) hydroxy acids spontaneously; thus γ-hydroxybutyric acid forms γ-butyrolactone. { 'lak‚tōn }

lactonitrile [ORGANIC CHEMISTRY] $CH_3CHOHCN$ A straw-colored liquid boiling at 183°C, soluble in water, insoluble in carbon disulfide and petroleum ether; used as a solvent, and as a chemical intermediate in making esters of lactic acid. Also known as acetaldehyde cyanohydrin. { ¦lak·tō'nī‚tril }

lactonization [ORGANIC CHEMISTRY] The process in which a lactone is formed by intramolecular attack of a hydroxyl group on an activated carbonyl group. { ‚lak·tə·nə'zā·shən }

lambda sulfur [CHEMISTRY] One of the two components of plastic (or gamma) sulfur, soluble in carbon disulfide. { 'lam·də ‚səl·fər }

Lambert-Beer law [ANALYTICAL CHEMISTRY] *See* Bouguer-Lambert-Beer law. { 'lam·bərt 'bir ‚lȯ }

Langmuir-Blodgett film [PHYSICAL CHEMISTRY] A highly ordered monomolecular film that results from compressing a surface layer of amphiphilic molecules into a floating monolayer and transferring it to a substrate by dipping. { 'laŋ‚myůr 'bläj·ət ‚film }

Langmuir isotherm equation [PHYSICAL CHEMISTRY] An equation, useful chiefly for gaseous systems, for the amount of material adsorbed on a surface as a function of pressure, while the temperature is held constant, assuming that a single layer of molecules is adsorbed; it is $f = ap/(1 + ap)$, where f is the fraction of surface covered, p is the pressure, and a is a constant. { 'laŋ‚myůr 'ī·sə‚thərm i‚kwā·zhən }

lanthana [INORGANIC CHEMISTRY] *See* lanthanum oxide. { 'lan·thə·nə }

lanthanide series [CHEMISTRY] Rare-earth elements of atomic numbers 57 through 71; their chemical properties are similar to those of lanthanum, atomic number 57. { 'lan·thə‚nīd ‚sir·ēz }

lanthanum [CHEMISTRY] A chemical element, symbol La, atomic number 57, atomic weight 138.91; it is the second most abundant element in the rare-earth group. { 'lan·thə·nəm }

lanthanum nitrate [INORGANIC CHEMISTRY] $La(NO_3)_3 \cdot 6H_2O$ Hygroscopic white crystals melting at 40°C; soluble in alcohol and water; used as an antiseptic and in gas mantles. { 'lan·thə·nəm 'nī,trāt }

lanthanum oxide [INORGANIC CHEMISTRY] La_2O_3 A white powder melting at about 2000°C; soluble in acid, insoluble in water; used to replace lime in calcium lights and in optical glass. Also known as lanthana; lanthanum sesquioxide; lanthanum trioxide. { 'lan·thə·nəm 'äk,sīd }

lanthanum sesquioxide [INORGANIC CHEMISTRY] *See* lanthanum oxide. { 'lan·thə·nəm ,ses·kwē'äk,sīd }

lanthanum sulfate [INORGANIC CHEMISTRY] $La_2(SO_4)_3 \cdot 9H_2O$ White crystals; slightly soluble in water, soluble in alcohol; used for atomic weight determinations for lanthanum. { 'lan·thə·nəm 'səl,fāt }

lanthanum trioxide [INORGANIC CHEMISTRY] *See* lanthanum oxide. { 'lan·thə·nəm trī'äk,sīd }

larixinic acid [ORGANIC CHEMISTRY] *See* maltol. { ¦lar·ik¦sin·ik 'as·əd }

laser heterodyne spectroscopy [SPECTROSCOPY] A high-resolution spectroscopic technique, used in astronomical and atmospheric observations, in which the signal to be measured is mixed with a laser signal in a solid-state diode, producing a difference-frequency signal in the radio-frequency range. { 'lā·zər ¦hed·ə·rə,dīn spek'träs·kə·pē }

laser spectroscopy [SPECTROSCOPY] A branch of spectroscopy in which a laser is used as an intense, monochromatic light source; in particular, it includes saturation spectroscopy, as well as the application of laser sources to Raman spectroscopy and other techniques. { 'lā·zər spek'träs·kə,pē }

laudanidine [ORGANIC CHEMISTRY] $C_{20}H_{25}NO_4$ An optically active alkaloid found in opium that crystallizes as prisms from an alcohol solution, and melts at 185°C. Also known as *l*-laudanine; tritopine. { lȯ'dan·ə,dēn }

laudanine [ORGANIC CHEMISTRY] $C_{20}H_{25}NO_4$ An optically inactive alkaloid derived from alkaline mother liquors from morphine extraction; it crystallizes in orthorhombic prisms from alcohol and chloroform; the prisms melt at 167°C, and are soluble in hot alcohol, benzene, and chloroform. Also known as *dl*-laudanidine. { lȯ'dan·ə,nēn }

laudanosine [ORGANIC CHEMISTRY] $C_{21}H_{27}NO_4$ An alkaloid that is the methyl ether of laudanine; the optically inactive form crystallizes from dilute alcohol and melts at about 115°C; the levorotatory active form crystallizes from light petroleum solution and melts at 89°C. { lȯ'dan·ə,sēn }

laughing gas [INORGANIC CHEMISTRY] *See* nitrous oxide. { 'laf·iŋ ,gas }

lauric acid [ORGANIC CHEMISTRY] $CH_3(CH_2)_{10}COOH$ A fatty acid melting at 44°C, boiling at 225°C (100 mmHg; 13,332 pascals); colorless needles soluble in alcohol and ether, insoluble in water; found as the glyceride in vegetable fats, such as coconut and laurel oils; used for wetting agents, in cosmetics, soaps, resins, and insecticides, and as a chemical intermediate. { 'lȯr·ik 'as·əd }

lauryl alcohol [ORGANIC CHEMISTRY] $CH_3(CH_2)_{11}OH$ A colorless solid which is obtained from coconut oil fatty acids, has a floral odor, and boils at 259°C; used in detergents, lubricating oils, and pharmaceuticals. { 'lȯr·əl 'al·kə,hōl }

lauryl aldehyde [ORGANIC CHEMISTRY] $CH_3(CH_2)_{10}CHO$ A constituent of an essential oil from the silver fir; a colorless solid or a liquid, with a floral odor, that is soluble in 90% alcohol; used in perfumes. { 'lȯr·əl 'al·də,hīd }

lauryl mercaptan [ORGANIC CHEMISTRY] $C_{12}H_{25}SH$ Pale-yellow or water-white liquid with mild odor; insoluble in water, soluble in organic solvents; used to manufacture plastics, pharmaceuticals, insecticides, fungicides, and elastomers. { 'lȯr·əl mər'kap,tan }

law of constant heat summation [PHYSICAL CHEMISTRY] *See* Hess's law. { 'lȯ əv ¦kän·stənt 'hēt sə,mā·shən }

law of corresponding states [CHEMISTRY] The law that when, for two substances, any two ratios of pressure, temperature, or volume to their respective critical properties are equal, the third ratio must equal the other two. { 'lȯ əv ¦kär·ə¦spän·diŋ 'stāts }

law of definite composition [CHEMISTRY] *See* law of definite proportion. { 'lȯ əv ¦def·ə·nət ˌkäm·pə'zish·ən }

law of definite proportion [CHEMISTRY] The law that a given chemical compound always contains the same elements in the same fixed proportion by weight. Also known as law of definite composition. { 'lȯ əv ¦def·ə·nət prə'pȯr·shən }

law of mass action [CHEMISTRY] The law stating that the rate at which a chemical reaction proceeds is directly proportional to the molecular concentrations of the reacting compounds. { 'lȯ əv ¦mas 'ak·shən }

lawrencium [CHEMISTRY] A chemical element, symbol Lr, atomic number 103; two isotopes have been discovered, mass number 257 or 258 and mass number 256. { 'lȯ'ren·sē·əm }

LDPE [ORGANIC CHEMISTRY] *See* low-density polyethylene.

leachate [CHEMISTRY] A solution formed by leaching. { 'lēˌchāt }

lead [CHEMISTRY] A chemical element, symbol Pb, atomic number 82, atomic weight 207.19. { led }

lead acetate [ORGANIC CHEMISTRY] $Pb(C_2H_3O_2)_2 \cdot 3H_2O$ Poisonous, water-soluble white crystals decomposing at 280°C; loses water at 75°C; used in hair dyes, medicines, and textile mordants, for waterproofing, for manufacture of varnishes and pigments, and as an analytical reagent. Also known as sugar of lead. { 'led 'as·əˌtāt }

lead antimonite [INORGANIC CHEMISTRY] $Pb_3(SbO_4)_2$ Poisonous, water-insoluble orange-yellow powder; used as a paint pigment and to stain glass and ceramics. Also known as antimony yellow; Naples yellow. { 'led an'tim·əˌnīt }

lead arsenate [INORGANIC CHEMISTRY] $Pb_3(AsO_4)_2$ Poisonous, water-insoluble white crystals; soluble in nitric acid; used as an insecticide. { 'led 'ärs·ənˌāt }

lead azide [INORGANIC CHEMISTRY] $Pb(N_3)_2$ Unstable, colorless needles that explode at 350°C; lead azide is shipped submerged in water to reduce sensitivity; used as a detonator for high explosives. { 'led 'āˌzīd }

lead borate [INORGANIC CHEMISTRY] $Pb(BO_2)_2 \cdot H_2O$ Poisonous, water-insoluble white powder; soluble in dilute nitric acid; used as varnish and paint drier, for galvanoplastic work, in lead glass, and in waterproofing paints. { 'led 'bȯrˌāt }

lead bromide [INORGANIC CHEMISTRY] $PbBr_2$ An alcohol-insoluble white powder melting at 373°C, boiling at 916°C; slightly soluble in hot water. { 'led 'brōˌmīd }

lead carbonate [INORGANIC CHEMISTRY] $PbCO_3$ Poisonous, acid-soluble white crystals decomposing at 315°C; insoluble in alcohol and water; used as a paint pigment. { 'led 'kär·bəˌnāt }

lead chloride [INORGANIC CHEMISTRY] $PbCl_2$ Poisonous white crystals melting at 498°C, boiling at 950°C; slightly soluble in hot water, insoluble in alcohol and cold water; used to make lead salts and lead chromate pigments and as an analytical reagent. { 'led 'klȯrˌīd }

lead chromate [INORGANIC CHEMISTRY] $PbCrO_4$ Poisonous, water-insoluble yellow crystals melting at 844°C; soluble in acids; used as a paint pigment. { 'led 'krōˌmāt }

lead cyanide [INORGANIC CHEMISTRY] $Pb(CN)_2$ Poisonous white to yellow powder; slightly soluble in water, decomposed by acids; used in metallurgy. { 'led 'sī·əˌnīd }

lead dioxide [INORGANIC CHEMISTRY] PbO_2 Poisonous brown crystals that decompose when heated; insoluble in water and alcohol, soluble in glacial acetic acid; used as an oxidizing agent, in electrodes, batteries, matches, and explosives, as a textile mordant, in dye manufacture, and as an analytical reagent. Also known as anhydrous plumbic acid; brown lead oxide; lead peroxide. { 'led dī'äkˌsīd }

lead fluoride [INORGANIC CHEMISTRY] PbF_2 A crystalline solid with a melting point of 824°C; used for laser crystals and electronic and optical applications. { 'led 'flu̇rˌīd }

lead formate [ORGANIC CHEMISTRY] $Pb(CHO_2)_2$ Poisonous, water-soluble brownish-white crystals that decompose at 190°C; used as an analytical reagent. { 'led 'fȯr,māt }

lead halide [INORGANIC CHEMISTRY] PbX_2, where X is a halogen (such as F, Br, Cl, or I). { 'led 'ha,līd }

lead hexafluorosilicate [INORGANIC CHEMISTRY] $PbSiF_6 \cdot 2H_2O$ Poisonous, colorless, water-soluble crystals; used in the electrolytic method for refining lead. { 'led ¦hek·sə,flu̇r·ə'sil·ə,kāt }

lead iodide [INORGANIC CHEMISTRY] PbI_2 Poisonous, water- and alcohol-insoluble golden-yellow crystals melting at 402°C, boiling at 954°C; used in photography, medicine, printing, mosaic gold, and bronzing. { 'led 'ī·ə,dīd }

lead metasilicate [INORGANIC CHEMISTRY] *See* lead silicate. { 'led ,med·ə'sil·ə,kāt }

lead molybdate [INORGANIC CHEMISTRY] $PbMoO_4$ Poisonous, acid-soluble yellow powder; insoluble in water and alcohol; used in pigments and as an analytical reagent. { 'led mə'lib,dāt }

lead monoxide [INORGANIC CHEMISTRY] PbO Yellow, tetragonal crystals that melt at 888°C and are soluble in alkalies and acids; used in storage batteries, ceramics, pigments, and paints. Also known as litharge; plumbous oxide; yellow lead oxide. { 'led mə'näk,sīd }

lead nitrate [INORGANIC CHEMISTRY] $Pb(NO_3)_2$ Strongly oxidizing, poisonous, water- and alcohol-soluble white crystals that decompose at 205-223°C; used as a textile mordant, paint pigment, and photographic sensitizer and in medicines, matches, explosives, tanning, and engraving. { 'led 'nī,trāt }

lead oleate [ORGANIC CHEMISTRY] $Pb(C_{18}H_{33}O_2)_2$ Poisonous, water-insoluble, white, ointmentlike material; soluble in alcohol, benzene, and ether; used in varnishes, lacquers, and high-pressure lubricants, and as a paint drier. { 'led 'ō·lē,āt }

lead orthoplumbate [INORGANIC CHEMISTRY] *See* lead tetroxide. { 'led ,ȯr·thō 'pləm,bāt }

lead oxide red [INORGANIC CHEMISTRY] *See* lead tetroxide. { 'led ¦äk,sīd 'red }

lead peroxide [INORGANIC CHEMISTRY] *See* lead dioxide. { 'led pə'räk,sīd }

lead phosphate [INORGANIC CHEMISTRY] Pb_3PO_4 A poisonous, white powder that melts at 1014°C; soluble in nitric acid and in fixed alkali hydroxide; used as a stabilizer in plastics. { 'led 'fäs,fāt }

lead pigments [CHEMISTRY] Chemical compounds of lead used in paints to give color; examples are white lead; basic lead carbonate; lead carbonate; lead thiosulfate; lead sulfide; basic lead sulfate (sublimed white lead); silicate white lead; basic lead silicate; lead chromate; basic lead chromate; lead oxychloride; and lead oxide (monoxide and dioxide). { 'led 'pig·məns }

lead resinate [ORGANIC CHEMISTRY] $Pb(C_{20}H_{29}O_2)_2$ Poisonous, insoluble, brown, lustrous, translucent lumps; used as a paint and varnish drier and for textile waterproofing. { 'led 'rez·ən,āt }

lead silicate [INORGANIC CHEMISTRY] $PbSiO_3$ Toxic, insoluble white crystals; used in ceramics, paints, and enamels, and to fireproof fabrics. Also known as lead metasilicate. { 'led 'sil·ə,kāt }

lead sodium hyposulfate [INORGANIC CHEMISTRY] *See* lead sodium thiosulfate. { 'led 'sōd·ē·əm ,hī·pō'səl,fāt }

lead sodium thiosulfate [INORGANIC CHEMISTRY] $Na_4Pb(S_2O_3)_3$ Poisonous, small, white, heavy crystals that are soluble in thiosulfate solutions; used in the manufacture of matches. Also known as lead sodium hyposulfate; sodium lead hyposulfate; sodium lead thiosulfate. { 'led 'sōd·ē·əm ,thī·ə'səl,fāt }

lead stearate [ORGANIC CHEMISTRY] $Pb(C_{18}H_{35}O_2)_2$ Poisonous white powder; soluble in alcohol and ether, insoluble in water; used as a lacquer and varnish drier and in high-pressure lubricants. { 'led 'stir,āt }

lead sulfate [INORGANIC CHEMISTRY] $PbSO_4$ Poisonous white crystals melting at 1170°C; slightly soluble in hot water, insoluble in alcohol; used in storage batteries and as a paint pigment. { 'led 'səl,fāt }

lead sulfide [INORGANIC CHEMISTRY] PbS Blue, metallic, cubic crystals that melt at

1120°C, derived from the mineral galena or by reacting hydrogen sulfide gas with a solution of lead nitrate; used in semiconductors and ceramics. Also known as plumbous sulfide. { 'led 'səl‚fīd }

lead telluride [INORGANIC CHEMISTRY] PbTe A crystalline solid that is very toxic if inhaled or ingested; melts at 902°C; used as a semiconductor and photoconductor in the form of single crystals. { 'led 'tel·yə‚rīd }

lead tetraacetate [ORGANIC CHEMISTRY] $Pb(CH_3COO)_4$ Crystals that are faintly pink or colorless; melts at 175°C; used as an oxidizing agent in organic chemistry, cleaving 1,2-diols to form aldehydes or ketones. { 'led ‚te·trə'as·ə‚tāt }

lead tetroxide [INORGANIC CHEMISTRY] Pb_3O_4 A poisonous, bright-red powder, soluble in excess glacial acetic acid and dilute hydrochloric acid; used in medicine, in cement for special applications, in manufacture of colorless glass, and in ship paint. Also known as lead orthoplumbate; lead oxide red; red lead. { 'led ‚te'träk‚sīd }

lead thiocyanate [INORGANIC CHEMISTRY] $Pb(SCN)_2$ Yellow, monoclinic crystals, soluble in potassium thiocyanate and slightly soluble in water; used in the powder mixture that primes small arm cartridges, in dyes, and in safety matches. { 'led ‚thī·ō'sī·ə‚nāt }

lead titanate [INORGANIC CHEMISTRY] $PbTiO_3$ A water-insoluble, pale-yellow solid; used as coloring matter in paints. { 'led 'tīt·ən‚āt }

lead tungstate [INORGANIC CHEMISTRY] $PbWO_4$ A yellowish powder, melting at 1130°C; insoluble in water, soluble in acid; used as a pigment. Also known as lead wolframate. { 'led 'təŋ·stə‚nāt }

lead vanadate [INORGANIC CHEMISTRY] $Pb(VO_3)_2$ A water-insoluble, yellow powder; used as a pigment and for the preparation of other vanadium compounds. { 'led 'van·ə‚dāt }

lead wolframate [INORGANIC CHEMISTRY] *See* lead tungstate. { 'led 'wu̇l·frə‚māt }

leakage [PHYSICAL CHEMISTRY] A phenomenon occurring in an ion-exchange process in which some influent ions are not adsorbed by the ion-exchange bed and appear in the effluent. { 'lēk·ij }

leaving group [ORGANIC CHEMISTRY] The group of charged or uncharged atoms that departs during a substitution or displacement reaction. Also known as nucleofuge. { 'lēv·iŋ ‚grüp }

Lennard-Jones potential [PHYSICAL CHEMISTRY] A semiempirical approximation to the potential of the force between two molecules, given by $\nu = (A/r^{12}) - (B/r^6)$, where r is the distance between the centers of the molecules, and A and B are constants. { 'len·ərd 'jōnz pə‚ten·chəl }

lepidine [ORGANIC CHEMISTRY] $C_9H_6NCH_3$ An alkaloid derived as an oily liquid from cinchona bark; boils at 266°C; soluble in ether, benzene, and alcohol; used in organic synthesis. { 'lep·ə‚dēn }

leptophos [ORGANIC CHEMISTRY] $C_{13}H_{10}BrCl_2O_2PS$ A white solid with a melting point of 70.2–70.6°C; slight solubility in water; used as an insecticide on vegetables, fruit, turf, and ornamentals. Also known as O-(4-bromo-2,5-dichlorophenyl) O-methyl phenylphosphorothioate. { 'lep·tə‚fäs }

leucaenine [ORGANIC CHEMISTRY] *See* mimosine. { 'lü·sə‚nēn }

leucaenol [ORGANIC CHEMISTRY] *See* mimosine. { 'lü·sə‚nól }

leucenine [ORGANIC CHEMISTRY] *See* mimosine. { 'lü·sə‚nēn }

leucenol [ORGANIC CHEMISTRY] *See* mimosine. { 'lü·sə‚nól }

leuco base [ORGANIC CHEMISTRY] Any group of colorless derivatives of triphenylmethane dyes that are produced by reducing the dye and are capable of being reconverted to the original dye by oxidation. Also known as leuco compound. { 'lü·kō ‚bās }

leuco compound [ORGANIC CHEMISTRY] *See* leuco base. { 'lü·ko ‚käm‚pau̇nd }

leucoline [ORGANIC CHEMISTRY] *See* quinoline. { 'lü·kə‚lēn }

leukol [ORGANIC CHEMISTRY] *See* quinoline. { 'lü‚kól }

levigate [CHEMISTRY] **1.** To separate a finely divided powder from a coarser material by suspending in a liquid in which both substances are insoluble. Also known as elutriation. **2.** To grind a moist solid to a fine powder. { 'lev·ə‚gāt }

levo form [PHYSICAL CHEMISTRY] An optical isomer which induces levorotation in a beam of plane polarized light. { 'lē·vō ˌfȯrm }

levulinic acid [ORGANIC CHEMISTRY] $CH_3COCH_2CH_2COOH$ Crystalline compound forming plates or leaflets that melt at 37°C; freely soluble in alcohol, ether, and chloroform; used in the manufacture of pharmaceuticals, plastics, rubber, and synthetic fibers. { ¦lev·yə¦lin·ik 'as·əd }

Lewis acid [CHEMISTRY] A substance that can accept an electron pair from a base; thus, $AlCl_3$, BF_3, and SO_3 are acids. { 'lü·əs ˌas·əd }

Lewis base [CHEMISTRY] A substance that can donate an electron pair; examples are the hydroxide ion, OH^-, and ammonia, NH_3. { 'lü·əs ˌbās }

Lewis formula [CHEMISTRY] *See* Lewis structure. { 'lü·is ˌfȯr·myə·lə }

lewisite [ORGANIC CHEMISTRY] $C_2H_2AsCl_3$ An oily liquid, colorless to brown or violet; forms a toxic gas, used in World War I. { 'lü·əˌsīt }

Lewis structure [CHEMISTRY] A structural formula in which electrons are represented by dots; two dots between atoms represent a covalent bond. Also known as electron-dot formula; Lewis formula. { 'lü·is ˌstrək·chər }

lidocaine [ORGANIC CHEMISTRY] $C_{14}H_{22}N_2O$ A crystalline compound, used as a local anesthetic. Also known as lignocaine. { 'līd·əˌkān }

ligand [CHEMISTRY] The molecule, ion, or group bound to the central atom in a chelate or a coordination compound; an example is the ammonia molecules in $[Co(NH_3)_6]^{3+}$. { 'lī·gənd }

ligand membrane [CHEMISTRY] A solvent immiscible with water and a reagent and acting as an extractant and complexing agent for an ion. { 'lī·gənd 'memˌbrān }

light-scattering photometry [ANALYTICAL CHEMISTRY] Use of optical methods to measure the extent of scattering of light by particles suspended in fluids or by macromolecules in solution. { ¦līt ˌskad·ər·iŋ fə'täm·ə·trē }

lignin plastic [ORGANIC CHEMISTRY] A plastic based on resins derived from lignin; used as a binder or extender. { 'lig·nən 'plas·tik }

lignocaine [ORGANIC CHEMISTRY] *See* lidocaine. { 'lī·nəˌkān }

lignosulfonate [ORGANIC CHEMISTRY] Any of several substances manufactured from waste liquor of the sulfate pulping process of soft wood; used in the petroleum industry to reduce the viscosity of oil well muds and slurries, and as extenders in glues, synthetic resins, and cements. { ¦lig·nō'səl·fəˌnāt }

limiting current density [PHYSICAL CHEMISTRY] The maximum current density to achieve a desired electrode reaction before hydrogen or other extraneous ions are discharged simultaneously. { 'lim·əd·iŋ ¦kə·rənt ˌden·səd·ē }

limiting density [PHYSICAL CHEMISTRY] The density of a gas when the ratio of density per unit pressure is extrapolated to zero pressure, the point at which a gas exhibits ideal-gas behavior. { 'lim·ət·iŋ ˌden·səd·ē }

limiting mean [ANALYTICAL CHEMISTRY] The value that the average approaches as the number of measurements made in a stable chemical measurement process increases indefinitely. { ¦lim·əd·iŋ 'mēn }

limiting reagent [CHEMISTRY] In a chemical reaction, the reagent that controls the quantity of product which can be formed. { 'lim·əd·iŋ rēˌā·jənt }

limit of detection [ANALYTICAL CHEMISTRY] The quantity or concentration that represents the smallest measure of an analyte that can be detected with reasonable certainty by a given analytical procedure. { ¦lim·ət əv di'tek·shən }

limiting viscosity number [PHYSICAL CHEMISTRY] *See* intrinsic viscosity. { 'lim·əd·iŋ vi'skäs·əd·ē ˌnəm·bər }

limonene [ORGANIC CHEMISTRY] $C_{10}H_{16}$ A terpene with a lemon odor that is optically active and is found in oils from citrus fruits and in oils from peppermint and spearmint; a colorless, water-insoluble liquid that boils at 176°C. { 'lim·nəˌlēn }

linalool [ORGANIC CHEMISTRY] $(CH_3)_2C{:}CH(CH_2)_2CCH_3OHCH{:}CH_2$ A terpene that is a colorless liquid, has a bergamot odor, boils at 195-196°C, and is found in many essential oils, particularly bergamot and rosewood; used as a flavoring agent and in perfumes. Also known as coriandrol. { lə'näl·əˌwȯl }

linalyl acetate [ORGANIC CHEMISTRY] $(CH_3)_2C{:}CH(CH_2)_2CCH_3(OCOCH_3)CH{:}CH_2$ The acetic acid ester of linalool, a colorless oily liquid with a bergamot odor that boils at 108-110°C; used in perfumes and as a flavoring agent. { 'lin·ə͵lil 'as·ə ͵tāt }

linear molecule [PHYSICAL CHEMISTRY] A molecule whose atoms are arranged so that the bond angle between each is 180°; an example is carbon dioxide, CO_2. { 'lin·ē·ər 'mäl·ə͵kyül }

linear polymer [ORGANIC CHEMISTRY] A polymer whose molecule is arranged in a chainlike fashion with few branches or bridges between the chains. { 'lin·ē·ər 'päl·ə·mər }

line broadening [SPECTROSCOPY] An increase in the range of wavelengths over which the characteristic absorption or emission of a spectral line takes place, due to a number of causes such as collision broadening and Doppler broadening. { 'līn ͵bród·ən·iŋ }

line-formula method [ORGANIC CHEMISTRY] A system of notation for hydrocarbons showing the chemical elements, functional groups, and ring systems in linear form; an example is acetone, CH_3COCH_3. { 'līn ͵fór·myə·lə ͵meth·əd }

line pair [SPECTROSCOPY] In spectrographic analysis, a particular spectral line and the internal standard line with which it is compared to determine the concentration of a substance. { 'līn ͵per }

line-segment formula [ORGANIC CHEMISTRY] *See* bond-line formula. { 'līn ͵seg·mənt ͵fór·myə·lə }

line spectrum [SPECTROSCOPY] **1.** A spectrum of radiation in which the quantity being studied, such as frequency or energy, takes on discrete values. **2.** Conventionally, the spectra of atoms, ions, and certain molecules in the gaseous phase at low pressures; distinguished from band spectra of molecules, which consist of a pattern of closely spaced spectral lines which could not be resolved by early spectroscopes. { 'līn ͵spek·trəm }

linolenyl alcohol [ORGANIC CHEMISTRY] $C_{18}H_{32}O$ A colorless, combustible solid used for paints, paper, leather, and flotation processes. Also known as octadecatrienol. { ¦lin·ə¦lēn·əl 'al·kə͵hȯl }

lipophilic [CHEMISTRY] **1.** Having a strong affinity for fats. **2.** Promoting the solubilization of lipids. { ¦lip·ə¦fil·ik }

lipophobic [CHEMISTRY] Lacking an affinity for, repelling, or failing to absorb or adsorb fats. { ͵lip·ə'fōb·ik }

liquid chromatography [ANALYTICAL CHEMISTRY] A form of chromatography employing a liquid as the moving phase and a solid or a liquid on a solid support as the stationary phase; techniques include column chromatography, gel permeation chromatography, and partition chromatography. { 'lik·wəd ͵krō·mə'täg·rə·fē }

liquid crystal [PHYSICAL CHEMISTRY] A liquid which is not isotropic; it is birefringent and exhibits interference patterns in polarized light; this behavior results from the orientation of molecules parallel to each other in large clusters. { 'lik·wəd 'krist·əl }

liquid dioxide [INORGANIC CHEMISTRY] *See* nitrogen dioxide. { 'lik·wəd dī'äk͵sīd }

liquid glass [INORGANIC CHEMISTRY] *See* sodium silicate. { 'lik·wəd 'glas }

liquid hydrocarbon [ORGANIC CHEMISTRY] A hydrocarbon that has been converted from a gas to a liquid by pressure or by reduction in temperature; usually limited to butanes, propane, ethane, and methane. { 'lik·wəd 'hī·drə͵kär·bən }

liquid junction emf [PHYSICAL CHEMISTRY] The emf (electromotive force) generated at the area of contact between the salt bridge and the test solution in a pH cell electrode. { 'lik·wəd ¦jəŋk·shən ¦ē¦em'ef }

liquid junction potential [PHYSICAL CHEMISTRY] *See* diffusion potential. { 'lik·wəd ¦jəŋk·shən pə'ten·chəl }

liquid-liquid chemical reaction [CHEMISTRY] Chemical reaction in which the reactants, two or more, are liquids. { 'lik·wəd 'lik·wəd ¦kem·ə·kəl rē'ak·shən }

liquid-liquid distribution [CHEMISTRY] The process in which a dissolved substance is transferred from one liquid phase to another, immiscible liquid phase. { 'lik·wəd 'lik·wəd ͵dis·trə'byü·shən }

liquid-solid chemical reaction [CHEMISTRY] Chemical reaction in which at least one of the reactants is a liquid, and another of the reactants is a solid. { 'lik·wəd 'säl·əd ¦kem·ə·kəl rē'ak·shən }

liquid-solid equilibrium [PHYSICAL CHEMISTRY] *See* solid-liquid equilibrium. { 'lik·wəd 'säl·əd ˌē·kwə'lib·rē·əm }

liquid-vapor chemical reaction [CHEMISTRY] Chemical reaction in which at least one of the reactants is a liquid, and another of the reactants is a vapor. { 'lik·wəd ¦vā·pər ¦kem·ə·kəl rē'ak·shən }

liquid-vapor equilibrium [PHYSICAL CHEMISTRY] The equilibrium relationship between the liquid and its vapor phase for a partially vaporized compound or mixture at specified conditions of pressure and temperature; for mixtures, it is expressed by $K = x/y$, where K is the equilibrium constant, x the mole fraction of a key component in the vapor, and y the mole fraction of the same key component in the liquid. Also known as vapor-liquid equilibrium. { 'lik·wəd ¦vā·pər ˌē·kwə'lib·rē·əm }

lithamide [INORGANIC CHEMISTRY] *See* lithium amide. { 'lith·əˌmīd }

litharge [INORGANIC CHEMISTRY] *See* lead monoxide. { 'liˌthärj }

lithium [CHEMISTRY] A chemical element, symbol Li, atomic number 3, atomic weight 6.939; an alkali metal. { 'lith·ē·əm }

lithium aluminum hydride [INORGANIC CHEMISTRY] $LiAlH_2$ A compound made by the reaction of lithium hydride and aluminum chloride; a powerful reducing agent for specific linkages in complex molecules; used in organic synthesis. { 'lith·ē·əm ə'lü·mə·nəm 'hīˌdrīd }

lithium amide [INORGANIC CHEMISTRY] $LiNH_2$ A compound crystallizing in the cubic form, and melting at 380-400°C; used in organic synthesis. Also known as lithamide. { 'lith·ē·əm 'amˌīd }

lithium bromide [INORGANIC CHEMISTRY] $LiBr \cdot H_2O$ A white, deliquescent, granular powder with a bitter taste, melting at 547°C; soluble in alcohol and glycol; used to add moisture to air-conditioning systems and as a sedative and hypnotic in medicine. { 'lith·ē·əm 'brōˌmīd }

lithium carbonate [INORGANIC CHEMISTRY] Li_2CO_3 A colorless, crystalline compound that melts at 700°C and has slight solubility in water; used in ceramic industries in the manufacture of powdered glass for porcelain enamel formulation. { 'lith·ē·əm 'kär·bəˌnāt }

lithium cell [CHEMISTRY] An electrolytic cell for the production of metallic lithium. { 'lith·ē·əm ˌsel }

lithium chloride [INORGANIC CHEMISTRY] $LiCl \cdot 2H_2O$ A colorless, water-soluble compound, forming octahedral crystals and melting at 614°C; used to form concentrated brine in commercial air-conditioning systems and as a pyrotechnic in welding and brazing fluxes. { 'lith·ē·əm 'klȯrˌīd }

lithium citrate [ORGANIC CHEMISTRY] $Li_3C_6H_5O_7 \cdot 4H_2O$ White powder that decomposes when heated; slightly soluble in alcohol; soluble in water; used in beverages and pharmaceuticals. { 'lith·ē·əm 'sīˌtrāt }

lithium fluoride [INORGANIC CHEMISTRY] LiF Poisonous, white powder melting at 870°C, boiling at 1670°C; insoluble in alcohol, slightly soluble in water, and soluble in acids; used as a heat-exchange medium, as a welding and soldering flux, in ceramics, and as crystals in infrared instruments. { 'lith·ē·əm 'flu̇rˌīd }

lithium halide [INORGANIC CHEMISTRY] A binary compound of lithium, LiX, where X is a halide; examples are lithium chloride, LiCl, and lithium fluoride, LiF. { 'lith·ē·əm 'halˌīd }

lithium hydride [INORGANIC CHEMISTRY] LiH Flammable, brittle, white, translucent crystals; decomposes in water; insoluble in ether, benzene, and toluene; used as a hydrogen source and desiccant, and to prepare lithium amide and double hydrides. { 'lith·ē·əm 'hīˌdrīd }

lithium hydroxide [INORGANIC CHEMISTRY] LiOH; $LiOH \cdot H_2O$ Colorless crystals; used as a storage-battery electrolyte, as a carbon dioxide absorbent, and in lubricating greases and ceramics. { 'lith·ē·əm hī'dräkˌsīd }

lithium iodide [INORGANIC CHEMISTRY] LiI; $LiI \cdot 3H_2O$ White, water- and alcohol-soluble crystals; LiI melts at 446°C; LiI · $3H_2O$ loses water at 72°C; used in medicine, photography, and mineral waters. { 'lith·ē·əm 'ī·ə,dīd }

lithium molybdate [INORGANIC CHEMISTRY] Li_2MoO_4 Water-soluble white crystals melting at 705°C; used as a catalytic cracking (petroleum) catalyst and as a mill additive for steel. { 'lith·ē·əm mə'lib,dāt }

lithium nitrate [INORGANIC CHEMISTRY] $LiNO_3$ Water- and alcohol-soluble colorless powder melting at 261°C; used as a heat-exchange medium and in ceramics, pyrotechnics, salt baths, and refrigeration systems. { 'lith·ē·əm 'nī,trāt }

lithium perchlorate [INORGANIC CHEMISTRY] $LiClO_4 \cdot 3H_2O$ A compound with high oxygen content (60% available oxygen), used as a source of oxygen in rockets and missiles. { 'lith·ē·əm pər'klȯr,at }

lithium stearate [ORGANIC CHEMISTRY] $LiC_{18}H_{35}O_2$ A white, crystalline compound with a melting point of 220°C; used in cosmetics, plastics, and greases, and as a corrosion inhibitor in petroleum. { 'lith·ē·əm 'stir,āt }

lithium tetraborate [INORGANIC CHEMISTRY] $Li_2B_4O_7 \cdot 5H_2O$ White crystals that lose water at 200°C; insoluble in alcohol, soluble in water; used in ceramics. { 'lith·ē·əm ,te·trə'bȯr,āt }

lithium titanate [INORGANIC CHEMISTRY] Li_2TiO_3 A water-insoluble white powder with strong fluxing ability when used in titanium-containing enamels; also used as a mill additive in vitreous and semivitreous glazes. { 'lith·ē·əm 'tī·tən,āt }

Littrow grating spectrograph [SPECTROSCOPY] A spectrograph having a plane grating at an angle to the axis of the instrument, and a lens in front of the grating which both collimates and focuses the light. { 'li,trō ¦grād·iŋ 'spek·trə,graf }

Littrow mounting [SPECTROSCOPY] The arrangement of the grating and other components of a Littrow grating spectrograph, which is analogous to that of a Littrow quartz spectrograph. { 'li,trō ,maủnt·iŋ }

Littrow quartz spectrograph [SPECTROSCOPY] A spectrograph in which dispersion is accomplished by a Littrow quartz prism with a rear reflecting surface that reverses the light; a lens in front of the prism acts as both collimator and focusing lens. { 'li,trō ¦kwȯrts 'spek·trə,graf }

Lobry de Bruyn-Ekenstein transformation [ORGANIC CHEMISTRY] The change in which an aldose sugar treated with dilute alkali results in a mixture of an epimeric pair and 2-keto-hexose due to the production of enolic forms in the presence of hydroxyl ions, followed by a rearrangement. { lō¦brē də¦brīn 'ā·kən,shtīn ,trans·fər,mā·shən }

locant [CHEMISTRY] The portion of a chemical name, usually a number or a letter, that designates the position of an atom or group of atoms in a formula unit. { 'lō,kant }

London dispersion force [PHYSICAL CHEMISTRY] *See* van der Waals force. { 'lən·dən di'spər·zhən ,fȯrs }

lone-pair electrons [PHYSICAL CHEMISTRY] A nonbonding pair of electrons in the valence shell of an atom. { ¦lōn ¦per i'lek,tränz }

Loomis-Wood diagram [SPECTROSCOPY] A graph used to assign lines in a molecular spectrum to the various branches of rotational bands when these branches overlap, in which the difference between observed wave numbers and wave numbers extrapolated from a few lines that apparently belong to one branch are plotted against arbitrary running numbers for that branch. { 'lü·məs 'wu̇d ,dī·ə,gram }

lophine [ORGANIC CHEMISTRY] $C_{21}H_{16}O_2$ A colorless, crystalline, water-insoluble compound that melts at 275°C, used as an indicator in fluorescent neutralization tests. { 'lō,fēn }

Lorentz unit [SPECTROSCOPY] A unit of reciprocal length used to measure the difference, in wave numbers, between a (zero field) spectrum line and its Zeeman components; equal to $eH/4\pi mc^2$, where H is the magnetic field strength, c is the speed of light, and e and m are the charge and mass of the electron respectively (gaussian units). { 'lȯr,ens ,yü·nət }

lot [ANALYTICAL CHEMISTRY] A specimen of bulk material that is to undergo chemical analysis. { lät }

lot sample [ANALYTICAL CHEMISTRY] *See* gross sample. { 'lät ,sam·pəl }

loturine [ORGANIC CHEMISTRY] *See* harman. { ,lä·chə,rēn }

low-boiling butene-2 [ORGANIC CHEMISTRY] *See* butene-2. { 'lō ,bȯil·iŋ ¦byü,tēn 'tü }

low-density polyethylene [ORGANIC CHEMISTRY] A thermoplastic polymer with a density of 0.910-0.940 gram per cubic centimeter (0.526-0.543 ounce per cubic inch). Abbreviated LDPE. { 'lō ˌden·səd·ē ˌpäl·ē'eth·əˌlēn }

low-frequency spectrum [SPECTROSCOPY] Spectrum of atoms and molecules in the microwave region, arising from such causes as the coupling of electronic and nuclear angular momenta, and the Lamb shift. { 'lō ˌfrē·kwən·sē 'spek·trəm }

Lr [CHEMISTRY] *See* lawrencium.

LSD [ORGANIC CHEMISTRY] *See* lysergic acid diethylamide.

LSD-25 [ORGANIC CHEMISTRY] *See* lysergic acid diethylamide.

Lu [CHEMISTRY] *See* lutetium.

Luggin probe [PHYSICAL CHEMISTRY] A device which transmits a significant current density on the surface of an electrode to measure its potential. { 'ləg·ən ˌprōb }

Lugol solution [CHEMISTRY] A solution of 5 grams of iodine and 10 grams of potassium iodide per 100 milliliters of water; used in medicine. { 'lüˌgȯl sə'lü·shən }

luminol [ORGANIC CHEMISTRY] $C_8H_7N_3O_2$ A white, water-soluble, crystalline compound that melts at 320°C; used in an alkaline solution for analytical testing in chemistry. Also known as 3-aminophthalic hydrazide. { 'lü·məˌnȯl }

Lundegardh vaporizer [ANALYTICAL CHEMISTRY] A device used for emission flame photometry in which a compressed air aspirator vaporizes the solution within a chamber; smaller droplets are carried into the fuel-gas stream and to the burner orifice where the solvent is evaporated, dissociated, and optically excited. { 'lu̇n·dəˌgard 'vā·pəˌrīz·ər }

lupinidine [ORGANIC CHEMISTRY] *See* sparteine. { lü'pin·əˌdēn }

lutetium [CHEMISTRY] A chemical element, symbol Lu, atomic number 71, atomic weight 174.97; a very rare metal and the heaviest member of the rare-earth group. { lü'tē·shəm }

lyate ion [CHEMISTRY] The anion that is produced when a solvent molecule loses a proton (hydrogen nucleus), for example, the hydroxide ion is the lyate ion of water. { 'līˌāt ˌī·ən }

lycine [ORGANIC CHEMISTRY] *See* betaine. { 'līˌsēn }

lye [INORGANIC CHEMISTRY] **1.** A solution of potassium hydroxide or sodium hydroxide used as a strong alkaline solution in industry. **2.** The alkaline solution that is obtained from the leaching of wood ashes. { lī }

Lyman-alpha radiation [SPECTROSCOPY] Radiation emitted by hydrogen associated with the spectral line in the Lyman series whose wavelength is 121.5 nanometers. { 'lī·mən 'al·fə ˌrād·ē'ā·shən }

Lyman band [SPECTROSCOPY] A band in the ultraviolet spectrum of molecular hydrogen, extending from 125 to 161 nanometers. { 'lī·mən ˌband }

Lyman continuum [SPECTROSCOPY] A continuous range of wavelengths (or wave numbers or frequencies) in the spectrum of hydrogen at wavelengths less than the Lyman limit, resulting from transitions between the ground state of hydrogen and states in which the single electron is freed from the atom. { 'lī·mən kən'tin·yə·wəm }

Lyman ghost [SPECTROSCOPY] A false line observed in a spectroscope as a result of a combination of periodicities in the ruling. { 'lī·mən ˌgōst }

Lyman limit [SPECTROSCOPY] The lower limit of wavelengths of spectral lines in the Lyman series (912 angstrom units), or the corresponding upper limit in frequency, energy of quanta, or wave number (equal to the Rydberg constant for hydrogen). { 'lī·mən ˌlim·ət }

Lyman series [SPECTROSCOPY] A group of lines in the ultraviolet spectrum of hydrogen covering the wavelengths of 121.5-91.2 nanometers. { 'lī·mən ˌsir·ēz }

lyonium ion [CHEMISTRY] The cation that is produced when a solvent molecule is protonated. { lī'än·ē·əm ˌī·ən }

lyophilic [CHEMISTRY] Referring to a substance which will readily go into colloidal suspension in a liquid. { ¦lī·ə¦fil·ik }

lyophobic [CHEMISTRY] Referring to a substance in a colloidal state that has a tendency to repel liquids. { ¦lī·ə¦fō·bik }

lyotopic series [CHEMISTRY] *See* Hofmeister series. { ¦lī·ə¦täp·ik 'sir·ēz }

lyotropic liquid crystal [PHYSICAL CHEMISTRY] A liquid crystal prepared by mixing two or more components, one of which is polar in character (for example, water). { ¦lī·ə¦träp·ik ¦lik·wəd 'krist·əl }

lysergic acid [ORGANIC CHEMISTRY] $C_{16}H_{16}N_2O_2$ A compound that crystallizes in the form of hexagonal plates that melt and decompose at 240°C; derived from ergot alkaloids; used as a psychotomimetic agent. { lə'sər·jīk 'as·əd }

lysergic acid diethylamide [ORGANIC CHEMISTRY] $C_{15}H_{15}N_2CON(C_2H_5)_2$ A psychotomimetic drug synthesized from compounds derived from ergot. Abbreviated LSD; LSD-25. { lə'sərjik ¦as·əd dī‚eth·əl'am·əd }

M

M [CHEMISTRY] *See* molarity.

MAA [ORGANIC CHEMISTRY] *See* methanearsonic acid.

M acid [ORGANIC CHEMISTRY] $NH_2C_{10}H_5(OH)SO_3H$ A sulfonic acid formed by alkaline fusion of a disulfonic acid of α-naphthylamine; used as a dye intermediate. { 'em ˌas·əd }

Macquer's salt [INORGANIC CHEMISTRY] *See* potassium arsenate. { mə'kerz ˌsȯlt }

macroanalysis [ANALYTICAL CHEMISTRY] Qualitative or quantitative analysis of chemicals that are in quantities of the order of grams. { 'mak·rō·ə'nal·ə·səs }

macrocycle [ORGANIC CHEMISTRY] An organic molecule with a large ring structure, usually containing over 15 atoms. { 'mak·rōˌsī·kəl }

macrolide [ORGANIC CHEMISTRY] A large ring molecule with many functional groups bonded to it. { 'mak·rəˌlīd }

macromolecular [ORGANIC CHEMISTRY] Composed of or characterized by large molecules. { ¦mak·rō·mə'lek·yə·lər }

macromolecule [ORGANIC CHEMISTRY] A large molecule in which there is a large number of one or several relatively simple structural units, each consisting of several atoms bonded together. { ¦mak·rō'mäl·əˌkyül }

macropore [CHEMISTRY] A pore in a catalytic material whose width is greater than 0.05 micrometer. { 'mak·rəˌpor }

macroporous resin [ORGANIC CHEMISTRY] A member of a class of very small, highly cross-linked polymer particles penetrated by channels through which solutions can flow; used as ion exchanger. Also known as macroreticular resin. { ¦mak·rə¦pȯr·əs 'rez·ən }

macroreticular resin [ORGANIC CHEMISTRY] *See* macroporous resin. { ¦mak·rō·rə'tik·yə·lər 'rez·ən }

magenta [ORGANIC CHEMISTRY] *See* fuchsin. { mə'jen·tə }

magic acid [INORGANIC CHEMISTRY] A superacid consisting of equal molar quantities of fluorosulfonic acid (HSO_3F) and antimony pentafluoride (SbF_5). { 'maj·ik ¦as·əd }

magic numbers [PHYSICAL CHEMISTRY] Numbers of atoms or molecules for which certain atom or molecular clusters have an unusually high abundance. { 'maj·ik 'nəm·bərz }

magister of sulfur [CHEMISTRY] Amorphous sulfur produced by acid precipitation from solutions of hyposulfites or polysulfides. { mə'jis·tər əv 'səl·fər }

magnesia [INORGANIC CHEMISTRY] Magnesium oxide that is processed for a particular purpose. { mag'nē·zhə }

magnesia mixture [ANALYTICAL CHEMISTRY] Reagent used to analyze for phosphorus; consists of the filtered liquor from an aqueous mixture of ammonium chloride, magnesium sulfate, and ammonia. { mag'nē·zhə ˌmiks·chər }

magnesium [CHEMISTRY] A metallic element, symbol Mg, atomic number 12, atomic weight 24.312. { mag'nē·zē·əm }

magnesium acetate [ORGANIC CHEMISTRY] $Mg(OOCCH_3)_2 \cdot 4H_2O$ or $Mg(OOCCH_3)_2$ A compound forming colorless crystals that are soluble in water and melt at 80°C; used in textile printing, in medicine as an antiseptic, and as a deodorant. { mag'nē·zē·əm 'as·əˌtāt }

magnesium arsenate [INORGANIC CHEMISTRY] $Mg_3(AsO_4)_2 \cdot xH_2O$ A white, poisonous, water-insoluble powder used as an insecticide. { mag'nē·zē·əm 'ärs·ən,āt }

magnesium benzoate [ORGANIC CHEMISTRY] $Mg(C_7H_5O_2) \cdot 3H_2O$ A crystalline white powder melting at 200°C; soluble in alcohol and hot water; used in medicine. { mag'nē·zē·əm 'ben·zə,wāt }

magnesium borate [INORGANIC CHEMISTRY] $3MgO \cdot B_2O_3$ Crystals that are white or colorless and transparent; soluble in alcohol and acids, slightly soluble in water; used as a fungicide, antiseptic, and preservative. { mag'nē·zē·əm 'bȯr,āt }

magnesium bromate [INORGANIC CHEMISTRY] $Mg(BrO_3)_2 \cdot 6H_2O$ A white crystalline compound, insoluble in alcohol, soluble in water; a fire hazard; used as an analytical reagent. { mag'nē·zē·əm 'brō,māt }

magnesium bromide [INORGANIC CHEMISTRY] $MgBr_2 \cdot 6H_2O$ Deliquescent, colorless, bitter-tasting crystals, melting at 172°C; soluble in water, slightly soluble in alcohol; used in medicine and in the synthesis of organic chemicals. { mag'nē·zē·əm 'brō,mīd }

magnesium carbonate [INORGANIC CHEMISTRY] $MgCO_3$ A water-insoluble, white powder, decomposing at about 350°C; used as a refractory material. { mag'nē·zē·əm 'kär·bə,nāt }

magnesium chlorate [INORGANIC CHEMISTRY] $Mg(ClO_3)_2 \cdot 6H_2O$ A white powder, bitter-tasting and hygroscopic; slightly soluble in alcohols, soluble in water; used in medicine. { mag'nē·zē·əm 'klȯr,āt }

magnesium chloride [INORGANIC CHEMISTRY] $MgCl_2 \cdot 6H_2O$ Deliquescent white crystals; soluble in water and alcohol; used in disinfectants and fire extinguishers, and in ceramics, textiles, and paper manufacture. { mag'nē·zē·əm 'klȯr,īd }

magnesium fluoride [INORGANIC CHEMISTRY] MgF_2 White, fluorescent crystals; insoluble in water and alcohol, soluble in nitric acid; melts at 1263°C; used in ceramics and glass. Also known as magnesium flux. { mag'nē·zē·əm 'flu̇r,īd }

magnesium fluosilicate [INORGANIC CHEMISTRY] $MgSiF_6 \cdot 6H_2O$ Water-soluble, efflorescent white crystals; used in ceramics, in mothproofing and waterproofing, and as a concrete hardener. Also known as magnesium silicofluoride. { mag'nē·zē·əm 'flu̇·ə'sil·ə,kāt }

magnesium flux [INORGANIC CHEMISTRY] *See* magnesium fluoride. { mag'nē·zē·əm 'fləks }

magnesium formate [ORGANIC CHEMISTRY] $Mg(CHO_2)_2 \cdot 2H_2O$ Colorless, water-soluble crystals; insoluble in alcohol and ether; used in analytical chemistry and medicine. { mag'nē·zē·əm 'fȯr,māt }

magnesium gluconate [ORGANIC CHEMISTRY] $Mg(C_6H_{11}O_7)_2 \cdot 2H_2O$ An odorless, tasteless, water-soluble powder; used in medicine. { mag'nē·zē·əm 'glü·kə,nāt }

magnesium halide [INORGANIC CHEMISTRY] A compound formed from the metal magnesium and any of the halide elements; an example is magnesium bromide. { mag'nē·zē·əm 'ha,līd }

magnesium hydrate [INORGANIC CHEMISTRY] *See* magnesium hydroxide. { mag'nē·zē·əm 'hī,drāt }

magnesium hydride [INORGANIC CHEMISTRY] MgH_2 A hydride compound formed from the metal magnesium; it decomposes violently in water, and in a vacuum at about 280°C. { mag'nē·zē·əm 'hī,drīd }

magnesium hydroxide [INORGANIC CHEMISTRY] $Mg(OH)_2$ A white powder, very slightly soluble in water, decomposing at 350°C; used as an intermediate in extraction of magnesium metal, and as a reagent in the sulfite wood pulp process. Also known as magnesium hydrate. { mag'nē·zē·əm hī'dräk,sīd }

magnesium hyposulfite [INORGANIC CHEMISTRY] *See* magnesium thiosulfate. { mag'nē·zē·əm ,hī·pō'səl,fīt }

magnesium iodide [INORGANIC CHEMISTRY] $MgI_2 \cdot 8H_2O$ Crystalline powder, white and deliquescent, discoloring in air; soluble in water, alcohol, and ether; used in medicine. { mag'nē·zē·əm 'ī·ə,dīd }

magnesium lactate [ORGANIC CHEMISTRY] $Mg(C_3H_5O_3)_2 \cdot 3H_2O$ Bitter-tasting, water-soluble white crystals; slightly soluble in alcohol; used in medicine. { mag'nē·ze·əm 'lak,tāt }

magnesium methoxide [ORGANIC CHEMISTRY] $(CH_3O)_2Mg$ Color-less crystals that decompose when heated; used as a catalyst, dielectric coating, and cross-linking agent, and to form gels. Also known as magnesium methylate. { mag'nē·zē·əm me'thäk,sīd }

magnesium methylate [ORGANIC CHEMISTRY] *See* magnesium methoxide. { mag'nē·zē·əm 'meth·ə,lāt }

magnesium nitrate [INORGANIC CHEMISTRY] $Mg(NO_3)_2 \cdot 6H_2O$ Deliquescent white crystals; soluble in alcohol and water; a fire hazard; used as an oxidizing material in pyrotechnics. { mag'nē·zē·əm 'nī,trāt }

magnesium oleate [ORGANIC CHEMISTRY] $Mg(C_{18}H_{33}O_2)_2$ Water-insoluble, yellowish mass; soluble in hydrocarbons, alcohol, and ether; used as a plasticizer lubricant and emulsifying agent, and in varnish driers and dry-cleaning solutions. { mag'nē·zē·əm 'ō·lē,āt }

magnesium oxide [INORGANIC CHEMISTRY] MgO A white powder that (depending on the method of preparation) may be light and fluffy, or dense; melting point 2800°C; insoluble in acids, slightly soluble in water; used in making refractories, and in cosmetics, pharmaceuticals, insulation, and medicine. { mag'nē·zē·əm 'äk,sīd }

magnesium perchlorate [INORGANIC CHEMISTRY] $Mg(ClO_4)_2 \cdot 6H_2O$ White, deliquescent crystals; soluble in water and alcohol; explosive when in contact with reducing materials; used as a drying agent for gases. { mag'nē·zē·əm pər'klȯr,āt }

magnesium peroxide [INORGANIC CHEMISTRY] MgO_2 A tasteless, odorless white powder; soluble in dilute acids, insoluble in water; a fire hazard; used as a bleaching and oxidizing agent, and in medicine. { mag'nē·zē·əm pə'räk,sīd }

magnesium phosphate [INORGANIC CHEMISTRY] A compound with three forms: monobasic, $MgH_4(PO_4)_2 \cdot 2H_2O$, used in medicine and wood fireproofing; dibasic, $MgHPO_4 \cdot 3H_2O$, used in medicine and as a plastics stabilizer, tribasic, $Mg_3(PO_4)_2 \cdot 8H_2O$, used in dentifrices, as an adsorbent, and in pharmaceuticals. { mag'nē·zē·əm 'fäs,fāt }

magnesium salicylate [ORGANIC CHEMISTRY] $Mg(C_7H_5O)_3 \cdot 4H_2O$ Efflorescent colorless crystals; soluble in water and alcohol; used in medicine. { mag'nē·zē·əm sə'lis·ə,lāt }

magnesium silicate [INORGANIC CHEMISTRY] $3MgSiO_3 \cdot 5H_2O$ White, water insoluble powder, containing variable proportions of water of hydration; used as a filler for rubber and in medicine. { mag'nē·zē·əm 'sil·ə,kāt }

magnesium silicofluoride [INORGANIC CHEMISTRY] *See* magnesium fluosilicate. { mag'nē·zē·əm ,sil·ə·kō'flu̇r,īd }

magnesium stearate [ORGANIC CHEMISTRY] $Mg(C_{18}H_{35}O_2)_2$ Tasteless, odorless white powder; soluble in hot alcohol, insoluble in water; melts at 89°C, used in paints and medicine, and as a plastics stabilizer and lubricant. Also known as dolomol. { mag'nē·zē·əm 'stir,āt }

magnesium sulfate [INORGANIC CHEMISTRY] $MgSO_4$ Colorless crystals with a bitter, saline taste; soluble in glycerol; used in fireproofing, textile processes, ceramics, cosmetics, and fertilizers. { mag'nē·zē·əm 'səl,fāt }

magnesium sulfite [INORGANIC CHEMISTRY] $MgSO_3 \cdot 6H_2O$ A white, crystalline powder; insoluble in alcohol, slightly soluble in water; used in medicine and paper pulp. { mag'nē·zē·əm 'səl,fīt }

magnesium thiosulfate [INORGANIC CHEMISTRY] $MgS_2O_3 \cdot 6H_2O$ Colorless crystals that lose water at 170°C; used in medicine. Also known as magnesium hyposulfite. { mag'nē·zē·əm ,thī·ə'səl,fāt }

magnesium trisilicate [INORGANIC CHEMISTRY] $Mg_2Si_3O_8 \cdot 5H_2O$ A white, odorless, tasteless powder; insoluble in water and alcohol; used as an industrial odor absorbent and in medicine. { mag'nē·zē·əm ,trī'sil·ə,kāt }

magnesium tungstate [INORGANIC CHEMISTRY] $MgWoO_4$ White crystals, insoluble in alcohol and water, soluble in acid; used in luminescent paint and for fluorescent x-ray screens. { mag'nē·zē·əm 'təŋ,stāt }

magneson [ORGANIC CHEMISTRY] $C_{12}H_9N_3O_4$ A brownish-red powder, soluble in dilute aqueous sodium hydroxide; used in the detection of magnesium and molybdenum. { 'mag·nə,sän }

magnetic scanning [SPECTROSCOPY] The magnetic field sorting of ions into their respective spectrums for analysis by mass spectroscopy; accomplished by varying the magnetic field strength while the electrostatic field is held constant. { mag'ned·ik 'skan·iŋ }

magnetochemistry [PHYSICAL CHEMISTRY] A branch of chemistry which studies the interrelationship between the bulk magnetic properties of a substance and its atomic and molecular structure. { mag¦nēd·ō'kem·ə·strē }

magnetofluid [PHYSICAL CHEMISTRY] A Newtonian or shear-thinning fluid whose flow properties become viscoplastic when it is modulated by a magnetic field. { ¦mag·nəd·ō'flü·əd }

malathion [ORGANIC CHEMISTRY] $C_{10}H_{19}O_6PS_2$ A yellow liquid, slightly soluble in water; malathion is the generic name for S-1,2-bis(ethoxycarbonyl)ethyl O,O-dimethylphosphorodithioate; used as an insecticide. { ˌmal·ə'thīˌän }

maleate [ORGANIC CHEMISTRY] An ester or salt of maleic acid. { mə'lēˌāt }

maleic acid [ORGANIC CHEMISTRY] HOOCCH:CHCOOH A colorless, crystalline dibasic acid; soluble in water, acetone, and alcohol; melting point 130-131°C; used in textile processing, and as an oil and fat preservative. { mə'lā·ik 'as·əd }

maleic anhydride [ORGANIC CHEMISTRY] $C_4H_2O_3$ Colorless crystals, soluble in acetone, hydrolyzing in water; used to form polyester resins. Also known as 2,5-furandione. { mə'lā·ik an'hīˌdrīd }

maleic hydrazide [ORGANIC CHEMISTRY] $C_4N_2H_4O_2$ Solid material, decomposing at 260°C; slightly soluble in alcohol and water; used as a weed killer and growth inhibitor. { mə'lā·ik 'hī·drəˌzīd }

malonamide nitrile [ORGANIC CHEMISTRY] *See* cyanoacetamide. { mə'län·ə·məd 'nī·trəl }

malonic acid [ORGANIC CHEMISTRY] $CH_2(COOH)_2$ A white, crystalline dicarboxylic acid, melting at 132-134°C; used to manufacture pharmaceuticals. { mə'län·ik 'as·əd }

malonic ester [ORGANIC CHEMISTRY] *See* ethyl malonate. { mə'län·ik 'es·tər }

malonic mononitrile [ORGANIC CHEMISTRY] *See* cyanoacetic acid. { mə'län·ik ¦män·ō'nī·trəl }

malonyl [ORGANIC CHEMISTRY] $CH_2(COO)_2$ A bivalent functional group formed from malonic acid. { 'mal·əˌnil }

maltol [ORGANIC CHEMISTRY] $C_6H_6O_3$ Crystalline substance with a melting point of 161-162°C and a fragrant caramellike odor; used as a flavoring agent in bread and cakes. Also known as larixinic acid. { 'mȯlˌtōl }

mandelic acid [ORGANIC CHEMISTRY] $C_6H_5CHOHCOOH$ A white, crystalline compound, melting at 117-119°C, darkening upon exposure to light; used in organic synthesis. { man'del·ik 'as·əd }

mandelic acid nitrile [ORGANIC CHEMISTRY] *See* mandelonitrile. { man'del·ik 'as·əd 'nī·trəl }

mandelonitrile [ORGANIC CHEMISTRY] $C_6H_5CH(OH)CN$ A liquid used to prepare bitter almond water. Also known as mandelic acid nitrile. { man¦del·ō'nī·trəl }

maneb [ORGANIC CHEMISTRY] $Mn[SSCH(CH_2)_2NHCSS]$ A generic term for manganese ethylene-1,2-bisdithiocarbamate; irritating to eyes, nose, skin, and throat; used as a fungicide. { 'maˌneb }

manganate [INORGANIC CHEMISTRY] **1.** Salts that have manganese in the anion. **2.** In particular, a salt of manganic acid formed by fusion of manganese dioxide with an alkali. { 'maŋ·gəˌnāt }

manganese [CHEMISTRY] A metallic element, symbol Mn, atomic weight 54.938, atomic number 25; a transition element whose properties fall between those of chromium and iron. { 'maŋ·gəˌnēs }

manganese acetate [ORGANIC CHEMISTRY] $Mn(C_2H_3O_2)_2 \cdot 4H_2O$ A pale-red crystalline compound melting at 80°C; soluble in water and alcohol; used in textile dyeing, as a catalyst, and for leather tanning. { 'maŋ·gəˌnēs 'as·əˌtāt }

manganese binoxide [INORGANIC CHEMISTRY] *See* manganese dioxide. { 'maŋ·gəˌnēs bi'näkˌsīd }

manganese black [INORGANIC CHEMISTRY] *See* manganese dioxide. { 'maŋ·gə,nēs 'blak }

manganese borate [INORGANIC CHEMISTRY] MnB_4O_7 Water-insoluble, reddish-white powder; used as a varnish and oil drier. { 'maŋ·gə,nēs 'bȯr,āt }

manganese bromide [INORGANIC CHEMISTRY] *See* manganous bromide. { 'maŋ·gə,nēs 'brō,mīd }

manganese carbonate [INORGANIC CHEMISTRY] $MnCO_3$ Rose-colored crystals found in nature as rhodocrosite; soluble in dilute acids, insoluble in water; used in medicine, in fertilizer, and as a paint pigment. { 'maŋ·gə,nēs 'kär·bə,nāt }

manganese citrate [ORGANIC CHEMISTRY] $Mn_3(C_6H_5O_7)_2$ A white powder, water-insoluble in the presence of sodium citrate; used in medicine. { 'maŋ·gə,nēs 'sī,trāt }

manganese dioxide [INORGANIC CHEMISTRY] MnO_2 A black, crystalline, water-insoluble compound, decomposing to manganese sesquioxide, Mn_2O_3, and oxygen when heated to 535°C; used as a depolarizer in certain dry-cell batteries, as a catalyst, and in dyeing of textiles. Also known as battery manganese; manganese binoxide; manganese black; manganese peroxide. { 'maŋ·gə,nēs dī'äk,sīd }

manganese fluoride [INORGANIC CHEMISTRY] *See* manganous fluoride. { 'maŋ·gə,nēs 'flu̇r,īd }

manganese gluconate [ORGANIC CHEMISTRY] $Mn(C_6H_{11}O_7)_2 \cdot 2H_2O$ A pinkish powder, insoluble in benzene and alcohol, soluble in water; used in medicine, in vitamin tablets, and as a feed additive and dietary supplement. { 'maŋ·gə,nēs 'glü·kə,nāt }

manganese green [INORGANIC CHEMISTRY] *See* barium manganate. { 'maŋ·gə,nēs 'grēn }

manganese halide [INORGANIC CHEMISTRY] Compound of manganese with a halide, such as chlorine, bromine, fluorine, or iodine. { 'maŋ·gə,nēs 'ha,līd }

manganese heptoxide [INORGANIC CHEMISTRY] Mn_2O_7 A compound formed as an explosive dark-green oil by the action of concentrated sulfuric acid on permanganate compounds. { 'maŋ·gə,nēs hep'täk,sīd }

manganese hydroxide [INORGANIC CHEMISTRY] *See* manganous hydroxide. { 'maŋ·gə,nēs hī'dräk,sīd }

manganese hypophosphite [INORGANIC CHEMISTRY] $Mn(H_2PO_2)_2 \cdot H_2O$ Odorless, tasteless pink crystals which explode if heated with oxidants; used in medicine. { 'maŋ·gə,nēs ,hī·pō'fäs,fīt }

manganese lactate [ORGANIC CHEMISTRY] $Mn(C_3H_5O_3)_2 \cdot 3H_2O$ Pale-red crystals; insoluble in water and alcohol; used in medicine. { 'maŋ·gə,nēs 'lak,tāt }

manganese linoleate [ORGANIC CHEMISTRY] $Mn(C_{18}H_{31}O_2)_2$ A dark-brown mass, soluble in linseed oil; used in pharmaceutical preparations and as a varnish and paint drier. { 'maŋ·gə,nēs lə'nō·lē,āt }

manganese monoxide [INORGANIC CHEMISTRY] *See* manganese oxide. { 'maŋ·gə,nēs mə'näk,sīd }

manganese naphthenate [ORGANIC CHEMISTRY] Hard brown resinous mass, soluble in mineral spirits; melts at 135°C; contains 6% manganese in commercial solutions; used as a paint and varnish drier. { 'maŋ·gə,nēs 'naf·thə,nāt }

manganese oleate [ORGANIC CHEMISTRY] $Mn(C_{18}H_{33}O_2)_2$ Granular brown mass, soluble in oleic acid and ether, insoluble in water; used in medicine and as a varnish drier. { 'maŋ·gə,nēs 'ō·lē,āt }

manganese oxalate [ORGANIC CHEMISTRY] $MnC_2O_4 \cdot 2H_2O$ A white crystalline compound, soluble in dilute acids, only slightly soluble in water; used as a paint and varnish drier. { 'maŋ·gə,nēs 'äk·sə,lāt }

manganese oxide [INORGANIC CHEMISTRY] MnO Green powder, soluble in acids, insoluble in water; melts at 1650°C; used in medicine, in textile printing, as a catalyst, in ceramics, and in dry batteries. Also known as manganese monoxide; manganous oxide. { 'maŋ·gə,nēs 'äk,sīd }

manganese peroxide [INORGANIC CHEMISTRY] *See* manganese dioxide. { 'maŋ·gə,nēs pə'räk,sīd }

manganese resinate [ORGANIC CHEMISTRY] $Mn(C_{20}H_{29}O_2)_2$ Water-insoluble mass, flesh-colored or brownish black; used as a varnish and oil drier. { 'maŋ·gə,nēs 'rez·ən,āt }

manganese silicate [INORGANIC CHEMISTRY] *See* manganous silicate. { 'maŋ·gə,nēs 'sil·ə,kāt }

manganese sulfate [INORGANIC CHEMISTRY] *See* manganous sulfate. { 'maŋ·gə,nēs 'səl,fāt }

manganese sulfide [INORGANIC CHEMISTRY] *See* manganous sulfide. { 'maŋ·gə,nēs 'səl,fīd }

manganic fluoride [INORGANIC CHEMISTRY] MnF_3 Poisonous red crystals, decomposed by heat and water; used as a fluorinating agent. { man'gan·ik 'flůr,īd }

manganic hydroxide [INORGANIC CHEMISTRY] $Mn(OH)_3$ A brown powder that rapidly loses water to form MnO(OH); used in ceramics and as a fabric pigment. Also known as hydrated manganic hydroxide. { man'gan·ik hī'dräk,sīd }

manganic oxide [INORGANIC CHEMISTRY] Mn_2O_3 Hard black powder, insoluble in water, soluble in cold hydrochloric acid, hot nitric acid, and sulfuric acid; occurs in nature as manganite. Also known as manganese sesquioxide. { man'gan·ik 'äk,sīd }

manganous bromide [INORGANIC CHEMISTRY] $MnBr_2 \cdot 4H_2O$ Water-soluble, deliquescent red crystals. Also known as manganese bromide. { 'maŋ·gə·nəs 'brō,mīd }

manganous chloride [INORGANIC CHEMISTRY] $MnCl_2 \cdot 4H_2O$ Water-soluble, deliquescent rose-colored crystals melting at 88°C; used as a catalyst and in paints, dyeing, and pharmaceutical preparations. { 'maŋ·gə·nəs 'klȯr,īd }

manganous fluoride [INORGANIC CHEMISTRY] MnF_2 Reddish powder, insoluble in water, soluble in acid. Also known as manganese fluoride. { 'maŋ·gə·nəs 'flůr,īd }

manganous hydroxide [INORGANIC CHEMISTRY] $Mn(OH)_2$ Heat-decomposable white-pink crystals; insoluble in water and alkali, soluble in acids; occurs in nature as pyrochroite. Also known as manganese hydroxide. { 'maŋ·gə·nəs hī'dräk,sīd }

manganous iodide [INORGANIC CHEMISTRY] $MnI_2 \cdot 4H_2O$ Water-soluble, deliquescent yellowish-brown crystals. Also known as manganese iodide. { 'maŋ·gə·nəs 'ī·ə,dīd }

manganous silicate [INORGANIC CHEMISTRY] $MnSiO_3$ Water-insoluble red crystals or yellowish-red powder; occurs in nature as rhodonite. Also known as manganese silicate. { 'maŋ·gə·nəs 'sil·ə,kāt }

manganous sulfate [INORGANIC CHEMISTRY] $MnSO_4 \cdot 4H_2O$ Water-soluble, translucent, efflorescent rose-red prisms; melts at 30°C; used in medicine, textile printing, and ceramics, as a fungicide and fertilizer, and in paint manufacture. Also known as manganese sulfate. { 'maŋ·gə·nəs 'səl,fāt }

manganous sulfide [INORGANIC CHEMISTRY] MnS An almost water-insoluble powder that decomposes on heating; used as a pigment and as an additive in making steel. Also known as manganese sulfide. { 'maŋ·gə·nəs 'səl,fīd }

manganous sulfite [INORGANIC CHEMISTRY] $MnSO_3$ Grayish-black or brownish-red powder, soluble in sulfur dioxide, insoluble in water. { 'maŋ·gə·nəs 'səl,fīt }

manna sugar [ORGANIC CHEMISTRY] *See* mannitol. { 'man·ə ,shůg·ər }

Mannich condensation reaction [ORGANIC CHEMISTRY] *See* Mannich reaction. { 'män·ik ,kän·dən'sā·shən rē,ak·shən }

Mannich reaction [ORGANIC CHEMISTRY] Condensation of a primary or secondary amine or ammonia (usually as the hydrochloride) with formaldehyde and a compound containing at least one reactive hydrogen atom, for example, acetophenone. Also known as Mannich condensation reaction. { 'män·ik rē,ak·shən }

mannite [ORGANIC CHEMISTRY] *See* mannitol. { 'ma,nīt }

mannitol [ORGANIC CHEMISTRY] $C_6H_8(OH)_6$ A straight-chain alcohol with six hydroxyl groups; a white, water-soluble, crystalline powder; used in medicine and as a dietary supplement. Also known as manna sugar; mannite. { 'man·ə,tȯl }

mannitol hexanitrate [ORGANIC CHEMISTRY] $C_6H_8(ONO_2)_6$ Explosive colorless crystals; soluble in alcohol, acetone, and ether, insoluble in water; melts at 112°C; used in explosives and medicine. { 'man·ə,tȯl ,hek·sə'nī,trāt }

manure salts [INORGANIC CHEMISTRY] Potash salts that have a high proportion of chloride and 20-30% potash; used in fertilizers. { mə'nůr ˌsȯlts }

margaric acid [ORGANIC CHEMISTRY] *See* *n*-heptadecanoic acid. { mär'gär·ik 'as·əd }

Mark-Houwink equation [PHYSICAL CHEMISTRY] The relationship between intrinsic viscosity and molecular weight for homogeneous linear polymers. { 'märk 'hau̇ˌwiŋk iˌkwā·zhən }

Markovnikoff's rule [ORGANIC CHEMISTRY] In an addition reaction, the additive molecule RH adds as H and R, with the R going to the carbon atom with the lesser number of hydrogen atoms bonded to it. { mär'kȯv·nəˌkȯfs ˌrül }

Marsh-Berzelius test [ANALYTICAL CHEMISTRY] *See* Marsh test. { 'märsh ber'zā·lē·əs ˌtest }

Marsh test [ANALYTICAL CHEMISTRY] A test for the presence of arsenic in a compound; the substance to be tested is mixed with granular zinc, and dilute hydrochloric acid is added to the mixture; gaseous arsine forms, which decomposes to a black deposit of arsenic, when the gas is passed through a heated glass tube. Also known as Marsh-Berzelius test. { 'märsh ˌtest }

Mars pigments [INORGANIC CHEMISTRY] A group of five pigments produced when milk of lime is added to a ferrous sulfate solution, and the precipitate is calcined; color is controlled by calcination temperature to give yellow, orange, brown, red, or violet. { 'märz ˌpig·məns }

masking agent [ANALYTICAL CHEMISTRY] *See* masking reagent. { 'mask·iŋ ˌā·jənt }

masking reagent [ANALYTICAL CHEMISTRY] A substance that decreases the concentration of a free metal ion or ligand by conversion into an essentially unreactive form, thus preventing undesirable chemical reactions that would interfere with the determination. Also known as masking agent. { 'mask·iŋ rēˌā·jənt }

mass action law [PHYSICAL CHEMISTRY] The law that the rate of a chemical reaction for a uniform system at constant temperature is proportional to the concentrations of the substances reacting. Also known as Guldberg and Waage law. { 'mas ¦ak·shən ˌlȯ }

mass-analyzed ion kinetic energy spectrometry [SPECTROSCOPY] A type of ion kinetic energy spectrometry in which the ionic products undergo mass analysis followed by energy analysis. Abbreviated MIKES. { 'mas ¦an·əˌlīzd 'īˌän kə¦ned·ik 'en·ər·jē spek'träm·ə·trē }

Massenfilter [ANALYTICAL CHEMISTRY] *See* quadrupole spectrometer. { 'mäs·ənˌfil·tər }

mass spectrometry [ANALYTICAL CHEMISTRY] An analytical technique for identification of chemical structures, determination of mixtures, and quantitative elemental analysis, based on application of the mass spectrometer. { 'mas spek'träm·ə·trē }

mass susceptibility [PHYSICAL CHEMISTRY] Magnetic susceptibility of a compound per gram. Also known as specific susceptibility. { 'mas səˌsep·tə'bil·əd·ē }

mass-to-charge ratio [ANALYTICAL CHEMISTRY] In analysis by mass spectroscopy, the measurement of the sample mass as a ratio to its ionic charge. { ¦mas tə 'chärj ˌrā·shō }

matrix [ANALYTICAL CHEMISTRY] The analyte as considered in terms of its being an assemblage of constituents, each with its own properties. { 'mā·triks }

matrix effects [ANALYTICAL CHEMISTRY] **1.** The enhancement or suppression of minor element spectral lines from metallic oxides during emission spectroscopy by the matrix element (such as graphite) used to hold the sample. **2.** The combined effect exerted by the various constituents of the matrix on the measurements of the analysis. { 'mā·triks iˌfeks }

matrix isolation [SPECTROSCOPY] A spectroscopic technique in which reactive species can be characterized by maintaining them in a very cold, inert environment while they are examined by an absorption, electron-spin resonance, or laser excitation spectroscope. { 'mā·triks ˌī·sə'lā·shən }

matrix spectrophotometry [SPECTROSCOPY] Spectrophotometric analysis in which the specimen is irradiated in sequence at more than one wavelength, with the visible spectrum evaluated for the energy leaving for each wavelength of irradiation. { 'mā·triks ˌspek·trō·fə'täm·ə·trē }

MBT [ORGANIC CHEMISTRY] *See* mercaptobenzothiazole.

Md [CHEMISTRY] *See* mendelevium.

measured spectrum [SPECTRUM] *See* spectrogram. { 'mezh·ərd 'spek·trəm }

mechanism [CHEMISTRY] A detailed description of the course of a chemical reaction as it proceeds from the reactants to the products, with as complete a characterization as possible of the reaction steps and intermediate species. Also known as reaction path. { 'mek·ə,niz·əm }

mechanochemical effect [PHYSICAL CHEMISTRY] Changes in the dimensions of certain polymers, particularly photoelectrolytic gels and crystalline polymers, in response to changes in their chemical environment. { ¦mek·ə·nō'kem·ə·kəl i,fekt }

mechanochemistry [PHYSICAL CHEMISTRY] The study of the conversion of mechanical energy into chemical energy in polymers. { ¦mek·ə·nō'kem·ə·strē }

mechanophotochemistry [PHYSICAL CHEMISTRY] The study of changes in the dimensions of certain photoresponsive polymers upon exposure to light. { ¦mek·ə·nō,fō·dō'kem·ə·strē }

meconin [ORGANIC CHEMISTRY] $C_{10}H_{10}O_4$ A neutral principle of opium; white crystals, soluble in hot water and alcohol and melting at 102-103°C. Also known as opianyl. { 'mek·ə·nən }

MEK [ORGANIC CHEMISTRY] *See* methyl ethyl ketone.

melamine [ORGANIC CHEMISTRY] $C_3H_6N_4$ A white crystalline compound that is slightly soluble in water, melts at 354°C and is a cyclic trimer of cyanamide; used to make melamine resins and in tanning of leather. { 'mel·ə,mēn }

melaniline [ORGANIC CHEMISTRY] *See* diphenylguanidine. { mel'an·ə·lən }

melissic acid [ORGANIC CHEMISTRY] $CH_3(CH_2)_{28}COOH$ Fatty acid found in beeswax; soluble in benzene and hot alcohol; melts at 90°C; used in biochemical research. { mə'lis·ik 'as·əd }

mellitate [ORGANIC CHEMISTRY] An ester or salt of mellitic acid. { 'mel·ə,tāt }

mellitic acid [ORGANIC CHEMISTRY] $C_6(COOH)_6$ A water-soluble compound forming colorless needles that melt at 287°C. { mə'lid·ik 'as·əd }

melt [CHEMISTRY] **1.** To change a solid to a liquid by the application of heat. **2.** A melted material. { melt }

melting [PHYSICAL CHEMISTRY] *See* fusion. { 'melt·iŋ }

membrane mimetic chemistry [ORGANIC CHEMISTRY] The study of processes and reactions that have been developed by using information obtained from biological membrane systems. { ¦mem,brān mi¦med·ik 'kem·ə·strē }

MEMC [ORGANIC CHEMISTRY] *See* methoxyethylmercury chloride.

mendelevium [CHEMISTRY] Synthetic radioactive element, symbol Md, with atomic number 101; made by bombarding lighter elements with light nuclei accelerated in cyclotrons. { ,men·də'lē·vē·əm }

menthane [ORGANIC CHEMISTRY] $C_{10}H_{20}$ A colorless, water-insoluble liquid hydrocarbon; used in organic synthesis. { 'men,thān }

menthene [ORGANIC CHEMISTRY] $C_{10}H_{18}$ A colorless, water-insoluble, liquid hydrocarbon; used in organic synthesis. { 'men,thēn }

menthol [ORGANIC CHEMISTRY] $CH_3C_6H_9(C_3H_7)OH$ An alcohol-soluble, white crystalline compound that may exist in levo form or a mixture of dextro and levo isomers; used in medicines and perfumes, and as a flavoring agent. Also known as peppermint camphor. { 'men,thȯl }

menthone [ORGANIC CHEMISTRY] $C_{10}H_{18}O$ Oily, colorless ketonic liquid with slight peppermint odor; slightly soluble in water, soluble in organic solvents. { 'men,thōn }

menthyl [ORGANIC CHEMISTRY] $C_{10}H_{19}$ A univalent radical that is derived from menthol by removal of the hydroxyl group. { 'men·thəl }

meq [CHEMISTRY] *See* milliequivalent. { mek }

-mer [ORGANIC CHEMISTRY] A combining form denoting the repeating structure unit of any high polymer. { mər }

merbromin [ORGANIC CHEMISTRY] $C_{20}H_8O_6Na_2Br_2Hg$ A green crystalline powder that gives a deep-red solution in water; used as an antiseptic. { mər'brō·mən }

mercaptal [ORGANIC CHEMISTRY] A group of organosulfur compounds that contain the group =C(SR)$_2$. { mər'kap,tal }

mercaptan [ORGANIC CHEMISTRY] A group of organosulfur compounds that are derivatives of hydrogen sulfide in the same way that alcohols are derivatives of water; have a characteristically disagreeable odor, and are found with other sulfur compounds in crude petroleum; an example is methyl mercaptan. Also known as thiol. { mər'kap,tan }

mercaptide [ORGANIC CHEMISTRY] A compound consisting of a metal and a mercaptan. { mər'kap,tīd }

mercapt-, mercapto- [CHEMISTRY] A combining form denoting the presence of the thiol (SH) group. { mər'kap·tō }

mercaptoacetic acid [ORGANIC CHEMISTRY] *See* thioglycolic acid. { mər¦kap·tō·ə¦sēd·ik 'as·əd }

2-mercaptobenzoic acid [ORGANIC CHEMISTRY] *See* thiosalicylic acid. { ¦tü mər¦kap·tō·ben'zō·ik 'as·əd }

mercaptobenzothiazole [ORGANIC CHEMISTRY] C_7H_5NS A yellow powder, melting at 164-174°C; used in rubber as a vulcanization accelerator with stearic acid. Abbreviated MBT. { mər¦kap·tō,ben·zō'thī·ə,zōl }

mercapto compound [CHEMISTRY] *See* sulfhydryl compound. { mər'kap·tō ,käm,paund }

mercaptoethanol [ORGANIC CHEMISTRY] $HSCH_2CH_2OH$ Mobile liquid, water-white; soluble in water, benzene, ether, and most organic solvents; boils at 157°C; used as a solvent, chemical intermediate, and reducing agent. { mər,kap·tō'eth·ə,nól }

mercaptol [ORGANIC CHEMISTRY] A compound formed by combining a mercaptal and a ketone. { mər'kap,tól }

mercaptosuccinic acid [ORGANIC CHEMISTRY] *See* thiomalic acid. { mər,kap·tō·sək'sin·ik 'as·əd }

mercuric [INORGANIC CHEMISTRY] The mercury ion with a 2+ oxidation state, for example $Hg(NO_3)_2$. { mər'kyůr·ik }

mercuric acetate [ORGANIC CHEMISTRY] $Hg(C_2H_3O_2)_2$ Poisonous, light-sensitive white crystals; soluble in alcohol and water; used in medicine and as a catalyst in organic synthesis. Also known as mercury acetate. { mər'kyůr·ik 'as·ə,tāt }

mercuric arsenate [INORGANIC CHEMISTRY] $HgHAsO_4$ A poisonous yellow powder; soluble in hydrochloric acid, insoluble in water; used in antifouling and waterproof paints and in medicine. Also known as mercury arsenate; mercury arseniate. { mər'kyůr·ik 'ärs·ən,āt }

mercuric barium iodide [INORGANIC CHEMISTRY] $HgI_2 \cdot BaI_2 \cdot 5H_2O$ Crystals that are yellow or reddish and deliquescent; soluble in alcohol and water; used in aqueous solution as Rohrbach's solution for mineral separation on the basis of density. Also known as barium mercury iodide; mercury barium iodide. { mər'kyůr·ik 'bar·ē·əm 'ī·ə,dīd }

mercuric benzoate [ORGANIC CHEMISTRY] $Hg(C_7H_5O_2)_2 \cdot H_2O$ Poisonous white crystals, sensitive to light, melting at 165°C; slightly soluble in alcohol and water; used in medicine. Also known as mercury benzoate. { mər'kyůr·ik 'ben·zə,wāt }

mercuric bromide [INORGANIC CHEMISTRY] $HgBr_2$ Poisonous white crystals, sensitive to light, melting at 235°C; soluble in alcohol and ether; used in medicine. Also known as mercury bromide. { mər'kyůr·ik 'brō,mīd }

mercuric chloride [INORGANIC CHEMISTRY] $HgCl_2$ An extremely toxic compound that forms white, rhombic crystals which sublime at 300°C and are soluble in alcohol or benzene; used for the manufacture of other mercuric compounds, as a fungicide, and in medicine and photography. Also known as bichloride of mercury; corrosive sublimate. { mər'kyůr·ik 'klór,īd }

mercuric cyanate [ORGANIC CHEMISTRY] *See* mercury fulminate. { mər'kyůr·ik 'sī·ə,nāt }

mercuric cyanide [INORGANIC CHEMISTRY] $Hg(CN)_2$ Poisonous, colorless, transparent crystals that darken in light, decompose when heated; soluble in water and alcohol; used in photography, medicine, and germicidal soaps. Also known as mercury cyanide. { mər'kyůr·ik 'sī·ə,nīd }

mercuric fluoride [INORGANIC CHEMISTRY] HgF_2 Poisonous, transparent crystals that decompose when heated; moderately soluble in alcohol and water; used to synthesize organic fluorides. { mər'kyu̇r·ik 'flu̇r‚īd }

mercuric iodide [INORGANIC CHEMISTRY] HgI_2 Poisonous red crystals that turn yellow when heated to 150°C; soluble in boiling alcohol; used in medicine and in Nessler's and Mayer's reagents. { mər'kyu̇r·ik 'ī·ə‚dīd }

mercuric lactate [ORGANIC CHEMISTRY] $Hg(C_3H_5O_3)_2$ A poisonous white powder that decomposes when heated; soluble in water; used in medicine. { mər'kyu̇r·ik 'lak‚tāt }

mercuric nitrate [INORGANIC CHEMISTRY] $Hg(NO_3)_2 \cdot H_2O$ Poisonous, colorless crystals that decompose when heated; soluble in water and nitric acid, insoluble in alcohol; a fire hazard; used in medicine, in nitrating organic aromatics, and in felt manufacture. Also known as mercury nitrate; mercury pernitrate. { mər'kyu̇r·ik 'nī‚trāt }

mercuric oleate [ORGANIC CHEMISTRY] $Hg(C_{18}H_{33}O_2)_2$ A poisonous yellowish-to-red liquid or solid mass; insoluble in water; used in medicine and antifouling paints, and as an antiseptic. Also known as mercury oleate. { mər'kyu̇r·ik 'ōl·ē‚āt }

mercuric oxide [INORGANIC CHEMISTRY] HgO A compound of mercury that exists in two forms, red mercuric oxide and yellow mercuric oxide; the red form decomposes upon heating, is insoluble in water, and is used in pigments and paints, and in ceramics; the yellow form is insoluble in water, decomposes upon heating, and is used in medicine. Also known as mercury oxide; red precipitate; yellow precipitate. { mər'kyu̇r·ik 'äk‚sīd }

mercuric phosphate [INORGANIC CHEMISTRY] $Hg_3(PO_4)_2$ Poisonous yellowish or white powder; insoluble in alcohol and water, soluble in acids; used in medicine. Also known as mercury phosphate; trimercuric orthophosphate. { mər'kyu̇r·ik 'fäs‚fāt }

mercuric salicylate [ORGANIC CHEMISTRY] $Hg(C_7H_5O_3)_2$ Poisonous, white powder; odorless and tasteless; almost insoluble in water and alcohol; variable composition; used in medicine. Also known as salicylated mercury. { mər'kyu̇r·ik sə'lis·ə‚lāt }

mercuric stearate [ORGANIC CHEMISTRY] $Hg(C_{17}H_{35}CO_2)_2$ Poisonous yellow powder; soluble in fatty acids, slightly soluble in alcohol; used as a germicide and in medicine. Also known as mercury stearate. { mər'kyu̇r·ik 'stir‚āt }

mercuric sulfate [INORGANIC CHEMISTRY] $HgSO_4$ A toxic, white, crystalline powder, soluble in acids; used in medicine, as a catalyst, and for galvanic batteries. Also known as mercury persulfate; mercury sulfate. { mər'kyu̇r·ik 'səl‚fāt }

mercuric sulfide [INORGANIC CHEMISTRY] HgS **1.** The black variety is a poisonous powder; insoluble in water, alcohol, and nitric acid, soluble in sodium sulfide solution; sublimes at 583°C; used as a pigment. Also known as black mercury sulfide; ethiops mineral. **2.** The red variety is a poisonous powder; insoluble in water and alcohol; sublimes at 446°C; used as a medicine and pigment. Also known as Chinese vermilion; quicksilver vermilion; red mercury sulfide; vermilion. { mər'kyu̇r·ik 'səl‚fīd }

mercuric thiocyanate [INORGANIC CHEMISTRY] $Hg(SCN)_2$ Poisonous white powder; soluble in alcohol, slightly soluble in water; decomposes when heated; used in photography. Also known as mercury thiocyanate. { mər'kyu̇r·ik ‚thī·ə'sī·ə‚nāt }

mercurous [INORGANIC CHEMISTRY] Referring to mercury with a valence of 1; for example, mercurous chloride, Hg_2Cl_2, where the mercury is covalently bonded, as Cl—Hg—Hg—Cl. { mər'kyu̇r·əs }

mercurous acetate [ORGANIC CHEMISTRY] $HgC_2H_3O_2$ Poisonous colorless plates or scales; decomposed by boiling water and by light; soluble in dilute nitric acids, slightly soluble in water. Also known as mercury acetate; mercury protoacetate. { mər'kyu̇r·əs 'as·ə‚tāt }

mercurous bromide [INORGANIC CHEMISTRY] HgBr Poisonous white powder, crystals, or fibrous mass; odorless and tasteless; darkens in light; soluble in hot sulfuric acid and fuming nitric acid, insoluble in alcohol and ether; used in medicine. Also known as mercury bromide. { mər'kyu̇r·əs 'brō‚mīd }

mercurous chlorate [INORGANIC CHEMISTRY] $Hg_2(ClO_3)_2$ Poisonous white crystals that decompose at 250°C; soluble in alcohol and water; explodes in contact with combustible substances. Also known as mercury chlorate. { mər'kyu̇r·əs 'klȯr,āt }

mercurous chloride [INORGANIC CHEMISTRY] Hg_2Cl_2 Odorless, nonpoisonous white crystals that darken in light; insoluble in water, alcohol, and ether; melts at 302°C; used in medicine and pyrotechnics. Also known as mercury monochloride; mercury protochloride; mild mercury chloride. { mər'kyu̇r·əs 'klȯr,īd }

mercurous chromate [INORGANIC CHEMISTRY] Hg_2CrO_4 Red powder with variable composition; decomposes when heated; soluble in nitric acid, insoluble in water and alcohol; used to color ceramics green. Also known as mercury chromate. { mər'kyu̇r·əs 'krō,māt }

mercurous iodide [INORGANIC CHEMISTRY] Hg_2I_2 Odorless, tasteless, poisonous yellow powder; darkens when heated; insoluble in water, alcohol, and ether; sublimes at 140°C; used as external medicine. Also known as mercury protoiodide. { mər'kyu̇r·əs 'ī·ə,dīd }

mercurous oxide [INORGANIC CHEMISTRY] Hg_2O A poisonous black powder; insoluble in water, soluble in acids; decomposes at 100°C. { mər'kyu̇r·əs 'äk,sīd }

mercurous phosphate [INORGANIC CHEMISTRY] Hg_3PO_4 Light-sensitive white powder with variable composition; insoluble in alcohol and water, soluble in nitric acids; used in medicine. Also known as mercury phosphate; trimercurous orthophosphate. { mər'kyu̇r·əs 'fäs,fāt }

mercurous sulfate [INORGANIC CHEMISTRY] Hg_2SO_4 Poisonous yellow-to-white powder; soluble in hot sulfuric acid or dilute nitric acid, insoluble in water; used as a catalyst and in laboratory batteries. { mər'kyu̇r·əs 'səl,fāt }

mercury [CHEMISTRY] A metallic element, symbol Hg, atomic number 80, atomic weight 200.59, existing at room temperature as a silvery, heavy liquid. Also known as quicksilver. { 'mər·kyə·rē }

mercury acetate [ORGANIC CHEMISTRY] *See* mercuric acetate; mercurous acetate. { 'mər·kyə·rē 'as·ə,tāt }

mercury arsenate [INORGANIC CHEMISTRY] *See* mercuric arsenate. { 'mər·kyə·rē 'ärs·ən,āt }

mercury arseniate [INORGANIC CHEMISTRY] *See* mercuric arsenate. { 'mər·kyə·rē är'sē·nē,āt }

mercury barium iodide [INORGANIC CHEMISTRY] *See* mercuric barium iodide. { 'mər·kyə·rē ba·rē·əm 'ī·ō·dīd }

mercury benzoate [ORGANIC CHEMISTRY] *See* mercuric benzoate. { 'mər·kyə·rē 'ben·zə,wāt }

mercury bromide [INORGANIC CHEMISTRY] *See* mercuric bromide; mercurous bromide. { 'mər·kyə·rē 'brō,mīd }

mercury chlorate [INORGANIC CHEMISTRY] *See* mercurous chlorate. { 'mər·kyə·rē 'klȯr,āt }

mercury chromate [INORGANIC CHEMISTRY] *See* mercurous chromate. { 'mər·kyə·rē 'krō,māt }

mercury cyanide [INORGANIC CHEMISTRY] *See* mercuric cyanide. { 'mər·kyə·rē 'sī·ə,nīd }

mercury fulminate [ORGANIC CHEMISTRY] $Hg(CNO)_2$ A gray, crystalline powder, explodes at the melting point; soluble in alcohol, ammonium hydroxide, and hot water; used for explosive caps and detonators. Also known as mercuric cyanate. { 'mər·kyə·rē 'fu̇l·mə,nāt }

mercury monochloride [INORGANIC CHEMISTRY] *See* mercurous chloride. { 'mər·kyə·rē ,män·ə'klȯr,īd }

mercury naphthenate [ORGANIC CHEMISTRY] Poisonous dark-amber liquid; soluble in mineral oils; used in gasoline antiknock compounds and as a paint antimildew promoter. { 'mər·kyə·rē 'naf·thə,nāt }

mercury nitrate [INORGANIC CHEMISTRY] *See* mercuric nitrate. { 'mər·kyə·rē 'nī,trāt }

mercury oleate [ORGANIC CHEMISTRY] *See* mercuric oleate. { 'mər·kyə·rē 'ōl·ē,āt }

mercury oxide [INORGANIC CHEMISTRY] *See* mercuric oxide. { 'mər·kyə·rē 'äk,sīd }

mercury pernitrate [INORGANIC CHEMISTRY] *See* mercuric nitrate. { 'mər·kyə·rē pər'nī,trāt }

mercury persulfate [INORGANIC CHEMISTRY] *See* mercuric sulfate. { 'mər·kyə·rē pər'səl‚fāt }

mercury phosphate [INORGANIC CHEMISTRY] *See* mercuric phosphate; mercurous phosphate. { 'mər·kyə·rē 'fäs‚fāt }

mercury protoacetate [ORGANIC CHEMISTRY] *See* mercurous acetate. { 'mər·kyə·rē ‚prō·dō'as·ə‚tāt }

mercury protochloride [INORGANIC CHEMISTRY] *See* mercurous chloride. { 'mər·kyə·rē ‚prō·dō'klȯr‚īd }

mercury protoiodide [INORGANIC CHEMISTRY] *See* mercurous iodide. { 'mər·kyə·rē ‚prō·dō'ī·ə‚dīd }

mercury stearate [ORGANIC CHEMISTRY] *See* mercuric stearate. { 'mər·kyə·rē 'stir‚āt }

mercury sulfate [INORGANIC CHEMISTRY] *See* mercuric sulfate. { 'mər·kyə·rē 'səl‚fāt }

mercury thiocyanate [INORGANIC CHEMISTRY] *See* mercuric thiocyanate. { 'mər·kyə·rē ‚thī·ə'sī·ə‚nāt }

mesaconic acid [ORGANIC CHEMISTRY] $C_5H_6O_4$ An unsaturated dibasic acid, an isomer of citraconic acid, that melts at 202°C. Also known as methyl fumaric acid. { ‚mes·ə'kän·ik 'as·əd }

mescaline [ORGANIC CHEMISTRY] $C_{11}H_{17}NO_3$ The alkaloid 3,4,5-trimethoxyphenethylamine, found in mescal buttons; produces unusual psychic effects and visual hallucinations. { 'mes·kə‚lēn }

mesityl oxide [ORGANIC CHEMISTRY] $(CH_3)_2C{=}CHCOCH_3$ A colorless, oily liquid with a honeylike odor; solidifies at -41.5°C; used as a solvent for resins, particularly vinyl resins, many gums, and nitrocellulose; also used in lacquers, paints, and varnishes. { 'mez·ə‚til 'äk‚sīd }

meso- [CHEMISTRY] A prefix meaning intermediate or middle, as in denoting inactive optical isomers, the form of intermediate inorganic acid, the middle position in cyclic organic compounds, or a ring system with middle ring positions. { 'me·zō }

mesogenic unit [PHYSICAL CHEMISTRY] A component of a molecule that induces a mesomorphic or liquid crystalline phase. Also known as mesagen. { ¦mez·ə¦jen·ik 'yü·nət }

meso-ionic compound [ORGANIC CHEMISTRY] Any of a class of five-membered ring heterocycles and their benzo derivatives which possess a sextet of pi electrons in association with the atoms composing the ring but which cannot be represented satisfactorily by any one covalent or polar structure. { ¦mez·ō·ī¦än·ik 'käm‚paůnd }

mesomerism [PHYSICAL CHEMISTRY] *See* resonance. { mə'säm·ə‚riz·əm }

mesomorphism [PHYSICAL CHEMISTRY] A state of matter intermediate between a crystalline solid and a normal isotropic liquid, in which long rod-shaped organic molecules contain dipolar and polarizable groups. { ¦mez·ə¦mȯr‚fiz·əm }

mesopore [CHEMISTRY] A pore in a catalytic material whose width ranges from 2 nanometers to 0.05 micrometer. { 'mez·ə‚pȯr }

mesyl [ORGANIC CHEMISTRY] *See* methylsulfonyl. { 'mes·əl }

meta- [ORGANIC CHEMISTRY] A prefix for benzene-ring compounds when two side chains are connected to carbon atoms with an unsubstituted carbon atom between them. { 'med·ə }

metachromasia [CHEMISTRY] **1.** The property exhibited by certain pure dyestuffs, chiefly basic dyes, of coloring certain tissue elements in a different color, usually of a shorter wavelength absorption maximum, than most other tissue elements. **2.** The assumption of different colors or shades by different substances when stained by the same dye. Also known as metachromatism. { ‚med‚ə·krō'mā·zhə }

metachromatism [CHEMISTRY] *See* metachromasia. { ¦med·ə'krō·mə ‚tiz·əm }

metatitanic acid [INORGANIC CHEMISTRY] *See* titanic acid. { ¦med·ə·tī'tan·ik 'as·əd }

metahydrate sodium carbonate [INORGANIC CHEMISTRY] $Na_2CO_3 \cdot H_2O$ Water-soluble, white crystals with an alkaline taste, loses water at 109°C, melts at 851°C; used in medicine, photography, and water pH control, and as a food additive. Also known as crystal carbonate; soda crystals. { ‚med·ə'hī‚drāt ¦sōd·ē·əm 'kär·bə‚nāt }

metal alkyl [ORGANIC CHEMISTRY] One of the family of organometallic compounds, a combination of an alkyl organic radical with a metal atom or atoms. { 'med·əl 'al·kəl }

metal cluster compound [CHEMISTRY] A compound in which two or more metal atoms aggregate so as to be within bonding distance of one another and each metal atom is bonded to at least two other metal atoms; some nonmetal atoms may be associated with the cluster. { 'med·əl ¦kləs·tər 'käm‚pau̇nd }

metaldehyde [ORGANIC CHEMISTRY] $(CH_3CHO)_n$ White acetaldehyde-polymer prisms; soluble in organic solvents, insoluble in water; used as a pesticide or fuel. { me 'tal·də‚hīd }

metal ion indicator [ANALYTICAL CHEMISTRY] A substance, usually a dyestuff, that changes color after forming a metal ion complex with a color different from that of the uncomplexed indicator. Also known as complexation indicator. { ¦med·əl ¦ī‚än ‚in·də‚kād·ər }

metallation [ORGANIC CHEMISTRY] The direct replacement of a hydrogen atom by a metal atom in an organic molecule to form a carbon-metal bond. { ‚med·ə'la·shən }

metallic bond [PHYSICAL CHEMISTRY] The type of chemical bond that is present in all metals, and may be thought of as resulting from a sea of valence electrons which are free to move throughout the metal lattice. { mə'tal·ik 'bänd }

metallic element [CHEMISTRY] An element generally distinguished (from a nonmetallic one) by its luster, electrical conductivity, malleability, and ability to form positive ions. { mə'tal·ik 'el·ə·mənt }

metallic hydrogen [PHYSICAL CHEMISTRY] **1.** A phase of hydrogen believed to occur at extremely high pressures, in which the material transforms to a conducting molecular solid. **2.** A phase of hydrogen believed to occur at still higher pressures, in which the molecular bonds that exist at lower pressures are broken and an atomic solid with the structure of an alkali metal is formed. { mə'tal·ik 'hī·drə·jən }

metallic soap [ORGANIC CHEMISTRY] A salt of stearic, oleic, palmitic, lauric, or erucic acid with a heavy metal such as cobalt or copper; used as a drier in paints and inks, in fungicides, decolorizing varnish, and waterproofing. { mə'tal·ik 'sōp }

metallo-carbohedrene [CHEMISTRY] A member of a class of molecular clusters in which atoms of an early transition metal (scandium through nickel in the third period of the periodic table) are bonded with carbon atoms in a cagelike network. { mə¦ta·lō‚kär·bə'hed‚rēn }

metallocene [ORGANIC CHEMISTRY] Organometallic coordination compound which is obtained as a cyclopentadienyl derivative of a transition metal or a metal halide. { mə'tal·ə‚sēn }

metallocycle [ORGANIC CHEMISTRY] A compound whose structure consists of a cyclic array of atoms of which one is a metal atom; frequently the ring contains three or four carbon atoms and one transition-metal atom. { mə'tal·ə‚sī·kəl }

metalloid [CHEMISTRY] An element whose properties are intermediate between those of metals and nonmetals. Also known as semimetal. { 'med·ə‚lȯid }

metamer [ORGANIC CHEMISTRY] One of two or more chemical compounds that exhibits isomerism with the others. { 'med·ə·mər }

metanilic acid [ORGANIC CHEMISTRY] $C_6H_4(NH_2)SO_3H$ A water-soluble, crystalline compound, isomeric with sulfanilic acid; used in medicines and dyes. { ¦med·ə¦nil·ik 'as·əd }

metastable equilibrium [PHYSICAL CHEMISTRY] A state of pseudo-equilibrium having higher free energy than the true equilibrium state. { ¦med·ə'stā·bəl ‚ē·kwə'lib·rē·əm }

metastable ion [ANALYTICAL CHEMISTRY] In mass spectroscopy, an ion formed by a secondary dissociation process in the analyzer tube (formed after the parent or initial ion has passed through the accelerating field). { ¦med·ə'stā·bəl 'ī‚än }

metastable phase [PHYSICAL CHEMISTRY] Existence of a substance as either a liquid, solid, or vapor under conditions in which it is normally unstable in that state. { ¦med·ə'sta·bəl ¦fāz }

metathesis [CHEMISTRY] A reaction involving the exchange of elements or groups as in the general equation $AX + BY \rightarrow AY + BX$. { mə'tath·ə·səs }

metathetical salts [CHEMISTRY] Salts that form a four-component, ternary equilibrium system in which there are four possible binary systems, resulting in two quadruple points. { ¦med·ə¦thed·ə·kəl 'sȯls }

metatitanic acid [INORGANIC CHEMISTRY] *See* titanic acid. { ¦med·ə·tī'tan·ik 'as·əd }

methacrolein [ORGANIC CHEMISTRY] $CH_2C(CH_3)CHO$ Liquid with 68°C boiling point; slightly soluble in water; used to make resins and copolymers. { mə'thak·rə·lən }

methacrylate ester [ORGANIC CHEMISTRY] $CH_2{:}C(CH_3)COOR$ Methacrylic acid ester in which R can be methyl, ethyl, isobutyl, or 50-50 *n*-butyl-isobutyl groups; used to make thermoplastic polymers or copolymers. { meth'ak·rə‚lāt 'es·tər }

methacrylic acid [ORGANIC CHEMISTRY] $CH_2C(CH_3)COOH$ Easily polymerized, colorless liquid melting at 15-16°C; soluble in water and most organic solvents; used to make water-soluble polymers and as a chemical intermediate. { ¦meth·ə¦kril·ik 'as·əd }

methacrylic polymer [ORGANIC CHEMISTRY] A polymer whose monomer is a methacrylic ester with the general formula $H_2C{=}C(CH_3)COOR$. { 'meth·ə‚kril·ik 'päl·ə·mər }

methacrylonitrile [ORGANIC CHEMISTRY] $CH_2{:}C(CH_3)CN$ Clear, colorless liquid boiling at 90°C; used to make solvent-resistant thermoplastic polymers and copolymers. { ‚meth·ə·krə'län·ə‚tril }

methallyl alcohol [ORGANIC CHEMISTRY] $H_2C{:}C(CH_3)CH_2OH$ Flammable, toxic, water-soluble, colorless liquid boiling at 115°C; has pungent aroma; soluble in most organic solvents; used as a chemical intermediate. Also spelled methyl allyl alcohol. { meth'al·əl 'al·kə‚hȯl }

methanal [ORGANIC CHEMISTRY] *See* formaldehyde. { 'meth·ə‚nal }

methanamide [ORGANIC CHEMISTRY] *See* formamide. { meth'an·ə·mid }

methane [ORGANIC CHEMISTRY] CH_4 A colorless, odorless, and tasteless gas, lighter than air and reacting violently with chlorine and bromine in sunlight, a chief component of natural gas; used as a source of methanol, acetylene, and carbon monoxide. Also known as methyl hydride. { 'meth‚ān }

methanearsonic acid [ORGANIC CHEMISTRY] $CH_3AsO(OH)_2$ A white solid with a melting point of 161°C; very soluble in water; used as an herbicide for cotton crops and for noncrop areas. Abbreviated MAA. { ¦meth‚ān·är'sän·ik 'as·əd }

methane hydrate [CHEMISTRY] Methane gas trapped or dissolved in ice formed in deep-sea sediments. { 'mēth‚ān 'hī‚drāt }

methanesulfonic acid [ORGANIC CHEMISTRY] CH_3SO_2OH A solid with a melting point of 20°C; used as a catalyst in polymerization, esterification, and alkylation reactions, and as a solvent. Also known as methysulfonic acid. { ¦meth‚ān·səl'fän·ik 'as·əd }

methanoic acid [ORGANIC CHEMISTRY] *See* formic acid. { ‚meth·ə'nō·ik 'as·əd }

methanol [ORGANIC CHEMISTRY] *See* methyl alcohol. { 'meth·ə‚nȯl }

methenyl [ORGANIC CHEMISTRY] *See* methine group. { 'meth·ə‚nil }

methidathion [ORGANIC CHEMISTRY] $C_4H_{11}O_4N_2PS_3$ A colorless, crystalline compound with a melting point of 39-40°C; used as an insecticide and miticide for pests on alfalfa, citrus, and cotton. { mə‚thid·ə'thī‚än }

methide [ORGANIC CHEMISTRY] A binary compound consisting of methyl and, most commonly, a metal, such as sodium (sodium methide, $NaCH_3$). { 'me‚thīd }

methine group [ORGANIC CHEMISTRY] $HC{\equiv}$ A radical consisting of a single carbon and a single hydrogen. Also known as methenyl; methylidyne. { 'me‚thēn 'grüp }

methionic acid [ORGANIC CHEMISTRY] $CH_2(SO_3H)_2$ An acid that exists as hygroscopic crystals; used in organic synthesis. { ¦meth·ē¦än·ik 'as·əd }

methoxide [ORGANIC CHEMISTRY] A compound formed from a metal and the methoxy radical; an example is sodium methoxide. Also known as methylate. { mə'thäk‚sīd }

methoxy- [ORGANIC CHEMISTRY] $OCH_3{-}$ A combining form indicating the oxygen-containing methane radical, found in many organic solvents, insecticides, and plasticizer intermediates. { mə'thäk·sē }

methoxychlor [ORGANIC CHEMISTRY] $Cl_3CCH(C_6H_4OCH_3)_2$ White, water-insoluble crystals melting at 89°C; used as an insecticide. Also known as DMDT; methoxy DDT. { me'thäk·si‚klȯr }

methoxy DDT [ORGANIC CHEMISTRY] *See* methoxychlor. { me'thäk·sē ¦de¦dē'tē }

2-methoxyethanol [ORGANIC CHEMISTRY] $CH_3OCH_2CH_2OH$ A poisonous liquid, used as a solvent for low-viscosity cellulose acetate, natural and some synthetic resins, and alcohol-soluble dyes, and also used in dyeing leather. { ¦tü mə‚thäk·sē′eth·ə‚nól }

methoxyethylmercury chloride [ORGANIC CHEMISTRY] $CH_3OCH_2CH_2HgCl$ A white, crystalline compound with a melting point of 65°C; used as a fungicide in diseases of sugarcane, pineapples, seed potatoes, and flower bulbs, and as seed dressings for cereals, legumes, and root crops. Abbreviated MEMC. { mə¦thäk·sē¦eth əl¦mər·kyə·rē ′klór‚īd }

4-methoxy-2-hydroxybenzophenone [ORGANIC CHEMISTRY] *See* oxybenzone. { ¦fór mə¦thäk·sē ¦tü ‚hī¦dräk·sē·ben′zä·fə‚nōn }

methoxyl [ORGANIC CHEMISTRY] $CH_3O—$ A functional group which is univalent. { mə′thäk·səl }

methyl [ORGANIC CHEMISTRY] The alkyl group derived from methane and usually written $CH_3—$. Also known as carbinyl. { ′meth·əl }

methyl abietate [ORGANIC CHEMISTRY] $C_{19}H_{29}COOCH_3$ Colorless to yellow liquid boiling at 365°C; miscible with most organic solvents; used as a solvent and plasticizer for lacquers, varnishes, and coatings. { ′meth·əl ′ab·ē·ə‚tāt }

methyl acetate [ORGANIC CHEMISTRY] $CH_3CO_2CH_3I$ Flammable, colorless liquid with fragrant odor; boils at 54°C; partially soluble in water, miscible with hydrocarbon solvents; used as a solvent and extractant. { ′meth·əl ′as·ə‚tāt }

methylacetic acid [ORGANIC CHEMISTRY] *See* propionic acid. { ¦meth·əl·ə¦sēd·ik ‚as·əd }

methyl acetoacetate [ORGANIC CHEMISTRY] $CH_3COCH_2CO_2CH_3$ Alcohol-soluble, colorless liquid boiling at 172°C; used as a chemical intermediate and as a solvent for cellulosics. { ′meth·əl ¦as·əd·ō¦as·ə‚tāt }

methyl acetophenone [ORGANIC CHEMISTRY] $CH_3C_6H_4COCH_3$ Fragrant (coumarin aroma), colorless or pale-yellow liquid, soluble in alcohol; used in perfumery. { ′meth·əl ‚as·ə′täf·ə‚nōn }

methyl acrylate [ORGANIC CHEMISTRY] $CH_2{:}CHCOOCH_3$ A readily polymerized, volatile, colorless liquid boiling at 80°C; slightly soluble in water; used as a chemical intermediate and in making polymers. { ′meth·əl ′ak·rə‚lāt }

methylal [ORGANIC CHEMISTRY] $CH_2OCH_2OCH_3$ Flammable, volatile, colorless liquid boiling at 42°C, soluble in ether, hydrocarbons, and alcohol, partially soluble in water; used as a solvent and chemical intermediate, and in perfumes, adhesives, coatings. Also known as formal. { ′meth·ə‚lal }

methyl alcohol [ORGANIC CHEMISTRY] CH_3OH A colorless, toxic, flammable liquid, boiling at 64.5°C, miscible with water, ether, alcohol; used in manufacture of formaldehyde, chemical synthesis, antifreeze for autos, and as a solvent. Also known as methanol, wood alcohol. { ′meth·əl ′al·kə‚hól }

methyl allyl alcohol [ORGANIC CHEMISTRY] *See* methallyl alcohol. { ′meth·əl ′al·əl ′al·kə‚hól }

methyl allyl chloride [ORGANIC CHEMISTRY] $CH_2{:}C(CH_3)CH_2Cl$ Volatile, flammable, colorless liquid boiling at 72°C; has disagreeable odor; used as an insecticide and fumigant, and for chemical synthesis. { ′meth·əl ′al·əl ′klór‚īd }

methylamine [ORGANIC CHEMISTRY] CH_3NH_2 A colorless gas that is highly toxic and flammable; used to prepare dyes, and as a chemical intermediate. { ¦meth·ə·lə¦mēn }

***N*-methyl-*para*-aminophenol** [ORGANIC CHEMISTRY] $CH_3NHC_6H_4OH$ Colorless, combustible needles with a melting point of 87°C; soluble in water, alcohol, and ether; used as a photographic developer. { ¦en ¦meth·əl ¦par·ə ¦am·ə·nō′fē‚nól }

methyl amyl acetate [ORGANIC CHEMISTRY] $CH_3COOCH(CH_3)CH_2CH(CH_3)_2$ Toxic, flammable, colorless liquid with mild, agreeable odor; boils at 146°C; used as nitrocellulose lacquer solvent. Also known as methyl isobutyl carbinol acetate. { ′meth·əl ′am·əl ′as·ə‚tāt }

methyl amyl alcohol [ORGANIC CHEMISTRY] $(CH_3)_2CHCH_2CHOHCH_3$ Toxic, flammable, colorless liquid; boils at 132°C; miscible with water and most organic solvents; used as a solvent and as a chemical intermediate. Also known as methyl isobutyl carbinol (MIBC). { ′meth·əl ′am·əl ′al·kə ‚hól }

methyl amyl carbinol [ORGANIC CHEMISTRY] $CH_3(CH_2)_4CHOHCH_3$ Colorless liquid with mild aroma; boils at 160°C; miscible with most organic liquids; used as an ore-flotation frothing agent and as a synthetic-resin solvent. { 'meth·əl 'am·əl 'kär·bə,nȯl }

methyl *n*-amyl ketone [ORGANIC CHEMISTRY] $CH_3(CH_2)_4COCH_3$ Stable, water-white liquid; miscible with organic lacquer solvents, slightly soluble in water; used as an inert reaction medium and as a solvent for nitrocellulose lacquers. Also known as 2-heptanone. { 'meth·əl ¦en ¦am·əl 'kē,tōn }

***N*-methylaniline** [ORGANIC CHEMISTRY] $C_6H_5NH(CH_3)$ Oily liquid, colorless to reddish-brown; soluble in water and organic solvents; boils at 190°C; used as an acid acceptor, solvent, and chemical intermediate. { ¦en ¦meth·əl'an·ə·lən }

α-methylanisalacetone [ORGANIC CHEMISTRY] $CH_3OC_6H_4CH{:}CHCOCH_2CH_3$ A white to pale yellow, combustible solid with a melting point of 60°C; used as a flavoring. { ¦al·fə ¦meth·əl,an·ə·sə'las·ə,tōn }

methyl anisole [ORGANIC CHEMISTRY] *See* methyl *para*-cresol. { 'meth·əl 'an·ə,sōl }

methyl anthranilate [ORGANIC CHEMISTRY] $H_2NC_6H_4CO_2CH_3$ A yellowish to colorless liquid, slightly soluble in water; used in flavoring and in perfumery. Also known as artificial neroli oil. { 'meth·əl an'thran·ə,lāt }

2-methyl anthraquinone [ORGANIC CHEMISTRY] *See* tectoquinone. { ¦tü ¦meth·əl ,an·thrə,kwē'nōn }

methyl arachidate [ORGANIC CHEMISTRY] $CH_3(CH_2)_{18}COOCH_3$ A waxlike solid with a melting point of 45.8°C; soluble in alcohol and ether; used in medical research and as a reference standard for gas chromatography. Also known as methyl eicosanoate. { 'meth·əl ə'rak·ə,dāt }

methylarsinic sulfide [ORGANIC CHEMISTRY] CH_3AsS A colorless compound whose flakes melt at 110°C; insoluble in water; used as a fungicide in treating cotton seeds. Also known as rhizoctol. { ¦meth·əl·är¦sin·ik 'səl,fīd }

methylate [ORGANIC CHEMISTRY] *See* methoxide. { 'meth·ə,lāt }

methylation [ORGANIC CHEMISTRY] A chemical process for introducing a methyl group (CH_3—) into an organic compound. { ,meth·ə'lā·shən }

methyl behenate [ORGANIC CHEMISTRY] $CH_3(CH_2)_{20}COOCH_3$ A combustible, waxlike solid with a melting point of 53.2°C; soluble in alcohol and ether; used in medical and biochemical research and as a reference standard for gas chromatography. Also known as methyl docosanoate. { 'meth·əl bə'he,nāt }

methylbenzene [ORGANIC CHEMISTRY] *See* toluene. { ¦meth·əl¦ben,zēn }

methylbenzethonium chloride [ORGANIC CHEMISTRY] $C_{27}H_{44}O_2Cl \cdot H_2O$ Colorless crystals with a melting point of 161-163°C; soluble in alcohol, hot benzene, chloroform, and water; used as a bactericide. { ,meth·əl,ben·zə'thō·nē·əm 'klȯr,īd }

methyl benzoate [ORGANIC CHEMISTRY] $C_6H_5CO_2CH_3$ Colorless, fragrant liquid boiling at 199°C; slightly soluble in alcohol and water, soluble in ether; used in perfumery and as a solvent. Also known as niobe oil. { 'meth·əl 'ben·zə,wāt }

methyl *ortho*-benzoylbenzoate [ORGANIC CHEMISTRY] $C_6H_5COC_6H_4COOCH_3$ A colorless, combustible liquid with a boiling point of 351°C; slightly soluble in water; used as a plasticizer. { 'meth·əl ¦or·thō,ben·zə,wil'ben·zə,wāt }

α-methylbenzyl acetate [ORGANIC CHEMISTRY] $C_6H_5CH(CH_3)OOCCH_3$ A colorless, combustible liquid with a strong floral odor; soluble in glycerin, mineral oil, and 70% alcohol; used in perfumes and as a flavoring. { ¦al·fə ¦meth·əl¦ben·zəl 'as·ə,tāt }

α-methylbenzyl alcohol [ORGANIC CHEMISTRY] $C_6H_5CH(CH_3)OH$ A colorless, combustible liquid with a mild floral odor and a boiling point of 204°C; soluble in water; used in perfumes and dyes and as a flavoring agent. { ¦al·fə ¦meth·əl¦ben·zəl 'al·kə,hȯl }

α-methylbenzylamine [ORGANIC CHEMISTRY] $C_6H_5CH(CH_3)NH_2$ A colorless, combustible liquid with a boiling point of 188.5°C; soluble in most organic solvents; used as an emulsifying agent. { ¦al·fə ,meth·əl·ben'zal·ə,mēn }

α-methylbenzyl ether [ORGANIC CHEMISTRY] $C_6H_5CH(CH_3)OCH(CH_3)C_6H_5$ A straw-colored, combustible liquid with a boiling point of 286.3°C; at 760 mmHg (101,325 pascals); slightly soluble in water; used as a solvent and as a synthetic rubber softener. { ¦al·fə ¦meth·əl¦ben·zəl 'ē·thər }

methyl blue [ORGANIC CHEMISTRY] Dark-blue powder or dye; sodium triphenyl *para*-rosaniline sulfonate; used as a biological and bacteriological stain and as an antiseptic. { 'meth·əl 'blü }

methyl borate [ORGANIC CHEMISTRY] *See* trimethyl borate. { 'meth·əl 'bȯr,āt }

methyl bromide [ORGANIC CHEMISTRY] CH_3Br A toxic, colorless gas that forms a crystalline hydrate with cold water; used in synthesis of organic compounds, and as a fumigant. { 'meth·əl 'brō,mīd }

2-methylbutanal [ORGANIC CHEMISTRY] *See* 2-methylbutyraldehyde. { ¦tü ,meth·əl'byüt·ən,al }

2-methylbutane [ORGANIC CHEMISTRY] *See* isopentane. { ¦tü ,meth·əl'byü,tān }

2-methyl-1-butanol [ORGANIC CHEMISTRY] $C_5H_{12}O$ A liquid with a boiling point of 128°C, miscible with alcohol and with ether, slightly soluble in water; used as a solvent, in organic synthesis, and as an additive in oils and paints. { ¦tü ¦meth·əl ¦wən 'byüt·ən,ȯl }

methyl butene [ORGANIC CHEMISTRY] C_5H_{10} Either of two colorless, flammable, volatile liquid isomers; soluble in alcohol, insoluble in water: 3-methyl-1-butene boils at 20°C, is used as a chemical intermediate and in the manufacture of high-octane fuel, and is also known as isopropylethylene; 3-methyl-2-butene boils at 38°C, is used as an anesthetic and high-octane fuel and as a chemical intermediate, and is also known as trimethylethylene. { 'meth·əl 'byü,tēn }

2-methyl-2-butene [ORGANIC CHEMISTRY] *See* amylene. { ¦tü 'meth·əl ¦tü 'byü,tēn }

methyl butyl ketone [ORGANIC CHEMISTRY] $CH_3COC_4H_9$ A liquid boiling at 127°C; soluble in water, alcohol, and ether; used as a solvent. Also known as propylacetone. { 'meth·əl 'byüd·əl 'kē,tōn }

methylbutynol [ORGANIC CHEMISTRY] $HC{:}CCOH(CH_3)_2$ Water-miscible, colorless liquid boiling at 104°C; soluble in most organic solvents; used as a stabilizer for chlorinated organic compounds, as a solvent, and as a chemical intermediate. { ,meth·əl'byüt·ən,ȯl }

2-methylbutyraldehyde [ORGANIC CHEMISTRY] $CH_3CH_2CH(CH_3)CHO$ A combustible liquid with a boiling point of 92.93°C; soluble in alcohol and ether; used as a brightener in electroplating. Also known as 2-methylbutanal. { ¦tü ,meth·əl,byüd·ə'ral·də,hīd }

methyl butyrate [ORGANIC CHEMISTRY] $CH_3CH_2CH_2COOCH_3$ Liquid boiling at 102°C; used as a solvent for cellulosic materials. { 'meth·əl 'byüd·ə,rāt }

methyl caprate [ORGANIC CHEMISTRY] $CH_3(CH_2)_8COOCH_3$ A colorless, combustible liquid with a boiling point of 244°C; soluble in alcohol and ether; used in the manufacture of detergents, stabilizers, plasticizers, textiles, and lubricants. Also known as methyl decanoate. { 'meth·əl 'ka,prāt }

methyl caproate [ORGANIC CHEMISTRY] $CH_3(CH_2)_4COOCH_3$ Colorless liquid boiling at 150°C; soluble in alcohol and ether, insoluble in water; used as an intermediate to make caproic acid. Also known as methyl hexanoate. { 'meth·əl 'kap·rə,wāt }

methyl caprylate [ORGANIC CHEMISTRY] $CH_3(CH_2)_6COOCH_3$ Colorless liquid boiling at 193°C; soluble in ether and alcohol, insoluble in water; used as an intermediate to make caprylic acid. { 'meth·əl 'kap·rə,lāt }

methyl carbonate [ORGANIC CHEMISTRY] $CO(OCH_3)_2$ Water-insoluble, colorless liquid boiling at 91°C; has pleasant odor; miscible with acids and alkalies; used as a chemical intermediate. { 'meth·əl 'kär·bə,nāt }

methylcellulose [ORGANIC CHEMISTRY] A grayish-white powder derived from cellulose; swells in water to a colloidal solution; soluble in glacial acetic acid; used in water-based paints and ceramic glazes, for leather tanning, and as a thickening and sizing agent, adhesive, and food additive. Also known as cellulose methyl ether. { 'meth·əl'sel·yə,lōs }

methyl chloride [ORGANIC CHEMISTRY] *See* chloromethane. { 'meth·əl 'klȯr,īd }

methyl chloroacetate [ORGANIC CHEMISTRY] $ClHC_2COOCH_3$ Colorless liquid boiling at 131°C; miscible with ether and alcohol, slightly soluble in water; used as a solvent. { 'meth·əl ,klȯr·ō'as·ə,tāt }

methyl chlorocarbonate [ORGANIC CHEMISTRY] *See* methyl chloroformate. { 'meth·əl ,klȯr·ō'kär·bə,nāt }

methyl chloroformate [ORGANIC CHEMISTRY] $ClCOOCH_3$ A toxic, corrosive, colorless liquid with a boiling point of 71.4°C; soluble in benzene, ether, and methanol; used as a lacrimator in military poison gas and for insecticides. Also known as methyl chlorocarbonate. { 'meth·əl ¦klȯrə¦fȯr‚māt }

methyl cinnamate [ORGANIC CHEMISTRY] $C_6H_5CH{:}CHCO_2CH_3$ A white crystalline compound with strawberry aroma; soluble in ether and alcohol, insoluble in water; boils at 260°C; used to flavor confectioneries and in perfumes. { 'meth·əl 'sin·ə‚māt }

methyl *para*-cresol [ORGANIC CHEMISTRY] $CH_3C_6H_4OCH_3$ Colorless liquid with floral aroma; used in perfumery. Also known as methyl anisole. { 'meth·əl ¦par·ə 'krē‚sȯl }

methyl cyanoacetate [ORGANIC CHEMISTRY] $CNCH_2COOCH_3$ A toxic, combustible, colorless liquid with a boiling point of 203°C; soluble in water, ether, and alcohol; used in pharmaceuticals and dyes. { 'meth·əl ¦sī·ə‚nō'as·ə‚tāt }

methyl cyclohexane [ORGANIC CHEMISTRY] C_7H_{14} Colorless liquid boiling at 101°C; used as a cellulosic solvent and as a chemical intermediate. Also known as hexahydrotoluene. { 'meth·əl ¦sī·klō'hek‚sān }

methyl cyclohexanol [ORGANIC CHEMISTRY] $CH_3C_6H_{10}OH$ A toxic, colorless liquid with menthol aroma; a mixture of three isomers; used as a solvent for lacquer and cellulosics, as a lubricant antioxidant, and in detergents and textile soaps. { 'meth·əl ¦sī·klō'hek·sə‚nȯl }

methyl cyclohexanone [ORGANIC CHEMISTRY] $CH_3C_5H_9CO$ A toxic, clear to pale-yellow liquid with acetonelike aroma; a mixture of cyclic ketones; used as a solvent and in lacquers. { 'meth·əl ¦sī·klō'hek·sə‚nōn }

methylcyclopentadiene dimer [ORGANIC CHEMISTRY] $C_{12}H_{16}$ A flammable, colorless liquid with a boiling range of 78-183°C; soluble in alcohol, benzene, and ether; used in high-energy fuels, plasticizers, dyes, and pharmaceuticals. { ¦meth·əl‚sī·klō‚pen·tə'dī‚ē n 'dī·mər }

methyl cyclopentane [ORGANIC CHEMISTRY] $C_5H_9CH_3$ Flammable, colorless liquid boiling at 72°C; used as a chemical intermediate. { 'meth·əl ¦sī·klō'pen‚tān }

methyl decanoate [ORGANIC CHEMISTRY] *See* methyl caprate. { 'meth·əl də'kan·ə ‚wāt }

methyl-*N*-(3,4-dichlorophenyl)carbamate [ORGANIC CHEMISTRY] *See* swep. { 'meth·əl ¦en ¦thrē ¦fȯr dī‚klȯr·ō'fen·əl 'kär·bə‚māt }

methyl diethanolamine [ORGANIC CHEMISTRY] $CH_3N(C_2H_4OH)_2$ A colorless liquid miscible with water and benzene; has amine aroma; boils at 247°C; used as a chemical intermediate and as an acid-gas absorbent. { 'meth·əl dī‚eth·ə'näl·ə‚mēn }

methyl dioxolane [ORGANIC CHEMISTRY] $C_4H_7O_2$ Water-soluble, clear liquid boiling at 81°C; used as a solvent and extractant. { 'meth·əl dī'äk·sə‚lān }

methyl docosanoate [ORGANIC CHEMISTRY] *See* methyl behenate. { 'meth·əl ‚dō·kə'san·ə‚wāt }

methyl eicosanoate [ORGANIC CHEMISTRY] *See* methyl arachidate. { 'meth·əl ‚ī·kə'san·ə‚wāt }

methylene [ORGANIC CHEMISTRY] $—CH_2—$ A radical that contains a bivalent carbon. { 'meth·ə‚lēn }

methylene blue [ORGANIC CHEMISTRY] Dark green crystals or powder; soluble in water (deep blue solution), alcohol, and chloroform; $C_{16}H_{18}N_3SCl \cdot 3H_2O$ used in medicine; $(C_{16}H_{18}N_3SCl)_2 \cdot ZnCl_2 \cdot H_2O$ used as a textile dye, biological stain, and indicator. Also known as methylthionine chloride. { 'meth·ə‚lēn 'blü }

methylene bromide [ORGANIC CHEMISTRY] CH_2Br_2 Colorless, clear liquid boiling at 97°C; miscible with organic solvents, slightly soluble in water; used as a solvent and chemical intermediate. Also known as dibromomethane. { 'meth·ə‚lēn 'brō‚mīd }

methylene chloride [ORGANIC CHEMISTRY] CH_2Cl_2 A colorless liquid, practically nonflammable and nonexplosive; used as a refrigerant in centrifugal compressors, a solvent for organic materials, and a component in nonflammable paint-remover mixtures. { 'meth·ə‚lēn 'klȯr‚īd }

methylene iodide [ORGANIC CHEMISTRY] CH_2I_2 Yellow liquid boiling at 180°C; soluble in ether and alcohol, insoluble in water; used as a chemical intermediate and to separate mineral mixtures. Also known as diiodomethane. { 'meth·ə‚lēn 'ī·ə‚dīd }

methylene oxide [ORGANIC CHEMISTRY] *See* formaldehyde. { 'meth·ə,lēn 'äk,sīd }

methyl ester [ORGANIC CHEMISTRY] An ester that forms methanol when hydrolyzed. { 'meth·əl 'es·tər }

methyl ether [ORGANIC CHEMISTRY] *See* dimethyl ether. { 'meth·əl 'ē·thər }

methylethylcellulose [ORGANIC CHEMISTRY] A combustible, white to cream-colored, fibrous solid or powder; disperses in cold water, forming solutions which undergo reversible transformation from sol to gel; used as an emulsifier and foaming agent. { ¦meth·əl¦eth·əl'sel·yə,lōs }

methyl ethylene [ORGANIC CHEMISTRY] *See* propylene. { 'meth·əl 'eth·ə,lēn }

methyl ethyl ketone [ORGANIC CHEMISTRY] $CH_3COC_2H_5$ A water-soluble, colorless liquid that is miscible in oil; used as a solvent in vinyl films and nitrocellulose coatings, and as a reagent in organic synthesis. Also known as ethyl methyl ketone; MEK. { ¦meth·əl ¦eth·əl 'kē,tōn }

methyl formate [ORGANIC CHEMISTRY] $HCOOCH_3$ A flammable, colorless liquid with a boiling point of 31.8°C; soluble in ether, water, and alcohol; used in military poison gases and larvicides, and as a fumigant. { 'meth·əl 'fȯr,māt }

methyl fumaric acid [ORGANIC CHEMISTRY] *See* mesaconic acid. { 'meth·əl fyü'mar·ik 'as·əd }

2-methylfuran [ORGANIC CHEMISTRY] $C_4H_3OCH_3$ A colorless liquid with ether flike aroma; boils at 64°C; used as a chemical intermediate. { ¦tü ,meth·ə'fyür,än }

methyl furoate [ORGANIC CHEMISTRY] $C_4H_3OCO_2CH_3$ Colorless liquid that turns yellow in light; soluble in ether and alcohol, insoluble in water; used as a solvent and chemical intermediate. { 'meth·əl 'fyür·ə,wāt }

methyl glucoside [ORGANIC CHEMISTRY] $C_7H_{14}O_6$ Odorless, water-soluble white crystals; used to make resins, drying oils, plasticizers, and surfactants. { 'meth·əl 'glü·kə,sīd }

methyl glycocol [ORGANIC CHEMISTRY] *See* sarcosine. { 'meth·əl 'glī·kə,kȯl }

methyl heptane [ORGANIC CHEMISTRY] C_8H_{18} Either of two colorless, water-insoluble liquids, soluble in alcohol and ether, used as chemical intermediates: 2-methylheptane boils at 118°C, is flammable; 4-methylheptane boils at 122°C. { ,meth·əl 'hep,tān }

methylheptenone [ORGANIC CHEMISTRY] $(CH_3)_2C{:}CH(CH_2)_2COCH_3$ A combustible, colorless liquid with a boiling point of 173-174°C; a constituent of many essential oils; used in perfumes and for flavoring. { ,meth·əl'hep·tə,non }

2-methylhexane [ORGANIC CHEMISTRY] C_7H_{16} Colorless liquid boiling at 90°C; insoluble in alcohol and water; used as a chemical intermediate. Also known as ethyl isobutylmethane. { ¦tü ¦meth·əl'hek,san }

methyl hexanoate [ORGANIC CHEMISTRY] *See* methyl caproate. { 'meth·əl ,hek'san·ə,wat }

methyl hexyl ketone [ORGANIC CHEMISTRY] $CH_3COC_6H_{13}$ A combustible, colorless liquid with a boiling point of 173.5°C; soluble in alcohol, hydrocarbons, ether, and esters; used in perfumes and as a flavoring and odorant. { ¦meth·əl ¦hek·səl 'kē,ton }

methyl hydride [ORGANIC CHEMISTRY] *See* methane. { 'meth·əl 'hī,drīd }

methyl hydroxystearate [ORGANIC CHEMISTRY] $C_{19}H_{38}O_3$ A white, waxy material; slightly soluble in organic solvents, insoluble in water; used in cosmetics, inks, and adhesives. { 'meth·əl hī,dräk·sē'stir,āt }

methylidyne [ORGANIC CHEMISTRY] *See* methine group. { me'thil·ə,dīn }

methyl iodide [ORGANIC CHEMISTRY] CH_3I Flammable colorless liquid that turns brown in light; boils at 42°C; soluble in ether and alcohol, insoluble in water; used as a chemical intermediate, in medicine, and in analytical chemistry. Also known as iodomethane. { 'meth·əl 'ī·ə,dīd }

3-methylindole [ORGANIC CHEMISTRY] *See* skatole. { ¦thrē meth·ə'lin,dōl }

methyl isobutyl carbinol [ORGANIC CHEMISTRY] *See* methyl amyl alcohol. { 'meth·əl ,ī·sō'byüd·əl 'kär·bə,nȯl }

methyl isobutyl carbinol acetate [ORGANIC CHEMISTRY] *See* methyl amyl acetate. { 'meth·əl ,ī·sō'byüd·əl 'kär·bə,nȯl 'as·ə,tāt }

methyl isobutyl ketone [ORGANIC CHEMISTRY] $(CH_3)_2CHCH_2COCH_3$ Flammable colorless liquid with pleasant aroma; boils at 116°C, miscible with most organic solvents; used as a solvent, extractant, and chemical intermediate. Also known as hexone. { 'meth·əl ,ī·sō'byüd·əl 'kē,tōn }

methylisothiocyanate [ORGANIC CHEMISTRY] C_2H_3NS A crystalline compound, with a melting point of 35-36°C; soluble in alcohol and ether; used as a pesticide and in amino acid sequence analysis. Also known as methyl mustard oil. { ¦meth·əl¦ī·sō,thī·ə'sī·ə,nāt }

methyl lactate [ORGANIC CHEMISTRY] $CH_3CHCHCOOCH_3$ Liquid boiling at 145°C; miscible with water and most organic liquids; used as a solvent for lacquers, stains, and cellulosic materials. { 'meth·əl 'lak,tāt }

methyl laurate [ORGANIC CHEMISTRY] $CH_3(CH_2)_{10}COOCH_3$ Water-insoluble, clear, colorless liquid boiling at 262°C; used as a chemical intermediate to make rust removers, and for leather treatment. { 'meth·əl 'lȯ,rāt }

methyl linoleate [ORGANIC CHEMISTRY] $C_{19}H_{34}O_2$ A combustible, colorless liquid with a boiling point of 212°C; soluble in alcohol and ether; used in the manufacture of detergents, emulsifiers, lubricants, and textiles, and in medical research. { 'meth·əl lə'nō·lē,āt }

methyl mercaptan [ORGANIC CHEMISTRY] CH_3SH Colorless, toxic, flammable gas with unpleasant odor; boils at 6.2°C; insoluble in water, soluble in organic solvents; used as a chemical intermediate. { 'meth·əl mər'kap,tan }

methylmercury compound [ORGANIC CHEMISTRY] Any member of a class of toxic compounds containing the methyl-mercury group, CH_3Hg. { ¦meth·əl'mər·kyə·rē ,käm,paùnd }

methylmercury cyanide [ORGANIC CHEMISTRY] *See* methylmercury nitrile. { ¦meth·əl'mər·kyə·rē 'sī·ə,nīd }

methylmercury nitrile [ORGANIC CHEMISTRY] CH_3HgCN A crystalline solid with a melting point of 95°C; soluble in water; used as a fungicide to treat seeds of cereals, flax, and cotton. Also known as methylmercury cyanide. { ¦meth·əl'mər·kyə·rē 'nī·trəl }

methyl methacrylate [ORGANIC CHEMISTRY] $CH_2C(CH_3)COOCH_3$ A flammable, colorless liquid, soluble in most organic solvents but insoluble in water; used as a monomer for polymethacrylate resins. { ¦meth·əl mə'thak·rə,lāt }

methyl mustard oil [ORGANIC CHEMISTRY] *See* methylisothiocyanate. { 'meth·əl 'məs·tərd ,ȯil }

methyl myristate [ORGANIC CHEMISTRY] $CH_3(CH_2)_{12}COOCH_3$ A colorless liquid with a boiling point of 186.8°C; used in the manufacture of detergents, plasticizers, resins, textiles, and animal feeds, and as a flavoring. Also known as methyl tetradecanoate. { 'meth·əl mə'ri,stāt }

methylnaphthalene [ORGANIC CHEMISTRY] $C_{10}H_7CH_3$ A solid melting at 34°C; used in insecticides and organic synthesis. { ,meth·əl'naf·thə,lēn }

methyl nitrate [ORGANIC CHEMISTRY] CH_3NO_3 Explosive liquid boiling at 60°C; slightly soluble in water, soluble in ether and alcohol; used as a rocket propellant. { 'meth·əl 'nī,trāt }

methyl nonanoate [ORGANIC CHEMISTRY] $CH_3(CH_2)_7COOCH_3$ A colorless liquid with a fruity odor and a boiling point of 213.5°C; soluble in alcohol and ether; used in perfumes and flavors, and for medical research. Also known as methyl pelargonate. { 'meth·əl nə'nan·ə,wāt }

methyl nonyl ketone [ORGANIC CHEMISTRY] $CH_3COC_9H_{19}$ An oily liquid with a boiling point of 225°C; soluble in two parts of 70% alcohol; used in perfumes and flavoring. Also known as 2-undecanone. { ¦'meth·əl ¦nō·nəl 'kē,tōn }

methyl oleate [ORGANIC CHEMISTRY] $C_{17}H_{33}COOCH_3$ Amber liquid with faint fatty odor; soluble in organic liquids, mineral spirits, and vegetable oil, insoluble in water; used as a plasticizer and softener. { 'meth·əl 'ōl·ē,āt }

methylol riboflavin [ORGANIC CHEMISTRY] An orange to yellow powder, soluble in water; used as a nutrient and in medicine. { 'meth·ə,lȯl 'rī·bə,flā·vən }

methylol urea [ORGANIC CHEMISTRY] $H_2NCONHCH_2OH$ Water-soluble, colorless crystals melting at 111°C; used to treat textiles and wood, and in the manufacture of resins and adhesives. { 'meth·ə,lȯl yü'rē·ə }

methyl palmitate [ORGANIC CHEMISTRY] $CH_3(CH_2)_{14}COOCH_3$ A colorless liquid with a boiling point of 211.5°C; soluble in alcohol and ether; used in the manufacture of detergents, resins, plasticizers, lubricants, and animal feed. { 'meth·əl 'pal·mə,tāt }

methyl pelargonate [ORGANIC CHEMISTRY] *See* methyl nonanoate. { 'meth·əl pə'lär·gə,nāt }

3-methylpentane [ORGANIC CHEMISTRY] C_6H_{14} Flammable, colorless liquid; insoluble in water, soluble in alcohol; boils at 64°C; used as a chemical intermediate. { ¦thrē ¦meth·əl'pen,tān }

2-methylpentanoic acid [ORGANIC CHEMISTRY] $(CH_3)_2CH(CH_2)_2COOH$ A colorless liquid with a boiling point of 197°C; soluble in alcohol, benzene, and acetone; used for plasticizers, vinyl stabilizers, and metallic salts. { ¦tü ¦meth·əl,pen·tə¦nō·ik 'as·əd }

methylpentene polymer [ORGANIC CHEMISTRY] Thermoplastic material based on 4-methylpentene-1; has low gravity, excellent electrical properties, and 90% optical transmission. { ¦meth·əl'pen,tēn 'päl·ə·mər }

methyl pentose [ORGANIC CHEMISTRY] **1.** Any compound that is a methyl derivative of a five-carbon sugar. **2.** In particular, the compound $CH_3(CHOH)_4CHO$. { 'meth·əl 'pen,tōs }

methyl phenyl acetate [ORGANIC CHEMISTRY] $C_6H_5CH_2COOCH_3$ A colorless liquid with honey odor; used to flavor tobacco and in perfumery. { ¦meth·əl ¦fen·əl 'as·ə,tāt }

methylphosphoric acid [ORGANIC CHEMISTRY] $CH_3H_2PO_4$ A straw-colored liquid used for textile- and paper-processing compounds, as a rust remover, and in soldering flux. { ,meth·əl,fäs'fȯr·ik 'as·əd }

methyl propionate [ORGANIC CHEMISTRY] $CH_3CH_2COOCH_3$ A flammable, colorless liquid with a boiling range of 78.0-79.5°C; soluble in most organic solvents; used as a solvent for cellulose nitrate, in lacquers, varnishes, and paints, and for flavoring. { 'meth·əl 'prō·pē·ə,nāt }

methyl propyl carbinol [ORGANIC CHEMISTRY] $CH_3CHOHC_3H_7$ Colorless liquid boiling at 119°C; miscible with ether and alcohol, slightly soluble in water; used as a pharmaceuticals intermediate and as a paint and lacquer solvent. Also known as *sec-n*-amyl alcohol; 2-pentanol. { ¦meth·əl ¦prō·pəl 'kär·bə,nȯl }

***N*-methyl-2-pyrrolidone** [ORGANIC CHEMISTRY] C_5H_9NO A liquid boiling at 202°C, miscible with water, castor oil, and organic solvents; used as a chemical intermediate and as a solvent for petroleum and resins, and in PVC spinning. { ¦en 'meth·əl ¦tü pə'räl·ə,dōn }

methyl red [ORGANIC CHEMISTRY] $(CH_3)_2NC_6H_4NNC_6H_4COOH$ A dark red powder or violet crystals with a melting point of 180°C; soluble in alcohol, ether, and glacial acetic acid; used as an acid-base indicator (pH 4.2-6.2). { 'meth·əl 'red }

methyl ricinoleate [ORGANIC CHEMISTRY] $C_{19}H_{36}O_3$ Clear, low-viscosity fluid used as a wetting agent, cutting oil additive, lubricant, and plasticizer. { 'meth·əl ,ris·ən'ōl·ē,at }

methyl salicylate [ORGANIC CHEMISTRY] $C_6H_4OHCOOCH_3$ A colorless, yellow, or reddish liquid, slightly soluble in water, boiling at 222.2°C, with an odor of wintergreen; used in medicine and perfumery, and as a solvent for cellulose derivatives. Also known as betula oil; gaultheria oil; wintergreen oil. { 'meth·əl sə'lis·ə,lāt }

3-methylsalicylic acid [ORGANIC CHEMISTRY] $C_8H_8O_3$ A white to reddish, crystalline compound with a melting point of 165-166°C; soluble in chloroform, alcohol, ether, and alkali hydroxides; used to make dyes. { ¦thrē ¦meth·əl¦sal·ə¦sil·ik 'as·əd }

methyl silicone [ORGANIC CHEMISTRY] $[(CH_3)_2SiO]_x$, $[C(CH_3)_2Si_2O_3]_y$, etc. The common varieties of silicones with properties of oil, resin, or rubber, depending on molecular size and arrangement. { 'meth·əl 'sil·ə,kōn }

methyl stearate [ORGANIC CHEMISTRY] $C_{17}H_{35}COOCH_3$ Colorless crystals melting at 39°C; soluble in alcohol and ether, insoluble in water; used as an intermediate for stearic acid manufacture. { 'meth·əl 'stir,āt }

methyl styrene [ORGANIC CHEMISTRY] *See* vinyltoluene. { 'meth·əl 'stī,rēn }

α-methyl styrene [ORGANIC CHEMISTRY] $C_6H_5C(CH_3){:}CH_2$ Colorless, toxic, polymerizable liquid boiling at 165°C; used to produce polystyrene resins. { ¦al·fə ¦meth·əl 'stī,rēn }

methyl sulfate [ORGANIC CHEMISTRY] *See* dimethyl sulfate. { 'meth·əl 'səl‚fāt }

methyl sulfide [ORGANIC CHEMISTRY] $(CH_3)_2S$ Flammable, colorless liquid with disagreeable aroma; soluble in ether and alcohol, insoluble in water; boils at 38°C; used as a chemical intermediate. Also known as dimethyl sulfide. { 'meth·əl 'səl‚fīd }

methylsulfonic acid [ORGANIC CHEMISTRY] *See* methanesulfonic acid. { ¦meth·əl·səl¦fän·ik 'as·əd }

methylsulfonyl [ORGANIC CHEMISTRY] A functional group with the formula CH_3SO_2—. Also known as mesyl. { ¦meth·əl 'səl·fə‚nil }

methyl tetradecanoate [ORGANIC CHEMISTRY] *See* methyl myristate. { 'meth·əl ‚te·trə·də'kan·ə‚wāt }

4-methyl-5-thiazoleethanol [ORGANIC CHEMISTRY] C_6H_9NOS A viscous, oily liquid; soluble in alcohol, ether, benzene, chloroform, and water; used as an intermediate in the synthesis of vitamin B_1 and as a sedative and hypnotic. { ¦fȯr 'meth·əl ¦fīv thī ‚az·ə·lō'eth·ə‚nȯl }

methylthionine chloride [ORGANIC CHEMISTRY] *See* methylene blue. { ¦meth·əl'thī·ə‚nēn 'klȯr‚īd }

α-methyl-*para*-tyrosine [ORGANIC CHEMISTRY] $C_{10}H_{13}NO_3$ A crystalline compound which acts as the inhibitor of the first and rate-limiting reaction in the biosynthesis of catecholamine; used as an inhibitor of tyrosine hydroxylase. { ¦al·fə 'meth·əl ¦par·ə 'tī·rə‚sēn }

methyl violet [ORGANIC CHEMISTRY] A derivative of pararosaniline, used as an antiallergen and bactericide, acid-base indicator, biological stain, and textile dye. Also known as crystal violet; gentian violet. { 'meth·əl 'vī·lət }

mevalonic acid [ORGANIC CHEMISTRY] $HO_2C_5H_9COOH$ A dihydroxy acid used in organic synthesis. { ¦mev·ə¦lan·ik 'as·əd }

mexacarbate [ORGANIC CHEMISTRY] $C_{12}H_{18}N_2O_2$ A tan solid with a melting point of 85°C; used to control insect pests of trees, flowers, and shrubs. { ¦mek·sə'kär‚bāt }

Mg [CHEMISTRY] *See* magnesium.

micellar catalysis [CHEMISTRY] Enhancement of the rate of a chemical reaction in solution by the addition of a surfactant, so that the reaction proceeds in the environment of surfactant aggregates. { mī¦sel·ər kə'tal·ə·səs }

micelle [PHYSICAL CHEMISTRY] A colloidal aggregate of a unique number (between 50 and 100) of amphipathic molecules, which occurs at a well-defined concentration known as the critical micelle concentration. { mī'sel }

Michler's ketone [ORGANIC CHEMISTRY] *See* tetramethyldiaminobenzophenone. { 'mik·lərz 'kē‚tōn }

micril [CHEMISTRY] *See* gammil. { 'mī·krəl }

microanalysis [ANALYTICAL CHEMISTRY] Identification and chemical analysis of material on a small scale so that specialized instruments such as the microscope are needed; the material analyzed may be on the scale of 1 microgram. { ¦mī·krō·ə'nal·ə·səs }

microchemistry [CHEMISTRY] The study of chemical reactions, using small quantities of materials, frequently less than 1 milligram or 1 milliliter, and often requiring special small apparatus and microscopical observation. { ¦mī·krō'kem·ə·strē }

microdensitometer [SPECTROSCOPY] A high-sensitivity densitometer used in spectroscopy to detect spectrum lines too faint on a negative to be seen by the human eye. { ¦mī·krō‚den·sə'täm·əd·ər }

microelectrolysis [PHYSICAL CHEMISTRY] Electrolysis of small quantities of material. { ¦mī·krō·i‚lek'träl·ə·səs }

microelectrophoresis [ANALYTICAL CHEMISTRY] Direct microscopic observation and measurement of the velocity of migration of ions or other charged bodies through a solution toward oppositely charged electrodes. Also known as optical cytopherometry. { ¦mī·krō·i‚lek·trə·fə'rē·səs }

microgammil [CHEMISTRY] *See* gammil. { ¦mī·krō'gam·əl }

microheterogeneity [CHEMISTRY] A small variation in the chemical structure of a molecule that does not result in a significant change in properties. { ‚mī·krō‚hed·ə·rə·jə'nē·əd·ē }

microincineration [CHEMISTRY] Reduction of small quantities of organic substances to ash by application of heat. { ¦mī·krō·in‚sin·ə'rā·shən }

micropore [CHEMISTRY] A pore in a catalytic material whose diameter is less than 2 nanometers. { 'mī·krə‚pȯr }

microprobe [SPECTROSCOPY] An instrument for chemical microanalysis of a sample, in which a beam of electrons is focused on an area less than a micrometer in diameter, and the characteristic x-rays emitted as a result are dispersed and analyzed in a crystal spectrometer to provide a qualitative and quantitative evaluation of chemical composition. Also known as x-ray microprobe. { 'mī·krə‚prōb }

microprobe spectrometry [SPECTROSCOPY] Microanalysis of a sample, using a microprobe. { 'mī·krə‚prōb spek'träm·ə·trē }

microradiography [ANALYTICAL CHEMISTRY] Technique for the study of surfaces of solids by monochromatic-radiation (such as x-ray) contrast effects shown via projection or enlargement of a contact radiograph. { ¦mī·krō‚rād·ē'äg·rə·fē }

microspectrograph [SPECTROSCOPY] A microspectroscope provided with a photographic camera or other device for recording the spectrum. { ¦mī·krō'spek·trə‚graf }

microspectrophotometer [SPECTROSCOPY] A split-beam or double-beam spectrophotometer including a microscope for the localization of the object under study, and capable of carrying out spectral analyses within the dimensions of a single cell. { ¦mī·krō¦spek·trə·fə'täm·əd·ər }

microspectroscope [SPECTROSCOPY] An instrument for analyzing the spectra of microscopic objects, such as living cells, in which light passing through the sample is focused by a compound microscope system, and both this light and the light which has passed through a reference sample are dispersed by a prism spectroscope, so that the spectra of both can be viewed simultaneously. { ¦mī·krō'spek·trə‚skōp }

microthrowing power [PHYSICAL CHEMISTRY] Relative ability of an electroplating solution to deposit metal in a small, shallow aperture or crevice not exceeding a few thousandths of an inch in dimensions. { ¦mī·krō'thrō·iŋ ‚pau̇·ər }

microwave spectrometer [SPECTROSCOPY] An instrument which makes a graphical record of the intensity of microwave radiation emitted or absorbed by a substance as a function of frequency, wavelength, or some related variable. { 'mī·krə‚wāv spek'träm·əd·ər }

microwave spectroscope [SPECTROSCOPY] An instrument used to observe the intensity of microwave radiation emitted or absorbed by a substance as a function of frequency, wavelength, or some related variable. { 'mī·krə‚wāv 'spek·trə‚skōp }

microwave spectroscopy [SPECTROSCOPY] The methods and techniques of observing and the theory for interpreting the selective absorption and emission of microwaves at various frequencies by solids, liquids, and gases. { 'mī·krə‚wāv spek'träs·kə·pē }

microwave spectrum [SPECTROSCOPY] A display, photograph, or plot of the intensity of microwave radiation emitted or absorbed by a substance as a function of frequency, wavelength, or some related variable. { 'mī·krə‚wāv 'spek·trəm }

migration [CHEMISTRY] The movement of an atom or group of atoms to new positions during the course of a molecular rearrangement. { mī'grā·shən }

migration current [PHYSICAL CHEMISTRY] Additional current produced by electrostatic attraction of cations to the surface of a dropping electrode; an unpredictable and undesirable effect to be avoided during analytical voltammetry. { mī'grā·shən ‚kə·rənt }

MIKES [SPECTROSCOPY] *See* mass-analyzed ion kinetic energy spectrometry. { mīks }

mild mercury chloride [INORGANIC CHEMISTRY] *See* mercurous chloride. { 'mīld 'mər·kyə·rē 'klȯr‚īd }

milk [CHEMISTRY] A suspension of certain metallic oxides, as milk of magnesia, iron, or bismuth. { milk }

milliequivalent [CHEMISTRY] One-thousandth of a compound's or an element's equivalent weight. Abbreviated meq. { ¦mil·ē·ə¦kwiv·ə·lənt }

Millon's reagent [CHEMISTRY] Reagent used to test for proteins; made by dissolving mercury in nitric acid, diluting, then decanting the liquid from the precipitate. { mē'lȯnz re‚ā·jənt }

mimosine [ORGANIC CHEMISTRY] $C_8H_{10}N_2O_4$ A crystalline compound with a melting point of 235-236°C; soluble in dilute acids or bases; used as a depilatory agent. Also known as leucaenine; leucaenol; leucenine; leucenol. { məˈmōˌsēn }

mineral acid [INORGANIC CHEMISTRY] Any one of the major inorganic acids, such as sulfuric, nitric, or hydrochloric acids. { ˈmin·rəl ˌas·əd }

mineral green [INORGANIC CHEMISTRY] *See* copper carbonate. { ˈmin·rəl ¦grēn }

mineralize [CHEMISTRY] To convert organic compounds to simpler inorganic compounds, namely, carbon dioxide and water (and halogen acids, if the organic substances are halogenated). { ˈmin·rəˌlīz }

mineralogy [INORGANIC CHEMISTRY] The science which concerns the study of natural inorganic substances called minerals. { ˌmin·əˈräl·ə·jē }

MIPC [ORGANIC CHEMISTRY] *See ortho*-isopropylphenyl-methylcarbamate.

miscibility [CHEMISTRY] The tendency or capacity of two or more liquids to form a uniform blend, that is, to dissolve in each other; degrees are total miscibility, partial miscibility, and immiscibility. { ˌmis·əˈbil·əd·ē }

misfire [CHEMISTRY] Failure of fuel or an explosive charge to ignite properly. { ˈmisˌfīr }

Mitscherlich law of isomorphism [CHEMISTRY] Substances which have similar chemical properties and crystalline forms usually have similar chemical formulas. { ˈmich·ərˌlik ¦lȯ əv ¦ī·sōˈmȯrˌfiz·əm }

mixed acid [INORGANIC CHEMISTRY] *See* nitrating acid. { ˈmikst ˈas·əd }

mixed aniline point [PHYSICAL CHEMISTRY] The minimum temperature at which a mixture of aniline, heptane, and hydrocarbon will form a solution; related to the aromatic character of the hydrocarbon. { ˈmikst ˈan·ə·lən ˌpȯint }

mixed indicator [ANALYTICAL CHEMISTRY] Color-change indicator for acid-base titration end points in which a mixture of two indicator substances is used to give sharper end-point color changes. { ˈmikst ˈin·dəˌkād·ər }

mixed potential [PHYSICAL CHEMISTRY] The electrode potential of a material while more than one electrochemical reaction is occurring simultaneously. { ˈmikst pəˈten·chəl }

Mn [CHEMISTRY] *See* manganese.

Mo [CHEMISTRY] *See* molybdenum.

mobile electron [PHYSICAL CHEMISTRY] An electron that can move readily from one atom to another within a chemical structure in response to changes in the external chemical environment. { ˌmō·bəl əˈlekˌträn }

mobile phase [ANALYTICAL CHEMISTRY] **1.** In liquid chromatography, the phase that is moving in the bed, including the fraction of the sample held by this phase. **2.** The carrier gas in a gas chromatography procedure. { ˈmō·bəl ˌfāz }

mobility coefficient [PHYSICAL CHEMISTRY] The average speed of motion of molecules in a solution in the direction of the concentration gradient, at unit concentration and unit osmotic pressure gradient. { mōˈbil·əd·ē ˌkō·əˌfish·ənt }

modified Lewis acid [PHYSICAL CHEMISTRY] An acid that is a halide ion acceptor. { ˈmäd·əˌfīd ˈlü·əs ¦as·əd }

modulated Raman scattering [SPECTROSCOPY] Application of modulation spectroscopy to the study of Raman scattering; in particular, use of external perturbations to lower the symmetry of certain crystals and permit symmetry-forbidden modes, and the use of wavelength modulation to analyze second-order Raman spectra. { ˈmäj·əˌlād·əd ˈrä·mən ˌskad·ə·riŋ }

modulation spectroscopy [SPECTROSCOPY] A branch of spectroscopy concerned with the measurement and interpretation of changes in transmission or reflection spectra induced (usually) by externally applied perturbation, such as temperature or pressure change, or an electric or magnetic field. { ˌmäj·əˈlā·shən spekˈträs·kə·pē }

Mohr's salt [INORGANIC CHEMISTRY] *See* ferrous ammonium sulfate. { ˈmȯrz ˌsȯlt }

Mohr titration [ANALYTICAL CHEMISTRY] Titration with silver nitrate to determine the concentration of chlorides in a solution; silver chromate precipitation is the end-point indicator. { ˈmȯr tīˈtrā·shən }

moiety [CHEMISTRY] A part or portion of a molecule, generally complex, having a characteristic chemical or pharmacological property. { ˈmȯi·əd·ē }

moisture [PHYSICAL CHEMISTRY] Water that is dispersed through a gas in the form of water vapor or small droplets, dispersed through a solid, or condensed on the surface of a solid. { 'mȯis·chər }

mol [CHEMISTRY] *See* mole. { mōl }

molal average boiling point [PHYSICAL CHEMISTRY] A pseudo boiling point for a mixture calculated as the summation of individual mole fraction-boiling point (in degrees Rankine) products. { 'mō·ləl 'av·rij 'bȯil·iŋ ‚pȯint }

molal elevation of the boiling point [PHYSICAL CHEMISTRY] *See* ebullioscopic constant. { ¦mō·ləl ‚el·ə¦vā·shən əv thə 'bȯil·iŋ ‚pȯint }

molal heat capacity [PHYSICAL CHEMISTRY] *See* molar heat capacity. { 'mō·ləl 'hēt kə‚pas·əd·ē }

molality [CHEMISTRY] Concentration given as moles per 1000 grams of solvent. { mō'lal·əd·ē }

molal quantity [CHEMISTRY] The number of moles (gram-molecular weights) present, expressed with weight in pounds, grams, or such units, numerically equal to the molecular weight; for example, pound-mole, gram-mole. { 'mō·ləl 'kwän·əd·ē }

molal solution [CHEMISTRY] Concentration of a solution expressed in moles of solute divided by 1000 grams of solvent. { 'mō·ləl sə‚lü·shən }

molal specific heat [PHYSICAL CHEMISTRY] *See* molar specific heat. { 'mō·ləl spə¦sif ik 'hēt }

molal volume [PHYSICAL CHEMISTRY] *See* molar volume. { 'mō·ləl 'väl·yəm }

molar [PHYSICAL CHEMISTRY] Denoting a physical quantity divided by the amount of substance expressed in moles. { 'mō·lər }

molar conductivity [PHYSICAL CHEMISTRY] The ratio of the conductivity of an electrolytic solution to the concentration of electrolyte in moles per unit volume. { 'mō·lər ‚kän‚dək'tiv·əd·ē }

molar heat capacity [PHYSICAL CHEMISTRY] The amount of heat required to raise 1 mole of a substance 1° in temperature. Also known as molal heat capacity; molecular heat capacity. { 'mō·lər 'hēt kə‚pas·əd·ē }

molarity [CHEMISTRY] Measure of the number of gram-molecular weights of a compound present (dissolved) in 1 liter of solution; it is indicated by M, preceded by a number to show solute concentration. { mō'lar·əd·ē }

molar solution [CHEMISTRY] Aqueous solution that contains 1 mole (gram-molecular weight) of solute in 1 liter of the solution. { 'mō·lər sə‚lü·shən }

molar specific heat [PHYSICAL CHEMISTRY] The ratio of the amount of heat required to raise the temperature of 1 mole of a compound 1°, to the amount of heat required to raise the temperature of 1 mole of a reference substance, such as water, 1° at a specified temperature. Also known as molal specific heat; molecular specific heat. { 'mō·lər spə‚sif·ik ‚hēt }

molar susceptibility [PHYSICAL CHEMISTRY] Magnetic susceptibility of a compound per gram-mole of that compound. { 'mō·lər sə‚sep·tə'bil·əd·ē }

molar volume [PHYSICAL CHEMISTRY] The volume occupied by one mole of a substance in the form of a solid, liquid, or gas. Also known as molal volume; mole volume. { 'mō·lər 'väl·yəm }

mole [CHEMISTRY] An amount of substance of a system which contains as many elementary units as there are atoms of carbon in 0.012 kilogram of the pure nuclide carbon-12; the elementary unit must be specified and may be an atom, molecule, ion, electron, photon, or even a specified group of such units. Symbolized mol. { mōl }

molecular adhesion [PHYSICAL CHEMISTRY] A particular manifestation of intermolecular forces which causes solids or liquids to adhere to each other; usually used with reference to adhesion of two different materials, in contrast to cohesion. { mə'lek·yə·lər ad'hē·zhən }

molecular amplitude [ANALYTICAL CHEMISTRY] The difference between the molecular rotation at the extreme (peak or trough) value caused by the longer light wavelength and the molecular rotation at the extreme value caused by the shorter wavelength. { mə'lek·yə·lər 'am·plə‚tüd }

molecular association [PHYSICAL CHEMISTRY] The formation of double molecules or polymolecules from a single species as a result of specific and moderately strong intermolecular forces. { mə'lek·yə·lər ə‚sō·sē'ā·shən }

molecular asymmetry [PHYSICAL CHEMISTRY] *See* asymmetry. { mə'lek·yə·lər ‚ā'sim·ə·trē }

molecular attraction [PHYSICAL CHEMISTRY] A force which pulls molecules toward each other. { mə'lek·yə·lər ə'trak·shən }

molecular cluster [PHYSICAL CHEMISTRY] An assembly of molecules that are weakly bound together and display properties intermediate between those of isolated gas-phase molecules and bulk condensed media. { mə'lek·yə·lər 'kləs·tər }

molecular conductivity [PHYSICAL CHEMISTRY] The conductivity of a volume of electrolyte containing 1 mole of dissolved substance. { mə'lek·yə·lər ‚kän‚dək'tiv·əd·ē }

molecular diamagnetism [PHYSICAL CHEMISTRY] Diamagnetism of compounds, especially organic compounds whose susceptibilities can often be calculated from the atoms and chemical bonds of which they are composed. { mə'lek·yə·lər ‚dī·ə'mag·nə‚tiz·əm }

molecular diameter [PHYSICAL CHEMISTRY] The diameter of a molecule, assuming it to be spherical; has a numerical value of 10^{-8} centimeter multiplied by a factor dependent on the compound or element. { mə'lek·yə·lər dī'am·əd·ər }

molecular dipole [PHYSICAL CHEMISTRY] A molecule having an electric dipole moment, whether it is permanent or produced by an external field. { mə'lek·yə·lər 'dī‚pōl }

molecular distillation [CHEMISTRY] A process by which substances are distilled in high vacuum at the lowest possible temperature and with least damage to their composition. { mə'lek·yə·lər ‚dis·tə'lā·shən }

molecular dynamics [PHYSICAL CHEMISTRY] A branch of physical chemistry concerned with molecular mechanisms of the elementary physical and chemical processes that control rates of reaction. { mə'lek·yə·lər di'nam·iks }

molecular energy level [PHYSICAL CHEMISTRY] One of the states of motion of nuclei and electrons in a molecule, having a definite energy, which is allowed by quantum mechanics. { mə'lek·yə·lər 'en·ər·jē ‚lev·əl }

molecular entity [CHEMISTRY] A chemically or isotopically distinct atom, molecule, ion, complex, free radical, or similar unit that can be distinguished from other kinds of units. { mə'lek·yə·lər 'en·təd·ē }

molecular exclusion chromatography [ANALYTICAL CHEMISTRY] *See* gel filtration. { mə'lek·yə·lər ik¦sklü·zhən ‚krō·mə'täg·rə·fē }

molecular formula [CHEMISTRY] A chemical formula that indicates the actual numbers and kinds of atoms in a molecule, but not the chemical structure. { mə'lek·yə·lər 'fȯr·myə·lə }

molecular gas [CHEMISTRY] A gas composed of a single species, such as oxygen, chlorine, or neon. { mə'lek·yə·lər 'gas }

molecular graphics [PHYSICAL CHEMISTRY] The use of computer graphics to display and manipulate chemical structures with sufficient accuracy that bond distances and angles may be displayed and reported and it is possible to dock or fit together two or more molecules. Also known as graphics-based molecular modeling. { mə'lek·yə·lər 'graf·iks }

molecular heat capacity [PHYSICAL CHEMISTRY] *See* molar heat capacity. { mə'lek·yə·lər ¦hēt kə‚pas·əd·ē }

molecular ion [ORGANIC CHEMISTRY] An ion that results from the loss of an electron by an organic molecule following bombardment with high-energy electrons during mass spectrometry. { mə'lek·yə·lər 'ī‚än }

molecularity [PHYSICAL CHEMISTRY] In a chemical reaction, the number of molecules which come together and form the activated complex. { mə‚lek·yə'lar·əd·ē }

molecular magnet [PHYSICAL CHEMISTRY] A molecule having a nonvanishing magnetic dipole moment, whether it is permanent or produced by an external field. { mə'lek·yə·lər 'mag·nət }

molecular mechanics [PHYSICAL CHEMISTRY] An empirical method of calculating the dynamics of molecules, in which bonds between atoms are represented by springs obeying Hooke's law, and additional terms representing bond angle bending, torsional interactions, and van der Waals-type interactions are included. Also known as force-field method. { mə'lek·yə·lər mi'kan·iks }

molecular modeling [CHEMISTRY] The use of computers for the simulation of chemical entities and processes. { mə'lek·yə·lər 'mäd·liŋ }

molecular orbital [PHYSICAL CHEMISTRY] A wave function describing an electron in a molecule. { mə'lek·yə·lər 'ȯr·bəd·əl }

molecular paramagnetism [PHYSICAL CHEMISTRY] Paramagnetism of molecules, such as oxygen, some other molecules, and a large number of organic compounds. { mə'lek·yə·lər ¦par·ə'mag·nə,tiz·əm }

molecular polarizability [PHYSICAL CHEMISTRY] The electric dipole moment induced in a molecule by an external electric field, divided by the magnitude of the field. { mə'lek·yə·lər ,pō·lə,rīz·ə'bil·əd·ē }

molecular rearrangement [ORGANIC CHEMISTRY] *See* rearrangement reaction. { mə'lek·yə·lər ,rē·ə'rānj·mənt }

molecular receptor [ORGANIC CHEMISTRY] A species that can select one of many possible binding partners and form a complex that is stabilized by interactions such as hydrogen bonding or changes in solvation. { mə'lek·yər·lər ri'sep·tər }

molecular relaxation [PHYSICAL CHEMISTRY] Transition of a molecule from an excited energy level to another excited level of lower energy or to the ground state. { mə'lek·yə·lər ,rē,lak'sā·shən }

molecular self-assembly [ORGANIC CHEMISTRY] The spontaneous aggregation of molecules into well-defined, stable, noncovalently bonded assemblies that are held together by intermolecular forces. { mə'lek·yə·lər ,self ə'sem·blē }

molecular sieve [CHEMISTRY] A naturally occurring or synthetic zeolite characterized by the ability to undergo dehydration with little or no change in crystal structure, thereby offering a very high surface area for adsorption of foreign molecules. { mə'lek·yə·lər 'siv }

molecular-sieve chromatography [ANALYTICAL CHEMISTRY] *See* gel filtration. { mə'lek·yə·lər ¦siv ,krō·mə'täg·rə·fē }

molecular specific heat [PHYSICAL CHEMISTRY] *See* molar specific heat. { mə'lek·yə·lər spə¦sif·ik 'hēt }

molecular spectroscopy [SPECTROSCOPY] The production, measurement, and interpretation of molecular spectra. { mə'lek·yə·lər spek'träs·kə·pē }

molecular spectrum [SPECTROSCOPY] The intensity of electromagnetic radiation emitted or absorbed by a collection of molecules as a function of frequency, wave number, or some related quantity. { mə'lek·yə·lər 'spek·trəm }

molecular still [CHEMISTRY] An apparatus used to conduct molecular distillation. { mə'lek·yə·lər 'stil }

molecular structure [PHYSICAL CHEMISTRY] The manner in which electrons and nuclei interact to form a molecule, as elucidated by quantum mechanics and a study of molecular spectra. { mə'lek·yə·lər 'strək·chər }

molecular velocity [PHYSICAL CHEMISTRY] The velocity of an individual molecule in a given sample of gas; the vector quantity is symbolized **u**, and the magnitude is symbolized *u*. { mə'lek·yə·lər və'läs·əd·ē }

molecular vibration [PHYSICAL CHEMISTRY] The theory that all atoms within a molecule are in continuous motion, vibrating at definite frequencies specific to the molecular structure as a whole as well as to groups of atoms within the molecule; the basis of spectroscopic analysis. { mə'lek·yə·lər vī'brā·shən }

molecular volume [CHEMISTRY] The volume that is occupied by 1 mole (gram-molecular weight) of an element or compound; equals the molecular weight divided by the density. { mə'lek·yə·lər 'väl·yəm }

molecular weight [CHEMISTRY] The sum of the atomic weights of all the atoms in a molecule. Also known as relative molecular mass. { mə'lek·yə·lər 'wāt }

molecular-weight distribution [ORGANIC CHEMISTRY] Frequency of occurrence of the different molecular-weight chains in a homologous polymeric system. { mə'lek·yə·lər ¦wāt ,di·strə'byü·shən }

molecule [CHEMISTRY] A group of atoms held together by chemical forces; the atoms in the molecule may be identical as in H_2, S_2, and S_8, or different as in H_2O and CO_2;

a molecule is the smallest unit of matter which can exist by itself and retain all its chemical properties. { 'mäl·ə‚kyül }

mole fraction [CHEMISTRY] The ratio of the number of moles of a substance in a mixture or solution to the total number of moles of all the components in the mixture or solution. { 'mōl ‚frak·shən }

mole percent [CHEMISTRY] Percentage calculation expressed in terms of moles rather than weight. { 'mōl pər‚sent }

mole volume [PHYSICAL CHEMISTRY] *See* molar volume. { 'mōl 'väl·yəm }

molinate [ORGANIC CHEMISTRY] $C_9H_{17}NOS$ A light yellow liquid with limited solubility in water; used as a herbicide to control watergrass in rice. { 'mäl·ə‚nāt }

molybdate [INORGANIC CHEMISTRY] A salt derived from a molybdic acid. { mə'lib‚dāt }

molybdenum [CHEMISTRY] A chemical element, symbol Mo, atomic number 42, and atomic weight 95.95. { mə'lib·de·nəm }

molybdenum dioxide [INORGANIC CHEMISTRY] MoO_2 Lead-gray powder; insoluble in hydrochloric and hydrofluoric acids; used in pigment for textiles. { mə'lib·də·nəm dī'äk‚sīd }

molybdenum disilicide [INORGANIC CHEMISTRY] $MoSi_2$ A dark gray, crystalline powder with a melting range of 1870-2030°C; soluble in hydrofluoric and nitric acids; used in electrical resistors and for protective coatings for high-temperature conditions. { mə'lib·də·nəm dī'sil·ə‚sīd }

molybdenum disulfide [INORGANIC CHEMISTRY] MoS_2 A black lustrous powder, melting at 1185°C, insoluble in water, soluble in aqua regia and concentrated sulfuric acid; used as a dry lubricant and an additive for greases and oils. Also known as molybdenum sulfide; molybdic sulfide. { mə'lib·də·nəm dī'səl‚fīd }

molybdenum pentachloride [INORGANIC CHEMISTRY] $MoCl_5$ Hygroscopic gray-black needles melting at 194°C; reacts with water and air; soluble in anhydrous organic solvents; used as a catalyst and as raw material to make molybdenum hexacarbonyl. { mə'lib·də·nəm ‚pen·tə'klȯr‚īd }

molybdenum sesquioxide [INORGANIC CHEMISTRY] MoO_3 Water-insoluble, gray-black powder with slight solubility in acids; used as a catalyst and as a coating for metal articles. { mə'lib·də·nəm ‚ses·kwē'äk‚sīd }

molybdenum sulfide [INORGANIC CHEMISTRY] *See* molybdenum disulfide. { mə'lib·də·nəm 'səl‚fīd }

molybdenum trioxide [INORGANIC CHEMISTRY] MoO_3 A white solid at room temperature, with a melting point of 795°C; soluble in concentrated mixtures of nitric and sulfuric acids and nitric and hydrochloric acids; used as a corrosion inhibitor, in enamels and ceramic glazes, in medicine and agriculture, and as a catalyst in the petroleum industry. { mə'lib·də·nəm trī'äk‚sīd }

molybdic acid [INORGANIC CHEMISTRY] Any acid derived from molybdenum trioxide, especially the simplest acid H_2MoO_4, obtained as white crystals. { mə'lib·dik 'as·əd }

molybdic sulfide [INORGANIC CHEMISTRY] *See* molybdenum disulfide. { mə'lib·dik 'səl‚fīd }

monatomic [CHEMISTRY] Composed of one atom. { ¦män·ə'täm·ik }

monatomic gas [CHEMISTRY] A gas whose molecules have only one atom; the inert gases are examples. { ¦män·ə¦täm·ik 'gas }

mono- [CHEMISTRY] A prefix for chemical compounds to show a single radical; for example, monoglyceride, a glycol ester on which a single acid group is attached to the glycerol group. { 'män·ō }

monoacetate [ORGANIC CHEMISTRY] A compound such as a salt or ester that contains one acetate group. { ¦män·ō'as·ə‚tāt }

monoacid [CHEMISTRY] **1.** An acid that has only one replaceable hydrogen. **2.** A base or an alcohol that has a single hydroxyl (—OH) group which can be replaced by an atom or a functional group to form a salt or ester. { ¦män·ō'as·əd }

monoamine [ORGANIC CHEMISTRY] An amine compound that has only one amino group. { ¦män·ō'am‚ēn }

monoammonium tartrate [ORGANIC CHEMISTRY] *See* ammonium bitartrate. { ¦män·ō·ə'mō·nē·əm 'tär,trāt }

monobasic [CHEMISTRY] Pertaining to an acid with one displaceable hydrogen atom, such as hydrochloric acid, HCl. { ¦män·ō'bās·ik }

monobasic sodium phosphate [INORGANIC CHEMISTRY] NaH_2PO_4 White crystals that are slightly hygroscopic, soluble in water, insoluble in alcohol; used in baking powders and acid cleansers, and as a cattle-food supplement. { män·ō'bās·ik 'sod·ē·əm 'fäs,fāt }

monochromator [SPECTROSCOPY] A spectrograph in which a detector is replaced by a second slit, placed in the focal plane, to isolate a particular narrow band of wavelengths for refocusing on a detector or experimental object. { ¦män·ə¦krō,mād·ər }

monodisperse colloidal system [CHEMISTRY] A colloidal system in which the suspended particles have identical size, shape, and interaction. { ¦män·ō·di¦spərs kə'lȯid·əl 'sis·təm }

monodispersity [ORGANIC CHEMISTRY] Polymer system that is homogeneous in molecular weight, that is, it does not have a distribution of different molecular-weight chains within the total mass. { ¦män·ō·di'spər·səd·ē }

monoester [ORGANIC CHEMISTRY] An ester that has only one ester group. { ¦män·ō'es·tər }

monofunctional compound [ORGANIC CHEMISTRY] An organic compound whose chemical structure possesses a single highly reactive site. { ,män·ō¦fəŋk·shən·əl 'käm,paund }

monoglyceride [ORGANIC CHEMISTRY] Any of the fatty-acid glycerol esters where only one acid group is attached to the glycerol group, for example, $RCOOCH_2CHOHCH_2OH$; examples are glycerol monostearate and monolaurate; used as emulsifiers in cosmetics and lubricants. { ¦män·ō'glis·ə,rīd }

monolayer [PHYSICAL CHEMISTRY] *See* monomolecular film. { 'män·ō,lā·ər }

monolayer capacity [CHEMISTRY] **1.** In chemisorption, the amount of adsorbate required to occupy all adsorption sites on the solid surface. **2.** In physisorption, the amount of material required to cover the solid surface with a complete monolayer of the adsorbate in a close-packed array. { 'män·ə,lā·ər kə,pas·əd·ē }

monomer [CHEMISTRY] A simple molecule which is capable of combining with a number of like or unlike molecules to form a polymer; it is a repeating structure unit within a polymer. Also known as repeating unit. { 'män·ə·mər }

monomeric unit [ORGANIC CHEMISTRY] *See* repeating unit. { ,män·ə¦mer·ik 'yü·nət }

monomolecular film [PHYSICAL CHEMISTRY] A film one molecule thick. Also known as monolayer. { ¦män·ō·mə¦lek·yə·lər 'film }

monopotassium L glutamate [ORGANIC CHEMISTRY] *See* potassium glutamate. { ¦män·ō·pə'tas·ē·əm ¦el'glüd·ə,māt }

monoprotic acid [CHEMISTRY] An acid that has only one ionizable hydrogen atom in each molecule. { ,män·ə¦präd·ik 'as·əd }

monosodium acid methanearsonate [ORGANIC CHEMISTRY] CH_4AsNaO_3 A white, crystalline solid; melting point is 132-139°C; soluble in water; used as an herbicide for grassy weeds on rights-of-way, storage areas, and noncrop areas, and as preplant treatment for cotton, citrus trees, and turf. Abbreviated MSMA. { ¦män·ə'sōd·ē·əm ¦as·əd ¦meth,ān'ärs·ən,āt }

monosodium glutamate [ORGANIC CHEMISTRY] *See* sodium glutamate. { ¦män·ə'sōd·ē·əm 'glüd·ə,māt }

monosubstituted alkene [ORGANIC CHEMISTRY] An alkene with the general formula $RHC{=}CH_2$, where R is any organic group; only one carbon atom is bonded directly to one of the carbons of the carbon-to-carbon double bond. { ,män·ō¦səb·stə,tüd·əd 'al·kēn }

monoterpene [ORGANIC CHEMISTRY] **1.** A class of terpenes with molecular formula $C_{10}H_{16}$; the members of the class contain two isoprene units. **2.** A derivative of a member of such a class. { ¦män·ō'tər,pēn }

monovalent [CHEMISTRY] A radical or atom whose valency is 1. { ¦män·ō'vā·lənt }

monoxide [CHEMISTRY] A compound that contains a single oxygen atom, such as carbon monoxide, CO. { mə'näk,sīd }

mordant [CHEMISTRY] An agent, such as alum, phenol, or aniline, that fixes dyes to tissues, cells, textiles, and other materials by combining with the dye to form an insoluble compound. Also known as dye mordant. { 'mȯrd·ənt }

morin [ORGANIC CHEMISTRY] $C_{15}H_{10}O_7 \cdot 2H_2O$ Colorless needles soluble in boiling alcohol, slightly soluble in water; used as a mordant dye and analytical reagent. { 'mȯr·ən }

morpholine [ORGANIC CHEMISTRY] C_4H_8ONH A hygroscopic liquid, soluble in water; used as a solvent and rubber accelerator. { 'mȯr·fə‚lēn }

morphosan [ORGANIC CHEMISTRY] $C_{17}H_{19}NO_3 \cdot CH_3Br$ A solid morphine derivative without morphine's disagreeable aftereffects; used in medicine. { 'mȯr·fə‚san }

Morse equation [PHYSICAL CHEMISTRY] An equation according to which the potential energy of a diatomic molecule in a given electronic state is given by a Morse potential. { 'mȯrs i‚kwā·zhən }

Morse potential [PHYSICAL CHEMISTRY] An approximate potential associated with the distance r between the nuclei of a diatomic molecule in a given electronic state; it is $V(r) = D\{1 - \exp[-a(r - r_e)]\}^2$, where r_e is the equilibrium distance, D is the dissociation energy, and a is a constant. { 'mȯrs pə‚ten·chəl }

mosaic gold [INORGANIC CHEMISTRY] *See* stannic sulfide. { mō'zā·ik ‚gōld }

Moseley's law [SPECTROSCOPY] The law that the square-root of the frequency of an x-ray spectral line belonging to a particular series is proportional to the difference between the atomic number and a constant which depends only on the series. { 'mōz·lēz ‚lȯ }

Mössbauer spectroscopy [SPECTROSCOPY] The study of Mössbauer spectra, for example, for nuclear hyperfine structure, chemical shifts, and chemical analysis. { 'mu̇s‚bau̇·ər spek'träs·kə·pē }

Mössbauer spectrum [SPECTROSCOPY] A plot of the absorption, by nuclei bound in a crystal lattice, of gamma rays emitted by similar nuclei in a second crystal, as a function of the relative velocity of the two crystals. { 'mu̇s‚bau̇·ər ‚spek·trəm }

mountain blue [INORGANIC CHEMISTRY] $2CuCO_3 \cdot Cu(OH)_2$ Ground azurite used as a paint pigment. Also known as copper blue. { 'mau̇nt·ən 'blü }

moving-boundary electrophoresis [ANALYTICAL CHEMISTRY] A U-tube variation of electrophoresis analysis that uses buffered solution so that all ions of a given species move at the same rate to maintain a sharp, moving front (boundary). { 'müv·iŋ ¦bau̇n·drē i¦lek·trə·fə'rē·səs }

MSMA [ORGANIC CHEMISTRY] *See* monosodium acid methanearsonate.

mucic acid [ORGANIC CHEMISTRY] $HOOC(CHOH)_4COOH$ A white, crystalline powder with a melting point of 210°C; soluble in water; used as a metal ion sequestrant and to retard concrete hardening. Also known as glactaric acid; saccharolactic acid; tetrahydroxyadipic acid. { 'myü·sik 'as·əd }

mull technique [SPECTROSCOPY] Method for obtaining infrared spectra of materials in the solid state; material to be scanned is first pulverized, then mulled with mineral oil. { 'məl tek‚nēk }

multident [ORGANIC CHEMISTRY] *See* polydent. { 'məl·tə‚dent }

multidentate ligand [CHEMISTRY] A ligand capable of donating two or more pairs of electrons in a complexation reaction to form coordinate bonds. { ‚məl·tē¦den‚tāt 'lī·gənd }

multiplet [SPECTROSCOPY] A collection of relatively closely spaced spectral lines resulting from transitions to or from the members of a multiplet (as in the quantum-mechanics definition). { 'məl·tə·plət }

multiplet intensity rules [SPECTROSCOPY] Rules for the relative intensities of spectral lines in a spin-orbit multiplet, stating that the sum of the intensities of all lines which start from a common initial level, or end on a common final level, is proportional to $2J + 1$, where J is the total angular momentum of the initial level or final level respectively. { 'məl·tə·plət in'ten·səd·ē ‚rülz }

multivalent [CHEMISTRY] *See* polyvalent. { ¦məl·tə'vā·lənt }

muriatic acid [INORGANIC CHEMISTRY] *See* hydrochloric acid. { myu̇r·ē'ad·ik 'as·əd }

muscarine [ORGANIC CHEMISTRY] $C_8H_{19}NO_3$ A quaternary ammonium compound, the toxic ingredient of certain mushrooms, as *Amanita muscaria*. Also known as hydroxycholine. { 'məs·kə‚rēn }

musk ambrette [ORGANIC CHEMISTRY] $C_{12}H_{16}N_2O_5$ White to yellow powder with heavy musky aroma; soluble in various oils and phthalates, insoluble in water; congeals at 83°C; used as a perfume fixative. Also known as 2,6-dinitro-3-methoxy-4-tert-butyltoluene. { 'məsk ‚am‚bret }

musk ketone [ORGANIC CHEMISTRY] $C_{14}H_{18}N_2O_5$ White to yellow crystals with sweet musk aroma; soluble in various oils and phthalates, insoluble in water; used as a perfume fixative. Also known as 3,5-dinitro-2,6-dimethyl-4-tert-butylacetophenone. { 'məsk 'kē‚tōn }

musk xylene [ORGANIC CHEMISTRY] *See* musk xylol. { 'məsk 'zī‚lēn }

musk xylol [ORGANIC CHEMISTRY] $(NO_2)_3C_6(CH_3)_2C(CH_3)_3$ White to yellow crystals with powerful musk aroma; soluble in various oils and phthalates, insoluble in water; congeals at 105°C; used as a perfume fixative. Also known as musk xylene; 2,4,6-trinitro-1,3-dimethyl-5-tert-butylbenzene. { 'məsk 'zī‚lòl }

mustard gas [ORGANIC CHEMISTRY] $HS(CH_2ClCH_2)_2S$ An oil, density 1.28, boiling point 215°C; used in chemical warfare. Also known as dichlorodiethylsulfide. { 'məs·tərd ‚gas }

mustard oil [ORGANIC CHEMISTRY] *See* allyl isothiocyanate. { 'məs·tərd ‚òil }

mutarotation [CHEMISTRY] A change in the optical rotation of light that takes place in the solutions of freshly prepared sugars. { ¦myüd·ə·rō'tā·shən }

mutual exclusion rule [PHYSICAL CHEMISTRY] The rule that if a molecule has a center of symmetry, then no transition is allowed in both its Raman scattering and infrared emission (and absorption), but only in one or the other. { 'myü·chə·wəl ik'sklü·zhən ‚rül }

mutuality of phases [CHEMISTRY] The rule that if two phases, with respect to a reaction, are in equilibrium with a third phase at a certain temperature, then they are in equilibrium with respect to each other at that temperature. { ‚myü·chə'wal·əd·ē əv 'fāz·əz }

***β*-myrcene** [ORGANIC CHEMISTRY] $C_{10}H_{16}$ An oily liquid with a pleasant odor; soluble in alcohol, chloroform, ether, and glacial acetic acid; used as an intermediate in the preparation of perfume chemicals. { ¦bād·ə 'mər‚sēn }

myricetin [ORGANIC CHEMISTRY] $C_{15}H_{10}O_8$ A yellow, crystalline compound with a melting point of 357°C; soluble in alcohol; used as an inhibitor of adenosinetriphosphatase. Also known as cannabiscetin; delphidenolon. { mə'ris·ə·tən }

myristic acid [ORGANIC CHEMISTRY] $CH_3(CH_2)_{12}COOH$ Oily white crystals melting at 58°C; soluble in ether and alcohol, insoluble in water; used to synthesize flavor and perfume esters, and in soaps and cosmetics. { mə'ris·tik 'as·əd }

myristyl alcohol [ORGANIC CHEMISTRY] $C_{14}H_{29}OH$ Liquid boiling at 264°C; soluble in ether and alcohol, insoluble in water; used as a chemical intermediate, plasticizer, and perfume fixative. { mə'rist·əl 'al·kə‚hol }

myristyl mercaptan [ORGANIC CHEMISTRY] *See* tetradecyl mercaptan. { mə'rist·əl mər'kap‚tan }

N

N [CHEMISTRY] *See* nitrogen; normality.

n- [ORGANIC CHEMISTRY] Chemical prefix for normal (straight-carbon-chain) hydrocarbon compounds.

Na [CHEMISTRY] *See* sodium.

NAA [ORGANIC CHEMISTRY] *See* naphthaleneacetic acid.

naled [ORGANIC CHEMISTRY] $C_4H_7Br_2Cl_2O_4$ A white solid with a melting point of 27°C; slight solubility in water; used as an insecticide and miticide for crops, farm buildings, and kennels, and for mosquito control. { 'nal·əd }

naphthacene [ORGANIC CHEMISTRY] $C_{18}H_{12}$ A hydrocarbon molecule that may be considered to be four benzene rings fused together; it is explosive when shocked; used in organic synthesis. Also known as rubene; tetracene. { 'naf·thə,sēn }

naphthalene [ORGANIC CHEMISTRY] $C_{10}H_8$ White, volatile crystals with coal tar aroma, insoluble in water, soluble in organic solvents; structurally it is represented as two benzenoid rings fused together; boiling point 218°C, melting point 80.1°C; used for moth repellents, fungicides, lubricants, and resins, and as a solvent. Also known as naphthalin; tar camphor. { 'naf·thə,lēn }

naphthaleneacetamide [ORGANIC CHEMISTRY] $C_{12}H_{11}NO$ A colorless solid with a melting point of 183°C; used as a growth regulator for root cuttings and for thinning of apples and pears. { ¦naf·thə,lēn·ə'sed·ə·məd }

naphthaleneacetic acid [ORGANIC CHEMISTRY] $C_{10}H_7CH_2COOH$ White, odorless crystals, melting at 132–135°C, soluble in organic solvents, slightly soluble in water; used as an agricultural spray. Abbreviated NAA. Also known as 1-naphthylacetic acid. { ¦naf·thə,lēn·ə¦sēd·ik 'as·əd }

naphthalene-1,5-disulfonic acid [ORGANIC CHEMISTRY] $C_{10}H_6(SO_3H)_2$ White crystals, decomposing when heated; used to make dyes. Also known as Armstrong's acid. { 'naf·thə,lēn ¦wən ¦fīv ,dī səl'fän·ik 'as·əd }

1-naphthalenesulfonic acid [ORGANIC CHEMISTRY] $C_{10}H_8O_3S$ A crystalline compound with a melting point of 90°C (dihydrate); soluble in water or alcohol, used to make *α*-naphthol. { ¦wən ¦naf·thə,lēn səl'fän·ik 'as·əd }

naphthalin [ORGANIC CHEMISTRY] *See* naphthalene. { 'naf·thə·lən }

naphthene [ORGANIC CHEMISTRY] Any of the cycloparaffin derivatives of cyclopentane (C_5H_{10}) or cyclohexane (C_6H_{12}) found in crude petroleum. { 'naf,thēn }

naphthenic acid [ORGANIC CHEMISTRY] Any of the derivatives of cyclopentane, cyclohexane, cycloheptane, or other naphthenic homologs derived from petroleum, molecular weights 180 to 350; soluble in organic solvents and hydrocarbons, slightly soluble in water; used as a paint drier and wood preservative, and in metals production. { naf'thēn·ik 'as·əd }

naphthionic acid [ORGANIC CHEMISTRY] $C_{10}H_6(NH_2)SO_3H$ White powder or crystals that decompose when heated; used to manufacture dyes. { ¦naf·thē¦än·ik 'as·əd }

***α*-naphthol** [ORGANIC CHEMISTRY] $C_{10}H_7OH$ Colorless to yellow powder, melting at 96°C; used to make dyes and perfumes, and in synthesis of organic molecules. { ¦al·fə 'naf,thȯl }

***β*-naphthol** [ORGANIC CHEMISTRY] $C_{10}H_7OH$ White crystals that melt at 121.6°C; insoluble in water; used to make pigments, dyes, and antioxidants. { ¦bād·ə 'naf,thȯl }

1,2-naphthoquinone [ORGANIC CHEMISTRY] $C_{10}H_6O_2$ A golden yellow, crystalline compound that decomposes at 145-147°C; soluble in benzene and ether; used as a reagent for resorcinol and thalline. { ¦wən ¦tü ,naf·thə·kwə'nōn }

1,4-naphthoquinone [ORGANIC CHEMISTRY] $C_{10}H_6O_2$ Greenish-yellow powder soluble in organic solvents, slightly soluble in water; melts at 123-126°C; used as an antimycotic agent, in synthesis, and as a rubber polymerization regulator. { ¦wən ¦fór ,naf·thə·kwə'nōn }

naphthoresorcinol [ORGANIC CHEMISTRY] $C_{10}H_6(OH)_2$ Crystals with a melting point of 124-125°C; soluble in ether, alcohol, and water; used as a reagent for sugars and oils, and to determine glucuronic acid in urine. { ,naf·thə·ri'sȯrs·ən,ȯl }

β-naphthoxyacetic acid [ORGANIC CHEMISTRY] $C_{12}H_{10}O_3$ A crystalline compound soluble in water, with a melting point of 156°C; used as a growth regulator to set blossoms and regulate growth for pineapples, strawberries, and tomatoes. Also known as O-(2-naphthyl)glycolic acid. Abbreviated BNOA. { ¦bād·ə naf¦thäk·sē·ə'sēd·ik 'as·əd }

2-(α-naphthoxy)-*N,N*-diethylpropionamide [ORGANIC CHEMISTRY] *See* devrinol. { ¦tü ¦al·fə naf'thäk·sē ¦en ¦en dī,eth·əl,pro·pē'än·ə·məd }

naphthylamine [ORGANIC CHEMISTRY] $C_{10}H_7NH_2$ White, toxic crystals, soluble in alcohol and ether; used in dyes; the two forms are α-naphthylamine, boiling at 301°C, and β-naphthylamine, boiling at 306°C. { naf'thil·ə,mēn }

2,5-naphthylamine sulfonic acid [ORGANIC CHEMISTRY] *See* gamma acid. { ¦tü ¦fīv naf'thil·ə,mēn səl'fän·ik 'as·əd }

β-naphthylmethyl ether [ORGANIC CHEMISTRY] $C_{10}H_7OCH_3$ White, crystalline scales with a melting point of 72°C; soluble in alcohol and ether; used for soap perfumes. { ¦bād·ə ¦naf·thil¦meth·əl 'ē·thər }

***N*-1-naphthylphthalamic acid** [ORGANIC CHEMISTRY] $C_{10}H_7NHCOC_6H_4COOH$ A crystalline solid with a melting point of 185°C; used as a preemergence herbicide. { ¦en ¦wən ¦naf·thil·thə'lam·ik 'as·əd }

1-(1-naphthyl)-2-thiourea [ORGANIC CHEMISTRY] $C_{10}H_7NHCSNH_2$ A crystalline compound with a melting point of 198°C; soluble in water, acetone, triethylene glycol, and hot alcohol; used as a poison to control the adult Norway rat. { ¦wən ¦wən 'naf·thil ¦tü ,thī·ə·yü'rē·ə }

Naples yellow [INORGANIC CHEMISTRY] *See* lead antimonite. { 'nā·pəlz 'yel·ō }

narceine [ORGANIC CHEMISTRY] $C_{23}H_{27}O_8N \cdot 3H_2O$ White, odorless crystals with bitter taste; soluble in alcohol and water, insoluble in ether; melts at 170°C; used in medicine. { 'när·sē,ēn }

narcissistic reaction [CHEMISTRY] A chemical reaction in which a reactant is converted into a product whose structure is the mirror image of the reactant molecule. { ,när·sə'sis·tik rē'ak·shən }

naringin [ORGANIC CHEMISTRY] $C_{27}H_{32}O_{14}$ A crystalline bioflavonoid with a melting point of 171°C; soluble in acetone and alcohol; used as a food supplement. Also known as aurantiin. { nə'rin·jən }

nascent [CHEMISTRY] Pertaining to an atom or simple compound at the moment of its liberation from chemical combination, when it may have greater activity than in its usual state. { 'nā·sənt }

natrium [CHEMISTRY] Latin name for sodium; source of the symbol Na. { 'nā·trē·əm }

natural linewidth [SPECTROSCOPY] The part of the linewidth of an absorption or emission line that results from the finite lifetimes of one or both of the energy levels between which the transition takes place. { 'nach·rəl 'līn,width }

natural red [ORGANIC CHEMISTRY] *See* purpurin. { 'nach·rəl 'red }

Nb [CHEMISTRY] *See* niobium.

Nd [CHEMISTRY] *See* neodymium.

NDGA [ORGANIC CHEMISTRY] *See* nordihydroguaiaretic acid.

Ne [CHEMISTRY] *See* neon.

near-infrared spectrophotometry [ANALYTICAL CHEMISTRY] Spectrophotometry at wavelengths in the near-infrared region, generally using instruments with quartz prisms in the monochromators and lead sulfide photoconductor cells as detectors to observe absorption bands which are harmonics of bands at longer wavelengths. { 'nir ˌin·frə'red ˌspek·trō·fə'täm·ə·trē }

negative catalysis [CHEMISTRY] A catalytic reaction such that the reaction is slowed down by the presence of the catalyst. { 'neg·əd·iv kə'tal·ə·səs }

negative ion [CHEMISTRY] An atom or group of atoms which by gain of one or more electrons has acquired a negative electric charge. { 'neg·əd·iv 'īˌän }

neighboring-group participation [ORGANIC CHEMISTRY] *See* anchimeric assistance. { 'nā·bər·iŋ ˌgrüp pärˌtis·ə'pā·shən }

nematic phase [PHYSICAL CHEMISTRY] A phase of a liquid crystal in the mesomorphic state, in which the liquid has a single optical axis in the direction of the applied magnetic field, appears to be turbid and to have mobile threadlike structures, can flow readily, has low viscosity, and lacks a diffraction pattern. { nə'mad·ik ˌfāz }

nematogenic solid [PHYSICAL CHEMISTRY] A solid which will form a nematic liquid crystal when heated. { nə'mad·əˌjen·ik 'säl·əd }

neodymium [CHEMISTRY] A metallic element, symbol Nd, with atomic weight 144.24, atomic number 60; a member of the rare-earth group of elements. { ˌnē·ō'dim·ē·əm }

neodymium chloride [INORGANIC CHEMISTRY] $NdCl_3 \cdot xH_2O$ Water- and acid-soluble, pink lumps; used to prepare metallic neodymium. { ˌnē·ō'dim·ē·əm 'klȯrˌīd }

neodymium oxide [INORGANIC CHEMISTRY] Nd_2O_3 A hygroscopic, blue-gray powder; insoluble in water, soluble in acids; used to color glass and in ceramic capacitors. { ˌnē·ō'dim·ē·əm 'äkˌsīd }

neohexane [ORGANIC CHEMISTRY] C_6H_{14} Volatile, flammable, colorless liquid boiling at 50°C; used as high-octane component of motor and aviation gasolines. { ¦nē·ō'hekˌsān }

neon [CHEMISTRY] A gaseous element, symbol Ne, atomic number 10, atomic weight 20.183; a member of the family of noble gases in the zero group of the periodic table. { 'nēˌän }

neo-, ne- [ORGANIC CHEMISTRY] Prefix indicating hydrocarbons where a carbon is bonded directly to at least four other carbon atoms, such as neopentane. { 'nē·ō }

neopentane [ORGANIC CHEMISTRY] C_5H_{12} Colorless liquid boiling at 10°C; soluble in alcohol, insoluble in water; a hydrocarbon found as a minor component of natural gasoline. { ¦nē·ō'penˌtān }

neptunium [CHEMISTRY] A chemical element, symbol Np, atomic number 93, atomic weight 237.0482; a member of the actinide series of elements. { nep'tü·nē·əm }

neptunium decay series [CHEMISTRY] Little-known radioactive elements with short lives; produced as successive series of decreasing atomic weight when uranium-237 and plutonium-241 decay radioactively through neptunium-237 to bismuth-209. { nep'tü·nē·əm di'kā ˌsir·ēz }

Nernst equation [PHYSICAL CHEMISTRY] The relationship showing that the electromotive force developed by a dry cell is determined by the activities of the reacting species, the temperature of the reaction, and the standard free-energy change of the overall reaction. { 'nernst iˌkwā·zhən }

Nernst-Thomson rule [PHYSICAL CHEMISTRY] The rule that in a solvent having a high dielectric constant the attraction between anions and cations is small so that dissociation is favored, while the reverse is true in solvents with a low dielectric constant. { 'nernst 'täm·sən ˌrül }

Nernst zero of potential [PHYSICAL CHEMISTRY] An electrode potential corresponding to the reversible equilibrium between hydrogen gas at a pressure of 1 standard atmosphere and hydrogen ions at unit activity. { 'nernst 'zir·ō əv pə'ten·chəl }

nerol [ORGANIC CHEMISTRY] $C_{10}H_{17}OH$ Colorless liquid with rose-neroli odor; derived from geraniol (a trans isomer); used in perfumery. { 'neˌrȯl }

nerolidol [ORGANIC CHEMISTRY] $C_{15}H_{26}O$ A straw-colored sesquiterpene alcohol; liquid with rose and apple aroma derived from cabreuva oil, oils of orange flower, and ylang ylang; soluble in alcohol; used in perfumery. { nə'räl·əˌdȯl }

Nessler's reagent [ANALYTICAL CHEMISTRY] Mercuric iodide-potassium iodide solution, used to analyze for small amounts of ammonia. { 'nes·lərz rē,ā·jənt }

Nessler tubes [ANALYTICAL CHEMISTRY] Standardized glass tubes for filling with standard solution colors for visual color comparison with similar tubes filled with solution samples. { 'nes·lər ,tübz }

neutral [CHEMISTRY] Property of a solution which is neither acidic nor basic, having the same concentration of hydrogen ions as water. { 'nü·trəl }

neutral flame [CHEMISTRY] Gas flame produced by a mixture of fuel and oxygen so as to be neither oxidizing nor reducing. { 'nü·trəl 'flām }

neutral granulation [CHEMISTRY] Propellant granulation in which the surface area of a grain remains constant during burning. { 'nü·trəl ,gran·yə'lā·shən }

neutralization [CHEMISTRY] The process of making a solution neutral (pH = 7) by adding a base to an acid solution, or adding an acid to an alkaline (basic) solution. Also known as neutralization reaction. { ,nü·trə·lə'zā·shən }

neutralization equivalent [CHEMISTRY] For an acid or base, the same as equivalent weight; multiplication of the neutralization equivalent by the number of acidic or basic groups in the molecule gives the molecular weight. { ,nü·trə·lə'zā·shən i,kwiv·ə·lənt }

neutralization number [ANALYTICAL CHEMISTRY] Petroleum product test; it is the milligrams of potassium hydroxide required to neutralize the acid in 1 gram of oil; used as an indication of oil acidity. { ,nü·trə·lə'zā·shən ,nəm·bər }

neutralization reaction [CHEMISTRY] *See* neutralization. { ,nü·trə·lə'zā·shən rē,ak·shən }

neutralize [CHEMISTRY] To make a solution neutral (neither acidic nor basic, pH of 7) by adding a base to an acidic solution, or an acid to a basic solution. { 'nü·trə,līz }

neutral molecule [PHYSICAL CHEMISTRY] A molecule in which the number of electrons surrounding the nuclei is the same as the total number of protons in the nuclei, so that there is no net electric charge. { 'nü·trəl 'mäl·ə,kyül }

neutral potassium phosphate [INORGANIC CHEMISTRY] *See* potassium phosphate. { 'nü·trəl pə'tas·ē·əm 'fäs,fāt }

neutral red [ORGANIC CHEMISTRY] $(CH_3)_2NC_6H_3N_2C_6H_2CH_3NH_2 \cdot ClH$ Water- and alcohol-soluble green powder; used as pH 6.8-8.0 acid-base indicator, and as a dye to test stomach function. Also known as dimethyl diaminophenazine chloride; toluylene red. { 'nü·trəl 'red }

neutral species [CHEMISTRY] *See* uncharged species. { 'nü·trəl 'spē·shēz }

Newland's law of octaves [CHEMISTRY] An arrangement of the elements that predated Mendeleev's periodic table; Newland's arrangement was a grouping of the elements in increasing atomic weights (starting with lithium) in horizontal rows of eight elements, with each new row directly beneath the previous one. { 'nü·lənz 'lȯ əv 'äk·tivz }

Newman projection [ORGANIC CHEMISTRY] A representation of the conformation of a molecule in which the viewer's eye is considered to be sighting down a carbon-carbon bond; the front carbon is represented by a point and the back carbon by a circle. { 'nü·mən prə,jek·shən }

Ni [CHEMISTRY] *See* nickel.

nickel [CHEMISTRY] A chemical element, symbol Ni, atomic number 28, atomic weight 58.71. { 'nik·əl }

nickel acetate [ORGANIC CHEMISTRY] $Ni(OOCCH_3)_2 \cdot 4H_2O$ Efflorescent green crystals that decompose upon heating; soluble in alcohol and water; used as textile dyeing mordant. { 'nik·əl 'as·ə,tāt }

nickel ammonium sulfate [INORGANIC CHEMISTRY] $NiSO_4 \cdot (NH_4)_2SO_4 \cdot 6H_2O$ A green, crystalline compound, soluble in water; used as a nickel electrolyte for electroplating. Also known as ammonium nickel sulfate; double nickel salt. { 'nik·əl ə'mō·nē·əm 'səl,fāt }

nickel arsenate [INORGANIC CHEMISTRY] $Ni_3(AsO_4)_2 \cdot H_2O$ Poisonous yellow-green powder; soluble in acids, insoluble in water; used as a fat-hardening catalyst in soapmaking. { 'nik·əl 'ars·ən,āt }

nickel carbonate [INORGANIC CHEMISTRY] $NiCO_3$ Light-green crystals that decompose upon heating; soluble in acid, insoluble in water; used in electroplating. { 'nik·əl 'kär·bə,nāt }

nickel carbonyl [INORGANIC CHEMISTRY] $Ni(CO)_4$ Colorless, flammable, poisonous liquid boiling at 43°C; soluble in alcohol and concentrated nitric acid, insoluble in water; used in gas plating (vapor decomposes at 60°C) and to produce metallic nickel. { 'nik·əl 'kär·bə,nil }

nickel cyanide [INORGANIC CHEMISTRY] $Ni(CN)_2 \cdot 4H_2O$ Poisonous, water-insoluble apple-green powder; melts and loses water at 200°C, decomposes at higher temperatures; used for electroplating and metallurgy. { 'nik·əl 'sī·ə,nīd }

nickel formate [ORGANIC CHEMISTRY] $Ni(HCOO)_2 \cdot 2H_2O$ Water-soluble green crystals; used in hydrogenation catalysts. { 'nik·əl 'fór,māt }

nickel iodide [INORGANIC CHEMISTRY] NiI_2 or $NiI_2 \cdot 6H_2O$ Hygroscopic black or blue-green solid, soluble in water and alcohol; sublimes when heated. { 'nik·əl 'ī·ə,dīd }

nickel nitrate [INORGANIC CHEMISTRY] $Ni(NO_3)_2 \cdot 6H_2O$ Fire-hazardous oxidant; deliquescent, green, water- and alcohol-soluble crystals; used for nickel plating and brown ceramic colors, and in nickel catalysts. { 'nik·əl 'nī,trāt }

nickelocene [ORGANIC CHEMISTRY] $(C_5H_5)_2Ni$ Dark green crystals with a melting point of 171-173°C; soluble in most organic solvents; used as an antiknock agent. { nə 'kəl·ə,sēn }

nickel oxide [INORGANIC CHEMISTRY] NiO Green powder; soluble in acids and ammonium hydroxide; insoluble in water; used to make nickel salts and for porcelain paints. Also known as green nickel oxide. { 'nik·əl 'äk,sīd }

nickel phosphate [INORGANIC CHEMISTRY] $Ni_3(PO_4)_2 \cdot 7H_2O$ A light-green powder; soluble in acids and ammonium hydroxide, insoluble in water; used for electroplating and production of yellow nickel. { 'nik·əl 'fäs,fāt }

nicotine [ORGANIC CHEMISTRY] $C_{10}H_{14}N_2$ A colorless liquid with a boiling point of 247.3°C; miscible with water; used as a contact insecticide fumigant in closed spaces. { 'nik·ə,tēn }

ninhydrin [ORGANIC CHEMISTRY] $C_9H_4O_3 \cdot H_2O$ White crystals or powder with a melting point of 240-245°C; soluble in water and alcohol; used for the detection and assay of peptides, amines, amino acids, and amino sugars. Also known as triketohydrindene hydrate. { nin'hī·drən }

niobe oil [ORGANIC CHEMISTRY] *See* methyl benzoate. { 'nī·ə,bē,óil }

niobic acid [INORGANIC CHEMISTRY] $Nb_2O_5 \cdot nH_2O$ Family of hydrates, white precipitate, soluble in inorganic acids and bases, insoluble in water; its formation is part of the analytical determination of niobium. { nī'ō·bik 'as·əd }

niobium [CHEMISTRY] A chemical element, symbol Nb, atomic number 41, atomic weight 92.906. { nī'ō·bē·əm }

niobium carbide [INORGANIC CHEMISTRY] NbC A lavender gray powder with a melting point of 3500°C; used for carbide-tipped tools and special steels. { nī'ō·bē·əm 'kär,bīd }

niter [INORGANIC CHEMISTRY] *See* potassium nitrate. { 'nīd·ər }

niter cake [INORGANIC CHEMISTRY] *See* sodium bisulfate. { 'nīd·ər ,kāk }

nitrate [CHEMISTRY] **1.** A salt or ester of nitric acid. **2.** Any compound containing the ion NO_3^-. { 'nī,trāt }

nitrating acid [INORGANIC CHEMISTRY] Sulfuric-nitric acid mix used to nitrate cellulosics and aromatic chemicals. Also known as mixed acid. { 'nī,trād·iŋ 'as·əd }

nitration [ORGANIC CHEMISTRY] Introduction of an NO_2^- group into an organic compound. { nī'trā·shən }

nitrene [ORGANIC CHEMISTRY] A molecular fragment that is an uncharged, electron-deficient species containing a monocovalent nitrogen. { 'nī,trēn }

nitric acid [INORGANIC CHEMISTRY] HNO_3 Strong oxidant that is fire-hazardous; colorless or yellowish liquid, miscible with water; boils at 86°C; used for chemical synthesis, explosives, and fertilizer manufacture, and in metallurgy, etching, engraving, and ore flotation. Also known as aqua fortis. { 'nī·trik 'as·əd }

nitric oxide [INORGANIC CHEMISTRY] NO A colorless gas that, at room temperature, reacts with oxygen to form nitrogen dioxide (NO_2, a reddish-brown gas); may be used to form other compounds. { 'nī·trik 'äk‚sīd }

nitride [INORGANIC CHEMISTRY] Compound of nitrogen and a metal, such as Mg_3N_2. { 'nī‚trīd }

nitrile [ORGANIC CHEMISTRY] $RC{\equiv}N$ Cyanide derived by removal of water from an acid amide. { 'nī‚trīl }

nitrile resin [ORGANIC CHEMISTRY] Any one of a family of polymers produced from acrylonitrile, various esters, butadiene, and styrene. { 'nī‚trīl ‚rez·ən }

nitrilotriacetic acid [ORGANIC CHEMISTRY] $N(CH_2COOH)_3$ A white powder, melting point 240°C, with some decomposition; soluble in water; it is toxic, and birth abnormalities may result from ingestion; may be used as a chelating agent in the laboratory. Also known as NTA; TGA. { ¦nī·trə·lō‚trī·ə'sēd·ik 'as·əd }

nitrite [CHEMISTRY] A compound containing the radical NO_2^-; can be organic or inorganic. { 'nī‚trīt }

nitro- [CHEMISTRY] Chemical prefix showing the presence of the NO_2^- radical. { 'nī·trō }

nitroalkane [ORGANIC CHEMISTRY] *See* nitroparaffin. { ¦nī·trō'al‚kān }

***meta*-nitroaniline** [ORGANIC CHEMISTRY] $NO_2C_6H_4NH_2$ Yellow crystals that melt at 112.5°C; a toxic material; used as a dye intermediate. { ¦med·ə ‚nī·trō'an·ə·lən }

***ortho*-nitroaniline** [ORGANIC CHEMISTRY] $NO_2C_6H_4NH_2$ Orange-red crystals that melt at 69.7°C, soluble in ethanol; a toxic material; used to manufacture dyes. { ¦ór·thō ‚nī·trō'an·ə·lən }

***para*-nitroaniline** [ORGANIC CHEMISTRY] $NO_2C_6H_4NH_2$ Yellow crystals that melt at 148°C; insoluble in water, soluble in ethanol; a toxic material; used to make dyes, and as a corrosion inhibitor. { ¦par·ə ‚nī·trō'an·ə·lən }

nitroaromatic [ORGANIC CHEMISTRY] A nitrated benzene or benzene derivative, such as nitrobenzene, $C_6H_5NO_2$, or nitrobenzoic acid, $NO_2 \cdot C_6H_4 \cdot COOH$. { ¦nī·trō‚ar·ə'mad·ik }

nitrobarite [INORGANIC CHEMISTRY] *See* barium nitrate. { ¦nī·trō'ba‚rīt }

nitrobenzene [ORGANIC CHEMISTRY] $C_6H_5NO_2$ Greenish crystals or a yellowish liquid, melting point 5.70°C; a toxic material; used in aniline manufacture. Also known as oil of mirbane. { ¦nī·trō'ben‚zēn }

***ortho*-nitrobiphenyl** [ORGANIC CHEMISTRY] $C_{12}H_9NO_2$ A crystalline compound with a sweetish odor; melting point is 36.7°C; used as a plasticizer for resins, cellulose acetate and nitrate, and polystyrenes, and as a fungicide for textiles. Abbreviated ONB. { ¦ór·thō ‚nī·trō·bī'fen·əl }

nitrobromoform [ORGANIC CHEMISTRY] *See* bromopicrin. { ¦nī·trō'brō·mə‚fórm }

nitrocalcite [INORGANIC CHEMISTRY] *See* calcium nitrate. { ¦nī·trō'kal‚sīt }

nitrocellulose [ORGANIC CHEMISTRY] *See* cellulose nitrate. { ¦nī·trō'sel·yə‚lōs }

nitrocotton [ORGANIC CHEMISTRY] *See* cellulose nitrate. { ¦nī·trō'kät·ən }

nitro dye [ORGANIC CHEMISTRY] A dye with the NO_2 chromophore group in the molecules. { 'nī·trō ‚dī }

nitroethane [ORGANIC CHEMISTRY] $CH_3CH_2NO_2$ A colorless liquid, slightly soluble in water; boils at 114°C; used as a solvent for cellulosics, resins, waxes, fats, and dyestuffs, and as a chemical intermediate. { ¦nī·trō'eth‚ān }

nitro explosive [ORGANIC CHEMISTRY] Explosive compound containing one or more NO_2^- groups, such as nitroglycerine, $C_3H_5(ONO_2)_3$, or trinitrotoluene, $C_6H_2(CH_3)(NO_2)_3$. { 'nī·trō ik'splō·siv }

nitrogen [CHEMISTRY] A chemical element, symbol N, atomic number 7, atomic weight 14.0067; it is a gas, diatomic (N_2) under normal conditions; about 78% of the atmosphere is N_2; in the combined form the element is a constituent of all proteins. { 'nī·trə·jən }

nitrogen acid anhydride [INORGANIC CHEMISTRY] *See* nitrogen pentoxide. { 'nī·trə·jən ¦as·əd an'hī‚drīd }

nitrogen dioxide [INORGANIC CHEMISTRY] NO_2 A reddish-brown gas; it exists in varying degrees of concentration in equilibrium with other nitrogen oxides; used to produce nitric acid. Also known as dinitrogen tetroxide, liquid dioxide; nitrogen peroxide; nitrogen tetroxide. { 'nī·trə·jən dī'äk‚sīd }

nitrogen monoxide [INORGANIC CHEMISTRY] *See* nitrous oxide. { 'nī·trə·jən mə'näk ˌsīd }

nitrogen mustard [ORGANIC CHEMISTRY] Any of the substituted mustard gases in which the sulfur is replaced by an amino nitrogen, such as for methyl bis(2-chlorethyl)amine, $(CH_2ClCH_2)_2NCH_3$; useful in cancer research. { 'nī·trə·jən 'məs·tərd }

nitrogen pentoxide [INORGANIC CHEMISTRY] N_2O_5 Colorless crystals, soluble in water (forms HNO_3); decomposes at 46°C. Also known as nitrogen acid anhydride. { 'nī·trə·jən pen 'täkˌsīd }

nitrogen peroxide [INORGANIC CHEMISTRY] *See* nitrogen dioxide. { 'nī·trə·jən pə 'räkˌsīd }

nitrogen solution [INORGANIC CHEMISTRY] Mixture used to neutralize super-phosphate in fertilizer manufacture; consists of 60% ammonium nitrate, and the balance a 50% aqua ammonia solution. { 'nī·trə·jən səˌlü·shən }

nitrogen tetroxide [INORGANIC CHEMISTRY] *See* nitrogen dioxide. { 'nī·trə·jən te'träkˌsīd }

nitrogen trifluoride [INORGANIC CHEMISTRY] NF_3 A colorless gas that has a melting point of −206.6°C and a boiling point of −128.8°C; used as an oxidizer for high-energy fuels. { 'nī·trə·jən trī'flu̇rˌīd }

nitrogen trioxide [INORGANIC CHEMISTRY] N_2O_3 Green, water-soluble liquid; boils at 3.5°C. { 'nī·trə·jən trī'äkˌsīd }

nitroglycerin [ORGANIC CHEMISTRY] $CH_2NO_3CHNO_3CH_2NO_3$ Highly unstable, explosive, flammable pale-yellow liquid; soluble in alcohol; freezes at 13°C and explodes at 260°C; used as an explosive, to make dynamite, and in medicine. { ¦nī·trə'glis ə·rən }

nitroguanidine [ORGANIC CHEMISTRY] $H_2NC(NH)NHNO_2$ Explosive yellow solid, soluble in alcohol; melts at 246°C; used in explosives and smokeless powder. { ¦nī·trō'gwän·əˌdēn }

nitrometer [ANALYTICAL CHEMISTRY] Glass apparatus used to collect and measure nitrogen and other gases evolved by a chemical reaction. Also known as azotometer. { nī'träm·əd·ər }

nitromethane [ORGANIC CHEMISTRY] CH_3NO_2 A liquid nitroparaffin compound; oily and colorless; boils at 101°C; used as a monopropellant for rockets, in chemical synthesis, and as an industrial solvent for cellulosics, resins, waxes, fats, and dyestuffs. { ¦nī·trō'methˌān }

nitron [ORGANIC CHEMISTRY] $CN_4(C_6H_5)_3CH$ Yellow crystals, soluble in chloroform and acetone; used as reagent to detect NO_3 ion in dilute solutions. { 'nīˌträn }

nitronium [CHEMISTRY] Positively charged NO_2 ion, believed to be formed from HNO_3. Also known as nitryl ion. { nī'trō·nē·əm }

nitroparaffin [ORGANIC CHEMISTRY] Any organic compound in which one or more hydrogens of an alkane are replaced by a nitro, or NO_2^-, group, such as nitromethane, CH_3NO_2, or nitroethane, $C_2H_5NO_2$. Also known as nitroalkane. { ¦nī trō'par ə·fən }

***ortho*-nitrophenol** [ORGANIC CHEMISTRY] $C_6H_5NO_3$ A yellow, crystalline compound; melting point is 44-45°C; soluble in hot water, alcohol, benzene, ether, carbon disulfide, and alkali hydroxides; used in the commercial preparation of many compounds. { ¦ór·thō ¦nī·trō'feˌnól }

***para*-nitrophenylhydrazine** [ORGANIC CHEMISTRY] $C_6H_7N_3O_2$ An orange-red, crystalline compound with a melting point of about 157°C; soluble in hot water or hot benzene; used as a reagent for aliphatic aldehydes and ketones. { ¦par·ə ¦nī·trō¦fen·əl'hī·drəˌzēn }

1-nitropropane [ORGANIC CHEMISTRY] $CH_3CH_2CH_2NO_2$ A colorless liquid with a boiling point of 132°C; used as a rocket propellant and gasoline additive. { ¦wən ¦nī·trō'prōˌpān }

2-nitropropane [ORGANIC CHEMISTRY] $CH_3CHNO_2CH_3$ A colorless liquid with a boiling point of 120°C; used as a solvent for vinyl coatings, as a rocket propellant, and as a gasoline additive. { ¦tü ¦nī·trō'prōˌpān }

nitroso [CHEMISTRY] The radical NO— with trivalent nitrogen. Also known as hydroximino; oximido. { nī'trō·sō }

nitrostarch [ORGANIC CHEMISTRY] $C_{12}H_{12}(NO_2)_8O_{10}$ Orange powder, soluble in ethyl alcohol; used in explosives. Also known as starch nitrate. { 'nī·trə‚stärch }

***meta*-nitrotoluene** [ORGANIC CHEMISTRY] $NO_2C_6H_4CH_3$ Yellow powder that melts at 15°C; insoluble in water; used in organic synthesis. { ¦med·ə ¦nī·trō'täl·yə‚wēn }

***ortho*-nitrotoluene** [ORGANIC CHEMISTRY] $NO_2C_6H_4CH_3$ A yellow liquid boiling at 220.4°C; insoluble in water; used to produce toluidine and dyes. { ¦ór·thō ¦nī·trō 'täl·yə‚wēn }

***para*-nitrotoluene** [ORGANIC CHEMISTRY] $NO_2C_6H_4CH_3$ Yellow crystals that melt at 51.7°C; insoluble in water, soluble in ethanol; used to produce toluidine and to manufacture dyes. { ¦par·ə ¦nī·trō'täl·yə‚wēn }

nitrourea [ORGANIC CHEMISTRY] $NH_2CONHNO_2$ Highly explosive white crystals, melting at 159°C; soluble in ether and alcohol, slightly soluble in water; used as a chemical intermediate. { ¦nī·trō·yu̇'rē·ə }

nitrous acid [INORGANIC CHEMISTRY] HNO_2 Aqueous solution of nitrogen trioxide, N_2O_3. { 'nī·trəs 'as·əd }

nitrous oxide [INORGANIC CHEMISTRY] N_2O Colorless, sweet-tasting gas, boiling at −90°C; slightly soluble in water, soluble in alcohol; used as a food aerosol, and as an anesthetic in dentistry and surgery. Also known as laughing gas; nitrogen monoxide. { 'nī·trəs 'äk‚sīd }

nitroxylene [ORGANIC CHEMISTRY] $C_6H_3(CH_3)_2NO_2$ Any of three isomers occurring either as a yellow liquid or as crystalline needles with a melting point of 2°C and boiling point of 246°C; soluble in alcohol and ether; used in gelatinizing accelerators for pyroxylin. { nī'träk·sə‚lēn }

nitryl halide [INORGANIC CHEMISTRY] NO_2X Compound containing a halide (X) and a nitro group (NO_2). { 'nī‚tril 'ha‚līd }

nitryl ion [CHEMISTRY] *See* nitronium. { 'nī‚tril ‚ī‚än }

N line [SPECTROSCOPY] One of the characteristic lines in an atom's x-ray spectrum, produced by excitation of an N electron. { 'en ‚līn }

No [CHEMISTRY] *See* nobelium.

nobelium [CHEMISTRY] A chemical element, symbol No, atomic number 102, atomic weight 254 when the element is produced in the laboratory; a synthetic element, in the actinium series. { nō'bel·ē·əm }

noble gas [CHEMISTRY] A gas in group 0 of the periodic table of the elements; it is monatomic and, with limited exceptions, chemically inert. Also known as inert gas; rare gas. { 'nō·bəl 'gas }

noble-gas electron configuration [CHEMISTRY] An electron structure of an atom or ion in which the outer electron shell contains eight electrons, corresponding to the electron configuration of a noble gas, such as neon or argon. { ¦nō·bəl ¦gas i'lek‚trän kən‚fig·yə‚rā·shən }

noble potential [PHYSICAL CHEMISTRY] A potential equaling or approaching that of the noble elements, such as gold, silver, or copper, of the electromotive series. { 'nō·bəl pə'ten·chəl }

NODA [ORGANIC CHEMISTRY] *See* *n*-octyl *n*-decyl adipate. { 'nō·də }

noise [SPECTROSCOPY] Random fluctuations of electronic signals appearing in a recorded spectrum. { nȯiz }{ nȯiz }

nonacosane [ORGANIC CHEMISTRY] $C_{29}H_{60}$ Colorless hydrocarbon, melting at 63°C; found in beeswax and the fat of cabbage leaves. { ‚nō·nə'kō‚sān }

nonadecane [ORGANIC CHEMISTRY] $CH_3(CH_2)_{17}CH_3$ Flammable crystals, soluble in ether and alcohol, insoluble in water; melts at 32°C; used as a chemical intermediate. { ‚nō·nə'de‚kān }

nonanal [ORGANIC CHEMISTRY] $C_8H_{17}CHO$ A colorless liquid with an orange rose odor; used in perfumes and for flavoring. { 'nän·ə‚näl }

nonane [ORGANIC CHEMISTRY] $CH_3(CH_2)_7CH_3$ Flammable, colorless liquid, boiling at 151°C; soluble in alcohol, insoluble in water; used as a chemical intermediate. Also known as nonyl hydride. { 'nō‚nān }

nonaqueous [CHEMISTRY] Pertaining to a liquid or solution containing no water. { ¦nän'ā·kwē·əs }

nonbenzenoid aromatic compound [ORGANIC CHEMISTRY] A compound exhibiting aromatic character but not containing a benzene nucleus, or having one or more rings in a fused ring system that are not benzene rings. { ¦nän'ben·zə‚nȯid ¦ar·ə¦mad·ik 'käm‚pau̇nd }

nonbonded distance [PHYSICAL CHEMISTRY] The distance between atoms in a molecule that are not bonded to each other. { ¦nän‚bän·dəd 'dis·təns }

noncrossing rule [PHYSICAL CHEMISTRY] The rule that when the potential energies of two electronic states of a diatomic molecule are plotted as a function of distance between the nuclei, the resulting curves do not cross, unless the states have different symmetry. { ¦nän'krȯs·iŋ 'rül }

nonfaradaic path [PHYSICAL CHEMISTRY] One of the two available paths for transfer of energy across an electrolyte-metal interface, in which energy is carried by capacitive transfer, that is, by charging and discharging the double-layer capacitance. { ¦nän‚far·ə'dā·ik 'path }

nonhypergolic [CHEMISTRY] Not capable of igniting spontaneously upon contact; used especially with reference to rocket fuels. { ‚nän‚hī·pər'gäl·ik }

nonideal solution [PHYSICAL CHEMISTRY] A solution whose behavior does not conform to that of an ideal solution; that is, the behavior is not predictable over a wide range of concentrations and temperatures by the use of Raoult's law. { 'nän·ī‚dēl sə·lü·shən }

nonlinear molecule [ORGANIC CHEMISTRY] A branched-chain molecule, that is, one whose atoms do not all lie along a straight line. Also known as isomolecule. { 'nän‚lin·ē·ər 'mäl·ə‚kyül }

nonlinear spectroscopy [SPECTROSCOPY] The study of energy levels not normally accessible with optical spectroscopy, through the use of nonlinear effects such as multiphoton absorption and ionization. { 'nän‚lin·ē·ər spek'träs·kə·pē }

nonlocalized bond [CHEMISTRY] *See* delocalized bond. { ¦nän'lō·kə‚līzd 'bänd }

nonoic acid [ORGANIC CHEMISTRY] $C_8H_{17}COOH$ Any of a family of acids which are mixed isomers produced in the Fischer-Tropsch process; pelargonic acid is the straight-chain member; used as a chemical intermediate. { nō'nō·ik 'as·əd }

nonpolar [CHEMISTRY] Pertaining to an element or compound which has no permanent electric dipole moment. { ¦nän'pō·lər }

nonpolar bond [PHYSICAL CHEMISTRY] A type of covalent bond in which both atoms attract the bonding electrons equally or nearly equally. { ¦nän¦pōl·ər 'bänd }

nonpolar covalent bond [PHYSICAL CHEMISTRY] A bond in which a pair of electrons is distributed or shared equally between two atoms. { ¦nän‚pō·lər ¦kō‚vā·lənt 'bänd }

nonpolar molecule [PHYSICAL CHEMISTRY] A molecule with equal distribution of electrons among its atoms. { ¦nän‚pō·lər 'mäl·ə‚kyül }

nonprotic solvent [CHEMISTRY] A solvent that does not contain a hydrogen ion source. { ¦nän¦prōd·ik 'säl·vənt }

nonyl acetate [ORGANIC CHEMISTRY] $C_9H_{19}OOCCH_3$ Any of a family of isomers, such as *n*-nonyl acetate and diisobutyl carbinyl acetate, which are products of Fischer-Tropsch and oxo syntheses. { 'nä‚nil 'as·ə‚tāt }

***n*-nonyl acetate** [ORGANIC CHEMISTRY] $CH_3COO(CH_2)_8CH_3$ Alcohol soluble, colorless liquid with pungent odor; boiling point 208-212°C; used in perfumery. { ¦en 'nä‚nil 'as·ə‚tāt }

***n*-nonyl alcohol** [ORGANIC CHEMISTRY] $CH_3(CH_2)_7CH_2OH$ One of a family of $C_9H_{19}OH$ isomers; a colorless liquid with rose aroma; boils at 215°C; insoluble in water, soluble in alcohol; used in perfumery and flavorings. { ¦en 'nä‚nil 'al·kə‚hȯl }

nonyl benzene [ORGANIC CHEMISTRY] $C_9H_{19}C_6H_5$ Liquid boiling at 245-252°C; straw-colored with aromatic aroma; used to make surface-active agents. { 'nä‚nil 'ben‚zēn }

1-nonylene [ORGANIC CHEMISTRY] C_9H_{18} Colorless liquid boiling at 150°C; soluble in alcohol, insoluble in water; used as a chemical intermediate. Also known as nonene. { ¦wən 'nän·ə‚lēn }

nonyl hydride [ORGANIC CHEMISTRY] *See* nonane. { 'nän·əl 'hī‚drīd }

nonyl phenol [ORGANIC CHEMISTRY] $C_9H_{19}C_6H_4OH$ Pale-yellow liquid boiling at 283-302°C; soluble in organic solvents, insoluble in water; a mixture of monoalkyl phenol isomers, mostly para-substituted; used to make surface-active agents, resins, and plasticizers. { 'nä‚nil 'fē‚nȯl }

nonylphenoxyacetic acid [ORGANIC CHEMISTRY] $C_9H_{19}C_6H_4OCH_2COOH$ A viscous, amber-colored liquid, soluble in alkali; used in turbine oils, lubricants, greases, and other materials as a corrosion inhibitor. { ¦nä‚nil·fe¦näk·sē·ə'sēd·ik 'as·əd }

nonyne [ORGANIC CHEMISTRY] $CH_3(CH_2)_6{\equiv}CCH$ Water-insoluble, colorless liquid boiling at 160°C. { 'nō‚nīn }

nopinene [ORGANIC CHEMISTRY] *See* pinene. { 'nä·pə‚nēn }

nor- [CHEMISTRY] Chemical formula prefix for normal; indicates a parent for another compound to be formed by removal of one or more carbons and associated hydrogens. { nȯr }

nordihydroguaiaretic acid [ORGANIC CHEMISTRY] $C_{18}H_{22}O_4$ A crystalline compound with a melting point of 184-185°C; soluble in alcohols, ether, acetone, glycerol, and propyleneglycol; used as an antioxidant in fats and oils. Abbreviated NDGA. { ¦nȯr·dī¦hī·drō¦gwī·ə¦red·ik 'as·əd }

normal bonded-phase chromatography [ANALYTICAL CHEMISTRY] A technique of bonded-phase chromatography in which the stationary phase is polar and the mobile phase is nonpolar. { 'nȯr·məl ¦ban·dəd ¦fāz ‚krō·mə'täg·rə·fē }

normality [CHEMISTRY] Measure of the number of gram-equivalent weights of a compound per liter of solution. Abbreviated N. { nȯr'mal·əd·ē }

normal potassium pyrophosphate [INORGANIC CHEMISTRY] *See* potassium pyrophosphate. { 'nȯr·məl pə'tas·ē·əm ‚pī·rō'fäs‚fāt }

normal salt [CHEMISTRY] A salt in which all of the acid hydrogen atoms have been replaced by a metal, or the hydroxide radicals of a base are replaced by an acid radical; for example, Na_2CO_3. { 'nȯr·məl 'sȯlt }

normal silver sulfate [INORGANIC CHEMISTRY] *See* silver sulfate. { 'nȯr·məl 'sil·vər 'səl‚fāt }

normal solution [CHEMISTRY] An aqueous solution containing one equivalent of the active reagent in grams in 1 liter of the solution. { 'nȯr·məl sə'lü·shən }

normal thorium sulfate [INORGANIC CHEMISTRY] *See* thorium sulfate. { 'nȯr·məl 'thȯr·ē·əm 'səl‚fāt }

norphytane [ORGANIC CHEMISTRY] *See* pristane. { nȯr'fī‚tān }

Np [CHEMISTRY] *See* neptunium.

NRS [SPECTROSCOPY] *See* nuclear reaction spectrometry.

NTA [ORGANIC CHEMISTRY] *See* nitrilotriacetic acid.

nuclear [CHEMISTRY] Pertaining to a group of atoms joined directly to the central group of atoms or central ring of a molecule. { 'nü·klē·ər }

nuclear atom [CHEMISTRY] An atomic structure consisting of dense, positively charged nucleus (neutrons and protons) surrounded by a corresponding set of negatively charged electrons. { 'nü·klē·ər 'ad·əm }

nuclear magnetic resonance spectrometer [SPECTROSCOPY] A spectrometer in which nuclear magnetic resonance is used for the analysis of protons and nuclei and for the study of changes in chemical and physical quantities over wide frequency ranges. { 'nü·klē·ər mag'ned·ik 'rez·ən·əns spek'träm·əd·ər }

nuclear reaction spectrometry [SPECTROSCOPY] A method of determining the concentration of a given element as a function of depth beneath the surface of a sample, by measuring the yield of characteristic gamma rays from a resonance reaction occurring when the surface is bombarded by a beam of ions. Abbreviated NRS. { 'nü·klē·ər rē'ak·shən spek'träm·ə·trē }

nuclear Zeeman effect [SPECTROSCOPY] A splitting of atomic spectral lines resulting from the interaction of the magnetic moment of the nucleus with an applied magnetic field. { 'nü·klē·ər 'zē·mən i‚fekt }

nucleation [CHEMISTRY] In crystallization processes, the formation of new crystal nuclei in supersaturated solutions. { ‚nü·klē'ā·shən }

nucleofuge [ORGANIC CHEMISTRY] *See* leaving group. { 'nü·klē·ə‚fyüj }

nucleophile [PHYSICAL CHEMISTRY] A species possessing one or more electron-rich sites, such as an unshared pair of electrons, the negative end of a polar bond, or pi electrons. Also known as electron donor. { 'nü·klē·ə‚fīl }

nucleophilic displacement [ORGANIC CHEMISTRY] *See* nucleophilic substitution. { ˌnü·klē·ə¦fil·ik di'splās·mənt }

nucleophilic reagent [PHYSICAL CHEMISTRY] A reactant that gives up electrons, or a share in electrons, to other molecules or ions in the course of a chemical reaction. { ¦nü·klē·ō¦fil·ik rē'ā·jənt }

nucleophilic substitution [ORGANIC CHEMISTRY] A reaction in which a nucleophile bonds to a carbon atom in a molecule, displacing a leaving group. Also known as nucleophilic displacement. { ˌnü·klē·ə¦fil·ik ˌsəb·stə'tü·shən }

Nylander reagent [CHEMISTRY] A solution of Rochelle salt (potassium sodium tartrate), potassium or sodium hydroxide, and bismuth subnitrate in water; used to test for sugar in urine. { 'nī·lən·dər rēˌā·jənt }

O

O [CHEMISTRY] *See* oxygen.

Obermayer's reagent [CHEMISTRY] A 0.4% solution of ferric chloride in concentrated hydrochloric acid; used to test for indican in urine, with a pale-blue or deep-violet color indicating positive. { 'ō·bər,mī·ərz rē,ā·jənt }

***n*-octadecane** [ORGANIC CHEMISTRY] $C_{18}H_{38}$ Colorless liquid boiling at 318°C; soluble in alcohol, acetone, ether, and petroleum, insoluble in water; used as a solvent and chemical intermediate. { ¦en ,äk·tə'de,kān }

1-octadecene [ORGANIC CHEMISTRY] $C_{18}H_{36}$ Colorless liquid boiling at 180°C; soluble in alcohol, acetone, ether, and petroleum, insoluble in water; used as a chemical intermediate. { ¦wən ,äk·tə'de,sēn }

octadecenyl aldehyde [ORGANIC CHEMISTRY] $C_{17}H_{35}CHO$ A flammable liquid with a boiling point of 167°C; used in the manufacture of vulcanization accelerators, rubber antioxidants, and pesticides. { ,äk·tə'des·ə,nəl 'al·də,hīd }

octafluorocyclobutane [ORGANIC CHEMISTRY] C_4F_8 A colorless gas or liquid with a boiling point of −4°C and a freezing point of −41.4°C; soluble in ether; used as a dielectric, refrigerant, and aerosol propellant. { ¦äk·tə¦flùr·ō,sī·klō'byü,tān }

octafluoropropane [ORGANIC CHEMISTRY] C_3F_8 A colorless gas with a boiling point of −36.7°C and a freezing point of approximately −160°C; used as a refrigerant and gaseous insulator. { ¦äk·tə,flùr·ō'prō,pān }

octahedral molecule [CHEMISTRY] A molecule whose structure forms an octahedron in which a central atom possesses six valence bonds that are directed to the points of the octahedron, for example, sulfur hexafluoride (SF_6). { ¦äk·tə¦hē·drəl 'mäl·ə,kyül }

octanal [ORGANIC CHEMISTRY] *See* octyl aldehyde. { 'äk·tə,nal }

***n*-octane** [ORGANIC CHEMISTRY] C_8H_{18} Colorless liquid boiling at 126°C; soluble in alcohol, acetone, and ether, insoluble in water; used as a solvent and chemical intermediate. { ¦en 'äk,tān }

octanedioic acid [ORGANIC CHEMISTRY] *See* suberic acid. { ¦äk,tan·dī'ō·ik 'as·əd }

1-octene [ORGANIC CHEMISTRY] $CH_3(CH_2)_5CHCH_2$ A colorless, flammable liquid; used as a plasticizer and in synthesis of organic compounds. Also known as 1-caprylene, 1-octylene. { ¦wən 'äk,ten }

2-octene [ORGANIC CHEMISTRY] $CH_3(CH_2)_4CHCHCH_3$ A colorless, flammable liquid, with trans and cis forms; used to manufacture lubricants and to synthesize organic materials. { ¦tü 'äk,tēn }

octet rule [CHEMISTRY] A concept of chemical bonding theory based on the assumption that in the formation of compounds, atoms exhibit a tendency for their valence shells either to be empty or to have a full complement of eight electrons (octet); for some elements there are more than the usual eight valence electrons in some of their compounds. { äk,tet ,rül }

octomethylene [ORGANIC CHEMISTRY] *See* cyclooctane. { ¦äk·tō'meth·ə,lēn }

octyl- [ORGANIC CHEMISTRY] Prefix indicating the eight-carbon hydrocarbon radical (C_8H_{17}—). { 'äkt·əl }

***n*-octyl acetate** [ORGANIC CHEMISTRY] $CH_3COO(CH_2)_7CH_3$ A colorless liquid with a fruity odor and a boiling point of 199°C; soluble in alcohol and other organic liquids; used for perfumes and flavoring. { ¦en 'äkt·əl 'as·ə,tāt }

octyl alcohol [ORGANIC CHEMISTRY] *See* 2-ethylhexyl alcohol. { 'äkt·əl 'al·kə,hȯl }

octyl aldehyde [ORGANIC CHEMISTRY] $C_8H_{16}O$ A liquid aldehyde boiling at 172°C; found in essential oils of many plants; used in perfume compositions. Also known as octanal. { 'äkt·əl 'al·də,hīd }

***n*-octyl *n*-decyl adipate** [ORGANIC CHEMISTRY] A liquid with a boiling range of 250-254°C; used as a low-temperature plasticizer. Abbreviated NODA. { ¦en 'äkt·əl ¦en 'des·əl 'ad·ə,pāt }

***n*-octyl *n*-decyl phthalate** [ORGANIC CHEMISTRY] A clear liquid with a boiling range of 232-267°C; used as a plasticizer for vinyl resins. { ¦en 'äkt·əl ¦en 'des·əl 'tha,lāt }

1-octylene [ORGANIC CHEMISTRY] *See* 1-octene. { ¦wən 'äk·tə,lēn }

octyl formate [ORGANIC CHEMISTRY] $C_8H_{17}OOCH$ A colorless liquid with a fruity odor; soluble in mineral oil; used for flavoring. { 'äkt·əl 'fȯr,māt }

***n*-octyl mercaptan** [ORGANIC CHEMISTRY] $C_8H_{17}SH$ Clear, colorless liquid boiling at 199°C; used as a chemical intermediate and polymerization conditioner. { ¦en 'äkt·əl mər'kap,tan }

octyl phenol [ORGANIC CHEMISTRY] $C_8H_{17}C_6H_4OH$ White flakes, congealing at 73°C; soluble in organic solvents, insoluble in water; used to make surfactants, plasticizers, and antioxidants. { 'äkt·əl 'fē,nȯl }

octyne [ORGANIC CHEMISTRY] $CHC(CH_2)_5CH_3$ Colorless hydrocarbon liquid, boiling at 125°C. { 'äk,tīn }

-oic [ORGANIC CHEMISTRY] A suffix indicating the presence of a —COOH group, as in ethyloic (—CH_2—COOH). { 'ō·ik }

oil blue [INORGANIC CHEMISTRY] Violet-blue copper sulfide pigment used in varnishes. { 'ȯil 'blü }

oil of mirbane [ORGANIC CHEMISTRY] *See* nitrobenzene. { 'ȯil əv 'mər,bān }

oil of vitriol [INORGANIC CHEMISTRY] *See* sulfuric acid. { 'ȯil əv 'vit·rē,ōl }

-ol [ORGANIC CHEMISTRY] Chemical suffix for an —OH group in organic compounds, such as phenol (C_6H_5OH). { ,ȯl }

oleate [ORGANIC CHEMISTRY] Salt made up of a metal or alkaloid with oleic acid; used for external medicines and in soaps and paints. { 'ō·lē,āt }

olefiant gas [ORGANIC CHEMISTRY] *See* ethylene. { ¦ō·lə¦fī·ənt 'gas }

olefin [ORGANIC CHEMISTRY] C_nH_{2n} A family of unsaturated, chemically active hydrocarbons with one carbon-carbon double bond; includes ethylene and propylene. { 'ō·lə·fən }

olefin copolymer [ORGANIC CHEMISTRY] Polymer made by the interreaction of two or more kinds of olefin monomers, such as butylene and propylene. { 'ō·lə·fən kō 'päl·ə·mər }

olefin resin [ORGANIC CHEMISTRY] Long-chain polymeric material produced by the chain reaction of olefinic monomers, such as polyethylene from ethylene, or polypropylene from propylene. { 'ō·lə·fən 'rez·ən }

oleic acid [ORGANIC CHEMISTRY] $C_{17}H_{33}COOH$ Yellowish, unsaturated fatty acid with lardlike aroma; soluble in organic solvents, slightly soluble in water; boils at 286°C (100 mmHg); the main component of olive and cooking oils; used in soaps, ointments, cosmetics, and ore beneficiation. Also known as red oil. { ō'lā·ik 'as·əd }

olein [ORGANIC CHEMISTRY] $(C_{17}H_{33}COO)_3C_3H_5$ Oleic acid triglyceride; yellow liquid melting at −5°C; slightly soluble in alcohol, soluble in chloroform, ether, and carbon tetrachloride; found in most fats and oils; used in textile lubrication. { 'ō·lē·ən }

oleum [CHEMISTRY] **1.** Latin name for oil. **2.** *See* fuming sulfuric acid. { 'ō·lē·əm }

oleyl alcohol [ORGANIC CHEMISTRY] $C_{18}H_{35}OH$ Clear liquid, boiling at 282-349°C; fatty alcohol derived from oleic acid; commercial grade 80-90% pure; used to make resins and surface-active agents, and as a chemical intermediate. { ō'lē·əl 'al·kə,hȯl }

oligomer [ORGANIC CHEMISTRY] A polymer made up of two, three, or four monomer units. { ə'lig·ə·mər }

oligopeptide [ORGANIC CHEMISTRY] A peptide composed of no more than 10 amino acids. { ,äl·ə·gō'pep,tīd }

omethioate [ORGANIC CHEMISTRY] *See* folimat. { ,ō·mə'thī·ə,wāt }

ONB [ORGANIC CHEMISTRY] *See* *ortho*-nitrobiphenyl.

-one [ORGANIC CHEMISTRY] Chemical suffix indicating a ketone, a substance related to starches and sugars, or an alkone. { ōn }

-onium [CHEMISTRY] Chemical suffix indicating a complex cation, as for hydronium, $(H_3O)^+$. { 'ō·nē·əm }

Onsager equation [PHYSICAL CHEMISTRY] An equation which relates the measured equivalent conductance of a solution at a certain concentration to that of the pure solvent. { 'ȯn‚säg·ər i‚kwā·zhən }

open-circuit potential [PHYSICAL CHEMISTRY] Steady-state or equilibrium potential of an electrode in absence of external current flow to or from the electrode. { 'ō·pən ¦sər·kət pə'ten·chəl }

open tubular column [ANALYTICAL CHEMISTRY] *See* capillary column. { 'ō·pən 'tü·byə·lər ‚käl·əm }

opianyl [ORGANIC CHEMISTRY] *See* meconin. { 'ō pē·ə‚nil }

Oppenauer oxidation [ORGANIC CHEMISTRY] The oxidation of a primary or secondary hydroxyl compound to form the corresponding carbonyl compound; aluminum alkoxide and an excess amount of a carbonyl hydrogen acceptor, such as benzophenone or acetone, are required. { 'äp·ə‚nau̇·ər ‚äk·sə'dā·shən }

optical anomaly [PHYSICAL CHEMISTRY] The phenomenon in which an organic compound has a molar refraction which does not agree with the value calculated from the equivalents of atoms and other structural units composing it. { 'äp·tə·kəl ə'näm·ə·lē }

optical antipode [CHEMISTRY] *See* enantiomorph. { 'äp·tə·kəl 'ant·i‚pōd }

optical cytopherometry [ANALYTICAL CHEMISTRY] *See* microelectrophoresis. { 'äp·tə·kəl ‚sī·dō‚fə'räm·ə·trē }

optical exaltation [PHYSICAL CHEMISTRY] Optical anomaly in which the observed molar refraction exceeds the calculated one; most cases of optical anomaly are in this category. { 'äp·tə·kəl ‚ek·səl'tā·shən }

optical isomer [CHEMISTRY] *See* enantiomorph. { 'äp·tə·kəl 'ī·sə·mər }

optical isomerism [PHYSICAL CHEMISTRY] Existence of two forms of a molecule such that one is a mirror image of the other; the two molecules differ in that rotation of light is equal but in opposite directions. { 'äp·tə·kəl ī'säm·ə‚riz·ən }

optical monochromator [SPECTROSCOPY] A monochromator used to observe the intensity of radiation at wavelengths in the visible, infrared, or ultraviolet regions. { 'äp·tə·kəl ¦män·ə'kräm·əd·ər }

optical null method [SPECTROSCOPY] In infrared spectrometry, the adjustment of a reference beam's energy transmission to match that of a beam that has been passed through a sample being analyzed. { 'äp·tə·kəl ¦nəl ‚meth·əd }

optical spectra [SPECTROSCOPY] Electromagnetic spectra for wavelengths in the ultraviolet, visible and infrared regions, ranging from about 10 nanometers to 1 millimeter, associated with excitations of valence electrons of atoms and molecules, and vibrations and rotations of molecules. { 'äp·tə·kəl 'spek·trə }

optical spectrograph [SPECTROSCOPY] An optical spectroscope provided with a photographic camera or other device for recording the spectrum made by the spectroscope. { 'äp·tə·kəl 'spek·trə‚graf }

optical spectrometer [SPECTROSCOPY] An optical spectroscope that is provided with a calibrated scale either for measurement of wavelength or for measurement of refractive indices of transparent prism materials. { 'äp·tə·kəl spek'träm·əd·ər }

optical spectroscope [SPECTROSCOPY] An optical instrument, consisting of a slit, collimator lens, prism or grating, and a telescope or objective lens, which produces an optical spectrum arising from emission or absorption of radiant energy by a substance, for visual observation. { 'äp·tə·kəl 'spek·trə‚skōp }

optical spectroscopy [SPECTROSCOPY] The production, measurement, and interpretation of optical spectra arising from either emission or absorption of radiant energy by various substances. { 'äp·tə·kəl spek'träs·kə·pē }

optoacoustic detection method [ANALYTICAL CHEMISTRY] A method of detecting trace inpurities in a gas, in which the absorption of a sample of the gas at various light frequencies is measured by directing a periodically interrupted laser beam through the sample in a spectrophone and measuring the sound generated by the optoacoustic effect at the frequency of interruption of the beam. { ¦äp·tō·ə¦küs·tik di'tek·shən ,meth·əd }

optogalvanic spectroscopy [SPECTROSCOPY] A method of obtaining absorption spectra of atomic and molecular species in flames and electrical discharges by measuring voltage and current changes upon laser irradiation. { ¦äp·tō·gal¦van·ik spek'träs·kə·pē }

orange oxide [INORGANIC CHEMISTRY] *See* uranium trioxide. { 'är·inj 'äk,sīd }

orange spectrometer [SPECTROSCOPY] A type of beta-ray spectrometer that consists of a number of modified double-focusing spectrometers employing a common source and a common detector, and has exceptionally high transmission. { 'är·inj spek'träm·əd·ər }

orange toner [ORGANIC CHEMISTRY] A diazo dyestuff coupled to diacetoacetic acid anhydride; contains no sulfonic or carboxylic groups; used for printing inks. { 'är·inj 'tōn·ər }

orbital overlap [PHYSICAL CHEMISTRY] The overlapping of two electron orbitals, one from each of two different atoms, such that each orbital obtains a share in the electron of the other atom, forming a chemical bond. { 'ȯr·bəd·əl 'ō·vər,lap }

orbital symmetry [PHYSICAL CHEMISTRY] The property of certain molecular orbitals of being carried into themselves or into the negative of themselves by certain geometrical operations, such as a rotation of 180° about an axis in the plane of the molecule, or reflection through this plane. { 'ȯr·bəd·əl 'sim·ə·trē }

orcin [ORGANIC CHEMISTRY] $CH_3C_6H_3(OH)_2 \cdot H_2O$ White crystals with strong, sweet, unpleasant taste; soluble in water, alcohol, and ether; extracted from lichens; used in medicine and as an analytical reagent. { 'ȯr·sən }

order [CHEMISTRY] A classification of chemical reactions, in which the order is described as first, second, third, or higher, according to the number of molecules (one, two, three, or more) which appear to enter into the reaction; decomposition of H_2O_2 to form water and oxygen is a first-order reaction. { 'ȯrd·ər }

organic [ORGANIC CHEMISTRY] Of chemical compounds, based on carbon chains or rings and also containing hydrogen with or without oxygen, nitrogen, or other elements. { ȯr'gan·ik }

organic acid [ORGANIC CHEMISTRY] A chemical compound with one or more carboxyl radicals (COOH) in its structure; examples are butyric acid, $CH_3(CH_2)_2COOH$, maleic acid, HOOCCHCHCOOH, and benzoic acid, C_6H_5COOH. { ȯr'gan·ik 'as·əd }

organic chemistry [CHEMISTRY] The study of the composition, reactions, and properties of carbon-chain or carbon-ring compounds or mixtures thereof. { ȯr'gan·ik 'kem·ə·strē }

organic pigment [ORGANIC CHEMISTRY] Any of the materials with organic-chemical bases used to add color to dyes, plastics, linoleum, tones, and lakes. { ȯr'gan·ik 'pig·mənt }

organic quantitative analysis [ANALYTICAL CHEMISTRY] Quantitative determination of elements, functional groups, or molecules in organic materials. { ȯr'gan·ik 'kwän·ə,tād·iv ə,nal·ə·səs }

organic reaction mechanism [ORGANIC CHEMISTRY] A pathway of chemical states traversed by an organic chemical system in its passage from reactants to products. { ȯr'gan·ik rē'ak·shən ,mek·ə,niz·əm }

organic salt [ORGANIC CHEMISTRY] The reaction product of an organic acid and an inorganic base, for example, sodium acetate (CH_3COONa) from the reaction of acetic acid (CH_3COOH) and sodium hydroxide (NaOH). { ȯr'gan·ik 'sȯlt }

organic solvent [ORGANIC CHEMISTRY] Liquid organic compound with the power to dissolve solids, gases, or liquids (miscibility); examples are methanol (methyl alcohol), CH_3OH, and benzene, C_6H_6. { ȯr'gan·ik 'säl·vənt }

organoborane [ORGANIC CHEMISTRY] A derivative of a borane (boron hydride) in which one or more hydrogen atoms have been replaced by functional groups. { ȯr¦gan·ə'bȯr,ān }

organometallic compound [ORGANIC CHEMISTRY] Molecules containing carbon-metal linkage; a compound containing an alkyl or aryl radical bonded to a metal, such as tetraethyllead, $Pb(C_2H_5)_4$. { ȯrˌgan·ə·məˈtal·ik ˈkämˌpau̇nd }

organophosphate [ORGANIC CHEMISTRY] A soluble fertilizer material made up of organic phosphate esters such as glucose, glycol, or sorbitol; useful for providing phosphorus to deep-root systems. { ȯrˌgan·əˈfäsˌfāt }

organophosphorus compound [ORGANIC CHEMISTRY] An organic compound that contains phosphorus in its chemical structure. { ȯrˌgan·əˌfäs·fə·rəs ˈkämˌpau̇nd }

organoselenium compound [ORGANIC CHEMISTRY] An organic compound that contains both selenium and carbon, and frequently other elements, such as the halogens, oxygen, sulfur, or nitrogen. { ȯrˌgan·ə·səˌlē·nē·əm ˈkämˌpau̇nd }

organosilicon compound [ORGANIC CHEMISTRY] A compound in which silicon is bonded to an organic functional group, either directly or indirectly via another atom. { ȯrˌgan·ōˈsil·ə·kən ˈkämˌpau̇nd }

organosulfur compound [ORGANIC CHEMISTRY] One of a group of substances which contain both carbon and sulfur. { ȯrˌgan·əˌsəl·fər ˈkämˌpau̇nd }

orientation [PHYSICAL CHEMISTRY] The arrangement of radicals in an organic compound in relation to each other and to the parent compound. { ˌȯr·ē·ənˈtā·shən }

orientation effect [PHYSICAL CHEMISTRY] A method of determining attractive forces among molecules, or components of these forces, from the interaction energy associated with the relative orientation of molecular dipoles. { ˌȯr·ē·ənˈtā·shən iˌfekt }

orientation force [PHYSICAL CHEMISTRY] A type of van der Waals force, resulting from interaction of the dipole moments of two polar molecules. Also known as dipole-dipole force; Keesom force. { ˌȯr·ē·ənˈtā·shən ˌfȯrs }

Orsat analyzer [ANALYTICAL CHEMISTRY] Gas analysis apparatus in which various gases are absorbed selectively (volumetric basis) by passing them through a series of preselected solvents. { ˈȯrˌsat ˈan·əˌlīz·ər }

ortho acid [ORGANIC CHEMISTRY] **1.** Aromatic acid with a carboxyl group in the ortho position (1,2 position). **2.** Organic acid with one added molecule of water in chemical combination; for example, $HC(OH)_3$, orthoformic acid, in contrast to HCOOH, formic acid; $H_3PO_4(P_2O_5 \cdot 3H_2O)$, orthophosphoric acid, in contrast to the less hydrated form, metaphosphoric acid, $HPO_3(P_2O_5 \cdot H_2O)$. { ˈȯr·thō ˈas·əd }

orthoarsenic acid [INORGANIC CHEMISTRY] *See* arsenic acid. { ˌȯr·thō·ärˌsen·ik ˈas·əd }

orthoboric acid [INORGANIC CHEMISTRY] *See* boric acid. { ˌȯr·thəˌbȯr·ik ˈas·əd }

orthophosphate [INORGANIC CHEMISTRY] One of the possible salts of orthophosphoric acid; the general formula is M_3PO_4, where M may be potassium as in potassium orthophosphate, K_3PO_4. { ˌȯr·thəˈfäsˌfāt }

orthophosphoric acid [INORGANIC CHEMISTRY] *See* phosphoric acid. { ˌȯr·thə·fäsˈfȯr·ik ˈas·əd }

orthotungstic acid [INORGANIC CHEMISTRY] *See* tungstic acid. { ˌȯr thōˈtəŋ·stik ˈas·əd }

Os [CHEMISTRY] *See* osmium.

oscillatory reaction [CHEMISTRY] A chemical reaction in which a variable of a chemical system exhibits regular periodic changes in time or in space. { ˈäs·ə·ləˌtȯr·ē rēˈak·shən }

oscillographic polarography [PHYSICAL CHEMISTRY] A type of voltammetry using a dropping mercury electrode with oscillographic scanning of the applied potential; used to measure the concentration of electroactive species in solutions. { ˌäs·ə·ləˌgraf·ik ˌpō·ləˈräg·rə·fē }

oscillometric titration [PHYSICAL CHEMISTRY] Radio-frequency technique used for conductometric and dielectrometric titrations; the changes in conductance or dielectric properties changes the solution capacity and thus the frequency of the connected oscillator circuit. { ˌäs·ə·lōˌme·trik tīˈtrā·shən }

oscillometry [PHYSICAL CHEMISTRY] Electrode measurement of oscillation-frequency changes to detect the progress of a titration of electrolytic solutions. { ˌäs·əˈläm·ə·trē }

oscine [CHEMISTRY] *See* scopoline. { ˈäˌsīn }

osmate [INORGANIC CHEMISTRY] A salt or ester of osmic acid, containing the osmate radical, OsO_4^{2-}; for example, potassium osmate (K_2OsO_4). { 'äz·mət }

osmic acid anhydride [INORGANIC CHEMISTRY] OsO_4 Poisonous yellow crystals with disagreeable odor; melts at 40°C; soluble in water, alcohol, and ether; used in medicine, photography, and catalysis. Also known as osmium oxide; osmium tetroxide. { 'äz·mik 'as·əd an'hī‚drīd }

osmium [CHEMISTRY] A chemical element, symbol Os, atomic number 76, atomic weight 190.2. { 'äz·mē·əm }

osmium oxide [INORGANIC CHEMISTRY] *See* osmic acid anhydride. { 'äz·mē·əm 'äk‚sīd }

osmium tetroxide [INORGANIC CHEMISTRY] *See* osmic acid anhydride. { 'äz·mē·əm te'träk‚sīd }

osmolality [CHEMISTRY] The molality of an ideal solution of a nondissociating substance that exerts the same osmotic pressure as the solution being considered. { ‚äz·mə'lal·əd·ē }

osmolarity [CHEMISTRY] The molarity of an ideal solution of a nondissociating substance that exerts the same osmotic pressure as the solution being considered. { ‚äz·mə'lar·əd·ē }

osmole [CHEMISTRY] **1.** The unit of osmolarity equal to the osmolarity of a solution that exerts an osmotic pressure equal to that of an ideal solution of a nondissociating substance that has a concentration of 1 mole of solute per liter of solution. **2.** The unit of osmolality equal to the osmolality of a solution that exerts an osmotic pressure equal to that of an ideal solution of a nondissociating substance that has a concentration of 1 mole of solute per kilogram of solvent. { 'äz‚mōl }

osmometer [ANALYTICAL CHEMISTRY] A device for measuring molecular weights by measuring the osmotic pressure exerted by solvent molecules diffusing through a semipermeable membrane. { äz'mäm·əd·ər }

osmosis [PHYSICAL CHEMISTRY] The transport of a solvent through a semipermeable membrane separating two solutions of different solute concentration, from the solution that is dilute in solute to the solution that is concentrated. { ä'smō·səs }

osmotic gradient [PHYSICAL CHEMISTRY] *See* osmotic pressure. { äz'mäd·ik 'grād·ē·ənt }

osmotic pressure [PHYSICAL CHEMISTRY] **1.** The applied pressure required to prevent the flow of a solvent across a membrane which offers no obstruction to passage of the solvent, but does not allow passage of the solute, and which separates a solution from the pure solvent. **2.** The applied pressure required to prevent passage of a solvent across a membrane which separates solutions of different concentration, and which allows passage of the solute, but may also allow limited passage of the solvent. Also known as osmotic gradient. { äz'mäd·ik 'presh·ər }

Ostwald dilution law [PHYSICAL CHEMISTRY] The law that for a sufficiently dilute solution of univalent electrolyte, the dissociation constant approximates $a^2c/(1 - a)$, where c is the concentration of electrolyte and a is the degree of dissociation. { 'ȯst‚vält di'lü·shən ‚lȯ }

Ostwald ripening [CHEMISTRY] Solution-crystallizer phenomenon in which small crystals, more soluble than large ones, dissolve and reprecipitate onto larger particles. { 'ȯst‚vält ¦rīp·ə·niŋ }

Oswald diagram [ANALYTICAL CHEMISTRY] Diagram used in fuel Orsat analyses by plotting percent by volume CO_2 (carbon dioxide) maximum in the fuel [ordinate] versus percent by volume O_2 (oxygen) in air [abscissa]; O_2 and CO_2 Orsat readings should fall on a line connecting these maximum values if the analysis is proceeding properly. { 'äz‚wȯld 'dī·ə‚gram }

ouabain [ORGANIC CHEMISTRY] $C_{29}H_{44}O_{12} \cdot 8H_2O$ White crystals that melt with decomposition at 190°C, soluble in water and ethanol; used in medicine. { wä'bī·ən }

Oudeman law [PHYSICAL CHEMISTRY] The law that the molecular rotations of the various salts of an acid or base tend toward an identical limiting value as the concentration of the solution is reduced to zero. { 'ȯd·ə·mən ‚lȯ }

outer orbital complex [PHYSICAL CHEMISTRY] A metal coordination compound in which the *d* orbital used in forming the coordinate bond is at the same energy level as the *s* and *p* orbitals. { 'au̇d·ər 'ȯrb·əd·əl 'käm,pleks }

overall stability constant [ANALYTICAL CHEMISTRY] Reaction equilibrium constant for the reaction forming soluble complexes during compleximetric titration. { ¦ō·vər¦ȯl stə'bil·əd·ē ,kän·stənt }

overpoint [CHEMISTRY] The initial boiling point in a distillation process; specifically, the temperature at which the first drop falls from the tip of the condenser into the condensate flask. { 'ō·vər,pȯint }

overpotential [PHYSICAL CHEMISTRY] *See* overvoltage. { ¦ō·vər·pə'ten·chəl }

overtone band [SPECTROSCOPY] The spectral band associated with transitions of a molecule in which the vibrational quantum number changes by 2 or more. { 'ō·vər,tōn ,band }

overvoltage [PHYSICAL CHEMISTRY] The difference between electrode potential under electrolysis conditions and the thermodynamic value of the electrode potential in the absence of electrolysis for the same experimental conditions. Also known as overpotential. { ¦ō·vər¦vōl·tij }

ovex [ORGANIC CHEMISTRY] $ClC_6H_4OSO_2C_6H_4Cl$ A white, crystalline solid with a melting point of 86.5°C; soluble in acetone and aromatic solvents; used as an insecticide and acaricide. { 'ō,veks }

oxadiazon [ORGANIC CHEMISTRY] $C_{13}H_{18}Cl_2N_2O_3$ A white solid with a melting point of 88-90°C; slight solubility in water; used as a pre- and postemergence herbicide to control weeds in rice, turf, soybeans, peanuts, and orchards. { ,äk·sə'dī·ə,zän }

oxalate [ORGANIC CHEMISTRY] Salt of oxalic acid; contains the $(COO)_2$ radical; examples are sodium oxalate, $Na_2C_2O_4$, ammonium oxalate, $(NH_4)_2C_2O_4 \cdot H_2O$, and ethyl oxalate, $C_2H_5(C_2O_4)C_2H_5$. { 'äk·sə,lāt }

oxalic acid [ORGANIC CHEMISTRY] $HOOCCOOH \cdot 2H_2O$ Poisonous, transparent, colorless crystals melting at 187°C; soluble in water, alcohol, and ether; used as a chemical intermediate and a bleach, and in polishes and rust removers. { äk'sal·ik 'as·əd }

oxalyl chloride [INORGANIC CHEMISTRY] $(COCl)_2$ Toxic, colorless liquid boiling at 64°C, soluble in ether, benzene, and chloroform; used as a chlorinating agent and for military poison gas. { 'äk·sə,lil 'klȯr,īd }

oxamide [ORGANIC CHEMISTRY] $NH_2COCONH_2$ Water-insoluble white powder, melting at 419°C; used as a stabilizer for nitrocellulose products. { 'äk·sə·məd }

oxamyl [ORGANIC CHEMISTRY] $C_7H_{13}N_3O_3S$ A white, crystalline compound with a melting point of 100-102°C; used to control pests of tobacco, ornamentals, fruits, and crops. { 'äk·sə,mil }

oxazole [ORGANIC CHEMISTRY] C_3H_3ON A structure that consists of a five-membered ring containing oxygen and nitrogen in the 1 and 3 position; a colorless liquid (boiling point 69-70°C) that is miscible with organic solvents and water; used to prepare other organic compounds. { 'äk·sə,zōl }

oxidant [CHEMISTRY] *See* oxidizing agent. { 'äk·səd·ənt }

oxidation [CHEMISTRY] **1.** A chemical reaction that increases the oxygen content of a compound. **2.** A chemical reaction in which a compound or radical loses electrons, that is in which the positive valence is increased. { ,äk·sə'dā·shən }

oxidation number [CHEMISTRY] **1.** Numerical charge on the ions of an element. **2.** *See* oxidation state. { ,äk·sə'dā·shən ,nəm·bər }

oxidation potential [PHYSICAL CHEMISTRY] The difference in potential between an atom or ion and the state in which an electron has been removed to an infinite distance from this atom or ion. { ,äk·sə'dā·shən pə,ten·chəl }

oxidation-reduction indicator [ANALYTICAL CHEMISTRY] A compound whose color in the oxidized state differs from that in the reduced state. { ,äk·sə'dā·shən ri'dək·shən ,in·də,kād·ər }

oxidation-reduction potential [PHYSICAL CHEMISTRY] *See* redox potential. { ,äk·sə'dā·shən ri'dək·shən pə,ten·chəl }

oxidation-reduction reaction [CHEMISTRY] An oxidizing chemical change, where an element's positive valence is increased (electron loss), accompanied by a simultaneous reduction of an associated element (electron gain). { ˌäk·səˈdā·shən riˈdək·shən rēˌak·shən }

oxidation state [CHEMISTRY] The number of electrons to be added (or subtracted) from an atom in a combined state to convert it to elemental form. Also known as oxidation number. { ˌäk·səˈdā·shən ˌstāt }

oxide [CHEMISTRY] Binary chemical compound in which oxygen is combined with a metal (such as Na_2O; basic) or nonmetal (such as NO_2; acidic). { ˈäkˌsīd }

oxidizing agent [CHEMISTRY] Compound that gives up oxygen easily, removes hydrogen from another compound, or attracts negative electrons. Also known as oxidant. { ˈäk·səˌdīz·iŋ ˌā·jənt }

oxidizing atmosphere [CHEMISTRY] Gaseous atmosphere in which an oxidation reaction occurs; usually refers to the oxidation of solids. { ˈäk·səˌdīz·iŋ ˈat·məˌsfir }

oxidizing flame [CHEMISTRY] A flame, or the portion of it, that contains an excess of oxygen. { ˈäk·səˌdīz·iŋ ˌflām }

oxime [ORGANIC CHEMISTRY] Compound containing the CH(:NOH) radical; condensation product of hydroxylamine with aldehydes or ketones. { ˈäkˌsēm }

oximido [CHEMISTRY] *See* nitroso. { äkˈsim·ə·dō }

oxine [ORGANIC CHEMISTRY] C_9H_6NOH White powder that darkens when exposed to light; slightly soluble in water, dissolves in ethanol, acetone, and benzene; used to prepare fungicides and to separate metals by precipitation. Also known as 8-hydroxyquinoline; oxyquinoline; 8-quinolinol. { ˈäkˌsēn }

oxirane [ORGANIC CHEMISTRY] *See* epoxide; ethylene oxide. { ˈäk·sə·rān }

oxo- [ORGANIC CHEMISTRY] Chemical prefix designating the keto group, C:O. { ˈäk·sō }

***para*-oxon** [ORGANIC CHEMISTRY] $(C_2H_5O)_2P(O)C_6H_4NO_2$ A reddish-yellow oil with a boiling point of 148-151°C; soluble in most organic solvents; used as an insecticide. Also known as diethyl *para*-nitrophenyl phosphate. { ¦par·ə ˈäkˌsän }

oxonium ion [CHEMISTRY] R_3O^+ A cation in which an oxygen atom is covalently bound to three atoms or groups of atoms. { äkˈsō·nē·əm ˈīˌän }

oxosilane [ORGANIC CHEMISTRY] *See* siloxane. { ¦äk·sōˈsiˌlān }

oxyacanthine [ORGANIC CHEMISTRY] $C_{37}H_{40}N_2O_6$ An alkaloid obtained from the root of *Berberis vulgaris*; a white, crystalline powder with a melting point of 202-214°C; soluble in water, chloroform, benzene, alcohol, and ether; used in medicine. Also known as vinetine. { ¦äk·sē·əˈkanˌthēn }

oxybenzone [ORGANIC CHEMISTRY] $C_{14}H_{12}O_3$ A crystalline substance with a melting point of 66°C; used as a sunscreen agent. Also known as 4-methoxy-2-hydroxybenzophenone. { ¦äk·sēˈbenˌzōn }

oxycarboxin [ORGANIC CHEMISTRY] $C_{12}H_{13}NO_4S$ An off-white, crystalline compound with a melting point of 127.5-130°C; used to control rust disease in greenhouse carnations. Also known as 5,6-dihydro-2-methyl- 1,4-oxathiin-3-carboxanilide-4,4-dioxide. { ¦äk·sē·kärˈbäk·sən }

oxy compound [CHEMISTRY] A compound containing two or more oxygen atoms that are not joined to each other but are covalently bound to other atoms in the structure. { ˈäk·sē ˌkämˌpau̇nd }

oxygen [CHEMISTRY] A gaseous chemical element, symbol O, atomic number 8, and atomic weight 15.9994; an essential element in cellular respiration and in combustion processes; the most abundant element in the earth's crust, and about 20% of the air by volume. { ˈäk·sə·jən }

oxygen absorbent [CHEMISTRY] Any material that will absorb (dissolve) oxygen into its body without reacting with it. { ˈäk·sə·jən əbˈsȯr·bənt }

oxygenate [CHEMISTRY] To treat, infuse, or combine with oxygen. { ˈäk·sə·jəˌnāt }

oxygen cell [PHYSICAL CHEMISTRY] *See* aeration cell. { ˈäk·sə·jən ˌsel }

oxygen concentration cell [PHYSICAL CHEMISTRY] *See* differential aeration cell. { ¦äk·sə·jən ˌkän·sənˈtrā·shən ˌsel }

oxygen-flask method [ANALYTICAL CHEMISTRY] Technique to determine the presence of combustible elements; the sample is burned with oxygen in a closed flask, and combustion products are absorbed in water of dilute alkali with subsequent analysis of the solution. { 'äk·sə·jən ¦flask ˌmeth·əd }

oxyhydrogen flame [CHEMISTRY] A flame obtained from the combustion of a mixture of oxygen and hydrogen. { ¦äk·sē'hī·drə·jən ¦flām }

oxyneurine [ORGANIC CHEMISTRY] *See* betaine. { ˌäk·sē'nu̇ˌrēn }

oxyquinoline [ORGANIC CHEMISTRY] *See* oxine. { ˌäk·sə'kwin·ə·lən }

ozone [CHEMISTRY] O_3 Unstable blue gas with pungent odor; an allotropic form of oxygen; a powerful oxidant boiling at −112°C; used as an oxidant, bleach, and water purifier, and to treat industrial wastes. { 'ōˌzōn }

ozonide [ORGANIC CHEMISTRY] Any of the oily, thick, unstable compounds formed by reaction of ozone with unsaturated compounds; an example is oleic ozonide from the reaction of oleic acid and ozone. { 'äz·əˌnīd }

ozonization [CHEMISTRY] The process of treating, impregnating, or combining with ozone. { ˌōˌzō·nə'zā·shən }

ozonolysis [ANALYTICAL CHEMISTRY] The use of ozone to locate double bonds. [ORGANIC CHEMISTRY] Oxidation of an organic substance by means of ozone. { ˌō·zə'näl·ə·səs }

P

P [CHEMISTRY] *See* phosphorus.

p- [ORGANIC CHEMISTRY] *See* para-.

Pa [CHEMISTRY] *See* protactinium.

Paal-Knorr synthesis [ORGANIC CHEMISTRY] A method of converting a 1,4-dicarbonyl compound by cyclization with ammonia or a primary amine to a pyrrole. { 'pȯl kə'nȯr 'sin·thə·səs }

Paar turbidimeter [ANALYTICAL CHEMISTRY] A visual-extinction device for measurement of solution turbidity; the length of the column of liquid suspension is adjusted until the light filament can no longer be seen. { pär ˌtər·bə'dim·əd·ər }

paired electron [PHYSICAL CHEMISTRY] One of two electrons that form a valence bond between two atoms. { ¦perd i'lekˌträn }

palladium [CHEMISTRY] A chemical element, symbol Pd, atomic number 46, atomic weight 106.4. { pə'lād·ē·əm }

palladium chloride [INORGANIC CHEMISTRY] $PdCl_2$ or $PdCl_2 \cdot 2H_2O$ Dark-brown, deliquescent powder that decomposes at 501°C; soluble in water, alcohol, acetone, and hydrochloric acid; used in medicine, analytical chemisty, photographic chemicals, and indelible inks. { pə'lād·ē·əm 'klȯrˌīd }

palladium iodide [INORGANIC CHEMISTRY] PdI_2 Black powder that decomposes above 100°C; soluble in potassium iodide solution, insoluble in water and alcohol. { pə'lād·ē·əm 'ī·əˌdīd }

palladium nitrate [INORGANIC CHEMISTRY] $Pd(NO_3)_2$ Brown, water-soluble, deliquescent salt; used as an analytical reagent. { pə'lād·ē·əm 'nīˌtrāt }

palladium oxide [INORGANIC CHEMISTRY] PdO Amber or black-green powder that decomposes at 750°C; soluble in dilute acids, used in chemical synthesis as a reduction catalyst. { pə'lād·ē·əm 'äkˌsīd }

palmitate [ORGANIC CHEMISTRY] A derivative ester or salt of palmitic acid. { 'päm·əˌtāt }

palmitic acid [ORGANIC CHEMISTRY] $C_{15}H_{31}COOH$ A fatty acid; white crystals, soluble in alcohol and ether, insoluble in water; melts at 63.4°C, boils at 271.5°C (100 mmHg); derived from spermaceti, used to make metallic palmitates and in soaps, waterproofing, and lubricating oils. { pal'mid·ik 'as·əd }

palmitoleic acid [ORGANIC CHEMISTRY] $C_{16}H_{30}O_2$ An unsaturated fatty acid, found in marine animal oils; it is a clear liquid used as a standard in chromatography. { ¦päl·məd·ō¦lē·ik 'as·əd }

Paneth's adsorption rule [PHYSICAL CHEMISTRY] The rule that an element is strongly absorbed on a precipitate which has a surface charge opposite in sign to that carried by the element, provided that the resulting adsorbed compound is very sparingly soluble in the solvent. { 'pan·əths ad'sȯrp·shən ˌrül }

Papanicolaou's stains [CHEMISTRY] A group of stains used on exfoliated cells, particularly those from the vagina, for examination and diagnosis. { ˌpä·pə'nēk·əˌlaủz ˌstān }

papaverine [ORGANIC CHEMISTRY] $C_{20}H_{21}O_4N$ A white, crystalline alkaloid, melting at 147°C; soluble in acetone and chloroform, insoluble in water; used as a smooth muscle relaxant and weak analgesic, usually as the water-soluble hydrochloride salt. { pə'pav·əˌrēn }

paper chromatography [ANALYTICAL CHEMISTRY] Procedure for analysis of complex chemical mixtures by the progressive absorption of the components of the unknown sample (in a solvent) on a special grade of paper. { 'pā·pər ˌkrō·mə'täg·rə·fē }

paper electrochromatography [ANALYTICAL CHEMISTRY] Variation of paper electrophoresis in which the electrolyte-impregnated absorbent paper is suspended vertically and the electrodes are connected to the sides of the paper, producing a current at right angles to the downward movement of the unknown sample. { 'pā·pər iˌlek·trōˌkrō·mə'täg·rə·fē }

paper electrophoresis [ANALYTICAL CHEMISTRY] A variation of paper chromatography in which an electric current is applied to the ends of the electrolyte-impregnated absorbent paper, thus moving chargeable molecules of the unknown sample toward the appropriate electrode. { 'pā·pər iˌlek·trə·fə'rē·səs }

paper-tape chemical analyzer [ANALYTICAL CHEMISTRY] Chemically treated paper tape that is continuously unreeled, exposed to the sample, and viewed by a phototube to measure the color change that is empirically related to changes in the sample's chemical composition. { 'pā·pər ¦tāp 'kem·ə·kəl 'an·əˌlīz·ər }

para- [ORGANIC CHEMISTRY] Chemical prefix designating the positions of substituting radicals on the opposite ends of a benzene nucleus, for example, paraxylene, $CH_3C_6H_4CH_3$. Abbreviated *p*-. { 'par·ə }

parabanic acid [ORGANIC CHEMISTRY] $C_3H_2O_3N_2$ A water-soluble cyclic compound that decomposes when heated to about 227°C; used in organic synthesis. { ¦par·ə¦ban·ik 'as·əd }

paracyanogen [INORGANIC CHEMISTRY] $(CN)_x$ A white solid produced by polymerization of cyanogen gas when heated to 400°C. { ¦par·ə·sī'an·ə·jən }

paraffinic hydrocarbon [ORGANIC CHEMISTRY] *See* alkane. { ˌpar·ə¦fin·ik 'hī·drəˌkär·bən }

paraffinicity [ORGANIC CHEMISTRY] The paraffinic nature or composition of crude petroleum or its products. { ˌpar·ə·fə'nis·əd·ē }

paraformaldehyde [ORGANIC CHEMISTRY] $(HCHO)_n$ Polymer of formaldehyde where *n* is greater than 6; white, alkali-soluble solid, insoluble in alcohol, ether, and water; used as a disinfectant, fumigant, and fungicide, and to make resins. { ¦par·ə·fȯr'mal·dəˌhīd }

paraldehyde [ORGANIC CHEMISTRY] $C_6H_{12}O_3$ Acetaldehyde polymer; colorless, flammable, toxic liquid, miscible with most organic solvents, soluble in water; melts at 12.6°C, boils at 124.5°C; used as a chemical intermediate, in medicine, and as a solvent. { pə'ral·dəˌhīd }

paraldol [ORGANIC CHEMISTRY] $(CH_3CHOHCH_2CHO)_2$ Water-soluble, white crystals, boiling at 90-100°C; used as a chemical intermediate, to make resins, and in cadmium plating baths. { par'alˌdȯl }

paramagnetic analytical method [ANALYTICAL CHEMISTRY] A method for analyzing fluid mixtures by measurement of the paramagnetic (versus diamagnetic) susceptibilities of materials when exposed to a magnetic field. { ¦par·ə·mag'ned·ik ˌan·ə'lid·ə·kəl 'meth·əd }

paramagnetic spectra [SPECTROSCOPY] Spectra associated with the coupling of the electronic magnetic moments of atoms or ions in paramagnetic substances, or in paramagnetic centers of diamagnetic substances, to the surrounding liquid or crystal environment, generally at microwave frequencies. { ¦par·ə·mag'ned·ik 'spek·trə }

paranitraniline red [ORGANIC CHEMISTRY] *See* para red. { ¦par·ə·nə'tran·ə·lən 'red }

paraquat [ORGANIC CHEMISTRY] $[CH_3(C_5H_4N)_2CH_3] \cdot 2CH_3SO_4$ A yellow, water-soluble solid, used as a herbicide. { 'par·əˌkwät }

para red [ORGANIC CHEMISTRY] $C_{10}H_6(OH)NNC_6H_4NO_2$ Red pigment derived from the coupling of β-naphthol with diazotized paranitroaniline. Also known as paranitraniline red. { 'par·ə 'red }

pararosaniline [ORGANIC CHEMISTRY] $HOC(C_6H_4NH_2)_3$ Red to colorless crystals, melting at 205°C; soluble in ethanol, in the hydrochloride salt; used as a dye. { ¦par·ə·rō'san·ə·lən }

parent compound [CHEMISTRY] A chemical compound that is the basis for one or more derivatives; for example, ethane is the parent compound for ethyl alcohol and ethyl acetate. { 'per·ənt ,käm,paůnd }

parent name [CHEMISTRY] That part of a chemical compound's name from which the name of a derivative comes; for example, ethane is the parent name for ethanol. { 'per·ənt ,nām }

partially ionic bond [PHYSICAL CHEMISTRY] A chemical bond that is neither wholly ionic nor wholly covalent in character. { 'pär·shə·lē ī¦än·ik 'bänd }

partial molal quantity [PHYSICAL CHEMISTRY] The molal concentration of one component of a mixture of components as related to total molal concentration for all components in the mixture. { 'pär·shəl 'mō·ləl 'kwän·əd·ē }

partial molar volume [PHYSICAL CHEMISTRY] That portion of the volume of a solution or mixture related to the molar content of one of the components within the solution or mixture. { 'pär·shəl 'mō·lər 'val·yəm }

particle counting [ANALYTICAL CHEMISTRY] Microscopic or photomicrographic technique for the visual counting of the numbers of particles in a known quantity of a solid-liquid suspension. { 'pärd·ə·kəl ,kaůnt·iŋ }

particle electrophoresis [PHYSICAL CHEMISTRY] Electrophoresis in which the particles undergoing analysis are of sufficient size to be viewed either with the naked eye or with the assistance of an optical microscope. { 'pärd·ə·kəl i,lek trō fə'rē·səs }

particle-induced x-ray emission [ANALYTICAL CHEMISTRY] A method of trace analysis in which a beam of ions is directed at a thin foil on which the sample to be analyzed has been deposited, and the energy spectrum of the resulting x-rays is measured. { 'pärd·ə·kəl in¦düst 'eks,rā i,mish·ən }

particle-scattering factor [ANALYTICAL CHEMISTRY] Factor in light-scattering equations used to compensate for the loss in scattered light intensity caused by destructive interference during the analysis of macromolecular compounds. { 'pärd·ə·kəl ¦skad·ə·riŋ ,fak·tər }

particle-thickness technique [PHYSICAL CHEMISTRY] Microscopic technique for visual measurement of the thickness of a fine particle (in the 3-100 micrometer range). { 'pärd·ə·kəl ¦thik·nəs tek,nēk }

partition chromatography [ANALYTICAL CHEMISTRY] Chromatographic procedure in which the stationary phase is a high-boiling liquid spread as a thin film on an inert support, and the mobile phase is a vaporous mixture of the components to be separated in an inert carrier gas. { pär'tish·ən ,krō mə'täg·rə·fē }

partition coefficient [ANALYTICAL CHEMISTRY] In the equilibrium distribution of a solute between two liquid phases, the constant ratio of the solute's concentration in the upper phase to its concentration in the lower phase. Symbolized K. { pär'tish·ən ,kō·i,fish·ənt }

parylene [ORGANIC CHEMISTRY] Polyparaxylylene, used in ultrathin plastic films for capacitor dielectrics, and as a pore-free coating. { 'par·ə,lēn }

PAS [SPECTROSCOPY] *See para*-aminosalicylic acid; photoacoustic spectroscopy.

Pascal rules [PHYSICAL CHEMISTRY] Rules which give the diamagnetic susceptiblity of a complex molecule in terms of the sum of the susceptibilities of its constituent atoms, and a correction factor which depends on the type of bonds linking the atoms. { pa'skal ,rülz }

Paschen-Back effect [SPECTROSCOPY] An effect on spectral lines obtained when the light source is placed in a very strong magnetic field; the anomalous Zeeman effect obtained with weaker fields changes over to what is, in a first approximation, the normal Zeeman effect. { 'päsh·ən 'bäk i,fekt }

Paschen-Runge mounting [SPECTROSCOPY] A diffraction grating mounting in which the slit and grating are fixed, and photographic plates are clamped to a fixed track running along the corresponding Rowland circle. { 'päsh·ən 'rüŋ·ə ,maůnt·iŋ }

Paschen series [SPECTROSCOPY] A series of lines in the infrared spectrum of atomic hydrogen whose wave numbers are given by $R_H [(1/9) - (1/n^2)]$, where R_H is the Rydberg constant for hydrogen, and n is any integer greater than 3. { 'päsh·ən ,sir·ēz }

passiflorin [ORGANIC CHEMISTRY] *See* harman. { ¦pas·ə¦flór·ən }

passivation potential [PHYSICAL CHEMISTRY] The potential corresponding to the critical anodic current density of an electrode which behaves in an active-passive manner. { ˌpas·əˈvā·shən pəˌten·chəl }

passivity [CHEMISTRY] A state of chemical inactivity, especially of a metal that is relatively resistant to corrosion due to loss of chemical activity. { pəˈsiv·əd·ē }

Pasteur's salt solution [ANALYTICAL CHEMISTRY] Laboratory reagent consisting of potassium phosphate and calcium phosphate, magnesium sulfate, and ammonium tartrate in distilled water. { paˈstərz ˈsȯlt səˌlü·shən }

Pauling scale [PHYSICAL CHEMISTRY] A numerical scale of electronegativities based on bond-energy calculations for different elements joined by covalent bonds. { ˈpȯl·iŋ ˌskāl }

Pavy's solution [ANALYTICAL CHEMISTRY] Laboratory reagent used to determine the concentration of sugars in solution by color titration; contains copper sulfate, sodium potassium tartrate, sodium hydroxide, and ammonia in water solution. { ˈpā·vēz səˌlü·shən }

Pb [CHEMISTRY] *See* lead.

p-block elements [CHEMISTRY] Elements of the main groups III-VII and 0 in the periodic table whose outer electronic configurations have occupied *p* levels. { ¦pē ˈbläk ˈel·əˌmənts }

P-branch [SPECTROSCOPY] A series of lines in molecular spectra that correspond, in the case of absorption, to a unit decrease in the rotational quantum number J. { ˈpē ˌbranch }

PCB [ORGANIC CHEMISTRY] *See* polychlorinated biphenyl.

PCNB [ORGANIC CHEMISTRY] *See* pentachloronitrobenzene.

Pd [CHEMISTRY] *See* palladium.

PDMS [SPECTROSCOPY] *See* plasma desorption mass spectrometry.

peacock blue [ORGANIC CHEMISTRY] $HSO_3C_6H_4COH[C_6H_4N(C_2H_5)CH_2C_6H_4SO_3Na]_2$ Blue pigment used in inks for multicolor printing. { ˈpēˌkäk ˈblü }

peak analysis [SPECTROSCOPY] Determination of the relevant peak parameters, such as position or area, from a spectogram. { ˈpēk əˌnal·ə·səs }

peak area [ANALYTICAL CHEMISTRY] The area enclosed between the peak and the base line on a spectrogram or chromatogram. { ˈpēk ˌer·ē·ə }

peak enthalpimetry [ANALYTICAL CHEMISTRY] A thermochemical analytical procedure applicable to biochemical and chemical analyses; the salient feature is rapid mixing of a reagent stream with an isothermal solvent stream into which discrete samples are intermittently injected; peak enthalpograms result which exhibit the response characteristics of genuine differential detectors. { ˈpēk ¦en·thəlˈpim·ə·trē }

peak width [ANALYTICAL CHEMISTRY] In a gas chromatogram (plot of eluent rise and fall versus time), the width of the base (time duration) of a symmetrical peak (rise and fall) of eluent. { ˈpēk ˈwidth }

pearl hardening [INORGANIC CHEMISTRY] Commercial name for a crystallized grade of calcium sulfate; used as a paper filler. { ˈpərl ¦härd·ən·iŋ }

pelargonic acid [ORGANIC CHEMISTRY] $CH_3(CH_2)_7CO_2H$ A colorless or yellowish oil, boiling at 254°C; soluble in ether and alcohol, insoluble in water; used as a chemical intermediate and flotation agent, in lacquers, pharmaceuticals, synthetic flavors and aromas, and plastics. { ¦peˌlärˈgän·ik ˈas·əd }

pellet technique [ANALYTICAL CHEMISTRY] *See* potassium bromide-disk technique. { ˈpel·ət tekˌnēk }

pellicular resins [ANALYTICAL CHEMISTRY] Glass spheres coated with a thin layer of ion-exchange resin, used in liquid chromatography. { pəˈlik·yə·lər ˈrez·ənz }

pellotine [ORGANIC CHEMISTRY] $C_{13}H_{19}O_3N$ A colorless, crystalline alkaloid, derived from the dried cactus pellote, *Lophophora williamsi* (Mexico), slightly soluble in water; used as a hypnotic. { ˈpel·əˌtēn }

pentabasic [CHEMISTRY] A description of a molecule that has five hydrogen atoms that may be replaced by metals or bases. { ¦pen·təˈbā·sik }

pentaborane [INORGANIC CHEMISTRY] B_5H_9 Flammable liquid boiling at 48°C; ignites spontaneously in air; proposed as high-energy fuel for aircraft and missiles. { ¦pen·təˈbȯrˌān }

pentachloride [CHEMISTRY] A molecule containing five atoms of chlorine in its structure. { ¦pen·tə'klȯr,īd }

pentachloroethane [ORGANIC CHEMISTRY] $CHCl_2CCl_3$ Colorless, water-insoluble liquid, boiling at 159°C; used as a solvent to degrease metals. Also known as pentalin. { ¦pen·tə,klȯr·ō'eth,ān }

pentachloronitrobenzene [ORGANIC CHEMISTRY] $C_6Cl_5NO_2$ Cream-colored crystals with a melting point of 142-145°C; slightly soluble in alcohols; used as a fungicide and herbicide. Abbreviated PCNB. Also known as quintozene; terrachlor. { ¦pen·tə¦klȯr·ō,nī·trə'ben,zēn }

pentachlorophenol [ORGANIC CHEMISTRY] C_6Cl_5OH A toxic white powder, decomposing at 310°C, melting at 190°C; soluble in alcohol, acetone, ether, and benzene; used as a fungicide, bactericide, algicide, herbicide, and chemical intermediate. { ¦pen·tə¦klȯr·ō·'fē,nȯl }

pentacosane [ORGANIC CHEMISTRY] $C_{25}H_{52}$ A water-insoluble hydrocarbon derived from beeswax. { ,pen·tə'kō,sān }

pentadecane [ORGANIC CHEMISTRY] $C_{15}H_{32}$ A colorless, water-insoluble liquid, boiling at 270.5°C; soluble in alcohol; used as a chemical intermediate. { ¦pen·tə¦de,kān }

pentadecanolide [ORGANIC CHEMISTRY] $C_{15}H_{18}O_2$ A colorless liquid with a musky odor extracted from angelica oil; soluble in 90% ethyl alcohol in equal volume; used in perfumes. { ¦pen·tə·də'kan·əl,īd }

pentadentate ligand [INORGANIC CHEMISTRY] A chelating agent having five groups capable of attachment to a metal ion. Also known as quinquidentate ligand. { ¦pen·tə¦den,tāt 'lī·gənd }

pentadiene [ORGANIC CHEMISTRY] C_5H_8 Any of several straight-chain liquid diolefins: $CH_3CH_2CH{=}C{=}CH_2$, a colorless liquid boiling at 45°C, also known as ethylallene; $CH_3CH{=}CHCH{=}CH_2$, a colorless liquid boiling at 43°C; $CH_2{=}CHCH_2CH{=}CH_2$, a colorless liquid boiling at 26°C. { ¦pen·tə'dī,ēn }

pentaerythritol [ORGANIC CHEMISTRY] $(CH_2OH)_4C$ A white crystalline solid, melting at 261-262°C; moderately soluble in cold water, freely soluble in hot water; used to make the explosive pentaerythritol tetranitrate (PETN) and in the manufacture of alkyol resins and other coating compounds. { ¦pen·tə·ə'rith·rə,tȯl }

pentaerythritol tetranitrate [ORGANIC CHEMISTRY] $C(CH_2ONO_2)_4$ A white crystalline compound, melting at 139°C; explodes at 205-215°C; soluble in acetone, insoluble in water; used in medicines and explosives. Also known as penthrite; PETN. { ¦pen·tə·ə'rith·rə,tȯl ,te·trə'nī,trāt }

pentaerythritol tetrastearate [ORGANIC CHEMISTRY] $C(CH_2OOCC_{17}H_{35})_4$ A hard, ivory-colored wax with a softening point of 67°C; used in polishes and textile finishes. { ¦pen·tə·ə'rith·rə,tȯl ,te·trə'stir,āt }

pentafluoride [CHEMISTRY] A chemical compound onto which five fluoride atoms are bonded. { ,pen·tə'flu̇r,īd }

pentalin [ORGANIC CHEMISTRY] *See* pentachloroethane. { 'pent·əl·ən }

***n*-pentane** [ORGANIC CHEMISTRY] $CH_3(CH_2)_3CH_3$ A colorless, flammable, water-insoluble hydrocarbon liquid, freezing at −130°C, boiling at 36°C; soluble in hydrocarbons and ethers; used as a chemical intermediate, solvent, and anesthetic. { ¦en 'pen,tān }

1,5-pentanediol [ORGANIC CHEMISTRY] $HOCH_2(CH_2)_3CH_2OH$ Colorless, water-miscible liquid boiling at 242.5°C; used as a hydraulic fluid, lube-oil additive, and antifreeze, and in manufacture of polyester and polyurethane resins. { ¦wən ¦fīv ,pen,tān'dī,ȯl }

pentane insolubles [ANALYTICAL CHEMISTRY] Insoluble matter that can be separated from used lubricating oil in solution in *n*-pentane; may include resinous bitumens produced from the oxidation of oil and fuel; used in an American Society for Testing and Material test. { 'pen,tān in'säl·yə·bəlz }

pentanol [ORGANIC CHEMISTRY] $C_5H_{11}OH$ A toxic organic alcohol; 1-pentanol is *n*-amyl alcohol, primary; 2-pentanol is methylpropylcarbinol; 3-pentanol is diethylcarbinol; *tert*-pentanol is *tert*-amyl alcohol; pentanols are used in pharmaceuticals, as chemical intermediates, and as solvents. { 'pen·tə,nȯl }

pentanone [ORGANIC CHEMISTRY] Either of two isomeric ketones derived from pentane:

$CH_3COC_3H_7$ is a flammable, colorless, clear liquid, a mixture of methyl propyl and diethyl ketones; insoluble in water, soluble in ether and alcohol; used as a solvent. $C_2H_5COC_2H_5$ is a colorless, flammable liquid with acetone aroma, boiling at 101°C; soluble in alcohol and ether; used in medicine and organic synthesis. { 'pen·tə,nōn }

pentavalent [CHEMISTRY] An atom or radical that exhibits a valency of 5. { ¦pen·tə 'vā·lənt }

pentene [ORGANIC CHEMISTRY] C_5H_{12} Colorless, flammable liquids derived from natural gasoline; isomeric forms are α-*n*-amylene and β-*n*-amylene. { 'pen,tēn }

penthrite [ORGANIC CHEMISTRY] *See* pentaerythritol tetranitrate. { 'pen,thrīt }

pentoglycerine [ORGANIC CHEMISTRY] *See* trimethylolethane. { ¦pen·tō'glis·ə·rən }

pentoxide [INORGANIC CHEMISTRY] A compound that is binary and has five atoms of oxygen; for example, phosphorus pentoxide, P_2O_5. { pen'täk,sīd }

pentyl [ORGANIC CHEMISTRY] *See* amyl. { 'pent·əl }

***para*-pentyloxyphenol** [ORGANIC CHEMISTRY] $C_{11}H_{16}O_2$ Compound melting at 49-50°C; used as a bactericide. { ¦par·ə ¦pent·əl¦äk·sē'fē,nól }

pentyne [ORGANIC CHEMISTRY] C_5H_8 Either of two normal isometric acetylene hydrocarbons: $HC{\equiv}C(CH_2)_2CH_3$, colorless liquid boiling at 40°C, also known as pentine, propylacetylene; $CH_3C{\equiv}CC_2H_5$, liquid boiling at 56°C. { 'pen,tēn }

peppermint camphor [ORGANIC CHEMISTRY] *See* menthol. { 'pep·ər,mint 'kam·fər }

peptide bond [ORGANIC CHEMISTRY] A bond in which the carboxyl group of one amino acid is condensed with the amino group of another to form a —CO · NH— linkage. Also known as peptide linkage. { 'pep,tīd ,bänd }

peptide linkage [ORGANIC CHEMISTRY] *See* peptide bond. { 'pep,tīd ,liŋ·kij }

peptization [CHEMISTRY] **1.** Aggregation in which a hydrophobic colloidal sol is stabilized by the addition of electrolytes (peptizing agents) which are adsorbed on the particle surfaces. **2.** Liquefaction of a substance by trace amounts of another substance. { ,pep·tə,zā·shən }

per- [CHEMISTRY] Prefix meaning: **1.** Complete, as in hydrogen peroxide. **2.** Extreme, or the presence of the peroxy (—O—O—) group. **3.** Exhaustive (complete) substitution, as in perchloroethylene. { pər *or* per }

peracetic acid [ORGANIC CHEMISTRY] CH_3COOOH A toxic, colorless liquid with strong aroma; boils at 105°C; explodes at 110°C; miscible with water, alcohol, glycerin, and ether; used as an oxidizer, bleach, catalyst, bactericide, fungicide, epoxy-resin precursor, and chemical intermediate. Also known as peroxyacetic acid. { ¦pər·ə¦sēd·ik 'as·əd }

peracid [CHEMISTRY] Acid containing the peroxy (—O—O—) group, such as peracetic acid or perchloric acid. { ¦pər'as·əd }

peralcohol [ORGANIC CHEMISTRY] Chemical compound containing the peroxy group (—O—O—), such as peracetic acid and perchromic acid. { pər'al·kə,hól }

perbenzoic acid [ORGANIC CHEMISTRY] $C_6H_5CO_2OH$ A crystalline compound forming leaflets from benzene solution, melting at 41-45°C, freely soluble in organic solvents; used in analysis of unsaturated compounds and to change ethylinic compounds into oxides. { ¦pər·ben¦zō·ik 'as·əd }

perchlorate [INORGANIC CHEMISTRY] A salt of perchloric acid containing the ClO_4^- radical; for example, potassium perchlorate, $KClO_4$. { pər'klór,āt }

perchloric acid [INORGANIC CHEMISTRY] $HClO_4$ Strongly oxidizing, corrosive, colorless, hygroscopic liquid, boiling at 16°C (8 mmHg, or 1067 pascals); soluble in water; unstable in pure form, but stable when diluted in water; used in medicine, electrolytic baths, electropolishing, explosives, and analytical chemistry, and as a chemical intermediate. Also known as Fraude's reagent. { pər'klór·ik 'as·əd }

perchloroethylene [ORGANIC CHEMISTRY] CCl_2CCl_2 Stable, colorless liquid, boiling at 121°C; nonflammable and nonexplosive, with low toxicity; used as a dry-cleaning and industrial solvent, in pharmaceuticals and medicine, and for metal cleaning. Also known as tetrachloroethylene. { pər¦klór·ō'eth·ə,lēn }

perchloromethyl mercaptan [ORGANIC CHEMISTRY] $ClSCCl_3$ Poisonous, yellow oil with disagreeable aroma; decomposes at 148°C; used as a chemical intermediate, granary fumigant, and military poison gas. { pər¦klór·ō'meth·əl mər'kap,tan }

perchloryl fluoride [INORGANIC CHEMISTRY] $ClFO_3$ A colorless gas with a sweet odor; boiling point is −46.8°C and melting point is −146°C; used as an oxidant in rocket fuels. { pər'klȯr·əl 'flu̇r‚īd }

perfect fractionation path [PHYSICAL CHEMISTRY] On a phase diagram, a line or a path representing a crystallization sequence in which any crystal that has been formed remains inert, that is, its composition is not altered. { 'pər·fikt ‚frak·shə'nā·shən ‚path }

perfectly mobile component [PHYSICAL CHEMISTRY] A component whose quantity in a system is determined by its externally imposed chemical potential rather than by its initial quantity in the system. Also known as boundary value component. { 'pər·fik·lē 'mō·bəl kəm'pō·nənt }

perfect solution [PHYSICAL CHEMISTRY] A solution that is ideal throughout its entire compositional range. { 'pər·fikt sə'lü·shən }

perfluorochemical [ORGANIC CHEMISTRY] A hydrocarbon in which all the hydrogen atoms have been replaced by fluorine. { pər¦flu̇r·ō'kem·ə·kəl }

perhydro- [ORGANIC CHEMISTRY] Prefix designating a completely saturated aromatic compound, as for decalin ($C_{10}H_{18}$), also known as perhydronaphthalene. { ¦pər'hī·drō }

pericondensed polycyclic [ORGANIC CHEMISTRY] Referring to an aromatic compound in which three or more rings share common carbon atoms. { ¦per·ə·kən'denst ¦päl·ē¦sī·klik }

pericyclic reaction [ORGANIC CHEMISTRY] Any one of a group of reactions that involve conjugated polyenes and proceed by single-step (concerted) reaction mechanisms. { ‚per·ə¦sīk·lik rē'ak·shən }

period [CHEMISTRY] A family of elements with consecutive atomic numbers in the periodic table and with closely related properties; for example, chromium through copper. { 'pir·ē·əd }

periodate [INORGANIC CHEMISTRY] A salt of periodic acid, HIO_4, for example, potassium periodate, KIO_4. { ¦pər'ī·ə‚dāt }

periodic acid [INORGANIC CHEMISTRY] $HIO_4 \cdot 2H_2O$ Water- and alcohol-soluble white crystals; loses water at 100°C; used as an oxidant. { ¦pir·ē¦äd·ik 'as·əd }

periodic law [CHEMISTRY] The law that the properties of the chemical elements and their compounds are a periodic function of their atomic weights. { ¦pir·ē¦äd·ik ¦lȯ }

periodic table [CHEMISTRY] A table of the elements, written in sequence in the order of atomic number or atomic weight and arranged in horizontal rows (periods) and vertical columns (groups) to illustrate the occurrence of similarities in the properties of the elements as a periodic function of the sequence. { ¦pir·ē¦äd·ik 'tā·bəl }

peritectic [PHYSICAL CHEMISTRY] An isothermal reversible reaction in which a liquid phase reacts with a solid phase during cooling to produce a second solid phase. { ¦per·ə¦tek·tik }

peritectic point [PHYSICAL CHEMISTRY] In a binary two-phase heteroazeotropic system at constant pressure, that point up to which the boiling point has remained constant until one of the phases has boiled away. { ¦per·ə¦tek·tik ‚pȯint }

peritectoid [PHYSICAL CHEMISTRY] An isothermal reversible reaction in which a solid phase on cooling reacts with another solid phase to form a third solid phase. { ‚per·ə'tek‚tȯid }

Perkin reaction [ORGANIC CHEMISTRY] The formation of unsaturated cinnamic-type acids by the condensation of aromatic aldehydes with fatty acids in the presence of acetic anhydride. { 'pər·kən rē‚ak·shən }

permanent hardness [CHEMISTRY] The hardness of water persisting after boiling. { 'pər·mə·nənt 'härd·nəs }

permanent-press resin [ORGANIC CHEMISTRY] A thermosetting resin, based on chemicals such as formaldehyde and maleic anhydride, which is used to impart crease resistance to textiles and fibers. Also known as durable-press resin. { 'pər·mə·nənt ¦pres 'rez·ən }

permanganate [INORGANIC CHEMISTRY] A purple salt of permanganic acid containing the MnO_4^- radical; used as an oxidizing agent and a disinfectant. { ¦pər'maŋ·gə,nāt }

permanganic acid [INORGANIC CHEMISTRY] $HMnO_4$ An unstable acid that exists only in dilute solutions; decomposes to manganese dioxide and oxygen. { ¦pər·man 'gan·ik 'as·əd }

permeable membrane [CHEMISTRY] A thin sheet or membrane of material through which selected liquid or gas molecules or ions will pass, either through capillary pores in the membrane or by ion exchange; used in dialysis, electrodialysis, and reverse osmosis. { 'pər·mē·ə·bəl 'mem,brān }

permeametry [ANALYTICAL CHEMISTRY] Determination of the average size of fine particles in a fluid (gas or liquid) by passing the mixture through a powder bed of known dimensions and recording the pressure drop and flow rate through the bed. { ,pər·mē'äm·ə·trē }

permeant [CHEMISTRY] A material that permeates another material. { 'pər·mē·ənt }

permeation [CHEMISTRY] The movement of atoms, molecules, or ions into or through a porous or permeable substance (such as zeolite or a membrane). { ,pər·mē'ā·shən }

permselective membrane [PHYSICAL CHEMISTRY] An ion-exchange material that allows ions of one electrical sign to enter and pass through. { ¦pərm·si¦lek·tiv 'mem,brān }

peroxide [CHEMISTRY] **1.** A compound containing the peroxy (—O—O—) group, as in hydrogen peroxide. **2.** *See* hydrogen peroxide. { pə'räk,sīd }

peroxide number [ANALYTICAL CHEMISTRY] Measure of millimoles of peroxide (or milliequivalents of oxygen) taken up by 1000 grams of fat or oil; used to measure rancidity. Also known as peroxide value. { pə'räk,sīd ,nəm·bər }

peroxide value [ANALYTICAL CHEMISTRY] *See* peroxide number. { pə'räk,sīd ,val·yü }

peroxyacetic acid [ORGANIC CHEMISTRY] *See* peracetic acid. { pə¦räk·sē·ə¦sēd·ik 'as·əd }

peroxydol [INORGANIC CHEMISTRY] *See* sodium perborate. { pə'räk·sə,dòl }

Persian red [INORGANIC CHEMISTRY] Red pigment made from basic lead chromate or ferric oxide. { 'pər·zhən 'red }

persulfate [INORGANIC CHEMISTRY] Salt derived from persulfuric acid and containing the radical $S_2O_8^{2-}$; made by electrolysis of sulfate solutions. { ¦pər'səl,fāt }

persulfuric acid [INORGANIC CHEMISTRY] $H_2S_2O_8$ Acid formed in lead-cell batteries by electrolyzing sulfuric acid; strong oxidizing agent. { ¦pər·səl'fyu̇r·ik 'as·əd }

pervaporation [CHEMISTRY] A chemical separation technique in which a solution is placed in contact with a heated semipermeable membrane that selectively retains one of the components of a solution. { pər,vap·ə'rā·shən }

PET [ORGANIC CHEMISTRY] *See* polyethylene terephthalate. { ¦pē¦ē'tē or pet }

PETN [ORGANIC CHEMISTRY] *See* pentaerythritol tetranitrate.

petrochemicals [ORGANIC CHEMISTRY] Chemicals made from feedstocks derived from petroleum or natural gas; examples are ethylene, butadiene, most large-scale plastics and resins, and petrochemical sulfur. Also known as petroleum chemicals. { ¦pe·trō'kem·ə·kəlz }

petrochemistry [ORGANIC CHEMISTRY] The chemistry and reactions of materials derived from petroleum, natural gas, or asphalt deposits. { ¦pe·trō'kem·ə·strē }

petroleum chemicals [ORGANIC CHEMISTRY] *See* petrochemicals. { pə'trō·lē·əm ,kem·ə·kəlz }

petroleum resin [ORGANIC CHEMISTRY] Any one of a family of polymers produced from mixed unsaturated monomers recovered from petroleum processing streams. { pə'trō·lē·əm ,rez·ən }

Pfund series [SPECTROSCOPY] A series of lines in the infrared spectrum of atomic hydrogen whose wave numbers are given by $R_H[(1/25) - (1/n^2)]$, where R_H is the Rydberg constant for hydrogen, and n is any integer greater than 5. { 'fu̇nt ,sir·ēz }

pH [CHEMISTRY] A term used to describe the hydrogen-ion activity of a system; it is equal to $-\log a_{H^+}$; here a_{H^+} is the activity of the hydrogen ion; in dilute solution, activity is essentially equal to concentration and pH is defined as $-\log_{10}[H^+]$, where

H^+ is hydrogen-ion concentration in moles per liter; a solution of pH 0 to 7 is acid, pH of 7 is neutral, pH over 7 to 14 is alkaline. { pē'āch }

pharmacology [CHEMISTRY] The science dealing with the nature and properties of drugs, particularly their actions. { ˌfär·mə'käl·ə·jē }

phase [CHEMISTRY] Portion of a physical system (liquid, gas, solid) that is homogeneous throughout, has definable boundaries, and can be separated physically from other phases. { fāz }

phase diagram [PHYSICAL CHEMISTRY] A graphical representation of the equilibrium relationships between phases (such as vapor-liquid, liquid-solid) of a chemical compound, mixture of compounds, or solution. { 'fāz ˌdī·əˌgram }

phase equilibria [PHYSICAL CHEMISTRY] The equilibrium relationships between phases (such as vapor, liquid, solid) of a chemical compound or mixture under various conditions of temperature, pressure, and composition. { 'fāz ˌē·kwə'lib·rē·ə }

phase ratio [ANALYTICAL CHEMISTRY] In chromatography, the ratio of the volume of the mobile phase to that of the stationary phase in a chromatographic column. { 'fāz ˌrā·shō }

phase rule [PHYSICAL CHEMISTRY] *See* Gibbs phase rule. { 'fāz rül }

phase solubility [PHYSICAL CHEMISTRY] The different solubilities of a sample's solid constituents (phases) in a selected solvent. { 'fāz ˌsäl·yəˌbil·əd·ē }

phase-solubility analysis [ANALYTICAL CHEMISTRY] Solvent technique used to determine the amount and number of components in a solid substance; the weight of sample added to the solvent is plotted against the weight of sample dissolved, with breakpoints in the curve occurring with each progressive saturation of the solvent with respect to each of the components; can be combined with extraction and recrystallization procedures. { 'fāz ˌsäl·yəˌbil·əd·ē əˌnal·ə·səs }

phase titration [ANALYTICAL CHEMISTRY] Analysis of a binary mixture of miscible liquids by titrating with a third liquid that is miscible with only one of the components, using the ternary phase diagram to determine the end point. { 'fāz tīˌtrā·shən }

phase transfer catalysis [ORGANIC CHEMISTRY] Enhancement of the reaction rate of a two-phase organic-water system by addition of a catalyst which alters the rate of transfer of water-soluble reactant across the interface to the organic phase. { 'fāz ¦transˌfər kə'tal·ə·səs }

pH electrode [ANALYTICAL CHEMISTRY] Membrane-type glass electrode used as the hydrogen-ion sensor of most pH meters; the pH-response electrode surface is a thin membrane made of a special glass. { ˌpē'āch i'lekˌtrōd }

α-phellandrene [ORGANIC CHEMISTRY] $C_{10}H_{16}$ A colorless oil soluble in ether; boiling point of *d*-optical isomer is 66-68°C, of *l*-optical isomer is 58-59°C; used in flavoring and perfumes. { ¦al·fə fə'lanˌdrēn }

phenanthrene [ORGANIC CHEMISTRY] $C_{14}H_{10}$ A colorless, crystalline hydrocarbon; melts at about 100°C; the nucleus is produced by the degradation of certain alkaloids; used in the synthesis of dyes and drugs. { fə'nanˌthrēn }

phenanthroline [ORGANIC CHEMISTRY] $C_{12}H_8N_2$ Any of three nitrogen bases related to phenanthrene; the ortho form is an oxidation-reduction indicator, turning faint blue when oxidized. { fə'nan·thrəˌlēn }

phenanthroline indicator [ANALYTICAL CHEMISTRY] A sensitive, red-colored specific reagent for iron. { fə'nan·thrəˌlēn 'in·dəˌkād·ər }

phenarsazine chloride [ORGANIC CHEMISTRY] $C_{12}H_9AsClN$ A yellow, crystalline compound obtained as a precipitate from carbon tetrachloride solutions; it sublimes readily, and is slightly soluble in xylene, benzene, and carbon tetrachloride; used as a war gas. Also known as adamsite. { fə'när·səˌzēn 'klōrˌīd }

phenazine [ORGANIC CHEMISTRY] $C_6H_4N_2C_6H_4$ Yellow crystals, melting at 170°C; slightly soluble in water, soluble in alcohol and ether; used as chemical intermediate and to make dyes. { 'fen·əˌzēn }

phenethyl acetate [ORGANIC CHEMISTRY] $C_6H_5CH_2CH_2OOCCH_3$ A colorless liquid with a peachlike odor and a boiling point of 226°C; soluble in alcohol, ether, and fixed oils; used in perfumes. Also known as phenylethyl acetate. { fen'eth·əl 'as·əˌtāt }

phenethyl alcohol [ORGANIC CHEMISTRY] $C_8H_{10}O$ A liquid with a floral odor found in many natural essential oils; soluble in 50% alcohol; used in perfumes and flavors, and in medicine as an antibacterial agent in diseases of the eye. { fen′eth·əl ′al·kə‚hȯl }

phenethyl isobutyrate [ORGANIC CHEMISTRY] $(CH_3)_2CHCOOC_2H_4C_6H_5$ A colorless liquid, soluble in alcohol and ether; used in perfumes and flavoring. { fen′eth·əl ‚ī·sō′byüd·ə‚rāt }

phenetidine [ORGANIC CHEMISTRY] $NH_2C_6H_4OC_2H_5$ Either of two toxic, oily liquids that darken when exposed to light and air; soluble in alcohol, insoluble in water; the ortho form boils at 228-230°C, is used to make dyes, and is also known at 2-aminophenetole; the para form boils at 253-255°C, is used to make dyes and in pharmaceuticals. { fə′ned·ə‚dēn }

phenol [ORGANIC CHEMISTRY] **1.** C_6H_5OH White, poisonous, corrosive crystals with sharp, burning taste; melts at 43°C, boils at 182°C; soluble in alcohol, water, ether, carbon disulfide, and other solvents; used to make resins and weed killers, and as a solvent and chemical intermediate. Also known as carbolic acid; phenylic acid. **2.** A chemical compound based on the substitution product of phenol, for example, ethylphenol ($C_2H_4C_4H_5OH$), the ethyl substitute of phenol. { ′fē‚nȯl }

phenol coefficient [ANALYTICAL CHEMISTRY] Number scale for comparison of antiseptics, using the efficacy of phenol as unity. { ′fē‚nȯl ‚kō·i‚fish·ənt }

phenol-coefficient method [CHEMISTRY] A method for evaluating water-miscible disinfectants in which a test organism is added to a series of dilutions of the disinfectant; the phenol coefficient is the number obtained by dividing the greatest dilution of the disinfectant killing the test organism by the greatest dilution of phenol showing the same result. { ′fē‚nȯl ‚kō·i‚fish·ənt ‚meth·əd }

phenol-formaldehyde resin [ORGANIC CHEMISTRY] Thermosetting resin made by the reaction of phenol and formaldehyde; has good strength and chemical resistance and low cost; used as a molding material for mechanical and electrical parts. Originally known as Bakelite. { ′fē‚nȯl fər′mal·də‚hīd ‚rez·ən }

phenol-furfural resin [ORGANIC CHEMISTRY] A phenolic resin characterized by the ability to be fabricated by injection molding since it hardens after curing conditions are reached. { ′fē‚nȯl ′fər·fə‚ral ‚rez·ən }

phenolphthalein [ORGANIC CHEMISTRY] $(C_6H_4OH)_2COC_6H_4CO$ Pale-yellow crystals; soluble in alcohol, ether, and alkalies, insoluble in water; used as an acid-base indicator (carmine-colored to alkalies, colorless to acids) for titrations, as a laxative and dye, and in medicine. { ¦fē‚nȯl′thal·ē·ən }

phenol red [ORGANIC CHEMISTRY] *See* phenolsulfonphthalein. { ′fē‚nȯl ′red }

phenolsulfonic acid [ORGANIC CHEMISTRY] $C_6H_5SO_3H$ Water- and alcohol-soluble mixture of *ortho-* and *para-*phenolsulfonic acids; yellowish liquid that turns brown when exposed to air; used as a chemical intermediate and in water analysis. Also known as sulfocarbolic acid. { ¦fē‚nȯl·səl′fän·ik ′as·əd }

phenolsulfonphthalein [ORGANIC CHEMISTRY] $C_{19}H_{14}O_5S$ A bright-red, crystalline compound, soluble in water, alcohol, and acetone; used as a pH indicator, and to test for kidney function in dogs. Also known as phenol red. { ¦fē‚nȯl¦səl‚fōn·′thal·ē·ən }

phenothiazine [ORGANIC CHEMISTRY] $C_{12}H_9N$ A yellow, crystalline compound, forming rhomboid leaflets or diamond-shaped plates, obtained from toluene or butanol solution; soluble in hot acetic acid, benzene, and ether; used as an insecticide and in pharmaceutical manufacture. { ′fē·nə′thī·ə‚zēn }

phenotole [ORGANIC CHEMISTRY] $C_6H_5OC_2H_5$ Combustible, colorless liquid, boiling at 172°C; soluble in alcohol and ether, insoluble in water. { ′fē·nə‚tōl }

phenoxyacetic acid [ORGANIC CHEMISTRY] $C_6H_5OCH_2COOH$ A light tan powder with a melting point of 98°C; soluble in ether, water, carbon disulfide, methanol, and glacial acetic acid; used in the manufacture of pharmaceuticals, pesticides, fungicides, and dyes. { fə¦näk·sē·ə¦sēd·ik ′as·əd }

phenoxybenzamine hydrochloride [ORGANIC CHEMISTRY] $C_{18}H_{22}ONCl \cdot HCl$ White crystals, slightly soluble in water, melting at 139°C; used in medicine. { fə¦näk·sē′ben·zə‚mēn ‚hī·drə′klȯr‚īd }

2-phenoxyethanol [ORGANIC CHEMISTRY] $C_6H_5OCH_2CH_2OH$ An oily liquid with a faint aromatic odor; melting point is 14°C; soluble in water; used in perfumes as a fixative, in organic synthesis, as an insect repellent, and as a topical anesthetic. { ¦tü fə¦näk·sē'eth·ə‚nól }

phenoxypropanediol [ORGANIC CHEMISTRY] $C_9H_{12}O_3$ A white, crystalline solid with a melting point of 53°C; soluble in water, alcohol, glycerin, and carbon tetrachloride; used in medicine and as a plasticizer. { fə¦näk·sē‚prō·pān'dī‚ól }

phenoxy resin [ORGANIC CHEMISTRY] A high-molecular-weight thermoplastic polyether resin based on bisphenol-A and epichlorohydrin with bisphenol-A terminal groups; used for injection molding, extrusion, coatings, and adhesives. { fə'näk·sē 'rez·ən }

phentolamine hydrochloride [ORGANIC CHEMISTRY] $C_{17}H_{19}ON_2 \cdot HCl$ White, water-soluble crystals, melting at 240°C; a sympatholytic; used in medicine. { fen'täl·ə‚mēn ‚hī·drə'klór‚īd }

phenyl [ORGANIC CHEMISTRY] $C_6H_5—$ A functional group consisting of a benzene ring from which a hydrogen has been removed. { 'fen·əl }

phenylacetaldehyde [ORGANIC CHEMISTRY] C_8H_8O A colorless liquid with a boiling point of 193-194°C; soluble in ether and fixed oils; used in perfumes and flavoring. Also known as α-toluic aldehyde. { 'fen·əl‚as·ə'täl·də‚hīd }

phenylacetic acid [ORGANIC CHEMISTRY] $C_8H_8O_2$ White crystals with a boiling point of 262°C; soluble in alcohol and ether; used in perfumes, medicine, and flavoring and in the manufacture of penicillin. Also known as α-toluic acid. { ¦fen·əl·ə'sēd·ik 'as·əd }

phenylaniline [ORGANIC CHEMISTRY] *See* diphenylamine. { ¦fen·əl'an·ə·lən }

***N*-phenylanthranilic acid** [ORGANIC CHEMISTRY] $(C_6H_5NH)C_6H_4COOH$ A crystalline compound, soluble in hot alcohol; decomposes at 183-184°C; used to detect vanadium in steel. { ¦en ¦fen·əl‚an·thrə¦nil·ik 'as·əd }

phenylbenzene [ORGANIC CHEMISTRY] *See* biphenyl. { ¦fen·əl'ben‚zēn }

phenylbutazone [ORGANIC CHEMISTRY] $C_{19}H_{20}O_2N_2$ White or light-yellow powder with aromatic aroma and bitter taste; melts at 107°C; slightly soluble in water, soluble in acetone; used in medicine as an analgesic and antipyretic. Also known as butazolidine. { ¦fen·əl'byüd·ə‚zōn }

phenyl cyanide [ORGANIC CHEMISTRY] *See* benzonitrile. { ¦fen·əl 'sī·ə‚nīd }

phenylcyclohexane [ORGANIC CHEMISTRY] $C_{12}H_{16}$ A colorless, oily liquid with a boiling point of 237.5°C; soluble in alcohol, benzene, castor oil, carbon tetrachloride, xylene, and hexane; used as a high-boiling solvent and a penetrating agent. { ¦fen·əl‚sī·klō'hek‚sān }

phenyldichloroarsine [ORGANIC CHEMISTRY] $C_6H_5AsCl_2$ A liquid which becomes a microcrystalline mass at −20°C (melting point) and decomposes in water; soluble in alcohol, ether, and benzene; used as a poison gas. { ¦fen·əl dī‚klór·ō'är‚sēn }

phenyl diglycol carbonate [ORGANIC CHEMISTRY] $C_{18}H_{18}O_7$ A colorless solid with a melting point of 40°C; soluble in organic solvents; used as a plasticizer. { ¦fen·əl dī'glī‚kól 'kär·bə‚nāt }

phenylene blue [ORGANIC CHEMISTRY] *See* indamine. { 'fən·əl‚ēn 'blü }

phenylenediamine [ORGANIC CHEMISTRY] $C_6H_4(NH_2)_2$ Also known as diaminobenzene. Any of three toxic isomeric crystalline compounds that are diamino derivatives of benzene; the ortho form, toxic colorless crystals melting at 102-104°C and soluble in alcohol, ether, water, and chloroform, is used to manufacture dyes, in photographic developers, and as a chemical intermediate; the meta form, colorless crystals unstable in air, melting at 63°C, and soluble in alcohol, ether, and water, is used to manufacture dyes, in textile dyeing, and as a nitrous acid detector; the para form, white to purple crystals melting at 147°C, soluble in alcohol and ether, and irritating to the skin, is used to manufacture dyes, in chemical analysis, and in photographic developers. { ¦fen·əl‚ēn'dī·ə‚mēn }

phenyl ether [ORGANIC CHEMISTRY] *See* diphenyl oxide. { 'fen·əl 'eth·ər }

phenylethyl acetate [ORGANIC CHEMISTRY] *See* phenethyl acetate. { ¦fen·əl¦eth·əl 'as·ə‚tāt }

phenylethylene [ORGANIC CHEMISTRY] *See* styrene. { ¦fen·əl'eth·ə‚lēn }

phenyl fluoride [ORGANIC CHEMISTRY] *See* fluorobenzene. { 'fen·əl 'flür‚īd }

***N*-phenylglycine** [ORGANIC CHEMISTRY] $C_6H_5NHCH_2COOH$ A crystalline compound, moderately soluble in water, melting at 127-128°C; used in dye manufacture (indigo). { ¦en ¦fen·əl'glī,sēn }

phenylglyoxylonitriloxime *O,O*-diethyl phosphorothioate [ORGANIC CHEMISTRY] $(H_5C_2O)_2PSONCCNC_6H_5$ A yellow liquid with a boiling point of 102°C at 0.01 mmHg (1.333 pascals); solubility in water is 7 parts per million at 20°C; used as an insecticide for stored products. Also known as phoxim. { ¦fen·əl·glī¦äk·sē¦län·ə·trəl'äk,sēm ¦ō¦ō dī'eth·əl ,fäs·fə·rō'thī·ə,wāt }

phenylhydrazine [ORGANIC CHEMISTRY] $C_6H_5NHNH_2$ Poisonous, oily liquid, boiling at 244°C; soluble in alcohol, ether, chloroform, and benzene, slightly soluble in water; used in analytical chemistry to detect sugars and aldehydes, and as a chemical intermediate. Also known as hydrazinobenzene. { ¦fen·əl'hī·drə,zēn }

phenyl ketone [ORGANIC CHEMISTRY] *See* benzophenone. { 'fen·əl 'kē,tōn }

phenyl mercaptan [ORGANIC CHEMISTRY] *See* thiophenol. { 'fen·əl mər'kap,tan }

phenylmercuric acetate [ORGANIC CHEMISTRY] $C_8H_8O_2Hg$ White to cream-colored prisms with a melting point of 148-150°C; soluble in alcohol, benzene, and glacial acetic acid; used as an antiseptic, fungicide, herbicide, and mildewcide. { ¦fen·əl·mər'kyu̇r·ik 'as·ə,tāt }

phenylmercuric chloride [ORGANIC CHEMISTRY] C_6H_5HgCl White crystals with a melting point of 251°C; soluble in benzene and ether; used as an antiseptic and fungicide. { ¦fen·əl·mər'kyu̇r·ik 'klȯr,īd }

phenylmercuric hydroxide [ORGANIC CHEMISTRY] C_6H_5HgOH White to cream-colored crystals with a melting point of 197-205°C; soluble in acetic acid and alcohol; used as a fungicide, germicide, and alcohol denaturant. { ¦fen·əl·mər'kyu̇r·ik hi'dräk,sīd }

phenylmercuric oleate [ORGANIC CHEMISTRY] $C_{41}H_{21}O_2Hg$ A white, crystalline powder with a melting point of 45°C; soluble in organic solvents; used in paints as a mildewproofing agent, and as a fungicide. { ¦fen·əl·mər'kyu̇r·ik 'ō·lē,āt }

phenylmercuric propionate [ORGANIC CHEMISTRY] $C_9H_{10}O_2Hg$ A white, waxlike powder with a melting point of 65-70°C; used in paints as a fungicide and bactericide. { ¦fen·əl·mər'kyu̇r·ik 'prō·pē·ə,nāt }

phenylmercuriethanolammonium acetate [ORGANIC CHEMISTRY] $C_{10}H_{15}O_3NHg$ A white, water-soluble, crystalline solid; used as an insecticide and fungicide. { ¦fen·əl·mər¦kyu̇r·ik¦eth·ə,nȯl·ə'mō·nē·əm 'as·ə,tāt }

phenylmethane [ORGANIC CHEMISTRY] *See* toluene. { ¦fen·əl'meth,ān }

phenylmethanol [ORGANIC CHEMISTRY] *See* benzyl alcohol. { ¦fen·əl'meth·ə,nȯl }

phenylmethyl acetate [ORGANIC CHEMISTRY] *See* benzyl acetate. { ¦fen·əl¦meth·əl 'as·ə,tāt }

***N*-phenylmorpholine** [ORGANIC CHEMISTRY] $C_{10}H_{13}NO$ A white, water-soluble solid with a melting point of 57°C; used in the manufacture of dyestuffs, corrosion inhibitors, and photographic developers, and as an insecticide. { ¦en ¦fen·əl'mȯr·fə,lēn }

phenyl mustard oil [ORGANIC CHEMISTRY] C_6H_5NCS A pale yellow or colorless liquid with a boiling point of 221°C; soluble in alcohol and ether; used in medicine. { 'fen·əl 'məs·tərd ,ȯil }

phenylphenol [ORGANIC CHEMISTRY] $C_6H_5C_6H_4OH$ Almost white crystals, soluble in alcohol, insoluble in water; the ortho form, melting at 56-58°C, is used to manufacture dyes, as germicide and fungicide, and in the rubber industry, and is also known as 2-hydroxybiphenyl, *ortho*-xenol; the para form, melting at 164-165°C, is used to manufacture dyes, resins, and rubber chemicals, and as a fungicide. { ¦fen·əl'fē,nȯl }

***N*-phenylpiperazine** [ORGANIC CHEMISTRY] $C_{10}H_{14}N_2$ A pale yellow oil with a boiling point of 286.5°C; soluble in alcohol and ether; used for pharmaceuticals and in the manufacture of synthetic fibers. { ¦en ¦fen·əl·pī'per·ə,zēn }

phenylpropane [ORGANIC CHEMISTRY] *See* propyl benzene. { ¦fen·əl'prō,pān }

1-phenyl-1-propanol [ORGANIC CHEMISTRY] $C_6H_5CH(OH)CH_2CH_3$ An oily liquid that has a weak esterlike odor; miscible with methanol, ethanol, ether, benzene, and toluene; used in industry as a heat transfer medium, in the manufacture of perfumes, and as a choleretic in medicine. { ¦wən ¦fen·əl ¦wən 'prō·pə,nȯl }

phenylpropyl alcohol [ORGANIC CHEMISTRY] $C_9H_{12}O$ A colorless liquid with a floral odor and a boiling point of 219°C; soluble in 70% alcohol; used in perfumes and flavoring. Also known as hydrocinnamic alcohol. { ¦fen·əl¦prō·pəl 'al·kə,hȯl }

phenylpropyl aldehyde [ORGANIC CHEMISTRY] $C_9H_{10}O$ A colorless liquid with a floral odor; soluble in 50% alcohol; used in perfumes and flavoring. Also known as hydrocinnamic aldehyde. { ¦fen·əl¦prō·pəl 'al·də,hīd }

1-phenyl-3-pyrazolidinone [ORGANIC CHEMISTRY] $C_9H_{10}N_2O$ A crystalline compound soluble in dilute aqueous solutions of acids and alkalies; melting point is 121°C; used as a high-contrast photographic developer. { ¦wən ¦fen·əl ¦thrē ,pī·rə·zə'lid·ən,ōn }

phenyl salicylate [ORGANIC CHEMISTRY] *See* salol. { ¦fen·əl sə'lis·ə,lāt }

phenylthiourea [ORGANIC CHEMISTRY] $C_6H_5NHCSNH_2$ A crystalline compound that has either a bitter taste or is tasteless, depending on the heredity of the taster; used in human genetics studies. { ¦fen·əl¦thī·ō·yu̇'rē·ə }

***ortho*-phenyl tolyl ketone** [ORGANIC CHEMISTRY] $CH_3C_6H_4COC_6H_5$ An oily liquid with a boiling point of 309-311°C; soluble in alcohol, oils, and organic solvents; used as a fixative in perfumery. { ¦ȯr·thō ¦fen·əl 'tä,lil 'kē,tōn }

phloridzin [ORGANIC CHEMISTRY] $C_{21}H_{24}O_{10} \cdot 2H_2O$ A glycoside extracted from the root bark of apple, plum, and pear trees; white needles with a melting point of 109°C, soluble in alcohol and hot water; used in medicine. { flə'rid·zən }

phloroglucinol [ORGANIC CHEMISTRY] $C_6H_3(OH)_3 \cdot 2H_2O$ White to yellow crystals with a melting point of 212-217°C when heated rapidly and 200-209°C when heated slowly; soluble in alcohol and ether; used as a bone decalcifying agent, as a floral preservative, and in the manufacture of pharmaceuticals. { ¦flȯr·ə'glüs·ən,ȯl }

pH measurement [ANALYTICAL CHEMISTRY] Determination of the hydrogen-ion concentration in an ionized solution by means of an indicator solution (such as phenolphthalein) or a pH meter. { ,pē'āch ,mezh·ər·mənt }

phorate [ORGANIC CHEMISTRY] $C_7H_{17}O_2PS_2$ A clear liquid with slight solubility in water; used as an insecticide for a wide range of insects on a wide range of crops. { 'fȯr,āt }

phosgene [ORGANIC CHEMISTRY] $COCl_2$ A highly toxic, colorless gas that condenses at 0°C to a fuming liquid; used as a war gas and in manufacture of organic compounds. { 'fäz,jēn }

phosphate [CHEMISTRY] **1.** Generic term for any compound containing a phosphate group (PO_4^{3-}), such as potassium phosphate, K_3PO_4. **2.** Generic term for a phosphate-containing fertilizer material. { 'fä,sfāt }

phosphate anion [INORGANIC CHEMISTRY] PO_4^{3-} The negative ion of phosphoric acid. { 'fä,sfāt 'an,ī·ən }

phosphate buffer [ANALYTICAL CHEMISTRY] Laboratory pH reference solution made of KH_2PO_4 and Na_2HPO_4; when 0.025 molal (equimolal of the potassium and sodium salts), the pH is 6.865 at 25°C. { 'fä,sfāt 'bəf·ər }

phosphide [INORGANIC CHEMISTRY] Binary compound of trivalent phosphorus, as in Na_3P. { 'fä,sfīd }

phosphine [INORGANIC CHEMISTRY] PH_3 Poisonous, colorless, spontaneously flammable gas with garlic aroma; soluble in alcohol, slightly soluble in cold water; boils at −85°C; used in organic reactions. Also known as hydrogen phosphide; phosphoretted hydrogen. { 'fä,sfēn }

phosphinic acid [ORGANIC CHEMISTRY] Organic derivative of hypophosphorous acid; contains the radical $—H_2PO_2$ or $=HPO_2$; examples are methylphosphinic acid, CH_3HPOOH, and dimethyl phosphinic acid, $(CH_3)_2POOH$. { fä'sfin·ik 'as·əd }

phosphite [INORGANIC CHEMISTRY] Salt of phosphorous acid; contains the radical PO_3^{3-}; an example is normal sodium phosphite, Na_3PO_3. { 'fä,sfīt }

phospholan [ORGANIC CHEMISTRY] $C_6H_{14}O_3PNS_2$ A colorless to yellow solid with a melting point of 37-45°C; used as an insecticide and miticide for cotton. { 'fä·sfə,lan }

phosphomolybdic acid [INORGANIC CHEMISTRY] $H_3PO_4 \cdot 12MoO_3 \cdot xH_2O$ Yellow crystals; soluble in alcohol, ether, and water; used as an alkaloid reagent and a pigment. Abbreviated PMA. { ¦fä·sfō·mə'lib·dik 'as·əd }

phosphonic acid [ORGANIC CHEMISTRY] $ROP(OH)_2$, where R is an organic radical such as $C_6H_5^-$, as in phenylphosphonic acid. { fä'sfän·ik 'as·əd }

phosphoric acid [INORGANIC CHEMISTRY] H_3PO_4 Water-soluble, transparent crystals, melting at 42°C; used as a fertilizer, in soft drinks and flavor syrups, pharmaceuticals, water treatment, and animal feeds and to pickle and rust-proof metals. Also known as orthophosphoric acid. { fä'sfór·ik 'as·əd }

phosphoric anhydride [INORGANIC CHEMISTRY] P_2O_5 A flammable, dangerous, soft-white deliquescent powder; used as a dehydrating agent, in medicine and sugar refining, and as a chemical intermediate and analytical reagent. Also known as anhydrous phosphoric acid; phosphoric oxide; phosphorus pentoxide. { fä'sfór·ik an'hī‚drīd }

phosphoric oxide [INORGANIC CHEMISTRY] *See* phosphoric anhydride. { fä'sfór·ik 'äk‚sīd }

phosphorimetry [ANALYTICAL CHEMISTRY] Low-temperature, analytical procedure related to fluorometry; based on the nature and intensity of the phosphorescent light emitted by an appropriately excited molecule. { ‚fäs·fə'rim·ə·trē }

phosphorous acid [INORGANIC CHEMISTRY] H_3PO_3 Alcohol- and water-soluble deliquescent white or yellowish crystals; decomposes at 200°C; used as an analytical reagent and reducing agent. { 'fäs·fə·rəs 'as·əd }

phosphorus [CHEMISTRY] A nonmetallic element, symbol P, atomic number 15, atomic weight 30.98; used to manufacture phosphoric acid, in phosphor bronzes, incendiaries, pyrotechnics, matches, and rat poisons; the white (or yellow) allotrope is a soft waxy solid melting at 44.5°C, is soluble in carbon disulfide, insoluble in water and alcohol, and is poisonous and self-igniting in air; the red allotrope is an amorphous powder subliming at 416°C, igniting at 260°C, is insoluble in all solvents, and is nonpoisonous; the black allotrope comprises lustrous crystals similar to graphite, and is insoluble in most solvents. { 'fäs·fə·rəs }

phosphorus nitride [INORGANIC CHEMISTRY] P_3N_5 Amorphous white solid that decomposes in hot water; insoluble in cold water, soluble in organic solvents; used to dope semiconductors. { 'fäs·fə·rəs 'nī‚trīd }

phosphorus oxide [INORGANIC CHEMISTRY] An oxygen compound of phosphorus, examples are phosphorus monoxide (P_2O), phosphorus trioxide (P_2O_3), phosphorus suboxide (P_4O). { 'fäs·fə·rəs 'äk‚sīd }

phosphorus oxychloride [INORGANIC CHEMISTRY] $POCl_3$ Toxic, colorless, fuming liquid with pungent aroma; boils at 107°C; decomposes in water or alcohol; causes skin burns; used as a catalyst, chlorinating agent, and in manufacture of various anhydrides. Also known as phosphoryl chloride. { 'fäs·fə·rəs ¦äk·sē'klór‚īd }

phosphorus pentabromide [INORGANIC CHEMISTRY] PBr_5 Yellow crystals, decomposing at 106°C and in water; used in organic synthesis. { 'fäs·fə·rəs ¦pen·tə'brō‚mīd }

phosphorus pentachloride [INORGANIC CHEMISTRY] PCl_5 Toxic, yellowish crystals with irritating aroma; an eye irritant; sublimes on heating, but will melt at 148°C under pressure; soluble in carbon disulfide; decomposes in water; used as a catalyst and chlorinating agent. { 'fäs·fə·rəs ¦pen·tə'klór‚īd }

phosphorus pentasulfide [INORGANIC CHEMISTRY] P_2S_5 Flammable, hygroscopic, yellow crystals, melting at 281°C; decomposes in moist air; soluble in alkali hydroxides; used to make lube-oil additives, rubber additives, and flotation agents. { 'fäs·fə·rəs ¦pen·tə'səl‚fīd }

phosphorus pentoxide [INORGANIC CHEMISTRY] *See* phosphoric anhydride. { 'fäs·fə·rəs pen'täk‚sīd }

phosphorus sesquisulfide [INORGANIC CHEMISTRY] P_4S_3 Flammable, yellow crystals, melting at 172°C; decomposed by hot water, insoluble in water, soluble in carbon disulfide; used as chemical intermediate and to make matches. Also known as tetraphosphorus trisulfide. { 'fäs·fə·rəs ¦ses·kwē'səl‚fīd }

phosphorus thiochloride [INORGANIC CHEMISTRY] $PSCl_3$ Yellow liquid, boiling at 125°C; used to make insecticides and oil additives. { 'fäs·fə·rəs ‚thī·ə'klór‚īd }

phosphorus tribromide [INORGANIC CHEMISTRY] PBr_3 A corrosive, fuming, colorless liquid with penetrating aroma; soluble in acetone, alcohol, carbon disulfide, and hydrogen sulfide; decomposes in water; used as an analytical reagent to test for sugar and oxygen. { ′fäs·fə·rəs trī′brō‚mīd }

phosphorus trichloride [INORGANIC CHEMISTRY] PCl_3 A colorless, fuming liquid that decomposes rapidly in moist air and water; soluble in ether, benzene, carbon disulfide, and carbon tetrachloride; boils at 76°C; used as a chlorinating agent, phosphorus solvent, and in saccharin manufacture. { ′fäs·fə·rəs trī′klȯr‚īd }

phosphorus triiodide [INORGANIC CHEMISTRY] PI_3 Hygroscopic, red crystals, melting at 61°C; soluble in alcohol and carbon disulfide; decomposes in water; used in organic syntheses. { ′fäs·fə·rəs trī′ī·ə‚dīd }

phosphorus trisulfide [INORGANIC CHEMISTRY] P_2S_3 or P_4S_6 Grayish-yellow, tasteless, odorless solid that burns in air; soluble in alcohol, carbon disulfide, and ether; melts at 290°C; used as an analytical reagent. { ′fäs·fə·rəs trī′səl‚fīd }

phosphorylation [ORGANIC CHEMISTRY] The esterification of compounds with phosphoric acid. { ‚fäs‚fȯr·ə′la·shən }

phosphoryl chloride [INORGANIC CHEMISTRY] *See* phosphorus oxychloride. { ′fäs·fə·rəl ′klȯr‚īd }

phosphotungstic acid [INORGANIC CHEMISTRY] $H_3PO_4 \cdot 12WO_3 \cdot xH_2O$ Heavy-greenish, water- and alcohol-soluble crystals; used as an analytical reagent and in the manufacture of organic pigments. Also known as heavy acid; phosphowolframic acid; PTA. { ¦fä·sfō′təŋ·stik ′as·əd }

phosphotungstic pigment [ORGANIC CHEMISTRY] A green or blue pigment prepared by precipitating solutions of phosphotungstic or phosphomolybdic acid with malachite green, Victoria blue, and other basic dyestuffs; used in printing inks, paints, and enamels. Also known as tungsten lake. { ¦fä·sfō′təŋ·stik ′pig·mənt }

phosphowolframic acid [INORGANIC CHEMISTRY] *See* phosphotungstic acid. { ¦fä·sfō·wu̇l′fram·ik ′as·əd }

photoacoustic spectroscopy [SPECTROSCOPY] A spectroscopic technique for investigating solid and semisolid materials, in which the sample is placed in a closed chamber filled with a gas such as air and illuminated with monochromatic radiation of any desired wavelength, with intensity modulated at some suitable acoustic frequency; absorption of radiation results in a periodic heat flow from the sample, which generates sound that is detected by a sensitive microphone attached to the chamber. Abbreviated PAS. Also known as optoacoustic spectroscopy. { ¦fōd·ō·ə¦kü·stik spek′träs·kə·pē }

photoaddition [PHYSICAL CHEMISTRY] A bimolecular photochemical process in which a single product is formed by electronically excited unsaturated molecules. { ¦fōd·ō·ə′dish·ən }

photochemical oxidant [CHEMISTRY] Any of the chemicals which enter into oxidation reactions in the presence of light or other radiant energy. { ¦fōd·ō′kem·ə·kəl ′äk·sə·dənt }

photochemical reaction [PHYSICAL CHEMISTRY] A chemical reaction influenced or initiated by light, particularly ultraviolet light, as in the chlorination of benzene to produce benzene hexachloride. { ¦fōd·ō′kem·ə·kəl rē′ak·shən }

photochemical reduction [CHEMISTRY] *See* photoreduction. { ‚fōd·ō¦kem·ə·kəl ri′dək·shən }

photochemistry [PHYSICAL CHEMISTRY] The study of the effects of light on chemical reactions. { ¦fōd·ō′kem·ə·strē }

photochromic compound [CHEMISTRY] A chemical compound that changes in color when exposed to visible or near-visible radiant energy; the effect is reversible; used to produce very-high-density microimages. { ¦fōd·ō¦krō·mik ′käm‚pau̇nd }

photochromic reaction [CHEMISTRY] A chemical reaction that produces a color change. { ¦fōd·ō¦krō·mik rē′ak·shən }

photochromism [CHEMISTRY] The ability of a chemically treated plastic or other transparent material to darken reversibly in strong light. { ¦fōd·ō′krō‚miz·əm }

photocurrent [PHYSICAL CHEMISTRY] An electric current induced at an electrode by radiant energy. { ¦fōd·ō′kə·rənt }

photodegradation [PHYSICAL CHEMISTRY] Decomposition of a compound by radiant energy. { ¦fōd·ō‚deg·rə′dā·shən }

photodetachment [PHYSICAL CHEMISTRY] The removal of an electron from a negative ion by absorption of a photon, resulting in a neutral atom or molecule. { ¦fōd·ō·di′tach·mənt }

photodimerization [PHYSICAL CHEMISTRY] A bimolecular photochemical process involving an electronically excited unsaturated molecule that undergoes addition with an unexcited molecule of the same species. { ¦fōd·ō‚dī·mə·rə′zā·shən }

photodissociation [PHYSICAL CHEMISTRY] The removal of one or more atoms from a molecule by the absorption of a quantum of electromagnetic energy. { ¦fōd·ō·di‚sō·shē′ā·shən }

photoelectric absorption analysis [ANALYTICAL CHEMISTRY] Type of activation analysis in which the γ-photon gives all of its energy to an electron in the crystal under analysis, generating a maximum-sized pulse for that particular γ-energy. { ¦fōd·ō·i′lek·trik əb′sȯrp·shən ə‚nal·ə·səs }

photoelectric color comparator [ANALYTICAL CHEMISTRY] *See* color comparator. { ¦fōd·ō·i′lek·trik ′kəl·ər kəm‚par·əd·ər }

photoelectric colorimetry [ANALYTICAL CHEMISTRY] Measurement of the colorant concentration in a solution by means of the tristimulus values of three primary light filter-photocell combinations. { ¦fōd·ō·i′lek·trik ‚kəl·ə′rim·ə·trē }

photoelectrolysis [PHYSICAL CHEMISTRY] The process of using optical energy to assist or effect electrolytic processes that ordinarily require the use of electrical energy. { ¦fōd·ō‚i‚lek′träl·ə·səs }

photoelectron spectroscopy [SPECTROSCOPY] The branch of electron spectroscopy concerned with the energy analysis of photoelectrons ejected from a substance as the direct result of bombardment by ultraviolet radiation or x-radiation. { ¦fōd·ō·i′lek‚trän spek′träs·kə·pē }

photoglycine [ORGANIC CHEMISTRY] *See* glycin. { ¦fōd·ō′glī‚sēn }

photographic photometry [SPECTROSCOPY] The use of a comparator-densitometer to analyze a photographed spectrograph spectrum by emulsion density measurements. { ¦fōd·ə¦graf·ik fə′täm·ə·trē }

photohomolysis [PHYSICAL CHEMISTRY] A homolysis reaction in which bond breaking is caused by radiant energy. { ¦fōd·ō·hə′mäl·ə·səs }

photoionization [PHYSICAL CHEMISTRY] The removal of one or more electrons from an atom or molecule by absorption of a photon of visible or ultraviolet light. Also known as atomic photoelectric effect. { ¦fōd·ō‚ī·ə·nə′zā·shən }

photoisomer [PHYSICAL CHEMISTRY] An isomer produced by photolysis. { ¦fōd·ō′ī·sə·mər }

photolysis [PHYSICAL CHEMISTRY] The use of radiant energy to produce chemical changes. { fō′täl·ə·səs }

photomechanochemistry [PHYSICAL CHEMISTRY] A branch of polymer sciences that deals with photochemical conversion of chemical energy into mechanical energy. { ¦fōd·ō·mə‚kan·ō′kem·ə·strē }

photometric titration [ANALYTICAL CHEMISTRY] A titration in which the titrant and solution cause the formation of a metal complex accompanied by an observable change in light absorbance by the titrated solution. { ¦fōd·ə¦me·trik tī′trā·shən }

photopolymer [PHYSICAL CHEMISTRY] Any of a class of light-sensitive polymers which undergo a spontaneous and permanent change in physical properties on exposure to light. { ‚fōd·ō′päl·ə·mər }

photoreduction [CHEMISTRY] A chemical reduction that is produced by electromagnetic radiation. Also known as photochemical reduction. { ‚fōd·ō·ri′dək·shən }

phoxim [ORGANIC CHEMISTRY] *See* phenylglyoxylonitriloxime O,O-diethyl phosphorothioate. { ′fäk‚sim }

pH standard [ANALYTICAL CHEMISTRY] Five standard laboratory solutions available from the U.S. National Bureau of Standards, each solution having a known pH value; the standards cover pH ranges from 3.557 to 8.833. Abbreviated pH(S). { ‚pē′āch ¦stan·dərd }

phthalate [ORGANIC CHEMISTRY] A salt of phthalic acid; contains the radical $C_6H_4(COO)_2^{2-}$; an example is dibutylphthalate, $C_{16}H_{22}O_4$; used as a plasticizer in plastics, and as a buffer in standard laboratory solutions. { ′tha‚lāt }

phthalate buffer [ANALYTICAL CHEMISTRY] Laboratory pH reference solution made of potassium hydrogen phthalate, $KHC_8H_4O_4$; at 0.05 molal, the pH is 4.008 at 25°C. { 'tha,lāt 'bəf·ər }

phthalate ester [ORGANIC CHEMISTRY] Any of a group of plastics plasticizers made by the direct action of alcohol on phthalic anhydride; generally characterized by moderate cost, good stability, and good general properties. { 'tha,lāt 'es·tər }

phthalazine [ORGANIC CHEMISTRY] $C_6H_4CHN_2CH$ Colorless crystals, melting at 91°C; soluble in alcohol. { 'thal·ə,zēn }

phthalic acid [ORGANIC CHEMISTRY] $C_6H_4(CO_2H)_2$ Any of three isomeric benzene dicarboxylic acids; the ortho form is usually called phthalic acid, comprises alcohol-soluble, colorless crystals decomposing at 191°C, slightly soluble in water and ether, is used to make dyes, medicine, and synthetic perfumes, and as a chemical intermediate, and is also known as benzene orthodicarboxylic acid; the para form, known as terephthalic acid, is used to make polyester resins (Dacron) and as poultry feed additives; the meta form is isophthalic acid. { 'thal·ik 'as·əd }

***meta*-phthalic acid** [ORGANIC CHEMISTRY] *See* isophthalic acid. { ¦med·ə 'thal·ik 'as·əd }

***para*-phthalic acid** [ORGANIC CHEMISTRY] *See* terephthalic acid. { ¦par·ə 'thal·ik 'as·əd }

phthalic anhydride [ORGANIC CHEMISTRY] $C_6H_4(CO)_2O$ White crystals, melting at 131°C; sublimes when heated; slightly soluble in ether and hot water, soluble in alcohol; used to make dyes, resins, plasticizers, and insect repellents. { 'thal·ik an'hī,drīd }

phthalocyanine pigments [ORGANIC CHEMISTRY] A group of light-fast organic pigments with four isoindole groups, $(C_6H_4)C_2N$, linked by four nitrogen atoms to form a conjugated chain; included are phthalocyanine (blue-green), copper phthalocyanine (blue), chlorinated copper phthalocyanine (green), and sulfonated copper phthalocyanine (green); used in enamels, plastics, linoleum, inks, wallpaper, and rubber goods. { ¦thal·ō'sī·ə·nən 'pig·məns }

phthalonitrile [ORGANIC CHEMISTRY] $C_6H_4(CN)_2$ Buff-colored crystals with a melting point of 138°C; soluble in acetone and benzene; used in organic synthesis and as an insecticide. { ¦thal·ō'nī·trəl }

physical adsorption [PHYSICAL CHEMISTRY] Reversible adsorption in which the adsorbate is held by weak physical forces. { 'fiz·ə·kəl ad'sȯrp·shən }

physical chemistry [CHEMISTRY] The branch of chemistry that deals with the interpretation of chemical phenomena and properties in terms of the underlying physical processes, and with the development of techniques for their investigation. { 'fiz·ə·kəl 'kem·ə·strē }

physical organic chemistry [ORGANIC CHEMISTRY] The study of the scope and limitations of the various rules, effects, and generalizations in use in organic chemistry by application of physical and mathematical means. { ¦fiz·ə·kəl ȯr¦gan·ik 'kem·ə·strē }

physical property [CHEMISTRY] Property of a compound that can change without involving a change in chemical composition; examples are the melting point and boiling point. { 'fiz·ə·kəl 'präp·ərd·ē }

physisorption [PHYSICAL CHEMISTRY] A physical adsorption process in which there are van der Waals forces of interaction between gas or liquid molecules and a solid surface. { ¦fiz·ə'sȯrp·shən }

physostigmine [ORGANIC CHEMISTRY] $C_{15}H_{21}O_2N_3$ An alkaloid; poisonous, colorless-to-pinkish crystals; soluble in alcohol and dilute acids; melts at 86°C; used as a source of salicylate and sulfate forms. Also known as calabarine; eserine. { ,fī·sə'stig·mēn }

physostigmine salicylate [ORGANIC CHEMISTRY] $C_{15}H_{21}O_2N_3 \cdot C_7H_6O_3$ Poisonous, colorless-to-yellow crystals; soluble in water, alcohol, and chloroform; melts at 182°C; used for medicines. { ,fī·sə'stig,mēn sə'lis·ə,lāt }

physostigmine sulfate [ORGANIC CHEMISTRY] $(C_{15}H_{21}O_2N_3)_2 \cdot H_2SO_4$ Poisonous, white crystals; soluble in water, alcohol, and chloroform; melts at 150°C; used for medicines. { ,fī·sə'stig,mēn 'səl,fāt }

phytane [ORGANIC CHEMISTRY] $C_{20}H_{42}$ A hydrocarbon derivative of chlorophyll that is found in rock specimens 2.5-3 × 10^9 years old; frequently associated with Precambrian fossil plant matter. { 'fī,tān }

phytic acid [ORGANIC CHEMISTRY] $C_6H_6[OPO(OH)_2]_6$ An acid found in seeds of plants as the insoluble calcium magnesium salt (phytin); derived from corn steep liquor; inhibits calcium absorption in intestine; used to treat hard water, to remove iron and copper from wines, and to inactivate trace-metal contaminants in animal and vegetable oils. { 'fīd·ik 'as·əd }

phytol [ORGANIC CHEMISTRY] $C_{20}H_{40}O$ A liquid with a boiling point of 202-204°C; soluble in organic solvents; used in the synthesis of vitamins E and K. { 'fī,tȯl }

phytonadione [ORGANIC CHEMISTRY] $C_{31}H_{46}O_2$ A yellow, viscous liquid soluble in benzene, chloroform, and vegetable oils; used in medicine and as a food supplement. Also known as vitamin K_1. { fī,tän·ə'dī,ōn }

pi bonding [PHYSICAL CHEMISTRY] Covalent bonding in which the greatest overlap between atomic orbitals is along a plane perpendicular to the line joining the nuclei of the two atoms. { 'pī ,bänd·iŋ }

Pickering series [SPECTROSCOPY] A series of spectral lines of singly ionized helium, observed in very hot O-type stars, associated with transitions between the level with principal quantum number $n = 4$ and higher energy levels. { 'pik·riŋ ,sir·ēz }

pickling acid [CHEMISTRY] Any of the acids used in pickling solutions, such as hydrochloric, sulfuric, nitric, phosphoric, or hydrofluoric acid. { 'pik·liŋ ,as·əd }

picoline [ORGANIC CHEMISTRY] $C_5H_4N(CH_3)$ Family of colorless liquid isomers, soluble in water and alcohol; the alpha form, boiling at 129°C, is used as a solvent and chemical intermediate, and is also known as 2-methyl pyridine; the beta form, boiling at 143.5°C, is used as a solvent for chemical synthesis reactions, to make nicotinic acid, and in fabric waterproofing, and is also known as 3-methyl pyridine; the gamma form, boiling at 143.1°C, is used as a solvent for chemical synthesis reactions and in fabric waterproofing. { 'pik·ə,lēn }

picolinic acid [ORGANIC CHEMISTRY] $C_{10}H_8N_4O_5$ An alcohol-soluble crystalline compound, forming yellow leaflets that melt at 116-117°C; used as a reagent in phenylalanine, tryptophan, and alkaloids production, and for the quantitative detection of calcium. { ¦pik·ə¦lin·ik 'as·əd }

picramic acid [ORGANIC CHEMISTRY] $C_6H_5N_3O_5$ A crystalline acid, forming dark red needles from alcohol solutions, melting at 169-170°C; used in dye manufacture and as a reagent in tests for albumin. { pi'kram·ik 'as·əd }

picric acid [ORGANIC CHEMISTRY] $C_6H_2(NO_2)_3OH$ Poisonous, explosive, highly oxidative yellow crystals with bitter taste; soluble in water, alcohol, chloroform, benzene, and ether; melts at 122°C; used in explosives, in external medicines; to make dyes, matches, and batteries, and to etch copper. { 'pik·rik 'as·əd }

pi electron [PHYSICAL CHEMISTRY] An electron which participates in pi bonding. { 'pī i'lek,trän }

piezochemistry [CHEMISTRY] The field of chemical reactions under high pressures. { pē¦ā·zō'kem·ə·strē }

piezoelectric polymer [ORGANIC CHEMISTRY] *See* piezopolymer. { pē¦ā·zō·ə'lek·trik 'päl·ə·mər }

piezopolymer [ORGANIC CHEMISTRY] A polymeric film that has the ability to reversibly convert heat and pressure to electricity. Also known as piezoelectric polymer. { pē¦ā·zō'päl·ə·mər }

pilocarpine [ORGANIC CHEMISTRY] $C_{11}H_{16}N_2O_2$ An alkaloid, in either oil or crystal form, melting at 34°C; soluble in chloroform, water, and alcohol; used in medicine. { ,pī·lə'kär,pēn }

pimaricin [ORGANIC CHEMISTRY] $C_{33}H_{47}NO_{13}$ A compound crystallizing from a methanol-water solution, decomposing at about 200°C; soluble in water and organic solvents; used in medicine as an antifungal agent for *Candida albicans* vaginitis. { pə'mar·ə·sən }

pimelic acid [ORGANIC CHEMISTRY] $HOOC(CH_2)_5COOH$ Crystals melting at 105°C; slightly soluble in water, soluble in alcohol and ether; used in biochemical research. { pə'mel·ik 'as·əd }

pinene [ORGANIC CHEMISTRY] $C_{10}H_{16}$ Either of two colorless isomeric unsaturated bicyclic terpene hydrocarbon liquids derived from sulfate wood turpentine; 95% of the alpha form boils in the range 156-160°C, and of the beta form boils in the range 164-169°C; used as solvents for coatings and wax formulations, as chemical intermediates for resins, and as lube-oil additives. Also known as nopinene. { 'pī‚nēn }

pinene hydrochloride [ORGANIC CHEMISTRY] *See* terpene hydrochloride. { 'pī‚nēn ‚hī·drə'klór‚īd }

pinic acid [ORGANIC CHEMISTRY] $C_9H_{14}O_4$ A crystalline dicarboxylic acid derived from α-pinene; used to make diesters for plasticizers and lubricants. { 'pī·nik 'as·əd }

piperazine [ORGANIC CHEMISTRY] $C_4H_{10}N_2$ A cyclic compound; colorless, deliquescent crystals, melting at 104-107°C; soluble in water, alcohol, glycerol, and glycols; absorbs carbon dioxide from air; used in medicine. { pī'par·ə‚zēn }

piperazine dihydrochloride [ORGANIC CHEMISTRY] $C_4H_{10}N_2 \cdot 2HCl$ White, water-soluble needles; used for insecticides and pharmaceuticals. { pī'par·ə‚zēn dī‚hī·drə'klór‚īd }

piperazine hexahydrate [ORGANIC CHEMISTRY] $C_4H_{10}N_2 \cdot 6H_2O$ White crystals with a melting point of 44°C; soluble in alcohol and water; used for pharmaceuticals and insecticides. { pī'par·ə‚zēn ¦hek·sə'hī‚drāt }

piperidine [ORGANIC CHEMISTRY] $C_5H_{11}N$ A cyclic compound, and strong base; colorless liquid with pepper aroma; boils at 106°C; soluble in water, alcohol, and ether; used as a chemical intermediate and rubber accelerator, and in medicine. { pī'per·ə‚dēn }

piperine [ORGANIC CHEMISTRY] $C_{17}H_{19}NO_3$ A crystalline compound that is found in black pepper, melting point is 130°C, soluble in benzene and acetic acid; used to give a pungent taste to brandy and as an insecticide. { 'pip·ə‚rēn }

piperocaine hydrochloride [ORGANIC CHEMISTRY] $C_{16}H_{23}NO_2 \cdot HCl$ A white, crystalline powder with a bitter taste and a melting point of 172-175°C; soluble in water, chloroform, and alcohol; used in medicine. { pī'per·ə‚kān ‚hī·drə'klór‚īd }

piperonal [ORGANIC CHEMISTRY] $C_8H_6O_3$ White crystals with a floral odor and a melting point of 35.5-37°C; soluble in alcohol and ether; used in medicine, perfumes, suntan preparations, and mosquito repellents. Also known as heliotropin. { pə'per·ə‚nal }

pipet [CHEMISTRY] Graduated or calibrated tube which may have a center reservoir (bulb), used to transfer known volumes of liquids from one vessel to another; types are volumetric or transfer, graduated, and micro. { pī'pet }

pirimiphosethyl [ORGANIC CHEMISTRY] $C_{13}H_{24}N_3O_3PS$ A straw-colored liquid which decomposes at 130°C; used as an insecticide for the control of soil insects in vegetables and other crops. { ‚pir·əm·fäs'eth·əl }

Pitzer equation [PHYSICAL CHEMISTRY] Equation for the approximation of data for heats of vaporization for organic and simple inorganic compounds; derived from temperature and reduced temperature relationships. { 'pit·sər i‚kwā·zhən }

PIXE [ANALYTICAL CHEMISTRY] *See* proton-induced x-ray emission. { 'pik‚sē }

pK [CHEMISTRY] The logarithm (to base 10) of the reciprocal of the equilibrium constant for a specified reaction under specified conditions.

plait point [CHEMISTRY] Composition conditions in which the three coexisting phases of partially soluble components of a three-phase liquid system approach each other in composition. { 'plāt ‚póint }

planocaine base [ORGANIC CHEMISTRY] *See* procaine base. { 'plan·ə‚kān ‚bās }

plasma desorption mass spectrometry [SPECTROSCOPY] A technique for analysis of nonvolatile molecules, particularly heavy molecules with atomic weight over 2000, in which heavy ions with energies on the order of 100 MeV penetrate and deposit energy in thin films, giving rise to chemical reactions that result in the formation of molecular ions and shock waves that result in the ejection of these ions from the surface; the ions are then analyzed in a mass spectrometer. Abbreviated PDMS. { 'plaz·mə dē¦sórp·shən 'mas spek'träm·ə·trē }

plasma-jet excitation [SPECTROSCOPY] The use of a high-temperature plasma jet to excite an element to provide measurable spectra with many ion lines similar to those from spark-excited spectra. { 'plaz·mə ‚jet ‚ek·sə'tā·shən }

plaster of paris [INORGANIC CHEMISTRY] White powder consisting essentially of the hemihydrate of calcium sulfate ($CaSO_4 \cdot {}^1/_2H_2O$ or $2CaSO_4 \cdot H_2O$), produced by calcining gypsum until it is partially dehydrated; forms with water a paste that quickly sets; used for casts and molds, building materials, and surgical bandages. Also known as calcined gypsum. { 'plas·tər əv 'par·əs }
plate theory [ANALYTICAL CHEMISTRY] In gas chromatography, the theory that the column operates similarly to a distillation column; for example, chromatographic columns are considered as consisting of a number of theoretical plates, each performing a partial separation of components. { 'plāt ‚thē·ə·rē }
platinic chloride [INORGANIC CHEMISTRY] *See* chloroplatinic acid. { plə'tin·ik 'klȯr‚īd }
platinic sodium chloride [INORGANIC CHEMISTRY] *See* sodium chloroplatinate. { plə'tin·ik 'sōd·ē·əm 'klȯr‚īd }
platinic sulfate [INORGANIC CHEMISTRY] *See* platinum sulfate. { plə'tin·ik 'səl‚fāt }
platinochloride [INORGANIC CHEMISTRY] *See* chloroplatinate. { ¦plat·ən·ō'klȯr‚īd }
platinocyanide [INORGANIC CHEMISTRY] A double salt of platinous cyanide and another cyanide, such as $K_2Pt(CN)_4$; used in photography and fluorescent x-ray screens. Also known as cyanoplatinate. { ¦plat·ən·ō'sī·ə‚nīd }
platinous chloride [INORGANIC CHEMISTRY] *See* platinum dichloride. { 'plat·ən·əs 'klȯr‚īd }
platinous iodide [INORGANIC CHEMISTRY] *See* platinum iodide. { 'plat·ən·əs 'ī·ə‚dīd }
platinum [CHEMISTRY] A chemical element, symbol Pt, atomic number 78, atomic weight 195.09. { 'plat·ən·əm }
platinum bichloride [INORGANIC CHEMISTRY] *See* platinum dichloride. { 'plat·ən·əm bī'klȯr‚īd }
platinum chloride [INORGANIC CHEMISTRY] $PtCl_4$ or $PtCl_4 \cdot 5H_2O$ A brown solid or red crystals; soluble in alcohol and water; decomposes when heated (loses $4H_2O$ at 100°C); used as an analytical reagent. { 'plat·ən·əm 'klȯr‚īd }
platinum dichloride [INORGANIC CHEMISTRY] $PtCl_2$ Water-insoluble, green-gray powder; decomposes to platinum at red heat; used to make platinum salts. Also known as platinous chloride; platinum bichloride. { 'plat·ən·əm dī'klȯr‚īd }
platinum diiodide [INORGANIC CHEMISTRY] *See* platinum iodide. { 'plat·ən·əm dī'ī·ə‚dīd }
platinum electrode [PHYSICAL CHEMISTRY] A solid platinum wire electrode used during voltammetric analyses of electrolytes. { 'plat·ən·əm i'lek‚trōd }
platinum iodide [INORGANIC CHEMISTRY] PtI_2 Water- and alkali-insoluble black powder; slightly soluble in hydrochloric acid; decomposes at 300-350°C. Also known as platinous iodide; platinum diiodide. { 'plat·ən·əm 'ī·ə‚dīd }
platinum metal [CHEMISTRY] A group of transition metals that includes ruthenium, osmium, rhodium, iridium, palladium, and platinum. { 'plat·ən·əm 'med·əl }
platinum oxide [INORGANIC CHEMISTRY] An oxide of platinum; examples are platinum monoxide (or platinous oxide), PtO, and platinum dioxide (or platinic oxide), PtO_2. { 'plat·ən·əm 'äk‚sīd }
platinum potassium chloride [INORGANIC CHEMISTRY] *See* potassium chloroplatinate. { 'plat·ən·əm pə'tas·ē·əm 'klȯr‚īd }
platinum sodium chloride [INORGANIC CHEMISTRY] *See* sodium chloroplatinate. { 'plat·ən·əm 'sōd·ē·əm 'klȯr‚īd }
platinum sulfate [INORGANIC CHEMISTRY] $Pt(SO_4)_2$ A hygroscopic, dark mass; soluble in alcohol, ether, water, and dilute acids; used in microanalysis for halogens. Also known as platinic sulfate. { 'plat·ən·əm 'səl‚fāt }
Plessy's green [INORGANIC CHEMISTRY] $CrPO_4 \cdot xH_2O$ Deep-green pigment made of chromium phosphate mixed with chromium oxide and calcium phosphate. { ple'sēz 'grēn }
plumbous oxide [INORGANIC CHEMISTRY] *See* lead monoxide. { 'pləm·bəs 'äk‚sīd }
plumbous sulfide [INORGANIC CHEMISTRY] *See* lead sulfide. { 'pləm·bəs 'səl‚fīd }
plumbum [CHEMISTRY] Latin name for lead; source of the element symbol Pb. { 'pləm·bəm }

plutonium [CHEMISTRY] A reactive metallic element, symbol Pu, atomic number 94, in the transuranium series of elements; the first isotope to be identified was plutonium-239; used as a nuclear fuel, to produce radioactive isotopes for research, and as the fissile agent in nuclear weapons. { plü'tō·nē·əm }

plutonium oxide [INORGANIC CHEMISTRY] PuO_2 A radioactively poisonous pyrophoric oxide of plutonium; particles may be easily airborne. { plü'tō·nē·əm 'äk‚sīd }

Pm [CHEMISTRY] *See* promethium.

PMA [INORGANIC CHEMISTRY] *See* phosphomolybdic acid. [ORGANIC CHEMISTRY] *See* pyromellitic acid.

PMDA [ORGANIC CHEMISTRY] *See* pyromellitic dianhydride.

pNa [CHEMISTRY] Logarithm of the sodium-ion concentration in a solution; that is, $pNa = -\log a_{Na^+}$, where a_{Na^+} is the sodium-ion concentration.

pnicogen [CHEMISTRY] Any member of the nitrogen family of elements, group V in the periodic table. { 'nī·kə·jən }

pnictide [CHEMISTRY] A simple compound of a pnicogen and an electropositive element. { 'nik‚tīd }

Po [CHEMISTRY] *See* polonium.

POD analysis [ANALYTICAL CHEMISTRY] A precision laboratory distillation procedure used to separate low-boiling hydrocarbon fractions quantitatively for analytical purposes. Also known as Podbielniak analysis. { ¦pē¦ō'dē ə‚nal·ə·səs }

Podbielniak analysis [ANALYTICAL CHEMISTRY] *See* POD analysis. { päd'bēl·nē‚ak ə‚nal·ə·səs }

poison [CHEMISTRY] A substance that exerts inhibitive effects on catalysts, even when present only in small amounts; for example, traces of sulfur or lead will poison platinum-based catalysts. { 'pȯiz·ən }

polar compound [CHEMISTRY] Molecules which contain polar covalent bonds; they can ionize when dissolved or fused; polar compounds include inorganic acids, bases, and salts. { 'pō·lər 'käm‚pau̇nd }

polar covalent bond [PHYSICAL CHEMISTRY] A bond in which a pair of electrons is shared in common between two atoms, but the pair is held more closely by one of the atoms. { 'pō·lər kō'vā·lənt 'bänd }

polarimetric analysis [ANALYTICAL CHEMISTRY] A method of chemical analysis based on the optical activity of the substance being determined; the measurement of the extent of the optical rotation of the substance is used to identify the substance or determine its quantity. { pō¦lar·ə¦me·trik ə'nal·ə·səs }

polarization potential [PHYSICAL CHEMISTRY] The reverse potential of an electrolytic cell which opposes the direct electrolytic potential of the cell. { ‚pō·lə·rə'zā·shən pə‚ten·chəl }

polarization spectroscopy [SPECTROSCOPY] A type of saturation spectroscopy in which a circularly polarized saturating laser beam depletes molecules with a certain orientation preferentially, leaving the remaining ones polarized; the latter are detected through their induction of elliptical polarization in a probe beam, allowing the beam to pass through crossed linear polarizers. { ‚pō·lə·rə'zā·shən spek'träs·kə·pē }

polarized scattering [PHYSICAL CHEMISTRY] In a quasielastic light scattering experiment performed with polarizers, the type of scattering produced when the polarizers select both the incident and final polarizations perpendicular to the scattering plane. { 'pō·lə‚rīzd 'skad·ə·riŋ }

polar molecule [PHYSICAL CHEMISTRY] A molecule having a permanent electric dipole moment. { 'pō·lər 'mäl·ə‚kyül }

polarogram [ANALYTICAL CHEMISTRY] Plotted output (current versus electrode voltage) for polarographic analysis of an electrolyte. { pə'lar·ə‚gram }

polarographic analysis [ANALYTICAL CHEMISTRY] An electroanalytical technique in which the current through an electrolysis cell is measured as a function of the applied potential; the apparatus consists of a potentiometer for adjusting the potential, a galvanometer for measuring current, and a cell which contains two electrodes, a reference electrode whose potential is constant and an indicator electrode which is commonly the dropping mercury electrode. Also known as polarography. { pō¦lar·ə¦graf·ik ə'nal·ə·səs }

polarographic cell [ANALYTICAL CHEMISTRY] Device for polarographic (voltammetric) analysis of an electrolyte solution; a known voltage is applied to the solution, and the ensuing current that passes through the cell (to an electrode) is measured. { pō¦lar·ə¦graf·ik 'sel }

polarographic maximum [ANALYTICAL CHEMISTRY] A deceptively high voltage buildup on an electrode during polarographic analysis of an electrolyte; caused by a reduction or oxidation process at the electrode. { pō¦lar·ə¦graf·ik 'mak·sə·məm }

polarography [ANALYTICAL CHEMISTRY] *See* polarographic analysis. { ˌpō·lə'räg·rə·fē }

polonium [CHEMISTRY] A chemical element, symbol Po, atomic number 84; all polonium isotopes are radioactive; polonium-210 is the naturally occurring isotope found in pitchblende. { pə'lō·nē·əm }

poly- [ORGANIC CHEMISTRY] A chemical prefix meaning many; for example, a polymer is made of a number of single molecules known as monomers, as polyethylene is made from ethylene. { 'päl·ē, 'päl·ə, 'päl·i }

polyacetals [ORGANIC CHEMISTRY] *See* acetal resins. { ¦päl·ē'as·əˌtalz }

polyacrylamide [ORGANIC CHEMISTRY] $(CH_2CHCONH_2)_x$ A white, water-soluble high polymer based on acrylamide; used as a thickening or suspending agent in water-base formulations. { ¦päl·ē·ə'kril·ə·məd }

polyacrylate [ORGANIC CHEMISTRY] A polymer of an ester or salt of acrylic acid. { ¦päl·ē'ak·rəˌlāt }

polyacrylic acid [ORGANIC CHEMISTRY] $(CH_2CHCOOH)_x$ An acrylic or acrylate resin formed by the polymerization of acrylic acid; water-soluble; used as a suspending and textile-sizing agent, and in adhesives, paints, and hydraulic fluids. { ¦päl·ē·ə¦kril·ik 'as·əd }

polyacrylic fiber [ORGANIC CHEMISTRY] Continuous-strand fiber extruded from an acrylate resin. { ¦päl·ē·ə¦kril·ik 'fī·bər }

polyacrylonitrile [ORGANIC CHEMISTRY] Polymer of acrylonitrile; semiconductive; used like an inorganic oxide catalyst to dehydrogenate *tert*-butyl alcohol to produce isobutylene and water. { ¦päl·ē¦ak·rə·lō'nī·trəl }

polyalcohol [ORGANIC CHEMISTRY] *See* polyhydric alcohol. { ¦päl·ē'al·kəˌhȯl }

polyallomer [ORGANIC CHEMISTRY] A copolymer of propylene with other olefins. { ¦päl·ē'al·ə·mər }

polyamide resin [ORGANIC CHEMISTRY] Product of polymerization of amino acid or the condensation of a polyamine with a polycarboxylic acid; an example is the nylons. { ¦päl·ē'am·əd 'rez·ən }

polyatomic ion [CHEMISTRY] An electrically charged species formed by covalent bonding of atoms of two or more different elements, usually nonmetals, for example, the ammonium ion (NH_4^+). { ˌpäl·ē·ə¦täm·ik 'ī·ən }

polyatomic molecule [CHEMISTRY] A chemical molecule with three or more atoms. { ¦päl·ē·ə¦täm·ik 'mäl·əˌkyül }

polybasic [CHEMISTRY] A chemical compound in solution that yields two or more H^- ions per molecule, such as sulfuric acid, H_2SO_4. { ¦päl·iˌbā·sik }

polybutadiene [ORGANIC CHEMISTRY] Oil-extendable synthetic elastomer polymer made from butadiene; resilience is similar to natural rubber; it is blended with natural rubber for use in tire and other rubber products. Also known as butadiene rubber. { ¦päl·iˌbyüd·ə'dīˌēn }

polybutene [ORGANIC CHEMISTRY] A polymer of isobutene, $(CH_3)_2CCH_2$; made in varying chain lengths to give a wide range of properties from oily to solid; used as a lube-oil additive, in adhesives, and in rubber products. { ¦päl·i'byüˌtēn }

polybutylene [ORGANIC CHEMISTRY] A polymer of one or more butylenes whose consistency ranges from a viscous liquid to a rubbery solid. { ¦päl·i'byüd·əˌlēn }

polycarbonate [ORGANIC CHEMISTRY] $[OC_6H_4C(CH_3)_2C_6H_4OCO]_x$ A linear polymer of carbonic acid which is a thermoplastic synthetic resin made from bisphenol and phosgene; used in emulsion coatings with glass fiber reinforcement. { ¦päl·i'kär·bə·nət }

polycarboxylic [ORGANIC CHEMISTRY] Prefix for a compound containing two or more carboxyl (—COOH) groups. { ¦päl·i¦kärˌbäk¦sil·ik }

polychlorinated biphenyl [ORGANIC CHEMISTRY] Any member of the group of chlorinated isomers of biphenyl. Abbreviated PCB. { ¦päl·i'klȯr·ə,nād·əd bī'fen·əl }

polycondensation [ORGANIC CHEMISTRY] A chemical condensation leading to the formation of a polymer by the linking together of molecules of a monomer and the releasing of water or a similar simple substance. { ¦päl·i,kän·dən'sā·shən }

polycyclic [ORGANIC CHEMISTRY] A molecule that contains two or more closed atomic rings; can be aromatic (such as DDT), aliphatic (bianthryl), or mixed (dicarbazyl). { ¦päl·i'sī·klik }

polycyclic hydrocarbon [ORGANIC CHEMISTRY] *See* polynuclear hydrocarbon. { ¦päl·i'sī·klik 'hī·drə,kär·bən }

polydent [ORGANIC CHEMISTRY] Pertaining to a chemical species whose molecules possess more than two reactive sites. Also known as multident. { 'päl·ə,dent }

polydisperse colloidal system [CHEMISTRY] A colloidal system in which the suspended particles have various sizes and shapes. { ¦päl·i·di'spərs kə¦lȯid·əl 'sis·təm }

polydispersity [CHEMISTRY] Molecular-weight nonhomogeneity in a polymer system; that is, there is some molecular-weight distribution throughout the body of the polymer. { ¦päl·i·di'spər·səd·ē }

polyelectrolyte [ORGANIC CHEMISTRY] A natural or synthetic electrolyte with high molecular weight, such as proteins, polysaccharides, and alkyl addition products of polyvinyl pyridine; can be a weak or strong electrolyte; when dissociated in solution, it does not give uniform distribution of positive and negative ions (the ions of one sign are bound to the polymer chain while the ions of the other sign diffuse through the solution). { ¦päl·ē·ə'lek·trə,līt }

polyene [ORGANIC CHEMISTRY] Compound containing many double bonds, such as the carotenoids. { 'päl·ē,ēn }

polyester resin [ORGANIC CHEMISTRY] A thermosetting or thermoplastic synthetic resin made by esterification of polybasic organic acids with polyhydric acids; examples are Dacron and Mylar; the resin has high strength and excellent resistance to moisture and chemicals when cured. { 'päl·ē,es·tər 'rez·ən }

polyester rubber [ORGANIC CHEMISTRY] *See* polyurethane rubber. { 'päl·ē,es·tər 'rəb·ər }

polyether [ORGANIC CHEMISTRY] Any compound whose molecular structure contains linked ethers, R—O—R', where R and R' represent functional groups. { 'päl·ē,ē·thər }

polyether resin [ORGANIC CHEMISTRY] Any member of a large group of thermoplastic or thermosetting polymers that contain the typical polyether linkages in the polymer chain. { 'päl·ē,ē·thər 'rez·ən }

polyethylene glycol [ORGANIC CHEMISTRY] Any of a family of colorless, water-soluble liquids with molecular weights from 200 to 6000; soluble also in aromatic hydrocarbons (not aliphatics) and many organic solvents; used to make emulsifying agents and detergents, and as plasticizers, humectants, and water-soluble textile lubricants. { ¦päl·ē'eth·ə,lēn 'glī,kȯl }

polyethylene glycol distearate [ORGANIC CHEMISTRY] *See* polyglycol distearate. { ¦päl·ē'eth·ə,lēn 'glī,kȯl dī'stir,āt }

polyethylene resin [ORGANIC CHEMISTRY] *See* ethylene resin. { ¦päl·ē'eth·ə,lēn 'rez·ən }

polyethylene terephthalate [ORGANIC CHEMISTRY] A thermoplastic polyester resin made from ethylene glycol and terephthalic acid; melts at 265°C; used to make films or fibers. Abbreviated PET. { ¦päl·ē'eth·ə,lēn ,ter·ə'tha,lāt }

polygen [CHEMISTRY] *See* polyvalent. { 'päl·i·jən }

polyglycol [ORGANIC CHEMISTRY] A dihydroxy ether derived from the dehydration (removal of a water molecule) of two or more glycol molecules; an example is diethylene glycol, $CH_2OHCH_2OCH_2CH_2OH$. { 'päl·i,glī,kȯl }

polyglycol distearate [ORGANIC CHEMISTRY] $(C_{17}H_{35})_2CO_2CO(CH_2CH_2O)_x$ An off-white, soft solid with a melting point of 43°C; soluble in chlorinated solvents, acetone, and light esters; used as a resin plasticizer. Also known as polyethylene glycol distearate. { 'päl·i,glī,kȯl dī'stir,āt }

polyhaloalkane [ORGANIC CHEMISTRY] An alkane derivative in which two or more hydrogen atoms have been replaced by halogen atoms. { ˌpäl·ēˌha·lō'alˌkān }

polyhalogeno compound [ORGANIC CHEMISTRY] An organic compound containing more than one halogen atom. { ˌpäl·ē·hə'läj·ə·nō ˌkämˌpau̇nd }

polyhydric alcohol [ORGANIC CHEMISTRY] An alcohol with many hydroxyl (—OH) radicals, such as glycerol, $C_3H_5(OH_3)$. Also known as polyalcohol; polyol. { ¦päl·i¦hī·drək 'al·kəˌhȯl }

polyhydric phenol [ORGANIC CHEMISTRY] A phenolic compound containing two or more hydroxyl groups, such as diphenol, $C_6H_4(OH)_2$. { ¦päl·i¦hī·drək 'fēˌnȯl }

polyimide resin [ORGANIC CHEMISTRY] An aromatic polyimide made by reacting pyromellitic dianhydride with an aromatic diamine; has high resistance to thermal stresses; used to make components of internal combustion engines. { ¦päl·ē'iˌmīd ˌrez·ən }

polyisoprene [ORGANIC CHEMISTRY] $(C_5H_8)_x$ The basis of natural rubber, balata, guttapercha, and other rubberlike materials; can also be made synthetically; the stereospecific forms are *cis*-1,4- and *trans*-1,4-polyisoprene; the polymer is thermoplastic. { ¦päl·ē'īs·əˌprēn }

polylactic resin [ORGANIC CHEMISTRY] A soft, elastic resin made by the heat reaction of lactic acid with castor oil or other fatty oils; used to produce tough, water-resistant coatings. { ¦päl·i¦lak·tik ˌrez·ən }

polyligated atom [PHYSICAL CHEMISTRY] An atom that is bonded to more than one other atom. { ˌpäl·ē¦līˌgād·əd 'ad·əm }

polymer [ORGANIC CHEMISTRY] Substance made of giant molecules formed by the union of simple molecules (monomers); for example polymerization of ethylene forms a polyethylene chain, or condensation of phenol and formaldehyde (with production of water) forms phenol-formaldehyde resins. { 'päl·ə·mər }

polymerization [CHEMISTRY] **1.** The bonding of two or more monomers to produce a polymer. **2.** Any chemical reaction that produces such a bonding. { pəˌlim·ə·rə'zā·shən }

polymethyl methacrylate [ORGANIC CHEMISTRY] A thermoplastic polymer that is derived from methyl methacrylate, $CH_2{=}C(CH_3)COOCH_3$; transparent solid with excellent optical qualities and water resistance; used for aircraft domes, lighting fixtures, optical instruments, and surgical appliances. { 'päl·iˌmeth·əl mə'thak·rəˌlāt }

polynuclear hydrocarbon [ORGANIC CHEMISTRY] Hydrocarbon molecule with two or more closed rings; examples are naphthalene, $C_{10}H_8$, with two benzene rings side by side, or diphenyl, $(C_6H_5)_2$, with two bond-connected benzene rings. Also known as polycyclic hydrocarbon. { ¦päl·ə'nü·klē·ər 'hī·drəˌkär·bən }

polyol [ORGANIC CHEMISTRY] *See* polyhydric alcohol. { 'päl·ēˌȯl }

polyolefin [ORGANIC CHEMISTRY] A resinous material made by the polymerization of olefins, such as polyethylene from ethylene, polypropylene from propylene, or polybutene from butylene. { ¦päl·ē'ōl·ə·fən }

polyoxyalkylene resin [ORGANIC CHEMISTRY] Condensation polymer produced from an oxyalkene, such as polyethylene glycol from oxyethylene or ethylene glycol. { ¦päl·ē¦äk·sē'al·kəˌlēn 'rez·ən }

polyoxyethylene (8) stearate [ORGANIC CHEMISTRY] *See* polyoxyl (8) stearate. { ¦päl·ēˌäk·sē'eth·əˌlēn ¦āt 'stirˌāt }

polyoxyl (8) stearate [ORGANIC CHEMISTRY] A cream-colored, soft, waxy solid at 25°C; soluble in toluene, acetone, ether, and ethanol; used in bakery products as an emulsifier. Also known as polyoxyethylene (8) stearate. { ˌpäl·ē'äk·səl ¦āt 'stirˌāt }

polyphenyl [ORGANIC CHEMISTRY] Any of a group of direct colors used to dye cotton and wool. { 'päl·iˌfen·əl }

polyphenylene oxide [ORGANIC CHEMISTRY] A polyether resin of 2,6-dimethylphenol, $(CH_3)_2C_6H_3OH$; useful temperature range is −275 to 375°F (−168 to 191°C), with intermittent use possible up to 400°F (204°C). { ˌpäl·i'fen·əlˌēn 'äkˌsīd }

polyphosphazene [ORGANIC CHEMISTRY] A high-molecular-weight, essentially linear polymer with alternating phosphorus and nitrogen atoms in the skeleton and two side groups attached to each phosphorus. { ˌpäl·i'fa·sfəˌzēn }

polyphosphoric acid [INORGANIC CHEMISTRY] $H_6P_4O_{13}$ Viscous, water-soluble, hygroscopic, water-white liquid; used wherever concentrated phosphoric acid is needed. { ¦päl·i·fä'sfȯr·ik 'as·əd }

polypropylene [ORGANIC CHEMISTRY] $(C_3H_6)_x$ A crystalline, thermoplastic resin made by the polymerization of propylene, C_3H_6; the product is hard and tough, resists moisture, oils, and solvents, and withstands temperatures up to 170°C; used to make molded articles, fibers, film, rope, printing plates, and toys. { ˌpäl·ə'prō pəˌlēn }

polypropylene glycol [ORGANIC CHEMISTRY] $CH_3CHOH(CH_2OCHCH_3)_xCH_2OH$ Polymeric material similar to polyethylene glycol, but with greater oil solubility and less water solubility; used as a solvent for vegetable oils, waxes, and resins, in hydraulic fluids and as a chemical intermediate. { ˌpäl·ə'prō·pəˌlēn 'glīˌkȯl }

polysiloxane [ORGANIC CHEMISTRY] $(R_2SiO)_n$ A polymer in which the chain contains alternate silicon and oxygen atoms; in the formula, R can be H or an alkyl or aryl group; commercially, the R is usually CH_3 (the methylsiloxanes); properties vary with molecular weight, from oils to greases to rubbers to plastics. { ¦päl·i·si'läkˌsān }

polysorbate [ORGANIC CHEMISTRY] Any compound that is an ester of sorbitol. { ˌpäl·ē'sȯrˌbāt }

polystyrene [ORGANIC CHEMISTRY] $(C_6H_5CHCH_2)_x$ A water-white, tough synthetic resin made by polymerization of styrene; soluble in aromatic and chlorinated hydrocarbon solvents; used for injection molding, extrusion or casting for electrical insulation, fabric lamination, and molding of plastic objects. { ¦päl·i'stīˌrēn }

polysulfide rubber [ORGANIC CHEMISTRY] A synthetic polymer made by the reaction of sodium polysulfide with an organic dichloride; resistant to light, oxygen, oils, and solvents; impermeable to gases; poor tensile strength and abrasion resistance. { ¦päl·i'səlˌfīd 'rəb·ər }

polyterpene resin [ORGANIC CHEMISTRY] A thermoplastic resin or viscous liquid from polymerization of turpentine, used in paints, polishes, and rubber plasticizers, and to cure concrete and impregnate paper. { ¦päl·i'tərˌpēn ˌrez·ən }

polythene [ORGANIC CHEMISTRY] Common name for polyethlylene in Great Britain. { 'päl·iˌthēn }

polytrifluorochloroethylene resin [ORGANIC CHEMISTRY] *See* chlorotrifluoroethylene polymer. { ¦päl·i·trī¦flu̇r·ō¦klȯr·ō'eth·əˌlēn ˌrez·ən }

polyunsaturated acid [ORGANIC CHEMISTRY] A fatty acid with two or more double bonds per molecule, such as linoleic or linolenic acid. { ¦päl·ēˌən'sach·əˌrād·əd 'as·əd }

polyurethane resin [ORGANIC CHEMISTRY] Any resin resulting from the reaction of diisocyanates (such as toluene diisocyanate) with a phenol, amine, or hydroxylic or carboxylic compound to produce a polymer with free isocyanate groups; used as protective coatings, potting or casting resins, adhesives, rubbers, and foams, and in paints, varnishes, and adhesives. { ¦päl·ē'yu̇r·əˌthān 'rez·ən }

polyurethane rubber [ORGANIC CHEMISTRY] A synthetic polyurethane-resin elastomer made by the reaction of a diisocyanate to a polyester (such as the glycol-adipic acid ester); has high resistance to abrasion, oil, ozone, and high temperatures. Also known as polyester rubber. { ¦päl·ē'yu̇r·əˌthān 'rəb·ər }

polyvalent [CHEMISTRY] Pertaining to an ion with more than one valency, such as the sulfate ion, SO_4^{2-}. Also known as multivalent; polygen. { ¦päl·i'va·lənt }

polyvinyl acetal resin [ORGANIC CHEMISTRY] *See* vinyl acetal resin. { ¦päl·i'vīn·əl 'as·əˌtal 'rez·ən }

polyvinyl acetate [ORGANIC CHEMISTRY] $(H_2CCHOOCCH_3)_x$ A thermoplastic polymer; insoluble in water, gasoline, oils, and fats, soluble in ketones, alcohols, benzene, esters, and chlorinated hydrocarbons; used in adhesives, films, lacquers, inks, latex paints, and paper sizes. Abbreviated PVA; PVAc. { ¦päl·i'vīn·əl 'as·əˌtāt }

polyvinyl alcohol [ORGANIC CHEMISTRY] Water-soluble polymer made by hydrolysis of a polyvinyl ester (such as polyvinyl acetate); used in adhesives, as textile and paper sizes, and for emulsifying, suspending, and thickening of solutions. Abbreviated PVA. { ¦päl·i'vīn·əl 'al·kəˌhȯl }

polyvinyl carbazole [ORGANIC CHEMISTRY] Thermoplastic resin made by reaction of acetylene with carbazole; softens at 150°C; has good electrical properties and heat and chemical stabilities; used as a paper-capacitor impregnant and as a substitute for electrical mica. { ¦päl·i'vīn·əl 'kär·bə,zōl }

polyvinyl chloride [ORGANIC CHEMISTRY] $(H_2CCHCl)_x$ Polymer of vinyl chloride; tasteless, odorless; insoluble in most organic solvents; a member of the family of vinyl resins; used in soft flexible films for food packaging and in molded rigid products such as pipes, fibers, upholstery, and bristles. Abbreviated PVC. { ¦päl·i'vīn·əl 'klȯr,īd }

polyvinyl chloride acetate [ORGANIC CHEMISTRY] Thermoplastic copolymer of vinyl chloride, CH_2CHCl, and vinyl acetate, $CH_3COOCH{=}CH_2$; colorless solid with good resistance to water, concentrated acids, and alkalies; compounded with plasticizers, it yields a flexible material superior to rubber in aging properties; used for cable and wire coverings and protective garments. { ¦päl·i'vīn·əl 'klȯr,īd 'as·ə,tāt }

polyvinyl dichloride [ORGANIC CHEMISTRY] A high-strength polymer of chlorinated polyvinyl chloride; it is self-extinguishing and has superior chemical resistance; used for pipes carrying hot, corrosive materials. Abbreviated PVDC. { ¦päl·i'vīn·əl dī'klȯr,īd }

polyvinyl ether [ORGANIC CHEMISTRY] *See* polyvinyl ethyl ether. { ¦päl·i'vīn·əl 'ē·thər }

polyvinyl ethyl ether [ORGANIC CHEMISTRY] $[{-}CH(OC_2H_5)CH_2{-}]_x$ A viscous gum to rubbery solid, soluble in organic solvents; used for pressure-sensitive tape. Also known as polyvinyl ether. { ¦päl·i'vīn·əl 'eth·əl 'ē·thər }

polyvinyl fluoride [ORGANIC CHEMISTRY] $({-}H_2CCHF{-})_x$ Vinyl fluoride polymer; has superior resistance to weather, chemicals, oils, and stains, and has high strength; used for packaging (but not of food) and electrical equipment. { ¦päl·i'vīn·əl 'flu̇r,īd }

polyvinyl formate resin [ORGANIC CHEMISTRY] $(CH_2{=}CHOOCH)_x$ Clear-colored resin that is hard and solvent-resistant; used to make clear, hard plastics. { ¦päl·i'vīn·əl 'fȯr,māt ,rez·ən }

polyvinylidene chloride [ORGANIC CHEMISTRY] Thermoplastic polymer of vinylidene chloride, $H_2C{=}CCl_2$; white powder softening at 185-200°C; used to make soft-flexible to rigid products. { ¦päl·i·vī'nil·ə,dēn 'klȯr,īd }

polyvinylidene fluoride [ORGANIC CHEMISTRY] $H_2C{=}CF_2$ Fluorocarbon polymer made from vinylidene fluoride; has good tensile and compressive strength and high impact strength; used in chemical equipment for gaskets, impellers, and other pump parts, and for drum linings and protective coatings. { ¦päl·i·vī'nil·ə,dēn 'flu̇r,īd }

polyvinylidene resin [ORGANIC CHEMISTRY] *See* vinylidene resin. { ¦päl·i·vī'nil·ə,dēn 'rez·ən }

polyvinyl isobutyl ether [ORGANIC CHEMISTRY] $[{-}CH_2CHOCH_2CH(CH_3)_2{-}]_x$ An odorless synthetic resin; elastomer to viscous liquid depending on molecular weight; soluble in hydrocarbons, esters, ethers, and ketones, insoluble in water; used in adhesives, waxes, plasticizers, lubricating oils, and surface coatings. Abbreviated PVI. { ¦päl·i'vīn·əl ¦ī·sə¦byüd·əl 'e·thər }

polyvinyl methyl ether [ORGANIC CHEMISTRY] $({-}CH_2CHOCH_3{-})_x$ A colorless, tacky liquid, soluble in organic solvents, except aliphatic hydrocarbons, and in water below 32°C; used for pressure-sensitive adhesives, as a heat sensitizer for rubber latex, and as a pigment binder in inks and textile finishing. Abbreviated PVM. { ¦päl·i'vīn·əl 'meth·əl 'ē·thər }

polyvinyl pyrrolidone [ORGANIC CHEMISTRY] $(C_6H_9NO)_x$ A water-soluble, white, resinous solid; used in pharmaceuticals, cosmetics, detergents, and foods, and as a synthetic blood plasma. Abbreviated PVP. { ¦päl·i'vīn·əl pə'räl·ə,dōn }

polyvinyl resin [ORGANIC CHEMISTRY] Any resin or polymer derived from vinyl monomers. Also known as vinyl plastic. { ¦päl·i'vīn·əl 'rez·ən }

porous alum [INORGANIC CHEMISTRY] *See* aluminum sodium sulfate. { 'pȯr·əs 'al·əm }

positional isomer [CHEMISTRY] One of a set of structural isomers which differ only in the point at which a side-chain group is attached. [ORGANIC CHEMISTRY] Constitutional isomer having the same functional group located in different positions along a chain or in a ring. { pə'zish·ən·əl 'ī·sə·mər }

positive ion [CHEMISTRY] An atom or group of atoms which by loss of one or more electrons has acquired a positive electric charge; occurs on ionization of chemical compounds as H^+ from ionization of hydrochloric acid, HCl. { 'päz·əd·iv 'ī,än }

positron emission spectroscopy [SPECTROSCOPY] A technique in which a solid surface is bombarded with a low-energy monoenergetic positron beam and the energies of positrons emitted from the surface are measured to determine the amounts of energy lost to molecules adsorbed on the surface. { 'päz·ə,trän i¦mish·ən spek'träs·kə·pē }

positronium velocity spectroscopy [SPECTROSCOPY] A technique in which a solid surface is bombarded with a low-energy monoenergetic positron beam and the velocities of the emitted positronium atoms are measured to determine the energy and momentum spectrum of the density of electron states near the surface. { ,päz·ə'trō·nē·əm və'läs·əd·ē spek'träs·kə·pē }

postignition [CHEMISTRY] Surface ignition after the passage of the normal spark. { ,pōst·ig'nish·ən }

postprecipitation [CHEMISTRY] Precipitation of an impurity from a supersaturated solution onto the surface of an already present precipitate; used for analytical laboratory separations. { ¦pōs·pri,sip·ə'tā·shən }

potash [INORGANIC CHEMISTRY] *See* potassium carbonate. { 'päd,ash }

potash blue [INORGANIC CHEMISTRY] A pigment made by oxidizing ferrous ferrocyanide; used in making carbon paper. { 'päd,ash 'blü }

potassium [CHEMISTRY] A chemical element, symbol K, atomic number 19, atomic weight 39.102; an alkali metal. Also known as kalium. { pə'tas·ē·əm }

potassium acetate [ORGANIC CHEMISTRY] $KC_2H_3O_2$ White, deliquescent solid; soluble in water and alcohol, insoluble in ether; melts at 292°C; used as analytical reagent, dehydrating agent, in medicine, and in crystal glass manufacture. { pə'tas·ē·əm 'as·ə,tāt }

potassium acid carbonate [INORGANIC CHEMISTRY] *See* potassium bicarbonate. { pə'tas·ē·əm 'as·əd 'kär·bə,nāt }

potassium acid fluoride [INORGANIC CHEMISTRY] *See* potassium bifluoride. { pə'tas·ē·əm 'as·əd 'flůr,īd }

potassium acid oxalate [ORGANIC CHEMISTRY] *See* potassium binoxalate. { pə'tas·ē·əm 'as·əd 'äk·sə,lāt }

potassium acid phosphate [INORGANIC CHEMISTRY] *See* potassium phosphate. { pə'tas·ē·əm 'as·əd 'fäs,fat }

potassium acid phthalate [ORGANIC CHEMISTRY] *See* potassium biphthalate. { pə'tas·ē·əm 'as·əd 'tha,lāt }

potassium acid saccharate [ORGANIC CHEMISTRY] $HOOC—(CHOH)_4COOK$ An off-white powder, soluble in hot water, acid, or alkaline solutions, used in rubber formulations, soaps, and detergents, and for metal plating. { pə'tas·ē·əm 'as·əd 'sak·ə,rāt }

potassium acid sulfate [INORGANIC CHEMISTRY] *See* potassium bisulfate. { pə'tas·ē·əm 'as·əd 'səl,fāt }

potassium acid sulfite [INORGANIC CHEMISTRY] *See* potassium bisulfite. { pə'tas·ē·əm 'as·əd 'səl,fīt }

potassium acid tartrate [ORGANIC CHEMISTRY] *See* potassium bitartrate. { pə'tas·ē·əm 'as·əd 'tär,trāt }

potassium alginate [ORGANIC CHEMISTRY] $(C_6H_7O_6K)_n$ A hydrophilic colloid occurring as filaments, grains, granules, and powder; used in food processing as a thickener and stabilizer. Also known as potassium polymannuronate. { pə'tas·ē·əm 'al·jə,nāt }

potassium alum [INORGANIC CHEMISTRY] *See* potassium aluminum sulfate. { pə'tas·ē·əm 'al·əm }

potassium aluminate [INORGANIC CHEMISTRY] $K_2Al_2O_4 \cdot 3H_2O$ Water-soluble, alcohol-insoluble, lustrous crystals; used as a dyeing and printing mordant, and as a paper sizing. { pə'tas·ē·əm ə'lüm·ə,nāt }

potassium aluminum fluoride [INORGANIC CHEMISTRY] K_3AlF_6 A toxic, white powder used as an insecticide. { pə'tas·ē·əm ə'lüm·ə·nəm 'flůr,īd }

potassium aluminum sulfate [INORGANIC CHEMISTRY] $KAl(SO_4)_2 \cdot 12H_2O$ White, odorless crystals that are soluble in water; used in medicines and baking powder, in dyeing, papermaking, and tanning. Also known as alum; aluminum potassium sulfate; potassium alum. { pə'tas·ē·əm ə'lüm·ə·nəm 'səl,fāt }

potassium antimonate [INORGANIC CHEMISTRY] $KSbO_3$ White, water-soluble crystals. Also known as potassium stibnate. { pə'tas·ē·əm 'ant·ə·mə,nāt }

potassium antimonyl tartrate [ORGANIC CHEMISTRY] *See* tartar emetic. { pə'tas·ē·əm 'ant·ə·mə,nil 'tär,trāt }

potassium argentocyanide [ORGANIC CHEMISTRY] *See* silver potassium cyanide. { pə'tas·ē·əm ¦ar·jən·tō'sī·ə,nīd }

potassium arsenate [INORGANIC CHEMISTRY] K_3AsO_4 Poisonous, colorless crystals; soluble in water, insoluble in alcohol; used as an insecticide, analytical reagent, and in hide preservation and textile printing. Also known as Macquer's salt. { pə'tas·ē·əm 'ärs·ən,āt }

potassium arsenite [INORGANIC CHEMISTRY] $KH(AsO_2)_2$ Poisonous, hygroscopic, white powder; soluble in alcohol; decomposes slowly in air; used in medicine, on mirrors, and as an analytical reagent. Also known as potassium metarsenite. { pə'tas·ē·əm 'ärs·ən,īt }

potassium aurichloride [INORGANIC CHEMISTRY] *See* potassium gold chloride. { pə'tas·ē·əm ¦ȯr·ə'klȯr,īd }

potassium bicarbonate [INORGANIC CHEMISTRY] $KHCO_3$ A white powder or granules, or transparent colorless crystals; used in baking powder and in medicine as an antacid. Also known as potassium acid carbonate. { pə'tas·ē·əm bī'kär·bə,nāt }

potassium bichromate [INORGANIC CHEMISTRY] *See* potassium dichromate. { pə'tas·ē·əm bī'krō,māt }

potassium bifluoride [INORGANIC CHEMISTRY] KHF_2 Colorless, corrosive, poisonous crystals; soluble in water and dilute alcohol; used to etch glass and as a metallurgy flux. Also known as Fremy's salt; potassium acid fluoride. { pə'tas·ē·əm bī'flu̇r,īd }

potassium binoxalate [ORGANIC CHEMISTRY] $KHC_2O_4 \cdot H_2O$ A poisonous, white, odorless, crystalline compound; used to clean wood and remove ink stains, as a mordant in dyeing, and in photography. Also known as potassium acid oxalate; sal acetosella; salt of sorrel. { pə'tas·ē·əm bə'näk·sə,lāt }

potassium biphthalate [ORGANIC CHEMISTRY] $HOOCC_6H_4COOK$ A crystalline compound, soluble in 12 parts of water; used as a buffer in pH determinations and as a primary standard for preparation of volumetric alkali solutions. Also known as acid potassium phthalate; potassium acid phthalate; potassium hydrogen phthalate. { pə'tas·ē·əm bī'tha,lāt }

potassium bismuth tartrate [ORGANIC CHEMISTRY] A white, granular powder with a sweet taste; soluble in water; used in medicine. Also known as bismuth potassium tartrate. { pə'tas·ē·əm 'biz·məth 'tär,trāt }

potassium bisulfate [INORGANIC CHEMISTRY] $KHSO_4$ Water-soluble, colorless crystals, melting at 214°C; used in winemaking, fertilizer manufacture, and as a flux and food preservative. Also known as acid potassium sulfate; potassium acid sulfate. { pə'tas·ē·əm bī'səl,fāt }

potassium bisulfite [INORGANIC CHEMISTRY] $KHSO_3$ White, water-soluble powder with sulfur dioxide aroma; insoluble in alcohol; decomposes when heated; used as an antiseptic and reducing chemical, and in analytical chemistry, tanning, and bleaching. Also known as potassium acid sulfite. { pə'tas·ē·əm bī'səl,fīt }

potassium bitartrate [ORGANIC CHEMISTRY] $KHC_4H_4O_6$ White, water-soluble crystals or powder; used in baking powder, for medicine, and as an acid and buffer in foods. Also known as cream of tartar; potassium acid tartrate. { pə'tas·ē·əm bī'tär,trāt }

potassium borohydride [INORGANIC CHEMISTRY] KBH_4 A white, crystalline powder, soluble in water, alcohol, and ammonia; used as a hydrogen source and a reducing agent for aldehydes and ketones. { pə'tas·ē·əm ¦bȯr·ō'hī,drīd }

potassium bromate [INORGANIC CHEMISTRY] $KBrO_3$ Water-soluble, white crystals, melting at 434°C; insoluble in alcohol; strong oxidizer and a fire hazard; used in analytical chemistry and as an additive for permanent-wave compounds. { pə'tas·ē·əm 'brō,māt }

potassium bromide [INORGANIC CHEMISTRY] KBr White, hygroscopic crystals with bitter taste; soluble in water and glycerin, slightly soluble in alcohol and ether; melts at 730°C; used in medicine, soaps, photography, and lithography. { pə'tas·ē·əm 'brō,mīd }

potassium bromide-disk technique [ANALYTICAL CHEMISTRY] Method of preparing an infrared spectrometry sample by grinding it and mixing it with a dry powdered alkali halide (such as KBr), then compressing the mixture into a tablet or pellet. Also known as pellet technique; pressed-disk technique. { pə'tas·ē·əm 'brō,mīd ¦disk tek,nēk }

potassium cadmium iodide [INORGANIC CHEMISTRY] *See* potassium tetraiodocadmate. { pə'tas·ē·əm 'kad·mē·əm 'ī·ə,dīd }

potassium carbonate [INORGANIC CHEMISTRY] K_2CO_3 White, water-soluble, deliquescent powder, melting at 891°C; insoluble in alcohol; used in brewing, ceramics, explosives, fertilizers, and as a chemical intermediate. Also known as potash; salt of tartar. { pə'tas·ē·əm 'kär·bə,nāt }

potassium chlorate [INORGANIC CHEMISTRY] $KClO_3$ Transparent, colorless crystals or a white powder with a melting point of 356°C; soluble in water, alcohol, and alkalies; used as an oxidizing agent, for explosives and matches, and in textile printing and paper manufacture. { pə'tas·ē·əm 'klòr,āt }

potassium chloride [INORGANIC CHEMISTRY] KCl Colorless crystals with saline taste; soluble in water, insoluble in alcohol; melts at 776°C; used as a fertilizer and in photography and pharmaceutical preparations. Also known as potassium muriate. { pə'tas·ē·əm 'klòr,īd }

potassium chloroaurate [INORGANIC CHEMISTRY] *See* potassium gold chloride. { pə'tas·ē·əm ¦klòr·ō'òr,āt }

potassium chloroplatinate [INORGANIC CHEMISTRY] K_2PtCl_6 Orange-yellow crystals or powder which decomposes when heated (250°C); used in photography. Also known as platinum potassium chloride; potassium platinichloride. { pə'tas·ē·əm ¦klòr·ō'plat·ən,āt }

potassium chromate [INORGANIC CHEMISTRY] K_2CrO_4 Yellow crystals, melting at 971°C; soluble in water, insoluble in alcohol; used as an analytical reagent and textile mordant, in enamels, inks, and medicines, and as a chemical intermediate. { pə'tas·ē·əm 'krō,māt }

potassium chromium sulfate [INORGANIC CHEMISTRY] *See* chrome alum. { pə'tas·ē·əm 'krō·mē·əm 'səl,fāt }

potassium citrate [ORGANIC CHEMISTRY] $K_3C_6H_5O_7 \cdot H_2O$ Odorless crystals with saline taste; soluble in water and glycerol, deliquesent and insoluble in alcohol; decomposes about 230°C; used in medicine. { pə'tas·ē·əm 'sī,trāt }

potassium cobaltinitrite [INORGANIC CHEMISTRY] *See* cobalt potassium nitrite. { pə'tas·ē·əm kō¦bòl·tə'nī,trāt }

potassium cyanate [INORGANIC CHEMISTRY] KOCN Colorless, water-soluble crystals; used as an herbicide and for the manufacture of drugs and organic chemicals. { pə'tas·ē·əm 'sī·ə,nāt }

potassium cyanide [INORGANIC CHEMISTRY] KCN Poisonous, white, deliquescent crystals with bitter almond taste; soluble in water, alcohol, and glycerol, used for metal extraction, for electroplating, for heat-treating steel, and as an analytical reagent and insecticide. { pə'tas·ē·əm 'sī·ə,nīd }

potassium cyanoargentate [ORGANIC CHEMISTRY] *See* silver potassium cyanide. { pə'tas·ē·əm ¦sī·ə·nō'är·jən,tāt }

potassium cyanoaurite [INORGANIC CHEMISTRY] *See* potassium gold cyanide. { pə'tas·ē·əm ¦sī·ə·nō'òr,īt }

potassium dichloroisocyanurate [INORGANIC CHEMISTRY] White, crystalline powder or granules; strong oxidant used in dry household bleaches, detergents, and scouring powders. { pə'tas·ē·əm dī¦klòr·ō,ī'sō,sī'an·yùr,āt }

potassium dichromate [INORGANIC CHEMISTRY] $K_2Cr_2O_7$ Poisonous, yellowish-red crystals with metallic taste; soluble in water, insoluble in alcohol; melts at 396°C, decomposes at 500°C; used as an oxidizing agent and analytical reagent, and in explosives, matches, and electroplating. Also known as potassium bichromate; red potassium chromate. { pə'tas·ē·əm dī'krō,māt }

potassium dihydrogen phosphate [INORGANIC CHEMISTRY] *See* potassium phosphate. { pə'tas·ē·əm dī'hī·drə·jən 'fäs,fāt }

potassium diphosphate [INORGANIC CHEMISTRY] *See* potassium phosphate. { pə'tas·ē·əm dī'fäs,fāt }

potassium ferric oxalate [INORGANIC CHEMISTRY] $K_3Fe(C_2O_4)_3 \cdot 3H_2O$ Green crystals decomposing at 230°C, soluble in water and acetic acid; used in photography and blueprinting. { pə'tas·ē·əm 'fer·ik 'äk·sə,lāt }

potassium ferricyanide [INORGANIC CHEMISTRY] $K_3Fe(CN)_6$ Poisonous, water-soluble, bright-red crystals; decomposes when heated; used in calico printing and wool dyeing. Also known as red potassium prussiate; red prussiate of potash. { pə'tas·ē·əm ,fer·ə'sī·ə,nīd }

potassium ferrocyanide [INORGANIC CHEMISTRY] $K_4Fe(CN)_6 \cdot 3H_2O$ Yellow crystals with saline taste; soluble in water, insoluble in alcohol; loses water at 60°C; used in medicine, dry colors, explosives, and as an analytical reagent. Also known as yellow prussiate of potash. { pə'tas·ē·əm ,fer·ō'sī·ə,nīd }

potassium fluoborate [INORGANIC CHEMISTRY] KBF_4 White powder or gelatinous crystals that decompose at high temperatures; slightly soluble in water and hot alcohol; used as a sand agent to cast magnesium and aluminum, and in electrochemical processes. { pə'tas·ē·əm ¦flü·ə'bȯr,āt }

potassium fluoride [INORGANIC CHEMISTRY] KF or $KF \cdot 2H_2O$ Poisonous, white, deliquescent crystals with saline taste; soluble in water and hydrofluoric acid, insoluble in alcohol; melts at 846°C; used to etch glass and as a preservative and insecticide. { pə'tas·ē·əm 'flür,īd }

potassium fluosilicate [INORGANIC CHEMISTRY] K_2SiF_6 An odorless, white crystalline compound; slightly soluble in water; used in vitreous frits, synthetic mica, metallurgy, and ceramics. Also known as potassium silicofluoride. { pə'tas·ē·əm ¦flü·ə'sil·ə·kət }

potassium gluconate [ORGANIC CHEMISTRY] $KC_6H_{11}O_7$ An odorless, white crystalline compound with salty taste; soluble in water, insoluble in alcohol and benzene; used in medicine. { pə'tas·ē·əm 'glü·kə,nāt }

potassium glutamate [ORGANIC CHEMISTRY] $KOOC(CH_2)_2CH(NH_2)COOH \cdot H_2O$ White, hygroscopic, water-soluble powder; used as a flavor enhancer and salt substitute. Also known as monopotassium L-glutamate. { pə'tas·ē·əm 'glüd·ə,māt }

potassium glycerinophosphate [ORGANIC CHEMISTRY] *See* potassium glycerophosphate. { pə'tas·ē·əm ¦glis·ə·rə·nō'fäs,fāt }

potassium glycerophosphate [ORGANIC CHEMISTRY] $K_2C_3H_5O_2 \cdot H_2PO_4 \cdot 3H_2O$ Pale yellow, syrupy liquid, soluble in alcohol; used in medicine and as a dietary supplement. Also known as potassium glycerinophosphate. { pə'tas·ē·əm ¦glis·ə·rō'fäs,fāt }

potassium gold chloride [INORGANIC CHEMISTRY] $KAuCl_4 \cdot 2H_2O$ Yellow crystals, soluble in water, ether, and alcohol; used in photography and medicine. Also known as gold potassium chloride; potassium aurichloride; potassium chloroaurate. { pə'tas·ē·əm 'gōld 'klȯr,īd }

potassium gold cyanide [INORGANIC CHEMISTRY] $KAu(CN)_2$ A white, water-soluble, crystalline powder; used in medicine and for gold plating. Also known as gold potassium cyanide; potassium cyanoaurite. { pə'tas·ē·əm 'gōld 'sī·ə,nīd }

potassium hydrate [INORGANIC CHEMISTRY] *See* potassium hydroxide. { pə'tas·ē·əm 'hī,drāt }

potassium hydrogen phosphate [INORGANIC CHEMISTRY] *See* potassium phosphate. { pə'tas·ē·əm 'hī·drə·jən 'fäs,fāt }

potassium hydrogen phthalate [ORGANIC CHEMISTRY] *See* potassium biphthalate. { pə'tas·ē·əm 'hī·drə·jən 'tha,lāt }

potassium hydroxide [INORGANIC CHEMISTRY] KOH Toxic, corrosive, water-soluble, white solid, melting at 360°C; used to make soap and matches, and as an analytical reagent and chemical intermediate. Also known as caustic potash; potassium hydrate. { pə'tas·ē·əm hī'dräk,sīd }

potassium hyperchlorate [INORGANIC CHEMISTRY] *See* potassium perchlorate. { pə'tas·ē·əm ,hī·pər'klȯr,āt }

potassium hypophosphite [INORGANIC CHEMISTRY] KH_2PO_2 White, opaque crystals or powder, soluble in water and alcohol; used in medicine. { pə'tas·ē·əm ˌhī·pō'fäsˌfīt }

potassium iodate [INORGANIC CHEMISTRY] KIO_3 Odorless, white crystals; soluble in water, insoluble in alcohol; melts at 560°C; used as an analytical reagent and in medicine. { pə'tas·ē·əm 'ī·əˌdāt }

potassium iodide [INORGANIC CHEMISTRY] KI Water- and alcohol-soluble, white crystals with saline taste; melts at 686°C; used in medicine and photography, and as an analytical reagent. { pə'tas·ē·əm 'ī·əˌdīd }

potassium linoleate [ORGANIC CHEMISTRY] $C_{17}H_{31}COOK$ Light tan, water-soluble paste; used as an emulsifying agent. { pə'tas·ē·əm li'nō·lēˌāt }

potassium manganate [INORGANIC CHEMISTRY] K_2MnO_4 Water-soluble dark-green crystals, decomposing at 190°C; used as an analytical reagent, bleach, oxidizing agent, disinfectant, mordant for dyeing wool and in photography, printing, and water purification. { pə'tas·ē·əm 'maŋ·gəˌnāt }

potassium metabisulfite [INORGANIC CHEMISTRY] $K_2S_2O_5$ White granules or powder, decomposing at 150-190°C; used as an antiseptic, for winemaking, food preservation, and process engraving, and as a source for sulfurous acid. Also known as potassium pyrosulfite. { pə'tas·ē·əm ˌmed·ə·bī'səlˌfīt }

potassium metarsenite [INORGANIC CHEMISTRY] *See* potassium arsenite. { pə'tas·ē·əm ¦med·ə'ärs·ənˌīt }

potassium monophosphate [INORGANIC CHEMISTRY] *See* potassium phosphate. { pə'tas·ē·əm ˌmän·ō'fäsˌfāt }

potassium muriate [INORGANIC CHEMISTRY] *See* potassium chloride. { pə'tas·ē·əm 'myu̇r·ēˌāt }

potassium nitrate [INORGANIC CHEMISTRY] KNO_3 Flammable, water-soluble, white crystals with saline taste; melts at 337°C; used in pyrotechnics, explosives, and matches, as a fertilizer, and as an analytical reagent. Also known as niter. { pə'tas·ē·əm 'nīˌtrāt }

potassium nitrite [INORGANIC CHEMISTRY] KNO_2 White, deliquescent prisms, melting at 297-450°C; soluble in water, insoluble in alcohol; strong oxidizer, exploding at over 550°C, used as an analytical reagent, in medicine, organic synthesis, pyrotechnics, and explosives. { pə'tas·ē·əm 'nīˌtrīt }

potassium oxalate [ORGANIC CHEMISTRY] $K_2C_2O_4 \cdot H_2O$ Odorless, efflorescent, water soluble, colorless crystals; decomposes when heated; used in analytical chemistry and photography and as a bleach and oxalic acid source. { pə'tas·ē·əm 'äk·səˌlāt }

potassium oxide [INORGANIC CHEMISTRY] K_2O Gray, water-soluble crystals, melts at red heat; forms potassium hydroxide in water. { pə'tas·ē·əm 'äkˌsīd }

potassium percarbonate [INORGANIC CHEMISTRY] $K_2C_2O_6 \cdot H_2O$ White, granular, water-soluble mass with a melting point of 200-300°C; used in microscopy, photography, and textile printing. { pə'tas·ē·əm pər'kär·bəˌnāt }

potassium perchlorate [INORGANIC CHEMISTRY] $KClO_4$ Explosive, oxidative, colorless crystals; soluble in water, insoluble in alcohol; decomposes at 400°C; used in explosives, medicine, pyrotechnics, analysis, and as a reagent and oxidizing agent. Also known as potassium hyperchlorate. { pə'tas·ē·əm pər'klórˌāt }

potassium permanganate [INORGANIC CHEMISTRY] $KMnO_4$ Highly oxidative, water-soluble, purple crystals with sweet taste; decomposes at 240°C; and explodes in contact with oxidizable materials; used as a disinfectant and analytical reagent, in dyes, bleaches, and medicines, and as a chemical intermediate. Also known as purple salt. { pə'tas·ē·əm pər'man·gəˌnāt }

potassium peroxide [INORGANIC CHEMISTRY] K_2O_2 Yellow mass with a melting point of 490°C; decomposes with oxygen evolution in water; used as an oxidizing and bleaching agent. { pə'tas·ē·əm pə'räkˌsīd }

potassium peroxydisulfate [INORGANIC CHEMISTRY] *See* potassium persulfate. { pə'tas·ē·əm pə¦räk·sē·dī'səlˌfāt }

potassium persulfate [INORGANIC CHEMISTRY] $K_2S_2O_8$ White, water-soluble crystals, decomposing below 100°C; used for bleaching and textile desizing, as an oxidizing agent and antiseptic, and in the manufacture of soap and pharmaceuticals. Also known as potassium peroxydisulfate. { pə'tas·ē·əm pər'səl‚fāt }

potassium phosphate [INORGANIC CHEMISTRY] Any one of three orthophosphates of potassium.The monobasic form, KH_2PO_4, consists of colorless, water-soluble crystals melting at 253°C; used in sonar transducers, optical modulation, medicine, baking powders, and nutrient solutions; also known as potassium acid phosphate, potassium dihydrogen phosphate (KDP), potassium diphosphate, potassium orthophosphate. The dibasic form, K_2HOP_4, consists of white, water-soluble crystals; used in medicine, fermentation, and nutrient solutions; also known as potassium hydrogen phosphate, potassium monophosphate. The tribasic form, K_3PO_4, is a water-soluble, hygroscopic white powder, melting at 1340°C; used to purify gasoline, to soften water, and to make liquid soaps and fertilizers; also known as neutral potassium phosphate, tripotassium orthophosphate. { pə'tas·ē·əm 'fäs‚fāt }

potassium platinichloride [INORGANIC CHEMISTRY] *See* potassium chloroplatinate. { pə'tas·ē·əm ¦plat·ən·ə'klȯr‚īd }

potassium polymannuronate [ORGANIC CHEMISTRY] *See* potassium alginate. { pə'tas·ē·əm ¦päl·ē·man'yu̇r·ə‚nāt }

potassium polymetaphosphate [INORGANIC CHEMISTRY] $(KPO_3)_n$ White powder with a molecular weight up to 500,000; used in foods as a fat emulsifier and moisture-retaining agent. { pə'tas·ē·əm päl·e‚med·ə'fäs‚fāt }

potassium pyrophosphate [INORGANIC CHEMISTRY] $K_4P_2O_7 \cdot 3H_2O$ Water-soluble, colorless crystals; dehydrates below 300°C, melts at 1090°C; used in tin plating, china-clay purification, dyeing, oil-drilling muds, and synthetic rubber production. Also known as normal potassium pyrophosphate; tetrapotassium pyrophosphate. { pə'tas·ē·əm ‚pī·rō'fäs‚fāt }

potassium pyrosulfite [INORGANIC CHEMISTRY] *See* potassium metabisulfite. { pə'tas·ē·əm ‚pī·rō'səl‚fīt }

potassium silicate [INORGANIC CHEMISTRY] $SiO_2{=}K_2O$ A compound existing in two forms, solution and solid (glass); as a solution, it is colorless to turgid in water, and is used in paints and coatings, as an arc-electrode binder and catalyst and in detergents; as a solid, it is colorless and water-soluble solid, and is used in glass manufacture and for dyeing and bleaching. { pə'tas·ē·əm 'sil·ə‚kāt }

potassium silicofluoride [INORGANIC CHEMISTRY] *See* potassium fluosilicate. { pə'tas·ē·əm ¦sil·ə·kō'flu̇r‚īd }

potassium sodium ferricyanide [INORGANIC CHEMISTRY] $K_2NaFe(CN)_6$ Red, water-soluble crystals; used for blueprint paper and in photography. { pə'tas·ē·əm 'sōd·ē·əm ‚fer·ə'sī·ə‚nīd }

potassium sodium tartrate [INORGANIC CHEMISTRY] $KNaC_4H_4O_6 \cdot 4H_2O$ Colorless, water-soluble, efflorescent crystals or white powder with a melting point of 70-80°C; used in medicine and as a buffer and sequestrant in foods. Also known as Rochelle salt. { pə'tas·ē·əm 'sōd·ē·əm 'tär‚trāt }

potassium sorbate [ORGANIC CHEMISTRY] $C_6H_7KO_2$ A crystalline compound, more soluble in water than in alcohol; decomposes above 270°C; used to inhibit mold and yeast growth in food. { pə'tas·ē·əm 'sȯr‚bāt }

potassium stannate [INORGANIC CHEMISTRY] $K_2SnO_3 \cdot 3H_2O$ White crystals; soluble in water, insoluble in alcohol; used in textile printing and dyeing, and in tin-plating baths. { pə'tas·ē·əm 'stan‚āt }

potassium stibnate [INORGANIC CHEMISTRY] *See* potassium antimonate. { pə'tas·ē·əm 'stib‚nīt }

potassium sulfate [INORGANIC CHEMISTRY] K_2SO_4 Colorless crystals with bitter taste; soluble in water, insoluble in alcohol; melts at 1072°C; used as an analytical reagent, medicine, and fertilizer, and in aluminum and glass manufacture. Also known as salt of Lemery. { pə'tas·ē·əm 'səl‚fāt }

potassium sulfide [INORGANIC CHEMISTRY] K_2S Moderately flammable, water-soluble, deliquescent red crystals; melts at 840°C; used in analytical chemistry, medicine, and depilatories. Also known as fused potassium sulfide, hepar sulfuris; potassium sulfuret. { pə'tas·ē·əm 'səl‚fīd }

potassium sulfite [INORGANIC CHEMISTRY] $K_2SO_3 \cdot 2H_2O$ Water-soluble, white crystals; used in medicine and photography. { pə'tas·ē·əm 'səl,fīt }

potassium sulfuret [INORGANIC CHEMISTRY] *See* potassium sulfide. { pə'tas·ē·əm 'səl·fə,ret }

potassium tetraiodocadmate [INORGANIC CHEMISTRY] $K_2(CdI_4) \cdot 2H_2O$ A crystalline compound; used in analytical chemistry for alkaloids, amines, and other compounds. Also known as cadmium potassium iodide; potassium cadmium iodide. { pə'tas·ē·əm ¦te·trə¦ī·ə·dō'kad,māt }

potassium thiocyanate [INORGANIC CHEMISTRY] KCNS Water- and alcohol-soluble, colorless, odorless hygroscopic crystals with saline taste; decomposes at 500°C; used as an analytical reagent and in freezing mixtures, chemicals manufacture, textile printing and dyeing, and photographic chemicals. { pə'tas·ē·əm ¦thī·ō'sī·ə,nāt }

potassium undecylenate [ORGANIC CHEMISTRY] $CH_2{:}CH(CH_2)_8COOK$ A white, water-soluble powder, decomposing at about 250°C; used in pharmaceuticals and cosmetics as a fungistat and bacteriostat. { pə'tas·ē·əm ,ən,des·ə'le,nāt }

potassium xanthate [ORGANIC CHEMISTRY] KC_2H_5OCSS Water- and alcohol-soluble, yellow crystals; used as an analytical reagent and soil-treatment fungicide. { pə'tas·ē·əm 'zan,thāt }

potential electrolyte [PHYSICAL CHEMISTRY] A solid material composed of uncharged molecules that can react chemically with a solvent to yield some ions in solution. { pə'ten·chəl i'lek·trə,līt }

potentiometric cell [ANALYTICAL CHEMISTRY] Container for the two electrodes and the electrolytic solution being titrated potentiometrically. { pə¦ten·chē·ə¦me·trik 'sel }

potentiometric titration [ANALYTICAL CHEMISTRY] Solution titration in which the end point is read from the electrode-potential variations with the concentrations of potential-determining ions, following the Nernst concept. Also known as constant-current titration. { pə¦ten·chē·ə¦me·trik tī'trā·shən }

Pr [CHEMISTRY] *See* praseodymium.

praseodymium [CHEMISTRY] A chemical element, symbol Pr, atomic number 59, atomic weight 140.91; a metallic element of the rare-earth group. { ,prā·zē·ō'dim·ē·əm }

precipitant [CHEMISTRY] A chemical or chemicals that cause a precipitate to form when added to a solution. { prə'sip·ət·əns }

precipitate [CHEMISTRY] **1.** A substance separating, in solid particles, from a liquid as the result of a chemical or physical change. **2.** To form a precipitate. { prə'sip·ə,tat }

precipitation [CHEMISTRY] The process of producing a separable solid phase within a liquid medium; represents the formation of a new condensed phase, such as a vapor or gas condensing to liquid droplets; a new solid phase gradually precipitates within a solid alloy as a result of slow, inner chemical reaction; in analytical chemistry, precipitation is used to separate a solid phase in an aqueous solution. { prə,sip·ə'tā·shən }

precipitation indicator [ANALYTICAL CHEMISTRY] In a titration, a substance that precipitates from solution in a clearly visible form at the end point. { prə,sip·ə'tā·shən ,in·də,kād·ər }

precipitation number [ANALYTICAL CHEMISTRY] The number of milliliters of asphaltic precipitate formed when 10 milliliters of petroleum-lubricating oil is mixed with 90 milliliters of a special-quality petroleum naphtha, then centrifuged according to American Society for Testing and Materials test conditions; used to determine the quantity of asphalt in petroleum-lubricating oil. { prə,sip·ə'tā·shən ,nəm·bər }

precipitation titration [ANALYTICAL CHEMISTRY] Amperometric titration in which the potential of a suitable indicator electrode is measured during the titration. { prə,sip·ə'tā·shən tī,trā·shən }

predissociation [PHYSICAL CHEMISTRY] The dissociation of a molecule that has absorbed energy before it can lose energy by radiation. { ¦prē·di,sō·sē'ā·shən }

Pregl procedure [ANALYTICAL CHEMISTRY] Microanalysis technique in which the sample is decomposed thermally, with subsequent oxidation of decomposition products. { 'prā·gəl prə,sē·jər }

preparing salt [INORGANIC CHEMISTRY] *See* sodium stannate. { prə'per·iŋ ˌsȯlt }

prepolymer [ORGANIC CHEMISTRY] A plastic or resin intermediate whose molecular weight is between that of the original monomer or monomers and that of the final, cured polymer or resin. { prē'päl·i·mər }

pressed-disk technique [ANALYTICAL CHEMISTRY] *See* potassium bromide-disk technique. { 'prest ¦disk tekˌnēk }

pressure broadening [SPECTROSCOPY] A spreading of spectral lines when pressure is increased, due to an increase in collision broadening. { 'presh·ər ˌbrȯd·ən·iŋ }

pressure effect [SPECTROSCOPY] The effect of changes in pressure on spectral lines in the radiation emitted or absorbed by a substance; namely, pressure broadening and pressure shift. { 'presh·ər iˌfekt }

pressure shift [SPECTROSCOPY] An increase in the wavelength at which a spectral line has maximum intensity, which takes place when pressure is increased. { 'presh·ər ˌshift }

primary [CHEMISTRY] A term used to distinguish basic compounds from similar or isomeric forms; in organic compounds, for example, RCH_2OH is a primary alcohol, R_1R_2CHOH is a secondary alcohol, and $R_1R_2R_3COH$ is a tertiary alcohol; in inorganic compounds, for example, NaH_2PO_4 is primary sodium phosphate, Na_2HPO_4 is the secondary form, and Na_3PO_4 is the tertiary form. { 'prī ˌmer·ē }

primary alcohol [ORGANIC CHEMISTRY] An alcohol whose molecular structure may be written as RCH_2OH, rather than as R_1R_2CHOH (secondary) or $R_1R_2R_3COH$ (tertiary). { 'prīˌmer·ē 'al·kəˌhȯl }

primary amine [ORGANIC CHEMISTRY] An amine whose molecular structure may be written as RNH_2, instead of R_1R_2NH (secondary) or $R_1R_2R_3N$ (tertiary). { 'prīˌmer·ē 'amˌēn }

primary carbon atom [ORGANIC CHEMISTRY] A carbon atom in a molecule that is singly bonded to only one other carbon atom. { 'prīˌmer·ē 'kär·bən ˌad·əm }

primary hydrogen atom [ORGANIC CHEMISTRY] A hydrogen atom that is bonded to a primary carbon atom. { 'prīˌmer·ē 'hī·drə·jən ˌad·əm }

primary structure [ORGANIC CHEMISTRY] The chemical structure of a polymer chain. { 'prīˌmer·ē 'strək·chər }

principal line [SPECTROSCOPY] That spectral line which is most easily excited or observed. { 'prin·sə·pəl 'līn }

principal moments [PHYSICAL CHEMISTRY] The three moments of inertia of a rigid molecule calculated with respect to the principal axes. { 'prin·sə·pəl 'mō·məns }

principal series [SPECTROSCOPY] A series occurring in the line spectra of many atoms and ions with one, two, or three electrons in the outer shell, in which the total orbital angular momentum quantum number changes from 1 to 0. { 'prin·sə·pəl 'sir·ēz }

priscol [ORGANIC CHEMISTRY] *See* tolazoline hydrochloride. { 'prisˌkȯl }

prism spectrograph [SPECTROSCOPY] Analysis device in which a prism is used to give two different but simultaneous light wavelengths derived from a common light source; used for the analysis of materials by flame photometry. { 'priz·əm 'spek·trəˌgraf }

pristane [ORGANIC CHEMISTRY] $C_{19}H_{40}$ A liquid soluble in such organic solvents as ether, petroleum ether, benzene, chloroform, and carbon tetrachloride; used as a lubricant, as an oil in transformers, and as an anticorrosion agent. Also known as norphytane. { 'priˌstān }

procaine [ORGANIC CHEMISTRY] *See* procaine base. { 'prōˌkān }

procaine base [ORGANIC CHEMISTRY] $C_6H_4NH_2COOCH_2CH_2N(C_2H_5)_2$ Water-insoluble, light-sensitive, odorless, white powder, melting at 60°C; soluble in alcohol, ether, chloroform, and benzene; used in medicine as a local anesthetic. Also known as planocaine base; procaine. { 'prōˌkān ˌbās }

procaine penicillin G [ORGANIC CHEMISTRY] $C_{29}H_{38}N_4O_6S \cdot H_2O$ White crystals or powder, fairly soluble in chloroform; used as an antibiotic in animal feed. { 'prōˌkān ˌpen·ə'sil·ən 'jē }

process analytical chemistry [ANALYTICAL CHEMISTRY] A branch of analytical chemistry concerned with quantitative and qualitative information about a chemical process. { 'präˌsəs ˌan·əl'it·i·kəl 'kem·ə·strē }

prochirality [ORGANIC CHEMISTRY] The property displayed by a molecule or atom which contains (or is bonded to) two constitutionally identical ligands. Also known as prostereoisomerism. { ¦prō·kī'ral·əd·ē }

product [CHEMISTRY] A substance formed as a result of a chemical reaction. { 'präd·əkt }

proflavine sulfate [ORGANIC CHEMISTRY] $C_{13}H_{11}N_3 \cdot H_2SO_4$ A reddish-brown, crystalline powder, soluble in alcohol and water; used in medicine. { prō'flā,vēn 'səl,fāt }

promazine hydrochloride [ORGANIC CHEMISTRY] $C_{17}H_{20}N_2S \cdot HCl$ A white to slightly yellow, crystalline powder, melting at 172-182°C; used in medicine and as a food additive. { 'präm·ə,zēn ,hī·drə'klȯr,īd }

promethium [CHEMISTRY] A chemical element, symbol Pm, atomic number 61; atomic weight of the most abundant isotope is 147; a member of the rare-earth group of metals. { prə'mē·thē·əm }

promoter [CHEMISTRY] A chemical which itself is a feeble catalyst, but greatly increases the activity of a given catalyst. { prə'mōd·ər }

propadiene [ORGANIC CHEMISTRY] *See* allene. { ,präp·ə'dī,ēn }

propagation rate [CHEMISTRY] The speed at which a flame front progresses through the body of a flammable fuel-oxidizer mixture, such as gas and air. { ,präp·ə'gā·shən ,rāt }

propagation step [CHEMISTRY] In a chain reaction, one of the fundamental steps that take place repeatedly until the reaction is complete. { ,präp·ə'gā·shən ,step }

propane [ORGANIC CHEMISTRY] $CH_3CH_2CH_3$ A heavy, colorless, gaseous petroleum hydrocarbon gas of the paraffin series; boils at −44.5°C; used as a solvent, refrigerant, and chemical intermediate. { 'prō,pān }

propanoic acid [ORGANIC CHEMISTRY] *See* propionic acid. { ¦prō·pə¦nō·ik 'as·əd }

propanol [ORGANIC CHEMISTRY] *See* propyl alcohol. { 'prō·ə,nȯl }

2-propanone [ORGANIC CHEMISTRY] *See* acetone. { ¦tü 'prō·pə,nōn }

propargyl alcohol [ORGANIC CHEMISTRY] $HCCCH_2OH$ Colorless, water- and alcohol-soluble liquid, boiling at 114°C; used as a chemical intermediate, stabilizer, and corrosion inhibitor. Also known as 2-propyn-1-ol. { prō'pär·jəl 'al·kə,hȯl }

propargyl bromide [ORGANIC CHEMISTRY] C_3H_3Cl A flammable liquid with a boiling point range of 56.0-57.1°C; used as a soil fumigant. { prō'pär·jəl 'brō,mīd }

propargyl chloride [ORGANIC CHEMISTRY] C_3H_3Cl A liquid miscible with benzene, carbon tetrachloride, ethanol, and ethylene glycol; used as an intermediate in organic synthesis. { prō'pär·jəl 'klȯr,īd }

propellant 23 [ORGANIC CHEMISTRY] *See* fluoroform. { prə'pel·ənt ¦twen·tē'thrē }

propenyl guaethol [ORGANIC CHEMISTRY] $C_{11}H_{14}O_2$ A white powder with a vanilla flavor and a melting point of 85-86°C; soluble in fats, essential oils, and edible solvents; used for artificial vanilla flavoring. { 'prō·pə,nil 'gwē,thȯl }

propham [ORGANIC CHEMISTRY] $C_{10}H_{13}NO_2$ A light brown solid with a melting point of 87-88°C; slightly soluble in water; used as a pre- and postemergence herbicide for vegetable crops. Abbreviated IPC (isopropyl-N-phenylcarbamate). { 'prō,fam }

β-propiolactone [ORGANIC CHEMISTRY] $C_3H_4O_2$ Water-soluble liquid that decomposes rapidly at boiling point (155°C); miscible with ethanol, acetone, chloroform, and ether; reacts with alcohol; used as a chemical intermediate. { ¦bād·ə ,prō·pē·ə'lak,tōn }

propionaldehyde [ORGANIC CHEMISTRY] C_2H_5CHO Flammable, water-soluble, water-white liquid, with suffocating aroma; boils at 48.8°C; used to manufacture acetals, plastics, and rubber chemicals, and as a disinfectant and preservative. { ¦prō·pē,än'al·də,hīd }

propionate [ORGANIC CHEMISTRY] A salt of propionic acid, CH_3CH_2COOH; an example is sodium propionate, CH_3CH_2COONa. { 'prō·pē·ə,nāt }

propionic acid [ORGANIC CHEMISTRY] CH_3CH_2COOH Water- and alcohol-soluble, clear, colorless liquid with pungent aroma; boils at 140.7°C; used to manufacture various propionates, in nickel-electroplating solutions, for perfume esters and artificial flavors, for pharmaceuticals, and as a cellulosics solvent. Also known as methylacetic acid; propanoic acid. { ¦prō·pē¦än·ik 'as·əd }

propionic anhydride [ORGANIC CHEMISTRY] $(CH_3CH_2CO)_2O$ A colorless liquid with a boiling point of 167-169°C; soluble in ether, alcohol, and chloroform; used as an esterifying agent and for dyestuffs and pharmaceuticals. { ¦prō·pē¦än·ik an'hī‚drīd }

propionic ether [ORGANIC CHEMISTRY] *See* ethyl propionate. { ¦prō·pē¦än·ik 'ē·thər }

propyl- [ORGANIC CHEMISTRY] The $CH_3CH_2CH_2$— radical, derived from propane; found, for example, in 1-propanol. { 'prō·pəl }

***n*-propyl acetate** [ORGANIC CHEMISTRY] $C_3H_7OOCCH_3$ Colorless liquid with pleasant aroma; miscible with alcohols, ketones, esters, and hydrocarbons; boils at 96-102°C; used for flavors and perfumes, in organic synthesis, and as a solvent. { ¦en 'prō·pəl 'as·ə‚tāt }

propylacetone [ORGANIC CHEMISTRY] *See* methyl butyl ketone. { ¦prō·pəl'as·ə‚tōn }

propyl alcohol [ORGANIC CHEMISTRY] $CH_3CH_2CH_2OH$ A colorless liquid made by oxidation of aliphatic hydrocarbons; boils at 97°C; used as a solvent and chemical intermediate. Also known as ethyl carbinol; propanol. { 'prō·pəl 'al·kə‚hȯl }

***n*-propylamine** [ORGANIC CHEMISTRY] $C_3H_7NH_2$ Colorless, flammable liquid, boiling at 46-51°C; used as a sedative. { ¦en ‚prō'pil·ə‚mēn }

propyl benzene [ORGANIC CHEMISTRY] $C_6H_5C_3H_7$ Water-insoluble, colorless liquid, boiling at 158°C. Also known as phenylpropane. { 'prō·pəl 'ben‚zēn }

propylene [ORGANIC CHEMISTRY] $CH_3CH{=}CH_2$ Colorless unsaturated hydrocarbon gas, with boiling point of −47°C; used to manufacture plastics and as a chemical intermediate. Also known as methyl ethylene; propene. { 'prō·pə‚lēn }

propylene aldehyde [ORGANIC CHEMISTRY] *See* crotonaldehyde. { 'prō·pə‚lēn 'al·də‚hīd }

propylene carbonate [ORGANIC CHEMISTRY] $C_3H_6CO_3$ Odorless, colorless liquid, boiling at 242°C; miscible with acetone, benzene, and ether; used as a solvent, extractant, plasticizer, and chemical intermediate. { 'prō·pə‚lēn 'kär·bə‚nāt }

propylene dichloride [ORGANIC CHEMISTRY] $CH_3CHClCH_2Cl$ Water-insoluble, colorless, moderately flammable liquid, with chloroform aroma; boils at 96.3°C; miscible with most common solvents; used as a solvent, dry-cleaning fluid, metal degreaser, and fumigant. { 'prō·pə‚lēn dī'klȯr‚īd }

propylene glycol [ORGANIC CHEMISTRY] $CH_3CHOHCH_2OH$ A viscous, colorless liquid, miscible with water, alcohol, and many solvents; boils at 188°C; used as a chemical intermediate, antifreeze, solvent, lubricant, plasticizer, and bactericide. { 'prō·pə‚lēn 'glī‚kȯl }

propylene glycol alginate [ORGANIC CHEMISTRY] $C_9H_{14}O_7$ A white, water-soluble powder; used as a stabilizer, thickener, and emulsifier. { 'prō·pə‚lēn 'glī‚kȯl 'al·jə‚nāt }

propylene glycol monomethyl ether [ORGANIC CHEMISTRY] $C_4H_{10}O_2$ A colorless liquid with a boiling point of 120.1°C; soluble in water, methanol, and ether; used as a solvent for cellulose, dyes, and inks. { 'prō·pə‚lēn 'glī‚kȯl ¦män·ō'meth·əl 'ē·thər }

propylene glycol monoricinoleate [ORGANIC CHEMISTRY] $C_{21}H_{30}O_4$ A pale yellow, moderately viscous oily liquid, soluble in organic solvents; used as a plasticizer and lubricant and in dye solvents and cosmetics. { 'prō·pə‚lēn 'glī‚kȯl ¦män·ō‚ris·ən'ō·lē‚āt }

propyleneimine [ORGANIC CHEMISTRY] C_3H_7N A clear, colorless liquid with a boiling point of 66-67°C; soluble in water and organic solvents; used as an intermediate in organic synthesis. { ‚prō·pə'lēn·ə‚mēn }

propylene oxide [ORGANIC CHEMISTRY] C_3H_6O Colorless, flammable liquid, with etherlike aroma; soluble in water, alcohol, and ether; boils at 33.9°C; used as a solvent and fumigant, in lacquers, coatings, and plastics, and as a petrochemical intermediate. { 'prō·pə‚lēn 'äk‚sīd }

propylene tetramer [ORGANIC CHEMISTRY] *See* dodecane. { 'prō·pə‚lēn 'te·trə·mər }

propyl formate [ORGANIC CHEMISTRY] $C_4H_8O_2$ A flammable liquid with a boiling point of 81.3°C; used for flavoring. { 'prō·pəl 'fȯr‚māt }

***n*-propyl furoate** [ORGANIC CHEMISTRY] $C_8H_{10}O_3$ A colorless, fragrant liquid with a boiling point of 210.9°C; soluble in alcohol and ether; used for flavoring. { ¦en 'prō·pəl 'fyu̇r·ə‚wāt }

propyl gallate [ORGANIC CHEMISTRY] $C_3H_7OOCC_6H_2(OH)_3$ Colorless crystals with a melting point of 150°C; used to prevent or retard rancidity in edible fats and oils. { 'prō·pəl 'ga,lāt }

propyliodone [ORGANIC CHEMISTRY] $C_{10}H_{11}O_3NI_2$ A white, crystalline powder with a melting point of 187-190°C; soluble in alcohol, acetone, and ether; used in medicine as a radiopaque medium. { ,prō·pəl'ī·ə,dōn }

***n*-propyl mercaptan** [ORGANIC CHEMISTRY] C_3H_7SH A liquid with an offensive odor and a boiling range of 67-73°C; used as a herbicide. Also known as 1-propanethlol. { ¦en 'prō·pəl mər'kap,tan }

***N*-propyl nitrate** [ORGANIC CHEMISTRY] $C_3H_7NO_3$ A white to straw-colored liquid with a boiling range of 104-127°C; used as a monopropellant rocket fuel. { ¦en 'prō·pəl 'nī,trāt }

propylparaben [ORGANIC CHEMISTRY] $C_{10}H_{12}O_3$ Colorless crystals or white powder with a melting point of 95-98°C; soluble in acetone, ether, and alcohol; used in medicine and as a food preservative and fungicide. { ,prō·pəl'par·ə bən }

1-propylphosphonic acid [ORGANIC CHEMISTRY] $C_3H_9O_3P$ A white solid with a melting point of 68-69°C; soluble in water; used as a growth regulator for herbaceous and woody species. { ¦wən ¦prō·pəl·fä'sfän·ik 'as·əd }

propylthiopyrophosphate [ORGANIC CHEMISTRY] $C_{12}H_{28}P_2S_2O$ A straw-colored to dark amber liquid with a boiling point of 148°C; used as an insecticide for chinch bugs in lawns and turf. { ¦prō·pəl¦thī·ō,pī·rə'fä,sfāt }

2-propyn-1-ol [ORGANIC CHEMISTRY] *See* propargyl alcohol. { ¦tü 'prō·pən ,wən ,ȯl }

prostereoisomerism [ORGANIC CHEMISTRY] *See* prochirality. { prō¦ster·ē·ō·ī'säm·ə,riz·əm }

protactinium [CHEMISTRY] A chemical element, symbol Pa, atomic number 91; the third member of the actinide group of elements; all the isotopes are radioactive; the longest-lived isotope is protactinium-231. { ¦prōd,ak'tin·ē·əm }

protective colloid [PHYSICAL CHEMISTRY] A colloidal substance that protects other colloids from the coagulative effect of electrolytes and other agents. { prə¦tek·tiv 'kä,lȯid }

proteinometer [ANALYTICAL CHEMISTRY] *See* hand sugar refractometer. { ,prōt·ən'äm·əd·ər }

protogenic [CHEMISTRY] Strongly acidic. { ¦prōd·ə¦jen·ik }

proton acid [CHEMISTRY] *See* Brönsted acid. { 'prō,tän 'as·əd }

protonic acid [CHEMISTRY] *See* Brönsted acid. { prō'tän·ik 'as·əd }

protonate [CHEMISTRY] To add protons to a base by a proton source. { 'prōt·ən,at }

proton-induced x-ray emission [ANALYTICAL CHEMISTRY] A method of elemental analysis in which the energy of the characteristic x-rays emitted when a sample is bombarded with a beam of energetic protons is used to identify the elements present in the sample. Abbreviated PIXE. { 'prō,tän in,düst 'eks,rā i,mish·ən }

proton resonance [SPECTROSCOPY] A phenomenon in which protons absorb energy from an alternating magnetic field at certain characteristic frequencies when they are also subjected to a static magnetic field; this phenomenon is used in nuclear magnetic resonance quantitative analysis technique. { 'prō,tän 'rez·ən·əns }

proton stability constant [PHYSICAL CHEMISTRY] The reciprocal of the dissociation constant of a weak base in solution. { 'prō,tän stə'bil·əd·ē ,kän·stənt }

protophilic [CHEMISTRY] Strongly basic. { ¦prōd·ō¦fil·ik }

prototropy [ORGANIC CHEMISTRY] A reversible interconversion of structural isomers that involves the transfer of a proton. { prō'tä·trə·pē }

protropic [CHEMISTRY] Pertaining to chemical reactions that are influenced by protons. { prō'träp·ik }

Prout's hypothesis [PHYSICAL CHEMISTRY] The hypothesis that all atoms are built up from hydrogen atoms. { 'prauts hī,päth·ə·səs }

Prussian blue [INORGANIC CHEMISTRY] $Fe_4[Fe(CN)_6]_3$ Ferric ferrocyanide, used as a blue pigment and in the removal of hydrogen sulfide from gases. { 'prəsh·ən 'blü }

prussic acid [INORGANIC CHEMISTRY] *See* hydrocyanic acid. { 'prəs·ik 'as·əd }

pryrrolidine [ORGANIC CHEMISTRY] C_4H_9N A colorless to pale yellow liquid with a boiling point of 87°C; soluble in water and alcohol; used in the manufacture of pharmaceuticals, insecticides, and fungicides. { pə'räl·ə,dēn }

pseudocritical properties [CHEMISTRY] Effective (empirical) values for the critical properties (such as temperature, pressure, and volume) of a multicomponent chemical system. { ¦sü·dō'krid·ə·kəl 'präp·ərd·ēz }

pseudocumene [ORGANIC CHEMISTRY] C_9H_{12} Water-insoluble, hydrocarbon liquid, boiling at 168°C; soluble in alcohol, benzene, and ether; used to manufacture perfumes and dyes, and as a catgut sterilant. Also known as pseudocumol; uns-trimethylbenzene. { ¦sü·dō'kyü·mēn }

pseudocumol [ORGANIC CHEMISTRY] *See* pseudocumene. { ¦sü·dō'kyü,mȯl }

pseudohalogen [CHEMISTRY] Any one of a group of molecules that exhibit significant similarity to the halogens, for example, cyanogen (NCCN). { ,süd·ō'hal·ə·jən }

pseudoionone [ORGANIC CHEMISTRY] $C_{13}H_{20}O$ A pale yellow liquid with a boiling point of 143-145°C; soluble in alcohol and ether; used for perfumes and cosmetics. { ¦sü·dō'ī·ə,nōn }

pseudoreduced compressibility [CHEMISTRY] The compressibility factor for a multicomponent gaseous system, calculated at reduced conditions using the pseudoreduced properties of the mixture. { ,sü·dō·ri'düst kəm,pres·ə'bil·əd·ē }

pseudoreduced properties [CHEMISTRY] Reduced-state relationships (such as reduced pressure, reduced temperature, and reduced volume) calculated for multicomponent chemical systems by using pseudocritical properties. { ,sü·dō·ri'düst 'präp·ərd·ēz }

Pt [CHEMISTRY] *See* platinum.

PTA [INORGANIC CHEMISTRY] *See* phosphotungstic acid.

Pu [CHEMISTRY] *See* plutonium.

pulse radiolysis [PHYSICAL CHEMISTRY] A method of studying fast chemical reactions in which a sample is subjected to a pulse of ionizing radiation, and the products formed by the resulting reactions are studied spectroscopically. { 'pəls ,rād·ē'äl·ə·səs }

pure substance [CHEMISTRY] A sample of matter, either an element or a compound, that consists of only one component with definite physical and chemical properties and a definite composition. { 'pyu̇r 'səb·stəns }

purity [CHEMISTRY] The degree to which the content of impurity can be detected by an analytical procedure in a sample of matter that is classified as a pure substance; the grade of purity is in inverse proportion to the amount of impurity present. Also known as chemical purity. { 'pyu̇r·əd·ē }

purple of Cassius [ORGANIC CHEMISTRY] *See* gold tin purple. { 'pər·pəl əv 'kash·əs }

purple salt [INORGANIC CHEMISTRY] *See* potassium permanganate. { 'pər·pəl 'sȯlt }

purpurin [ORGANIC CHEMISTRY] $C_{14}H_8O_5$ A compound crystallizing as long orange needles from dilute alcohol solutions; used in the manufacture of dyes, and as a reagent for the detection of boron. Also known as natural red. { 'pər·pyə·rən }

purpurogallin [ORGANIC CHEMISTRY] $C_{11}H_8O_5$ A red, crystalline compound, the aglycon of several glycosides from nutgalls; decomposes at 274-275°C; soluble in boiling alcohol, methanol, and acetone; used as an antioxidant or to retard metal contamination in hydrocarbon fuels or lubricants. { ,pər·pyə·rō'gal·ən }

PVA [ORGANIC CHEMISTRY] *See* polyvinyl acetate; polyvinyl alcohol.

PVAc [ORGANIC CHEMISTRY] *See* polyvinyl acetate.

PVC [ORGANIC CHEMISTRY] *See* polyvinyl chloride.

PVDC [ORGANIC CHEMISTRY] *See* polyvinyl dichloride.

PVI [ORGANIC CHEMISTRY] *See* polyvinyl isobutyl ether.

PVM [ORGANIC CHEMISTRY] *See* polyvinyl methyl ether.

PVP [ORGANIC CHEMISTRY] *See* polyvinyl pyrrolidone.

pyracetic acid [ORGANIC CHEMISTRY] *See* pyroligneous acid. { ¦pī·rə¦sēd·ik 'as·əd }

pyramidal molecule [CHEMISTRY] A molecular structure in the shape of a pyramid in which the central atom at the peak possesses either three or four valence bonds that are directed to the other atoms, which form the base of the pyramid. { ¦pir·ə,mid·əl 'mäl·ə,kyül }

pyrazolone dye [ORGANIC CHEMISTRY] An acid dye containing both —N=N— and =C=C= chromophore groups, such as tartrazine; used for silk and wool. { pə'raz·ə‚lōn ‚dī }

pyridine [ORGANIC CHEMISTRY] C_5H_5N Organic base; flammable, toxic yellowish liquid, with penetrating aroma and burning taste; soluble in water, alcohol, ether, benzene, and fatty oils; boils at 116°C; used as an alcohol denaturant, solvent, in paints, medicine, and textile dyeing. { 'pir·ə‚dēn }

pyro- [CHEMISTRY] A chemical prefix for compounds formed by heat, such as pyrophosphoric acid, an inorganic acid formed by the loss of one water molecule from two molecules of an ortho acid. { 'pī·rō, 'pī·rə }

pyrocatechuic acid [ORGANIC CHEMISTRY] *See* catechol. { ¦pī·rō¦kad·ə¦chü·ik 'as·əd }

pyrocellulose [ORGANIC CHEMISTRY] Highly nitrated cellulose; used to make explosives; originally called guncotton in the United States, cordite in England. { ¦pī·rō'sel·yə‚lōs }

pyrogallic acid [ORGANIC CHEMISTRY] $C_6H_3(OH)_3$ Lustrous, light-sensitive white crystals, melting at 133°C; soluble in alcohol, ether, and water; used for photography, dyes, drugs, medicines, and process engravings, and as an analytical reagent and protective colloid. Also known as pyrogallol. { ¦pī·rō'gal·ik 'as·əd }

pyrogallol [ORGANIC CHEMISTRY] *See* pyrogallic acid. { ‚pī·rō'ga‚lȯl }

pyrogallolphthalein [ORGANIC CHEMISTRY] *See* gallein. { ¦pī·rō¦gal·ō'thal·ē·ən }

pyroligneous acid [ORGANIC CHEMISTRY] An impure acetic acid derived from destructive distillation of wood or pine tar. Also known as pyracetic acid; wood vinegar. { ¦pī·rō'lig·nē·əs 'as·əd }

pyrolithic acid [ORGANIC CHEMISTRY] *See* cyanuric acid. { ¦pī·rō¦lith·ik 'as·əd }

pyrolysate [CHEMISTRY] Any product of pyrolysis. { pī'räl·ə‚zāt }

pyrolysis [CHEMISTRY] The breaking apart of complex molecules into simpler units by the use of heat, as in the pyrolysis of heavy oil to make gasoline. { pə'räl·ə·səs }

pyromellitic acid [ORGANIC CHEMISTRY] $C_6H_2(COOH)_4$ A white powder with a melting point of 257-265°C; used as an intermediate for polyesters and polyamides. Abbreviated PMA. { ‚pī·rō·mə'lid·ik 'as·əd }

pyromellitic dianhydride [ORGANIC CHEMISTRY] $C_6H_2(C_2O_3)_2$ A white powder with a melting point of 286°C; soluble in some organic solvents; used for curing epoxy resins. Abbreviated PMDA. { ‚pī·rō·mə'lid·ik ¦dī·an'hī‚drīd }

pyromucic acid [ORGANIC CHEMISTRY] *See* furoic acid. { ¦pī·rō'myü·sik 'as·əd }

pyrophosphoric acid [INORGANIC CHEMISTRY] $H_4P_2O_7$ Water-soluble, syrupy liquid melting at 61°C; used as a catalyst and to make organic phosphate esters. { ¦pī·rō·fä'sfȯr·ik 'as·əd }

pyrosin [ORGANIC CHEMISTRY] *See* tetraiodofluorescein. { ‚pī·rə·sən }

pyroxylin [ORGANIC CHEMISTRY] $[C_{12}H_{16}O_6(NO_3)_4]_x$ Any member of the group of commercially available nitrocelluloses that are used for properties other than their combustibility; the term is commonly used to identify products that are principally made from nitrocellulose, such as pyroxylin plastic or pyroxylin lacquer. Also known as collodion cotton; soluble guncotton; soluble nitrocellulose. { pə'räk·sə·lən }

pyrrole [ORGANIC CHEMISTRY] C_4H_5N Water-insoluble, yellowish oil, with pungent taste; soluble in alcohol, ether, and dilute acids; boils at 130°C; polymerizes in light; used to make drugs. { 'pi‚rōl }

pyrrole ring [ORGANIC CHEMISTRY] A five-member heterocycle containing one nitrogen atom and four carbon atoms in the ring; frequently found in structures of natural products occurring as joined rings or attached to straight chains. { 'pī‚rōl ‚riŋ }

2-pyrrolidone [ORGANIC CHEMISTRY] C_4H_7ON Combustible, light-yellow liquid, boiling at 245°C; soluble in ethyl alcohol, water, chloroform, and carbon disulfide; used as a plasticizer and polymer solvent, in insecticides and specialty inks, and as a nylon-4 precursor. { ¦tü pə'räl·ə‚dēn }

pyrrone [ORGANIC CHEMISTRY] A polyimidazopyrrolone synthesized from dianhydrides and tetramines; soluble in sulfuric acid; resists temperatures to 600°C. { 'pi‚rōn }

Q

Q branch [SPECTROSCOPY] A series of lines in molecular spectra that correspond to changes in the vibrational quantum number with no change in the rotational quantum number. { 'kyü ,branch }

quadridentate ligand [CHEMISTRY] A group which forms a chelate and has four points of attachment. { ¦kwä·drə¦den,tāt 'līg·ənd }

quadruple point [PHYSICAL CHEMISTRY] Temperature at which four phases are in equilibrium, such as a saturated solution containing an excess of solute. { kwə'drüp·əl 'pȯint }

quadrupole spectrometer [ANALYTICAL CHEMISTRY] A type of mass spectroscope in which ions pass along a line of symmetry between four parallel cylindrical rods; an alternating potential superimposed on a steady potential between pairs of rods filters out all ions except those of a predetermined mass. Also known as Massenfilter. { 'kwä·drə,pōl spek'träm·əd·ər }

qualitative analysis [ANALYTICAL CHEMISTRY] The analysis of a gas, liquid, or solid sample or mixture to identify the elements, radicals, or compounds composing the sample. { 'kwäl·ə,tād·iv ə,nal·ə·səs }

quantitative analysis [ANALYTICAL CHEMISTRY] The analysis of a gas, liquid, or solid sample or mixture to determine the precise percentage composition of the sample in terms of elements, radicals, or compounds. { 'kwän·ə·tad·iv ə'nal·ə·səs }

quantum chemistry [PHYSICAL CHEMISTRY] A branch of physical chemistry concerned with the explanation of chemical phenomena by means of the laws of quantum mechanics. { 'kwăn·təm 'kem·ə·strē }

quantum limit [SPECTROSCOPY] The shortest wavelength present in a continuous x-ray spectrum. Also known as boundary wavelength, end radiation. { 'kwän·təm 'lim·ət }

quantum theory of valence [PHYSICAL CHEMISTRY] The theory of valence based on quantum mechanics; it accounts for many experimental facts, explains the stability of a chemical bond, and allows the correlation and prediction of many different properties of molecules not possible in earlier theories. { 'kwän·təm ,thē·ə·rē əv 'vā·ləns }

quantum yield [PHYSICAL CHEMISTRY] For a photochemical reaction, the number of moles of a stated reactant disappearing, or the number of moles of a stated product produced, per einstein of light of the stated wavelength absorbed. { 'kwăn·təm 'yēld }

quarterpolymer [CHEMISTRY] A polymer in which the repeating groups comprise four species of monomer. { ¦kwȯrd·ər¦päl·i·mər }

quaternary ammonium base [ORGANIC CHEMISTRY] Ammonium hydroxide (NH_4OH) with the ammonium hydrogens replaced by organic radicals, such as $(CH_3)_4NOH$. { 'kwät·ən,er·ē ə¦mō·nē·əm 'bās }

quaternary ammonium salt [ORGANIC CHEMISTRY] A nitrogen compound in which a central nitrogen atom is joined to four organic radicals and one acid radical, for example, hexamethonium chloride; used as an emulsifying agent, corrosion inhibitor, and antiseptic. { 'kwät·ən,er·ē ə¦mō·nē·əm 'sȯlt }

quaternary carbon atom [ORGANIC CHEMISTRY] A carbon atom bonded to four other carbon atoms with single bonds. { 'kwät·ən,er·ē ¦kär·bən 'ad·əm }

quaternary phase equilibria [PHYSICAL CHEMISTRY] The solubility relationships in any liquid system with four nonreactive components with varying degrees of mutual solubility. { 'kwät·ən,er·ē ¦fāz ,ē·kwə'lib·rē·ə }

quaternary system [PHYSICAL CHEMISTRY] An equilibrium relationship between a mixture of four (four phases, four components, and so on). { 'kwät·ən,er·ē 'sis·təm }

quercimelin [ORGANIC CHEMISTRY] *See* quercitrin. { ,kwer'sim·ə·lən }

quercitrin [ORGANIC CHEMISTRY] $C_{21}H_{20}O_{11}$ The 3-rhamnoside of quercitin, forming yellow crystals from dilute ethanol or methanol solution, melting at 176-179°C, soluble in alcohol; used as a textile dye. Also known as quercimelin; quercitroside. { 'kwer·sə·trən }

quercitroside [ORGANIC CHEMISTRY] *See* quercitrin. { kwer'si·trə,sīd }

Quevenne scale [CHEMISTRY] Arbitrary scale used with hydrometers or lactometers in the determination of the specific gravity of milk; degrees Quevenne = 1000 (specific gravity −1). { kə'ven ,skāl }

quicksilver [CHEMISTRY] *See* mercury. { 'kwik,sil·vər }

quicksilver vermilion [INORGANIC CHEMISTRY] *See* mercuric sulfide. { 'kwik,sil·vər vər'mil·yən }

quinaldine [ORGANIC CHEMISTRY] $C_9H_6NCH_3$ A colorless, oily liquid with a boiling point of 246-247°C; soluble in alcohol, chloroform, and ether; used in medicine as an antimalarial. Also known as chinaldine. { 'kwin·əl,dēn }

quinalizarin [ORGANIC CHEMISTRY] $C_{14}H_8O_6$ A red, crystalline compound, soluble in water solutions of alkalies, and in acetic and sulfuric acid; used to dye cottons. { ,kwin·ə'liz·ə·rən }

quinhydrone [ORGANIC CHEMISTRY] $C_6H_4O_2 \cdot C_6H_4(OH)_2$ Green, water-soluble powder, subliming at 171°C; a compound of quinone and hydroquinone dissociating in solution. { kwin'hī,drōn }

quinhydrone electrode [ANALYTICAL CHEMISTRY] A platinum wire in a saturated solution of quinhydrone; used as a reversible electrode standard in pH determinations. { kwin'hī,drōn i'lek,trōd }

quinic acid [ORGANIC CHEMISTRY] $C_6H_7(OH)_4COOH \cdot H_2O$ Ether-insoluble, white crystals with acid taste; melts at 162°C; soluble in alcohol, water, and glacial acetic acid; used in medicine. Also known as chinic acid; kinic acid. { 'kwin·ik 'as·əd }

quinidine [ORGANIC CHEMISTRY] $C_{20}H_{24}N_2O_2$ A crystalline alkaloid that melts at 174-175°C and that may be derived from the bark of cinchona; used as the salt in medicine. Also known as β-quinine. { 'kwin·ə,dēn }

quinine [ORGANIC CHEMISTRY] $C_{20}H_{24}N_2O_2 \cdot 3H_2O$ White powder or crystals, soluble in alcohol, ether, carbon disulfide, chloroform, and glycerol; an alkaloid derived from cinchona bark; used as an antimalarial drug and in beverages. { 'kwī,nīn }

β-quinine [ORGANIC CHEMISTRY] *See* quinidine. { ¦bād·ə 'kwī,nīn }

quinoidine [ORGANIC CHEMISTRY] A brownish-black mass consisting of a mixture of alkaloids which remain in solution after extracting crystallized alkaloids from cinchona bark; soluble in dilute acids, alcohol, and chloroform; used in medicine. Also known as chinoidine. { kwi'nō·ə,dēn }

quinol [ORGANIC CHEMISTRY] *See* hydroquinone. { 'kwi,nól }

quinoline [ORGANIC CHEMISTRY] C_9H_7N Water-soluble, aromatic nitrogen compound; colorless, hygroscopic liquid; also soluble in alcohol, ether, and carbon disulfide; boils at 238°C; used in medicine and as a chemical intermediate. Also known as chinoline; leucoline; leukol. { 'kwin·ə,lēn }

quinoline blue [ORGANIC CHEMISTRY] *See* cyanine dye. { 'kwin·ə,lēn 'blü }

8-quinolinol [ORGANIC CHEMISTRY] *See* 8-hydroxyquinoline; oxine. { ¦āt kwi'näl·ə,nól }

quinone [ORGANIC CHEMISTRY] $CO(CHCH)_2CO$ Yellow crystalline compound with irritating aroma; melts at 116°C; soluble in alcohol, alkalies, and ether; used to make dyes and hydroquinone. Also known as benzoquinone; chinone. { 'kwi,nōn }

quinoxaline [ORGANIC CHEMISTRY] $C_8H_6N_2$ Bicyclic organic base; colorless powder, soluble in water and organic solvents; melts at 30°C; used in organic synthesis. { kwi'näk·sə,lēn }

***N*'-2-quinoxalysulfanilimide** [ORGANIC CHEMISTRY] $C_{14}H_{12}N_4SO_2$ Crystals with a melting point of 247°C; almost insoluble in water; used as a rodenticide. Also known as sulfaquinoxaline. { ¦en,prīm ¦tü kwi¦näk·sə·lē,səl·fə'nil·ə,mīd }

quintozene [ORGANIC CHEMISTRY] *See* pentachloronitrobenzene. { 'kwin·tə,zēn }

R

Ra [CHEMISTRY] *See* radium.

racemate [ORGANIC CHEMISTRY] A compound whose individual crystals contain equal numbers of dextrorotatory and levorotatory molecules. { 'ras·ə‚māt }

racemic acid [ORGANIC CHEMISTRY] $C_2H_4O_2(COOH)_2 \cdot H_2O$ Colorless crystals, melting at 205°C; soluble in water, slightly soluble in alcohol; used as a chemical intermediate. Also known as inactive tartaric acid. { rə'sēm·ik 'as·əd }

racemic mixture [ORGANIC CHEMISTRY] A mixture of equal quantities of crystals of pure dextrorotatory and levorotatory isomers of the same compound, and therefore optically inactive. Also known as conglomerate; racemic modification. { rə'sēm·ik 'miks·chər }

racemic modification [ORGANIC CHEMISTRY] *See* racemic mixture. { rā¦sēm·ik ‚mäd·ə·fə'kā·shən }

racemization [ORGANIC CHEMISTRY] A process by which an optically active form of a substance is converted into a racemic mixture. { ‚rā·sə·mə'zā·shən }

radial chromatography [ANALYTICAL CHEMISTRY] A circular disk of absorbent paper which has a strip (wick) cut from edge to center to dip into a solvent; the solvent climbs the wick, touches the sample, and resolves it into concentric rings (the chromatogram). Also known as circular chromatography; radial paper chromatography. { 'rād·ē·əl ‚krō·mə'täg·rə·fē }

radial distribution function [PHYSICAL CHEMISTRY] A function $\rho(r)$ equal to the average over all directions of the number density of molecules at distance r from a given molecule in a liquid. { 'rād·ē·əl ‚dis·trə'byü·shən ‚fəŋk·shən }

radial paper chromatography [ANALYTICAL CHEMISTRY] *See* radial chromatography. { 'rād·ē·əl 'pā·pər ‚krō·mə'täg·rə·fē }

radiation catalysis [CHEMISTRY] The use of radiation (such as gamma, neutron, proton, electron, or x-ray) to activate or speed up a chemical or physical change; for example, radiation alone can initiate polymerization without heat, pressure, or chemical catalysts. { ‚rād·ē'ā·shən kə'tal·ə·səs }

radical [ORGANIC CHEMISTRY] *See* free radical. { 'rad·ə·kəl }

radical scavenger [CHEMISTRY] One of a group of molecules that combines with free radicals in a chemical or biochemical system to render them less active chemically. { 'rad·ə·kəl ‚skav·ən·jər }

radicofunctional name [ORGANIC CHEMISTRY] A name for an organic compound that uses two key words; the first word corresponds to the group or groups involved and the second word indicates the functional group—for example, alkyl halide. { ¦rad·ə·kō'fəŋk·shən·əl 'nām }

radioassay [ANALYTICAL CHEMISTRY] An assay procedure involving the measurement of the radiation intensity of a radioactive sample. { ¦rād·ē·ō'a‚sā }

radiochemical laboratory [CHEMISTRY] A specially equipped and shielded chemical laboratory designed for conducting radiochemical studies without danger to the laboratory personnel. { ¦rad·ē·ō'kem·ə·kəl 'lab·rə‚tór·ē }

radiochemistry [CHEMISTRY] That area of chemistry concerned with the study of radioactive substances. { ¦rad·ē·ō'kem·ə·strē }

radiochromatography [ANALYTICAL CHEMISTRY] An analytic process for quantitative or qualitative determination of radioactive substances in a mixture by measuring the radioactivity of various zones in the chromatogram. { ¦rād·ē·ō‚krō·mə'täg·rə·fē }

radiocolloid [CHEMISTRY] A colloid having a component that consists of radioactive atoms. { ‚rād·ē·ō'kä‚lȯid }

radio-frequency spectrometer [SPECTROSCOPY] An instrument which measures the intensity of radiation emitted or absorbed by atoms or molecules as a function of frequency at frequencies from 10^5 to 10^9 hertz; examples include the atomic-beam apparatus, and instruments for detecting magnetic resonance. { 'rād·ē·ō ¦frē·kwən·sē spek'träm·əd·ər }

radio-frequency spectroscopy [SPECTROSCOPY] The branch of spectroscopy concerned with the measurement of the intervals between atomic or molecular energy levels that are separated by frequencies from about 10^5 to 10^9 hertz, as compared to the frequencies that separate optical energy levels of about 6×10^{14} hertz. { 'rād·ē·ō ¦frē·kwən·sē spek'träs·kə·pē }

radioisotope assay [ANALYTICAL CHEMISTRY] An analytical technique including procedures for separating and reproducibly measuring a radioactive tracer. { ¦rād·ē·ō'ī·sə‚tōp 'a‚sā }

radiolysis [PHYSICAL CHEMISTRY] The dissociation of molecules by radiation; for example, a small amount of water in a reactor core dissociates into hydrogen and oxygen during operation. { ‚rād·ē'äl·ə·səs }

radiometric analysis [ANALYTICAL CHEMISTRY] Quantitative chemical analysis that is based on measurement of the absolute disintegration rate of a radioactive component having a known specific activity. { ¦rād·ē·ō¦me·trik ə'nal·ə·səs }

radiometric titration [ANALYTICAL CHEMISTRY] Use of radioactive indicator to track the transfer of material between two liquid phases in equilibrium, such as titration of $^{110}AgNO_3$ (silver nitrate, with the silver atom having mass number 110) against potassium chloride. { ¦rād·ē·ō¦me·trik tī'trā·shən }

radiomimetic substances [CHEMISTRY] Chemical substances which cause biological effects similar to those caused by ionizing radiation. { ¦rād·ē·ō·mi'med·ik 'səb·stəns·əz }

radio recombination line [SPECTROSCOPY] A radio-frequency spectral line that results from an electron transition between energy levels in an atom or ion having a large principal quantum number n, greater than 50. { 'rād·ē·ō rē‚käm·bə'nā·shən ‚līn }

radium [CHEMISTRY] **1.** A radioactive member of group II, symbol Ra, atomic number 88; the most abundant naturally occurring isotope has mass number 226 and a half-life of 1620 years. **2.** A highly toxic solid that forms water-soluble compounds; decays by emission of α, β, and γ-radiation; melts at 700°C, boils at 1140°C; turns black in air; used in medicine, in industrial radiography, and as a source of neutrons and radon. { 'rād·ē·əm }

radium bromide [INORGANIC CHEMISTRY] $RaBr_2$ Water-soluble, poisonous, radioactive white powder, corrosive to skin or flesh; melts at 728°C; used in medicine, physical research, and luminous paint. { 'rād·ē·əm 'brō‚mīd }

radium carbonate [INORGANIC CHEMISTRY] $RaCO_3$ Water-insoluble, poisonous, radioactive, white powder; used in medicine. { 'rād·ē·əm 'kär·bə‚nāt }

radium chloride [INORGANIC CHEMISTRY] $RaCl_2$ Water- and alcohol-soluble, poisonous, radioactive, yellow-white crystals; corrosive effect on skin and flesh; melts at 1000°C; used in medicine, physical research, and luminous paint. { 'rād·ē·əm 'klȯr‚īd }

radium sulfate [INORGANIC CHEMISTRY] $RaSO_4$ Water-insoluble, radioactive, poisonous, white crystals; used in medicine. { 'rād·ē·əm 'səl‚fāt }

radius ratio [PHYSICAL CHEMISTRY] The ratio of the radius of a cation to the radius of an ion; relative ionic radii are pertinent to crystal lattice structure, particularly the determination of coordination number. { 'rād·ē·əs ‚rā·shō }

radon [CHEMISTRY] A chemical element, symbol Rn, atomic number 86; all isotopes are radioactive, the longest half-life being 3.82 days for mass number 222; it is the heaviest element of the noble-gas group, produced as a gaseous emanation from the radioactive decay of radium. { 'rā‚dän }

Raman spectrophotometry [SPECTROSCOPY] The study of spectral-line patterns on a photograph taken at right angles through a substance illuminated with a quartz mercury lamp. { ′räm·ən ˌspek·trə·fə′täm·ə·trē }

Raman spectroscopy [SPECTROSCOPY] Analysis of the intensity of Raman scattering of monochromatic light as a function of frequency of the scattered light. { ′räm·ən spek′träs·kə·pē }

Raman spectrum [SPECTROSCOPY] A display, record, or graph of the intensity of Raman scattering of monochromatic light as a function of frequency of the scattered light. { ′räm·ən ˌspek·trəm }

random coil [PHYSICAL CHEMISTRY] Any of various irregularly coiled polymers that can occur in solution. Also known as cyclic coil. { ′ran·dəm ′kȯil }

random copolymer [ORGANIC CHEMISTRY] Resin copolymer in which the molecules of each monomer are randomly arranged in the polymer backbone. { ′ran·dəm kō′päl·i·mər }

Raoult's law [PHYSICAL CHEMISTRY] The law that the vapor pressure of a solution equals the product of the vapor pressure of the pure solvent and the mole fraction of solvent. { rä′ülz ˌlȯ }

rare-earth element [CHEMISTRY] The name given to any of the group of chemical elements with atomic numbers 58 to 71; the name is a misnomer since they are neither rare nor earths; examples are cerium, erbium, and gadolinium. { ′rer ˌərth ′el·ə·mənt }

rare-earth salts [INORGANIC CHEMISTRY] Salts derived from monazite, and with rare earths in similar proportions as in monazite; contains La, Ce, Pr, Nd, Sm, Gd, and Y as acetates, carbonates, chlorides, fluorides, nitrates, sulfates, and so on. { ′rer ˌərth ′sȯls }

rare gas [CHEMISTRY] *See* noble gas. { ′rer ′gas }

Rast method [ANALYTICAL CHEMISTRY] The melting-point depression method often used for the determination of the molecular weight of organic compounds. { ′rast ˌmeth·əd }

rate constant [PHYSICAL CHEMISTRY] Numerical constant in a rate-of-reaction equation; for example, $r_A = kC_A^a C_B^b C_C^c$, where C_A, C_B, and C_C are reactant concentrations, k is the rate constant (specific reaction rate constant), and a, b, and c are empirical constants. { ′rāt ˌkän·stənt }

rate-determining step [CHEMISTRY] In a multistep chemical reaction, the step with the lowest velocity, which determines the rate of the overall reaction. { ′rāt di′tər·mən·iŋ ˌstep }

rate of reaction [CHEMISTRY] A measurement based on the mass of reactant consumed in a chemical reaction during a given period of time. { ′rāt əv rē′ak·shən }

rational synthesis [CHEMISTRY] The production of a compound using a sequence of chemical reaction steps strategically chosen. { ′rash·ən·əl ′sin·thə·səs }

ratio of specific heats [PHYSICAL CHEMISTRY] The ratio of specific heat at constant pressure to specific heat at constant volume, $\gamma = C_p/C_v$. { ′rā·shō əv spə′sif·ik ′hēts }

Rayleigh line [SPECTROSCOPY] Spectrum line in scattered radiation which has the same frequency as the corresponding incident radiation. { ′rā·lē ˌlīn }

Rb [CHEMISTRY] *See* rubidium.

R-branch [SPECTROSCOPY] A series of lines in molecular spectra that correspond, in the case of absorption, to a unit increase in the rotational quantum number J. { ′är ˌbranch }

RDGE [ORGANIC CHEMISTRY] *See* resorcinol diglycidyl ether.

RDX [ORGANIC CHEMISTRY] *See* cyclonite.

Re [CHEMISTRY] *See* rhenium.

reactant [CHEMISTRY] A substance that reacts with another one to produce a new set of substances (products). { rē′ak·tənt }

reaction boundary [PHYSICAL CHEMISTRY] *See* reaction line. { rē′ak·shən ˌbaůn·drē }

reaction curve [PHYSICAL CHEMISTRY] *See* reaction line. { rē′ak·shən ˌkərv }

reaction enthalpy number [PHYSICAL CHEMISTRY] A dimensionless number used in the study of interphase transfer in chemical reactions, equal to the enthalpy of reaction per unit mass of a specified compound produced in a reaction, times the mass fraction of that compound, divided by the product of the specific heat at constant pressure and the temperature change during the reaction. { rē′ak·shən ′en‚thal·pē ‚nəm·bər }

reaction kinetics [PHYSICAL CHEMISTRY] *See* chemical kinetics. { rē′ak·shən ki ′ned·iks }

reaction line [PHYSICAL CHEMISTRY] In a ternary system, a special case of the boundary line along which one of the two crystalline phases present reacts with the liquid, as the temperature is decreased, to form the other crystalline phase. Also known as reaction boundary; reaction curve. { rē′ak·shən ‚līn }

reaction mechanism [CHEMISTRY] The sequence of steps during which a chemical reaction occurs, including the transition state during which the reactants are converted into products. { rē′ak·shən ‚mek·ə‚niz·əm }

reaction path [CHEMISTRY] *See* mechanism. { rē′ak·shən ‚path }

reactive bond [CHEMISTRY] A bond between atoms that is easily invaded (reacted to) by another atom or radical; for example, the double bond in $CH_2{=}CH_2$ (ethylene) is highly reactive to other ethylene molecules in the reaction known as polymerization to form polyethylene. { rē′ak·tiv ′bänd }

reactive intermediate [CHEMISTRY] An unstable compound formed as an intermediate during a chemical reaction. { rē′ak·tiv ‚in·tər′mē·dē·ət }

reactivity [CHEMISTRY] The relative capacity of an atom, molecule, or radical to combine chemically with another atom, molecule, or radical. { ‚rē·ak′tiv·əd·ē }

reagent [ANALYTICAL CHEMISTRY] A substance, chemical, or solution used in the laboratory to detect, measure, or otherwise examine other substances, chemicals, or solutions; grades include ACS (American Chemical Society standards), reagent (for analytical reagents), CP (chemically pure), USP (U.S. Pharmacopeia standards), NF (National Formulary standards), and purified, technical (for industrial use). [CHEMISTRY] The compound that supplies the molecule, ion, or free radical which is arbitrarily considered as the attacking species in a chemical reaction. { rē′ā·jənt }

reagent chemicals [ANALYTICAL CHEMISTRY] High-purity chemicals used for analytical reactions, for testing of new reactions where the effect of impurities are unknown, and, in general, for chemical work where impurities must either be absent or at a known concentration. { rē′ā·jənt ‚kem·ə·kəlz }

rearrangement reaction [ORGANIC CHEMISTRY] A chemical reaction involving a change in the bonding sequence within a molecule. Also known as molecular rearrangement. { ‚rē·ə′rānj·mənt rē‚ak·shən }

reconstructive processing [INORGANIC CHEMISTRY] The spinning of an inorganic compound of an organic support or binder subsequently removed by oxidation or volatilization to form an inorganic polymer. { ‚rē·kən′strək·tiv ′prä‚ses·iŋ }

recording balance [ANALYTICAL CHEMISTRY] An analytical balance equipped to record weight results by electromagnetic or servomotor-driven accessories. { ri′kȯrd·iŋ ‚bal·əns }

recrystallization [CHEMISTRY] Repeated crystallization of a material from fresh solvent to obtain an increasingly pure product. { rē‚krist·əl·ə′zā·shən }

red lead [INORGANIC CHEMISTRY] *See* lead tetroxide. { ′red ′led }

red mercury sulfide [INORGANIC CHEMISTRY] *See* mercuric sulfide. { ′red ′mər·kyə·rē ′səl‚fīd }

red ocher [INORGANIC CHEMISTRY] *See* ferric oxide. { ′red ′ō·kər }

redox polymer [ORGANIC CHEMISTRY] A polymer whose structure contains functional groups that can be reversibly reduced or oxidized. Also known as electron exchanger. { ′rē‚däks ‚päl·ə·mər }

redox potential [PHYSICAL CHEMISTRY] Voltage difference at an inert electrode immersed in a reversible oxidation-reduction system; measurement of the state of oxidation of the system. Also known as oxidation-reduction potential. { ′rē‚däks pə‚ten·chəl }

redox potentiometry [ANALYTICAL CHEMISTRY] Use of neutral electrode probes to measure the solution potential developed as the result of an oxidation or reduction reaction. { ′rē‚däks pə‚ten·chē′äm·ə·trē }

redox system [CHEMISTRY] A chemical system in which reduction and oxidation (redox) reactions occur. { 'rē,däks ,sis·təm }

redox titration [ANALYTICAL CHEMISTRY] A titration characterized by the transfer of electrons from one substance to another (from the reductant to the oxidant) with the end point determined colorimetrically or potentiometrically. { 'rē,däks tī'trā·shən }

red phosphorus [CHEMISTRY] An allotropic form of the element phosphorus; violet-red, amorphous powder subliming at 416°C, igniting at 260°C; insoluble in all solvents; nonpoisonous. { 'red 'fä·sfə·rəs }

red potassium chromate [INORGANIC CHEMISTRY] *See* potassium dichromate. { 'red pə'tas·ē·əm 'krō,māt }

red precipitate [INORGANIC CHEMISTRY] *See* mercuric oxide. { 'red prə'sip·ə,tāt }

red prussiate of soda [INORGANIC CHEMISTRY] *See* sodium ferricyanide. { 'red 'prəs·ē,āt əv 'sōd·ə }

red tetrazolium [ORGANIC CHEMISTRY] *See* triphenyltetrazolium chloride. { 'red ,te·trə'zäl·ē·əm }

reducer [CHEMISTRY] *See* reducing agent. { ri'dü·sər }

reducing agent [CHEMISTRY] Also known as reducer. **1.** A material that adds hydrogen to an element or compound. **2.** A material that adds an electron to an element or compound, that is, decreases the positiveness of its valence. { ri'düs·iŋ ,ā·jənt }

reducing atmosphere [CHEMISTRY] An atmosphere of hydrogen (or other substance that readily provides electrons) surrounding a chemical reaction or physical device; the effect is the opposite to that of an oxidizing atmosphere. { ri'düs·iŋ 'at·mə,sfir }

reducing flame [CHEMISTRY] A flame having excess fuel and being capable of chemical reduction, such as extracting oxygen from a metallic oxide. { ri'dus·iŋ ,flām }

reducing sugar [ORGANIC CHEMISTRY] Any of the sugars that because of their free or potentially free aldehyde or ketone groups, possess the property of readily reducing alkaline solutions of many metallic salts such as copper, silver, or bismuth; examples are the monosaccharides and most of the disaccharides, including maltose and lactose. { ri'düs·iŋ ,shu̇g·ər }

reduction [ANALYTICAL CHEMISTRY] Preparation of one or more subsamples from a sample of material that is to be analyzed chemically. [CHEMISTRY] **1.** Reaction of hydrogen with another substance. **2.** Chemical reaction in which an element gains an electron (has a decrease in positive valence). { ri'dək·shən }

reduction cell [CHEMISTRY] A vessel in which aqueous solutions of salts or fused salts are reduced electrolytically. { ri'dək·shən ,sel }

reduction potential [PHYSICAL CHEMISTRY] The potential drop involved in the reduction of a positively charged ion to a neutral form or to a less highly charged ion, or of a neutral atom to a negatively charged ion. { ri'dək·shən pə,ten chəl }

reference electrode [PHYSICAL CHEMISTRY] A nonpolarizable electrode that generates highly reproducible potentials; used for pH measurements and polarographic analyses; examples are the calomel electrode, silver-silver chloride electrode, and mercury pool. { 'ref rəns i'lek,trōd }

reference material [ANALYTICAL CHEMISTRY] A material or substance whose properties are sufficiently well established to be used in calibrating an apparatus, assessing a measurement method, or assigning values to other materials. { 'ref·rəns mə,tir·ē·əl }

reflectance spectrophotometry [SPECTROSCOPY] Measurement of the ratio of spectral radiant flux reflected from a light-diffusing specimen to that reflected from a light-diffusing standard substituted for the specimen. { ri'flek·təns ,spek·trə·fə'täm·ə·trē }

Reformatsky reaction [ORGANIC CHEMISTRY] A condensation-type reaction between ketones and α-bromoaliphatic acids in the presence of zinc or magnesium, such as $R_2CO + BrCH_2.\ COOR + Zn \rightarrow (ZnO \cdot HBr) + R_2C(OH)CH_2COOR$. { ,rif·ər'mat·skē rē,ak·shən }

refractory hard metals [CHEMISTRY] True chemical compounds composed of two or more metals in the crystalline form, and having a very high melting point and high hardness. { ri'frak·trē 'härd 'med·əlz }

refrigerant 23 [ORGANIC CHEMISTRY] *See* fluoroform. { ri'frij·ə·rənt ¦twen·tē'thrē }

regenerant [CHEMISTRY] A solution whose purpose is to restore the activity of an ion-exchange bed. { rē'jen·ə·rənt }

regeneration [CHEMISTRY] Restoration of the activity of a deactivated catalyst. { rē‚jen·ə'rā·shən }

regioselective [ORGANIC CHEMISTRY] Pertaining to a chemical reaction which favors a single positional or structural isomer, leading to its yield being greater than that of the other products in the reaction. Also known as regiospecific. { ¦rē·jē·ō·si'lek·tiv }

regiospecific [ORGANIC CHEMISTRY] *See* regioselective. { ¦rē·jē·ō·spə'sif·ik }

regular polymer [CHEMISTRY] A polymer whose molecules possess only one kind of constitutional unit in a single sequential structure. { 'reg·yə·lər 'päl·ə·mər }

Reichert-Meissl number [ANALYTICAL CHEMISTRY] An indicator of the measure of volatile soluble fatty acids. { 'rī·kərt 'mīs·əl ‚nəm·bər }

Reimer-Tiemann reaction [ORGANIC CHEMISTRY] Formation of phenolic aldehydes by reaction of phenol with chloroform in the presence of an alkali. { 'rīm·ər ¦tē·mən rē‚ak·shən }

Reinecke's salt [ANALYTICAL CHEMISTRY] $[(NH_3)_2Cr(SCN)_4]NH_4 \cdot H_2O$ A reagent to detect mercury (gives a red color or a precipitate), and to isolate organic bases (such as proline or histidine). { 'rīn·ə·kēz ‚sȯlt }

Reinsch test [ANALYTICAL CHEMISTRY] A test for detecting small amounts of arsenic, silver, bismuth, and mercury. { 'rīnsh ‚test }

relative atomic mass [CHEMISTRY] *See* atomic weight. { 'rel·əd·iv ə'täm·ik 'mas }

relative fugacity [PHYSICAL CHEMISTRY] The ratio of the fugacity in a given state to the fugacity in a defined standard state. { 'rel·əd·iv fyü'gas·əd·ē }

relative molecular mass [CHEMISTRY] *See* molecular weight. { 'rel·əd·iv mə'lek·yə·lər 'mas }

relative stability test [ANALYTICAL CHEMISTRY] A color test using methylene blue that indicates when the oxygen present in a sewage plant's effluent or polluted water is exhausted. { 'rel·əd·iv stə'bil·əd·ē ‚test }

relative volatility [CHEMISTRY] The volatility of a standard material whose relative volatility is by definition equal to unity. { 'rel·əd·iv ‚väl·ə'til·əd·ē }

relaxation kinetics [PHYSICAL CHEMISTRY] A branch of kinetics that studies chemical systems by disturbing their states of equilibrium and making observations as they return to equilibrium. { ‚rē‚lak'sā·shən ki‚ned·iks }

Renner-Teller effect [PHYSICAL CHEMISTRY] The splitting, into two, of the potential function along the bending coordinate in degenerate electronic states of linear triatomic or polyatomic molecules. { 'ren·ər 'tel·ər i‚fekt }

repeating unit [ORGANIC CHEMISTRY] The group of atoms that is derived from a monomer and repeats throughout a polymer. Also known as monomeric unit. { ri¦pēd·iŋ ‚yü·nət }

repellency [CHEMISTRY] Ability to repel water, or being hydrophobic; opposite to water wettability. { ri'pel·ən·sē }

replication [ANALYTICAL CHEMISTRY] The formation of a faithful mold or replica of a solid that is thin enough for penetration by an electron microscope beam; can use plastic (such as collodion) or vacuum deposition (such as of carbon or metals) to make the mold. { ‚rep·lə'kā·shən }

resbenzophenone [ORGANIC CHEMISTRY] *See* benzoresorcinol. { rez‚ben'zäf·ə‚nōn }

residual intensity [SPECTROSCOPY] The intensity of radiation at some wavelength in a spectral line divided by the intensity in the adjacent continuum. { rə'zij·yə·wəl in'ten·səd·ē }

resin [ORGANIC CHEMISTRY] Any of a class of solid or semisolid organic products of natural or synthetic origin with no definite melting point, generally of high molecular weight; most resins are polymers. { 'rez·ən }

resin of copper [INORGANIC CHEMISTRY] *See* cuprous chloride. { 'rez·ən əv 'käp·ər }

resin matrix [PHYSICAL CHEMISTRY] The molecular network of an ion exchange material that carries the ionogenic groups. { ′rez·ən ‚mā‚triks }

resinography [CHEMISTRY] Science of resins, polymers, plastics, and their products; includes study of morphology, structure, and other characteristics relatable to composition or treatment. { ‚rez·ən′äg·rə·fē }

resinoid [ORGANIC CHEMISTRY] A thermosetting synthetic resin either in its initial (temporarily fusible) or in its final (infusible) state. { ′rez·ən‚ȯid }

resite [ORGANIC CHEMISTRY] *See* C stage. { ′re‚zīt }

resolution [ORGANIC CHEMISTRY] The process of separating a racemic mixture into the two component optical isomers. [SPECTROSCOPY] *See* resolving power. { ‚rez·ə′lü·shən }

resolving power [SPECTROSCOPY] A measure of the ability of a spectroscope or interferometer to separate spectral lines of nearly equal wavelength, equal to the average wavelength of two equally strong spectral lines whose images can barely be separated, divided by the difference in wavelengths; for spectroscopes, the lines must be resolved according to the Rayleigh criterion; for interferometers, the wavelengths at which the lines have half of maximum intensity must be equal. Also known as resolution. { ri′zälv·iŋ ‚pau̇·ər }

resonance [PHYSICAL CHEMISTRY] A feature of the valence-bond method that accounts for the anomalies in certain molecules by representing their structures with approximate resonance hybrid formulas; no single electronic formula conforms both to the observed properties and to the octet rule. Also known as mesomerism. { ′rez·ən·əns }

resonance hybrid [CHEMISTRY] A molecule that may be considered an intermediate between two or more valence bond structures. { ′rez·ən·əns ‚hī·brəd }

resonance ionization spectroscopy [SPECTROSCOPY] A technique capable of detecting single atoms or molecules of a given element or compound in a gas, in which an atom or molecule in its ground state is excited to a bound state when a photon is absorbed from a laser beam at a very well-controlled wavelength that is resonant with the excitation energy; a second photon removes the excited electron from the atom or molecule, and this electron is then accelerated by an electric field and collides with the gas molecules creating additional ionization which is detected by a proportional counter. Abbreviated RIS. { ′rez·ən·əns ‚ī·ə·nə′zā·shən spek′träs·kə·pē }

resonance line [SPECTROSCOPY] The line of longest wavelength associated with a transition between the ground state and an excited state. { ′rez·ən·əns ‚līn }

resonance spectrum [SPECTROSCOPY] An emission spectrum resulting from illumination of a substance (usually a molecular gas) by radiation of a definite frequency or definite frequencies. { ′rez·ən·əns ‚spek·trəm }

resonance structure [ORGANIC CHEMISTRY] Any of two or more possible structures of the same compound that have identical geometry but different arrangements of their paired electrons; none of the structures has physical reality or adequately accounts for the properties of the compound, which exists as an intermediate form. { ′rez·ən·əns ‚strək·chər }

resorcin [ORGANIC CHEMISTRY] *See* resorcinol. { rə′zȯrs·ən }

resorcinol [ORGANIC CHEMISTRY] $C_6H_4(OH)_2$ Sweet-tasting, white, toxic crystals; soluble in water, alcohol, ether, benzene, and glycerol; melts at 111°C; used for resins, dyes, pharmaceuticals, and adhesives, and as a chemical intermediate. Also known as resorcin. { rə′zȯrs·ən‚ȯl }

resorcinol acetate [ORGANIC CHEMISTRY] $HOC_6H_4OCOCH_3$ A viscous, combustible, yellow to amber liquid with burning taste; soluble in alcohol; boils at 283°C; used in cosmetics and medicine. Also known as resorcinol monoacetate. { rə′zȯrs·ən‚ȯl ′as·ə‚tāt }

resorcinol diglycidyl ether [ORGANIC CHEMISTRY] $C_{12}H_{14}O_2$ A straw yellow liquid with a boiling point of 172°C (at 0.8 mmHg or 100 pascals); used for epoxy resins. Abbreviated RDGE. { rə′zȯrs·ən‚ȯl di′glis·ə‚dil ‚ē·thər }

resorcinol-formaldehyde resin [ORGANIC CHEMISTRY] A phenol-formaldehyde resin, soluble in water, ketones, and alchol; used to make fast-curing adhesives for wood gluing. { rə'zȯrs·ən‚ȯl fȯr'mal·də‚hīd 'rez·ən }

resorcinol monoacetate [ORGANIC CHEMISTRY] *See* resorcinol acetate. { rə'zȯrs·ən‚ȯl ¦män·ō'as·ə‚tāt }

β-resorcylic acid [ORGANIC CHEMISTRY] $(OH)_2C_6H_3COOH$ Combustible, white needles; soluble in alcohol and ether, very slightly soluble in water; decomposes at 220°C; used as a dyestuff and a pharmaceutical intermediate, and in the manufacture of fine chemicals. { ¦bād·ə ¦rē·zȯr¦sil·ik 'as·əd }

restricted internal rotation [PHYSICAL CHEMISTRY] Restrictions on the rotational motion of molecules or parts of molecules in some substances, such as solid methane, at certain temperatures. { ri'strik·təd in¦tərn·əl rō'tā·shən }

ret [CHEMISTRY] The reduction or digestion of fibers (usually linen) by enzymes. { ret }

retene [ORGANIC CHEMISTRY] $C_{18}H_{18}$ A cyclic hydrocarbon, melting at 100.5-101°C, soluble in benzene and hot ethanol; used in organic syntheses. { 'rē‚tēn }

retention index [ANALYTICAL CHEMISTRY] In gas chromatography, the relationship of retention volume with arbitrarily assigned numbers to the compound being analyzed; used to indicate the volume retention behavior during analysis. { ri'ten·chən ‚in‚deks }

retention time [ANALYTICAL CHEMISTRY] In gas chromatography, the time at which the center, or maximum, of a symmetrical peak occurs on a gas chromatogram. { ri'ten·chən ‚tīm }

retention volume [ANALYTICAL CHEMISTRY] In gas chromatography, the product of retention time and flow rate. { ri'ten·chən ‚väl·yəm }

rethrolone [ORGANIC CHEMISTRY] A generic name for the five-member ring portion of a pyrethrin. { 'reth·rə‚lōn }

retrogradation [CHEMISTRY] **1.** Generally, a process of deterioration; a reversal or retrogression to a simpler physical form. **2.** A chemical reaction involving vegetable adhesives, which revert to a simpler molecular structure. { ¦re·trō·grā'dā·shən }

retrograde condensation [ORGANIC CHEMISTRY] Phenomenon associated with the behavior of a hydrocarbon mixture in the critical region wherein, at constant temperature, the vapor phase in contact with the liquid may be condensed by a decrease in pressure; or at constant pressure, the vapor is condensed by an increase in temperature. { 're·trə‚grād ‚kän·dən'sā·shən }

retrograde evaporation [ORGANIC CHEMISTRY] Phenomenon associated with the behavior of a hydrocarbon mixture in the critical region wherein, at constant temperature, the liquid phase in contact with the vapor may be vaporized by an increase in pressure; or at constant pressure, the liquid is evaporated by a decrease in temperature. { 're·trə‚grād i‚vap·ə'rā·shən }

retrosynthetic analysis [ORGANIC CHEMISTRY] A method for planning an organic chemical synthesis in which the desired product molecule is considered first, and then steps are considered one at a time leading back to the appropriate starting materials. { ‚re·trō·sin¦thed·ik ə'nal·ə·səs }

reversal spectrum [SPECTROSCOPY] A spectrum which may be observed in intense white light which has traversed luminous gas, in which there are dark lines where there were bright lines in the emission spectrum of the gas. { ri'vər·səl ‚spek·trəm }

reversal temperature [SPECTROSCOPY] The temperature of a blackbody source such that, when light from this source is passed through a luminous gas and analyzed in a spectroscope, a given spectral line of the gas disappears, whereas it appears as a bright line at lower blackbody temperatures, and a dark line at higher temperatures. { ri'vər·səl ‚tem·prə·chər }

reverse bonded-phase chromatography [ANALYTICAL CHEMISTRY] A technique of bonded-phase chromatography in which the stationary phase is nonpolar and the mobile phase is polar. { ri'vərs 'bän·dəd ¦fāz ‚krō·mə'täg·rə·fē }

reverse deionization [CHEMISTRY] A process in which an ion-exchange unit and a cation-exchange unit are used in sequence to remove all ions from a solution. { ri'vərs dē‚ī·ə·nə'zā·shən }

reversed-phase partition chromatography [ANALYTICAL CHEMISTRY] Paper chromatography in which the low-polarity phase (such as paraffin, paraffin jelly, or grease) is put onto the support (paper) and the high-polarity phase (such as water, acids, or organic solvents) is allowed to flow over it. { ri'vərst ¦fāz pär'tish·ən ˌkrō·mə'täg·rə·fē }

reversible chemical reaction [CHEMISTRY] A chemical reaction that can be made to proceed in either direction by suitable variations in the temperature, volume, pressure, or quantities of reactants or products. { ri'vər·sə·bəl 'kem·ə·kəl rē'ak·shən }

reversible electrode [PHYSICAL CHEMISTRY] An electrode that owes its potential to unit charges of a reversible nature, in contrast to electrodes used in electroplating and destroyed during their use. { ri'vər·sə·bəl i'lekˌtrōd }

Reychler's acid [ORGANIC CHEMISTRY] *See* *d*-camphorsulfonic acid. { 'rī·klərz ˌas·əd }

Rf [CHEMISTRY] *See* rutherfordium.

Rh [CHEMISTRY] *See* rhodium.

rhenium [CHEMISTRY] A metallic element, symbol Re, atomic number 75, atomic weight 186.2; a transition element. { 'rē·ne·əm }

rhenium halide [INORGANIC CHEMISTRY] Halogen compound of rhenium; examples are $ReCl_3$, $ReCl_4$, ReF_4, and ReF_6. { 'rē·nē·əm 'haˌlīd }

rheopexy [PHYSICAL CHEMISTRY] A property of certain sols, having particles shaped like rods or plates, which set to gel form more quickly when mechanical means are used to hasten the orientation of the particles. { 'rē·əˌpek·sē }

rhizoctol [ORGANIC CHEMISTRY] *See* methylarsinic sulfide. { rī'zäkˌtȯl }

rhodamine B [ORGANIC CHEMISTRY] $C_{28}H_{31}ClN_2O_3$ Red, green, or reddish-violet powder, soluble in alcohol and water; forms bluish red, fluorescent solution in water; used as red dye for paper, wool, and silk, and as an analytical reagent and biological stain. { 'rōd·əˌmēn 'bē }

rhodanic acid [INORGANIC CHEMISTRY] *See* thiocyanic acid. { rō'dan·ik 'as·əd }

rhodanine [ORGANIC CHEMISTRY] $C_3H_3NOS_2$ A pale-yellow crystalline compound that may decompose violently when heated, giving off toxic by-products; used in organic synthesis. { 'rōd·ənˌīn }

rhodium [CHEMISTRY] A chemical element, symbol Rh, atomic number 45, atomic weight 102.905. { 'rōd·ē·əm }

rhodium chloride [INORGANIC CHEMISTRY] $RhCl_3$ Water-insoluble, brown-red powder, soluble in cyanides and alkalies, decomposes at 450-500°C. Also known as rhodium trichloride. { 'rōd·ē·əm 'klȯrˌīd }

rhodium trichloride [INORGANIC CHEMISTRY] *See* rhodium chloride. { 'rōd·ē·əm trī'klȯrˌīd }

rhombic sulfur [CHEMISTRY] Crystalline sulfur with three unequal axes, all at right angles. { 'räm·bik 'səl·fər }

Rice's bromine solution [ANALYTICAL CHEMISTRY] Analytical reagent for the quantitative analysis of urea; has 12.5% bromine and sodium bromide in aqueous solution. { 'rīs·əz 'brōˌmēn səˌlü·shən }

rich mixture [CHEMISTRY] An air-fuel mixture that is high in its concentration of combustible component. { 'rich ¦miks·chər }

ricinoleic acid [ORGANIC CHEMISTRY] $C_{18}H_{34}O_3$ Unsaturated fatty acid; a combustible, water-insoluble, viscous liquid, soluble in most organic solvents; boils at 226°C (10 mmHg); used as a chemical intermediate, in soaps and Turkey red oils, and for textile finishing. Also known as castor oil acid. { ¦ris·ən·ō¦lē·ik 'as·əd }

ricinoleyl alcohol [ORGANIC CHEMISTRY] $C_{18}H_{36}O_2$ Fatty alcohol of ricinoleic acid; a combustible, colorless, nondrying liquid, boiling at 170-328°C; used as a chemical intermediate, in protective coatings, surface-active agents, pharmaceuticals, and plasticizers. { ¦ris·ən·ō'lē·əl 'al·kəˌhōl }

Riegler's test [ANALYTICAL CHEMISTRY] Analytical technique for nitrous acid; uses sodium naphthionate and *β*-naphthol. { 'rēg·lərz ˌtest }

ring [ORGANIC CHEMISTRY] A closed loop of bonded atoms in a chemical structure, for example, benzene or cyclohexane. { riŋ }

ring closure [ORGANIC CHEMISTRY] A chemical reaction in which one part of an open chain of a molecule reacts with another part to form a ring. { 'riŋ ˌklō·zhər }

Ringer's solution [CHEMISTRY] A solution of 0.86 gram sodium chloride, 0.03 gram potassium chloride, and 0.033 gram calcium chloride in boiled, purified water, used topically as a physiological salt solution. { 'riŋ·ərz sə,lü·shən }

ring isomerism [ORGANIC CHEMISTRY] A type of geometrical isomerism in which bond lengths and bond angles prevent the existence of the trans structure if substituents are attached to alkenic carbons which are part of a cyclic system, the ring of which contains fewer than eight members; for example, 1,2-dichlorocyclohexene. { 'riŋ ī'säm·ə,riz·əm }

ring structure [ORGANIC CHEMISTRY] A cyclic chemical structure consisting of a chain whose ends are connected by bonds. { 'riŋ ,strək·chər }

ring system [ORGANIC CHEMISTRY] Arbitrary designation of certain compounds as closed, circular structures, as in the six-carbon benzene ring; common rings have four, five, and six members, either carbon or some combination of carbon, nitrogen, oxygen, sulfur, or other elements. { 'riŋ ,sis·təm }

ring whizzer [INORGANIC CHEMISTRY] A fluxional molecule frequently encountered in organometallic chemistry in which rapid rearrangements occur by migrations about unsaturated organic rings. { 'riŋ ,wiz·ər }

RIS [SPECTROSCOPY] *See* resonance ionization spectroscopy.

Ritter reaction [ORGANIC CHEMISTRY] A procedure for the preparation of amides by reacting alkenes or tertiary alcohols with nitriles in an acidic medium. { 'rid·ər rē,ak·shən }

Ritz's combination principle [SPECTROSCOPY] The empirical rule that sums and differences of the frequencies of spectral lines often equal other observed frequencies. Also known as combination principle. { 'rit·səz ,käm·bə'nā·shən ,prin·sə·pəl }

Rn [CHEMISTRY] *See* radon.

Rochelle salt [INORGANIC CHEMISTRY] *See* potassium sodium tartrate. { rō'shel ,sȯlt }

roentgen spectrometry [SPECTROSCOPY] *See* x-ray spectrometry. { 'rent·gən spek'träm·ə·trē }

Roese-Gottlieb method [ANALYTICAL CHEMISTRY] A solvent extraction method used to obtain an accurate determination of the fat content of milk. { 'rez·ə 'gät,lēb ,meth·əd }

rosaniline [ORGANIC CHEMISTRY] *See* fuchsin. { rōz'an·ə·lən }

Rosenmund reaction [ORGANIC CHEMISTRY] Catalytic hydrogenation of an acid chloride to form an aldehyde; the reaction is in the presence of sulfur to prevent the subsequent hydrogenation of the aldehyde. { 'rōz·ən,mu̇nd rē,ak·shən }

rosin ester [ORGANIC CHEMISTRY] *See* ester gum. { 'räz·ən 'es·tər }

rotating platinum electrode [ANALYTICAL CHEMISTRY] Platinum wire sealed in a soft-glass tubing and rotated by a constant-speed motor; used as the electrode in amperometric titrations. Abbreviated RPE. { 'rō,tād·iŋ 'plat·ən·əm i'lek,trōd }

rotational constant [PHYSICAL CHEMISTRY] That constant inversely proportioned to the moment of inertia of a linear molecule; used in calculations of microwave spectroscopy quantums. { rō'tā·shən·əl 'kän·stənt }

rotational energy [PHYSICAL CHEMISTRY] For a diatomic molecule, the difference between the energy of the actual molecule and that of an idealized molecule which is obtained by the hypothetical process of gradually stopping the relative rotation of the nuclei without placing any new constraint on their vibration, or on motions of electrons. { rō'tā·shən·əl 'en·ər·jē }

rotational level [PHYSICAL CHEMISTRY] An energy level of a diatomic or polyatomic molecule characterized by a particular value of the rotational energy and of the angular momentum associated with the motion of the nuclei. { rō'tā·shən·əl ,lev·əl }

rotational quantum number [PHYSICAL CHEMISTRY] A quantum number J characterizing the angular momentum associated with the motion of the nuclei of a molecule; the angular momentum is $(h/2\pi)\sqrt{J(J+1)}$ and the largest component is $(h/2\pi)J$, where h is Planck's constant. { rō'tā·shən·əl 'kwän·təm ,nəm·bər }

rotational spectrum [SPECTROSCOPY] The molecular spectrum resulting from transitions between rotational levels of a molecule which behaves as the quantum-mechanical analog of a rotating rigid body. { rō'tā·shən·əl 'spek·trəm }

rotational sum rule [SPECTROSCOPY] The rule that, for a molecule which behaves as a symmetric top, the sum of the line strengths corresponding to transitions to or from a given rotational level is proportional to the statistical weight of that level, that is, to 2J+1, where J is the total angular momentum quantum number of the level. { rō′tā·shən·əl ′səm ,rül }

rotational transition [PHYSICAL CHEMISTRY] A transition between two molecular energy levels which differ only in the energy associated with the molecule's rotation. { rō′tā·shən·əl tran′zish·ən }

rotation spectrum [PHYSICAL CHEMISTRY] Absorption-spectrum (absorbed electromagnetic energy) wavelengths produced if only the rotational energy of a molecule is affected during excitation. { rō′tā·shən ,spek·trəm }

rotation-vibration spectrum [PHYSICAL CHEMISTRY] Absorption-spectrum (absorbed electromagnetic energy) wavelengths produced when both the energy of vibration and energy of rotation of a molecule are affected by excitation. { rō′tā·shən vī′brā·shən ,spek·trəm }

rotatory power [PHYSICAL CHEMISTRY] The product of the specific rotation of an element or compound and its atomic or molecular weight. { ′rōd·ə,tȯr·ē ,pau̇·ər }

rotaxane [ORGANIC CHEMISTRY] A compound with two or more independent portions not bonded to each other but linked by a linear portion threaded through a ring and maintained in this position by bulky end groups. { rō′tak,sān }

rotenone [ORGANIC CHEMISTRY] $C_{23}H_{22}O_6$ White crystals with a melting point of 163°C; soluble in ether and acetone; used as an insecticide and in flea powders and fly sprays. Also known as tubatoxin. { ′rōt·ən,ōn }

rowland [SPECTROSCOPY] A unit of length, formerly used in spectroscopy, equal to 999.81/999.94 angstrom, or approximately 0.99987×10^{-10} meter. { ′rō,lənd }

Rowland circle [SPECTROSCOPY] A circle drawn tangent to the face of a concave diffraction grating at its midpoint, having a diameter equal to the radius of curvature of a grating surface; the slit and camera for the grating should lie on this circle. { ′rō,lənd ,sər·kəl }

Rowland grating [SPECTROSCOPY] *See* concave grating. { ′rō,lənd ,grād·iŋ }

Rowland ghost [SPECTROSCOPY] A false spectral line produced by a diffraction grating, arising from periodic errors in groove position. { ′rō,lənd ,gōst }

Rowland mounting [SPECTROSCOPY] A mounting for a concave grating spectrograph in which camera and grating are connected by a bar forming a diameter of the Rowland circle, and the two run on perpendicular tracks with the slit placed at their junction. { ′rō,lənd ,mau̇nt·iŋ }

roxarsone [ORGANIC CHEMISTRY] *See* 4-hydroxy-3-nitrobenzenearsonic acid. { ,räks′är,sōn }

RPE [ANALYTICAL CHEMISTRY] *See* rotating platinum electrode.

Ru [SPECTROSCOPY] *See* ruthenium.

rubber [ORGANIC CHEMISTRY] A natural, synthetic, or modified high polymer with elastic properties and, after vulcanization, elastic recovery. { ′rəb·ər }

rubber accelerator [ORGANIC CHEMISTRY] A substance that increases the speed of curing of rubber, such as thiocarbanilide. { ′rəb·ər ak′sel·ə,rād·ər }

rubber hydrochloride [ORGANIC CHEMISTRY] White, thermoplastic hydrochloric acid derivative of rubber; water-insoluble powder or clear film, soluble in aromatic hydrocarbons, softens at 110-120°C; used for protective coverings, food packaging, shower curtains, and rainwear. { ′rəb·ər ,hī·drə′klȯr,īd }

rubidium [CHEMISTRY] A chemical element, symbol Rb, atomic number 37, atomic weight 85.47; a reactive alkali metal; salts of the metal may be used in glass and ceramic manufacture. { rü′bid·ē·əm }

rubidium bromide [INORGANIC CHEMISTRY] RbBr Colorless, regular crystals, melting at 683°C; soluble in water; used as a nerve sedative. { rü′bid·ē·əm ′brō,mīd }

rubidium chloride [INORGANIC CHEMISTRY] RbCl A water-soluble, white, lustrous powder melting at 715°C; used as a source for rubidium metal, and as a laboratory reagent. { rü′bid·ē·əm ′klȯr,īd }

rubidium halide [INORGANIC CHEMISTRY] Any of the halogen compounds of rubidium; examples are RbBr, RbCl, RbF, RbIBrCl, $RbBr_2Cl$, and $RbIBr_2$. { rü′bid·ē·əm ′ha,līd }

rubidium halometallate [INORGANIC CHEMISTRY] Halogen-metal-containing compounds of rubidium; examples are Rb_2GeF_6 (rubidium hexafluorogermanate), Rb_2PtCl_6 (rubidium chloroplatinate), and Rb_2PdCl_5 (rubidium palladium chloride). { rü'bid·ē·əm ,ha·lō'med·əl,āt }

rubidium sulfate [INORGANIC CHEMISTRY] Rb_2SO_4 Colorless, water-soluble rhomboid crystals, melting at 1060°C; used as a cathartic. { rü'bid·ē·əm 'səl,fāt }

ruling engine [SPECTROSCOPY] A machine operated by a long micrometer screw which rules equally spaced lines on an optical diffraction grating. { 'rül·iŋ ,en·jən }

ruthenic chloride [INORGANIC CHEMISTRY] *See* ruthenium chloride. { rü'then·ik 'klȯr,īd }

ruthenium [CHEMISTRY] A chemical element, symbol Ru, atomic number 44, atomic weight 101.07. { rü'thē·nē·əm }

ruthenium chloride [INORGANIC CHEMISTRY] $RuCl_3$ Black, deliquescent, water-insoluble solid that decomposes in hot water and above 500°C; used as a laboratory reagent. Also known as ruthenic chloride; ruthenium sesquichloride. { rü'thē·nē·əm 'klȯr,īd }

ruthenium halide [INORGANIC CHEMISTRY] Halogen compound of ruthenium; examples are $RuCl_2$, $RuCl_3$, $RuCl_4$, $RuBr_3$, and RuF_5. { rü'thē·nē·əm 'ha,līd }

ruthenium red [INORGANIC CHEMISTRY] $Ru_2(OH)_2Cl_4 \cdot 7NH_3 \cdot 3H_2O$ A water-soluble, brownish-red powder; used as an analytical reagent and stain. { rü'thē·nē·əm 'red }

ruthenium sesquichloride [INORGANIC CHEMISTRY] *See* ruthenium chloride. { rü'thē·nē·əm ,ses·kwi'klȯr,īd }

ruthenium tetroxide [INORGANIC CHEMISTRY] RuO_4 A yellow, toxic solid, melting at 25°C; used as an oxidizing agent. { rü'thē·nē·əm te'träk,sīd }

Rutherford backscattering spectrometry [SPECTROSCOPY] A method of determining the concentrations of various elements as a function of depth beneath the surface of a sample, by measuring the energy spectrum of ions which are backscattered out of a beam directed at the surface. { 'rəth·ər·fərd 'bak,skad·ə·riŋ spek'träm·ə·trē }

rutherfordium [CHEMISTRY] The name suggested by workers in the United States for element 104. Symbolized Rf. { ,rəth·ər'fȯr·dē·əm }

rutin [ORGANIC CHEMISTRY] $C_{27}H_{32}O_{16}$ A hydroxyflavone glucorhamnoside derived from cowslip and other plants; yellow needles melting at 190°C; used to treat capillary disorders. { 'rüt·ən }

R_F value [ANALYTICAL CHEMISTRY] In chromatography, a measurement based on the relative distance traveled by a sample of a substance in a specific procedure; under standard conditions it is a characteristic property of the substance. { ,är'ef *or* ,är səb'ef ,val·yü }

rydberg [SPECTROSCOPY] *See* kayser. { 'rid,bərg }

Rydberg series formula [SPECTROSCOPY] An empirical formula for the wave numbers of various lines of certain spectral series such as neutral hydrogen and alkali metals; it states that the wave number of the *n*th member of the series is $\lambda_\infty - R/(n + a)^2$, where λ_∞ is the series limit, R is the Rydberg constant of the atom, and a is an empirical constant. { 'rid,bərg ¦sir·ēz ,fȯr·myə·lə }

Rydberg spectrum [SPECTROSCOPY] An ultraviolet absorption spectrum produced by transitions of atoms of a given element from the ground state to states in which a single electron occupies an orbital farther from the nucleus. { 'rīd,bərg ,spek·trəm }

S

S [CHEMISTRY] *See* secondary winding; siemens; stoke; sulfur.

saccharin [ORGANIC CHEMISTRY] $C_6H_4COSO_2NH$ A sweet-tasting, white powder, soluble in acetates, benzene, and alcohol; slightly soluble in water and ether; melts at 228°C; used as a sugar substitute for syrups, and in medicines, foods, and beverages. Also known as benzosulfimide; gluside. { 'sak·ə·rən }

saccharolactic acid [ORGANIC CHEMISTRY] *See* mucic acid. { ¦sak·ə·rō¦lak·tik 'as·əd }

saccharose [ORGANIC CHEMISTRY] *See* sucrose. { 'sak·ə‚rōs }

sacrificial anode [PHYSICAL CHEMISTRY] A protective coating applied to a metal surface to act as an anode and be consumed in an electrochemical reaction, thereby preventing electrolytic corrosion of the metal. { ¦sak·rə‚fish·əl 'an‚ōd }

sacrificial metal [PHYSICAL CHEMISTRY] A metal that can be used for a sacrificial anode. { ‚sak·rə'fish·əl 'med·əl }

saddle-point azeotrope [PHYSICAL CHEMISTRY] A rarely occurring azeotrope which is formed in ternary systems and for which the boiling point is intermediate between the highest and lowest boiling mixture in the system. { 'sad·əl ¦pȯint 'ā·zē·ə‚trōp }

safranine [ORGANIC CHEMISTRY] Any of a group of phenazine-based dyes; some are used as biological stains. { 'saf·rə‚nēn }

safrole [ORGANIC CHEMISTRY] $C_3H_5C_6H_3O_2CH_2$ A toxic, water-insoluble, colorless oil that boils at 233°C; found in sassafras and camphorwood oils, used in medicine, perfumes, insecticides, and soaps, and as a chemical intermediate. { 'sa‚frōl }

sal acetosella [ORGANIC CHEMISTRY] *See* potassium binoxalate. { 'sal ¦as·ə·dō ¦sel·ə }

salazosulfadimidine [ORGANIC CHEMISTRY] $C_{19}H_{17}N_5O_5S$ A brown crystalline compound that melts at 207°C, used in medicine in cases of ulcerative colitis. { ¦sal·ā·zō‚səl·fə'dī·mə‚dēn }

sal ethyl [ORGANIC CHEMISTRY] *See* ethyl salicylate. { 'sal 'eth·əl }

salicin [ORGANIC CHEMISTRY] $C_{13}H_{18}O_7$ A glucoside; colorless crystals, soluble in water, alcohol, alkalies, and glacial acetic acid; melts at 199°C; used in medicine and as an analytical reagent. { 'sal·ə·sən }

salicyl alcohol [ORGANIC CHEMISTRY] $C_7H_8O_2$ A crystalline alcohol that forms plates or powder, melting at 86-87°C; used in medicine as a local anesthetic. { 'sal·ə·səl 'al·kə‚hȯl }

salicylaldehyde [ORGANIC CHEMISTRY] C_6H_4OHCHO Clear to dark-red oily liquid, with burning taste and almond aroma; soluble in alcohol, benzene, and ether, very slightly soluble in water; boils at 196°C; used in analytical chemistry, in perfumery, and for synthesis of chemicals. Also known as helicin. { ¦sal·ə səl'al·də‚hīd }

salicylamide [ORGANIC CHEMISTRY] $C_6H_4(OH)CONH_2$ Pinkish or white crystals; soluble in alcohol, ether, chloroform, and hot water; melts at 193°C; used in medicine as an analgesic, antipyretic, and antirheumatic drug. { ‚sal·ə'sī·lə·məd }

salicylate [ORGANIC CHEMISTRY] A salt of salicylic acid with the formula $C_6H_4(OH)$-COOM, where M is a monovalent metal; for example, $NaC_7H_5O_3$, sodium salicylate. { sə'lis·ə‚lāt }

salicylated mercury [ORGANIC CHEMISTRY] *See* mercuric salicylate. { sə′lis·ə‚lād·əd ′mər·kyə·rē }

salicylic acid [ORGANIC CHEMISTRY] $C_6H_4(OH)COOH$ White crystals with sweetish taste; soluble in alcohol, acetone, ether, benzene, and turpentine, slightly soluble in water; discolored by light; melts at 158°C; used as a chemical intermediate and in medicine, dyes, perfumes, and preservatives. { ¦sal·ə¦sil·ik ′as·əd }

salicylic acid ethyl ether [ORGANIC CHEMISTRY] *See* ethyl salicylate. { ¦sal·ə¦sil·ik ′as·əd ′eth·əl ′ē·thər }

salicylic ether [ORGANIC CHEMISTRY] *See* ethyl salicylate. { ¦sal·ə¦sil·ik ′ē·thər }

salol [ORGANIC CHEMISTRY] $C_6H_4OHCOOC_6H_5$ White powder with aromatic taste and aroma; soluble in alcohol, ether, chloroform, and benzene; slightly soluble in water; melts at 42°C; used in medicinals and as a preservative. Also known as phenyl salicylate. { ′sa‚lól }

sal soda [INORGANIC CHEMISTRY] $Na_2CO_3 \cdot 10H_2O$ White, water-soluble crystals, insoluble in alcohol; melts and loses water at about 33°C; mild irritant to mucous membrane; used in cleansers and for washing textiles and bleaching linen and cotton. Also known as sodium carbonate decahydrate; washing soda. { ′sal ‚sō·də }

salt [CHEMISTRY] The reaction product when a metal displaces the hydrogen of an acid; for example, $H_2SO_4 + 2NaOH \rightarrow Na_2SO_4$ (a salt) $+ 2H_2O$. { sólt }

sal tartar [ORGANIC CHEMISTRY] *See* sodium tartrate. { ′sal ′tär·tər }

salt bridge [PHYSICAL CHEMISTRY] A bridge of a salt solution, usually potassium chloride, placed between the two half-cells of a galvanic cell, either to reduce to a minimum the potential of the liquid junction between the solutions of the two half-cells or to isolate a solution under study from a reference half-cell and prevent chemical precipitations. { ′sólt ‚brij }

salt cake [INORGANIC CHEMISTRY] Impure sodium sulfate; used in soaps, paper pulping, detergents, glass, ceramic glaze, and dyes. { ′sólt ‚kāk }

salt error [ANALYTICAL CHEMISTRY] An error introduced in an analytical determination of a saline liquid such as sea water; caused by the effect of the neutral ions in the solution on the color of the pH indicator, and hence upon the apparent pH. { ′sólt ‚er·ər }

salt fingers [CHEMISTRY] A close-packed array of rising and sinking columns of fluid that form in a liquid when a slow-diffusing solute is separated by an interface from another, lower solute that diffuses more rapidly. { ′sólt ‚fiŋ·gərz }

salt of Lemery [INORGANIC CHEMISTRY] *See* potassium sulfate. { ′sólt əv lem′rē }

salt pan [CHEMISTRY] A pool used for obtaining salt by the natural evaporation of seawater. { ′sólt ‚pan }

salt of sorrel [ORGANIC CHEMISTRY] *See* potassium binoxalate. { ′sólt əv ′sär·əl }

salt of tartar [INORGANIC CHEMISTRY] *See* potassium carbonate. { ′sólt əv ′tär·tər }

samarium [CHEMISTRY] A rare-earth metal, atomic number 62, symbol Sm; melts at 1350°C, tarnishes in air, ignites at 200-400°C. { sə′mar·ē·əm }

samarium oxide [INORGANIC CHEMISTRY] Sm_2O_3 A cream-colored powder with a melting point of 2300°C; soluble in acids; used for infrared-absorbing glass and as a neutron absorber. { sə′mar·ē·əm ′äk‚sīd }

SAN [ORGANIC CHEMISTRY] *See* styrene-acrylonitrile resin.

Sandmeyer's reaction [CHEMISTRY] Conversion of diazo compounds (in the presence of cuprous halogen salts) into halogen compounds. { ′san‚mī·ərz rē‚ak·shən }

Sanger's reagent [ORGANIC CHEMISTRY] *See* 1-fluoro-2,4-dinitrobenzene. { ′saŋ·ərz rē‚ā·jənt }

santalol [ORGANIC CHEMISTRY] $C_{15}H_{24}O$ A colorless liquid with a boiling point of 300°C; derived from sandalwood oil and used for perfumes. { ′san·tə‚lól }

santonin [ORGANIC CHEMISTRY] $C_{15}H_{18}O_3$ A white powder with a melting point of 170-173°C; soluble in chloroform and alcohol; used in medicine. { ′sant·ən·ən }

saponification [CHEMISTRY] The process of converting chemicals into soap; involves the alkaline hydrolysis of a fat or oil, or the neutralization of a fatty acid. { sə‚pän·ə·fə′kā·shən }

saponification equivalent [CHEMISTRY] The quantity of fat in grams that can be saponified by 1 liter of normal alkalies. { sə‚pän·ə·fə′kā·shən i‚kwiv·ə·lənt }

saponification number [ANALYTICAL CHEMISTRY] Milligrams of potassium hydroxide required to saponify the fat, oil, or wax in a 1-gram sample of a given material, using a specific American Society for Testing and Materials test method. { sə,pän·ə·fə 'kā·shən ,nəm·bər }

saponin [ORGANIC CHEMISTRY] Any of numerous plant glycosides characterized by foaming in water and by producing hemolysis when water solutions are injected into the bloodstream; used as beverage foam producer, textile detergent and sizing, soap substitute, and emulsifier. { 'sap·ə·nən }

sarcosine [ORGANIC CHEMISTRY] CH_3NHCH_2COOH Sweet-tasting, deliquescent crystals; soluble in water, slightly soluble in alcohol; decomposes at 210-215°C; used in toothpaste manufacture. Also known as methyl glycocol. { 'sär·kə,sēn }

SAS [INORGANIC CHEMISTRY] *See* aluminum sodium sulfate.

satellite infrared spectrometer [SPECTROSCOPY] A spectrometer carried aboard satellites in the Nimbus series which measures the radiation from carbon dioxide in the atmosphere at several different wavelengths in the infrared region, giving the vertical temperature structure of the atmosphere over a large part of the earth. Abbreviated SIRS. { 'sad·əl,īt ¦in·frə'red spek'träm·əd·ər }

saturated ammonia [CHEMISTRY] **1.** Liquid ammonia in a state in which adding heat at constant pressure causes the liquid to vaporize at constant temperature, and in which removing heat at constant pressure causes the temperature of the liquid to drop immediately. **2.** Ammonia vapor in a state in which adding heat at constant pressure causes an immediate temperature rise (superheating) and in which removing heat at constant pressure starts immediate condensation at constant temperature. { 'sach·ə,rād·əd ə'mō·nyə }

saturated calomel electrode [PHYSICAL CHEMISTRY] A reference electrode of mercury topped by a layer of mercury (I) chloride paste with potassium chloride solution placed above, easier to assemble than the normal and the one-tenth normal (referring to the concentration of KCl) calomel electrodes. { 'sach·ə,rād·əd 'kal·ə,mel i'lek,trōd }

saturated compound [ORGANIC CHEMISTRY] An organic compound with all carbon bonds satisfied; it does not contain double or triple bonds and thus cannot add elements or compounds. { 'sach·ə,rād·əd 'käm,paủnd }

saturated hydrocarbon [ORGANIC CHEMISTRY] A saturated carbon-hydrogen compound with all carbon bonds filled; that is, there are no double or triple bonds as in olefins and acetylenics. { 'sach·ə,rād·əd ¦hī·drə'kär·bən }

saturated interference spectroscopy [SPECTROSCOPY] A version of saturation spectroscopy in which the gas sample is placed inside an interferometer that splits a probe laser beam into parallel components in such a way that they cancel on recombination; intensity changes in the recombined probe beam resulting from changes in absorption or refractive index induced by a laser saturating beam are then measured. { 'sach·ə,rād·əd ,in·tər,fir·əns spek'träs·kə·pē }

saturated liquid [CHEMISTRY] A solution that contains enough of a dissolved solid, liquid, or gas so that no more will dissolve into the solution at a given temperature and pressure. { 'sach·ə,rād·əd 'lik·wəd }

saturation [PHYSICAL CHEMISTRY] The condition in which the partial pressure of any fluid constituent is equal to its maximum possible partial pressure under the existing environmental conditions, such that any increase in the amount of that constituent will initiate within it a change to a more condensed state. { ,sach·ə'rā·shən }

saturation spectroscopy [SPECTROSCOPY] A branch of spectroscopy in which the intense, monochromatic beam produced by a laser is used to alter the energy-level populations of a resonant medium over a narrow range of particle velocities, giving rise to extremely narrow spectral lines that are free from Doppler broadening; used to study atomic, molecular, and nuclear structure, and to establish accurate values for fundamental physical constants. { ,sach·ə'rā·shən spek'träs·kə·pē }

Sb [CHEMISTRY] *See* antimony.

s-block element [CHEMISTRY] A chemical element whose valence shell contains s-electrons only; found in groups 1 and 2 of the periodic table. { 'es ,bläk ,el·ə·mənt }

Sc [CHEMISTRY] *See* scandium.

scale [CHEMISTRY] *See* boiler scale. { skāl }

scandia [INORGANIC CHEMISTRY] *See* scandium oxide. { 'skan·dē·ə }

scandium [CHEMISTRY] A transition element, symbol Sc, atomic number 21; melts at 1200°C; found associated with rare-earth elements. { 'skan·dē·əm }

scandium halide [INORGANIC CHEMISTRY] A compound of scandium and a halogen; for example, scandium chloride, $ScCl_3$. { 'skan·dē·əm 'ha,līd }

scandium oxide [INORGANIC CHEMISTRY] Sc_2O_3 White powder, soluble in hot acids; used to prepare scandium. Also known as scandia. { 'skan·dē·əm 'äk,sīd }

scandium sulfate [INORGANIC CHEMISTRY] $Sc_2(SO_4)_3$ Water-soluble, colorless crystals. { 'skan·dē·əm 'səl,fāt }

scandium sulfide [INORGANIC CHEMISTRY] Sc_3S_3 Yellowish powder; decomposes in dilute acids and boiling water to give off hydrogen sulfide. { 'skan·dē·əm 'səl,fīd }

scarlet [ORGANIC CHEMISTRY] *See* scarlet red. { 'skär·lət }

scarlet red [ORGANIC CHEMISTRY] $CH_3C_6H_4H{:}NC_6H_3CH_3N{:}NC_{10}H_{15}OH$ A brown, water-insoluble powder, used as a dye in ointments. Also known as Biebrich red; scarlet. { 'skär·lət 'red }

scattering plane [PHYSICAL CHEMISTRY] In a quasielastic light-scattering experiment performed with the use of polarizers, the plane containing the incident and scattered beams. { 'skad·ə·riŋ ,plān }

scavenger [CHEMISTRY] A substance added to a mixture or other system to remove or inactivate impurities. Also known as getter. { 'skav·ən·jər }

Schaeffer's salt [ORGANIC CHEMISTRY] $HOC_{10}SO_3Na$ A light-yellow to pink, water-soluble powder; the sodium salt formed from 2-naphthol-6-sulfonic acid; used as an intermediate in synthesis of organic compounds. { 'shā·fərz ,sölt }

Scheele's green [INORGANIC CHEMISTRY] *See* copper arsenite. { 'shā·ləz 'grēn }

Schiff base [ORGANIC CHEMISTRY] $RR'C{=}NR''$ Any of a class of derivatives of the condensation of aldehydes or ketones with primary amines; colorless crystals, weakly basic; hydrolyzed by water and strong acids to form carbonyl compounds and amines; used as chemical intermediates and perfume bases, in dyes and rubber accelerators, and in liquid crystals for electronics. { 'shif ,bās }

Schiff's reagent [ANALYTICAL CHEMISTRY] An aqueous solution of rosaniline and sulfurous acid; used in the Schiff test. { 'shifs rē,ā·jənt }

Schiff test [ANALYTICAL CHEMISTRY] A test for aldehydes by using an aqueous solution of rosaniline and sulfurous acid. { 'shif ,test }

schiller layer [PHYSICAL CHEMISTRY] One of a series of layers formed by sedimenting particles that exhibit bright colors in reflected light, because the layers are separated by approximately equal distances, with the distances being of the same order of magnitude as the wavelength of visible light. Also known as iridescent layer. { 'shil·ər ,lā·ər }

Schmidt number 3 [PHYSICAL CHEMISTRY] A dimensionless number used in electrochemistry, equal to the product of the dielectric susceptibility and the dynamic viscosity of a fluid divided by the product of the fluid density, electrical conductivity, and the square of a characteristic length. Symbolized Sc_3. { 'shmit ,nəm·bər 'thrē }

Schoelkopf's acid [ORGANIC CHEMISTRY] A dye of the following types: 1-naphtho-4,8-disulfonic acid, 1-naphthylamine-4,8-disulfonic acid, and 1-naphthylamine-8-sulfonic acid; may be toxic. { 'shəl,köpfs ,as·əd }

Schotten-Baumann reaction [ORGANIC CHEMISTRY] An acylation reaction that uses an acid chloride in the presence of dilute alkali to acylate the hydroxyl and amino group of organic compounds. { 'shät·ən 'bau̇,män rē,ak·tən }

Schuermann series [CHEMISTRY] A list of metals so arranged that the sulfide of one is precipitated at the expense of the sulfide of any lower metal in the series. { 'shöi·ər,män ,sir·ēz }

Schultz-Hardy rule [CHEMISTRY] The sensitivity of lyophobic colloids to coagulating electrolytes is governed by the charge of the ion opposite that of the colloid, and the sensitivity increases more rapidly than the charge of the ion. { 'shu̇lts 'här·dē ˌrül }

Schulze's reagent [ANALYTICAL CHEMISTRY] An oxidizing mixture consisting of a saturated aqueous solution of $KClO_3$ and varying amounts of concentrated HNO_3; commonly used in palynologic macerations. { 'shu̇lt·səz rē'ā·jənt }

Schuster method [SPECTROSCOPY] A method for focusing a prism spectroscope without using a distant object or a Gauss eyepiece. { 'shüs·tər ˌmeth·əd }

Schweitzer's reagent [CHEMISTRY] An ammoniacal solution of cupric hydroxide; used to dissolve cellulose, silk, and linen, and to test for wool. { 'shvīt·sərz rē'ā·jənt }

scopoline [ORGANIC CHEMISTRY] $C_8H_{13}O_2H$ A white crystalline alkaloid that melts at 108-109°C, soluble in water and ethanol; used in medicine. Also known as oscine. { 'skō·pəˌlēn }

screening agent [ANALYTICAL CHEMISTRY] A nonchelating dye used to improve the colorimetric end point of a complexometric titration; a dye addition forms a complementary pair of colors with the metalized and unmetalized forms of the end-point indicator. { 'skrēn·iŋ ˌā·jənt }

SDDC [INORGANIC CHEMISTRY] *See* sodium dimethyldithiocarbamate.

Se [CHEMISTRY] *See* selenium.

sebacic acid [ORGANIC CHEMISTRY] $COOH(CH_2)_8COOH$ Combustible, white crystals; slightly soluble in water, soluble in alcohol and ether; melts at 133°C; used in perfumes, paints, and hydraulic fluids and to stabilize synthetic resins. { si'bas·ik 'as·əd }

secondary alcohol [ORGANIC CHEMISTRY] An organic alcohol with molecular structure R_1R_2CHOH, where R_1 and R_2 designate either identical or different groups. { 'sek·ənˌder·ē 'al·kəˌho̊l }

secondary amine [ORGANIC CHEMISTRY] An organic compound that may be written R_1R_2NH, where R_1 and R_2 designate either identical or different groups. { 'sek·ənˌder·ē 'amˌēn }

secondary carbon atom [ORGANIC CHEMISTRY] A carbon atom that is singly bonded to two other carbon atoms. { 'sek·ənˌder·ē 'kär·bən 'ad·əm }

secondary hydrogen atom [ORGANIC CHEMISTRY] A hydrogen atom that is bonded to a secondary carbon atom. { 'sek·ənˌder·ē 'hī·drə·jən ˌad·əm }

second boiling point [PHYSICAL CHEMISTRY] In certain mixtures, the temperature at which a gas phase develops from a liquid phase upon cooling. { 'sek·ənd 'bȯil·iŋ ˌpȯint }

second-order reaction [PHYSICAL CHEMISTRY] A reaction whose rate of reaction is determined by the concentration of two chemical species. { 'sek·ənd ¦ȯr·dər rē'ak·shən }

sedimentation [CHEMISTRY] The settling of suspended particles within a liquid under the action of gravity or a centrifuge. { ˌsed·ə·mən'tā·shən }

sedimentation balance [ANALYTICAL CHEMISTRY] A device to measure and record the weight of sediment (solid particles settled out of a liquid) versus time; used to determine particle sizes of fine solids. { ˌsed·ə·mən'tā·shən ˌbal·əns }

sedimentation coefficient [PHYSICAL CHEMISTRY] In the sedimentation of molecules in an accelerating field, such as that of a centrifuge, the velocity of the boundary between the solution containing the molecules and the solvent divided by the accelerating field. (In the case of a centrifuge, the accelerating field equals the distance of the boundary from the axis of rotation multiplied by the square of the angular velocity in radians per second.) { ˌsed·ə·mən'tā·shən ˌkō·i'fish·ənt }

sedimentation constant [PHYSICAL CHEMISTRY] A quantity used in studying the behavior of colloidal particles subject to forces, especially centrifugal forces; it is equal to $2r^2(\rho - \rho')/9\eta$, where r is the particle's radius, ρ and ρ' are reciprocals of partial specific volumes of particle and medium respectively, and η is the medium's viscosity. { ˌsed·ə·mən'tā·shən ˌkän·stənt }

sedimentation equilibrium [ANALYTICAL CHEMISTRY] The equilibrium between the forward movement of a sample's liquid-sediment boundary and reverse diffusion during centrifugation; used in molecular-weight determinations. { ˌsed·ə·mən'tā·shən ˌē·kwəˌlib·rē·əm }

sedimentation velocity [ANALYTICAL CHEMISTRY] The rate of movement of the liquid-sediment boundary in the sample holder during centrifugation; used in molecular-weight determinations. { ˌsed·ə·mən′tā·shən vəˈläs·əd·ē }

seed [CHEMISTRY] A small, single crystal of a desired substance added to a solution to induce crystallization. { sēd }

seed charge [CHEMISTRY] A small amount of material added to a supersaturated solution to initiate precipitation. { ′sēd ˌchärj }

seeding [CHEMISTRY] The adding of a seed charge to a supersaturated solution, or a single crystal of a desired substance to a solution of the substance to induce crystallization. { ′sēd·iŋ }

segment [ANALYTICAL CHEMISTRY] A specific, demarcated portion of a lot of a substance that is to be chemically analyzed. { ′seg·mənt }

selective inhibition [CHEMISTRY] *See* selective poisoning. { si′lek·tiv ˌin·ə′bish·ən }

selective poisoning [CHEMISTRY] Retardation of the rate of one catalyzed reaction more than that of another by the use of a catalyst poison. Also known as selective inhibition. { si′lek·tiv ′pȯiz·ən·iŋ }

selectivity [ANALYTICAL CHEMISTRY] The ability of a type of method or instrumentation to respond to a specified substance or constituent and not to others. { səˌlek′tiv·əd·ē }

selectivity coefficient [ANALYTICAL CHEMISTRY] Ion equilibria relationship formula for ion-exchange-resin systems. { səˌlek′tiv·əd·ē ˌkō·iˌfish·ənt }

selenic acid [INORGANIC CHEMISTRY] H_2SeO_4 A highly toxic, water-soluble, white solid, melting point 58°C, decomposing at 260°C. { sə′len·ik ′as·əd }

selenide [INORGANIC CHEMISTRY] M_2Se A binary compound of divalent selenium, such as Ag_2Se, silver selenide. [ORGANIC CHEMISTRY] An organic compound containing divalent selenium, such as $(C_2H_5)_2Se$, ethyl selenide. { ′sel·əˌnīd }

selenious acid [INORGANIC CHEMISTRY] *See* selenous acid. { sə′lē·nē·əs ′as·əd }

selenium [CHEMISTRY] A highly toxic, nonmetallic element in group VI, symbol Se, atomic number 34; steel-gray color; soluble in carbon disulfide, insoluble in water and alcohol; melts at 217°C; and boils at 690°C; used in analytical chemistry, metallurgy, and photoelectric cells, and as a lube-oil stabilizer and chemicals intermediate. { sə′lē·nē·əm }

selenium bromide [INORGANIC CHEMISTRY] Any of three compounds of selenium and bromine: Se_2Br_2, a red liquid that melts at −46°C, also known as selenium monobromide; $SeBr_2$, a brown liquid, also known as selenium dibromide; and $SeBr_4$, orange, carbon-disulfide-soluble crystals, also known as selenium tetrabromide. { sə′lē·nē·əm ′brōˌmīd }

selenium dibromide [INORGANIC CHEMISTRY] *See* selenium bromide. { sə′lē·nē·əm dī′brōˌmīd }

selenium dioxide [INORGANIC CHEMISTRY] SeO_2 Water- and alcohol-soluble, white to reddish, lustrous crystals; melts at 340°C; used in medicine, and as an oxidizing agent and catalyst. Also known as selenous acid anhydride; selenous anhydride; selenium oxide. { sə′lē·nē·əm dī′äkˌsīd }

selenium halide [INORGANIC CHEMISTRY] A compound of selenium and a halogen, for example, Se_2Br_2, $SeBr_2$, $SeBr_4$; Se_2Cl_2, $SeCl_2$, $SeCl_4$; Se_2I_2, SeI_4. { sə′lē·nē·əm ′haˌlīd }

selenium monobromide [INORGANIC CHEMISTRY] *See* selenium bromide. { sə′lē·nē·əm ˌmän·ō′brōˌmīd }

selenium nitride [INORGANIC CHEMISTRY] Se_2N_2 A water-insoluble, yellow solid that explodes at 200°C. { sə′lē·nē·əm ′nīˌtrīd }

selenium oxide [INORGANIC CHEMISTRY] *See* selenium dioxide. { sə′lē·nē·əm ′äkˌsīd }

selenium tetrabromide [INORGANIC CHEMISTRY] *See* selenium bromide. { sə′lē·nē·əm ˌte·trə′brōˌmīd }

selenone [ORGANIC CHEMISTRY] A group of organic selenium compounds with the general formula R_2SeO_2. { ′sel·əˌnōn }

selenonic acid [ORGANIC CHEMISTRY] Any organic acid containing the radical $—SeO_3H$; analogous to a sulfonic acid. { ˌsel·ə′nän·ik ′as·əd }

selenous acid [INORGANIC CHEMISTRY] H_2SeO_3 Colorless, transparent crystals; soluble in water and alcohol, insoluble in ammonia; decomposes when heated; used as an analytical reagent. Also spelled selenious acid. { sə'lē·nəs 'as·əd }

selenous acid anhydride [INORGANIC CHEMISTRY] *See* selenium dioxide. { sə'lē·nəs 'as·əd an'hī‚drīd }

selenous anhydride [INORGANIC CHEMISTRY] *See* selenium dioxide. { sə'lē·nəs an'hī‚drīd }

selenoxide [ORGANIC CHEMISTRY] A group of organic selenium compounds with the general formula R_2SeO. { ‚sel·ə'näk‚sīd }

self-absorption [SPECTROSCOPY] Reduction of the intensity of the center of an emission line caused by selective absorption by the cooler portions of the source of radiation. Also known as self-reduction; self-reversal. { ¦self əb¦sȯrp·shən }

self-poisoning [CHEMISTRY] Inhibition of a chemical reaction by a product of the reaction. Also known as autopoisoning. { ‚self 'pȯiz·ən·iŋ }

self-reduction [SPECTROSCOPY] *See* self-absorption. { ¦self ri¦dək·shən }

self-reversal [SPECTROSCOPY] *See* self-absorption. { ¦self ri¦vər·səl }

Seliwanoff's test [ANALYTICAL CHEMISTRY] A color test helpful in the identification of ketoses, which develop a red color with resorcinol in hydrochloric acid. { sə'liv·ə‚nȯfs ‚test }

sellite [INORGANIC CHEMISTRY] A solution of sodium sulfite (Na_2SO_3) used in the purification of 2,4,6-trinitrotoluene to remove unsymmetrical isomers. { 'se‚līt }

Semenov number 1 [PHYSICAL CHEMISTRY] A dimensionless number used in reaction kinetics, equal to a mass transfer constant divided by a reaction rate constant. Symbolized S_m. Formerly known as Schmidt number 2. { 'se·mə‚nȯf 'nəm·bər 'wən }

semicarbazide [ORGANIC CHEMISTRY] $H_2NNHCONH_2$ A reagent used to produce semicarbazones by reaction with aldehydes or ketones. { ¦sem·i'kär·bə‚zīd }

semicarbazide hydrochloride [ORGANIC CHEMISTRY] $CH_5ON_3 \cdot HCl$ Colorless prisms, soluble in water, decomposing at 175°C; used as an analytical reagent for aldehydes and ketones, and to recover constituents of essential oils. { ¦sem·i 'kär·bə‚zīd ‚hī·drə'klȯr‚īd }

semicarbazone [ORGANIC CHEMISTRY] $R_2C{:}N_2HCONH_2$ A condensation product of an aldehyde or ketone with semicarbazide. { ‚sem·i'kär·bə‚zōn }

semiempirical computation [PHYSICAL CHEMISTRY] Computation of the geometry of a molecule by using parameters that have been experimentally determined for similar molecules. { ‚sem·ē·em¦pir·ə·kəl ‚käm·pyə'tā·shən }

semiforbidden line [SPECTROSCOPY] A spectral line associated with a semiforbidden transition. { ¦sem·i fər'bid·ən 'līn }

semimetal [CHEMISTRY] *See* metalloid. { ¦sem·ē'med·əl }

semimicroanalysis [ANALYTICAL CHEMISTRY] A chemical analysis procedure in which the weight of the sample is between 10 and 100 milligrams. { ¦sem·i‚mī·krō·ə'nal·ə·səs }

semioxamazide [ORGANIC CHEMISTRY] $H_2NCOCONHNH_2$ A crystalline compound that decomposes at 220°C; soluble in hot water, acids, and alkalies, used as a reagent for aldehydes and ketones. { ¦sem·ē‚äk'sam·ə·zəd }

sensing zone technique [ANALYTICAL CHEMISTRY] Particle-size measurement in a dilute solution, with fine particles passed through a small zone (opening) so that individual particles may be observed and measured by electrolytic, photic, or sonic methods. { 'sens·iŋ ¦zōn tek‚nēk }

separatory funnel [CHEMISTRY] A funnel-shaped device used for the careful and accurate separation of two immiscible liquids; a stopcock on the funnel stem controls the rate and amount of outflow of the lower liquid. { 'sep·rə‚tȯr·ē 'fən·əl }

sequestering agent [CHEMISTRY] A substance that removes a metal ion from a solution system by forming a complex ion that does not have the chemical reactions of the ion that is removed; can be a chelating or a complexing agent. { si'kwes·tə·riŋ ‚ā·jənt }

series [ANALYTICAL CHEMISTRY] A group of results of repeated analyses completed by using a single analytical method on samples of a homogeneous substance. [SPECTROSCOPY] A collection of spectral lines of an atom or ion for a set of transitions, with the same selection rules, to a single final state; often the frequencies have the general formula $[R/(a + c_1)^2] - [R/(n + c_2)^2]$, where R is the Rydberg constant for the atom, a and c_1 and c_2 are constants, and n takes on the values of the integers greater than a for the various lines in the series. { 'sir·ēz }

sesquioxide [CHEMISTRY] A compound composed of a metal and oxygen in the ratio 2:3; for example, Al_2O_3. { ¦ses·kwē'äk‚sīd }

sesquiterpene [ORGANIC CHEMISTRY] Any terpene with the formula $C_{15}H_{24}$; that is, $1^1/_2$ times the terpene formula. { ¦ses·kwē'tər‚pēn }

set [CHEMISTRY] The hardening or solidifying of a plastic or liquid substance. { set }

SFS [ORGANIC CHEMISTRY] *See* sodium formaldehyde sulfoxylate.

sharp series [SPECTROSCOPY] A series occurring in the line spectra of many atoms and ions with one, two, or three electrons in the outer shell, in which the total orbital angular momentum quantum number changes from 0 to 1. { 'shärp 'sir·ēz }

shift [SPECTROSCOPY] A small change in the position of a spectral line that is due to a corresponding change in frequency which, in turn, results from one or more of several causes, such as the Doppler effect. { shift }

Shpol'skii effect [SPECTROSCOPY] The occurrence of very narrow fluorescent lines in the spectra of certain compounds from molecules frozen at low temperatures. { 'shpōl·skē i‚fekt }

Si [CHEMISTRY] *See* silicon.

side chain [ORGANIC CHEMISTRY] A grouping of similar atoms (two or more, generally carbons, as in the ethyl radical, C_2H_5—) that branches off from a straight-chain or cyclic (for example, benzene) molecule. Also known as branch; branched chain. { 'sīd ‚chān }

side reaction [CHEMISTRY] A secondary or subsidiary reaction that takes place simultaneously with the reaction of primary interest. { 'sīd rē‚ak·shən }

siderophile element [CHEMISTRY] An element with a weak affinity for oxygen and sulfur and that is readily soluble in molten iron; includes iron, nickel, cobalt, platinum, gold, tin, and tantalum. { 'sid·ə·rə‚fīl 'el·ə·mənt }

siegbahn [SPECTROSCOPY] A unit of length, formerly used to express wavelengths of x-rays, equal to 1/3029.45 of the spacing of the (200) planes of calcite at 18°C, or to $(1.00202 \pm 0.00003) \times 10^{-13}$ meter. Also known as x-ray unit; X-unit. Symbolized X; XU. { 'sēg‚bän }

sigma bond [PHYSICAL CHEMISTRY] The chemical bond resulting from the formation of a molecular orbital by the end-on overlap of atomic orbitals. { 'sig·mə ‚bänd }

sigmatropic shift [ORGANIC CHEMISTRY] A rearrangement reaction that consists of the migration of a sigma bond (that is, the sigma electrons) and the group of atoms that are attached to it from one position in a chain or ring into a new position. { ¦sig·mə¦träp·ik 'shift }

silane [INORGANIC CHEMISTRY] Si_nH_{2n+2} A class of silicon-based compounds analogous to alkanes, that is, straight-chain, saturated paraffin hydrocarbons; they can be gaseous or liquid. Also known as silicon hydride. { 'si‚lān }

silanol [CHEMISTRY] A member of the family of compounds whose structure contains a silicon atom that is bound directly to one or more hydroxyl groups. { 'sī·lə‚nòl }

silica gel [INORGANIC CHEMISTRY] A colloidal, highly absorbent silica used as a dehumidifying and dehydrating agent, as a catalyst carrier, and sometimes as a catalyst. { 'sil·ə·kə ¦jel }

silicate [INORGANIC CHEMISTRY] The generic term for a compound that contains silicon, oxygen, and one or more metals, and may contain hydrogen. { 'sil·ə·kət }

silicate of soda [INORGANIC CHEMISTRY] *See* sodium silicate. { 'sil·ə·kət əv 'sōd·ə }

silicic acid [INORGANIC CHEMISTRY] $SiO_2 \cdot nH_2O$ A white, amorphous precipitate; used to bleach fats, waxes, and oils. Also known as hydrated silica. { sə'lis·ik 'as·əd }

silicide [CHEMISTRY] A binary compound in which silicon is bonded with a more electropositive element. { 'sil·ə‚sīd }

silicon [CHEMISTRY] A group IV nonmetallic element, symbol Si, with atomic number 14, atomic weight 28.086; dark-brown crystals that burn in air when ignited; soluble in hydrofluoric acid and alkalies; melts at 1410°C; used to make silicon-containing alloys, as an intermediate for silicon-containing compounds, and in rectifiers and transistors. { 'sil·ə·kən }

silicon bromide [INORGANIC CHEMISTRY] *See* silicon tetrabromide. { 'sil·ə·kən 'brō ˌmīd }

silicon carbide [INORGANIC CHEMISTRY] SiC Water-insoluble, bluish-black crystals, very hard and iridescent; soluble in fused alkalies; sublimes at 2210°C; used as an abrasive and a heat refractory, and in light-emitting diodes to produce green or yellow light. { 'sil·ə·kən 'kärˌbīd }

silicon chloride [INORGANIC CHEMISTRY] *See* silicon tetrachloride. { 'sil·ə·kən 'klȯrˌīd }

silicon dioxide [INORGANIC CHEMISTRY] SiO_2 Colorless, transparent crystals, soluble in molten alkalies and hydrofluoric acid; melts at 1710°C; used to make glass, ceramic products, abrasives, foundry molds, and concrete. { 'sil·ə·kən dī'äkˌsīd }

silicon fluoride [INORGANIC CHEMISTRY] *See* silicon tetrafluoride. { 'sil·ə·kən 'flu̇rˌīd }

silicon halide [INORGANIC CHEMISTRY] A compound of silicon and a halogen; for example, $SiBr_4$, Si_2Br_6, $SiCl_4$, Si_2Cl_6, Si_3Cl_8, SiF_4, Si_2F_6, SiI_4, and Si_2F_6. { 'sil·ə·kən 'haˌlīd }

silicon hydride [INORGANIC CHEMISTRY] *See* silane. { 'sil·ə·kən 'hīˌdrīd }

silicon monoxide [INORGANIC CHEMISTRY] SiO A hard, abrasive, amorphous solid used as thin surface films to protect optical parts, mirrors, and aluminum coatings. { 'sil·ə·kən mə'näkˌsīd }

silicon nitride [INORGANIC CHEMISTRY] Si_3N_4 A white, water-insoluble powder, resistant to thermal shock and to chemical reagents; used as a catalyst support and for stator blades of high-temperature gas turbines. { 'sil·ə·kən 'nīˌtrīd }

silicon tetrabromide [INORGANIC CHEMISTRY] $SiBr_4$ A fuming, colorless liquid that yellows in air; disagreeable aroma; boils at 153°C. Also known as silicon bromide. { 'sil·ə·kən ¦te·trə'brōˌmīd }

silicon tetrachloride [INORGANIC CHEMISTRY] $SiCl_4$ A clear, corrosive, fuming liquid with suffocating aroma; decomposes in water and alcohol; boils at 57.6°C; used in warfare smoke screens, to make ethyl silicate and silicones, and as a source of pure silicon and silica. Also known as silicon chloride. { 'sil·ə·kən ¦te·trə'klȯrˌīd }

silicon tetrafluoride [INORGANIC CHEMISTRY] SiF_4 A colorless, suffocating gas absorbed readily by water, in which it decomposes; boiling point, −86°C; used in chemical analysis and to make fluosilicic acid. Also known as silicon fluoride. { 'sil·ə·kən ¦te·trə'flu̇rˌīd }

siloxane [ORGANIC CHEMISTRY] R_2SiO Any of a family of silica-based polymers in which R is an alkyl group, usually methyl; these polymers exist as oily liquids, greases, rubbers, resins, or plastics. Also known as oxosilane. { si'läkˌsān }

silver [CHEMISTRY] A white metallic transition element, symbol Ag, with atomic number 47; soluble in acids and alkalies, insoluble in water; melts at 961°C, boils at 2212°C; used in photographic chemicals, alloys, conductors, and plating. { 'sil·vər }

silver acetate [ORGANIC CHEMISTRY] CH_3COOAg A white powder, moderately soluble in water and nitric acid; used in medicine. { 'sil·vər 'as·əˌtāt }

silver acetylide [INORGANIC CHEMISTRY] Ag_2C_2 A white explosive powder used in detonators. { 'sil·vər ə'sed·əlˌīd }

silver arsenite [INORGANIC CHEMISTRY] Ag_3AsO_3 A poisonous, light-sensitive, yellow powder; soluble in acids and alkalies, insoluble in water and alcohol; decomposes at 150°C; used in medicine. { 'sil·vər 'ärs·ənˌīt }

silver bromate [INORGANIC CHEMISTRY] $AgBrO_3$ A poisonous, light- and heat-sensitive, white powder; soluble in ammonium hydroxide, slightly soluble in hot water; decomposed by heat. { 'sil·vər 'broˌmāt }

silver bromide [INORGANIC CHEMISTRY] AgBr Yellowish, light-sensitive crystals; soluble in potassium bromide and potassium cyanide, slightly soluble in water; melts at 432°C; used in photographic films and plates. { 'sil·vər 'brōˌmīd }

silver carbonate [INORGANIC CHEMISTRY] Ag_2CO_3 Yellowish, light-sensitive crystals; insoluble in water and alcohol, soluble in alkalies and acids; decomposes at 220°C; used as a reagent. { 'sil·vər 'kär·bə·nət }

silver chloride [INORGANIC CHEMISTRY] $AgCl$ A white, poisonous, light-sensitive powder; slightly soluble in water, soluble in alkalies and acids; melts at 445°C; used in photography, photometry, silver plating, and medicine. { 'sil·vər 'klȯr‚īd }

silver chromate [INORGANIC CHEMISTRY] Ag_2CrO_4 Dark-colored crystals insoluble in water, soluble in acids and in solutions of alkali chromates; used as an analytical reagent. { 'sil·vər 'krō‚māt }

silver cyanide [INORGANIC CHEMISTRY] $AgCN$ A poisonous, white, light-sensitive powder; insoluble in water, soluble in alkalies and acids; decomposes at 320°C; used in medicine and in silver plating. { 'sil·vər 'sī·ə‚nīd }

silver fluoride [INORGANIC CHEMISTRY] $AgF \cdot H_2O$ A light-sensitive, yellow or brownish solid, soluble in water; dehydrated form melts at 435°C; used in medicine. Also known as tachiol. { 'sil·vər 'flu̇r‚īd }

silver halide [INORGANIC CHEMISTRY] A compound of silver and a halogen; for example, silver bromide ($AgBr$), silver chloride ($AgCl$), silver fluoride (AgF), and silver iodide (AgI). { 'sil·vər 'ha‚līd }

silver iodate [INORGANIC CHEMISTRY] $AgIO_3$ A white powder, soluble in ammonium hydroxide and nitric acid, slightly soluble in water; melts above 200°C; used in medicine. { 'sil·vər 'ī·ə‚dāt }

silver iodide [INORGANIC CHEMISTRY] AgI A pale-yellow powder, insoluble in water, soluble in potassium iodide-sodium chloride solutions and ammonium hydroxide; melts at 556°C; used in medicine, photography, and artificial rainmaking. { 'sil·vər 'ī·ə‚dīd }

silver lactate [ORGANIC CHEMISTRY] $CH_3CHOHCOOAg \cdot H_2O$ Gray-to-white, light-sensitive crystals; slightly soluble in water and in alcohol; used in medicine. Also known as actol. { 'sil·vər 'lak‚tāt }

silver nitrate [INORGANIC CHEMISTRY] $AgNO_3$ Poisonous, corrosive, colorless crystals; soluble in glycerol, water, and hot alcohol; melts at 212°C; used in external medicine, photography, hair dyeing, silver plating, ink manufacture, and mirror silvering, and as a chemical reagent. { 'sil·vər 'nī‚trāt }

silver nitrite [INORGANIC CHEMISTRY] $AgNO_2$ Yellow or grayish-yellow needles which decompose at 140°C; soluble in hot water; used in organic synthesis and in testing for alcohols. { 'sil·vər 'nī‚trīt }

silver orthophosphate [INORGANIC CHEMISTRY] *See* silver phosphate. { 'sil·vər ¦ȯr·thō'fäs‚fāt }

silver oxide [INORGANIC CHEMISTRY] Ag_2O An odorless, dark-brown powder with a metallic taste; soluble in nitric acid and ammonium hydroxide, insoluble in alcohol; decomposes above 300°C; used in medicine and in glass polishing and coloring, as a catalyst, and to purify drinking water. { 'sil·vər 'äk‚sīd }

silver permanganate [INORGANIC CHEMISTRY] $AgMnO_4$ Water-soluble, violet crystals that decompose in alcohol; used in medicine and in gas masks. { 'sil·vər pər'maŋ·gə‚nāt }

silver phosphate [INORGANIC CHEMISTRY] Ag_3PO_4 A poisonous, yellow powder; darkens when heated or exposed to light; soluble in acids and in ammonium carbonate, very slightly soluble in water; melts at 849°C; used in photographic emulsions and in pharmaceuticals, and as a catalyst. Also known as silver orthophosphate. { 'sil·vər 'fä‚sfāt }

silver picrate [ORGANIC CHEMISTRY] $C_6H_2O(NO_2)_3Ag \cdot H_2O$ Yellow crystals, soluble in water, insoluble in ether and chloroform; used in medicine. { 'sil·vər 'pi‚krāt }

silver potassium cyanide [ORGANIC CHEMISTRY] $KAg(CN)_2$ Toxic, white crystals soluble in water and alcohol; used in silver plating and as a bactericide and antiseptic. Also known as potassium argentocyanide; potassium cyanoargentate. { 'sil·vər pə'tas·ē·əm 'sī·ə‚nīd }

silver protein [ORGANIC CHEMISTRY] A brown, hygroscopic powder containing 7.5-8.5% silver; made by reaction of a silver compound with gelatin in the presence of an alkali; used as an antibacterial. { 'sil·vər 'prō‚tēn }

silver selenide [INORGANIC CHEMISTRY] Ag_2Se A gray powder, insoluble in water, soluble in ammonium hydroxide; melts at 880°C. { 'sil·vər 'sel·ə‚nīd }

silver suboxide [INORGANIC CHEMISTRY] AgO A charcoal-gray powder that crystallizes in the cubic or orthorhombic system, and has diamagnetic properties; used in making silver oxide-zinc alkali batteries. Also known as argentic oxide. { 'sil·vər səb 'äk‚sīd }

silver sulfate [INORGANIC CHEMISTRY] Ag_2SO_4 Light-sensitive, colorless, lustrous crystals; soluble in alkalies and acids, insoluble in alcohol; melts at 652°C; used as an analytical reagent. Also known as normal silver sulfate. { 'sil·vər 'səl‚fāt }

silver sulfide [INORGANIC CHEMISTRY] Ag_2S A dark, heavy powder, insoluble in water, soluble in concentrated sulfuric and nitric acids; melts at 825°C; used in ceramics and in inlay metalwork. { 'sil·vər 'səl‚fīd }

simple salt [CHEMISTRY] One of four classes of salts in a classification system that depends on the character of completeness of the ionization; examples are NaCl, $NaHCO_3$, and Pb(OH)Cl. { 'sim·pəl 'sölt }

single-replacement reaction [CHEMISTRY] A chemical reaction in which an element replaces one element in a compound. { ¦siŋ·gəl ri'plās·mənt rē‚ak·shən }

singlet [SPECTROSCOPY] A spectral line that cannot be resolved into components at even the highest resolution. { 'siŋ·glət }

SIRS [SPECTROSCOPY] *See* satellite infrared spectrometer. { sərz }

skatole [ORGANIC CHEMISTRY] C_9H_9N A white, crystalline compound that melts at 93-95°C, dissolves in hot water, and has an unpleasant feceslike odor. Also known as 3-methylindole. { 'ska‚tōl }

Skraup synthesis [ORGANIC CHEMISTRY] A method for the preparation of commercial synthetic quinoline by heating aniline and glycerol in the presence of sulfuric acid and an oxidizing agent to form pyridine unsubstituted quinolines. { skraup 'sin·thə·səs }

Sm [CHEMISTRY] *See* samarium.

smectic-A [PHYSICAL CHEMISTRY] A subclass of smectic liquid crystals in which molecules are free to move within layers and are oriented perpendicular to the layers. { 'smek·dik 'ā }

smectic-B [PHYSICAL CHEMISTRY] A subclass of smectic liquid crystals in which molecules in each layer are arranged in a close-packed lattice and are oriented perpendicular to the layers. { 'smek·dik 'bē }

smectic-C [PHYSICAL CHEMISTRY] A subclass of smectic liquid crystals in which molecules are free to move within layers and are oriented with their axes tilted with respect to the normal to the layers. { 'smek·dik 'sē }

smectic phase [PHYSICAL CHEMISTRY] A form of the liquid crystal (mesomorphic) state in which molecules are arranged in layers that are free to glide over each other with relatively small viscosity. { 'smek·dik ‚fāz }

smectogenic solid [PHYSICAL CHEMISTRY] A solid which will form a smectic liquid crystal when heated. { ¦smek·tə¦jen·ik 'säl·əd }

smoldering [CHEMISTRY] Combustion of a solid without a flame, often with emission of smoke. { 'smōl·driŋ }

Sn [CHEMISTRY] *See* tin.

snow point [PHYSICAL CHEMISTRY] Referring to a gas mixture, the temperature at which the vapor pressure of the sublimable component is equal to the actual partial pressure of that component in the gas mixture; analogous to dew point. { 'snō ‚póint }

soda [INORGANIC CHEMISTRY] *See* sodium carbonate.

soda alum [INORGANIC CHEMISTRY] *See* aluminum sodium sulfate. { 'sōd·ə 'al·əm }

soda ash [INORGANIC CHEMISTRY] Na_2CO_3 The commercial grade of sodium carbonate; a powder soluble in water, insoluble in alcohol; used in glass manufacture and petroleum refining, and for soaps and detergents. Also known as anhydrous sodium carbonate; calcined soda. { 'sōd·ə 'ash }

soda crystals [INORGANIC CHEMISTRY] *See* metahydrate sodium carbonate. { 'sōd·ə ‚krist·əlz }

sodamide [INORGANIC CHEMISTRY] *See* sodium amide. { 'sōd·ə‚mīd }

sodide [INORGANIC CHEMISTRY] A member of a class of alkalides in which the metal anion is sodium (Na^-). { 'sä‚dīd }

sodium [CHEMISTRY] A metallic element of group I, symbol Na, with atomic number 11, atomic weight 22.9898; silver-white, soft, and malleable; oxidizes in air; melts at

97.6°C; used as a chemical intermediate and in pharmaceuticals, petroleum refining, and metallurgy; the source of the symbol Na is natrium. { 'sōd·ē·əm }

sodium acetate [ORGANIC CHEMISTRY] $NaC_2H_3O_2$ Colorless, efflorescent crystals, soluble in water and ether; melts at 324°C; used as a chemical intermediate and for pharmaceuticals, dyes, and dry colors. { 'sōd·ē·əm 'as·ə‚tāt }

sodium acid carbonate [INORGANIC CHEMISTRY] *See* sodium bicarbonate. { 'sōd·ē·əm 'as·əd 'kär·bə·nət }

sodium acid chromate [INORGANIC CHEMISTRY] *See* sodium dichromate. { 'sōd·ē·əm 'as·əd 'krō‚māt }

sodium acid fluoride [INORGANIC CHEMISTRY] *See* sodium bifluoride. { 'sōd·ē·əm 'as·əd 'flu̇r‚īd }

sodium acid sulfate [INORGANIC CHEMISTRY] *See* sodium bisulfate. { 'sōd·ē·əm 'as·əd 'səl‚fāt }

sodium acid sulfite [INORGANIC CHEMISTRY] *See* sodium bisulfite. { 'sōd·ē·əm 'as·əd 'səl‚fīt }

sodium alginate [ORGANIC CHEMISTRY] $C_6H_7O_6Na$ Colorless or light yellow filaments, granules, or powder which forms a viscous colloid in water; used in food thickeners and stabilizers, in medicine and textile printing, and for paper coating and water-base paint. Also known as algin; alginic acid sodium salt; sodium polymannuronate. { 'sōd·ē·əm 'al·jə‚nāt }

sodium aluminate [INORGANIC CHEMISTRY] $Na_2Al_2O_4$ A white powder soluble in water, insoluble in alcohol; melts at 1800°C; used as a zeolite-type of material and a mordant, and in water purification, milkglass manufacture, and cleaning compounds. { 'sōd·ē·əm ə'lü·mə‚nāt }

sodium aluminosilicate [INORGANIC CHEMISTRY] White, amorphous powder or beads of variable stoichiometry, partially soluble in strong acids and alkali hydroxide solutions between 80 and 100°C; used in food as an anticaking agent. Also known as sodium silicoaluminate. { 'sōd·ē·əm ə¦lü·mə·nō'sil·ə‚kāt }

sodium aluminum phosphate [INORGANIC CHEMISTRY] $NaAl_3H_{14}(PO_4)_8 \cdot 4H_2O$ or $Na_3Al_2H_{15}(PO_4)_8$ White powder, soluble in hydrochloric acid; used as a food additive for baked products. { 'sōd·ē·əm ə'lü·mə·nəm 'fä‚sfāt }

sodium aluminum silicofluoride [INORGANIC CHEMISTRY] $Na_5Al(SiF_6)_4$ A toxic, white powder, used for mothproofing and in insecticides. { 'sōd·ē·əm ə'lü·mə·nəm ‚sil·ə·kō'flu̇r‚īd }

sodium aluminum sulfate [INORGANIC CHEMISTRY] *See* aluminum sodium sulfate. { 'sōd·ē·əm ə'lü·mə·nəm 'səl‚fāt }

sodium amalgam [INORGANIC CHEMISTRY] Na_xHg_x A fire-hazardous, silver-white crystal mass that decomposes in water; used to make hydrogen and as an analytical reagent. { 'sōd·ē·əm ə'mal·gəm }

sodium amide [INORGANIC CHEMISTRY] $NaNH_2$ White crystals that decompose in water; melts at 210°C; a fire hazard; used to make sodium cyanide. Also known as sodamide. { 'sōd·ē·əm 'am‚īd }

sodium antimonate [INORGANIC CHEMISTRY] $NaSbO_3$ A white, granular powder, used as an enamel opacifier and high-temperature oxidizing agent. Also known as antimony sodiate. { 'sōd·ē·əm an'tim·ə‚nāt }

sodium arsanilate [ORGANIC CHEMISTRY] $C_6H_4NH_2(AsO \cdot OH \cdot ONa)$ A white, water-soluble, poisonous powder with a faint saline taste, used in medicine and as a chemical intermediate. { 'sōd·ē·əm är'san·əl‚āt }

sodium arsenate [INORGANIC CHEMISTRY] $Na_3AsO_4 \cdot 12H_2O$ Water-soluble, poisonous, clear, colorless crystals with a mild alkaline taste; melts at 86°C; used in medicine, insecticides, dry colors, and textiles, and as a germicide and a chemical intermediate. { 'sōd·ē·əm 'ärs·ən‚āt }

sodium arsenite [INORGANIC CHEMISTRY] $NaAsO_2$ A poisonous, water-soluble, grayish powder; used in antiseptics, dyeing, insecticides, and soaps for taxidermy. { 'sōd·ē·əm 'ärs·ən‚īt }

sodium ascorbate [ORGANIC CHEMISTRY] $CH_2OH(CHOH)_2COHCOHCOONa$ White, odorless crystals; soluble in water, insoluble in alcohol; decomposes at 218°C; used in therapy for vitamin C deficiency. { 'sōd·ē·əm ə'skȯr‚bāt }

sodium azide [INORGANIC CHEMISTRY] NaN_3 Poisonous, colorless crystals; soluble in water and liquid ammonia; decomposes at 300°C; used in medicine and to make lead azide explosives. { 'sōd·ē·əm 'ā‚zīd }

sodium barbiturate [ORGANIC CHEMISTRY] $C_4H_3N_2O_3Na$ White to slightly yellow powder, soluble in water and dilute mineral acid; used in wood-impregnating solutions. { 'sōd·ē·əm bär'bich·ə·rət }

sodium benzoate [ORGANIC CHEMISTRY] $NaC_7H_5O_2$ Water- and alcohol-soluble, white, amorphous crystals with a sweetish taste; used as a food preservative and an antiseptic and in tobacco, pharmaceuticals, and medicine. { 'sōd·ē·əm 'ben·zə‚wāt }

sodium benzoylacetone dihydrate [ORGANIC CHEMISTRY] A metal chelate with low melting point (115°C) and slight solubility in acetone. { 'sōd·ē·əm ¦ben·zə‚wil'as·ə‚tōn dī'hī‚drāt }

sodium bicarbonate [INORGANIC CHEMISTRY] $NaHCO_3$ White, water-soluble crystals with an alkaline taste; loses carbon dioxide at 270°C; used as a medicine and a butter preservative, in food preparation, in effervescent salts and beverages, in ceramics, and to prevent timber mold. Also known as baking soda; bicarbonate of soda; sodium acid carbonate. { 'sōd·ē·əm bī'kär·bə‚nət }

sodium bichromate [INORGANIC CHEMISTRY] *See* sodium dichromate. { 'sōd·ē·əm bī'krō‚māt }

sodium bifluoride [INORGANIC CHEMISTRY] $NaHF_2$ Poisonous, water-soluble, white crystals; decomposes when heated; used as a laundry-rinse neutralizer, preservative, and antiseptic, and in glass etching and tinplating. Also known as sodium acid fluoride. { 'sōd·ē·əm bi'flu̇r·īd }

sodium bismuthate [INORGANIC CHEMISTRY] $NaBiO_3$ A yellow to brown amorphous powder, used as an analytical reagent and in pharmaceuticals. { 'sōd·ē·əm 'biz·mə‚thāt }

sodium bisulfate [INORGANIC CHEMISTRY] $NaHSO_4$ Colorless crystals, soluble in water; the aqueous solution is strongly acidic; decomposes at 315°C; used for flux to decompose minerals, as a disinfectant, and in dyeing and manufacture of magnesia, cements, perfumes, brick, and glue. Also known as niter cake; sodium acid sulfate. { 'sōd·ē·əm bī'səl‚fāt }

sodium bisulfide [INORGANIC CHEMISTRY] *See* sodium hydrosulfide. { 'sōd·ē·əm bī'səl‚fīd }

sodium bisulfite [INORGANIC CHEMISTRY] $NaHSO_3$ A colorless, water-soluble solid, decomposes when heated. Also known as sodium acid sulfite. { 'sōd·ē·əm bī'səl‚fīt }

sodium bisulfite test [ANALYTICAL CHEMISTRY] A test for aldehydes in which aldehydes form a crystalline salt upon addition of a 40% aqueous solution of sodium bisulfite. { 'sōd·ē·əm bī'səl‚fīt ‚test }

sodium bitartrate [ORGANIC CHEMISTRY] $NaHC_4H_4O_6 \cdot H_2O$ A white, combustible, water-soluble powder that loses water at 100°C, decomposes at 219°C; used in effervescing mixtures and as an analytical reagent. Also known as acid sodium tartrate. { 'sōd·ē·əm bī'tär‚trāt }

sodium borate [INORGANIC CHEMISTRY] $Na_2B_4O_7 \cdot 10H_2O$ A water-soluble, odorless, white powder; melts between 75 and 200°C; used in glass, ceramics, starch and adhesives, detergents, agricultural chemicals, pharmaceuticals, and photography; the impure form is known as borax. Also known as sodium pyroborate; sodium tetraborate. { 'sōd·ē·əm 'bȯ‚rāt }

sodium boroformate [ORGANIC CHEMISTRY] $NaH_2BO_3 \cdot 2HCOOH \cdot 2H_2O$ Water-soluble, white crystals, melting at 110°C; used in textile treating and in tanning, and as a buffering agent. { 'sōd·ē·əm ‚bȯr·ō'fȯr‚māt }

sodium borohydride [INORGANIC CHEMISTRY] $NaBH_4$ A flammable, hygroscopic, white to gray powder; soluble in water, insoluble in ether and hydrocarbons; decomposes in damp air; used as a hydrogen source, a chemical reagent, and a rubber foaming agent. { 'sōd·ē·əm ‚bȯr·ō'hī‚drīd }

sodium bromate [INORGANIC CHEMISTRY] $NaBrO_3$ Odorless, white crystals; soluble in water, insoluble in alcohol; decomposes at 381°C; a fire hazard, used as an analytical reagent. { 'sōd·ē·əm 'brō‚māt }

sodium bromide [INORGANIC CHEMISTRY] NaBr White, water-soluble, crystals with a bitter, saline taste; absorbs moisture from air; melts at 758°C; used in photography and medicine, as a chemical intermediate, and to make bromides. { 'sōd·ē·əm 'brō‚mīd }

sodium carbolate [ORGANIC CHEMISTRY] *See* sodium phenate. { 'sōd·ē·əm 'kär·bə‚lāt }

sodium carbonate [INORGANIC CHEMISTRY] Na_2CO_3 A white, water-soluble powder that decomposes when heated to about 852°C; used as a reagent; forms a monohydrate compound, $Na_2CO_3 \cdot H_2O$, and a decahydrate compound, $Na_2CO_3 \cdot 10H_2O$. Also known as soda. { 'sōd·ē·əm 'kär·bə·nət }

sodium carbonate decahydrate [INORGANIC CHEMISTRY] *See* sal soda. { 'sōd·ē·əm 'kär·bə·nət ‚dek·ə'hī‚drāt }

sodium carbonate peroxide [INORGANIC CHEMISTRY] $2Na_2CO_3 \cdot 3H_2O$ A white, crystalline powder; used in household detergents, in dental cleansers, and for bleaching and dyeing. { 'sōd·ē·əm 'kär·bə·nət pə'räk‚sīd }

sodium caseinate [ORGANIC CHEMISTRY] A tasteless, odorless, water-soluble, white powder; used in medicine, foods, emulsification, and stabilization; formed by dissolving casein in sodium hydroxide and then evaporating. Also known as casein sodium; nutrose. { 'sōd·ē·əm 'kā·sē·ə‚nāt }

sodium chlorate [INORGANIC CHEMISTRY] $NaClO_3$ Water- and alcohol-soluble, colorless crystals with a saline taste; melts at 255°C; used as a medicine, weed killer, defoliant, and oxidizing agent, and in matches, explosives, and bleaching. { 'sōd·ē·əm 'klȯr‚āt }

sodium chloride [INORGANIC CHEMISTRY] NaCl Colorless or white crystals; soluble in water and glycerol, slightly soluble in alcohol; melts at 804°C; used in foods and as a chemical intermediate and an analytical reagent. Also known as common salt; table salt. { 'sōd·ē·əm 'klȯr‚īd }

sodium chlorite [INORGANIC CHEMISTRY] $NaClO_2$ An explosive, white, mildly hygroscopic, water-soluble powder; decomposes at 175°C; used as an analytical reagent and oxidizing agent. { 'sōd·ē·əm 'klȯr‚īt }

sodium chloroacetate [ORGANIC CHEMISTRY] $ClCH_2COONa$ A white, water-soluble powder; used as a defoliant and in the manufacture of weed killers, dyes, and pharmaceuticals. { 'sōd·ē·əm ‚klȯr·ō'as·ə‚tāt }

sodium chloroplatinate [INORGANIC CHEMISTRY] $Na_2PtCl_6 \cdot 4H_2O$ A yellow powder, soluble in alcohol and water; used for zinc etching, indelible ink, plating, and mirrors, and in photography and medicine. Also known as platinic sodium chloride; platinum sodium chloride; sodium platinichloride. { 'sōd·ē·əm ‚klȯr·ō'plat·ən‚āt }

sodium chromate [INORGANIC CHEMISTRY] $Na_2CrO_4 \cdot 10H_2O$ Water-soluble, translucent, yellow, efflorescent crystals that melt at 20°C; used as a rust preventive and in inks, dyeing, and leather tanning. { 'sōd·ē·əm 'krō‚māt }

sodium citrate [ORGANIC CHEMISTRY] $C_6H_5Na_3O_7 \cdot 2H_2O$ A white powder with the taste of salt; soluble in water, slightly soluble in alcohol; has an acid taste; loses water at 150°C; decomposes at red heat; used in medicine as an anticoagulant, in soft drinks, cheesemaking, and electroplating. Also known as trisodium citrate. { 'sōd·ē·əm 'sī‚trāt }

sodium cobaltinitrite [INORGANIC CHEMISTRY] $Na_3Co(NO_2)_6 \cdot 1/2H_2O$ Purple, water-soluble, hygroscopic crystals; used as a reagent for analysis of potassium. { 'sōd·ē·əm ¦kō‚bȯl·tə'nī‚trīt }

sodium cyanate [INORGANIC CHEMISTRY] NaOCN A poisonous, white powder; soluble in water, insoluble in alcohol and ether; used as a chemical intermediate and for the manufacture of medicine and the heat-treating of steels. { 'sōd·ē·əm 'sī·ə‚nāt }

sodium cyanide [INORGANIC CHEMISTRY] NaCN A poisonous, water-soluble, white powder melting at 563°C; decomposes rapidly when standing; used to manufacture pigments, in heat treatment of metals, and as a silver- and gold-ore extractant. { 'sōd·ē·əm 'sī·ə‚nīd }

sodium cyclamate [ORGANIC CHEMISTRY] $C_6H_{11}NHSO_3Na$ White, water-soluble crystals; sweetness 30 times that of sucrose; formerly used as an artificial sweetener for foods, but now prohibited. { 'sōd·ē·əm 'sī·klə‚māt }

sodium dehydroacetate [ORGANIC CHEMISTRY] $C_8H_7NaO_4 \cdot H_2O$ A tasteless, white powder, soluble in water and propylene glycol; used as a fungicide and plasticizer, in toothpaste, and for pharmaceuticals. { 'sōd·ē·əm dē,hī·drō'as·ə,tāt }

sodium diacetate [ORGANIC CHEMISTRY] $CH_3COONa \cdot x(CH_3COOH)$ Combustible, white, water-soluble crystals with an acetic acid aroma; decomposes above 150°C; used to inhibit mold, and as a buffer, varnish hardener, sequestrant, and food preservative, and in mordants. { 'sōd·ē·əm dī'as·ə,tāt }

sodium diatrizoate [ORGANIC CHEMISTRY] $C_{11}H_8NO_4I_3Na$ White, water-soluble crystals which give a radiopaque solution; used in medicine as a radiopaque medium. { 'sōd·ē·əm ,dī·ə'trī·zə,wāt }

sodium dichloroisocyanate [ORGANIC CHEMISTRY] $HC_3N_3O_3NaCl$ A white, crystalline compound, soluble in water; used as a bactericide and algicide in swimming pools. { 'sōd·ē·əm dī¦klȯr·ō¦ī·sō'sī·ə,nāt }

sodium dichloroisocyanurate [ORGANIC CHEMISTRY] $C_3N_3O_3Cl_2Na$ A white, crystalline powder; used in dry bleaches, detergents, and cleaning compounds, and for water and sewage treatment. { 'sōd·ē·əm dī¦klȯr·ō¦ī·sō,sī·ə'nu̇r,āt }

sodium dichromate [INORGANIC CHEMISTRY] $Na_2Cr_2O_7 \cdot 2H_2O$ Poisonous, red to orange deliquescent crystals; soluble in water, insoluble in alcohol; melts at 320°C; loses water of hydration upon prolonged heating at 105°C; used as a chemical intermediate and corrosion inhibitor and in the manufacture of pigments, leather tanning, and electroplating. Also known as bichromate of soda; sodium acid chromate; sodium bichromate. { 'sōd·ē·əm dī'krō,māt }

sodium diethyldithiocarbamate [ORGANIC CHEMISTRY] $(C_2H_5)_2NCS_2Na$ A solid that is soluble in water and in alcohol, the trihydrate is used to determine small amounts of copper and to separate copper from other metals. { 'sōd·ē·əm dī¦eth·əl dī,thī·ō'kär·bə,māt }

sodium dimethyldithiocarbamate [ORGANIC CHEMISTRY] $(CH_3)_2NCS_2Na$ Amber to light green liquid; used as a fungicide, corrosion inhibitor, and rubber accelerator. Abbreviated SDDC. { 'sōd·ē·əm dī¦meth·əl·dī,thī·ō'kär·bə,māt }

sodium dinitro-*ortho*-cresylate [ORGANIC CHEMISTRY] $CH_3C_6H_2(NO_2)_2ONa$ A toxic, orange-yellow dye, used as a herbicide and fungicide. { 'sōd·ē·əm dī¦nī,trō ¦ȯr·thō 'kres·ə,lat }

sodium dithionite [INORGANIC CHEMISTRY] *See* sodium hydrosulfite. { 'sōd·ē·əm dī'thī·ə,nīt }

sodium diuranate [ORGANIC CHEMISTRY] $Na_2U_2O_7 \cdot 6H_2O$ A yellow-orange solid, soluble in dilute acids; used for colored glazes on ceramics and in the manufacture of fluorescent uranium glass. Also known as uranium yellow. { 'sōd·ē·əm dī'yu̇r·ə,nat }

sodium dodecylbenzenesulfonate [ORGANIC CHEMISTRY] $C_{18}H_{29}SO_3Na$ Biodegradable, white to yellow flakes, granules, or powder, used as a synthetic detergent. { 'sōd·ē·əm ¦dō·də,sil¦ben,zēn'səl·fə,nat }

sodium ethoxide [ORGANIC CHEMISTRY] *See* sodium ethylate. { 'sōd·ē·əm e'thäk,sīd }

sodium ethylate [ORGANIC CHEMISTRY] C_2H_5ONa A white powder formed from ethanol by replacement of the hydroxyl groups' hydrogen by monovalent sodium; used in organic synthesis. Also known as caustic alcohol; sodium ethoxide. { 'sōd·ē·əm 'eth·ə,lāt }

sodium 2-ethylhexyl sulfoacetate [ORGANIC CHEMISTRY] $C_{10}H_{19}O_2SO_3Na$ Cream-colored, water-soluble flakes, used as a stabilizing agent in soapless shampoos. { 'sōd·ē·əm ¦tü ¦eth·əl'hek·səl ,səl·fō'as·ə,tāt }

sodium ethylxanthate [ORGANIC CHEMISTRY] $C_2H_5OC(S)SNa$ A yellowish powder, soluble in water and alcohol; used as an ore flotation agent. Also known as sodium xanthate; sodium xanthogenate. { 'sōd·ē·əm ,eth·əl'zan,thāt }

sodium ferricyanide [INORGANIC CHEMISTRY] $Na_3Fe(CN)_6 \cdot H_2O$ A poisonous, deliquescent, red powder; soluble in water, insoluble in alcohol; used in printing and for the manufacture of pigments. Also known as red prussiate of soda. { 'sōd·ē·əm ,fer·ə'sī·ə,nīd }

sodium ferrocyanide [INORGANIC CHEMISTRY] $Na_4Fe(CN)_6 \cdot 10H_2O$ Semitransparent crystals, soluble in water; insoluble in alcohol; used in photography, dyes, tanning, and blueprint paper. Also known as yellow prussiate of soda. { 'sōd·ē·əm ˌfer·ə'sī·əˌnīd }

sodium fluoborate [INORGANIC CHEMISTRY] $NaBF_4$ A white powder with a bitter taste; soluble in water, slightly soluble in alcohol; decomposes when heated, fuses below 500°C; used in electrochemical processes, as flux for nonferrous metals refining, and as an oxidation inhibitor. { 'sōd·ē·əm ˌflü·ə'bȯrˌāt }

sodium fluorescein [ORGANIC CHEMISTRY] *See* uranine. { 'sōd·ē·əm flu̇'res·ē·ən }

sodium fluoride [INORGANIC CHEMISTRY] NaF A poisonous, water-soluble, white powder, melting at 988°C; used as an insecticide and a wood and adhesive preservative, and in fungicides, vitreous enamels, and dentistry. { 'sōd·ē·əm 'flu̇rˌīd }

sodium fluoroacetate [ORGANIC CHEMISTRY] $C_2H_2FO_2Na$ A white powder, hygroscopic and nonvolatile; decomposes at 200°C; very soluble in water; used as a repellent for rodents and predatory animals. { 'sōd·ē·əm ˌflu̇r·ō'as·əˌtāt }

sodium fluosilicate [INORGANIC CHEMISTRY] Na_2SiF_6 A poisonous, white, amorphous powder; slightly soluble in water; decomposes at red heat; used to fluoridate drinking water and to kill rodents and insects. Also known as sodium silicofluoride. { 'sōd·ē·əm ˌflü·ə'sil·ə·kət }

sodium folate [ORGANIC CHEMISTRY] $C_{19}H_{18}N_7NaO_6$ A yellow to yellow-orange liquid; used in medicine for folic acid deficiency. Also known as folic acid sodium salt. { 'sōd·ē·əm 'fōˌlāt }

sodium formaldehyde sulfoxylate [ORGANIC CHEMISTRY] $HCHO \cdot HSO_2Na \cdot 2H_2O$ A white solid with a melting point of 64°C; soluble in water and alcohol; used as a textile stripping agent and a bleaching agent for soap and molasses. Abbreviated SFS. { 'sōd·ē·əm fȯr'mal·dəˌhīd səl'fäk·səˌlāt }

sodium formate [ORGANIC CHEMISTRY] HCOONa A mildly hygroscopic, white powder, soluble in water; has a formic acid aroma; melts at 245°C; used in medicine and as a chemical intermediate and reducing agent. { 'sōd·ē·əm 'fȯrˌmāt }

sodium glucoheptonate [ORGANIC CHEMISTRY] $C_7H_{13}O_8Na$ A light tan, crystalline powder; used for cleaning metal, mercerizing, paint stripping, and aluminum etching. { 'sōd·ē·əm ¦glü·kō'hep·təˌnāt }

sodium gluconate [ORGANIC CHEMISTRY] $C_6H_{11}NaO_7$ A water-soluble, yellow to white, crystalline powder, produced by fermentation; used in food and pharmaceutical industries, and as a metal cleaner. Also known as gluconic acid sodium salt. { 'sōd·ē·əm 'glü·kəˌnāt }

sodium glutamate [ORGANIC CHEMISTRY] $COOH(CH_2)_2CH(NH_2)COONa$ A salt of an amino acid; a white powder, soluble in water and alcohol; used as a taste enhancer. Also known as monosodium glutamate (MSG). { 'sōd·ē·əm 'glüd·əˌmāt }

sodium gold chloride [INORGANIC CHEMISTRY] $NaAuCl_4 \cdot 2H_2O$ Yellow crystals, soluble in water and alcohol; used in photography, fine glass staining, porcelain decorating, and medicine. Also known as gold salt; gold sodium chloride. { 'sōd·ē·əm 'gōld 'klȯrˌīd }

sodium gold cyanide [ORGANIC CHEMISTRY] $NaAu(CN)_2$ A yellow, water-soluble powder; used for gold plating radar and electric parts, jewelry, and tableware. Also known as gold sodium cyanide. { 'sōd·ē·əm 'gōld 'sī·əˌnīd }

sodium halide [INORGANIC CHEMISTRY] A compound of sodium with a halogen; for example, sodium bromide (NaBr), sodium chloride (NaCl), sodium iodide (NaI), and sodium fluoride (NaF). { 'sōd·ē·əm 'haˌlīd }

sodium halometallate [INORGANIC CHEMISTRY] A compound of sodium with halogen and a metal; for example, sodium platinichloride, $Na_2PtCl_6 \cdot 6H_2O$. { 'sōd·ē·əm ˌhal·ō'med·əlˌāt }

sodium hexylene glycol monoborate [ORGANIC CHEMISTRY] $C_6H_{12}O_3BNa$ An amorphous, white solid with a melting point of 426°C; used as a corrosion inhibitor, flame retardant, and lubricating-oil additive. { 'sōd·ē·əm 'hek·səˌlēn 'glīˌkȯl ¦män·ə'bȯrˌāt }

sodium hydrate [INORGANIC CHEMISTRY] *See* sodium hydroxide. { 'sōd·ē·əm 'hīˌdrāt }

sodium hydride [INORGANIC CHEMISTRY] NaH A white powder, decomposed by water, and igniting in moist air; used to make sodium borohydride and as a drying agent and a reagent. { 'sōd·ē·əm 'hī‚drīd }

sodium hydrogen phosphate [INORGANIC CHEMISTRY] $NaH_2PO_4 \cdot H_2O$ Hygroscopic, transparent, water-soluble crystals; used as a purgative, reagent, and buffer. { 'sōd·ē·əm 'hī·drə·jən 'fä‚sfāt }

sodium hydrogen sulfide [INORGANIC CHEMISTRY] *See* sodium hydrosulfide. { 'sōd·ē·əm 'hī·drə·jən 'səl‚fīd }

sodium hydrosulfide [INORGANIC CHEMISTRY] $NaSH \cdot 2H_2O$ Toxic, colorless, water-soluble needles, melting at 55°C; used in pulping of paper, processing dyestuffs, hide dehairing, and bleaching. Also known as sodium bisulfide; sodium hydrogen sulfide; sodium sulfhydrate. { 'sōd·ē·əm ‚hī·drə'səl‚fīd }

sodium hydrosulfite [INORGANIC CHEMISTRY] $Na_2S_2O_4$ A fire-hazardous, lemon to whitish-gray powder; soluble in water, insoluble in alcohol; melts at 55°C; used as a chemical intermediate and catalyst and in ore flotation. Also known as sodium dithionite. { 'sōd·ē·əm ‚hī·drə'səl‚fīt }

sodium hydroxide [INORGANIC CHEMISTRY] NaOH White, deliquescent crystals; absorbs carbon dioxide and water from air; soluble in water, alcohol, and glycerol; melts at 318°C; used as an analytical reagent and chemical intermediate, in rubber reclaiming and petroleum refining, and in detergents. Also known as sodium hydrate. { 'sōd·ē·əm hī'dräk‚sīd }

sodium hypochlorite [INORGANIC CHEMISTRY] NaOCl Air-unstable, pale-green crystals with sweet aroma; soluble in cold water, decomposes in hot water; used as a bleaching agent for paper pulp and textiles, as a chemical intermediate, and in medicine. { 'sōd·ē·əm ¦hī·pō'klȯr‚īt }

sodium hypophosphite [INORGANIC CHEMISTRY] $NaH_2PO_2 \cdot H_2O$ Colorless, pearly, water-soluble crystalline plates or a white, granular powder; used in medicine and electroless nickel plating of plastic and metal. { 'sōd·ē·əm ¦hī·pō'fä‚sfīt }

sodium hyposulfite [INORGANIC CHEMISTRY] *See* sodium thiosulfate. { 'sōd·ē·əm ¦hī·pō'səl‚fīt }

sodium iodate [INORGANIC CHEMISTRY] $NaIO_3$ A white, water- and acetone-soluble powder; used as a disinfectant and in medicine. { 'sōd·ē·əm 'ī·ə‚dāt }

sodium iodide [INORGANIC CHEMISTRY] NaI A white, air-sensitive powder, deliquescent, with bitter taste; soluble in water, alcohol, and glycerin; melts at 653°C; used in photography and in medicine and as an analytical reagent. { 'sōd·ē·əm 'ī·ə‚dīd }

sodium isopropylxanthate [ORGANIC CHEMISTRY] $C_4H_7ONaS_2$ Light yellow, crystalline compound that decomposes at 150°C, soluble in water; used as a postemergence herbicide and as a flotation agent for ores. { 'sōd·ē·əm ¦ī·sə¦prō·pəl'zan‚thāt }

sodium lactate [ORGANIC CHEMISTRY] $CH_3CHOHCOONa$ A water-soluble, hygroscopic, yellow to colorless, syrupy liquid; solidifies at 17°C; used in medicine, as a corrosion inhibitor in antifreeze, and a hygroscopic agent. { 'sōd·ē·əm 'lak‚tāt }

sodium lauryl sulfate [ORGANIC CHEMISTRY] $CH_3(CH_2)_{10}CH_2OSO_3Na$ A water-soluble salt, produced as a white or cream powder, crystals, or flakes; used in the textile industry as a wetting agent and detergent. Also known as dodecyl sodium sulfate. { 'sōd·ē·əm 'lȯr·əl 'səl‚fāt }

sodium lead hyposulfate [INORGANIC CHEMISTRY] *See* lead sodium thiosulfate. { 'sōd·ē·əm 'led ¦hī·pō'səl‚fāt }

sodium lead thiosulfate [INORGANIC CHEMISTRY] *See* lead sodium thiosulfate. { 'sōd·ē·əm 'led ¦thī·ə'səl‚fāt }

sodium metaborate [INORGANIC CHEMISTRY] $NaBO_2$ Water-soluble, white crystals, melting at 966°C; the aqueous solution is alkaline; made by fusing sodium carbonate with borax; used as an herbicide. { 'sōd·ē·əm ¦med·ə'bȯr‚āt }

sodium metaphosphate [INORGANIC CHEMISTRY] $(NaPO_3)_x$ Sodium phosphate groupings; cyclic forms range from $x = 3$ for the trimetaphosphate, to $x = 10$ for the decametaphosphate; sodium hexametaphosphate with $x = 10$ to 20 is probably a polymer; used for dental polishing, building detergents, and water softening, and as a sequestrant, emulsifier, and food additive. { 'sōd·ē·əm ¦med·ə'fä‚sfāt }

sodium metasilicate [INORGANIC CHEMISTRY] *See* sodium silicate. { 'sōd·ē·əm ¦med·ə'sil·ə‚kāt }

sodium metavanadate [INORGANIC CHEMISTRY] $NaVO_3$ Colorless crystals or a pale green, crystalline powder with a melting point of 630°C; soluble in water; used in inks, fur dyeing, and photography, and as a corrosion inhibitor in gas scrubbers. { 'sōd·ē·əm ¦med·ə'van·ə‚dāt }

sodium methiodal [ORGANIC CHEMISTRY] ICH_2SO_3Na A white, crystalline powder, soluble in water and methanol; used in medicine as a radiopaque medium. { 'sōd·ē·əm me'thī·ə‚dal }

sodium methoxide [ORGANIC CHEMISTRY] CH_3ONa A salt produced as a free-flowing powder, soluble in methanol and ethanol; used as an intermediate in organic synthesis. Also known as sodium methylate. { 'sōd·ē·əm me'thäk‚sīd }

sodium methylate [ORGANIC CHEMISTRY] *See* sodium methoxide. { 'sōd·ē·əm 'meth·ə‚lāt }

sodium *N*-methyldithiocarbamate dihydrate [ORGANIC CHEMISTRY] $CH_3NHC(S)SNa \cdot 2H_2O$ A white, water-soluble, crystalline solid; used as a fungicide, insecticide, nematicide, and weed killer. { 'sōd·ē·əm ¦en ¦meth·əl·dī¦thī·ə'kär·bə‚māt dī'hī‚drāt }

sodium molybdate [INORGANIC CHEMISTRY] Na_2MoO_4 Water-soluble crystals, melting at 687°C; used as an analytical reagent, corrosion inhibitor, catalyst, and zinc-plating brightening agent, and in medicine. { 'sōd·ē·əm mə'lib‚dāt }

sodium 12-molybdophosphate [INORGANIC CHEMISTRY] $Na_3PMo_{12}O_{40}$ Yellow, water-soluble crystals; used in neuromicroscopy and photography, and as a water-resisting agent in plastic adhesives and cements. { 'sōd·ē·əm ¦twelv mə‚lib·dō'fä‚sfāt }

sodium monoxide [INORGANIC CHEMISTRY] Na_2O A strong basic white powder soluble in molten caustic soda; forms sodium hydroxide in water; used as a dehydrating and polymerization agent. Also known as sodium oxide. { 'sōd·ē·əm mə'näk‚sīd }

sodium naphthalenesulfonate [ORGANIC CHEMISTRY] $C_{10}H_7SO_3Na$ Yellow, water-soluble crystalline plates or white scales; used as a liquefying agent in animal glue. { 'sōd·ē·əm ¦naf·thə‚lēn'səl·fə‚nāt }

sodium naphthionate [ORGANIC CHEMISTRY] $NaC_{10}H_6(NH_2)SO_3 \cdot 4H_2O$ White, light-sensitive crystals, soluble in water and insoluble in ether; used in analysis (Riegler's reagent) for nitrous acid. { 'sōd·ē·əm naf'thī·ə‚nāt }

sodium nitrate [INORGANIC CHEMISTRY] $NaNO_3$ Fire-hazardous, transparent, colorless crystals with bitter taste; soluble in glycerol and water; melts at 308°C; decomposes when heated; used in manufacture of glass and pottery enamel and as a fertilizer and food preservative. { 'sōd·ē·əm 'nī‚trāt }

sodium nitrite [INORGANIC CHEMISTRY] $NaNO_2$ A fire-hazardous, air-sensitive, yellowish powder, soluble in water; decomposes above 320°C; used as an intermediate for dyestuffs and for pickling meat, textiles dyeing, and rust-proofing, and in medicine. { 'sōd·ē·əm 'nī‚trīt }

sodium nitroferricyanide [INORGANIC CHEMISTRY] $Na_2Fe(CN)_5NO \cdot 2H_2O$ Water-soluble, transparent, reddish crystals; slowly decomposes in water; used as an analytical reagent. { 'sōd·ē·əm ¦nī·trō‚fer·ə'sī·ə‚nīd }

sodium oleate [ORGANIC CHEMISTRY] $C_{17}H_{33}COONa$ A white powder with a tallow aroma; soluble in alcohol and water, with partial decomposition; used in medicine and textile waterproofing. { 'sōd·ē·əm 'ō·lē‚āt }

sodium oxalate [ORGANIC CHEMISTRY] $Na_2C_2O_4$ A poisonous, white powder; soluble in water, insoluble in alcohol; used for leather tanning and as an analytical reagent. { 'sōd·ē·əm 'äk·sə‚lāt }

sodium oxide [INORGANIC CHEMISTRY] *See* sodium monoxide. { 'sōd·ē·əm 'äk‚sīd }

sodium paraperiodate [INORGANIC CHEMISTRY] $Na_3H_2IO_6$ White, crystalline solid, soluble in concentrated sodium hydroxide solutions; used to wet-strengthen paper and to aid in tobacco combustion. { 'sōd·ē·əm ¦par·ə·pər'ī·ə‚dāt }

sodium pentaborate [INORGANIC CHEMISTRY] $Na_2B_{10}O_{16} \cdot 10H_2O$ A white, water-soluble powder; used in glassmaking, weed killers, and fireproofing compositions. { 'sōd·ē·əm ‚pen·tə'bȯr‚āt }

sodium pentachlorophenate [ORGANIC CHEMISTRY] C_6Cl_5ONa A white or tan powder, soluble in water, ethanol, and acetone; used as a fungicide and herbicide. { 'sōd·ē·əm ¦pen·tə‚klór·ə'fe‚nāt }

sodium perborate [INORGANIC CHEMISTRY] $NaBO_2 \cdot H_2O_2 \cdot 3H_2O$ A white powder with a saline taste; slightly soluble in water, decomposes in moist air; used in deodorants, in dental compositions, and as a germicide. Also known as peroxydol. { 'sōd·ē·əm pər'bór‚āt }

sodium perchlorate [INORGANIC CHEMISTRY] $NaClO_4$ Fire-hazardous, white, deliquescent crystals; soluble in water and alcohol; melts at 482°C; explosive when in contact with concentrated sulfuric acid; used in jet fuel, as an analytical reagent, and for explosives. { 'sōd·ē·əm pər'klór‚āt }

sodium permanganate [INORGANIC CHEMISTRY] $NaMnO_4 \cdot 3H_2O$ A fire-hazardous, water-soluble, purple powder; decomposes when heated; used to make saccharin, as a disinfectant, and as an oxidizing agent. { 'sōd·ē·əm pər'maŋ·gə‚nāt }

sodium peroxide [INORGANIC CHEMISTRY] Na_2O_2 A fire-hazardous, white powder that yellows with heating; decomposes when heated; causes ignition when in contact with water; used as an oxidizing agent and a bleach, and in medicinal soap. { 'sōd·ē·əm pə'räk‚sīd }

sodium persulfate [INORGANIC CHEMISTRY] $Na_2S_2O_8$ A white, water-soluble, crystalline powder; used as a bleaching agent and in medicine. { 'sōd·ē·əm pər'səl‚fāt }

sodium phenate [ORGANIC CHEMISTRY] C_6H_5ONa White, deliquescent crystals, soluble in water and alcohol; decomposed by carbon dioxide in air; used as a chemical intermediate, antiseptic, and military gas absorbent. Also known as sodium carbolate; sodium phenolate. { 'sōd·ē·əm 'fe‚nāt }

sodium phenolate [ORGANIC CHEMISTRY] *See* sodium phenate. { 'sōd·ē·əm 'fen·əl‚āt }

sodium phenylacetate [ORGANIC CHEMISTRY] $C_6H_5CH_2 \cdot COONa$ Pale yellow, 50% aqueous solution which crystallizes at 15°C; used in the manufacture of penicillin G. { 'sōd·ē·əm ‚fen·əl'as·ə‚tāt }

sodium phenylphosphinate [ORGANIC CHEMISTRY] $C_6H_5PH(O)(ONa)$ Crystals with a melting point of 355°C; used as an antioxidant and as a heat and light stabilizer. { 'sōd·ē·əm ‚fen·əl'fä·sfə‚nāt }

sodium phosphate [INORGANIC CHEMISTRY] A general term encompassing the following compounds: sodium hexametaphosphate, sodium metaphosphate, dibasic sodium phosphate, hemibasic sodium phosphate, monobasic sodium phosphate, tribasic sodium phosphate, sodium pyrophosphate, and acid sodium pyrophosphate. { 'sōd·ē·əm 'fä‚sfāt }

sodium phosphite [INORGANIC CHEMISTRY] $Na_2HPO_3 \cdot 5H_2O$ White, hygroscopic crystals, melting at 53°C; soluble in water, insoluble in alcohol; used in medicine. { 'sōd·ē·əm 'fä‚sfīt }

sodium phosphotungstate [INORGANIC CHEMISTRY] *See* sodium tungstophosphate. { 'sōd·ē·əm ‚fä·sfō'təŋ‚stāt }

sodium phytate [ORGANIC CHEMISTRY] $C_6H_9O_{24}P_6Na_9$ A hygroscopic, water-soluble powder; used as a chelating agent for trace metals and in medicine. { 'sōd·ē·əm 'fī‚tāt }

sodium picramate [ORGANIC CHEMISTRY] $NaOC_6H_2(NO_2)_2NH_2$ A yellow salt, soluble in water; used in the manufacture of dye intermediates. { 'sōd·ē·əm 'pik·rə‚māt }

sodium platinichloride [INORGANIC CHEMISTRY] *See* sodium chloroplatinate. { 'sōd·ē·əm ¦plat·ən·ə'klór‚īd }

sodium plumbite [INORGANIC CHEMISTRY] $Na_2PbO_2 \cdot 3H_2O$ A toxic, corrosive solution of lead oxide (litharge) in sodium hydroxide; used (as doctor solution) to sweeten gasoline. { 'sōd·ē·əm 'pləm‚bīt }

sodium polymannuronate [ORGANIC CHEMISTRY] *See* sodium alginate. { 'sōd·ē·əm ¦päl·i·mə'nyùr·ə‚nāt }

sodium polysulfide [INORGANIC CHEMISTRY] Na_2S_x Yellow-brown granules, used to make dyes and colors, and insecticides, as a petroleum additive, and in electroplating. { 'sōd·ē·əm ¦päl·i'səl‚fīd }

sodium propionate [ORGANIC CHEMISTRY] CH_3CH_2COONa Deliquescent, transparent crystals; soluble in water, slightly soluble in alcohol; used as a fungicide, and mold preventive. { 'sōd·ē·əm 'prō·pē·ə‚nāt }

sodium pyroborate [INORGANIC CHEMISTRY] *See* sodium borate. { 'sōd·ē·əm ¦pī·rō'bȯr‚āt }

sodium pyrophosphate [INORGANIC CHEMISTRY] $Na_4P_2O_7$ A white powder; soluble in water, insoluble in alcohol and ammonia; melts at 880°C; used as a water softener and newsprint deinker, and to control drilling-mud viscosity. Also known as normal sodium pyrophosphate; tetrasodium pyrophosphate (TSPP). { 'sōd·ē·əm ¦pī·rō'fä‚sfāt }

sodium saccharin [ORGANIC CHEMISTRY] $C_7H_4NNaO_3S \cdot 2H_2O$ White crystals or a crystalline powder, soluble in water and slightly soluble in alcohol; used in medicine and as a nonnutritive food sweetener. { 'sōd·ē·əm 'sak·ə·rən }

sodium salicylate [ORGANIC CHEMISTRY] HOC_6H_4COONa A shiny, white powder with sweetish taste and mild aromatic aroma; soluble in water, glycerol, and alcohol; used in medicine and as a preservative. { 'sōd·ē·əm sə'lis·ə‚lāt }

sodium selenate [INORGANIC CHEMISTRY] $Na_2SeO_4 \cdot 10H_2O$ White, poisonous, water-soluble crystals; used as an insecticide. { 'sōd·ē·əm 'sel·ə‚nāt }

sodium selenite [INORGANIC CHEMISTRY] $Na_2SeO_3 \cdot 5H_2O$ White, water-soluble crystals; used in glass manufacture, as a bacteriological reagent, and for decorating porcelain. { 'sōd·ē·əm 'sel·ə‚nīt }

sodium sesquicarbonate [INORGANIC CHEMISTRY] $Na_2CO_3 \cdot NaHCO_3 \cdot 2H_2O$ White, water-soluble, needle-shaped crystals; used as a detergent, an alkaline agent for water softening and leather tanning, and a food additive. { 'sōd·ē·əm ¦ses·kwē 'kär·bə‚nāt }

sodium sesquisilicate [INORGANIC CHEMISTRY] $Na_6Si_2O_7$ A white, water-soluble powder; used for metals cleaning and textile processing. { 'sōd·ē·əm ¦ses·kwē'sil·ə‚kāt }

sodium silicate [INORGANIC CHEMISTRY] Na_2SiO_3 A gray-white powder; soluble in alkalies and water, insoluble in alcohol and acids; used to fireproof textiles, in petroleum refining and corrugated paperboard manufacture, and as an egg preservative. Also known as liquid glass; silicate of soda; sodium metasilicate; soluble glass; water glass. { 'sōd·ē·əm 'sil·ə‚kāt }

sodium silicoaluminate [INORGANIC CHEMISTRY] *See* sodium aluminosilicate. { 'sōd·ē·əm ¦sil·ə·kō·ə'lü·mə‚nāt }

sodium silicofluoride [INORGANIC CHEMISTRY] *See* sodium fluosilicate. { 'sōd·ē·əm ¦sil·ə·kō'flu̇r‚īd }

sodium stannate [INORGANIC CHEMISTRY] $Na_2SnO_3 \cdot 3H_2O$ Water- and alcohol-insoluble, whitish crystals; used in ceramics, dyeing, and textile fireproofing, and as a mordant. Also known as preparing salt. { 'sōd·ē·əm 'sta‚nāt }

sodium stearate [ORGANIC CHEMISTRY] $NaC_{18}H_{35}O_2$ A white powder with a fatty aroma; soluble in hot water and alcohol; used in medicine and toothpaste and as a waterproofing agent. { 'sōd·ē·əm 'stir‚āt }

sodium subsulfite [INORGANIC CHEMISTRY] *See* sodium thiosulfate. { 'sōd·ē·əm ¦səb'səl‚fīt }

sodium succinate [ORGANIC CHEMISTRY] $Na_2C_4H_4O_4 \cdot 6H_2O$ Water-soluble, white crystals; loses water at 120°C; used in medicine. { 'sōd·ē·əm 'sək·sə‚nāt }

sodium sulfate [INORGANIC CHEMISTRY] Na_2SO_4 Crystalline compound, melts at 888°C, soluble in water; used to make paperboard, kraft paper, glass, and freezing mixtures. { 'sōd·ē·əm 'səl‚fāt }

sodium sulfhydrate [INORGANIC CHEMISTRY] *See* sodium hydrosulfide. { 'sōd·ē·əm ¦səlf'hī‚drāt }

sodium sulfide [INORGANIC CHEMISTRY] Na_2S An irritating, water-soluble, yellow to red, deliquescent powder; melts at 1180°C; used as a chemical intermediate, solvent, photographic reagent, and analytical reagent. Also known as sodium sulfuret. { 'sōd·ē·əm 'səl‚fīd }

sodium sulfite [INORGANIC CHEMISTRY] Na_2SO_3 White, water-soluble, crystals with a sulfurous, salty taste; decomposes when heated; used as a chemical intermediate and food preservative, in medicine and paper manufacturing, and for dyes and photographic developing. { 'sōd·ē·əm 'səl‚fīt }

sodium sulfocyanate [INORGANIC CHEMISTRY] *See* sodium thiocyanate. { 'sōd·ē·əm ¦səl·fō'sī·ə‚nāt }

sodium sulfuret [INORGANIC CHEMISTRY] *See* sodium sulfide. { 'sōd·ē·əm 'səl·fyə‚ret }

sodium tartrate [ORGANIC CHEMISTRY] $Na_2C_4H_4O_6 \cdot 2H_2O$ White, water-soluble crystals or granules; loses water at 150°C; used in medicine and as a food stabilizer and sequestrant. Also known as disodium tartrate; sal tartar. { 'sōd·ē·əm 'tär‚trāt }

sodium TCA [INORGANIC CHEMISTRY] *See* sodium trichloroacetate. { 'sōd·ē·əm ¦tē¦sē'ā }

sodium tetraborate [INORGANIC CHEMISTRY] *See* sodium borate. { 'sōd·ē·əm ¦te·trə'bȯr‚āt }

sodium tetrafluorescein [ORGANIC CHEMISTRY] *See* easin. { 'sōd·ē·əm ¦te·trə'flu̇r·ə‚sēn }

sodium tetraphenylborate [ORGANIC CHEMISTRY] $[(C_6H_5)_4B]Na$ A snow-white, crystalline compound, soluble in water and acetone; used as a reagent in the determination of the following ions: potassium, ammonium, rubidium, and cesium. { 'sōd·ē·əm ¦tre·trə¦fen·əl'bȯr‚at }

sodium tetrasulfide [INORGANIC CHEMISTRY] Na_2S_4 Hygroscopic, yellow or dark-red crystals, melting at 275°C; used for insecticides and fungicides, ore flotation, and dye manufacture, and as a reducing agent. { 'sōd·ē·əm ¦tet·rə'səl‚fīd }

sodium thiocyanate [INORGANIC CHEMISTRY] NaSCN A poisonous, water- and alcohol-soluble, deliquescent, white powder; melts at 287°C; used as an analytical reagent, solvent, and chemical intermediate, and for rubber treatment and textile dyeing and printing. Also known as sodium sulfocyanate. { 'sōd·ē·əm ‚thī·ə'sī·ə‚nāt }

sodium thioglycolate [ORGANIC CHEMISTRY] $C_2H_3NaO_3S$ A water-soluble compound produced as hygroscopic crystals; used as an ingredient in bacteriology media, and in hair-waving solutions. { 'sōd·ē·əm ‚thī·ə'glī·kə‚lāt }

sodium thiosulfate [INORGANIC CHEMISTRY] $Na_2S_2O_3 \cdot 5H_2O$ White, translucent crystals or powder with a melting point of 48°C; soluble in water and oil of turpentine; used as a fixing agent in photography, for extracting silver from ore, in medicine, and as a sequestrant in food. Also known as sodium hyposulfite; sodium subsulfite. { 'sōd·ē·əm ‚thī·ə'səl‚fāt }

sodium trichloroacetate [ORGANIC CHEMISTRY] CCl_3COONa A toxic material, used in herbicides and pesticides. Abbreviated sodium TCA. { 'sōd·ē·əm trī¦klȯr·ō'as·ə‚tat }

sodium 2,4,5-trichlorophenate [ORGANIC CHEMISTRY] $C_6H_2Cl_3ONa \cdot 1^1/_2H_2O$ Buff to light brown flakes, soluble in water, methanol, and acetone, used as a bactericide and fungicide. { 'sōd·ē·əm ¦tü ¦fȯr ¦fīv trī¦klȯr·ō'fe‚nāt }

sodium tripolyphosphate [INORGANIC CHEMISTRY] $Na_5P_3O_{10}$ A white powder with a melting point of 622°C; used for water softening and as a food additive and texturizer. Abbreviated STPP. { 'sōd·ē·əm trī‚pal·i'fä‚sfāt }

sodium tungstate [INORGANIC CHEMISTRY] $Na_2WO_4 \cdot 2H_2O$ Water-soluble, colorless crystals; lose water at 100°C, melts at 692°C; used as a chemical intermediate analytical reagent, and for fireproofing. Also known as sodium wolframate. { 'sōd·ē·əm 'təŋ‚stat }

sodium tungstophosphate [INORGANIC CHEMISTRY] Approximately $2Na_2O \cdot P_2O_5 \cdot 12WO_3 \cdot 18H_2O$ A yellowish-white powder, soluble in water and alcohols; used to manufacture organic pigments, as an antistatic agent for textiles, in leather tanning, and as a water-resistant agent in plastic films, adhesives, and cements. Also known as sodium phosphotungstate. { 'sōd·ē·əm ¦twelv ¦təŋ·stō'fä‚sfāt }

sodium undecylenate [ORGANIC CHEMISTRY] $C_{11}H_{19}O_2Na$ A white, water-soluble powder that decomposes above 200°C; used in cosmetics and pharmaceuticals as a bacteriostat and fungistat. { 'sōd·ē·əm ‚ən‚de·sə'le‚nāt }

sodium wolframate [INORGANIC CHEMISTRY] *See* sodium tungstate. { 'sōd·ē·əm 'wu̇l·frə‚mīt }

sodium xanthate [ORGANIC CHEMISTRY] *See* sodium ethylxanthate. { 'sōd·ē·əm 'zan‚thāt }

sodium xanthogenate [ORGANIC CHEMISTRY] *See* sodium ethylxanthate. { 'sōd·ē·əm zan'thä·jə‚nāt }

soft electrophile [PHYSICAL CHEMISTRY] A molecule that readily accepts electrons during a primary reaction step. { 'sȯft i'lek·trə‚fīl }

soft water [CHEMISTRY] Water that is free of magnesium or calcium salts. { 'sȯft 'wȯd·ər }

soft-x-ray absorption spectroscopy [SPECTROSCOPY] A spectroscopic technique which is used to get information about unoccupied states above the Fermi level in a metal or about empty conduction bands in an inoculator. { 'sȯft ¦eks‚ra əb'sōrp·shən spek'träs·kə·pē }

soft-x-ray appearance potential spectroscopy [SPECTROSCOPY] A branch of electron spectroscopy in which a solid surface is bombarded with monochromatic electrons, and small but abrupt changes in the resulting total x-ray emission intensity are detected as the energy of the electrons is varied. Abbreviated SXAPS. { 'sȯft ¦eks‚ra ə'pir·əns pə¦ten·chəl spek'träs·kə·pē }

sol [CHEMISTRY] A colloidal solution consisting of a suitable dispersion medium, which may be gas, liquid, or solid, and the colloidal substance, the disperse phase, which is distributed throughout the dispersion medium. { säl }

solation [PHYSICAL CHEMISTRY] The change of a substance from a gel to a sol. { sə 'lā·shən }

sol-gel glass [PHYSICAL CHEMISTRY] An optically transparent amorphous silica or silicate material produced by forming interconnections in a network of colloidal, submicrometer particles under increasing viscosity until the network becomes completely rigid, with about one-half the density of glass. { 'säl 'jel 'glas }

solid-liquid equilibrium [PHYSICAL CHEMISTRY] **1.** The interrelation of a solid material and its melt at constant vapor pressure. **2.** The concentration relationship of a solid with a solvent liquid other than its melt. Also known as liquid-solid equilibrium. { 'säl·əd 'lik·wəd ‚ē·kwə'lib·rē·əm }

solidus [PHYSICAL CHEMISTRY] In a constitution or equilibrium diagram, the locus of points representing the temperature below which the various compositions finish freezing on cooling, or begin to melt on heating. { 'säl·əd·əs }

solidus curve [PHYSICAL CHEMISTRY] A curve on the phase diagram of a system with two components which represents the equilibrium between the liquid phase and the solid phase. { 'säl·əd·əs ‚kərv }

soliquid [PHYSICAL CHEMISTRY] A system in which solid particles are dispersed in a liquid. { ¦sä'lik·wəd }

solubility [PHYSICAL CHEMISTRY] The ability of a substance to form a solution with another substance. { ‚säl·yə'bil·əd·ē }

solubility coefficient [PHYSICAL CHEMISTRY] The volume of a gas that can be dissolved by a unit volume of solvent at a specified pressure and temperature. { ‚säl·yə'bil·əd·ē ‚kō·i‚fish·ənt }

solubility curve [PHYSICAL CHEMISTRY] A graph showing the concentration of a substance in its saturated solution in a solvent as a function of temperature. { ‚säl·yə'bil·əd·ē ‚kərv }

solubility product constant [PHYSICAL CHEMISTRY] A type of simplified equilibrium constant, K_{sp}, defined for and useful for equilibria between solids and their respective ions in solution; for example, the equilibrium

$$AgCl(s) \rightleftharpoons Ag^+ + Cl^-, \quad [Ag^+][Cl^-] \approx K_{sp}$$

where $[Ag^+]$ and $[Cl^-]$ are molar concentrations of silver ions and chloride ions. { ‚säl·yə'bil·əd·ē ¦präd·əkt ‚kän·stənt }

solubility test [ANALYTICAL CHEMISTRY] **1.** A test for the degree of solubility of asphalts and other bituminous materials in solvents, such as carbon tetrachloride, carbon disulfide, or petroleum ether. **2.** Any test made to show the solubility of one material in another (such as liquid-liquid, solid-liquid, gas-liquid, or solid-solid). { ‚säl·yə'bil·əd·ē ‚test }

soluble [CHEMISTRY] Capable of being dissolved. { 'säl·yə·bəl }

soluble glass [INORGANIC CHEMISTRY] *See* sodium silicate. { 'säl·yə·bəl 'glas }

soluble guncotton [ORGANIC CHEMISTRY] *See* pyroxylin. { 'säl·yə·bəl 'gən‚kat·ən }

soluble indigo blue [ORGANIC CHEMISTRY] *See* indigo carmine. { 'säl·yə·bəl 'in·də·gō 'blü }

soluble nitrocellulose [ORGANIC CHEMISTRY] *See* pyroxylin. { 'säl·yə·bəl ¦nī·trō 'sel·yə‚lōs }

solute [CHEMISTRY] The substance dissolved in a solvent. { 'säl·yüt }

solution [CHEMISTRY] A single, homogeneous liquid, solid, or gas phase that is a mixture in which the components (liquid, gas, solid, or combinations thereof) are uniformly distributed throughout the mixture. { sə'lü·shən }

solution pressure [PHYSICAL CHEMISTRY] **1.** A measure of the tendency of molecules or atoms to cross a bounding surface between phases and to enter into a solution. **2.** A measure of the tendency of hydrogen, metals, and certain nonmetals to pass into solution as ions. { sə'lü·shən ‚presh·ər }

solutrope [CHEMISTRY] A ternary mixture with two liquid phases and a third component distributed between the phases, or selectively dissolved in one or the other of the phases; analogous to an azeotrope. { 'säl·yə‚trōp }

solvation [CHEMISTRY] The process of swelling, gelling, or dissolving of a material by a solvent; for resins, the solvent can be a plasticizer. { säl'vā·shən }

solvent [CHEMISTRY] That part of a solution that is present in the largest amount, or the compound that is normally liquid in the pure state (as for solutions of solids or gases in liquids). { 'säl·vənt }

solvent front [ANALYTICAL CHEMISTRY] In paper chromatography, the wet moving edge of the solvent that progresses along the surface where the separation of the mixture is occurring. { 'säl·vənt ‚frənt }

solvolysis [CHEMISTRY] A reaction in which a solvent reacts with the solute to form a new substance. { säl'väl·ə·səs }

solvus [PHYSICAL CHEMISTRY] In a phase or equilibrium diagram, the locus of points representing the solid-solubility temperatures of various compositions of the solid phase. { 'säl·vəs }

Sommelet process [ORGANIC CHEMISTRY] The preparation of thiophene aldehydes by treatment of thiophene with hexamethylenetetramine. { ‚sȯ·məl'yā ‚prä·səs }

Sonnenschein's reagent [ANALYTICAL CHEMISTRY] A solution of phosphomolybdic acid that forms a yellow precipitate with alkaloid sulfates. { 'zȯn·ən‚shīnz rē‚ā·jənt }

sonocatalysis [CHEMISTRY] **1.** Initiation of a catalytic reaction by irradiation with sound or ultrasound. **2.** Use of sound to impart catalytic activity to a chemical compound. { ¦sän·ə·kə'tal·ə·səs }

sonochemistry [CHEMISTRY] Any chemical change, such as in reaction type or rate, that occurs in response to sound or ultrasound. { ¦sän·ə'kem·ə·strē }

sorbate [CHEMISTRY] A substance that has been either adsorbed or absorbed. [ORGANIC CHEMISTRY] A salt or an ester of sorbic acid. { 'sȯr‚bāt }

sorbic acid [ORGANIC CHEMISTRY] $CH_3CH{=}CHCH{=}CHCOOH$ A white, crystalline compound; soluble in most organic solvents, slightly soluble in water; melts at 135°C; used as a fungicide and food preservative, and in the manufacture of plasticizers and lubricants. { 'sȯr·bik 'as·əd }

sorbide [CHEMISTRY] The generic term for anhydrides derived from sorbitol. { 'sȯr‚bīd }

sorbitol [ORGANIC CHEMISTRY] $C_6H_8(OH)_6$ Combustible, white, water-soluble, hygroscopic crystals with a sweet taste; melt at 93 to 97.5°C (depending on the form); used in cosmetic creams and lotions, toothpaste, and resins; as a food additive; and for ascorbic acid fermentation. { 'sȯr·bə‚tȯl }

Sörensen titration [ANALYTICAL CHEMISTRY] Titration with one of the Sörensen hydrogen-ion-concentration indicators. { 'sȯr·ən·sən tī‚trā·shən }

sorption [PHYSICAL CHEMISTRY] A general term used to encompass the processes of adsorption, absorption, desorption, ion exchange, ion exclusion, ion retardation, chemisorption, and dialysis. { 'sȯrp·shən }

sosoloid [PHYSICAL CHEMISTRY] A system consisting of particles of a solid dispersed in another solid. { 'säs·ə,lȯid }

sour [CHEMISTRY] Containing large amounts of malodorous sulfur compounds (such as mercaptans or hydrogen sulfide), as in crude oils, naphthas, or gasoline. { saür }

source [SPECTROSCOPY] The arc or spark that supplies light for a spectroscope. { sȯrs }

Soxhlet extractor [CHEMISTRY] A flask and condenser device for the continuous extraction of alcohol- or ether-soluble materials. { 'säks·lət ik,strak·tər }

spark excitation [SPECTROSCOPY] The use of an electric spark (10,000 to 30,000 volts) to excite spectral line emissions from otherwise hard-to-excite samples; used in emission spectroscopy. { 'spärk ,ek,sī'tā·shən }

spark explosion method [ANALYTICAL CHEMISTRY] A technique for the analysis of hydrogen; the sample is mixed with an oxidant and exploded by a spark or hot wire, and the combustion products are then analyzed. { 'spärk ik'splō·zhən ,meth·əd }

spark spectrum [SPECTROSCOPY] The spectrum produced by a spark discharging through a gas or vapor; with metal electrodes, a spectrum of the metallic vapor is obtained. { 'spärk ,spek·trəm }

sparteine [ORGANIC CHEMISTRY] $C_{15}H_{26}N_2$ A poisonous, colorless, oily alkaloid; soluble in alcohol and ether, slightly soluble in water; boils at 173°C; used in medicine. Also known as lupinidine. { 'spärd·ē,ēn }

special cause [ANALYTICAL CHEMISTRY] A cause of variance or bias in a measurement process that is external to the system. { ¦spesh·əl ¦kȯz }

species [CHEMISTRY] A chemical entity or molecular particle, such as a radical, ion, molecule, or atom. Also known as chemical species. { 'spē·shēz }

specific catalysis [CHEMISTRY] The acceleration of a given chemical reaction by a unique catalyst rather than by a family of related substances. { spə¦sif·ik kə'tal·ə·səs }

specificity [CHEMISTRY] The selective reactivity that occurs between substances, such as between an antigen and its corresponding antibody. { ,spes·ə'fis·əd·ē }

specific retention volume [ANALYTICAL CHEMISTRY] The relationship among retention volume, void volume, and adsorbent weight, used to standardize gas chromatography adsorbents by the elution of a standard solute by a standard eluent from the adsorbent under test. { spə'sif·ik ri'ten·chən ,väl·yəm }

specific susceptibility [PHYSICAL CHEMISTRY] *See* mass susceptibility. { spə'sif·ik sə,sep·tə'bil·əd·ē }

spectator ion [CHEMISTRY] An ion that serves to balance the electrical charges in a reaction environment without participating in product formation. { 'spek,tād·ər 'ī,än }

spectral bandwidth [SPECTROSCOPY] The minimum radiant-energy bandwidth to which a spectrophotometer is accurate; that is, 1-5 nanometers for better models. { 'spek·trəl 'band,width }

spectral directional reflectance factor [ANALYTICAL CHEMISTRY] In spectrophotometric colorimetry, the ratio of the energy diffused in any desired direction by the object under analysis to that energy diffused in the same direction by an ideal perfect (energy) diffuser. { 'spek·trəl di'rek·shən·əl ri'flek·təns ,fak·tər }

spectral line [SPECTROSCOPY] A discrete value of a quantity, such as frequency, wavelength, energy, or mass, whose spectrum is being investigated; one may observe a finite spread of values resulting from such factors as level width, Doppler broadening, and instrument imperfections. Also known as spectrum line. { 'spek·trəl ,līn }

spectral radiance factor [ANALYTICAL CHEMISTRY] A situation when the desired directions for analysis of energy diffused from (reflected from) an object under spectrophotometric colorimetric analysis are all substantially the same (a solid angle of nearly zero steradians). { 'spek·trəl 'rād·ē·əns ,fak·tər }

spectral reflectance [ANALYTICAL CHEMISTRY] Situation when the desired directions for analysis of energy from (reflected from) an object under spectrophotometric colorimetric analysis is diffused in all directions (not directed as a single beam). { 'spek·trəl ri'flek·təns }

spectral regions [SPECTROSCOPY] Arbitrary ranges of wavelength, some of them overlapping, into which the electromagnetic spectrum is divided, according to the types of sources that are required to produce and detect the various wavelengths, such as x-ray, ultraviolet, visible, infrared, or radio-frequency. { 'spek·trəl ˌrē·jənz }

spectral series [SPECTROSCOPY] Spectral lines or groups of lines that occur in sequence. { 'spek·trəl ˌsir·ēz }

spectrobolometer [SPECTROSCOPY] An instrument that measures radiation from stars; measurement can be made in a narrow band of wavelengths in the electromagnetic spectrum; the instrument itself is a combination spectrometer and bolometer. { ¦spek·trō·bō'läm·əd·ər }

spectrofluorometer [SPECTROSCOPY] A device used in fluorescence spectroscopy to increase the selectivity of fluorometry by passing emitted fluorescent light through a monochromator to record the fluorescence emission spectrum. { ¦spek·trō·flü'räm·əd·ər }

spectrogram [SPECTROSCOPY] The record of a spectrum produced by a spectrograph. Also known as measured spectrum. { 'spek·trəˌgram }

spectrograph [SPECTROSCOPY] A spectroscope provided with a photographic camera or other device for recording the spectrum. { 'spek·trəˌgraf }

spectrography [SPECTROSCOPY] The use of photography to record the electromagnetic spectrum displayed in a spectroscope. { spek'träg·rə·fē }

spectrometer [SPECTROSCOPY] **1.** A spectroscope that is provided with a calibrated scale either for measurement of wavelength or for measurement of refractive indices of transparent prism materials. **2.** A spectroscope equipped with a photoelectric photometer to measure radiant intensities at various wavelengths. { spek'träm·əd·ər }

spectrometry [SPECTROSCOPY] The use of spectrographic techniques for deriving the physical constants of materials. { spek'träm·ə·trē }

spectrophone [ANALYTICAL CHEMISTRY] A cell containing the sample in the optoacoustic detection method, equipped with windows through which the laser beam enters the cell and a microphone for detecting sound. { 'spek·trəˌfōn }

spectrophotometer [SPECTROSCOPY] An instrument that measures transmission or apparent reflectance of visible light as a function of wavelength, permitting accurate analysis of color or accurate comparison of luminous intensities of two sources or specific wavelengths. { ¦spek·trō fə'täm·əd·ər }

spectrophotometric titration [ANALYTICAL CHEMISTRY] An analytical method in which the radiant-energy absorption of a solution is measured spectrophotometrically after each increment of titrant is added. { ¦spek·trōˌfōd·ə'me·trik tī'trā·shən }

spectrophotometry [SPECTROSCOPY] A procedure to measure photometrically the wavelength range of radiant energy absorbed by a sample under analysis; can be by visible light, ultraviolet light, or x-rays. { ¦spek·trō·fə'täm·ə·trē }

spectropyrheliometer [SPECTROSCOPY] An astronomical instrument used to measure distribution of radiant energy from the sun in the ultraviolet and visible wavelengths. { ¦spek·trō¦pīrˌhē·le'äm·əd·ər }

spectroscope [SPECTROSCOPY] An optical instrument consisting of a slit, collimator lens, prism or grating, and a telescope or objective lens which produces a spectrum for visual observation. { 'spek·trəˌskōp }

spectroscopic displacement law [SPECTROSCOPY] The spectrum of an un-ionized atom resembles that of a singly ionized atom of the element one place higher in the periodic table, and that of a doubly ionized atom two places higher in the table, and so forth. { ¦spek·trə¦skäp·ik di'splās·mənt ˌlȯ }

spectrum line [SPECTROSCOPY] *See* spectral line. { 'spek·trəm ˌlīn }

sphere of attraction [PHYSICAL CHEMISTRY] The distance within which the potential energy arising from mutual attraction of two molecules is not negligible with respect to the molecules' average thermal energy at room temperature. { 'sfir əv ə'trak·shən }

spin label [PHYSICAL CHEMISTRY] A molecule which contains an atom or group of atoms exhibiting an unpaired electron spin that can be detected by electron spin resonance (ESR) spectroscopy and can be bonded to another molecule. { 'spin ˌlā·bəl }

spinning-band column [ANALYTICAL CHEMISTRY] An analytical distillation column inside of which is a series of driven, spinning bands; centrifugal action of the bands throws a layer of liquid onto the inner surface of the column; used as an aid in liquid-vapor contact. { ˈspin·iŋ ¦band ˌkäl·əm }

spin-polarized atomic hydrogen [PHYSICAL CHEMISTRY] A system of hydrogen atoms cooled to a very low temperature in a very high magnetic field so that electron spins in almost all the atoms are antiparallel to the magnetic field, with the result that the atoms interact only through the weak triplet-state interaction so that no hydrogen molecules are formed. { ˈspin ¦pō·ləˌrīzd əˈtäm·ik ˈhī·drə·jən }

spiral wire column [ANALYTICAL CHEMISTRY] An analytical rectification (distillation) column with a wire spiral the length of the inside of the column to serve as a liquid-vapor contact surface. { ˈspī·rəl ¦wīr ˈkäl·əm }

spiran [ORGANIC CHEMISTRY] A polycyclic compound containing a carbon atom which is a member of two rings. { ˈspīˌran }

spirit [ORGANIC CHEMISTRY] A solution of alcohol and a volatile substance, such as an essential oil. { ˈspir·ət }

spiro atom [ORGANIC CHEMISTRY] A single atom that is the only common member of two ring structures. { ˈspir·ō ˌad·əm }

spiro ring system [ORGANIC CHEMISTRY] A molecular structure with two ring structures having one atom in common; for example, spiropentane. { ˈspī·rō ˈriŋ ˌsis·təm }

spontaneous combustion [CHEMISTRY] Ignition that can occur when certain materials such as tung oil are stored in bulk, resulting from the generation of heat, which cannot be readily dissipated; often heat is generated by microbial action. Also known as spontaneous ignition. { spänˈtā·nē·əs kəmˈbəs·chən }

spontaneous heating [CHEMISTRY] The slow reaction of material with atmospheric oxygen at ambient temperatures; liberated heat, if undissipated, accumulates so that in the presence of combustible substances a fire will result. { spänˈtā·nē·əs ˈhēd·iŋ }

spontaneous ignition [CHEMISTRY] *See* spontaneous combustion. { spänˈtā·nē·əs igˈnish·ən }

spot test [ANALYTICAL CHEMISTRY] The addition of a drop of reagent to a drop or two of sample solution to obtain distinctive colors or precipitates; used in qualitative analysis. { ˈspät ˌtest }

square planar molecule [CHEMISTRY] A molecule in which a central atom possesses four valence bonds directed to the corners of a square, with all atoms lying in the same plane. { ˈskwer ¦plā·nər ˌmäl·əˌkyül }

Sr [CHEMISTRY] *See* strontium.

SRMS [SPECTROSCOPY] *See* structure resonance modulation spectroscopy.

SSD [ANALYTICAL CHEMISTRY] *See* steady-state distribution.

stability [CHEMISTRY] The property of a chemical compound which is not readily decomposed and does not react with other compounds. { stəˈbil·əd·ē }

stability constant [CHEMISTRY] Refers to the equilibrium reaction of a metal cation and a ligand to form a chelating mononuclear complex; the absolute-stability constant is expressed by the product of the concentration of products divided by the product of the concentrations of the reactants; the apparent-stability constant (also known as the conditional- or effective-stability constant) allows for the nonideality of the system because of the combination of the ligand with other complexing agents present in the solution. { stəˈbil·əd·ē ˌkän·stənt }

standard calomel electrode [PHYSICAL CHEMISTRY] A mercury-mercurous chloride electrode used as a reference (standard) measurement in polarographic determinations. { ˈstan·dərd ˈkal·ə·məl iˈlekˌtrōd }

standard electrode potential [PHYSICAL CHEMISTRY] The reversible or equilibrium potential of an electrode in an environment where reactants and products are at unit activity. { ˈstan·dərd iˈlekˌtrōd pəˌten·chəl }

standardization [ANALYTICAL CHEMISTRY] A process in which the value of a potential standard is fixed by a measurement made with respect to a standard whose value is known. { ˌstan·dər·dəˈzā·shən }

standard potential [PHYSICAL CHEMISTRY] The potential of an electrode composed of a substance in its standard state, in equilibrium with ions in their standard states compared to a hydrogen electrode. { ˈstan·dərd pəˈten·chəl }

standard reference material [ANALYTICAL CHEMISTRY] A reference material distributed and certified by the appropriate national institute for standardization. { 'stan·dərd 'ref·rəns mə'tir·ē·əl }

standard solution [ANALYTICAL CHEMISTRY] *See* titrant. { ¦stan·dərd sə'lü·shən }

stannane [INORGANIC CHEMISTRY] *See* tin hydride. { 'sta,nān }

stannic acid [INORGANIC CHEMISTRY] *See* stannic oxide. { 'stan·ik 'as·əd }

stannic anhydride [INORGANIC CHEMISTRY] *See* stannic oxide. { 'stan·ik an'hī,drīd }

stannic bromide [INORGANIC CHEMISTRY] $SnBr_4$ Water- and alcohol-soluble, white crystals that fume when exposed to air, and melt at 31°C; used in mineral separations. Also known as tin bromide; tin tetrabromide. { 'stan·ik 'brō,mīd }

stannic chloride [INORGANIC CHEMISTRY] $SnCl_4$ A colorless, fuming liquid; soluble in cold water, alcohol, carbon disulfide, and oil of turpentine; decomposed by hot water; boils at 114°C; used as a conductive coating and a sugar bleach, and in drugs, ceramics, soaps, and blueprinting. Also known as tin chloride; tin tetrachloride. { 'stan·ik 'klȯr,īd }

stannic chromate [INORGANIC CHEMISTRY] $Sn(CrO_4)_2$ Toxic, brownish-yellow crystals, slightly soluble in water; used to decorate porcelain and china. Also known as tin chromate. { 'stan·ik 'krō,māt }

stannic iodide [INORGANIC CHEMISTRY] SnI_4 Yellow-reddish crystals; insoluble in water, soluble in alcohol, ether, chloroform, carbon disulfide, and benzene; decomposed by water, melt at 144°C, sublime at 180°C. Also known as tin iodide; tin tetraiodide. { 'stan·ik 'ī·ə,dīd }

stannic oxide [INORGANIC CHEMISTRY] SnO_2 A white powder; insoluble in water, soluble in concentrated sulfuric acid; melts at 1127°C; used in ceramic glazes and colors, special glasses, putty, and cosmetics, and as a catalyst. Also known as flowers of tin; stannic acid; stannic anhydride; tin dioxide; tin oxide; tin peroxide. { 'stan·ik 'äk,sīd }

stannic sulfide [INORGANIC CHEMISTRY] SnS_2 A yellow-brown powder; insoluble in water, soluble in alkaline sulfides; decomposes at red heat; used as a pigment and for imitation gilding. Also known as artificial gold; mosaic gold; tin bisulfide. { 'stan·ik 'səl,fīd }

stannous bromide [INORGANIC CHEMISTRY] $SnBr_2$ A yellow powder, soluble in water, alcohol, acetone, ether, and dilute hydrochloric acid; browns in air; melts at 215°C. Also known as tin bromide. { 'stan·əs 'brō,mīd }

stannous chloride [INORGANIC CHEMISTRY] $SnCl_2$ White crystals, soluble in water, alcohol, and alkalies; oxidized in air to the oxychloride; melt at 247°C; used as a chemical intermediate, reducing agent, and ink-stain remover, and for silvering mirrors. Also known as tin chloride; tin crystals; tin dichloride; tin salts. { 'stan·əs 'klȯr,īd }

stannous chromate [INORGANIC CHEMISTRY] $SnCrO_4$ A brown powder; very slightly soluble in water; used to decorate porcelain. Also known as tin chromate. { 'stan·əs 'krō,māt }

stannous 2-ethylhexoate [ORGANIC CHEMISTRY] $Sn(C_8H_{15}O_2)_2$ A light yellow liquid, soluble in benzene, toluene, and petroleum ether; used as a lubricant, a vulcanizing agent, and a stabilizer for transformer oil. { 'stan·əs ¦tü ¦eth·əl'hek·sə,wāt }

stannous fluoride [INORGANIC CHEMISTRY] SnF_2 A white, lustrous powder, slightly soluble in water; used to fluoridate toothpaste and as a medicine. { 'stan·əs 'flůr,īd }

stannous oxalate [ORGANIC CHEMISTRY] SnC_2O_4 A white, crystalline powder that decomposes at about 280°C; soluble in acids; used in textile dyeing and printing. Also known as tin oxalate. { 'stan·əs 'äk·sə,lāt }

stannous oxide [INORGANIC CHEMISTRY] SnO An air-unstable, brown to black powder; insoluble in water, soluble in acids and strong bases; decomposes when heated; used as a reducing agent and chemical intermediate, and for glass plating. Also known as tin oxide; tin protoxide. { 'stan·əs 'äk,sīd }

stannous sulfate [INORGANIC CHEMISTRY] $SnSO_4$ Heavy light-colored crystals; decomposes rapidly in water, loses SO_2 at 360°C; used for dyeing and tin plating. Also known as tin sulfate. { 'stan·əs 'səl,fāt }

stannous sulfide [INORGANIC CHEMISTRY] SnS Dark crystals; insoluble in water, soluble (with decomposition) in concentrated hydrochloric acid; melts at 880°C; used as an analytical reagent and catalyst, and in bearing material. Also known as tin monosulfide; tin protosulfide; tin sulfide. { 'stan·əs 'səl‚fīd }

stannum [CHEMISTRY] The Latin name for tin, thus the symbol Sn for the element. { 'stan·əm }

starch nitrate [ORGANIC CHEMISTRY] *See* nitrostarch. { 'stärch 'nī‚trāt }

Stark effect [SPECTROSCOPY] The effect on spectrum lines of an electric field which is either externally applied or is an internal field caused by the presence of neighboring ions or atoms in a gas, liquid, or solid. Also known as electric field effect. { 'stärk i‚fekt }

Stark-Einstein law [PHYSICAL CHEMISTRY] *See* Einstein photochemical equivalence law. { 'stärk 'īn‚stīn ‚lȯ }

stationary phase [ANALYTICAL CHEMISTRY] In chromatography, the nonmobile phase contained in the chromatographic bed. { 'stā·shə‚ner·ē 'fāz }

statistical control [ANALYTICAL CHEMISTRY] In an analytical procedure, a state that exists when the means of a large number of individual values in the output of a measurement process tend to approach a limiting value known as the limiting mean. { stə'tis·tə·kəl kən'trōl }

steady-state distribution [ANALYTICAL CHEMISTRY] The equilibrium condition between phases in each step of a multistage, countercurrent liquid-liquid extraction. Abbreviated SSD. { 'sted·ē ¦stāt ‚dis·trə'byü·shən }

stearamide [ORGANIC CHEMISTRY] $CH_3(CH_2)_{16}CONH_2$ Colorless leaflets with a melting point of 109°C; used as a corrosion inhibitor in oil wells. { 'stir·ə·məd }

stearate [ORGANIC CHEMISTRY] $C_{17}H_{35}COOM$ A salt or ester of stearic acid where M is a monovalent radical, for example, sodium stearate, $C_{17}H_{35}COONa$. { 'stir‚āt }

stearic acid [ORGANIC CHEMISTRY] $CH_3(CH_2)_{16}COOH$ Nature's most common fatty acid, derived from natural animal and vegetable fats; colorless, waxlike solid, insoluble in water, soluble in alcohol, ether, and chloroform; melts at 70°C; used as a lubricant and in pharmaceuticals, cosmetics, and food packaging. { 'stir·ik 'as·əd }

stearin [ORGANIC CHEMISTRY] $C_3H_5(C_{18}H_{35}O_2)_3$ A colorless combustible powder; insoluble in water, soluble in alcohol, chloroform, and carbon disulfide; melts at 72°C; used in metal polishes, pastes, candies, candles, and soap, and to waterproof paper. Also known as glyceryl tristearate; tristearin. { 'stir·ən }

stearyl alcohol [ORGANIC CHEMISTRY] $CH_3(CH_2)_{16}CH_2OH$ Oily white, combustible flakes; insoluble in water, soluble in alcohol, acetone, and ether; melt at 59°C; used in lubricants, resins, perfumes, and cosmetics, and as a surface-active agent. { 'sti‚ril 'al·kə‚hȯl }

step [ORGANIC CHEMISTRY] *See* elementary reaction. { step }

stepwise reaction [CHEMISTRY] A chemical reaction in which at least one reactive intermediate is produced and at least two elementary reactions are involved. { ¦step‚wīz rē¦ak·shən }

sterane [ORGANIC CHEMISTRY] A cycloalkane derived from a sterol. { 'sti‚rān }

stereochemistry [PHYSICAL CHEMISTRY] The study of the spatial arrangement of atoms in molecules and the chemical and physical consequences of such arrangement. { ¦ster·ē·ə'kem·ə·strē }

stereogenic center [ORGANIC CHEMISTRY] *See* asymmetric carbon atom. { ‚ster·ē·ə¦jen·ik 'sen·tər }

stereoisomers [ORGANIC CHEMISTRY] Compounds whose molecules have the same number and kind of atoms and the same atomic arrangement, but differ in their spatial relationship. { ¦ster·ē·ō'ī·sə·mərz }

stereoregular polymer [ORGANIC CHEMISTRY] *See* stereospecific polymer. { ¦ster·ē·ə'reg·yə·lər 'päl·i·mər }

stereorubber [ORGANIC CHEMISTRY] Synthetic rubber, *cis*-polyisoprene, a polymer with stereospecificity. { 'ster·ō·ə‚rəb·ər }

stereoselective reaction [ORGANIC CHEMISTRY] A chemical reaction in which one stereoisomer is produced or decomposed more rapidly than another. Also known as enantioselective reaction. { ¦ster·ē·ə·si'lek·tiv rē'ak·shən }

stereospecificity [ORGANIC CHEMISTRY] The condition of a polymer whose molecular structure has a fixed spatial (geometric) arrangement of its constituent atoms, thus having crystalline properties; for example, synthetic natural rubber, *cis*-polyisoprene. { ¦ster·ē·ō·spə·sə'fis·əd·ē }

stereospecific polymer [ORGANIC CHEMISTRY] A polymer with specific or definite order of arrangement of molecules in space, as in isotactic polypropylene; permits close packing of molecules and leads to a high degree of polymer crystallinity. Also known as stereoregular polymer. { ¦ster·ē·ō·spə'sif·ik 'päl·i·mər }

stereospecific synthesis [ORGANIC CHEMISTRY] Catalytic polymerization of monomer molecules to produce stereospecific polymers, as with Ziegler or Natta catalysts (derived from a transition metal halide and a metal alkyl). { ¦ster·ē·ō·spə'sif·ik 'sin·thə·səs }

steric effect [PHYSICAL CHEMISTRY] The influence of the spatial configuration of reacting substances upon the rate, nature, and extent of reaction. { 'ster·ik i,fekt }

steric hindrance [ORGANIC CHEMISTRY] The prevention or retardation of chemical reaction because of neighboring groups on the same molecule; for example, ortho-substituted aromatic acids are more difficult to esterify than are the meta and para substitutions. { 'ster·ik 'hin·drəns }

stern layer [PHYSICAL CHEMISTRY] One of two electrically charged layers of electrolyte ions, the layer of ions immediately adjacent to the surface, in the neighborhood of a negatively charged surface. { 'stərn ,lā·ər }

stibide [INORGANIC CHEMISTRY] *See* antimonide. { 'sti,bīd }

stibium [CHEMISTRY] The Latin name for antimony, thus the symbol Sb for the element. { 'stib·ē·əm }

sticking coefficient [PHYSICAL CHEMISTRY] The fraction of all atoms incident on a surface that are adsorbed on the surface. { 'stik·iŋ ,kō·i,fish·ənt }

stilbene [ORGANIC CHEMISTRY] $C_6H_5CH{:}CHC_6H_5$ Colorless crystals soluble in ether and benzene, insoluble in water; melts at 124°C; used to make dyes and bleaches and as phosphors. Also known as diphenylethylene; toluylene. { 'stil,bēn }

Stobbe reaction [ORGANIC CHEMISTRY] A type of aldol condensation reaction represented by the reaction of benzophenone with dimethyl succinate and sodium methoxide to form monoesters of an α-alkylidene (or arylidene) succinic acid. { 'shtöb·ə rē,ak·shən }

stoichiometry [PHYSICAL CHEMISTRY] The numerical relationship of elements and compounds as reactants and products in chemical reactions. { ,stȯi·kē'äm·ə·trē }

Stokes' law [SPECTROSCOPY] The wavelength of luminescence excited by radiation is always greater than that of the exciting radiation. { 'stōks ,lȯ }

Stokes line [SPECTROSCOPY] A spectrum line in luminescent radiation whose wavelength is greater than that of the radiation which excited the luminescence, and thus obeys Stokes' law. { 'stōks ,līn }

Stokes shift [SPECTROSCOPY] The displacement of spectral lines or bands of luminescent radiation toward longer wavelengths than those of the absorption lines or bands. { 'stōks ,shift }

stopped-flow method [CHEMISTRY] A method for studying chemical reactions in which the reactants are rapidly mixed, then abruptly stopped after a very short time. { 'stäpt ¦flō ,meth·əd }

stratified film [PHYSICAL CHEMISTRY] A film in which two thicknesses are present in a fixed configuration for a significant period of time. { 'strad·ə,fīd 'film }

stripping analysis [ANALYTICAL CHEMISTRY] An analytic process of solutions or concentrations containing ions, in which the ions are electrodeposited onto an electrode, stripped (dissolved) from the material from the electrode, and weighed. { 'strip·iŋ ə,nal·ə·səs }

strong acid [CHEMISTRY] An acid with a high degree of dissociation in solution, for example, mineral acids, such as hydrochloric acid, HCl, sulfuric acid, H_2SO_4, or nitric acid, HNO_3. { 'strȯŋ 'as·əd }

strong base [CHEMISTRY] A base with a high degree of dissociation in solution, for example, sodium hydroxide, NaOH, potassium hydroxide, KOH. { 'strȯŋ 'bās }

strontia [INORGANIC CHEMISTRY] *See* strontium oxide. { 'strän·chə }

strontium [CHEMISTRY] A metallic element in group II, symbol Sr, with atomic number 38, atomic weight 87.62; flammable, soft, pale-yellow solid; soluble in alcohol and acids, decomposes in water; melts at 770°C, boils at 1380°C; chemistry is similar to that of calcium; used as electron-tube getter. { 'strän·tē·əm }

strontium acetate [ORGANIC CHEMISTRY] $Sr(C_2H_3O_2)_2 \cdot {}^1/_2H_2O$ White, water-soluble crystals, loses water at 150°C; used for catalysts, as a chemical intermediate, and in medicine. { 'strän·tē·əm 'as·ə,tāt }

strontium bromide [INORGANIC CHEMISTRY] $SrBr_2 \cdot 6H_2O$ A white, hygroscopic powder soluble in water and alcohol; loses water at 180°C, melts at 643°C; used in medicine and as an analytical reagent. { 'strän·tē·əm 'brō,mīd }

strontium carbonate [INORGANIC CHEMISTRY] $SrCO_3$ A white powder slightly soluble in water, decomposes at 1340°C; used to make TV-tube glass, strontium salts, and ceramic ferrites, and in pyrotechnics. { 'strän·tē·əm 'kär·bə,nāt }

strontium chlorate [INORGANIC CHEMISTRY] $Sr(ClO_3)_2$ Shock-sensitive, highly combustible, white, water-soluble crystals that decompose at 120°C; used in pyrotechnics and tracer bullets. { 'strän·tē·əm 'klȯr,āt }

strontium chloride [INORGANIC CHEMISTRY] $SrCl_2$ Water- and alcohol-soluble white crystals, melts at 872°C; used in medicine and pyrotechnics and to make strontium salts. { 'strän·tē·əm 'klȯr,īd }

strontium chromate [INORGANIC CHEMISTRY] $SrCrO_4$ A light yellow, rust- and corrosion-resistant pigment used in metal coatings and for pyrotechnics. { 'strän·tē·əm 'krō,māt }

strontium dioxide [INORGANIC CHEMISTRY] *See* strontium peroxide. { 'strän·tē·əm dī'äk,sīd }

strontium fluoride [INORGANIC CHEMISTRY] SrF_2 A white powder, soluble in hydrochloric acid and hydrofluoric acid; used in medicine and for single crystals for lasers. { 'strän·tē·əm 'flu̇r,īd }

strontium hydrate [INORGANIC CHEMISTRY] *See* strontium hydroxide. { 'strän·tē·əm 'hī,drāt }

strontium hydroxide [INORGANIC CHEMISTRY] $Sr(OH)_2$ Colorless deliquescent crystals that absorb carbon dioxide from air, soluble in hot water and acids, melts at 375°C; used by the sugar industry, in lubricants and soaps, and as a plastic stabilizer. Also known as strontium hydrate. { 'strän·tē·əm hī'dräk,sīd }

strontium iodide [INORGANIC CHEMISTRY] SrI_2 Air-yellowing, white crystals that decompose in moist air, melts at 515°C; used in medicine and as a chemicals intermediate. { 'strän·tē·əm 'ī·ə,dīd }

strontium monosulfide [INORGANIC CHEMISTRY] *See* strontium sulfide. { 'strän·tē·əm ¦män·ə'səl,fīd }

strontium nitrate [INORGANIC CHEMISTRY] $Sr(NO_3)_2$ A white, water-soluble powder melting at 570°C; used in pyrotechnics, signals and flares, medicine, and matches, and as a chemicals intermediate. { 'strän·tē·əm 'nī,trāt }

strontium oxalate [INORGANIC CHEMISTRY] $SrC_2O_4 \cdot H_2O$ A white powder that loses water at 150°C; used in pyrotechnics and tanning. { 'strän·tē·əm 'äk·sə,lāt }

strontium oxide [INORGANIC CHEMISTRY] SrO A grayish powder, melts at 2430°C, becomes the hydroxide in water; used in medicine, pyrotechnics, pigments, greases, soaps, and as a chemicals intermediate. Also known as strontia. { 'strän·tē·əm 'äk,sīd }

strontium peroxide [INORGANIC CHEMISTRY] SrO_2 A strongly oxidizing, fire-hazardous, white, alcohol-soluble powder that decomposes in hot water; used in medicine, bleaching, and fireworks. Also known as strontium dioxide. { 'strän·tē·əm pər'äk,sīd }

strontium salicylate [ORGANIC CHEMISTRY] $Sr(C_7H_5O_3)_2 \cdot 2H_2O$ White crystals or powder with a sweet saline taste; soluble in water and alcohol; used in medicine and manufacture of pharmaceuticals. { 'strän·tē·əm sə'lis·ə,lāt }

strontium sulfate [INORGANIC CHEMISTRY] $SrSO_4$ White crystals insoluble in alcohol, slightly soluble in water and concentrated acids, melts at 1605°C; used in paper manufacture, pyrotechnics, ceramics, and glass. { 'strän·tē·əm 'səl,fāt }

strontium sulfide [INORGANIC CHEMISTRY] SrS A gray powder with a hydrogen sulfide aroma in moist air, slightly soluble in water, soluble (with decomposition) in acids, melts above 2000°C; used in depilatories and luminous paints and as a chemicals intermediate. Also known as strontium monosulfide. { 'strän·tē·əm 'səl‚fīd }

strontium titanate [INORGANIC CHEMISTRY] $SrTiO_3$ A solid material, insoluble in water and melting at 2060°C; used in electronics and electrical insulation. { 'strän·tē·əm 'tīt·ən‚āt }

structural formula [CHEMISTRY] A system of notation used for organic compounds in which the exact structure, if it is known, is given in schematic representation. { 'strək·chə·rəl 'fȯr·myə·lə }

structural isomers [ORGANIC CHEMISTRY] *See* constitutional isomers. { 'strək·chə·rəl 'ī·sə·mərz }

structure resonance [SPECTROSCOPY] An extremely narrow resonance exhibited by a small aerosol particle at a natural electromagnetic frequency at which the dielectric sphere oscillates, observed in the particle's scattered light excitation spectrum. { 'strək·chər ‚rez·ən·əns }

structure resonance modulation spectroscopy [SPECTROSCOPY] The infrared modulation of visible scattered light near a structure resonance to determine the absorption spectrum of an aerosol particle. Abbreviated SRMS. { 'strək·chər ‚rez·ən·əns ‚mäj·ə'lā·shən spek'träs·kə·pē }

strychnine [ORGANIC CHEMISTRY] $C_{21}H_{22}O_2N_2$ An alkaloid obtained primarily from the plant nux vomica, formerly used for therapeutic stimulation of the central nervous system. { 'strik‚nīn }

styphnic acid [ORGANIC CHEMISTRY] $C_6H(OH)_2(HO_2)_3$ An explosive, yellow, crystalline compound, melting at 179-180°C, slightly soluble in water, used in explosives as a priming agent. { 'stif·nik 'as·əd }

styrene [ORGANIC CHEMISTRY] $C_6H_5CH{:}CH_2$ A colorless, toxic liquid with a strong aroma; insoluble in water, soluble in alcohol and ether; polymerizes rapidly, can become explosive; boils at 145°C; used to make polymers and copolymers, polystyrene plastics, and rubbers. Also known as phenylethylene; styrene monomer; vinylbenzene. { 'stī‚rēn }

styrene-acrylonitrile resin [ORGANIC CHEMISTRY] A thermoplastic copolymer of styrene and acrylonitrile with good stiffness and resistance to scratching, chemicals, and stress. Also known as SAN. { 'stī‚rēn ‚ak·rə'län·ə·trəl 'rez·ən }

styrene monomer [ORGANIC CHEMISTRY] *See* styrene. { 'stī‚rēn 'män·ə·mər }

styrene oxide [ORGANIC CHEMISTRY] C_8H_8O A moderately toxic, combustible, colorless or straw-colored liquid miscible in acetone, ether, and benzene, and melts at 195°C; used as a chemical intermediate. { 'stī‚rēn 'äk‚sīd }

styrene plastic [ORGANIC CHEMISTRY] A plastic made by the polymerization of styrene or the copolymerization of styrene with other unsaturated compounds. { 'stī‚rēn 'plas·tik }

subcompound [CHEMISTRY] A compound, generally in the vapor phase, in which an element exhibits a valency lower than that exhibited in its ordinary compounds. { ¦səb'käm‚pau̇nd }

suberic acid [ORGANIC CHEMISTRY] $HOOC(CH_2)_6COOH$ A colorless, crystalline compound that melts at 143°C, and dissolves slightly in cold water; used in organic synthesis. Also known as octanedioic acid. { sü'ber·ik 'as·əd }

sublimatography [ANALYTICAL CHEMISTRY] A procedure of fractional sublimation in which a solid mixture is separated into bands along a condensing tube with a temperature gradient. { ‚səb·lə·mə'täg·rə·fē }

sublimator [CHEMISTRY] Device used for the heating of solids (usually under vacuum) to the temperature at which the solid sublimes. { 'səb·lə‚mād·ər }

subsample [ANALYTICAL CHEMISTRY] A portion taken from a sample of material for which a chemical analysis has been specified. { ‚səb'sam·pəl }

subsolvus [PHYSICAL CHEMISTRY] A range of conditions in which two or more solid phases can form by exsolution from an original homogeneous phase. { ¦səb'säl·vəs }

substituent [ORGANIC CHEMISTRY] An atom or functional group substituted for another in a chemical structure. { səb'stich·ə·wənt }

substitution reaction [CHEMISTRY] Replacement of an atom or radical by another one in a chemical compound. { ˌsəb·stəˈtü·shən rēˌak·shən }

substitutive nomenclature [ORGANIC CHEMISTRY] A system in which the name of a compound is derived by using the functional group (the substituent) as a prefix or suffix to the name of the parent compound to which it is attached; for example, in 2-chloropropane a chlorine atom has replaced a hydrogen atom on the central carbon of the propane chain. { ˈsəb·stəˌtüd·iv ˈnō·mənˌklā·chər }

substrate [ORGANIC CHEMISTRY] A compound with which a reagent reacts. { ˈsəb ˌstrāt }

succinate [ORGANIC CHEMISTRY] A salt or ester of succinic acid; for example, sodium succinate, $Na_2C_4H_4O_4 \cdot 6H_2O$, the reaction product of succinic acid and sodium hydroxide. { ˈsək·səˌnāt }

succinic acid [ORGANIC CHEMISTRY] $CO_2H(CH_2)_2CO_2H$ Water-soluble, colorless crystals with an acid taste; melts at 185°C; used as a chemical intermediate, in medicine, and to make perfume esters. { səkˈsin·ik ˈas·əd }

succinic acid 2,2-dimethylhydrazide [ORGANIC CHEMISTRY] $C_6H_{12}O_3N_2$ White crystals with a melting point of 154-156°C; soluble in water; used as a growth regulator for many crops and ornamentals. Also known as aminocide. { səkˈsin·ik ˈas·əd ¦tü ¦tü dīˌmeth·əlˈhī·drəˌzīd }

succinic anhydride [ORGANIC CHEMISTRY] $C_4H_4O_3$ Colorless or pale needles soluble in alcohol and chloroform; converts to succinic acid in water; melts at 120°C; used as a chemical and pharmaceutical intermediate and a resin hardener. { səkˈsin·ik anˈhīˌdrīd }

succinimide [ORGANIC CHEMISTRY] $C_4H_5O_2N \cdot H_2O$ Colorless or tannish water-soluble crystals with a sweet taste; melts at 126°C; used to make plant growth stimulants and as a chemical intermediate. { səkˈsin·əˌmīd }

succinonitrite [ORGANIC CHEMISTRY] *See* ethylene cyanide. { ¦sək·sə·nōˈnīˌtrāt }

succinylcholine chloride [ORGANIC CHEMISTRY] $[Cl(CH_3)_3N(CH_2)_2OOCH_2]_2 \cdot 2H_2O$ Water-soluble white crystals with a bitter taste, melts at 162°C; used in medicine. { ¦sək·sən·əlˈkōˌlēn ˈklȯrˌīd }

sucrochemical [ORGANIC CHEMISTRY] A chemical made from a feedstock derived from sucrose extracted from sugarcane or sugarbeet. { ˈsü·krōˌkem·i·kəl }

sucrochemistry [ORGANIC CHEMISTRY] A type of chemistry based on sucrose as a starting point. { ¦sü·krōˈkem·i·strē }

sucrose [ORGANIC CHEMISTRY] $C_{12}H_{22}O_{11}$ Combustible, white crystals soluble in water, decomposes at 160 to 186°C; derived from sugarcane or sugarbeet; used as a sweetener in drinks and foods and to make syrups, preserves, and jams. Also known as saccharose; table sugar. { ˈsüˌkrōs }

sucrose octoacetate [ORGANIC CHEMISTRY] $C_{28}H_{38}O_{19}$ A bitter crystalline compound that forms needles from alcohol solution, melts at 89°C, and breaks down at 286°C or above; used as an adhesive, to impregnate and insulate paper, and in lacquers and plastics. { ˈsüˌkrōs ¦äk·tōˈas·əˌtāt }

sugar alcohol [ORGANIC CHEMISTRY] Any of the acyclic linear polyhydric alcohols; may be considered sugars in which the aldehydic group of the first carbon atom is reduced to a primary alcohol; classified according to the number of hydroxyl groups in the molecule; sorbitol (D-glucitol, sorbite) is one of the most widespread of all the naturally occurring sugar alcohols. { ˈshu̇g·ər ˈal·kəˌhȯl }

sugar of lead [ORGANIC CHEMISTRY] *See* lead acetate. { ˈshu̇g·ər əv ˈled }

sulfallate [ORGANIC CHEMISTRY] $C_8H_{14}NS_2Cl$ An oily liquid, used as a preemergence herbicide for vegetable crops and ornamentals. Also known as 2-chloroallyl diethyldithiocarbamate (CDEC). { səlˈfaˌlāt }

sulfamate [CHEMISTRY] A salt of sulfamic acid; for example, calcium sulfamate, $Ca(SO_3NH_2)_2 \cdot 4H_2O$. { ˈsəl·fəˌmāt }

sulfamic acid [INORGANIC CHEMISTRY] HSO_3NH_2 White, nonvolatile crystals slightly soluble in water and organic solvents, decomposes at 205°C; used to clean metals and ceramics, and as a plasticizer, fire retardant, chemical intermediate, and textile and paper bleach. { ¦səl¦fam·ik ˈas·əd }

sulfanilic acid [ORGANIC CHEMISTRY] $C_6H_4NH_2 \cdot SO_3H \cdot H_2O$ Combustible, grayish-white crystals slightly soluble in water, alcohol, and ether, soluble in fuming hydrochloric acid; chars at 280-300°C; used in medicine and dyestuffs and as a chemical intermediate. { ¦səl·fə¦nil·ik 'as·əd }

sulfaquinoxaline [ORGANIC CHEMISTRY] *See* N′-2-quinoxalysulfanilimide. { ¦səl·fə·kwə'näk·sə,lēn }

sulfate [CHEMISTRY] **1.** A compound containing the —SO_4 group, as in sodium sulfate, Na_2SO_4. **2.** A salt of sulfuric acid. { 'səl,fāt }

sulfation [CHEMISTRY] The conversion of a compound into a sulfate by the oxidation of sulfur, as in sodium sulfide, Na_2S, oxidized to sodium sulfate, Na_2SO_4; or the addition of a sulfate group, as in the reaction of sodium and sulfuric acid to form Na_2SO_4. { səl'fā·shən }

sulfenic acid [ORGANIC CHEMISTRY] An oxy acid of sulfur with the general formula RSOH, where R is an alkyl or aryl group such as CH_3, known as the esters and halides. { ¦səl¦fen·ik 'as·əd }

sulfenyl chloride [ORGANIC CHEMISTRY] Any of a group of well-known organosulfur compounds with the general formula RSCl; although highly reactive compounds, they can generally be synthesized and isolated; examples are trichloromethanesulfenyl chloride and 2,4-dinitrobenzenesulfenyl chloride. { ,səl'fen·əl 'klȯr,īd }

sulfhydryl compound [CHEMISTRY] A compound with a —SH group. Also known as a mercapto compound. { ,səlf'hī·drəl 'käm,paůnd }

sulfidation [CHEMISTRY] The chemical insertion of a sulfur atom into a compound. { ,səl·fə'dā·shən }

sulfide [CHEMISTRY] Any compound with one or more sulfur atoms in which the sulfur is connected directly to a carbon, metal, or other nonoxygen atom; for example, sodium sulfide, Na_2S. { 'səl,fīd }

sulfide dye [ORGANIC CHEMISTRY] A dye containing sulfur and soluble in a 0.25-0.50% sodium sulfide solution, and used to dye cotton; the dyes are manufactured from aromatic polyamines or hydroxy amines; the amine group is primary, secondary, or tertiary, or may be an equivalent nitro, nitroso, or imino group; an example is the dye sulfur blue. Also known as sulfur dye. { 'səl,fīd ,dī }

sulfinate [ORGANIC CHEMISTRY] **1.** A compound containing the R_2SX_2 grouping, where X is a halide. **2.** A salt of sulfinic acid having the general formula R · OH · S:O. { ,səl·fə'nāt }

sulfinic acid [ORGANIC CHEMISTRY] Any of the monobasic organic acids of sulfur with the general formula RS:O(OH); for example, ethanesulfinic acid, $C_2H_5SO_2H$. { ¦səl¦fin·ik 'as·əd }

sulfinyl bromide [ORGANIC CHEMISTRY] *See* thionyl bromide. { səl·fə,nil 'brō,mīd }

sulfite [INORGANIC CHEMISTRY] M_2SO_3 A salt of sulfurous acid, for example, sodium sulfite, Na_2SO_3. { 'səl,fīt }

sulfo- [CHEMISTRY] Prefix for a compound with either a divalent sulfur atom, or the presence of —SO_3H, the sulfo group in a compound. Also spelled sulpho-. { 'səl·fō *or* 'səl·fə }

sulfocarbanilide [ORGANIC CHEMISTRY] *See* thiocarbanilide. { ¦səl·fō·kär'ban·ə·ləd }

sulfocarbimide [ORGANIC CHEMISTRY] *See* isothiocyanate. { ¦səl·fō'kär·bə,mīd }

sulfocyanate [INORGANIC CHEMISTRY] *See* thiocyanate. { ¦səl·fō'sī·ə,nāt }

sulfocyanic acid [INORGANIC CHEMISTRY] *See* thiocyanic acid. { ¦səl·fō·sī¦an·ik 'as·əd }

sulfocyanide [INORGANIC CHEMISTRY] *See* thiocyanate. { ¦səl·fō'sī·ə,nīd }

sulfolane [ORGANIC CHEMISTRY] $C_4H_8SO_2$ A liquid with a boiling point of 285°C and outstanding solvent properties; used for extraction of aromatic hydrocarbons, fractionation of fatty acids, and textile finishing, and as a solvent and plasticizer. { 'səl·fə,lān }

sulfonamide [ORGANIC CHEMISTRY] One of a group of organosulfur compounds, RSO_2NH_2, prepared by the reaction of sulfonyl chloride and ammonia; used for sulfa drugs. { ,səl'fän·ə,mīd }

sulfonate [CHEMISTRY] A sulfuric acid derivative or a sulfonic acid ester containing a —SO_3— group. [ORGANIC CHEMISTRY] Any of a group of petroleum hydrocarbons derived from sulfuric-acid treatment of oils, used as synthetic detergents, emulsifying and wetting agents, and chemical intermediates. { 'səl·fə,nāt }

sulfonation [CHEMISTRY] Substitution of —SO_3H groups (from sulfuric acid) for hydrogen atoms, for example, conversion of benzene, C_6H_6, into benzenesulfonic acid, $C_6H_5SO_3H$. { ,səl·fə'nā·shən }

sulfone [ORGANIC CHEMISTRY] R_2SO_2 (or RSOOR) A compound formed by the oxidation of sulfides, for example, ethyl sulfone, $C_4H_{10}SO_2$, from ethyl sulfide, $C_4H_{10}S$; the use of sulfones, particularly 4,4′-sulfonyldianiline (dapsone) in the treatment of leprosy leads to apparent improvement; relapses associated with sulfone-resistant strains have been encountered. { 'səl,fōn }

sulfonic acid [ORGANIC CHEMISTRY] A compound with the radical —SO_2OH, derived by the sulfuric acid replacement of a hydrogen atom; for example, conversion of benzene, C_6H_6, to the water-soluble benzenesulfonic acid, $C_6H_5SO_3H$, by treatment with sulfuric acid; used to make dyes and drugs. { ¦səl¦fän·ik 'as·əd }

sulfonyl [CHEMISTRY] Also known as sulfuryl. **1.** A compound containing the radical —SO_2—. **2.** A prefix denoting the presence of a sulfone group. { 'səl·fə,nil }

sulfonyl chloride [INORGANIC CHEMISTRY] *See* sulfuryl chloride. { 'səl·fə,nil 'klȯr,īd }

sulfosalicylic acid [ORGANIC CHEMISTRY] $C_7H_6O_6S$ A trifunctional aromatic compound whose dihydrate is in the form of white crystals or crystalline powder; soluble in water and alcohol; melting point is 120°C; used as an indicator for albumin in urine and as a reagent for the determination of ferric ion; it also has industrial uses. { ¦səl·fō¦sal·ə¦sil·ik 'as·əd }

sulfoxide [ORGANIC CHEMISTRY] R_2SO A compound with the group =SO; derived from oxidation of sulfides, the proportion of oxidant, such as hydrogen peroxide, and temperature being set to avoid excessive oxidation; an example is dimethyl sulfoxide, $(CH_3)_2SO$. { ,səl'fäk,sīd }

sulfur [CHEMISTRY] A nonmetallic element in group VI, symbol S, atomic number 16, atomic weight 32.064, existing in a crystalline or amorphous form and in four stable isotopes; used as a chemical intermediate and fungicide, and in rubber vulcanization. { 'səl·fər }

sulfurated lime [INORGANIC CHEMISTRY] *See* calcium sulfide. { ,səl·fə'rād·əd 'līm }

sulfuration [CHEMISTRY] The chemical act of combining an element or compound with sulfur. { ,səl·fə'rā·shən }

sulfur bichloride [INORGANIC CHEMISTRY] *See* sulfur dichloride. { 'səl·fər bī'klȯr,īd }

sulfur bromide [INORGANIC CHEMISTRY] S_2Br_2 A toxic, irritating, yellow liquid that reddens in air, soluble in carbon disulfide, decomposes in water, boils at 54°C. Also known as sulfur monobromide. { 'səl·fər 'brō,mīd }

sulfur chloride [INORGANIC CHEMISTRY] S_2Cl_2 A combustible, water-soluble, oily, fuming, amber to yellow-red liquid with an irritating effect on the eyes and lungs, boils at 138°C; used to make military gas and insecticides, in rubber substitutes and cements, to purify sugar juices, and as a chemical intermediate. Also known as sulfur subchloride. { 'səl·fər 'klȯr,īd }

sulfur dichloride [INORGANIC CHEMISTRY] SCl_2 A red-brown liquid boiling (when heated rapidly) at 60°C, decomposes in water; used to make insecticides, for rubber vulcanization, and as a chemical intermediate and a solvent. Also known as sulfur bichloride. { 'səl·fər dī'klȯr,īd }

sulfur dioxide [INORGANIC CHEMISTRY] SO_2 A toxic, irritating, colorless gas soluble in water, alcohol, and ether; boils at −10°C; used as a chemical intermediate, in artificial ice, paper pulping, and ore refining, and as a solvent. Also known as sulfurous acid anhydride. { 'səl·fər dī'äk,sīd }

sulfur dye [ORGANIC CHEMISTRY] *See* sulfide dye. { 'səl·fər ,dī }

sulfur hexafluoride [INORGANIC CHEMISTRY] SF_6 A colorless gas soluble in alcohol and ether, slightly soluble in water, sublimes at −64°C; used as a dielectric in electronics. { 'səl·fər ¦hek·sə'flu̇r,īd }

sulfuric acid [INORGANIC CHEMISTRY] H_2SO_4 A toxic, corrosive, strongly acid, colorless liquid that is miscible with water and dissolves most metals, and melts at 10°C; used in industry in the manufacture of chemicals, fertilizers, and explosives, and in petroleum refining. Also known as dipping acid; oil of vitriol, vitriolic acid. { ¦səl¦fyůr·ik 'as·əd }

sulfuric chloride [INORGANIC CHEMISTRY] *See* sulfuryl chloride. { ¦səl¦fyůr·ik 'klȯr,īd }

sulfur iodide [INORGANIC CHEMISTRY] *See* sulfur iodine. { ¦səl¦fyůr·ik 'ī·ə,dīd }

sulfur iodine [INORGANIC CHEMISTRY] I_2S_2 A gray-black brittle mass with an iodine aroma and a metallic luster, insoluble in water, soluble in carbon disulfide; used in medicine. Also known as iodine bisulfide; iodine disulfide; sulfur iodide. { ¦səl¦fyůr·ik 'ī·ə,dīn }

sulfur monobromide [INORGANIC CHEMISTRY] *See* sulfur bromide. { 'səl·fər ¦män·ə'brō,mīd }

sulfur monoxide [INORGANIC CHEMISTRY] SO A gas at ordinary temperatures; produces an orange-red deposit when cooled to temperatures of liquid air; prepared by passing an electric discharge through a mixture of sulfur vapor and sulfur dioxide at low temperature. { 'səl·fər mə'näk,sīd }

sulfur number [ANALYTICAL CHEMISTRY] The number of milligrams of sulfur per 100 milliliters of sample, determined by electrometric titration; used in the petroleum industry for oils. { 'səl·fər ,nəm·bər }

sulfurous acid [INORGANIC CHEMISTRY] H_2SO_3 An unstable, water-soluble, colorless liquid with a strong sulfur aroma; derived from absorption of sulfur dioxide in water; used in the synthesis of medicine and chemicals, manufacture of paper and wine, brewing, metallurgy, and ore flotation, as a bleach and analytic reagent, and to refine petroleum products. { 'səl·fə·rəs 'as·əd }

sulfurous acid anhydride [INORGANIC CHEMISTRY] *See* sulfur dioxide. { 'səl·fə·rəs 'as·əd an'hī,drīd }

sulfurous oxychloride [INORGANIC CHEMISTRY] *See* thionyl chloride. { 'səl·fə·rəs ¦äk·sē'klȯr,īd }

sulfur oxide [INORGANIC CHEMISTRY] An oxide of sulfur, such as sulfur dioxide, SO_2, and sulfur trioxide, SO_3. { 'səl·fər 'äk,sīd }

sulfur oxychloride [INORGANIC CHEMISTRY] *See* thionyl chloride. { 'səl·fər ¦äk·sē'klȯr,īd }

sulfur subchloride [INORGANIC CHEMISTRY] *See* sulfur chloride. { 'səl·fər ¦səb 'klȯr,īd }

sulfur test [ANALYTICAL CHEMISTRY] **1.** Method to determine the sulfur content of a petroleum material by combustion in a bomb. **2.** Analysis of sulfur in petroleum products by lamp combustion in which combustion of the sample is controlled by varying the flow of carbon dioxide and oxygen to the burner. { 'səl·fər ,test }

sulfur trioxide [INORGANIC CHEMISTRY] SO_3 A toxic, irritating liquid in three forms, α, β, γ, with respective melting points of 62°C, 33°C, and 17°C; a strong oxidizing agent and fire hazard; used for sulfonation of organic chemicals. { 'səl·fər trī'äk,sīd }

sulfuryl [CHEMISTRY] *See* sulfonyl. { 'səl·fə,ril }

sulfuryl chloride [INORGANIC CHEMISTRY] SO_2Cl_2 A colorless liquid with a pungent aroma, boils at 69°C, decomposed by hot water and alkalies; used as a chlorinating agent and solvent and for pharmaceuticals, dyestuffs, rayon, and poison gas. Also known as sulfonyl chloride; sulfuric chloride. { 'səl·fə,ril 'klȯr,īd }

sulfuryl fluoride [INORGANIC CHEMISTRY] SO_2F_2 A colorless gas with a melting point of −136.7°C and a boiling point of 55.4°C; used as an insecticide and fumigant. { 'səl·fə,ril 'flůr,īd }

Sullivan reaction [ORGANIC CHEMISTRY] The formation of a red-brown color when cysteine is reacted with 1,2-naphthoquinone-4-sodium sulfate in a highly alkaline reducing medium. { 'səl·ə·vən rē,ak·shən }

sulpho- [CHEMISTRY] *See* sulfo-. { 'səl·fō }

superacid [CHEMISTRY] **1.** An acidic medium that has a proton-donating ability equal to or greater than 100% sulfuric acid. **2.** A solution of acetic or phosphoric acid. { ¦sü·pər'as·əd }

supercritical-fluid chromatography [ANALYTICAL CHEMISTRY] Any chemical separation technique using chromatography in which a supercritical fluid is used as the mobile phase. { ¦sü·pər,krid·ə·kəl ¦flü·əd ,krō·mə'täg·rə·fē }

superheavy element [INORGANIC CHEMISTRY] A chemical element with an atomic number of 110 or greater. { ¦sü·pər'hev·ē 'el·ə·mənt }
supermolecule [PHYSICAL CHEMISTRY] A single quantum-mechanical entity presumably formed by two reacting molecules and in existence only during the collision process; a concept in the hard-sphere collision theory of chemical kinetics. { ¦sü·pər'mäl·ə‚kyül }
superoxide ion [CHEMISTRY] O_2^- An ion formed by the combination of one molecule of dioxygen (O_2) and one electron (e^-). { ‚sü·pər'äk‚sīd ‚ī·ən }
supersaturation [PHYSICAL CHEMISTRY] The condition existing in a solution when it contains more solute than is needed to cause saturation. Also known as supersolubility. { ¦sü·pər‚sach·ə'rā·shən }
supersolubility [PHYSICAL CHEMISTRY] *See* supersaturation. { ¦sü·pər‚säl·yə'bil·əd·ē }
supertransuranics [INORGANIC CHEMISTRY] A group of relatively stable elements, with atomic numbers around 114 and mass numbers around 298, that are predicted to exist beyond the present periodic table of known elements. { ¦sü·pər‚tranz·yů'ran·iks }
support-coated capillary column [ANALYTICAL CHEMISTRY] A capillary column that utilizes a fine-granular solid support to disperse the stationary liquid. { sə'pòrt ¦kōd·əd 'kap·ə‚ler·ē ‚käl·əm }
suppressor [SPECTROSCOPY] In an analytical procedure, a substance added to the analyte to reduce the extraneous emission, absorption, or light scattering caused by the presence of an impurity. { sə'pres·ər }
surface chemistry [PHYSICAL CHEMISTRY] The study and measurement of the forces and processes that act on the surfaces of fluids (gases and liquids) and solids, or at an interface separating two phases; for example, surface tension. { 'sər·fəs ‚kem·ə·strē }
surface orientation [PHYSICAL CHEMISTRY] Arrangement of molecules on the surface of a liquid with one part of the molecule turned toward the liquid. { 'sər·fəs ‚òr·ē·ən'tā·shən }
surface reaction [CHEMISTRY] A chemical reaction carried out on a surface as on an adsorbent or solid catalyst. { 'sər·fəs rē‚ak·shən }
suspended solids [CHEMISTRY] *See* suspension. { sə'spen·dəd 'säl·ədz }
suspension [CHEMISTRY] A mixture of fine, nonsettling particles of any solid within a liquid or gas, the particles being the dispersed phase, while the suspending medium is the continuous phase. Also known as suspended solids. { sə'spen·shən }
svedberg [PHYSICAL CHEMISTRY] A unit of sedimentation coefficient, equal to 10^{-13} second. { 'sfed‚bərg }
Swarts reaction [ORGANIC CHEMISTRY] The reaction of chlorinated hydrocarbons with metallic fluorides to form chlorofluorohydrocarbons, such as CCl_2F_2, which is quite inert and nontoxic. { 'svärts rē‚ak·shən }
sweat [CHEMISTRY] Exudation of nitroglycerin from dynamite due to separation of nitroglycerin from its adsorbent. { swet }
sweet spirits of niter [ORGANIC CHEMISTRY] *See* ethyl nitrite. { 'swēt 'spir·əts əv 'nī·tər }
swep [ORGANIC CHEMISTRY] $C_8H_7Cl_2NO_2$ A white, crystalline compound with a melting point of 112-114°C; insoluble in water; used as a pre- and postemergence herbicide for rice, carrots, potatoes, and cotton. Also known as methyl-N-(3,4-dichlorophenyl)carbamate. { swep }
SXAPS [SPECTROSCOPY] *See* soft x-ray appearance potential spectroscopy.
sym- [ORGANIC CHEMISTRY] A chemical prefix; denotes structure of a compound in which substituents are symmetrical with respect to a functional group or to the carbon skeleton. { sim }
symbol [CHEMISTRY] Letter or combination of letters and numbers that represent various conditions or properties of an element, for example, a normal atom, O (oxygen); with its atomic weight, ^{16}O; its atomic number, $_8{}^{16}O$; as a molecule, O_2; as an ion, O^{2+}; in excited state, O*; or as an isotope, ^{18}O. { 'sim·bəl }
symclosene [ORGANIC CHEMISTRY] *See* trichloroisocyanuric acid. { 'sim·klə‚zēn }

symmetric top molecule [PHYSICAL CHEMISTRY] A nonlinear molecule which has one and only one axis of threefold or higher symmetry. { sə'me·trik ¦täp 'mäl·ə,kyül }

symmetry number [PHYSICAL CHEMISTRY] The number of indistinguishable orientations that a molecule can exhibit by being rotated around symmetry axes. { 'sim·ə,trē ,nəm·bər }

syndiotactic polymer [ORGANIC CHEMISTRY] A vinyl polymer in which the side chains alternate regularly above and below the plane of the backbone. { ¦sin·dē·ə¦tak·tik 'päl·i·mər }

syneresis [CHEMISTRY] Spontaneous separation of a liquid from a gel or colloidal suspension due to contraction of the gel. { sə'ner·ə·səs }

synthesis [CHEMISTRY] Any process or reaction for building up a complex compound by the union of simpler compounds or elements. { 'sin·thə·səs }

synthetic resin [ORGANIC CHEMISTRY] Amorphous, organic, semisolid, or solid material derived from the polymerization of unsaturated monomers such as ethylene, butylene, propylene, and styrene. { sin'thed·ik 'rez·ən }

systematic nomenclature [CHEMISTRY] A system for naming chemical compounds according to a specific set of rules, usually those developed by the International Union of Pure and Applied Chemistry. { ¦sis·tə,mad·ik 'nō·mən,klā·chər }

T

2,4,5-T [ORGANIC CHEMISTRY] *See* 2,4,5-trichlorophenoxyacetic acid.

2,4,6-T [ORGANIC CHEMISTRY] *See* trichlorophenol.

Tl [CHEMISTRY] *See* thallium.

Ta [CHEMISTRY] *See* tantalum.

table salt [INORGANIC CHEMISTRY] *See* sodium chloride. { 'tā·bəl ˌsȯlt }

table sugar [ORGANIC CHEMISTRY] *See* sucrose. { 'tā·bəl ˌshu̇g·ər }

tabun [ORGANIC CHEMISTRY] $(CH_3)_2NP(O)(C_2H_5O)(CN)$ A toxic liquid with a boiling point of 240°C; soluble in organic solvents; used as a nerve gas. { 'täˌbu̇n }

tachiol [INORGANIC CHEMISTRY] *See* silver fluoride. { 'tak·ēˌȯl }

tactic polymer [ORGANIC CHEMISTRY] A polymer with regularity or symmetry in the structural arrangement of its molecules, as in a stereospecific polymer such as some types of polypropylene. { 'tak·tə·kəl 'päl·i·mər }

Tag closed-cup tester [ANALYTICAL CHEMISTRY] A laboratory device used to determine the flash point of mobile petroleum liquids flashing below 175°F (79.4°C). Also known as Tagliabue closed tester. { 'tag ¦klōzd ¦kəp 'tes·tər }

tagged molecule [CHEMISTRY] A molecule having one or more atoms which are either radioactive or have a mass which differs from that of the atoms which normally make up the molecule. { 'tagd 'mäl·əˌkyül }

Tagliabue closed tester [ANALYTICAL CHEMISTRY] *See* Tag closed-cup tester. { ˌtäl·yə'bü·ē 'klōsd 'tes·tər }

tannic acid [ORGANIC CHEMISTRY] **1.** $C_{14}H_{10}O_9$ A yellowish powder with an astringent taste, soluble in water and alcohol, insoluble in acetone and ether; derived from nutgalls; decomposes at 210°C; used as an alcohol denaturant and a chemical intermediate, and in tanning and textiles. Also known as digallic acid; gallotannic acid; gallotannin; tannin. **2.** $C_{76}H_{52}O_{46}$ Yellowish-white to light-brown amorphous powder or flakes, decomposes at 210–215°C; very soluble in alcohol and acetone; used as a mordant in dyeing, in photography, as a reagent, and in clarifying wine or beer. Also known as pentadigalloylglucose. { 'tan·ik 'as·əd }

tannin [ORGANIC CHEMISTRY] *See* tannic acid. { 'tan·ən }

tantalic acid anhydride [INORGANIC CHEMISTRY] *See* tantalum oxide. { tan'tal·ik 'as·əd an'hīˌdrīd }

tantalic chloride [INORGANIC CHEMISTRY] *See* tantalum chloride. { tan'tal·ik 'klȯrˌīd }

tantalum [CHEMISTRY] A metallic transition element, symbol Ta, atomic number 73, atomic weight 180.948; black powder or steel-blue solid soluble in fused alkalies, insoluble in acids (except hydrofluoric and fuming sulfuric); melts about 3000°C. { 'tant·əl·əm }

tantalum carbide [INORGANIC CHEMISTRY] TaC Hard, chemical-resistant crystals melting at 3875°C; used in cutting tools and dies. { 'tant·əl·əm 'kärˌbīd }

tantalum chloride [INORGANIC CHEMISTRY] $TaCl_5$ A highly reactive, pale-yellow powder decomposing in moist air; soluble in alcohol and potassium hydroxide; melts at 221°C; used to produce tantalum and as a chemical intermediate. Also known as tantalic chloride; tantalum pentachloride. { 'tant·əl·əm 'klȯrˌīd }

tantalum nitride [INORGANIC CHEMISTRY] TaN A very hard, black, water-insoluble solid, melting at 3360°C. { 'tant·əl·əm 'nīˌtrīd }

tantalum oxide [INORGANIC CHEMISTRY] Ta_2O_5 Prisms insoluble in water and acids (except for hydrofluoric); melts at 1800°C; used to make tantalum, in optical glass and electronic equipment, and as a chemical intermediate. Also known as tantalic acid anhydride; tantalum pentoxide. { 'tant·əl·əm 'äk,sīd }

tantalum pentachloride [INORGANIC CHEMISTRY] *See* tantalum chloride. { 'tant·əl·əm ¦pen·tə'klör,īd }

tantalum pentoxide [INORGANIC CHEMISTRY] *See* tantalum oxide. { 'tant·əl·əm pen'täk,sīd }

tar base [CHEMISTRY] A basic nitrogen compound found in coal tar, for example, pyridine and quinoline. { 'tär ,bās }

tar camphor [ORGANIC CHEMISTRY] *See* naphthalene. { 'tär ,kam·fər }

tartar emetic [ORGANIC CHEMISTRY] $K(SbO)C_4H_4O_6 \cdot {}^1/_2H_2O$ A transparent crystalline compound, soluble in water; used to attract and kill moths, wasps, and yellow jackets. Also known as antimony potassium tartrate; potassium antimonyl tartrate. { 'tärd·ər i'med·ik }

tartaric acid [ORGANIC CHEMISTRY] $HOOC(CHOH)_2COOH$ Water- and alcohol-soluble colorless crystals with an acid taste, melts at 170°C; used as a chemical intermediate and a sequestrant and in tanning, effervescent beverages, baking power, ceramics, photography, textile processing, mirror silvering, and metal coloring. { tär'tar·ik 'as·əd }

tartrate [ORGANIC CHEMISTRY] A salt or ester of tartaric acid, for example, sodium tartrate, $Na_2C_4H_4O_6$. { 'tär,trāt }

tartrazine [ORGANIC CHEMISTRY] $C_{16}H_9N_4O_9S_2$ A bright orange-yellow, water-soluble powder, used as a food, drug, and cosmetic dye. { 'tär·trə,zēn }

Tauber test [ANALYTICAL CHEMISTRY] A color test for identification of pentose sugars; the sugars produce a cherry-red color when heated with a solution of benzidine in glacial acetic acid. { 'taů·bər ,test }

taurine [ORGANIC CHEMISTRY] $NH_2CH_2CH_2SO_3H$ A crystalline compound that decomposes at about 300°C; present in bile combined with cholic acid. { 'tö,rēn }

tautomerism [CHEMISTRY] The reversible interconversion of structural isomers of organic chemical compounds; such interconversions usually involve transfer of a proton. { tö'täm·ə,riz·əm }

Tb [CHEMISTRY] *See* terbium.

TBH [ORGANIC CHEMISTRY] *See* 1,2,3,4,5,6-hexachlorocyclohexane.

TBP [ORGANIC CHEMISTRY] *See* tributyl phosphate.

TBT [ORGANIC CHEMISTRY] *See* tetrabutyl titanate.

Tc [CHEMISTRY] *See* technetium.

TCA [ORGANIC CHEMISTRY] *See* trichloroacetic acid.

TCP [ORGANIC CHEMISTRY] *See* tricresyl phosphate.

TDE [ORGANIC CHEMISTRY] *See* 2,2-bis(*para*-chlorophenyl)-1,1-dichloroethane.

Te [CHEMISTRY] *See* tellurium.

TEA chloride [ORGANIC CHEMISTRY] *See* tetraethylammonium chloride. { ,tē,ē'ā 'klör,īd }

technetium [CHEMISTRY] A transition element, symbol Tc, atomic number 43; derived from uranium and plutonium fission products; chemically similar to rhenium and manganese; isotope ^{99}Tc has a half-life of 200,000 years; used to absorb slow neutrons in reactor technology. { tek'nē·shē·əm }

tectoquinone [ORGANIC CHEMISTRY] $C_{15}H_{10}O_2$ A white compound with needlelike crystals; sublimes at 177°C; insoluble in water; used as an insecticide to treat wood. Also known as 2-methyl anthraquinone. { ¦tek·tō·kwə'nōn }

TEG [ORGANIC CHEMISTRY] *See* tetraethylene glycol; triethylene glycol.

TEL [ORGANIC CHEMISTRY] *See* tetraethyllead.

Teller-Redlich rule [PHYSICAL CHEMISTRY] For two isotopic molecules, the product of the frequency ratio values of all vibrations of a given symmetry type depends only on the geometrical structure of the molecule and the masses of the atoms, and not on the potential constants. { 'tel·ər 'red·lik ,rül }

telluric acid [INORGANIC CHEMISTRY] H_6TeO_6 Toxic white crystals, slightly soluble in cold water, soluble in hot water and alkalies; melts at 136°C; used as an analytical reagent. Also known as hydrogen tellurate. { tə'lůr·ik 'as·əd }

telluric line [SPECTROSCOPY] Any of the spectral bands and lines in the spectrum of the sun and stars produced by the absorption of their light in the atmosphere of the earth. { tə'lůr·ik 'līn }

tellurinic acid [ORGANIC CHEMISTRY] A compound of tellurium with the general formula R_2TeOOH; an example is methanetellurinic acid, C_6H_5TeOOH. { ¦tel·yə¦rin·ik 'as·əd }

tellurium [CHEMISTRY] A member of group VI, symbol Te, atomic number 52, atomic weight 127.60; dark-gray crystals, insoluble in water, soluble in nitric and sulfuric acids and potassium hydroxide; melts at 452°C, boils at 1390°C; used in alloys (with lead or steel), glass, and ceramics. { tə'lůr·ē·əm }

tellurium dibromide [INORGANIC CHEMISTRY] $TeBr_2$ Toxic, hygroscopic, green- or gray-black crystals with violet vapor, soluble in ether, decomposes in water, and melts at 210°C. { tə'lůr·ē·əm dī'brō,mīd }

tellurium dichloride [INORGANIC CHEMISTRY] $TeCl_2$ A toxic, amorphous, black or green-yellow powder decomposing in water, melting at 209°C. { tə'lůr·ē·əm dī'klȯr,īd }

tellurium dioxide [INORGANIC CHEMISTRY] TeO_2 The most stable oxide of tellurium, formed when tellurium is burned in oxygen or air or by oxidation of tellurium with cold nitric acid; crystallizes as colorless, tetragonal, hexagonlike crystals that melt at 452°C. { tə'lůr·ē·əm dī'äk,sīd }

tellurium disulfide [INORGANIC CHEMISTRY] TeS_2 A toxic, red powder, insoluble in water and acids. Also known as tellurium sulfide. { tə'lůr·ē·əm dī'səl,fīd }

tellurium hexafluoride [INORGANIC CHEMISTRY] TeF_6 A colorless gas which is formed from the elements tellurium and fluorine; it is slowly hydrolyzed by water. { tə'lůr·ē·əm ¦hek·sə'flůr,īd }

tellurium monoxide [INORGANIC CHEMISTRY] TeO A black, amorphous powder, stable in cold dry air; formed by heating the mixed oxide $TeSO_3$. { tə'lůr·ē·əm mə'näk,sīd }

tellurium sulfide [INORGANIC CHEMISTRY] *See* tellurium disulfide. { tə'lůr·ē·əm 'səl,fīd }

tellurokotone [ORGANIC CHEMISTRY] One of a group of compounds with the general formula R_2CTe. { ¦tel·yə·rō'kē,tōn }

telluromercaptan [ORGANIC CHEMISTRY] One of a group of compounds with the general formula RTeH. { ¦tel·yə·rō·mər'kap,tan }

tellurous acid [INORGANIC CHEMISTRY] H_2TeO_3 Toxic, white crystals, soluble in alkalies and acids, slightly soluble in water and alcohol; decomposes at 40°C. { 'tel·yə·rəs 'as·əd }

telvar [ORGANIC CHEMISTRY] The common name for the herbicide 3-(*para*-chlorophenyl)-1,1-dimethylurea; used as a soil sterilant. { 'tel,vär }

TEM [ORGANIC CHEMISTRY] *See* triethylenemelamine.

temporary hardness [CHEMISTRY] The portion of the total hardness of water that can be removed by boiling whereby the soluble calcium and magnesium bicarbonate are precipitated as insoluble carbonates. { 'tem·pə,rer·ē 'härd·nəs }

TEP [ORGANIC CHEMISTRY] *See* triethyl phosphate.

terbacil [ORGANIC CHEMISTRY] $C_9H_{13}ClN_2O_2$ A colorless, crystalline compound with a melting point of 175–177°C; used as an herbicide to control weeds in sugarcane, apples, peaches, citrus, and mints. { 'tər·bə,sil }

terbia [INORGANIC CHEMISTRY] *See* terbium oxide. { tər·bē·ə }

terbium [CHEMISTRY] A rare-earth element, symbol Tb, in the yttrium subgroup of the transition elements, atomic number 65, atomic weight 158.924. { 'tər·bē·əm }

terbium chloride [INORGANIC CHEMISTRY] $TbCl_3 \cdot 6H_2O$ Water- and alcohol-soluble, hygroscopic, colorless, transparent prisms; anhydrous form melts at 588°C. { 'tər·bē·əm 'klȯr,īd }

terbium nitrate [INORGANIC CHEMISTRY] $Tb(NO_3)_3 \cdot 6H_2O$ A colorless, fire-hazardous (strong oxidant) powder, soluble in water; melts at 89°C. { 'tər·bē·əm 'nī,trāt }

terbium oxide [INORGANIC CHEMISTRY] Tb_2O_3 A slightly hygroscopic, dark-brown powder soluble in dilute acids, absorbs carbon dioxide from air. Also known as terbia. { 'tər·bē·əm 'äk‚sīd }

terbutol [ORGANIC CHEMISTRY] The common name for the herbicide 2,6-di-*tert*-butyl-*p*-tolylmethylcarbamate; used as a selective preemergence crabgrass herbicide for turf. { 'tər·byə‚tȯl }

terbutryn [ORGANIC CHEMISTRY] $C_{13}H_{19}N_5S$ A colorless powder with a melting point of 104-105°C; used for weed control for wheat, barley, and grain sorghum. { tər'byü·trən }

terbutylhylazine [ORGANIC CHEMISTRY] $C_9H_{16}N_5Cl$ A white solid with a melting point of 177-179°C; used as a preemergence herbicide. { ‚tər‚byüd·əl'hī·lə‚zēn }

terephthalic acid [ORGANIC CHEMISTRY] $C_6H_4(COOH)_2$ A combustible white powder, insoluble in water, soluble in alkalies, sublimes above 300°C; used to make polyester resins for fibers and films and as an analytical reagent and poultry-feed additive. Also known as *para*-phthalic acid; TPA. { ¦ter·əf¦thal·ik 'as·əd }

terephthaloyl chloride [ORGANIC CHEMISTRY] $C_6H_4(COCl)_2$ Colorless needles with a melting point of 82-84°C; soluble in ether; used in the manufacture of dyes, synthetic fibers, resins, and pharmaceuticals. { ‚ter·əf'thal·ə‚wil 'klȯr‚īd }

term [SPECTROSCOPY] A set of (2S+1)(2L+1) atomic states belonging to a definite configuration and to definite spin and orbital angular momentum quantum numbers S and L. { tərm }

termination [CHEMISTRY] The steps that end a chain reaction by destroying or rendering inactive the reactive intermediates. { ‚tər·mə'nā·shən }

termination step [CHEMISTRY] In a chain reaction, the mechanism that halts the reaction. { ‚tər·mə'nā·shən ‚steps }

ternary compound [CHEMISTRY] A molecule consisting of three different types of atoms; for example, sulfuric acid, H_2SO_4. { 'tər·nə·rē 'käm‚pau̇nd }

ternary system [CHEMISTRY] Any system with three nonreactive components; in liquid systems, the components may or may not be partially soluble. { 'tər·nə·rē 'sis·təm }

terpene [ORGANIC CHEMISTRY] **1.** $C_{10}H_{16}$ A moderately toxic, flammable, unsaturated hydrocarbon liquid found in essential oils and plant oleoresins; used as an intermediate for camphor, menthol, and terpineol. **2.** A class of naturally occurring compounds whose carbon skeletons are composed exclusively of isopentyl (isoprene) C_5 units. Also known as isoprenoid. { 'tər‚pēn }

terpene alcohol [ORGANIC CHEMISTRY] A generic name for an alcohol related to or derived from a terpene hydrocarbon, such as terpineol or borneol. { 'tər‚pēn 'al·kə‚hȯl }

terpene hydrochloride [ORGANIC CHEMISTRY] $C_{10}H_{16} \cdot HCl$ A solid, water-insoluble material melting at 125°C; used as an antiseptic. Also known as artificial camphor; dipentene hydrochloride; pinene hydrochloride; turpentine camphor. { 'tər‚pēn ¦hī·drə'klȯr‚īd }

terpenoid [ORGANIC CHEMISTRY] Any compound with an isoprenoid structure similar to that of the terpene hydrocarbons. { 'tər·pə‚nȯid }

***para*-terphenyl** [ORGANIC CHEMISTRY] $(C_6H_5)_2C_6H_4$ A combustible, toxic liquid boiling at 405°C; crystals are used for scintillation counters; polymerized with styrene to make plastic phosphor. { ¦par·ə ¦tər'fen·əl }

terpineol [ORGANIC CHEMISTRY] $C_{10}H_{17}OH$ A combustible, colorless liquid with a lilac scent, derived from pine oil, soluble in alcohol, slightly soluble in water, boils at 214-224°C; used in medicine, perfumes, soaps, and disinfectants, and as an antioxidant, a flavoring agent, and a solvent; isomeric forms are alpha-, beta- and gamma-. { tər'pin·ē‚ȯl }

terpin hydrate [ORGANIC CHEMISTRY] $CH_3(OH)C_6H_9C(CH_3)_2OH \cdot H_2O$ Combustible, efflorescent, lustrous white prisms soluble in alcohol and ether, slightly soluble in water; melts at 116°C; used for pharmaceuticals and to make terpineol. Also known as dipentene glycol. { 'tər·pən 'hī‚drāt }

terpinolene [ORGANIC CHEMISTRY] $C_{10}H_{16}$ A flammable, water-white liquid insoluble in water, soluble in alcohol, ether, and glycols, boils at 184°C; used as a solvent and as a chemical intermediate for resins and essential oils. { tər'pin·ə‚lēn }

terpinyl acetate [ORGANIC CHEMISTRY] $C_{10}H_{17}OOCCH_3$ A combustible, colorless, liquid slightly soluble in water and glycerol, soluble in water, boils at 220°C; used in perfumes and flavors. { 'tər·pən·əl 'as·ə,tāt }

terpolymer [ORGANIC CHEMISTRY] A polymer that contains three distinct monomers; for example, acrylonitrile-butadiene-styrene terpolymer, ABS. { ¦tər'päl·i·mər }

terrachlor [ORGANIC CHEMISTRY] *See* pentachloronitrobenzene. { 'ter·ə,klȯr }

tert- [ORGANIC CHEMISTRY] Abbreviation for tertiary; trisubstituted methyl radical with the central carbon attached to three other carbons ($R_1R_2R_3C$—). { tərt }

tertiary alcohol [ORGANIC CHEMISTRY] A trisubstituted alcohol in which the hydroxyl group is attached to a carbon that is joined to three carbons; for example, *tert*-butyl alcohol. { 'tər·shē ,er·ē 'al·kə,hȯl }

tertiary amine [ORGANIC CHEMISTRY] R_3N A trisubstituted amine in which the hydroxyl group is attached to a carbon that is joined to three carbons; for example, trimethylamine, $(CH_3)_3N$. { 'tər·shē,er·ē 'am,ēn }

tertiary carbon atom [ORGANIC CHEMISTRY] A carbon atom bonded to three other carbon atoms with single bonds. { 'tər·shē,er·ē 'kär·bən 'ad·əm }

tertiary hydrogen atom [ORGANIC CHEMISTRY] A hydrogen atom that is bonded to a tertiary carbon atom. { 'tər·shē,er·ē 'hī·drə·jən 'ad·əm }

tertiary sodium phosphate [INORGANIC CHEMISTRY] *See* trisodium phosphate. { 'tər·shē,er·ē 'sōd·ē·əm 'fä,sfāt }

tetraamylbenzene [ORGANIC CHEMISTRY] $(C_5H_{11})_4C_6H_2$ A colorless liquid with a boiling range of 320-350°C; used as a solvent. { ¦te·trə¦am·əl'ben,zēn }

tetrabromobisphenol A [ORGANIC CHEMISTRY] $(C_6H_2Br_2OH)_2C(CH_3)_2$ An off-white powder with a melting point of 180-184°C, soluble in methanol and ether; used as a flame retardant for plastics, paper, and textiles. { ¦te·trə¦brō·mō,bis'fē,nȯl 'ā }

tetrabromophthalic anhydride [ORGANIC CHEMISTRY] $C_6Br_4C_2O_3$ A pale yellow, crystalline solid with a melting point of 280°C; used as a flame retardant for paper, plastics, and textiles. { ¦te·trə,brōm·äf'thal·ik an'hī,drīd }

tetrabutylthiuram monosulfide [ORGANIC CHEMISTRY] $[(C_4H_9)_2NCS]_2S$ A brown liquid, soluble in acetone, benzene, gasoline, and ethylene dichloride; used as a rubber accelerator. { ¦te·trə¦byüd·əl'thī·yə,ram ¦män·ə'səl,fīd }

tetrabutyltin [ORGANIC CHEMISTRY] $(C_4H_9)_4Sn$ A colorless or slightly yellow, oily liquid with a boiling point of 145°C; soluble in most organic solvents; used as a stabilizing agent and rust inhibitor for silicones, and as a lubricant and fuel additive. { ,te·trə'byüd·əl·tən }

tetrabutyl titanate [ORGANIC CHEMISTRY] $Ti(OC_4H_9)_4$ A combustible, colorless to yellowish liquid soluble in many solvents, boils at 312°C, decomposes in water; used in paints, surface coatings, and heat-resistant paints. Abbreviated TBT. { ¦te·trə¦byüd·əl 'tīt ən,āl }

tetrabutyl urea [ORGANIC CHEMISTRY] $(C_4H_9)_4N_2CO$ A liquid with a boiling point of 305°C; used as a plasticizer. { te·trə¦byüd·əl yu'rē·ə }

tetracaine hydrochloride [ORGANIC CHEMISTRY] $C_{15}H_{24}O_2N_2 \cdot HCl$ Bitter-tasting, water-soluble crystals melting at 148°C; used as a local anesthetic. { 'te·trə,kān ,hī·drə'klȯr,īd }

tetracene [ORGANIC CHEMISTRY] *See* naphthacene. { 'te·trə,sēn }

tetrachlorobenzene [ORGANIC CHEMISTRY] $C_6H_2Cl_4$ Water-insoluble, combustible white crystals that appear in two forms: 1,2,3,4-tetrachlorobenzene which melts at 47°C and is used in chemical synthesis and in dielectric fluids; and 1,2,4,5-tetrachlorobenzene which melts at 138°C and is used to make herbicides, defoliants, and electrical insulation. { ¦te·trə¦klȯr·ə'ben,zēn }

***sym*-tetrachlorodifluoroethane** [ORGANIC CHEMISTRY] CCl_2FCCl_2F A white, toxic liquid with a camphor aroma, soluble in alcohol, insoluble in water, boils at 93°C; used for metal degreasing. { ¦sim ¦te·trə¦klȯr ə·dī¦flůr·ō'eth,ān }

***sym*-tetrachloroethane** [ORGANIC CHEMISTRY] $CHCl_2CHCl_2$ A colorless, corrosive, toxic liquid with a chloroform scent, soluble in alcohol and ether, slightly soluble in water, boils at 147°C; used as a solvent, metal cleaner, paint remover, and weed killer. { ¦sim ¦te·trə¦klȯr·ō'eth,ān }

tetrachloroethylene [ORGANIC CHEMISTRY] *See* perchloroethylene. { ¦te·trə¦klȯr·ō'eth·ə‚lēn }

tetrachlorophenol [ORGANIC CHEMISTRY] C_6HCl_4OH Either of two toxic compounds: 2,3,4,6-tetrachlorophenol comprises brown flakes, soluble in common solvents, melting at 70°C, and is used as a fungicide; 2,4,5,6-tetrachlorophenol is a brown solid, insoluble in water, soluble in sodium hydroxide, has a phenol scent, melts at about 50°C, and is used as a fungicide and for wood preservatives. { ¦te·trə¦klȯr·ə'fē‚nȯl }

tetrachlorophthalic acid [ORGANIC CHEMISTRY] $C_6Cl_4(CO_2H)_2$ Colorless plates, soluble in hot water; used in making dyes. { ¦te·trə¦klȯr·ə¦thal·ik 'as·əd }

tetrachlorophthalic anhydride [ORGANIC CHEMISTRY] $C_6Cl_4(CO)_2O$ A white powder with a melting point of 254-255°C; slightly soluble in water; used in the manufacture of dyes and pharmaceuticals and as a flame retardant for epoxy resins. { ¦te·trə¦klȯr·ə¦thal·ik an'hī‚drīd }

tetracosane [ORGANIC CHEMISTRY] $C_{24}H_{50}$ Combustible crystals insoluble in water, soluble in alcohol, melts at 52°C; used as a chemical intermediate. { ¦te·trə'kō‚sān }

tetracyanoethylene [ORGANIC CHEMISTRY] $(CN)_2C{:}C(CN)_2$ A member of the cyanocarbon compounds; colorless crystals with a melting point of 198-200°C; used in dye manufacture. { ¦te·trə¦sī·ə·nō'eth·ə‚lēn }

***n*-tetradecane** [ORGANIC CHEMISTRY] $C_{14}H_{30}$ A combustible, colorless, water-insoluble liquid boiling at 254°C; used as a solvent and distillation chaser and in organic synthesis. { ¦en ¦te·trə'de‚kān }

1-tetradecene [ORGANIC CHEMISTRY] $CH_2{:}CH(CH_2)_{11}CH_3$ A combustible, colorless, water-insoluble liquid boiling at 256°C; used as a solvent for perfumes and flavors and in medicine. { ¦wən ¦te·trə'de‚sēn }

tetradecylamine [ORGANIC CHEMISTRY] $C_{14}H_{29}NH_2$ A white solid with a melting point of 37°C; soluble in alcohol and ether; used in making germicides. { ¦te·trə·də'sil·ə‚mēn }

tetradecyl mercaptan [ORGANIC CHEMISTRY] $CH_3(CH_2)_{13}SH$ A combustible liquid with a boiling point of 176-180°C; used for processing synthetic rubber. Also known as myristyl mercaptan. { ¦te·trə'des·əl mər'kap‚tan }

tetradentate ligand [INORGANIC CHEMISTRY] A chelating agent which has four groups capable of attachment to a metal ion. Also known as quadridentate ligand. { ¦te·trə'den‚tāt 'līg·ənd }

tetraethanolammonium hydroxide [ORGANIC CHEMISTRY] $(HOCH_2CH_2)_4NOH$ A white, water-soluble, crystalline solid with a melting point of 123°C; used as a dye solvent and in metal-plating solutions. { ¦te·trə¦eth·ə‚nȯl·ə'mō·nē·əmhī'dräk‚sīd }

tetraethylammonium chloride [ORGANIC CHEMISTRY] $(C_2H_5)_4NCl$ Colorless, hygroscopic crystals with a melting point of 37.5°C; soluble in water, alcohol, acetone, and chloroform; used in medicine. Abbreviated TEAC. Also known as TEA chloride. { ¦te·trə¦eth·əl·ə'mō·nē·əm 'klȯr‚īd }

tetra-(2-ethylbutyl)silicate [ORGANIC CHEMISTRY] $[(C_2H_5)C_4H_8O]_4Si$ A colorless liquid with a boiling point of 238°C at 50 mmHg (6660 pascals); used as a lubricant and hydraulic fluid. { ¦te·trə ¦tü ¦eth·əl¦byüd·əl 'sil·ə‚kāt }

tetraethylene glycol [ORGANIC CHEMISTRY] $HO(C_2H_4O)_3C_2H_4OH$ A combustible, hygroscopic, colorless, water-soluble liquid, boils at 327°C; used as a nitrocellulose solvent and plasticizer and in lacquers and coatings. Abbreviated TEG. { ¦te·trə'eth·ə‚lēn 'glī‚kȯl }

tetraethylene glycol dimethacrylate [ORGANIC CHEMISTRY] A colorless to pale straw-colored liquid with a boiling point of 200°C at 1 mmHg (133.32 pascals); soluble in styrene and some esters and aromatics; used as a plasticizer. { ¦te·trə'eth·ə‚lēn 'glī‚kȯl ‚dī·mə'thak·rə‚lāt }

tetraethylenepentamine [ORGANIC CHEMISTRY] $C_8H_{23}N_2$ A toxic, viscous liquid with a boiling point of 333°C and a freezing point of −30°C; soluble in water and organic solvents, used as a motor oil additive, in the manufacture of synthetic rubber, and as a solvent for dyes, acid gases, and sulfur. { ¦te·trə¦eth·ə‚lēn'pent·ə‚mēn }

tetraethyllead [ORGANIC CHEMISTRY] $Pb(C_2H_5)_4$ A highly toxic lead compound that, when added in small proportions to gasoline, increases the fuel's antiknock quality. Abbreviated TEL. { ¦te·tre¦eth·əl′led }

tetraethylpyrophosphate [ORGANIC CHEMISTRY] $C_8H_{20}O_7P_2$ A hygroscopic corrosive liquid miscible with although decomposed by water, and miscible with many organic solvents; inhibits the enzyme acetylcholinesterase; used as an insecticide in place of nicotine sulfate. { ¦te·trə¦eth·əl¦pī·rō′fä‚sfāt }

tetrafluoroethylene [ORGANIC CHEMISTRY] $F_2C{:}CF_2$ A flammable, colorless, heavy gas, insoluble in water, boils at 78°C; used as a monomer to make poly(tetrafluoroethylene). Abbreviated TFE. { ¦te·trə¦flu̇r·ō′eth·ə‚lēn }

tetrafluorohydrazine [INORGANIC CHEMISTRY] F_2NNF_2 A colorless liquid or gas with a calculated boiling point of −73°C; used as an oxidizer in rocket fuels. { ¦te·trə¦flu̇r·ō′hī·drə‚zēn }

tetrafluoromethane [ORGANIC CHEMISTRY] *See* carbon tetrafluoride. { ¦te·trə¦flu̇r·ō′meth‚ān }

tetrafunctional molecule [ORGANIC CHEMISTRY] A chemical structure that possesses four highly reactive sites. { ‚te·trə¦fəŋk·shən·əl ′mäl·ə‚kyül }

tetrahedral molecule [CHEMISTRY] A molecule whose structure forms a tetrahedron with a central atom possessing four valence bonds that are directed toward the four points of the tetrahedron. { ‚te·trə¦hē·drəl ′mäl·ə‚kyül }

tetrahydrocannabinol [ORGANIC CHEMISTRY] $C_{21}H_{30}O_2$ Any member of a group of isomers that are active components of marijuana. Abbreviated THC. { ¦te·trə¦hī·drə·kə′nab·ə‚nȯl }

tetrahydrofuran [ORGANIC CHEMISTRY] C_4H_8O A clear, colorless liquid with a boiling point of 66°C; soluble in water and organic solvents; used as a solvent for resins and in adhesives, printing inks, and polymerizations. Abbreviated THF. { ¦te·trə‚hī drə′fyü‚rän }

tetrahydrofurfuryl acetate [ORGANIC CHEMISTRY] $C_7H_{12}O_3$ A colorless liquid with a boiling point of 194-195°C; soluble in water, alcohol, ether, and chloroform; used for flavoring. { ¦te·trə¦hī·drə′fər·fə‚ril ′as·ə‚tāt }

tetrahydrofurfuryl alcohol [ORGANIC CHEMISTRY] $C_4H_7OCH_2OH$ A hygroscopic, colorless liquid, miscible with water, boils at 178°C; used as a solvent for resins, in leather dyes, and in nylon. { ¦te·trə¦hī·drə′fər·fə‚ril ′al·kə‚hȯl }

tetrahydrofurfurylamine [ORGANIC CHEMISTRY] $C_4H_7OCH_2NH_2$ A colorless to light yellow liquid with a distillation range of −150 to −156°C; used for fine-grain photographic development and to accelerate vulcanization. { ¦te·trə¦hī·drə‚fər·fə′ril·ə‚mēn }

tetrahydrofurfuryl oleate [ORGANIC CHEMISTRY] $C_{23}H_{42}O_3$ A colorless liquid with a boiling point of 240°C at 2 mmHg (266.64 pascals); used as a plasticizer. { ¦te·trə¦hī·drə‚fər·fə′ril ′ōl·ē‚āt }

tetrahydrofurfuryl phthalate [ORGANIC CHEMISTRY] $C_6H_4(COOCH_2C_4H_7O)_2$ A colorless liquid with a melting point below 15°C; used as a plasticizer. { ¦te·trə¦hī·drə‚fər·fə′ril ′tha‚lāt }

tetrahydrolinalool [ORGANIC CHEMISTRY] $C_{10}H_{21}OH$ A colorless liquid with a floral odor, used in perfumery and flavoring. { ¦te·trə¦hī·drə·lə′nal·ō‚ȯl }

tetrahydronaphthalene [ORGANIC CHEMISTRY] $C_{10}H_{12}$ A colorless, oily liquid that boils at 206°C, and is miscible with organic solvents; used as an intermediate in chemical synthesis and as a solvent. { ¦te·trə¦hī·drə′naf·thə‚lēn }

tetrahydroxyadipic acid [ORGANIC CHEMISTRY] *See* mucic acid. { ¦te·trə·hī¦dräk·sē·ə¦dip·ik ′as·əd }

tetraiodoethylene [ORGANIC CHEMISTRY] $I_2C{:}CI_2$ Light yellow crystals with a melting point of 187°C; soluble in organic solvents; used in surgical dusting powder and antiseptic ointments, and as a fungicide. Also known as iodoethylene. { ¦te·trə¦ī·ə‚dō′eth·ə‚lēn }

tetraiodofluorescein [ORGANIC CHEMISTRY] $C_{20}H_8O_5I_4$ A yellow, water-insoluble, crystalline compound; used as a dye. Also known as pyrosin. { ¦te·trə¦ī·ə‚dō′flu̇r·ə‚sēn }

tetrakis(hydroxymethyl)phosphonium chloride [ORGANIC CHEMISTRY] $(HOCH_2)_4PCl$ A crystalline compound made from phosphine, formaldehyde, and hydrochloric acid;

used as a flame retardant for cotton fabrics. Abbreviated THPC. { ¦te·trə·kəs·hī¦dräk·sē¦meth·əl·f aua'sfō·nē·əm 'klȯr‚īd }

tetralite [ORGANIC CHEMISTRY] *See* tetryl. { 'te·trə‚līt }

tetramer [ORGANIC CHEMISTRY] A polymer that results from the union of four identical monomers; for example, the tetramer C_8H_8 forms from union of four molecules of C_2H_2. { 'te·trə·mər }

tetramethyldiaminobenzophenone [ORGANIC CHEMISTRY] $[(CH_3)_2NC_6H_4]_2CO$ White to greenish, crystalline leaflets with a melting point of 172°C; soluble in alcohol, ether, water, and warm benzene; used in the manufacture of dyes. Also known as Michler's ketone. { ¦te·trə¦meth·əl·dī¦am·ə·nō·ben'z äf·ə‚nōn }

tetramethylene [ORGANIC CHEMISTRY] *See* cyclobutane. { ¦te·trə'meth·ə‚lēn }

tetramethylethylenediamine [ORGANIC CHEMISTRY] $(CH_3)_4N_2(CH_2)_2$ A colorless liquid with a boiling point of 121-122°C; soluble in organic solvents and water; used in the formation of polyurethane, as a corrosion inhibitor, and for textile finishing agents. { ¦te·trə¦meth·əl¦eth·ə‚lēn'dī·ə‚mēn }

tetramethyllead [ORGANIC CHEMISTRY] $Pb(CH_3)_4$ An organic compound of lead that, when added in small amounts to motor gasoline, increases the antiknock quality of the fuel; not widely used. { ¦te·trə¦meth·əl'led }

tetramethylsilane [ORGANIC CHEMISTRY] $(CH_3)_4Si$ A colorless, volatile, toxic liquid with a boiling point of 26.5°C; soluble in organic solvents; used as an aviation fuel. { ¦te·trə¦meth·əl'sī‚lān }

tetramethylthiuram monosulfide [ORGANIC CHEMISTRY] $[(CH_3)_2NCS]_2S$ A yellow powder with a melting point of 104-107°C; soluble in acetone, benzene, and ethylene dichloride; used as a rubber accelerator, fungicide, and insecticide. { ¦te·trə¦meth·əl'thī·yə‚ram ¦män·ə'səl‚fīd }

tetramethylurea [ORGANIC CHEMISTRY] $C_5H_{12}N_2O$ A liquid that boils at 176.5°C, and is miscible in water and organic solvents; used as a reagent and solvent. { ¦te·trə¦meth·əl·yu̇'rē·ə }

tetranitromethane [ORGANIC CHEMISTRY] $C(NO_2)_4$ A powerful oxidant; toxic, colorless liquid with a pungent aroma, insoluble in water, soluble in alcohol and ether, boils at 126°C; used in rocket fuels and as an analytical reagent. { ¦te·trə¦nī·trō'meth‚ān }

tetraphenyltin [ORGANIC CHEMISTRY] $(C_6H_5)_4Sn$ A white powder with a melting point of 225-228°C; soluble in hot benzene, toluene, and xylene; used for mothproofing. { ‚te·trə'fen·əl·tən }

tetraphosphorus trisulfide [INORGANIC CHEMISTRY] *See* phosphorus sesquisulfide. { ¦te·trə'fä·sfə·rəs trī'səl‚fīd }

tetrapotassium pyrophosphate [INORGANIC CHEMISTRY] *See* potassium pyrophosphate. { ¦te·trə·pə'tas·ē·əm ¦pī·rō'fä‚sfāt }

tetrapropylene [ORGANIC CHEMISTRY] *See* dodecane. { ¦te·trə'prō·pə‚lēn }

tetrapyrrole [ORGANIC CHEMISTRY] A chemical structure in which four pyrrole rings are joined in straight chains, as in a phycobilin, or as joined rings, as in a chlorophyll. { ‚te·trə'pī‚rōl }

tetrasodium pyrophosphate [INORGANIC CHEMISTRY] *See* sodium pyrophosphate. { ¦te·trə¦sōd·ē·əm ¦pī·rō'fä‚sfāt }

tetraterpene [ORGANIC CHEMISTRY] A class of terpene compounds that contain isoprene units; best known are the carotenoid pigments from plants and animals, such as lycopene, the red coloring matter in tomatoes. { ¦te·trə'tər‚pēn }

tetrazene [ORGANIC CHEMISTRY] $H_2NC(NH)_3N_2C(NH)_3NO$ An explosive, colorless to yellowish solid practically insoluble in water and alcohol; used as an explosive initiator and in detonators. { 'te·trə‚zēn }

tetrol [ORGANIC CHEMISTRY] *See* furan. { 'te‚trȯl }

tetryl [ORGANIC CHEMISTRY] $(NO_2)_3C_6H_2N(NO_2)CH_3$ A yellow, water-insoluble, crystalline explosive material melting at 130°C; used in explosives and ammunition. Also known as tetralite. { 'te·trəl }

TFE [ORGANIC CHEMISTRY] *See* tetrafluoroethylene.

Th [CHEMISTRY] *See* thorium.

thalline [ORGANIC CHEMISTRY] $C_9H_6N(OCH_3)H_4$ Colorless rhomboids soluble in water and melting at 40°C. { 'tha‚lēn }

thallium [CHEMISTRY] A metallic element in group III, symbol Tl, atomic number 81, atomic weight 204.37; insoluble in water, soluble in nitric and sulfuric acids, melts at 302°C, boils at 1457°C. { 'thal·ē·əm }

thallium acetate [ORGANIC CHEMISTRY] $TlOCOCH_3$ Toxic, white, deliquescent crystals, soluble in water and alcohol, melts at 131°C, used as an ore-flotation solvent and in medicine. { 'thal·ē·əm 'as·ə‚tāt }

thallium bromide [INORGANIC CHEMISTRY] TlBr A toxic, yellowish powder soluble in alcohol, slightly soluble in water, melts at 460°C; used in infrared radiation transmitters and detectors. Also known as thallous bromide. { 'thal·ē·əm 'brō‚mīd }

thallium carbonate [INORGANIC CHEMISTRY] Tl_2CO_3 Toxic, shiny, colorless needles soluble in water, insoluble in alcohol, melts at 272°C; used as an analytical reagent and in artificial gems. Also known as thallous carbonate. { 'thal·ē·əm 'kär·bə‚nāt }

thallium chloride [INORGANIC CHEMISTRY] TlCl A white, toxic, light-sensitive powder, slightly soluble in water, insoluble in alcohol, melts at 430°C; used as a chlorination catalyst and in medicine and suntan lamps. Also known as thallous chloride. { 'thal·ē·əm 'klȯr‚īd }

thallium hydroxide [INORGANIC CHEMISTRY] $TlOH \cdot H_2O$ Toxic yellow, water- and alcohol-soluble needles, decomposes at 139°C; used as an analytical reagent. Also known as thallous hydroxide. { 'thal·ē·əm hī'dräk‚sīd }

thallium iodide [INORGANIC CHEMISTRY] TlI A toxic, yellow powder, insoluble in alcohol, slightly soluble in water, melts at 440°C; used in infrared radiation transmitters and in medicine. Also known as thallous iodide. { 'thal·ē·əm 'ī·ə‚dīd }

thallium monoxide [INORGANIC CHEMISTRY] Tl_2O A black, toxic, water- and alcohol-soluble powder, melts at 300°C; used as an analytical reagent and in artificial gems and optical glass. Also known as thallium oxide; thallous oxide. { 'thal·ē·əm mə'näk‚sīd }

thallium nitrate [INORGANIC CHEMISTRY] $TlNO_3$ Colorless, toxic, fire-hazardous crystals soluble in hot water, insoluble in alcohol, melts at 206°C, decomposes at 450°C; used as an analytical reagent and in pyrotechnics. Also known as thallous nitrate. { 'thal·ē·əm 'nī‚trāt }

thallium oxide [INORGANIC CHEMISTRY] *See* thallium monoxide. { 'thal·ē·əm 'äk‚sīd }

thallium sulfate [INORGANIC CHEMISTRY] Tl_2SO_4 Toxic, water-soluble, colorless crystals melting at 632°C; used as an analytical reagent and in medicine, rodenticides, and pesticides. Also known as thallous sulfate. { 'thal·ē·əm 'səl‚fāt }

thallium sulfide [INORGANIC CHEMISTRY] Tl_2S Lustrous, toxic, blue-black crystals insoluble in water, alcohol, and ether, soluble in mineral acids, melts at 448°C; used in infrared-sensitive devices. Also known as thallous sulfide. { 'thal·ē·əm 'səl‚fīd }

thallous bromide [INORGANIC CHEMISTRY] *See* thallium bromide. { 'thal·əs 'brō‚mīd }

thallous carbonate [INORGANIC CHEMISTRY] *See* thallium carbonate. { 'thal·əs 'kär·bə‚nāt }

thallous chloride [INORGANIC CHEMISTRY] *See* thallium chloride. { 'thal·əs 'klȯr‚īd }

thallous hydroxide [INORGANIC CHEMISTRY] *See* thallium hydroxide. { 'thal·əs hī'dräk‚sīd }

thallous iodide [INORGANIC CHEMISTRY] *See* thallium iodide. { 'thal·əs 'ī·ə‚dīd }

thallous nitrate [INORGANIC CHEMISTRY] *See* thallium nitrate. { 'thal·əs 'nī‚trāt }

thallous oxide [INORGANIC CHEMISTRY] *See* thallium monoxide. { 'thal·əs 'äk‚sīd }

thallous sulfate [INORGANIC CHEMISTRY] *See* thallium sulfate. { 'thal·əs 'səl‚fāt }

thallous sulfide [INORGANIC CHEMISTRY] *See* thallium sulfide. { 'thal·əs 'səl‚fīd }

THAM [ORGANIC CHEMISTRY] *See* tromethamine.

THC [ORGANIC CHEMISTRY] *See* tetrahydrocannabinol.

thenyl [ORGANIC CHEMISTRY] $C_4H_3SCH_2$— An organic radical based on methylthiophene; thus thenyl alcohol is also known as thiophenemethanol. { 'then·əl }

theobromine [ORGANIC CHEMISTRY] $C_7H_8N_4O_2$ A toxic alkaloid found in cocoa, chocolate products, tea, and cola nuts; closely related to caffeine. { ‚thē·ə'brō‚mēn }

theophylline [ORGANIC CHEMISTRY] $C_7H_8N_4O_2 \cdot H_2O$ Alkaloid from tea leaves; bitter-tasting white crystals slightly soluble in water and alcohol, melts at 272°C; used in medicine. { ˌthē·ə'fīˌlēn }

thermal analysis [ANALYTICAL CHEMISTRY] Any analysis of physical or thermodynamic properties of materials in which heat (or its removal) is directly involved; for example, boiling, freezing, solidification-point determinations, heat of fusion and heat of vaporization measurements, distillation, calorimetry, and differential thermal, thermogravimetric, thermometric, and thermometric titration analyses. Also known as thermoanalysis. { 'thər·məl ə'nal·ə·səs }

thermal black [CHEMISTRY] A type of carbon black made by a thermal process using natural gas; used in the rubber industry. { 'thər·məl 'blak }

thermal degradation [CHEMISTRY] Molecular deterioration of materials (usually organics) because of overheat; can be avoided by low-temperature or vacuum processing, as for foods and pharmaceuticals. { 'thər·məl ˌdeg·rə'dā·shən }

thermal diffusion [PHYSICAL CHEMISTRY] A phenomenon in which a temperature gradient in a mixture of fluids gives rise to a flow of one constituent relative to the mixture as a whole. Also known as thermodiffusion. { 'thər·məl di'fyü·zhən }

thermal titration [ANALYTICAL CHEMISTRY] *See* thermometric titration. { 'thər·məl tī'trā·shən }

thermoanalysis [ANALYTICAL CHEMISTRY] *See* thermal analysis. { ¦thərmo·ə'nal·ə·səs }

thermobalance [ANALYTICAL CHEMISTRY] An analytical balance modified for thermogravimetric analysis, involving the measurement of weight changes associated with the transformations of matter when heated. { ¦thər·mō'bal·əns }

thermochemistry [PHYSICAL CHEMISTRY] The measurement, interpretation, and analysis of heat changes accompanying chemical reactions and changes in state. { ¦thər·mō'kem·ə·strē }

thermodiffusion [PHYSICAL CHEMISTRY] *See* thermal diffusion. { ¦thər·mō·di'fyü·zhən }

thermoelectric diffusion potential [PHYSICAL CHEMISTRY] A potential difference across an electrolyte that results when a temperature gradient causes one constituent to attempt to flow relative to the other. { ˌthər·mō·i¦lek·trik də'fyü·zhən pəˌten·chəl }

thermogravimetric analysis [ANALYTICAL CHEMISTRY] Chemical analysis by the measurement of weight changes of a system or compound as a function of increasing temperature. { ¦thər·mōˌgrav·ə'me·trik ə'nal·ə·səs }

thermokinetic analysis [ANALYTICAL CHEMISTRY] A type of enthalpimetric analysis which uses kinetic titrimetry; involves rapid and continuous automatic delivery of a suitable titrant, under judiciously controlled experimental conditions with temperature measurement; the end points obtained are converted by mathematical procedures into valid stoichiometric equivalence points and used for determining reaction rate constants. { ¦thər·mō·ki¦ned·ik ə'nal·ə·səs }

thermometric analysis [PHYSICAL CHEMISTRY] A method for determination of the transformations a substance undergoes while being heated or cooled at an essentially constant rate, for example, freezing-point determinations. { ¦thər·mə¦me·trik ə'nal·ə·səs }

thermometric titration [ANALYTICAL CHEMISTRY] A titration in an adiabatic system, yielding a plot of temperature versus volume of titrant; used for neutralization, precipitation, redox, organic condensation, and complex-formation reactions. Also known as calorimetric titration; enthalpy titration; thermal titration. { ¦thər·mə¦me·trik tī'trā·shən }

thermoplastic elastomer [ORGANIC CHEMISTRY] A polymer that can be processed as a thermoplastic material but also possesses the properties of a conventional thermoset rubber. Abbreviated TPE. { ¦thər·məˌpla·stik i'las·tə·mər }

thermotropic liquid crystal [PHYSICAL CHEMISTRY] A liquid crystal prepared by heating the substance. { ¦thər·mō¦träp·ik 'lik·wəd 'krist·əl }

THF [ORGANIC CHEMISTRY] *See* tetrahydrofuran.

thiabendazole [ORGANIC CHEMISTRY] $C_{10}H_7N_3S$ A white powder with a melting point of 304-305°C; controls fungi on citrus fruits, sugarbeets, turf, and ornamentals, and roundworms of cattle and other animals. Also known as 2-(4-thiazolyl)benzimidazole. { ˌthī·ə'ben·dəˌzōl }

thiacetic acid [ORGANIC CHEMISTRY] *See* thioacetic acid. { ¦thī·ə¦sēd·ik 'as·əd }

thiamine hydrochloride [ORGANIC CHEMISTRY] $C_{12}H_{17} \cdot ON_4S \cdot HCl$ White, hygroscopic crystals soluble in water, insoluble in ether, with a yeasty aroma and a salty, nutlike taste, decomposes at 247°C; the form in which thiamine is generally employed. { 'thī·ə·mən ¦hī·drə'klórˌīd }

thianaphthene [ORGANIC CHEMISTRY] C_8H_6S A crystalline compound with a melting point of 32°C; soluble in organic solvents; used in the production of pharmaceuticals. Also known as benzothiofuran. { ˌthī·ə'nafˌthēn }

thiazole [ORGANIC CHEMISTRY] C_3H_3NS A colorless to yellowish liquid with a pyridinelike aroma, slightly soluble in water, soluble in alcohol and ether, used as an intermediate for fungicides, dyes, and rubber accelerators. { 'thī·əˌzōl }

thiazole dye [ORGANIC CHEMISTRY] One of a family of dyes in which the chromophore groups are ═C═N—, —S—C═, and used mainly for cotton; an example is primuline. { 'thī·əˌzōl 'dī }

2-(4-thiazolyl)benzimidazole [ORGANIC CHEMISTRY] *See* thiabendazole. { ¦tü ¦fór thī¦az·əˌwil ˌbenzˌim·ə'daˌzōl }

Thiele melting-point apparatus [ANALYTICAL CHEMISTRY] A stirred, specially shaped test-tube device used for the determination of the melting point of a crystalline chemical. { 'tēl·ə 'melt·iŋ ˌpóint ˌap·əˌrad·əs }

thin-layer chromatography [ANALYTICAL CHEMISTRY] Chromatographing on thin layers of adsorbents rather than in columns; adsorbent can be alumina, silica gel, silicates, charcoals, or cellulose. { 'thin ¦lā·ər ˌkrō·mə'täg·rə·fē }

thio- [CHEMISTRY] A chemical prefix derived from the Greek *theion*, meaning sulfur; indicates the replacement of an oxygen in an acid radical by sulfur with a negative valence of 2. { 'thī·ō }

thioacetamide [ORGANIC CHEMISTRY] C_2H_5NS A crystalline compound with a melting point of 113-114°C; soluble in water and ethanol; used in laboratories in place of hydrogen sulfide. { ¦thī·ō·ə'sed·ə·mīd }

thioacetic acid [ORGANIC CHEMISTRY] CH_3COSH A toxic, clear-yellow liquid with an unpleasant aroma, soluble in water, alcohol, and ether, boils at 82°C; used as an analytical reagent and a lacrimator. Also known as thiacetic acid. { ¦thī·ō·ə¦sēd·ik 'as·əd }

thioaldehyde [ORGANIC CHEMISTRY] An organic compound that contains the —CHS radical and has the suffix -thial; for example, ethanethial, CH_3CHS. { ¦thī·ō'al·dəˌhīd }

thiobarbituric acid [ORGANIC CHEMISTRY] $C_4H_4N_2O_2S$ Malonyl thiourea, the parent compound of the thiobarbiturates; represents barbituric acid in which the oxygen atom of the urea component has been replaced by sulfur. { ¦thī·ōˌbär·bə'túr·ik 'as·əd }

thiocarbamide [ORGANIC CHEMISTRY] *See* thiourea. { ¦thī·ō'kär·bəˌmīd }

thiocarbanilide [ORGANIC CHEMISTRY] $CS(NHC_6H_5)_2$ A gray powder with a melting point of 148°C; soluble in alcohol and ether; used for making dyes, and as a vulcanization accelerator and ore flotation agent. Also known as sulfocarbanilide. { ¦thī·ōˌkär·bə'niˌlīd }

thiocyanate [INORGANIC CHEMISTRY] A salt of thiocyanic acid that contains the —SCN radical, for example, sodium thiocyanate, NaSCN. Also known as sulfocyanate; sulfocyanide; thiocyanide. { ¦thī·ō'sī·əˌnāt }

thiocyanic acid [INORGANIC CHEMISTRY] HSC:N A colorless, water-soluble liquid decomposing at 200°C; used to inhibit paper deterioration due to the action of light, and (in the form of organic esters) as an insecticide. Also known as rhodanic acid; sulfocyanic acid. { ¦thī·ō·sī¦an·ik 'as·əd }

thiocyanide [INORGANIC CHEMISTRY] *See* thiocyanate. { ¦thī·ō'sī·əˌnīd }

thiocyanogen [INORGANIC CHEMISTRY] NCSSCN White, light-unstable rhombic crystals melting at −2°C. { ¦thī·ō·sī·'an·ə·jən }

thiodiglycol [ORGANIC CHEMISTRY] $(CH_2CH_2OH)_2S$ A combustible, colorless, syrupy liquid soluble in water, alcohol, acetone, and chloroform, boils at 283°C; used as a chemical intermediate, textile-dyeing solvent, and antioxidant. { ¦thī·ō·dī'glī‚kȯl }

thiodiglycolic acid [ORGANIC CHEMISTRY] $HOOCCH_2SCH_2COOH$ Combustible, colorless, water- and alcohol-soluble crystals melting at 128°C; used as an analytical reagent. { ¦thī·ō‚dī·glī'käl·ik 'as·əd }

3,3′-thiodiproprionic acid [ORGANIC CHEMISTRY] $(CH_2CH_2COOH)_2S$ A crystalline compound with a melting point of 134°C; soluble in hot water, acetone, and alcohol; used as an antioxidant for soap products and polymers of ethylene. { ¦thrē ¦thre‚prīm ¦thī·ō‚prō·prē'än·ik 'as·əd }

thioether [ORGANIC CHEMISTRY] RSR A general formula for colorless, volatile organic compounds obtained from alkyl halides and alkali sulfides; the R groups can be the same, or different as in methylthioethane ($CH_3SC_2H_5$). { ¦thī·ō'ē·thər }

thioethyl alcohol [ORGANIC CHEMISTRY] *See* ethyl mercaptan. { ¦thī·ō'eth·əl 'al·kə‚hȯl }

thioflavine T [ORGANIC CHEMISTRY] $C_{16}H_{17}N_2Cl$ A yellow basic dye, used for textile dyeing and fluorescent sign paints. { ¦thī·ō'flā·vən 'tē }

thiofuran [ORGANIC CHEMISTRY] *See* thiophene. { ¦thī·ō'fyu̇‚ran }

thioglycolic acid [ORGANIC CHEMISTRY] $HSCH_2COOH$ A liquid with a strong unpleasant odor; used as a reagent for metals such as iron, molybdenum, silver, and tin, and in bacteriology. Also known as mercaptoacetic acid. { ¦thī·ō·glī'käl·ik 'as·əd }

2-thiohydantoin [ORGANIC CHEMISTRY] $NHC(S)NHC(O)CH_2$ Crystals or a tan powder with a melting point of 230°C; used in the manufacture of pharmaceuticals, rubber accelerators, and copper-plating brighteners. Also known as glycolythiourea. { ¦tü ¦thī·ō·hī'dänt·ə‚win }

thiol [ORGANIC CHEMISTRY] *See* mercaptan. { 'thī‚ȯl }

thiolactic acid [ORGANIC CHEMISTRY] $CH_3CH(SH)COOH$ An oil with a disagreeable odor; used in toiletry preparation. Also known as 2-mercaptopropionic acid; 2-thiolpropionic acid. { ¦thī·ō¦lak·tik 'as·əd }

thiomalic acid [ORGANIC CHEMISTRY] $C_4H_6O_4S$ White crystals or powder with a melting point of 149-150°C; soluble in water, alcohol, and acetone; used as a sealer for fuel cells and machine and electrical parts, for caulking compounds, and as a propellant binder. Also known as mercaptosuccinic acid. { ¦thī·ō¦mal·ik 'as·əd }

thionic acid [INORGANIC CHEMISTRY] $H_2S_xO_6$, where x varies from 2 to 6. [ORGANIC CHEMISTRY] An organic acid with the radical —CSOH. { thī'än·ik 'as·əd }

thionyl bromide [ORGANIC CHEMISTRY] $SOBr_2$ A red liquid boiling at 68°C (40 mmHg). Also known as sulfinyl bromide. { 'thī·ən·əl 'brō‚mīd }

thionyl chloride [INORGANIC CHEMISTRY] $SOCl_2$ A toxic, yellowish to red liquid with a pungent aroma, soluble in benzene, decomposes in water and at 140°C; boils at 79°C; used as a chemical intermediate and catalyst. Also known as sulfur oxychloride; sulfurous oxychloride. { 'thī·ən·əl 'klȯr‚īd }

thiopental sodium [ORGANIC CHEMISTRY] $C_{11}H_{17}O_2N_2NaS$ Yellow, water-soluble crystals with a characteristic aroma; used in medicine as a short-acting anesthetic. Also known as thiopentone sodium. { ¦thī·ō'pen‚tal 'sōd·ē·əm }

thiopentone sodium [ORGANIC CHEMISTRY] *See* thiopental sodium. { ¦thī·ō'pen‚tēn 'sōd·ē·əm }

thiophanate [ORGANIC CHEMISTRY] $C_{14}H_{18}N_4O_4S_2$ A tan to colorless solid that decomposes at 195°C; slightly soluble in water; used to control fungus diseases of turf. { thī·ō'fa‚nāt }

thiophene [ORGANIC CHEMISTRY] C_4H_4S A toxic, flammable, highly reactive, colorless liquid, insoluble in water, soluble in alcohol and ether, boils at 84°C; used as a chemical intermediate and to make condensation copolymers. Also known as thiofuran. { 'thī·ə‚fēn }

thiophenol [ORGANIC CHEMISTRY] C_6H_5SH A toxic, fire-hazardous, water-white liquid with a disagreeable aroma, insoluble in water, soluble in alcohol and ether, boils at 168°C; used to make pharmaceuticals. Also known as phenyl mercaptan. { ¦thī·ō'fē‚nȯl }

thiosalicylic acid [ORGANIC CHEMISTRY] $HOOCC_6H_4SH$ A yellow solid with a melting point of 164-165°C; soluble in alcohol, ether, and acetic acid; used for making dyes. Also known as 2-mercaptobenzoic acid. { ¦thī·ō‚sal·ə¦sil·ik 'as·əd }

thiosemicarbazide [ORGANIC CHEMISTRY] $NH_2CSNHNH_2$ A white, water- and alcohol-soluble powder melting at 182°C; used as an analytical reagent and in photography and rodenticides. { ¦thī·ō‚sem·i'kär·bə‚zīd }

thiosulfate [INORGANIC CHEMISTRY] $M_2S_2O_3$ A salt of thiosulfuric acid and a base; for example, reaction of sodium hydroxide and thiosulfuric acid to produce sodium thiosulfate. { ¦thī·ə'səl‚fāt }

thiosulfonic acid [ORGANIC CHEMISTRY] Name for a group of oxy acids of sulfur, with the general formula RS_2O_2H; they are known as esters and salts. { ¦thī·ə‚səl'fän·ik 'as·əd }

thiosulfuric acid [INORGANIC CHEMISTRY] $H_2S_2O_3$ An unstable acid that decomposes readily to form sulfur and sulfurous acid. { ¦thī·ə‚səl'fyu̇r·ik 'as·əd }

thiourea [ORGANIC CHEMISTRY] $(NH_2)_2CS$ Bitter-tasting, white crystals with a melting point of 180-182°C, soluble in cold water and alcohol; used in photography and photocopying, as a rubber accelerator, and as an antithyroid drug in treating hyperthyroidism. Also known as thiocarbamide. { ¦thī·ō·yu̇'rē·ə }

third-order reaction [PHYSICAL CHEMISTRY] A chemical reaction in which the rate of reaction is determined by the concentration of three reactants. { 'thərd ¦ȯr·dər rē'ak·shən }

thiuram [ORGANIC CHEMISTRY] A chemical compound containing a R_2NCS radical; occurs mainly in disulfide compounds; the most common monosulfide compound is tetramethylthiuram monosulfide. { 'thī·yə‚ram }

thixotropy [PHYSICAL CHEMISTRY] Property of certain gels which liquefy when subjected to vibratory forces, such as ultrasonic waves or even simple shaking, and then solidify again when left standing. { thik'sä·trə·pē }

Thomson-Berthelot principle [PHYSICAL CHEMISTRY] The assumption that the heat released in a chemical reaction is directly related to the chemical affinity, and that, in the absence of the application of external energy, that chemical reaction which releases the greatest heat is favored over others; the principle is in general incorrect, but applies in certain special cases. { 'täm·sən ber·tə'lō ‚prin·sə·pəl }

thoria [INORGANIC CHEMISTRY] *See* thorium dioxide. { 'thȯr·ē·ə }

thorium [CHEMISTRY] An element of the actinium series, symbol Th, atomic number 90, atomic weight 232; soft, radioactive, insoluble in water and alkalies, soluble in acids, melts at 1750°C, boils at 4500°C. { 'thȯr·ē·əm }

thorium anhydride [INORGANIC CHEMISTRY] *See* thorium dioxide. { 'thȯr·ē·əm an'hī‚drīd }

thorium carbide [INORGANIC CHEMISTRY] ThC_2 A yellow solid melting at above 2630°C, decomposes in water; used in nuclear fuel. { 'thȯr·ē·əm 'kär‚bīd }

thorium chloride [INORGANIC CHEMISTRY] $ThCl_4$ Hygroscopic, toxic colorless crystal needles soluble in alcohol, melts at 820°C, decomposes at 928°C; used in incandescent lighting. Also known as thorium tetrachloride. { 'thȯr·ē·əm 'klȯr‚īd }

thorium dioxide [INORGANIC CHEMISTRY] ThO_2 A heavy, white powder soluble in sulfuric acid, insoluble in water, melts at 3300°C; used in medicine, ceramics, flame spraying, and electrodes. Also known as thoria; thorium anhydride; thorium oxide. { 'thȯr·ē·əm dī'äk‚sīd }

thorium fluoride [INORGANIC CHEMISTRY] ThF_4 A white, toxic powder, melts at 1111°C; used to make thorium metal and magnesium-thorium alloys and in high-temperature ceramics. { 'thȯr·ē·əm 'flu̇r‚īd }

thorium nitrate [INORGANIC CHEMISTRY] $Th(NO_3)_4 \cdot 4H_2O$ Explosive white crystals soluble in water and alcohol, strong oxidizer; the anhydrous form decomposes at 500°C; used in medicine and as an analytical reagent. { 'thȯr·ē·əm 'nī‚trāt }

thorium oxalate [ORGANIC CHEMISTRY] $Th(C_2O_4)_2 \cdot 2H_2O$ A white, toxic powder soluble in alkalies and ammonium oxalate, insoluble in water and most acids, decomposes to thorium dioxide, ThO_2, above 300-400°C; used in ceramics. { 'thȯr·ē·əm 'äk·sə‚lāt }

thorium oxide [INORGANIC CHEMISTRY] *See* thorium dioxide. { 'thȯr·ē·əm 'äk‚sīd }

thorium sulfate [INORGANIC CHEMISTRY] $Th(SO_4)_2 \cdot 8H_2O$ A white powder soluble in ice water, loses water at 42° and 400°C. Also known as normal thorium sulfate. { 'thȯr·ē·əm 'səl,fāt }

thorium tetrachloride [INORGANIC CHEMISTRY] *See* thorium chloride. { 'thȯr·ē·əm ¦te·trə'klȯr,īd }

Thorpe reaction [ORGANIC CHEMISTRY] The reaction by which, in presence of lithium amides, α,ω-dinitriles undergo base-catalyzed condensation to cyclic iminonitriles, which can be hydrolyzed and decarboxylated to cyclic ketones. { 'thȯrp rē,ak·shən }

THPC [ORGANIC CHEMISTRY] *See* tetrakis(hydroxymethyl)phosphonium chloride.

thulia [INORGANIC CHEMISTRY] *See* thulium oxide. { 'thü·lē·ə }

thulium [CHEMISTRY] A rare-earth element, symbol Tm, of the lanthanide group, atomic number 69, atomic weight 168.934; reacts slowly with water, soluble in dilute acids, melts at 1550°C, boils at 1727°C; the dust is a fire hazard; used as x-ray source and to make ferrites. { 'thü·lē·əm }

thulium chloride [INORGANIC CHEMISTRY] $TmCl_3 \cdot 7H_2O$ Green, deliquescent crystals soluble in water and alcohol; melts at 824°C. { 'thü·lē·əm 'klȯr,īd }

thulium oxalate [ORGANIC CHEMISTRY] $Tm_2(C_2O_4)_3 \cdot 6H_2O$ A toxic, greenish-white solid, soluble in aqueous alkali oxalates, loses one water at 50°C; used for analytical separation of thulium from common metals. { 'thü·lē·əm 'äk·sə,lāt }

thulium oxide [INORGANIC CHEMISTRY] Tm_2O_3 A white, slightly hygroscopic powder that absorbs water and carbon dioxide from the air, and is slowly soluble in strong acids; used to make thulium metal. Also known as thulia. { 'thü·lē·əm 'äk,sīd }

thyme camphor [ORGANIC CHEMISTRY] *See* thymol. { 'tīm ,kam·fər }

thymol [ORGANIC CHEMISTRY] $C_{10}H_{14}O$ A naturally occurring crystalline phenol obtained from thyme or thyme oil, melting at 515°C; used to kill parasites in herbaria, to preserve anatomical specimens, and in medicine as a topical antifungal agent. Also known as thyme camphor. { thī,mȯl }

thymol blue [ORGANIC CHEMISTRY] $C_6H_4SO_2OC[C_6H_2(CH_3)(OH)CH(CH_3)_2]_2$ Brown-green crystals soluble in alcohol and dilute alkalies, insoluble in water, decomposes at 223°C; used as acid-base pH indicator. { thī,mȯl 'blü }

thymol iodide [ORGANIC CHEMISTRY] $[C_6H_2(CH_3)(OI)(C_3H_7)]_2$ A red-brown, light-sensitive powder with an aromatic aroma, soluble in ether and chloroform, insoluble in water; used in medicine and as a feed additive. { thī,mȯl 'ī·ə,dīd }

thymolphthalein [ORGANIC CHEMISTRY] $C_6H_4COOC[C_6H_2(CH_3)(OH)CH(CH_3)_2]_2$ A white powder insoluble in water, soluble in alcohol and acetone, melts at 245°C; used in medicine and as an acid-base titration indicator. { ¦thī,mȯl¦thā,lēn }

Ti [CHEMISTRY] *See* titanium.

tiba [ORGANIC CHEMISTRY] $C_7H_3I_3O_2$ A colorless solid with a melting point of 226-228°C; insoluble in water; used as a growth regulator for fruit. { 'tī·bə }

tie line [PHYSICAL CHEMISTRY] A line on a phase diagram joining the two points which represent the composition of systems in equilibrium. Also known as conode. { 'tī ,līn }

tight ion pair [ORGANIC CHEMISTRY] An ion pair composed of individual ions which keep their stereochemical configuration; no solvent molecules separate the cation and anion. Also known as contact ion pair; intimate ion pair. { 'tīt 'ī,än ,per }

time-of-flight mass spectrometer [SPECTROSCOPY] A mass spectrometer in which all the positive ions of the material being analyzed are ejected into the drift region of the spectrometer tube with essentially the same energies, and spread out in accordance with their masses as they reach the cathode of a magnetic electron multiplier at the other end of the tube. { ¦tīm əv ¦flīt 'mas spek'träm·əd·ər }

time-resolved laser spectroscopy [SPECTROSCOPY] A method of studying transient phenomena in the interaction of light with matter through the exposure of samples to extremely short and intense pulses of laser light, down to subnanosecond or subpicosecond duration. { 'tīm ri¦zälvd 'lā·zər spek'träs·kə·pē }

tin [CHEMISTRY] Metallic element in group IV, symbol Sn, atomic number 50, atomic weight 118.69; insoluble in water, soluble in acids and hot potassium hydroxide solution; melts at 232°C, boils at 2260°C. { tin }

tin bisulfide [INORGANIC CHEMISTRY] *See* stannic sulfide. { 'tin bī'səl,fīd }

tin bromide [INORGANIC CHEMISTRY] *See* stannic bromide; stannous bromide.

tin chloride [INORGANIC CHEMISTRY] *See* stannic chloride; stannous chloride. { 'tin 'klȯr‚īd }

tin chromate [INORGANIC CHEMISTRY] *See* stannic chromate; stannous chromate. { 'tin 'krō‚māt }

tin crystals [INORGANIC CHEMISTRY] *See* stannous chloride. { 'tin ‚krist·əlz }

tin dichloride [INORGANIC CHEMISTRY] *See* stannous chloride. { 'tin dī'klȯr‚īd }

tin dioxide [INORGANIC CHEMISTRY] *See* stannic oxide. { 'tin dī'äk‚sīd }

tin hydride [INORGANIC CHEMISTRY] SnH_4 A gas boiling at −52°C. Also known as stannane. { 'tin 'hī‚drīd }

tin iodide [] *See* stannic iodide. { 'tin 'ī·ə‚dīd }

tin monosulfide [INORGANIC CHEMISTRY] *See* stannous sulfide. { 'tin ¦män·ə'səl‚fīd }

tin oxalate [ORGANIC CHEMISTRY] *See* stannous oxalate. { 'tin 'äk·sə‚lāt }

tin oxide [INORGANIC CHEMISTRY] *See* stannic oxide; stannous oxide. { 'tin 'äk‚sīd }

tin peroxide [INORGANIC CHEMISTRY] *See* stannic oxide. { 'tin pə'räk‚sīd }

tin protosulfide [INORGANIC CHEMISTRY] *See* stannous sulfide. { 'tin ¦prōd·ō'səl‚fīd }

tin protoxide [INORGANIC CHEMISTRY] *See* stannous oxide. { 'tin prə'täk‚sīd }

tin salts [INORGANIC CHEMISTRY] *See* stannous chloride. { 'tin ‚sȯlts }

tin sulfate [INORGANIC CHEMISTRY] *See* stannous sulfate. { 'tin 'səl‚fāt }

tin sulfide [INORGANIC CHEMISTRY] *See* stannous sulfide. { 'tin 'səl‚fīd }

tin tetrabromide [INORGANIC CHEMISTRY] *See* stannic bromide. { 'tin ¦te·trə'brō‚mīd }

tin tetrachloride [INORGANIC CHEMISTRY] *See* stannic chloride. { 'tin ¦te·trə'klȯr‚īd }

tin tetraiodide [INORGANIC CHEMISTRY] *See* stannic iodide. { 'tin ¦te·trə'ī·ə‚dīd }

Tischenko reaction [ORGANIC CHEMISTRY] The formation of an ester by the condensation of two molecules of aldehyde utilizing a catalyst of aluminum alkoxides in the presence of a halide. { ti'shəŋ·kō rē‚ak·shən }

titanate [INORGANIC CHEMISTRY] A salt of titanic acid; titanates of the M_2TiO_3 type are called metatitanates, those of the M_4TiO_4 type are called orthotitanates; an example is sodium titanate, $(Na_2O)_2Ti_2O_5$. { 'tīt·ən‚āt }

titanellow [INORGANIC CHEMISTRY] *See* titanium trioxide. { ¦tīt·ən'el·ō }

titania [INORGANIC CHEMISTRY] *See* titanium dioxide.

titanic acid [INORGANIC CHEMISTRY] H_2TiO_3 A white, water-insoluble powder; used as a dyeing mordant. Also known as metatitanic acid; titanic hydroxide. { tī'tan·ik 'as·əd }

titanic anhydride [INORGANIC CHEMISTRY] *See* titanium dioxide.

titanic chloride [INORGANIC CHEMISTRY] *See* titanium tetrachloride. { tī'tan·ik 'klȯr‚īd }

titanic hydroxide [INORGANIC CHEMISTRY] *See* titanic acid. { tī'tan·ik hī'dräk‚sīd }

titanic sulfate [INORGANIC SULFATE] *See* titanium sulfate.

titanium [CHEMISTRY] A metallic transition element, symbol Ti, atomic number 22, atomic weight 47.90; ninth most abundant element in the earth's crust; insoluble in water, melts at 1660°C, boils above 3000°C. { tī'ta·nē·əm }

titanium boride [INORGANIC CHEMISTRY] TiB_2 A hard solid that resists oxidation at elevated temperatures and melts at 2980°C, used as a refractory and in alloys, high-temperature electrical conductors, and cermets. { tī'tā·nē·əm 'bȯr‚īd }

titanium carbide [INORGANIC CHEMISTRY] TiC Very hard gray crystals insoluble in water, soluble in nitric acid and aqua regia, melts at about 3140°C; used in cermets, arc-melting electrodes, and tungsten-carbide tools. { tī'tā·nē·əm 'kär‚bīd }

titanium chloride [INORGANIC CHEMISTRY] *See* titanium dichloride. { tī'tā·nē·əm 'klȯr‚īd }

titanium dichloride [INORGANIC CHEMISTRY] $TiCl_2$ A flammable, alcohol-soluble, black powder that decomposes in water, and in vacuum at 475°C, and burns in air. Also known as titanium chloride. { tī'tā·nē·əm dī'klȯr‚īd }

titanium dioxide [INORGANIC CHEMISTRY] TiO_2 A white, water-insoluble powder that melts at 1560°C, and which is produced commercially from the titanium dioxide minerals ilmenite and rutile; used in paints and cosmetics. Also known as titania; titanic anhydride; titanium oxide; titanium white. { tī'tā·nē·əm dī'äk‚sīd }

titanium hydride [INORGANIC CHEMISTRY] TiH_2 A black metallic powder whose dust is an explosion hazard and which dissociates above 288°C; used in powder metallurgy, hydrogen production, foamed metals, glass solder, and refractories, and as an electronic gas getter. { tī'tā·nē·əm 'hī‚drīd }

titanium nitride [INORGANIC CHEMISTRY] TiN Golden-brown brittle crystals melting at 2927°C; used in refractories, alloys, cermets, and semiconductors. { tī'tā·nē·əm 'nī‚trīd }

titanium oxalate [ORGANIC CHEMISTRY] $Ti_2(C_2O_4)_3 \cdot 10H_2O$ Toxic, yellow prisms soluble in water, insoluble in alcohol; used to make titanic acid and titanium metal. Also known as titanous oxalate. { tī'tā·nē·əm 'äk·sə‚lāt }

titanium oxide [INORGANIC CHEMISTRY] *See* titanium dioxide; titanium trioxide. { tī'tā·nē·əm 'äk‚sīd }

titanium peroxide [INORGANIC CHEMISTRY] *See* titanium trioxide. { tī'tā·nē·əm pə'räk‚sīd }

titanium sesquisulfate [INORGANIC CHEMISTRY] *See* titanous sulfate. { tī'tā·nē·əm ¦ses·kwə'səl‚fāt }

titanium sulfate [INORGANIC CHEMISTRY] $Ti(SO_4)_2 \cdot 9H_2O$ Caked solid, soluble in water, toxic, highly acidic; used as a dye stripper, reducing agent, laundry chemical, and in treatment of chrome yellow colors. Also known as titanic sulfate; titanyl sulfate. { tī'tā·nē·əm 'səl‚fāt }

titanium tetrachloride [INORGANIC CHEMISTRY] $TiCl_4$ A colorless, toxic liquid soluble in water, fumes when exposed to moist air, boils at 136°C; used to make titanium and titanium salts, as a dye mordant and polymerization catalyst, and in smoke screens and pigments. Also known as titanic chloride. { tī'tā·nē·əm ¦te·trə 'klȯr‚īd }

titanium trichloride [INORGANIC CHEMISTRY] $TiCl_3$ Toxic, dark-violet, deliquescent crystals soluble in alcohol and some amines, decomposes in water with heat evolution, decomposes above 440°C; used as a reducing agent, chemical intermediate, polymerization catalyst, and laundry stripping agent. Also known as titanous chloride. { tī'tā·nē·əm trī'klȯr‚īd }

titanium trioxide [INORGANIC CHEMISTRY] TiO_3 Yellow titanium oxide used to make ivory shades in ceramics. Also known as titanellow; titanium oxide; titanium peroxide. { tī'tā·nē·əm trī'äk‚sīd }

titanium white [INORGANIC CHEMISTRY] *See* titanium dioxide. { tī'tā·nē·əm 'wīt }

titanous chloride [INORGANIC CHEMISTRY] *See* titanium trichloride. { tī'tan·əs 'klȯr‚īd }

titanous oxalate [ORGANIC CHEMISTRY] *See* titanium oxalate. { tī'tan·əs 'äk·sə‚lāt }

titanous sulfate [INORGANIC CHEMISTRY] $Ti_2(SO_4)_3$ Green crystals soluble in dilute hydrochloric and sulfuric acids, insoluble in water and alcohol; used as a textile reducing agent. Also known as titanium sesquisulfate. { tī'tan·əs 'səl‚fāt }

titanyl sulfate [INORGANIC CHEMISTRY] *See* titanium sulfate. { 'tīt·ən·əl 'səl‚fāt }

titer [CHEMISTRY] **1.** The concentration in a solution of a dissolved substance as shown by titration. **2.** The least amount or volume needed to give a desired result in titration. **3.** The solidification point of hydrolyzed fatty acids. { 'tī·tər }

titrand [ANALYTICAL CHEMISTRY] The substance that is analyzed in a titration procedure. { 'tī‚trand }

titrant [ANALYTICAL CHEMISTRY] A solution of known concentration and composition used for analytical titrations. Also known as standard solution. { 'tī·trənt }

titration [ANALYTICAL CHEMISTRY] A method of analyzing the composition of a solution by adding known amounts of a standardized solution until a given reaction (color change, precipitation, or conductivity change) is produced. { ti'trā·shən }

titrimetric analysis [ANALYTICAL CHEMISTRY] *See* volumetric analysis. { ¦tī·trə‚me·trik ə'nal·ə·səs }

Tm [CHEMISTRY] *See* thulium.

TMA [ORGANIC CHEMISTRY] *See* trimethylamine.

TNF [ORGANIC CHEMISTRY] *See* 2,4,7-trinitrofluorenone.

TNT [ORGANIC CHEMISTRY] *See* 2,4,6-trinitrotoluene.

tocopherol [ORGANIC CHEMISTRY] Any of several substances having vitamin E activity that occur naturally in certain oils; α-tocopherol possesses the highest biological activity. { tə'käf·ə‚röl }

tolazoline hydrochloride [ORGANIC CHEMISTRY] $C_{10}H_{12}N_2 \cdot HCl$ Water-soluble white crystals, melting at 173°C; used as a sympatholytic and vasodilator. Also known as priscol. { täl′az·ə‚lēn ¦hī·drə′klȯr‚īd }

tolerance interval [ANALYTICAL CHEMISTRY] That range of values within which it has been calculated that a specified percentage of individual values of measurements will lie with a stated confidence level. { ′täl·ə·rəns ‚in·tər·vəl }

***ortho*-tolidine** [ORGANIC CHEMISTRY] $[C_6H_3(CH_3)NH_2]$ Light-sensitive, combustible white to reddish crystals soluble in alcohol and ether, slightly soluble in water, melts at 130°C; used as an anlytical reagent and a curing agent for urethane resins. { ¦ȯr·thō ′täl·ə‚dēn }

Tollen's aldehyde test [ANALYTICAL CHEMISTRY] A test that uses an ammoniacal solution of silver oxides to test for aldehydes and ketones. { ′täl·ənz ′al·də‚hīd ‚test }

toluene [ORGANIC CHEMISTRY] $C_6H_5CH_3$ A colorless, aromatic liquid derived from coal tar or from the catalytic reforming of petroleum naphthas; insoluble in water, soluble in alcohol and ether, boils at 111°C; used as a chemical intermediate, for explosives, and in high-octane gasolines. Also known as methylbenzene; phenylmethane; toluol. { ′täl·yə‚wēn }

toluene 2,4-diisocyanate [ORGANIC CHEMISTRY] $CH_3C_6H_3(NCO)_2$ A liquid (at room temperature) with a sharp, pungent odor; miscible with ether, acetone, and benzene; used to make polyurethane foams and other elastomers, and also as a protein cross-linking agent. { ′täl·yə‚wēn ¦tü ¦fȯr dī¦ī·sō′sī·ə‚nāt }

***para*-toluenesulfonic acid** [ORGANIC CHEMISTRY] $C_6H_4(SO_3H)(CH_3)$ Toxic, colorless, combustible crystals soluble in water, alcohol, and ether; melts at 107°C; used in dyes and as a chemical intermediate and organic catalyst. { ¦par·ə ¦täl·yə‚wēn¦səl¦fän·ik ′as·əd }

α-toluic acid [ORGANIC CHEMISTRY] *See* phenylacetic acid. { ¦al·fə tə′lü·ik ′as·əd }

***meta*-toluic acid** [ORGANIC CHEMISTRY] $C_6H_4CH_3COOH$ White to yellow, combustible crystals soluble in alcohol and ether, slightly soluble in water, melts at 109°C; used as a chemical intermediate and base for insect repellants. Also known as *meta*-toluylic acid. { ¦med·ə tə′lü·ik ′as·əd }

***ortho*-toluic acid** [ORGANIC CHEMISTRY] $C_6H_4CH_3COOH$ White, combustible crystals soluble in alcohol and chloroform, slightly soluble in water, melts at 104°C; used as a bacteriostat. Also known as *ortho*-toluylic acid. { ¦ȯr·thō tə′lü·ik ′as·əd }

***para*-toluic acid** [ORGANIC CHEMISTRY] $C_6H_4CH_3COOH$ Transparent, combustible crystals soluble in alcohol and ether, slightly soluble in water, melts at 180°C; used in agricultural chemicals and as an animal feed supplement. Also known as *para*-toluylic acid. { ¦par·ə tə′lü·ik ′as·əd }

α-toluic aldehyde [ORGANIC CHEMISTRY] *See* phenylacetaldehyde. { ¦al·fə tə′lü·ik ′al·də‚hīd }

***meta*-toluidine** [ORGANIC CHEMISTRY] $CH_3C_6H_4NH_2$ A combustible, colorless, toxic liquid soluble in alcohol and ether, slightly soluble in water, boils at 203°C; used for dyes and as a chemical intermediate. { ¦med·ə tə′lü·ə‚dēn }

***ortho*-toluidine** [ORGANIC CHEMISTRY] $CH_3C_6H_4NH_2$ A light green, light-sensitive, combustible, toxic liquid soluble in alcohol and ether, very slightly soluble in water, boils at 200°C; used for dyes and textile printing and as a chemical intermediate. { ¦ȯr·thō tə′lü·ə‚dēn }

***para*-toluidine** [ORGANIC CHEMISTRY] $CH_3C_6H_4NH_2$ Toxic, combustible, white leaflets soluble in alcohol and ether, very slightly soluble in water, boils at 200°C; used as an analytical reagent and in dyes. { ¦par·ə tə′lü·ə‚dēn }

toluol [ORGANIC CHEMISTRY] *See* toluene. { ′täl·yə‚wȯl }

toluylene [ORGANIC CHEMISTRY] *See* stilbene. { ′täl·yə·wə‚lēn }

toluylene red [ORGANIC CHEMISTRY] *See* neutral red. { ′täl·yə·wə‚lēn ′red }

***meta*-toluylic acid** [ORGANIC CHEMISTRY] *See* *meta*-toluic acid. { ¦med·ə ¦täl·yə¦wil·ik ′as·əd }

***ortho*-toluylic acid** [ORGANIC CHEMISTRY] *See* *ortho*-toluic acid. { ¦ȯr·thō ¦täl·yə¦wil·ik ′as·əd }

para-toluylic acid [ORGANIC CHEMISTRY] *See para*-toluic acid. { ¦par·ə ¦täl·yə¦wil·ik 'as·əd }

para-tolylsulfonylmethylnitrosamide [ORGANIC CHEMISTRY] $C_8H_{10}N_2O_3S$ Yellow crystals with a melting point of 62°C; soluble in ether, petroleum ether, benzene, carbon tetrachloride, and chloroform; a precursor to diazomethane; a useful reagent for the preparation of a wide range of biologically active compounds for gas chromatography analysis. { ¦par·ə ¦tä,lil¦səl·fə,nil¦meth·əl,nī'trō·sə,mīd }

tomatine [ORGANIC CHEMISTRY] $C_{50}H_{83}NO_{21}$ A glycosidal alkaloid obtained from the leaves and stems from the tomato plant; the crude extract is known as tomatin: white, toxic crystals; used as a plant fungicide and as a precipitating agent for cholesterol. { 'täm·ə,tēn }

topochemical control [CHEMISTRY] In a chemical reaction, product formation that is determined by the orientation of molecules in the crystal. { ¦täp·ə'kem·ə·kəl kən'trōl }

torsional angle [PHYSICAL CHEMISTRY] The angle between bonds on adjacent atoms. { 'tȯr·shən·əl 'aŋ·gəl }

total heat of dilution [PHYSICAL CHEMISTRY] *See* heat of dilution. { 'tōd·əl 'hēt əv di'lü·shən }

total heat of solution [PHYSICAL CHEMISTRY] *See* heat of solution. { 'tōd·əl 'hēt əv sə'lü·shən }

total solids [CHEMISTRY] The total content of suspended and dissolved solids in water. { 'tōd·əl 'säl·ədz }

TPA [ORGANIC CHEMISTRY] *See* terephthalic acid.

TPE [ORGANIC CHEMISTRY] *See* thermoplastic elastomer.

trace analysis [ANALYTICAL CHEMISTRY] Analysis of a very small quantity of material of a sample by such techniques as polarography or spectroscopy. { 'trās ə,nal·ə·səs }

trace element [ANALYTICAL CHEMISTRY] An element in a sample that has an average concentration of less than 100 parts per million atoms or less than 100 micrograms per gram. { 'trās ,el·ə·mənt }

tracer [CHEMISTRY] A foreign substance, usually radioactive, that is mixed with or attached to a given substance so the distribution or location of the latter can later be determined; used to trace chemical behavior of a natural element in an organism. Also known as tracer element. { 'trā·sər }

tracer element [CHEMISTRY] *See* tracer. { 'trā·sər ,el·ə·mənt }

transamination [CHEMISTRY] **1.** The transfer of one or more amino groups from one compound to another. **2.** The transposition of an amino group within a single compound. { tran,sam·ə'nā·shən }

transesterification [ORGANIC CHEMISTRY] Conversion of an organic acid ester into another ester of that same acid. { ¦trans·e¦ster·ə·fə'kā·shən }

transition element [CHEMISTRY] One of a group of metallic elements in which the members have the filling of the outermost shell to 8 electrons interrupted to bring the penultimate shell from 8 to 18 or 32 electrons; includes elements 21 through 29 (scandium through copper), 39 through 47 (yttrium through silver), 57 through 79 (lanthanum through gold), and all known elements from 89 (actinium) on. { tran'zish·ən ,el·ə·mənt }

transition interval [ANALYTICAL CHEMISTRY] In a titrimetric analysis, the range in concentration of the species being determined over which a variation in a chemical indicator can be observed visually. { tran'zish·ən ,in·tər·vəl }

transition state [PHYSICAL CHEMISTRY] *See* activated complex. { tran'zish·ən ,stāt }

transition temperature [CHEMISTRY] The temperature at which an enantiotropic polymorph is converted into a different form. { tran'zish·ən ,tem·prə·chər }

transition time [ANALYTICAL CHEMISTRY] The time interval needed for a working (nonreference) electrode to become polarized during chronopotentiometry (time-measurement electrolysis of a sample). { tran'zish·ən ,tīm }

translational energy [PHYSICAL CHEMISTRY] The kinetic energy of gaseous or liquid molecules that is associated with their motion within their particular chemical systems. { tran¦slā·shən·əl 'en·ər·jē }

transmission diffraction [ANALYTICAL CHEMISTRY] A type of electron diffraction analysis in which the electron beam is transmitted through a thin film or powder whose smallest dimension is no greater than a few tenths of a micrometer. { tranz'mish·ən di͵frak·shən }

transmittance [ANALYTICAL CHEMISTRY] During absorption spectroscopy, the amount of radiant energy transmitted by the solution under analysis. { tranz'mid·əns }

transpassive region [PHYSICAL CHEMISTRY] That portion of an anodic polarization curve in which metal dissolution increases as the potential becomes noble. { tranz'pas·iv ¦rē·jən }

transplutonium element [INORGANIC CHEMISTRY] An element having an atomic number greater than that of plutonium (94). { ¦tranz·plə'tō·nē·əm 'el·ə·mənt }

transport number [PHYSICAL CHEMISTRY] The fraction of the total current carried by a given ion in an electrolyte. Also known as transference number. { 'tranz͵pȯrt ͵nəm·bər }

transuranic elements [CHEMISTRY] Elements that have atomic numbers greater than 92; all are radioactive, are products of artificial nuclear changes, and are members of the actinide group. Also known as transuranium elements. { ¦tranz·yu̇'ran·ik 'el·ə·məns }

transuranium elements [CHEMISTRY] *See* transuranic elements. { ¦tranz·yu̇'rā·nē·əm 'el·ə·məns }

trapping [CHEMISTRY] A method for intercepting a reactive intermediate or molecule and removing it from the system or converting it to a more stable form for further study and identification. { 'trap·iŋ }

Traube's rule [PHYSICAL CHEMISTRY] In dilute solutions, the concentration of a member of a homologous series at which a given lowering of surface tension is observed decreases threefold for each additional methylene group in a given series. { 'trau̇·bəz ͵rül }

tretamine [ORGANIC CHEMISTRY] *See* triethylenemelamine. { 'tred·ə͵mēn }

triacetin [ORGANIC CHEMISTRY] $C_3H_5(CO_2CH_3)_3$ A colorless, combustible oil with a bitter taste and a fatty aroma; found in cod liver and butter; soluble in alcohol and ether, slightly soluble in water; boils at 259°C; used in plasticizers, perfumery, cosmetics, and external medicine and as a solvent and food additive. { trī'as·əd·ən }

triamcinolone [ORGANIC CHEMISTRY] $C_{21}H_{27}FO_6$ White, toxic crystals; insoluble in water, soluble in dimethylformamide; melts at 266°C; used as an intermediate for ion-exchange resin, wetting and frothing agent, and photographic developer. { ͵trī·am'sin·əl͵än }

triamylamine [ORGANIC CHEMISTRY] $(C_5H_{11})_3N$ A combustible, colorless, toxic liquid; soluble in gasoline, insoluble in water; used to inhibit corrosion and in insecticides. { ͵trī·ə'mil·ə͵mēn }

triamyl borate [ORGANIC CHEMISTRY] $(C_5H_{11})_3BO_3$ A combustible, colorless liquid with an alcoholic aroma; soluble in alcohol and ether; boils at 220-280°C; used in varnishes. { trī'am·əl 'bȯr͵āt }

triatomic [CHEMISTRY] Consisting of three atoms. { ¦trī·ə¦täm·ik }

triazole [ORGANIC CHEMISTRY] A five-membered chemical ring compound with three nitrogens in the ring; for example, $C_2H_3N_3$; proposed for use as a photoconductor and for copying systems. { 'trī·ə͵zōl }

tribasic calcium phosphate [INORGANIC CHEMISTRY] *See* calcium phosphate. { trī'bā·sik 'kal·sē·əm 'fä͵sfāt }

tribasic zinc phosphate [INORGANIC CHEMISTRY] *See* zinc phosphate. { trī'bā·sik 'ziŋk 'fä͵sfāt }

tributoxyethyl phosphate [ORGANIC CHEMISTRY] $[CH_3(CH_2)_3O(CH_2)_2O]PO$ A light yellow, oily liquid with a boiling range of 215-228°C; soluble in organic solvents; used as a plasticizer and flame retardant, and in floor waxes. { ͵trī·byü¦täk·sē'eth·əl 'fä͵sfāt }

tributyl borate [ORGANIC CHEMISTRY] $(C_4H_9)_3BO_3$ A combustible, water-white liquid miscible with common organic liquids; boils at 232°C; used in welding fluxes and as a chemical intermediate and textile flame-retardant. { trī'byüd·əl 'bȯr͵āt }

tributyl phosphate [ORGANIC CHEMISTRY] $(C_4H_9)_3PO_4$ A combustible, toxic, stable liquid; soluble in most solvents, and very slightly soluble in water; boils at 292°C; used as a heat-exchange medium, pigment-grinding assistant, antifoam agent, and solvent. Abbreviated TBP. { trī'byüd·əl 'fä,sfāt }

tributyltin acetate [ORGANIC CHEMISTRY] $(C_4H_9)_3SnOOCCH_3$ An organic compound of tin, used as an antimicrobial agent in the paper, wood, plastics, leather, and textile industries. { trī'byüd·əl·tən 'as·ə,tāt }

tributyltin chloride [ORGANIC CHEMISTRY] $(C_4H_9)_3SnCl$ A colorless liquid with a boiling point of 145-147°C; soluble in alcohol, benzene, and other organic solvents; used as a rodenticide. { trī'byüd·əl·tən 'klȯr,īd }

tricaine [ORGANIC CHEMISTRY] $C_{10}H_{15}NO_5S$ Fine, needlelike crystals, soluble in water; used as an anesthetic for fish. { 'trī,kān }

tricalcium phosphate [INORGANIC CHEMISTRY] *See* calcium phosphate. { trī'kal·sē·əm 'fä,sfāt }

trichloroacetic acid [ORGANIC CHEMISTRY] CCl_3COOH Toxic, deliquescent, colorless crystals with a pungent aroma; soluble in water, alcohol, and ether; boils at 198°C; used as a chemical intermediate and laboratory reagent, and in medicine, pharmacy, and herbicides. Abbreviated TCA. { trī¦klȯr·ō·ə¦sed·ik 'as·əd }

trichloroacetic aldehyde [ORGANIC CHEMISTRY] *See* chloral. { trī¦klȯr·ō·ə¦sed·ik 'al·də,hīd }

trichlorobenzene [ORGANIC CHEMISTRY] $C_6H_3Cl_3$ Either of two toxic compounds: 1,2,3-trichlorobenzene forms white crystals, soluble in ether, insoluble in water, boiling at 221°C, and is used as a chemical intermediate; 1,2,4-trichlorobenzene is a combustible, colorless liquid, soluble in most organic solvents and oils, insoluble in water, boiling at 213°C, and is used as a solvent and in dielectric fluids, synthetic transformer oils, lubricants, and insecticides. { trī¦klȯr·ō'ben,zēn }

trichloroethane [ORGANIC CHEMISTRY] $C_2H_3Cl_3$ Either of two nonflammable, irritating liquid isomeric compounds: 1,1,1-trichloroethane (CH_3CCl_3) is toxic, soluble in alcohol and ether, insoluble in water, and boils at 75°C; it is used as a solvent, aerosol propellant, and pesticide and for metal degreasing, and is also known as methyl chloroform; 1,1,2-trichloroethane ($CHCl_2CH_2Cl$) is clear and colorless, is soluble in alcohols, ethers, esters, and ketones, insoluble in water, has a sweet aroma, and boils at 114°C; it is used as a chemical intermediate and solvent, and is also known as vinyl trichloride. { trī¦klȯr·ō'eth,ān }

trichloroethanal [ORGANIC CHEMISTRY] *See* chloral. { trī¦klȯr·ō'eth·ə,nal }

trichloroethylene [ORGANIC CHEMISTRY] $CHCl{:}CCl_2$ A heavy, stable, toxic liquid with a chloroform aroma; slightly soluble in water, soluble with greases and common organic solvents; boils at 87°C; used for metal degreasing, solvent extraction, and dry cleaning and as a fumigant and chemical intermediate. { trī¦klȯr·ō'eth·ə,lēn }

trichlorofluoromethane [ORGANIC CHEMISTRY] CCl_3F A toxic, noncombustible, colorless liquid boiling at 24°C; used as a chemical intermediate, solvent, refrigerant, aerosol prepellant, and blowing agent (plastic foams) and in fire extinguishers. Also known as fluorocarbon-11; fluorotrichloromethane. { trī¦klȯr·ō¦flür·ō'meth,ān }

trichloroiminocyanuric acid [ORGANIC CHEMISTRY] *See* trichloroisocyanuric acid. { trī¦klȯr·ō¦im·ə·nō¦sī·ə¦nü r·ik 'as·əd }

trichloroisocyanuric acid [ORGANIC CHEMISTRY] $C_3Cl_3N_3O_3$ A crystalline substance that releases hypochlorous acid on contact with water; melting point is 246-247°C; soluble in chlorinated and highly polar solvents; used as a chlorinating agent, disinfectant, and industrial deodorant. Also known as symclosene; trichloroiminocyanuric acid. { trī¦klȯr·ō¦i·sō¦sī·ə¦nür·ik 'as·əd }

trichloromethane [ORGANIC CHEMISTRY] *See* chloroform. { trī¦klȯr·ō'meth,ān }

trichloromethyl chloroformate [ORGANIC CHEMISTRY] $ClCOOCCl_3$ A toxic, colorless liquid with a boiling point of 127-128°C; soluble in alcohol, ether, and benzene; used in organic synthesis, and as a military poison gas during World War I. Also known as diphosgene. { trī¦klȯr·ō'meth·əl ¦klȯr·ə'fȯr,māt }

trichloronate [ORGANIC CHEMISTRY] *See ortho*-ethyl(O-2,4,5-trichlorophenyl)ethyl phosphonothioate. { trī'klȯr·ə,nāt }

trichloronitromethane [INORGANIC CHEMISTRY] *See* chloropicrin. { trī¦klȯr·ō¦nī·trō'meth,ān }

trichlorophenol [ORGANIC CHEMISTRY] $C_6H_2Cl_3OH$ Either of two toxic nonflammable compounds with a phenol aroma: 2,4,5-trichlorophenol is a gray solid, is soluble in alcohol, acetone, and ether, melts at 69°C, and is used as a fungicide and bactericide; 2,4,6-trichlorophenol forms yellow flakes, is soluble in alcohol, acetone, and ether, boils at 248°C, and is used as a fungicide, defoliant, and herbicide; it is also known as 2,4,6-T. { trī¦klȯr·ō'fē,nȯl }

2,4,5-trichlorophenoxyacetic acid [ORGANIC CHEMISTRY] $C_6H_2Cl_3OCH_2CO_2H$ A toxic, light-tan solid; soluble in alcohol, insoluble in water; melts at 152°C; used as a defoliant, plant hormone, and herbicide. Also known as 2,4,5-T. { ¦tü ¦fȯr ¦fīv trī¦klȯr·ō·fə¦näk·sē·ə¦sēd·ik 'as·əd }

1,2,3-trichloropropane [ORGANIC CHEMISTRY] $CH_2ClCHClCH_2Cl$ A toxic, colorless liquid with a boiling point of 156.17°C; used as a paint and varnish remover and degreasing agent. { ¦wən ¦tü ¦thrē trī¦klȯr·ō'prō,pan }

1,1,2-trichloro-1,2,2-trifluoroethane [ORGANIC CHEMISTRY] CCl_2FCClF_2 A colorless, volatile liquid with a boiling point of 47.6°C; used as a solvent for dry cleaning, as a refrigerant, and in fire extinguishers. Also known as trifluorotrichloroethane. { ¦wən ¦wən ¦tü trī¦klȯr·ō ¦wən ¦tü ¦tü trī¦flu̇r·ō'eth,ān }

tricosane [ORGANIC CHEMISTRY] $CH_3(CH_2)_{21}CH_3$ Combustible, glittering crystals; soluble in alcohol, insoluble in water; melts at 48°C; used as a chemical intermediate. { 'trī·kə,sān }

tricresyl phosphate [ORGANIC CHEMISTRY] $(CH_3C_6H_4O)_3PO$ A combustible, colorless liquid; insoluble in water, soluble in common solvents and vegetable oils; boils at 420°C; used as a plasticizer, plastics fire retardant, air filter medium, and gasoline and lubricant additive. Abbreviated TCP. { trī'kres·əl 'fä,sfāt }

tricyclic dibenzopyran [ORGANIC CHEMISTRY] *See* xanthene. { trī'sī·klik dī¦ben·zō'pī·rən }

***n*-tridecane** [ORGANIC CHEMISTRY] $CH_3(CH_2)_{11}CH_3$ A combustible liquid; soluble in alcohol, insoluble in water; boils at 226°C; used as a distillation chaser and chemical intermediate. { ¦en trī'de,kān }

tridecanol [ORGANIC CHEMISTRY] *See* tridecyl alcohol. { trī'dek·ə,nȯl }

tridecyl alcohol [ORGANIC CHEMISTRY] $C_{12}H_{25}CH_2OH$ An isomer mixture; a white, combustible solid with a pleasant aroma; melts at 31°C; used in detergents and perfumery and to make synthetic lubricants. Also known as tridecanol. { trī'des·əl 'al·kə,hȯl }

tridentate ligand [INORGANIC CHEMISTRY] A chelating agent having three groups capable of attachment to a metal ion. Also known as terdentate ligand. { trī'den,tāt 'lig·ənd }

triethanolamine [ORGANIC CHEMISTRY] $(HOCH_2CH_2)_3N$ A viscous, hygroscopic liquid with an ammonia aroma, soluble in chloroform, water, and alcohol, and boiling at 335°C; used in dry-cleaning soaps, cosmetics, household detergents, and textile processing, for wool scouring, and as a corrosion inhibitor. { trī,eth·ə'näl·ə,mēn }

triethanolamine stearate [ORGANIC CHEMISTRY] *See* trihydroxyethylamine stearate. { trī,eth·ə'näl·ə,mēn 'stir,āt }

triethylamine [ORGANIC CHEMISTRY] $(C_2H_5)_3N$ A colorless, toxic, flammable liquid with an ammonia aroma; soluble in water and alcohol; boils at 90°C; used as a solvent, rubber-accelerator activator, corrosion inhibitor, and propellant, and in penetrating and waterproofing agents. { trī¦eth·ə·lə¦mēn }

triethylborane [ORGANIC CHEMISTRY] $(C_2H_5)_3B$ A colorless liquid with a boiling point of 95°C; used as a jet fuel or igniter for jet engines and as a fuel additive. Also known as boron triethyl; triethylborine. { trī,eth·əl'bȯr,ān }

triethylborine [ORGANIC CHEMISTRY] *See* triethylborane. { trī,eth·əl'bȯr,ēn }

triethylene glycol [ORGANIC CHEMISTRY] $HO(C_2H_4O)_3H$ A colorless, combustible, hygroscopic, water-soluble liquid; boils at 287°C; used as a chemical intermediate, solvent, bactericide, humectant, and fungicide. Abbreviated TEG. { trī'eth·ə,lēn 'glī,kȯl }

triethylenemelamine [ORGANIC CHEMISTRY] $NC[N(CH_2)_2]NC[N(CH_2)_2]NC[N(CH_2)_2]$ White crystals, soluble in water, alcohol, acetone, chloroform, and methanol; polymerizes at 160°C; used in medicine and insecticides and as a chemosterilant. Abbreviated TEM. Also known as tretamine. { trī¦eth·ə,lēn'mel·ə,mēn }

triethylenetetramine [ORGANIC CHEMISTRY] $NH_2(C_2H_4NH)_2C_2H_4NH_2$ A yellow, water-soluble liquid with a boiling point of 277.5°C; used in detergents and in the manufacture of dyes and pharmaceuticals. { trī¦eth·ə,lēn'te·trə,mēn }

triethylic borate [ORGANIC CHEMISTRY] *See* ethyl borate. { ,trī·ə'thil·ik 'bȯr,āt }

triethyl phosphate [ORGANIC CHEMISTRY] $(C_2H_5)_3PO_4$ A toxic, colorless liquid that acts as a cholinesterase inhibitor; boiling point is 216°C; soluble in organic solvents; used as a solvent and plasticizer and for pesticides manufacture. Abbreviated TEP. { trī'eth·əl 'fä,sfāt }

trifluorochlorethylene resin [ORGANIC CHEMISTRY] A fluorocarbon used as a base for polychlorotrifluoroethylene resin. { trī¦flůr·ō¦klȯr·ō'eth·ə,lēn ,rez·ən }

trifluoromethane [ORGANIC CHEMISTRY] *See* fluoroform. { trī¦flůr·ō'meth,ān }

trifluorotrichloroethane [ORGANIC CHEMISTRY] *See* 1,1,2-trichloro-1,2,2-trifluoroethane. { trī¦flůr·ō·trī¦klȯr·ō'eth,ān }

triformal [ORGANIC CHEMISTRY] *See sym*-trioxane. { trī'fȯr·məl }

triglyceride [ORGANIC CHEMISTRY] $CH_2(OOCR_1)CH(OOCR_2)CH_2(OOCR_3)$ A naturally occurring ester of normal, fatty acids and glycerol; used in the manufacture of edible oils, fats, and monoglycerides. { trī'glis·ə,rīd }

trigonal planar molecule [CHEMISTRY] A molecule having a central atom that is bonded to three other atoms, with all four atoms lying in the same plane. { trī¦gōn·əl ¦plā,när 'mäl·ə,kyül }

trihydroxyethylamine stearate [ORGANIC CHEMISTRY] $C_{24}H_{49}NO_5$ A cream-colored, waxy solid with a melting point of 42-44°C; soluble in methanol, ethanol, mineral oil, and vegetable oil; used as an emulsifier in cosmetics and pharmaceuticals. Also known as triethanolamine stearate. { ,trī·hī¦dräk·sē¦eth·ə·lə¦mē n 'stir,āt }

triiodomethane [ORGANIC CHEMISTRY] *See* iodoform. { trī,ī¦ō·də'meth,ān }

triisobutylene [ORGANIC CHEMISTRY] $(C_4H_8)_3$ A mixture of isomers; combustible liquid boiling at 348-354°C; used as a chemical and resin intermediate, lubricating oil additive, and motor-fuel alkylation feedstock. { trī,ī·sō'byüd·əl,ēn }

trimer [CHEMISTRY] An oligomer whose molecule is composed of three identical monomers. { 'trī·mər }

trimercuric orthophosphate [INORGANIC CHEMISTRY] *See* mercuric phosphate. { ,trī·mər'kyůr·ik ¦ȯr·thō'fä,sfāt }

trimercurous orthophosphate [INORGANIC CHEMISTRY] *See* mercurous phosphate. { ,trə·mər'kyůr·əs ¦ȯr·thō'fä,sfāt }

trimethylamine [ORGANIC CHEMISTRY] $(CH_3)_3N$ A colorless, liquefied gas with a fishy odor and a boiling point of −4°C; soluble in water, ether, and alcohol; used as a warning agent for natural gas, a flotation agent, and insect attractant. Abbreviated TMA. { trī¦meth·ə·lə¦mēn }

***uns*-trimethylbenzene** [INORGANIC CHEMISTRY] *See* pseudocumene.

trimethyl borate [ORGANIC CHEMISTRY] $B(OCH_3)_3$ A water-white liquid, boiling at 67-68°C; used as a solvent for resins, waxes, and oils, and as a catalyst and a reagent in analysis of paint and varnish. Also known as methyl borate. { trī'meth·əl 'bȯr,āt }

trimethylchlorosilane [ORGANIC CHEMISTRY] $(CH_3)_3SiCl$ A colorless liquid with a boiling point of 57°C; soluble in ether and benzene; used as a water-repelling agent. { trī¦meth·əl¦klȯr·ō'si,lān }

trimethylethylene [ORGANIC CHEMISTRY] *See* methyl butene. { trī¦meth·əl'eth·ə,lēn }

trimethylol aminomethane [ORGANIC CHEMISTRY] *See* tromethamine. { trī'meth·ə,lȯl ə¦mē·nō'meth,ān }

trimethylolethane [ORGANIC CHEMISTRY] $CH_3C(CH_2OH)_3$ Colorless crystals, soluble in alcohol and water; used in the manufacture of varnishes and drying oils. Also known as methyltrimethylolmethane; pentoglycerine. { trī¦meth·ə,lȯl'eth,ān }

trinitrobenzene [ORGANIC CHEMISTRY] $C_6H_3(NO_2)_3$ A yellow crystalline compound, soluble in alcohol and ether; used as an explosive. { trī¦nī·trō'ben,zēn }

2,4,7-trinitrofluorenone [ORGANIC CHEMISTRY] $C_{13}H_5N_3O_7$ A yellow, crystalline compound with a melting point of 175.2-176°C; forms crystalline complexes with indoles for identification by mass spectroscopy. Abbreviated TNF. { ¦tü ¦fór ¦sev·ən trī¦nī·trō'flůr·ə,nōn }

trinitromethane [ORGANIC CHEMISTRY] $CH(NO_2)_3$ A crystalline compound, melting at 150°C, decomposing above 25°C; used to make explosives. Also known as nitroform. { trī¦nī·trō'meth,ān }

2,4,6-trinitrotoluene [ORGANIC CHEMISTRY] $CH_3C_6H_2(NO_2)_3$ Toxic, flammable, explosive, yellow crystals; soluble in alcohol and ether, insoluble in water; melts at 81°C; used as an explosive and chemical intermediate and in photographic chemicals. Abbreviated TNT. { ¦tü ¦fór ¦siks trī¦nī·trō'täl·yə,wēn }

***sym*-trioxane** [ORGANIC CHEMISTRY] $(CH_2O)_3$ White, flammable, explosive crystals; soluble in water, alcohol, and ether; melts at 62°C; used as a chemical intermediate, disinfectant, and fuel. Also known as metaformaldehyde; triformol, trioxin. { ¦sim trī'äk,sān }

trioxygen [CHEMISTRY] *See* ozone.

tripalmitin [ORGANIC CHEMISTRY] $C_3H_5(OOCC_{15}H_{31})_3$ A white, water-insoluble powder that melts at 65.5°C; used in the preparation of leather dressings and soaps. { trī'päm·ə·tən }

triphenylmethane dye [ORGANIC CHEMISTRY] A family of dyes with a molecular structure derived from $(C_6H_5)_3CH$, usually by NH_2, OH, or HSO_3 substitution for one of the C_6H_5 hydrogens; includes many coal tar dyes, for example, rosaniline and fuchsin. { trī¦fen·əl'meth,ān 'dī }

triphenylmethyl radical [ORGANIC CHEMISTRY] A free radical in which three phenyl rings are bonded to a single carbon. Also known as trityl radical. { trī¦fen·əl¦meth·əl 'rad·ə·kəl }

triphenyl phosphate [ORGANIC CHEMISTRY] $(C_6H_5O)_3PO$ A crystalline compound with a melting point of 49-50°C; soluble in benzene, chloroform, ether, and acetone; used as a substitute for camphor in celluloid, as a plasticizer in lacquers and varnishes, and to impregnate roofing paper. { trī'fen·əl 'fä,sfāt }

triphenylphosphine [ORGANIC CHEMISTRY] $(C_6H_5)_3P$ A crystalline compound with a melting point of 80.5°C; soluble in ether, benzene, chloroform, and glacial acetic acid; used as an initiator of polymerization and in organic synthesis. { trī¦fen·əl'fä,sfēn }

triphenyltetrazolium chloride [ORGANIC CHEMISTRY] $C_{19}H_{15}ClN_4$ A crystalline compound, soluble in water, alcohol, and acetone; used as a sensitive reagent for reducing sugars. Also known as red tetrazolium. { trī¦fen·əl¦te·trə'zō·lē·əm 'klór,īd }

triphenyltinacetate [ORGANIC CHEMISTRY] *See* fentinacetate. { trī¦fen·əl tə'nas·ə,tāt }

triple phosphate [CHEMISTRY] A phosphate containing magnesium, calcium, and ammonium ions. { 'trip·əl 'fäs,fāt }

triple point [PHYSICAL CHEMISTRY] A particular temperature and pressure at which three different phases of one substance can coexist in equilibrium. { 'trip·əl 'póint }

tripotassium orthophosphate [INORGANIC CHEMISTRY] *See* potassium phosphate. { ,trī·pə'tas·ē·əm ¦ór·thō'fä,sfāt }

triprene [ORGANIC CHEMISTRY] $C_{18}H_{32}O_2S$ An amber liquid used as a growth regulator for crops. { 'trī,prēn }

triprotic acid [CHEMISTRY] An acid that has three ionizable hydrogen atoms in each molecule. { trī¦präd·ik 'as·əd }

triptane [ORGANIC CHEMISTRY] C_7H_{16} A hydrocarbon compound made commercially in small quantities, but having one of the highest antiknock ratings known. { 'trip,tān }

TRIS [ORGANIC CHEMISTRY] *See* tromethamine. { tris }

trisamine [ORGANIC CHEMISTRY] *See* tromethamine. { 'tris·ə,mēn }

tris buffer [INORGANIC CHEMISTRY] *See* tromethamine. { 'tris ,bəf·ər }

tris[2-(2,4-dichlorophenoxy)ethyl]phosphite [ORGANIC CHEMISTRY] $C_{24}H_{21}Cl_6O_6P$ A dark liquid that boils above 200°C; used as a preemergence herbicide for corn, peanuts, and strawberries. Abbreviated 2,4-DEP. { ¦tris ¦tü ¦tü ¦fór dī¦klór·ō·fə¦näk·sē ¦eth·əl 'fä,sfīt }

trisodium citrate [ORGANIC CHEMISTRY] *See* sodium citrate. { trī'sōd·ē·əm 'sī,trāt }

trisodium orthophosphate [INORGANIC CHEMISTRY] *See* trisodium phosphate. { trī'sōd·ē·əm ¦ór·thō'fä,sfāt }

trisodium phosphate [INORGANIC CHEMISTRY] Na_3PO_4 A water-soluble crystalline compound; used as a cleaning compound and as a water softener. Abbreviated TSP. Also known as tertiary sodium phosphate; trisodium orthophosphate. { trī'sōd·ē·əm 'fä,sfāt }

tristearin [ORGANIC CHEMISTRY] *See* stearin. { ¦trī'stir·ən }

trisulfide [CHEMISTRY] A binary chemical compound that contains three sulfur atoms in its molecule, for example, iron trisulfude, Fe_2S_3. { trī'səl,fīd }

triterpene [ORGANIC CHEMISTRY] One of a class of compounds having molecular skeletons containing 30 carbon atoms, and theoretically composed of six isoprene units; numerous and widely distributed in nature, occurring principally in plant resins and sap; an example is ambrein. { trī'tər,pēn }

trithioacetaldehyde [ORGANIC CHEMISTRY] $(C_4H_4S_2)_3$ A colorless, water-insoluble, crystalline compound; used as a hypnotic. { trī¦thī·ō,as·ə'tal·də,hīd }

tritiated [CHEMISTRY] Pertaining to matter in which tritium atoms have replaced one or more atoms of ordinary hydrogen. { 'trish·ē,ād·əd }

trityl radical [ORGANIC CHEMISTRY] *See* triphenylmethyl radical. { 'trīd·əl ,rad·ə·kəl }

triuranium octoxide [INORGANIC CHEMISTRY] U_3O_8 Olive green to black crystals or granules, soluble in nitric acid and sulfuric acid; decomposes at 1300°C; used in nuclear technology and in the preparation of other uranium compounds. Also known as uranous-uranic oxide; uranyl uranate. { ,trī·yů'rā·nē·əm äk'täk,sīd }

trivial name [ORGANIC CHEMISTRY] A common name for a chemical compound derived from the names of the natural source of the compound at the time of its isolation and before anything is known about its molecular structure. { 'triv·ē·əl ¦nām }

tromethamine [ORGANIC CHEMISTRY] $C_4H_{11}NO_3$ A crystalline compound with a melting point of 171-172°C; soluble in water, ethylene glycol, methanol, and ethanol; used to make surface-active agents, vulcanization accelerators, and pharmaceuticals, and as a titrimetric standard. Also known as THAM; trimethylol aminomethane; TRIS; trisamine; tris buffers. { trō'meth·ə,mēn }

tropeoline 00 [ORGANIC CHEMISTRY] $NaSO_3C_6H_4NNC_6H_4NHC_6H_5$ An acid-base indicator with a pH range of 1.4-3.0, color change (from acid to base) red to yellow; used as a biological stain. { trō'pē·ə·lən 'zir·ō 'zir·ō }

Trouton's rule [PHYSICAL CHEMISTRY] An approximation rule for the derivation of molar heats of vaporization of normal liquids at their boiling points. { 'traůt·ənz ,rül }

true condensing point [PHYSICAL CHEMISTRY] *See* critical condensation temperature. { 'trü kən'dens·iŋ ,pȯint }

true electrolyte [PHYSICAL CHEMISTRY] A substance in the solid state that consists entirely of ions. { ¦trü i¦lek·trə,līt }

true freezing point [PHYSICAL CHEMISTRY] The temperature at which the liquid and solid forms of a substance exist in equilibrium at a given pressure (usually 1 standard atmosphere, or 101,325 pascals). { 'trü 'frēz·iŋ ,pȯint }

TSP [INORGANIC CHEMISTRY] *See* trisodium phosphate.

TSPP [INORGANIC CHEMISTRY] *See* sodium pyrophosphate.

tubatoxin [ORGANIC CHEMISTRY] *See* rotenone. { 'tü·bə,täk·sən }

tungstate [INORGANIC CHEMISTRY] M_2WO_4 A salt of tungstic acid; for example, sodium tungstate, Na_2WO_4. { 'təŋ,stāt }

tungstate white [INORGANIC CHEMISTRY] *See* barium tungstate. { 'təŋ,stāt 'wīt }

tungsten [CHEMISTRY] A metallic transition element, symbol W, atomic number 74, atomic weight 183.85; soluble in mixed nitric and hydrofluoric acids; melts at 3400°C. Also known as wolfram. { 'təŋ·stən }

tungsten boride [INORGANIC CHEMISTRY] WB_2 A silvery solid; insoluble in water, soluble in aqua regia and concentrated acids; melts at 2900°C; used as a refractory for furnaces and chemical process equipment. { 'təŋ·stən 'bȯr,īd }

tungsten carbide [INORGANIC CHEMISTRY] WC A hard, gray powder; insoluble in water; readily attacked by nitric-hydrofluoric acid mixture; melts at 2780°C; used in tools, dies, ceramics, cermets, and wear-resistant mechanical parts, and as an abrasive. { 'təŋ·stən 'kär,bīd }

tungsten carbonyl [INORGANIC CHEMISTRY] *See* tungsten hexacarbonyl. { 'təŋ·stən 'kär·bə,nil }

tungsten disulfide [INORGANIC CHEMISTRY] WS_2 A grayish-black solid with a melting point above 1480°C; used as a lubricant and aerosol. { 'təŋ·stən dī'səl,fīd }

tungsten hexacarbonyl [INORGANIC CHEMISTRY] $W(CO)_6$ A white, refractive, crystalline solid which decomposes at 150°C; used for tungsten coatings on base metals. Also known as tungsten carbonyl. { 'təŋ·stən ¦hek·sə'kär·bə,nil }

tungsten hexachloride [INORGANIC CHEMISTRY] WCl_6 Dark blue or violet crystals with a melting point of 275°C; soluble in organic solvents; used for tungsten coatings on base metals and as a catalyst for olefin polymers. { 'təŋ·stən ¦hek·sə'klȯr,īd }

tungsten lake [ORGANIC CHEMISTRY] *See* phosphotungstic pigment. { 'təŋ·stən 'lāk }

tungsten oxychloride [INORGANIC CHEMISTRY] $WOCl_4$ Dark red crystals with a melting point of approximately 211°C; soluble in carbon disulfide; used for incandescent lamps. { 'təŋ·stən ¦äk·sə'klȯr,īd }

tungstic acid [INORGANIC CHEMISTRY] H_2WO_4 A yellow powder; insoluble in water, soluble in alkalies; used as a color-resist mordant for textiles, as an ingredient in plastics, and for the manufacture of tungsten metal products. Also known as orthotungstic acid; wolframic acid. { 'təŋ·stik 'as·əd }

tungstic acid anhydride [INORGANIC CHEMISTRY] *See* tungstic oxide. { 'təŋ·stik 'as·əd an'hī,drīd }

tungstic anhydride [INORGANIC CHEMISTRY] *See* tungstic oxide. { 'təŋ·stik an'hī,drīd }

tungstic oxide [INORGANIC CHEMISTRY] WO_3 A heavy, canary-yellow powder; soluble in caustic, insoluble in water; melts at 1473°C; used in alloys, in fabric fireproofing, for ceramic pigments, and for the manufacture of tungsten metal. Also known as anhydrous wolframic acid; tungstic acid anhydride; tungstic anhydride; tungstic trioxide. { 'təŋ·stik 'äk,sīd }

tungstic trioxide [INORGANIC CHEMISTRY] *See* tungstic oxide. { 'təŋ·stik trī'äk,sīd }

turbidimetric analysis [ANALYTICAL CHEMISTRY] A scattered-light procedure for the determination of the weight concentration of particles in cloudy, dull, or muddy solutions; uses a device that measures the loss in intensity of a light beam as it passes through the solution. Also known as turbidimetry. { ¦tər·bə·də¦me·trik ə'nal·ə·səs }

turbidimetric titration [ANALYTICAL CHEMISTRY] Titration in which the end point is indicated by the developing turbidity of the titrated solution. { ¦tər·bə·də¦me·trik 'tī,trā·shən }

turbidimetry [ANALYTICAL CHEMISTRY] *See* turbidimetric analysis. { ,tər·bə'dim·ə·trē }

turbidity [ANALYTICAL CHEMISTRY] **1.** Measure of the clarity of an otherwise clear liquid by using colorimetric scales. **2.** Cloudy or hazy appearance in a naturally clear liquid caused by a suspension of colloidal liquid droplets or fine solids. { tər'bid·əd·ē }

Turnbull's blue [INORGANIC CHEMISTRY] A blue pigment that precipitates from the reaction of potassium ferricyanide with a ferrous salt. { 'tərn,bu̇lz 'blü }

turpentine camphor [ORGANIC CHEMISTRY] *See* terpene hydrochloride. { 'tər·pən,tīn 'kam·fər }

twinned double bonds [ORGANIC CHEMISTRY] *See* cumulative double bonds. { ¦twind ¦dəb·əl 'bänz }

Twitchell reagent [ORGANIC CHEMISTRY] A catalyst for the acid hydrolysis of fats; a sulfonated addition product of naphthalene and oleic acid, that is, a naphthalenestearosulfonic acid. { 'twich·əl rē,ā·jənt }

two-dimensional chromatography [ANALYTICAL CHEMISTRY] A paper chromatography technique in which the sample is resolved by standard procedures (ascending, descending, or horizontal solvent movement) and then turned at right angles in a second solvent and re-resolved. { 'tü ¦di¦men·shən·əl ,krō·mə'täg·rə·fē }

two-fluid cell [PHYSICAL CHEMISTRY] Cell having different electrolytes at the positive and negative electrodes. { 'tü ¦flü·əd 'sel }

U

U [CHEMISTRY] *See* uranium.

UDMH [ORGANIC CHEMISTRY] *See uns*-dimethylhydrazine.

Ullmann reaction [ORGANIC CHEMISTRY] A variation of the Fittig synthesis, using copper powder instead of sodium. { 'əl·mən rē‚ak·shən }

ultimate analysis [ANALYTICAL CHEMISTRY] The determination of the percentage of elements contained in a chemical substance. { 'əl·tə·mət ə'nal·ə·səs }

ultramarine blue [INORGANIC CHEMISTRY] A blue pigment; a powder with heat resistance, used for enamels on toys and machinery, white baking enamels, printing inks, and cosmetics, and in textile printing. { ¦əl'trə·mə'rēn 'blü }

ultrasensitive mass spectrometry [ANALYTICAL CHEMISTRY] A form of mass spectrometry in which the ions to be detected are accelerated to megaelectronvolt energies in a particle accelerator and passed through a thin gas cell or foil, stripping away outer electrons, so that contaminating molecules dissociate into lower-mass fragments, and isobars can be distinguished by particle detectors that measure ionization rate and total energy. { ¦əl·trə'sen·səd·iv 'mas spek'träm·ə·trē }

ultraviolet absorption spectrophotometry [SPECTROSCOPY] The study of the spectra produced by the absorption of ultraviolet radiant energy during the transformation of an electron from the ground state to an excited state as a function of the wavelength causing the transformation. { ¦əl·trə'vī·lət əb'sorp·shən ‚spek·trō·fə'täm·ə·trē }

ultraviolet densitometry [SPECTROSCOPY] An ultraviolet-spectrophotometry technique for measurement of the colors on thin-layer chromatography absorbents following elution. { ¦əl·trə'vī·lət ‚den·sə'täm·ə·trē }

ultraviolet photoemission spectroscopy [SPECTROSCOPY] A spectroscopic technique in which photons in the energy range 10-200 electronvolts bombard a surface and the energy spectrum of the emitted electrons gives information about the states of electrons in atoms and chemical bonding. Abbreviated UPS. { ¦əl·trə'vī·lət ¦fōd·ō·i'mish·ən spek'träs·kə·pē }

ultraviolet spectrometer [SPECTROSCOPY] A device which produces a spectrum of ultraviolet light and is provided with a calibrated scale for measurement of wavelength. { ¦əl·trə'vī·lət spek'träm·əd·ər }

ultraviolet spectrophotometry [SPECTROSCOPY] Determination of the spectra of ultraviolet absorption by specific molecules in gases or liquids (for example, Cl_2, SO_2, NO_2, CS_2, ozone, mercury vapor, and various unsaturated compounds). { ¦əl·trə'vī·lət ¦spek·trō·fə'täm·ə·trē }

ultraviolet spectroscopy [SPECTROSCOPY] Absorption spectroscopy involving electromagnetic wavelengths in the range 4-400 nanometers. { ¦əl·trə'vī·lət spek'träs·kə·pē }

ultraviolet stabilizer [CHEMISTRY] *See* UV stabilizer. { ¦əl·trə'vī·lət 'stā·bə‚līz·ər }

uncharged species [CHEMISTRY] A chemical entity with no net electric charge. Also known as neutral species. { ¦ən'chärjd 'spē·shēz }

uncoupling phenomena [SPECTROSCOPY] Deviations of observed spectra from those predicted in a diatomic molecule as the magnitude of the angular momentum increases, caused by interactions which could be neglected at low angular momenta. { ¦ən'kəp·liŋ fə‚näm·ə·nä }

undecanal [ORGANIC CHEMISTRY] $CH_3(CH_2)_9CHO$ A sweet-smelling, colorless liquid, soluble in oils and alcohol; used in perfumes and flavoring. Also known as hendecanal; *n*-undecyclic aldehyde. { ən'dek·ə,nal }

undecane [ORGANIC CHEMISTRY] $CH_3(CH_2)_9CH_3$ A colorless, combustible liquid, boiling at 385°F (196°C), flash point at 149°F (65°C); used as a chemical intermediate and in petroleum research. Also known as hendecane. { ,ən'de,kān }

2-undecanone [ORGANIC CHEMISTRY] *See* methyl nonyl ketone. { ¦tü ,ən'dek·ə,nōn }

undecanoic acid [ORGANIC CHEMISTRY] $CH_3(CH_2)_9COOH$ Colorless crystals, soluble in alcohol and ether, insoluble in water; melts at 29°C; used as a chemical intermediate. { ¦ən¦dek·ə¦nō·ik 'as·əd }

undecyl [ORGANIC CHEMISTRY] $C_{11}H_{23}$ The radical of undecane. Also known as hendecyl. { ¦ən¦des·əl }

undecylenic acid [ORGANIC CHEMISTRY] $CH_2 \cdot CH(CH_2)_8COOH$ A light-colored, combustible liquid with a fruity aroma; soluble in alcohol, ether, chloroform, and benzene, almost insoluble in water; used in medicine, perfumes, flavors, and plastics. { ¦ən¦des·ə¦lin·ik 'as·əd }

undecylenic alcohol [ORGANIC CHEMISTRY] $C_{11}H_{22}O$ A colorless liquid with a citrus odor, soluble in 70% alcohol; used in perfumes. { ¦ən¦des·ə¦lin·ik 'al·kə,hol }

undecylenyl acetate [ORGANIC CHEMISTRY] $C_{13}H_{24}O_2$ A colorless liquid with a floral-fruity odor, soluble in 80% alcohol; used in perfumes and for flavoring. { ¦ən¦des·ə¦len·əl 'as·ə,tāt }

undersaturated fluid [PHYSICAL CHEMISTRY] Any fluid (liquid or gas) capable of holding additional vapor or liquid components in solution at specified conditions of pressure and temperature. { ¦ən·dər¦sach·ə,rād·əd 'flü·əd }

unidentate ligand [CHEMISTRY] A ligand that donates one pair of electrons in a complexation reaction to form coordinate bonds. { ¦yü·nē¦den,tāt 'līg·ənd }

unimolecular reaction [PHYSICAL CHEMISTRY] A chemical reaction involving only one molecular species as a reactant; for example, $2H_2O \rightarrow 2H_2 + O_2$, as in the electrolytic dissociation of water. { ¦yü·nə·mə'lek·yə·lər rē'ak·shən }

unsaturated compound [CHEMISTRY] Any chemical compound with more than one bond between adjacent atoms, usually carbon, and thus reactive toward the addition of other atoms at that point; for example, olefins, diolefins, and unsaturated fatty acids. { ¦ən'sach·ə,rād·əd 'käm,paủnd }

unsaturated hydrocarbon [ORGANIC CHEMISTRY] One of a class of hydrocarbons that have at least one double or triple carbon-to-carbon bond that is not in an aromatic ring; examples are ethylene, propadiene, and acetylene. { ¦ən'sach·ə,rād·əd ¦hī·drə'kär·bən }

unsaturation [ORGANIC CHEMISTRY] A state in which the atomic bonds of an organic compound's chain or ring are not completely satisfied (that is, not saturated); usually applies to carbon, but can include other ring or chain atoms; unsaturation usually results in a double bond (as for olefins) or a triple bond (as for the acetylenes). { ¦ən,sach·ə'rā·shən }

uns-, unsym- [ORGANIC CHEMISTRY] A chemical prefix denoting that the substituents of an organic compound are structurally unsymmetrical with respect to the carbon skeleton, or with respect to a function group (for example, double or triple bond). { əns, ¦ən·sim }

upflow [CHEMISTRY] In an ion-exchange unit, an operation in which solutions enter at the bottom of the unit and leave at the top. { 'əp,flō }

upper explosive limit [CHEMISTRY] *See* upper flammable limit. { ¦əp·ər ik'splō·siv ,lim·ət }

upper flammable limit [CHEMISTRY] The maximum percentage of flammable gas or vapor in the air above which ignition cannot take place because the ratio of the gas to oxygen is too high. Also known as upper explosive limit. { ¦əp·ər 'flam·ə·bəl ,lim·ət }

UPS [SPECTROSCOPY] *See* ultraviolet photoemission spectroscopy.

uranate [INORGANIC CHEMISTRY] A salt of uranic acid; for example, sodium uranate, Na_2UO_4. { 'yủr·ə,nāt }

urania [INORGANIC CHEMISTRY] *See* uranium dioxide. { yə'rā·nē·ə }

uranic chloride [INORGANIC CHEMISTRY] *See* uranium tetrachloride. { yu̇'ran·ik 'klȯr,īd }

uranic oxide [INORGANIC CHEMISTRY] *See* uranium dioxide. { yu̇'ran·ik 'äk,sīd }

uranin [ORGANIC CHEMISTRY] *See* uranine. { 'yu̇r·ə·nən }

uranine [ORGANIC CHEMISTRY] $Na_2C_{20}H_{10}O_5$ A brown or orange-red hygroscopic powder soluble in water; used as a yellow dye for silk and wool, a marker in the ocean to facilitate air and sea rescues, and as an analytical reagent. Also known as sodium fluorescein; uranin; uranine yellow. { 'yu̇r·ə,nēn }

uranine yellow [ORGANIC CHEMISTRY] *See* uranine. { 'yu̇r·ə,nēn 'yel·ō }

uranium [CHEMISTRY] A metallic element in the actinide series, symbol U, atomic number 92, atomic weight 238.03; highly toxic and radioactive; ignites spontaneously in air and reacts with nearly all nonmetals; melts at 1132°C, boils at 3818°C; used in nuclear fuel and as the source of ^{235}U and plutonium. { yə'rā·nē·əm }

uranium acetate [INORGANIC CHEMISTRY] *See* uranyl acetate. { yə'rā·nē·əm 'as·ə,tāt }

uranium carbide [INORGANIC CHEMISTRY] One of the carbides of uranium, such as uranium monocarbide; used chiefly as a nuclear fuel. { yə'rā·nē·əm 'kär,bīd }

uranium dioxide [INORGANIC CHEMISTRY] UO_2 Black, highly toxic, spontaneously flammable, radioactive crystals; insoluble in water, soluble in nitric and sulfuric acids; melts at approximately 3000°C; used to pack nuclear fuel rods and in ceramics, pigments, and photographic chemicals. Also known as urania; uranic oxide; uranium oxide. { yə'rā·nē·əm dī'äk,sīd }

uranium hexafluoride [INORGANIC CHEMISTRY] UF_6 Highly toxic, radioactive, corrosive, colorless crystals; soluble in carbon tetrachloride, fluorocarbons, and liquid halogens; it reacts vigorously with alcohol, water, ether, and most metals, and it sublimes; used to separate uranium isotopes in the gaseous-diffusion process. { yə'rā·nē·əm ¦hek·sə'flu̇r,īd }

uranium hydride [INORGANIC CHEMISTRY] UH_3 A highly toxic, gray to black powder that ignites spontaneously in air, and that conducts electricity; used for making powdered uranium metal, for hydrogen-isotope separation, and as a reducing agent. { yə'rā·nē·əm 'hī,drīd }

uranium nitrate [INORGANIC CHEMISTRY] *See* uranyl nitrate. { yə'rā·nē·əm 'nī,trāt }

uranium oxide [INORGANIC CHEMISTRY] *See* uranium dioxide; uranium trioxide. { yə'rā·nē·əm 'äk,sīd }

uranium sulfate [INORGANIC CHEMISTRY] *See* uranyl sulfate. { yə'rā·nē·əm 'səl,fāt }

uranium tetrachloride [INORGANIC CHEMISTRY] UCl_4 Poisonous, radioactive, hygroscopic, dark-green crystals; soluble in alcohol and water; melts at 590°C, boils at 792°C. Also known as uranic chloride. { yə'rā·nē·əm ¦te·trə'klȯr,īd }

uranium tetrafluoride [INORGANIC CHEMISTRY] UF_4 Toxic, radioactive, corrosive green crystals; insoluble in water; melts at 1036°C; used in the manufacture of uranium metal. Also known as green salt. { yə'rā·nē·əm ¦te·trə'flu̇r,īd }

uranium trioxide [INORGANIC CHEMISTRY] UO_3 A poisonous, radioactive, red to yellow powder; soluble in nitric acid, insoluble in water; decomposes when heated; used in ceramics and pigments and for uranium refining. Also known as orange oxide; uranium oxide. { yə'rā·nē·əm trī'äk,sīd }

uranium yellow [ORGANIC CHEMISTRY] *See* sodium diuranate. { yə'rā·nē·əm yel·ō }

uranous-uranic oxide [INORGANIC CHEMISTRY] *See* triuranium octoxide. { 'yu̇r·ə·nəs yə'ran·ik 'äk,sīd }

uranyl acetate [INORGANIC CHEMISTRY] $UO_2(C_2H_3O_2)_2 \cdot 2H_2O$ Poisonous, radioactive yellow crystals, decomposed by light; soluble in cold water, decomposes in hot water; loses water of crystallization at 110°C, decomposes at 275°C; used in medicine and as an analytical reagent and bacterial oxidant. Also known as uranium acetate. { 'yu̇r·ə,nil 'as·ə,tāt }

uranyl nitrate [INORGANIC CHEMISTRY] $UO_2(NO_3)_2 \cdot 6H_2O$ Toxic, explosive, unstable yellow crystals; soluble in water, alcohol, and ether; melts at 60°C and boils at 118°C; used in photography, in medicine, and for uranium extraction and uranium glaze. Also known as uranium nitrate; yellow salt. { 'yu̇r·ə,nil 'nī,trāt }

uranyl salts [INORGANIC CHEMISTRY] Salts of UO_3 that ionize to form UO_2^{2+} and that are yellow in solution; for example, uranyl chloride, UO_2Cl_2. { 'yu̇r·ə,nil ,sȯls }

uranyl sulfate [INORGANIC CHEMISTRY] $UO_2SO_4 \cdot 3^1/_2H_2O$ and $UO_2SO_4 \cdot 3H_2O$ Poisonous, radioactive yellow crystals; soluble in water and concentrated hydrochloric acid; used as an analytical reagent. Also known as uranium sulfate. { 'yu̇r·ə,nil 'səl,fāt }

uranyl uranate [INORGANIC CHEMISTRY] *See* triuranium octoxide. { 'yu̇r·ə,nil 'yu̇r·ə,nāt }

urbacid [ORGANIC CHEMISTRY] $C_7H_{15}AsN_2S_3$ A colorless, crystalline compound with a melting point of 144°C; insoluble in water; used to control apple scale and diseases of coffee trees. { 'ər·bə,sid }

urea [ORGANIC CHEMISTRY] $CO(HN_2)_2$ A natural product of protein metabolism found in urine; synthesized as white crystals or powder with a melting point of 132.7°C; soluble in water, alcohol, and benzene; used as a fertilizer, in plastics, adhesives, and flameproofing agents, and in medicine. Also known as carbamide. { yu̇'rē·ə }

urea anhydride [INORGANIC CHEMISTRY] *See* cyanamide. { yu̇'rē·ə an'hī,drīd }

urea-formaldehyde resin [ORGANIC CHEMISTRY] A synthetic thermoset resin derived by the reaction of urea (carbamide) with formaldehyde or its polymers. Also known as urea resin. { yu̇'rē·ə fȯr'mal·də,hīd 'rez·ən }

urea nitrate [ORGANIC CHEMISTRY] $CO(NH_2)_2 \cdot HNO_3$ Colorless, explosive, fire-hazardous crystals; soluble in alcohol, slightly soluble in water; decomposes at 152°C; used in explosives and to make urethane. { yu̇'rē·ə 'nī,trāt }

urea peroxide [ORGANIC CHEMISTRY] $CO(NH_2)_2 \cdot H_2O_2$ An unstable, fire-hazardous white powder; soluble in water, alcohol, and ethylene glycol; decomposes at 75-85°C or by moisture; used as a source of water-free hydrogen peroxide, as a disinfectant, in cosmetics and pharmaceuticals, and for bleaching. { yu̇'rē·ə pə'räk,sīd }

urea resin [ORGANIC CHEMISTRY] *See* urea-formaldehyde resin. { yu̇'rē·ə 'rez·ən }

urethane [ORGANIC CHEMISTRY] $CO(NH_2)OC_2H_5$ A combustible, toxic, colorless powder; soluble in water and alcohol; melts at 49°C; used as a solvent and chemical intermediate and in biochemical research and veterinary medicine. Also known as ethyl carbamate; ethyl urethane. { 'yu̇r·ə,thān }

uronic acid [ORGANIC CHEMISTRY] One of the compounds that are similar to sugars, except that the terminal carbon has been oxidized from the alcohol to a carboxyl group; for example, galacturonic acid and glucuronic acid. { yə'rän·ik 'as·əd }

urotropin [ORGANIC CHEMISTRY] *See* cystamine. { yə'rä·trə·pən }

USP acid test [ANALYTICAL CHEMISTRY] A United States Pharmacopoeia test to determine the carbonizable substances present in petroleum white oils. { ¦yü¦es'pē 'as·əd ,test }

UV stabilizer [CHEMISTRY] Any chemical compound that, admixed with a thermoplastic resin, selectively absorbs ultraviolet rays; used to prevent ultraviolet degradation of polymers. Also known as ultraviolet stabilizer. { ¦yü¦vē 'stā·bə,līz·ər }

V

V [CHEMISTRY] *See* vanadium.

vacuum condensing point [CHEMISTRY] Temperature at which the sublimate (vaporized solid) condenses in a vacuum.Abbreviated vcp. { 'vak·yəm kən'dens·iŋ ‚pȯint }

vacuum thermobalance [ANALYTICAL CHEMISTRY] An instrument used in thermogravimetry consisting of a precision balance and furnace that have been adapted for continuously measuring or recording changes in weight of a substance as a function of temperature; used in many types of physicochemical reactions where rates of reaction and energies of activation for vaporization, sublimation, and chemical reaction can be obtained. { 'vak·yəm ¦thər·mō'bal·əns }

vacuum ultraviolet spectroscopy [SPECTROSCOPY] Absorption spectroscopy involving electromagnetic wavelengths shorter than 200 nanometers; so called because the interference of the high absorption of most gases necessitates work with evacuated equipment. { 'vak·yəm ¦əl·trə'vī·lət spek'träs·kə·pē }

valence [CHEMISTRY] A positive number that characterizes the combining power of an element for other elements, as measured by the number of bonds to other atoms which one atom of the given element forms upon chemical combination; hydrogen is assigned valence 1, and the valence is the number of hydrogen atoms, or their equivalent, with which an atom of the given element combines. { 'vā·ləns }

valence angle [PHYSICAL CHEMISTRY] *See* bond angle. { 'vā·ləns ‚aŋ·gəl }

valence bond [PHYSICAL CHEMISTRY] The bond formed between the electrons of two or more atoms. { 'vā·ləns ‚bänd }

valence-bond method [PHYSICAL CHEMISTRY] A method of calculating binding energies and other parameters of molecules by taking linear combinations of electronic wave functions, some of which represent covalent structures, others ionic structures; the coefficients in the linear combination are calculated by the variational method. Also known as valence-bond resonance method. { 'vā·ləns ¦bänd ‚meth·əd }

valence-bond resonance method [PHYSICAL CHEMISTRY] *See* valence-bond method. { 'vā·ləns ¦bänd 'rez·ən·əns ‚meth·əd }

valence-bond theory [CHEMISTRY] A theory of the structure of chemical compounds according to which the principal requirements for the formation of a covalent bond are a pair of electrons and suitably oriented electron orbitals on each of the atoms being bonded; the geometry of the atoms in the resulting coordination polyhedron is coordinated with the orientation of the orbitals on the central atom. { 'vā·ləns ¦bänd ‚thē·ə·rē }

valence number [CHEMISTRY] A number that is equal to the valence of an atom or ion multiplied by +1 or −1, depending on whether the ion is positive or negative, or equivalently on whether the atom in the molecule under consideration has lost or gained electrons from its free state. { 'vā·ləns ‚nəm·bər }

valence transition [PHYSICAL CHEMISTRY] A change in the electronic occupation of the 4*f* or 5*f* orbitals of the rare-earth or actinide atoms in certain substances at a certain temperature, pressure, or composition. { 'vā·ləns tran‚zish·ən }

***n*-valeraldehyde** [ORGANIC CHEMISTRY] $CH_3(CH_2)_3CHO$ A flammable liquid, soluble in ether and alcohol, slightly soluble in water; boils at 102°C; used in flavors and as a rubber accelerator. { ¦en ¦val·ər'al·də‚hīd }

valeramide [ORGANIC CHEMISTRY] $CH_3(CH_2)_3CONH_2$ Water-soluble, colorless crystals, melting at 127°C. Also known as pentanamide; valeric amide. { ¦val·ər′am·əd }

valerianic acid [ORGANIC CHEMISTRY] *See* valeric acid. { və¦lir·ē¦an·ik ′as·əd }

valeric acid [ORGANIC CHEMISTRY] $CH_3(CH_2)_3COOH$ A combustible, toxic, colorless liquid with a penetrating aroma; soluble in water, alcohol, and ether; boils at 185°C; used to make flavors, perfumes, lubricants, plasticizers, and pharmaceuticals. Also known as valerianic acid. { və′lir·ik ′as·əd }

γ-valerolactone [ORGANIC CHEMISTRY] $C_5H_8O_2$ A combustible, mostly immiscible, colorless liquid, boiling at 205°C; used as a dye-bath coupling agent, in brake fluids and cutting oils, and as a solvent for adhesives, lacquers, and insecticides. { ¦gam·ə və¦lir·ō′lak‚tōn }

value of isotope mixture [CHEMISTRY] A measure of the effort required to prepare a quantity of an isotope mixture; it is proportional to the amount of the mixture, and also depends on the composition of the mixture to be prepared and the composition of the original mixture. { ′val·yü əv ′ī·sə‚tōp ‚miks·chər }

vamidothion [ORGANIC CHEMISTRY] $C_7H_{16}NO_4PS_2$ A white wax with a melting point of 40°C; very soluble in water; used to control pests in orchards, vineyards, rice, cotton, and ornamentals. { ¦vam·əd·ō′thī‚än }

vanadic acid [INORGANIC CHEMISTRY] Any of various acids that do not exist in a pure state and are found in various alkali and other metal vanadates; forms are meta- (HVO_3), ortho- (H_3VO_4), and pyro- ($H_4V_2O_7$). { və′nād·ik ′as·əd }

vanadic acid anhydride [INORGANIC CHEMISTRY] *See* vanadium pentoxide. { və′nād·ik ′as·əd an′hī‚drīd }

vanadic sulfide [INORGANIC CHEMISTRY] *See* vanadium sulfide. { və′nād·ik ′səl‚fīd }

vanadium [CHEMISTRY] A metallic transition element, symbol V, atomic number 23; soluble in strong acids and alkalies; melts at 1900°C, boils about 3000°C; used as a catalyst. { və′nād·ē·əm }

vanadium carbide [INORGANIC CHEMISTRY] VC Hard, black crystals, melting at 2800°C, boiling at 3900°C; insoluble in acids, except nitric acid; used in cutting-tool alloys and as a steel additive. { və′nād·ē·əm ′kär‚bīd }

vanadium dichloride [INORGANIC CHEMISTRY] VCl_2 Toxic, green crystals, soluble in alcohol and ether; decomposes in hot water; used as a reducing agent. Also known as vanadous chloride. { və′nād·ē·əm dī′klȯr‚īd }

vanadium oxide [INORGANIC CHEMISTRY] A compound of vanadium with oxygen, for example, vanadium tetroxide (V_2O_4), vanadium trioxide or sesquioxide (V_2O_3), vanadium oxide (VO), and vanadium pentoxide (V_2O_5). { və′nād·ē·əm ′äk‚sīd }

vanadium oxytrichloride [INORGANIC CHEMISTRY] $VOCl_3$ A toxic, yellow liquid that dissolves or reacts with many organic substances; hydrolyzes in moisture; boils at 126°C; used as an olefin-polymerization catalyst and in organovanadium synthesis. { və′nād·ē·əm ¦äk·sē·trī′klȯr‚īd }

vanadium pentasulfide [INORGANIC CHEMISTRY] *See* vanadium sulfide. { və′nād·ē·əm ¦pen·tə′səl‚fīd }

vanadium pentoxide [INORGANIC CHEMISTRY] V_2O_5 A toxic, yellow to red powder, soluble in alkalies and acids, slightly soluble in water; melts at 690°C; used in medicine, as a catalyst, as a ceramics coloring, for ultraviolet-resistant glass, photographic developers, textiles dyeing, and nuclear reactors. Also known as vanadic acid anhydride. { və′nād·ē·əm ¦pen′täk‚sīd }

vanadium sesquioxide [INORGANIC CHEMISTRY] *See* vanadium trioxide. { və′nād·ē·əm ¦ses·kwē′äk‚sīd }

vanadium sulfate [INORGANIC CHEMISTRY] *See* vanadyl sulfate. { və′nād·ē·əm ′səl‚fāt }

vanadium sulfide [INORGANIC CHEMISTRY] V_2S_5 A toxic, black-green powder; insoluble in water, soluble in alkalies and acids; decomposes when heated; used to make vanadium compounds. Also known as vanadic sulfide; vanadium pentasulfide. { və′nād·ē·əm ′səl‚fīd }

vanadium tetrachloride [INORGANIC CHEMISTRY] VCl_4 A toxic, red liquid; soluble in ether and absolute alcohol; boils at 154°C; used in medicine and to manufacture vanadium and organovanadium compounds. { və′nād·ē·əm ¦te·trə′klȯr‚īd }

vanadium tetraoxide [INORGANIC CHEMISTRY] V_2O_4 A toxic blue-black powder; insoluble in water, soluble in alkalies and acids; melts at 1967°C; used as a catalyst. { və'nād·ē·əm ¦te·trə'äk‚sīd }

vanadium trichloride [INORGANIC CHEMISTRY] VCl_3 Toxic, deliquescent, pink crystals; soluble in ether and absolute alcohol; decomposes in water and when heated; used to prepare vanadium and organovanadium compounds. { və'nād·ē·əm ¦trī'klór‚īd }

vanadium trioxide [INORGANIC CHEMISTRY] V_2O_3 Toxic, black crystals; soluble in alkalies and hydrofluoric acid, slightly soluble in water; melts at 1970°C; used as a catalyst. Also known as vanadium sesquioxide. { və'nād·ē·əm trī'äk‚sīd }

vanadous chloride [INORGANIC CHEMISTRY] *See* vanadium dichloride. { və'nād·əs 'klór‚īd }

vanadyl chloride [INORGANIC CHEMISTRY] $V_2O_2Cl_4 \cdot 5H_2O$ Toxic, deliquescent, water- and alcohol-soluble green crystals; used to mordant textiles. { və'nād·əl 'klór‚īd }

vanadyl sulfate [INORGANIC CHEMISTRY] $VOSO_4 \cdot 2H_2O$ Blue, toxic, water-soluble crystals; used as a reducing agent, catalyst, glass and ceramics colorant, and mordant. Also known as vanadic sulfate; vanadium sulfate. { və'nād·əl 'səl‚fāt }

Van Deemter rate theory [ANALYTICAL CHEMISTRY] A theory that the sample phase in gas chromatography flows continuously, not stepwise. { van 'dām·tər 'rāt ‚thē·ə·rē }

van der Waals adsorption [PHYSICAL CHEMISTRY] Adsorption in which the cohesion between gas and solid arises from van der Waals forces. { 'van dər ‚wólz ad‚sórp·shən }

van der Waals attraction [PHYSICAL CHEMISTRY] *See* van der Waals force. { 'van dər ‚wólz ə‚trak·shən }

van der Waals covolume [PHYSICAL CHEMISTRY] The constant b in the van der Waals equation, which is approximately four times the volume of an atom of the gas in question multiplied by Avogadro's number. { 'van dər ‚wólz ¦kō'väl·yəm }

van der Waals equation [PHYSICAL CHEMISTRY] An empirical equation of state which takes into account the finite size of the molecules and the attractive forces between them: $p = [RT/(v - b)] - (a/v^2)$, where p is the pressure, v is the volume per mole, T is the absolute temperature, R is the gas constant, and a and b are constants. { 'van dər ‚wólz i‚kwā·zhən }

van der Waals force [PHYSICAL CHEMISTRY] An attractive force between two atoms or nonpolar molecules, which arises because a fluctuating dipole moment in one molecule induces a dipole moment in the other, and the two dipole moments then interact. Also known as dispersion force; London dispersion force; van der Waals attraction. { 'van dər ‚wólz ‚fórs }

van der Waals molecule [PHYSICAL CHEMISTRY] A molecule that is held together by van der Waals forces. { 'van·dər ‚wälz ‚mäl·ə‚kyül }

van der Waals radius [PHYSICAL CHEMISTRY] One-half the distance between two atoms of an element that are as close to each other as possible without being formally bonded to each other except for van der Waals forces. { ¦van·dər‚wälz 'rād·e·əs }

vanillic aldehyde [ORGANIC CHEMISTRY] *See* vanillin. { və'nil·ik 'al·də‚hīd }

vanillin [ORGANIC CHEMISTRY] $C_8H_8O_3$ A combustible solid, soluble in water, alcohol, ether, and chloroform; melts at 82°C; used in pharmaceuticals, perfumes, and flavors, and as an analytical reagent. Also known as vanillic aldehyde. { və'nil·ən }

van't Hoff equation [PHYSICAL CHEMISTRY] An equation for the variation with temperature T of the equilibrium constant K of a gaseous reaction in terms of the heat of reaction at constant pressure, ΔH: $d(\ln K)/dT = \Delta H/RT^2$, where R is the gas constant. Also known as van't Hoff isochore. { van'tóf i‚kwā·zhən }

van't Hoff formula [ORGANIC CHEMISTRY] The expression that the number of stereoisomers of a sugar molecule is equal to 2^n, where n is the number of asymmetric carbon atoms. { van'tóf ‚fór·myə·lə }

van't Hoff isochore [PHYSICAL CHEMISTRY] *See* van't Hoff equation. { van'tóf 'ī·sə‚kór }

van't Hoff isotherm [PHYSICAL CHEMISTRY] An equation for the change in free energy during a chemical reaction in terms of the reaction, the temperature, and the concentration and number of molecules of the reactants. { van'tóf 'ī·sə‚thərm }

vapor-liquid equilibrium [PHYSICAL CHEMISTRY] *See* liquid-vapor equilibrium. { 'vā·pər 'lik·wəd ‚e·kwə'lib·rē·əm }

vapor-pressure osmometer [ANALYTICAL CHEMISTRY] A device for the determination of molecular weights by the decrease of vapor pressure of a solvent upon addition of a soluble sample. { 'vā·pər ¦presh·ər äz'mäm·əd·ər }

V band [SPECTROSCOPY] Absorption bands that appear in the ultraviolet part of the spectrum due to color centers produced in potassium bromide by exposure of the crystal at temperature of liquid nitrogen (81 K) to intense penetrating x-rays. { 'vē ‚band }

vcp [CHEMISTRY] *See* vacuum condensing point.

Venetian red [INORGANIC CHEMISTRY] A pigment with a true red hue; contains 15-40% ferric oxide and 60-80% calcium sulfate. { və'nēsh·ən 'red }

verdigris [ORGANIC CHEMISTRY] *See* cupric acetate. { 'vərd·ə‚grēs }

vermilion [INORGANIC CHEMISTRY] *See* mercuric sulfide. { vər'mil·yən }

vernolate [ORGANIC CHEMISTRY] $C_{10}H_{21}NOS$ An amber liquid, used to control weeds in sweet potatoes, peanuts, soybeans, and tobacco. { 'vərn·əl‚āt }

vibration [PHYSICAL CHEMISTRY] Oscillation of atoms about their equilibrium positions within a molecular system. { vī'brā·shən }

vibrational energy [PHYSICAL CHEMISTRY] For a diatomic molecule, the difference between the energy of the molecule idealized by setting the rotational energy equal to zero, and that of a further idealized molecule which is obtained by gradually stopping the vibration of the nuclei without placing any new constraint on the motions of electrons. { vī'brā·shən·əl 'en·ər·jē }

vibrational level [PHYSICAL CHEMISTRY] An energy level of a diatomic or polyatomic molecule characterized by a particular value of the vibrational energy. { vī'brā·shən·əl 'lev·əl }

vibrational quantum number [PHYSICAL CHEMISTRY] A quantum number v characterizing the vibrational motion of nuclei in a molecule; in the approximation that the molecule behaves as a quantum-mechanical harmonic oscillator, the vibrational energy is $h(v + {}^1/_2)f$, where h is Planck's constant and f is the vibration frequency. { vī'brā·shən·əl 'kwän·təm ‚nəm·bər }

vibrational spectrum [SPECTROSCOPY] The molecular spectrum resulting from transitions between vibrational levels of a molecule which behaves like the quantum-mechanical harmonic oscillator. { vī'brā·shən·əl 'spek·trəm }

vibrational sum rule [SPECTROSCOPY] **1.** The rule that the sums of the band strengths of all emission bands with the same upper state is proportional to the number of molecules in the upper state, where the band strength is the emission intensity divided by the fourth power of the frequency. **2.** The sums of the band strengths of all absorption bands with the same lower state is proportional to the number of molecules in the lower state, where the band strength is the absorption intensity divided by the frequency. { vī'brā·shən·əl 'səm ‚rül }

vibrational transition [PHYSICAL CHEMISTRY] A transition between two quantized levels of a molecule that have different vibrational energies. Also known as vibronic transition. { vī'brā·shən·əl tran'zish·ən }

vibronic transition [PHYSICAL CHEMISTRY] *See* vibrational transition. { vī¦brän·ik tran'zish·ən }

vic- [ORGANIC CHEMISTRY] A chemical prefix indicating vicinal (neighboring or adjoining) positions on a carbon structure (ring or chain); used to identify the location of substituting groups when naming derivatives. { vik }

vicinal [ORGANIC CHEMISTRY] Referring to neighboring or adjoining positions on a carbon structure (ring or chain). { 'vis·ən·əl }

Victoria blue [ORGANIC CHEMISTRY] $C_{33}H_{31}N_3 \cdot HCl$ Bronze crystals, soluble in hot water, alcohol, and ether; used as a dye for silk, wool, and cotton, as a biological stain, and to make pigment toners. { vik'tór·ē·ə 'blü }

Vigreaux column [ANALYTICAL CHEMISTRY] An obsolete apparatus used in laboratory fractional distillation; it is a long glass tube with indentation in its walls; a thermometer is placed at the top of the tube and a side arm is attached to a condenser. { ve'grō ˌkäl·əm }

vinetine [ORGANIC CHEMISTRY] *See* oxyacanthine. { 'vin·əˌtēn }

vinyl acetal resin [ORGANIC CHEMISTRY] $[CH_2CH(OC_2H_5)]_x$ A colorless, odorless, light-stable thermoplastic that is unaffected by water, gasoline, or oils; soluble in lower alcohols, benzene, and chlorinated hydrocarbons; used in lacquers, coatings, and molded objects. Also known as polyvinyl acetal resin. { 'vīn·əl 'as·əˌtəl 'rez·ən }

vinyl acetate [ORGANIC CHEMISTRY] $CH_3COOCH{:}CH_2$ A colorless, water-insoluble, flammable liquid that boils at 73°C; used as a chemical intermediate and in the production of polymers and copolymers (for example, the polyvinyl resins). { 'vīn·əl 'as·əˌtāt }

vinyl acetate resin [ORGANIC CHEMISTRY] $(CH_2{:}CHOOCCH_3)_x$ An odorless thermoplastic formed by the polymerization of vinyl acetate; resists attack by water, gasoline, and oils; soluble in lower alcohols, benzene, and chlorinated hydrocarbons; used in lacquers, coatings, and molded products. { 'vīn·əl 'as·əˌtāt 'rez·ən }

vinylacetylene [ORGANIC CHEMISTRY] $H_2CCHCCH$ A combustible dimer of acetylene, boiling at 5°C; used for the manufacture of neoprene rubber and as a chemical intermediate. { ¦vīn·əl·ə'sed·əlˌēn }

vinyl alcohol [ORGANIC CHEMISTRY] $CH_2{:}CHOH$ A flammable, unstable liquid found only in ester or polymer form. Also known as ethenol. { 'vīn·əl 'al·kəˌhȯl }

vinylation [CHEMISTRY] Formation of a vinyl-derived product by reaction with acetylene; for example, vinylation of alcohols gives vinyl ethers, such as vinyl ethyl ether. { ˌvīn·əl'ā·shən }

vinylbenzene [ORGANIC CHEMISTRY] *See* styrene. { ¦vīn·əl'benˌzēn }

vinyl chloride [ORGANIC CHEMISTRY] $CH_2{:}CHCl$ A flammable, explosive gas with an ethereal aroma; soluble in alcohol and ether, slightly soluble in water; boils at −14°C; an important monomer for polyvinyl chloride and its copolymers; used in organic synthesis and in adhesives. Also known as chloroethene; chloroethylene. { 'vīn·əl 'klȯrˌīd }

vinyl chloride resin [ORGANIC CHEMISTRY] $(CH_2CHCl)_x$ A white-power polymer made by the polymerization of vinyl chloride; used to make chemical-resistant pipe (when unplasticized) or bottles and parts (when plasticized). { 'vīn·əl 'klȯrˌīd ˌrez·ən }

vinylcyanide [ORGANIC CHEMISTRY] *See* acrylonitrile. { ¦vīn·əl'sī·əˌnīd }

vinyl ether [ORGANIC CHEMISTRY] $CH_2{:}CHOCH{:}CH_2$ A colorless, light-sensitive, flammable, explosive liquid; soluble in alcohol, acetone, ether, and chloroform, slightly soluble in water; boils at 39°C; used as an anesthetic and a comonomer in polyvinyl chloride polymers. Also known as divinyl ether; divinyl oxide. { 'vīn·əl 'ē·thər }

vinyl ether resin [ORGANIC CHEMISTRY] Any of a group of vinyl ether polymers; for example, polyvinyl methyl ether, polyvinyl ethyl ether, and polyvinyl butyl ether. { 'vīn·əl 'ē·thər 'rez·ən }

vinyl group [ORGANIC CHEMISTRY] $CH_2{=}CH{-}$ A group of atoms derived when one hydrogen atom is removed from ethylene. { 'vīn·əl ˌgrüp }

vinylidene chloride [ORGANIC CHEMISTRY] $CH_2{:}CCl_2$ A colorless, flammable, explosive liquid, insoluble in water; boils at 37°C; used to make polymers copolymerized with vinyl chloride or acrylonitrile (Saran). { vī'nil·əˌdēn 'klȯrˌīd }

vinylidene resin [ORGANIC CHEMISTRY] A polymer made up of the $({-}H_2CCX_2{-})$ unit, with X usually a chloride, fluoride, or cyanide radical. Also known as polyvinylidene resin. { vī'nil·əˌdēn 'rez·ən }

vinylog [ORGANIC CHEMISTRY] Any of the organic compounds that differ from each other by a vinylene linkage $({-}CH{=}CH{-})$; for example, ethyl crotonate is a vinylog of ethyl acetate and of the next higher vinylog, ethyl sorbate. { 'vīn·əlˌäg }

vinyl plastic [ORGANIC CHEMISTRY] *See* polyvinyl resin. { 'vīn·əl 'plas·tik }

vinyl polymerization [ORGANIC CHEMISTRY] Addition polymerization where the unsaturated monomer contains a $CH_2{=}C{-}$ group. { 'vīn·əl pəˌlim·ə·rə'zā·shən }

vinylpyridine [ORGANIC CHEMISTRY] $C_5H_4NCH{:}CH_2$ A toxic, combustible liquid; soluble in water, alcohol, hydrocarbons, esters, ketones, and dilute acids; used to manufacture elastomers and pharmaceuticals. { ¦vīn·əl'pir·əˌdēn }

N-vinyl-2-pyrrolidone [ORGANIC CHEMISTRY] C_6H_9ON A colorless, toxic, combustible liquid, boiling at 148°C (100 mmHg); used as a chemical intermediate and to make polyvinyl pyrrolidone. { ¦en ¦vīn·əl ¦tü pə'räl·ə‚dōn }

vinylstyrene [ORGANIC CHEMISTRY] *See* divinylbenzene. { ¦vīn·əl'stī‚rēn }

vinyltoluene [ORGANIC CHEMISTRY] $CH_2{:}CHC_6H_4CH_3$ A colorless, flammable, moderately toxic liquid; soluble in ether and methanol, slightly soluble in water; boils at 170°C; used as a chemical intermediate and solvent. Also known as methyl styrene. { ¦vīn·əl'täl·yə‚wēn }

vinyl trichloride [ORGANIC CHEMISTRY] *See* trichloroethane. { 'vīn·əl trī'klȯr‚īd }

vinyl trichlorosilone [ORGANIC CHEMISTRY] $CH_2CH_3SiCl_3$ A liquid that boils at 90.6°C and is soluble in organic solvents; used in silicones and adhesives. { 'vīn·əl trī¦klȯr·ō'si‚lōn }

viologen [CHEMISTRY] Any member of a group of chlorides of certain quaternary bases derived from γ,γ'-dipyridyl that are used as oxidation-reduction indicators; color is exhibited in the reduced form. { vī'äl·ə·jən }

virtual orbital [PHYSICAL CHEMISTRY] An orbital that is either empty or unoccupied while in the ground state. { 'vər·chə·wəl 'ȯr·bəd·əl }

visible absorption spectrophotometry [SPECTROSCOPY] Study of the spectra produced by the absorption of visible-light energy during the transformation of an electron from the ground state to an excited state as a function of the wavelength causing the transformation. { 'viz·ə·bəl əb'sȯrp·shən ¦spek·trō·fə'täm·ə·trē }

visible spectrophotometry [SPECTROSCOPY] In spectrophotometric analysis, the use of a spectrophotometer with a tungsten lamp that has an electromagnetic spectrum of 380-780 nanometers as a light source, glass or quartz prisms or gratings in the monochromator, and a photomultiplier cell as a detector. { 'viz·ə·bəl ¦spek·trō·fə'täm·ə·trē }

visible spectrum [SPECTROSCOPY] **1.** The range of wavelengths of visible radiation. **2.** A display or graph of the intensity of visible radiation emitted or absorbed by a material as a function of wavelength or some related parameter. { 'viz·ə·bəl 'spek·trəm }

visual colorimetry [ANALYTICAL CHEMISTRY] A procedure for the determination of the color of an unknown solution by visual comparison to color standards (solutions or color-tinted disks). { 'vizh·ə·wəl ‚kəl·ə'rim·ə·trē }

vitamin K$_1$ [ORGANIC CHEMISTRY] *See* phytonadione. { 'vīd·ə·mən ¦kā¦wən }

vitriolic acid [INORGANIC CHEMISTRY] *See* sulfuric acid. { 'vi·trē‚äl·ik 'əs·ed }

volatile [CHEMISTRY] Readily passing off by evaporation. { 'väl·əd·əl }

volatile fluid [CHEMISTRY] A liquid with the tendency to become vapor at specified conditions of temperature and pressure. { 'väl·əd·əl 'flü·əd }

volatility product [CHEMISTRY] The product of the concentrations of two or more molecules or ions that react to form a volatile substance. { ‚väl·ə'til·əd·ē ‚präd·əkt }

Volhard titration [ANALYTICAL CHEMISTRY] Determination of the halogen content of a solution by titration with a standard thiocyanate solution. { 'fōl‚härt tī'trā·shən }

voltameter [PHYSICAL CHEMISTRY] *See* coulometer. { väl'tam·əd·ər }

voltammetry [PHYSICAL CHEMISTRY] Any electrochemical technique in which a faradaic current passing through the electrolysis solution is measured while an appropriate potential is applied to the polarizable or indicator electrode; for example, polarography. { väl'täm·ə·trē }

Volta series [CHEMISTRY] *See* displacement series. { 'vōl·tə ‚sir·ēz }

volume susceptibility [PHYSICAL CHEMISTRY] The magnetic susceptibility of a specified volume (for example, 1 cubic centimeter) of a magnetically susceptible material. { 'väl·yəm sə‚sep·tə'bil·əd·ē }

volumetric analysis [ANALYTICAL CHEMISTRY] Quantitative analysis of solutions of known volume but unknown strength by adding reagents of known concentration until a reaction end point (color change or precipitation) is reached; the most common technique is by titration. Also known as titrimetric analysis. { ¦väl·yə¦me·trik ə'nal·ə·səs }

volumetric flask [ANALYTICAL CHEMISTRY] A laboratory flask primarily intended for the preparation of definite, fixed volumes of solutions, and therefore calibrated for a single volume only. { ¦väl·yə¦me·trik 'flask }

volumetric pipet [ANALYTICAL CHEMISTRY] A graduated glass tubing used to measure quantities of a solution; the tube is open at the top and bottom, and a slight vacuum (suction) at the top pulls liquid into the calibrated section; breaking the vacuum allows liquid to leave the tube. { ¦väl·yə¦me·trik pī'pet }

W [CHEMISTRY] *See* tungsten.

Wagner's reagent [ANALYTICAL CHEMISTRY] An aqueous solution of iodine and potassium iodide; used for microchemical analysis of alkaloids. Also known as Wagner's solution. { 'väg·nərz rē,ā·jənt }

Wagner's solution [ANALYTICAL CHEMISTRY] *See* Wagner's reagent. { 'väg·nərz sə,lü·shən }

Walden's rule [PHYSICAL CHEMISTRY] A rule which states that the product of the viscosity and the equivalent ionic conductance at infinite dilution in electrolytic solutions is a constant, independent of the solvent, it is only approximately correct. { 'wȯl·dənz ,rül }

Wallach transformation [ORGANIC CHEMISTRY] By the use of concentrated sulfuric acid, an azoxybenzene is converted into a *para*-hydroxyazobenzene. { 'wäl·ək ,tranz·fər,mā·shən }

wall-coated capillary column [ANALYTICAL CHEMISTRY] A capillary column characterized by a layer of stationary liquid coated directly on the inner wall of a coiled capillary tube. { 'wȯl ¦kōd·əd 'kap·ə,ler·ē ,käl·əm }

washing [ANALYTICAL CHEMISTRY] **1.** In the purification of a laboratory sample, the cleaning of residual liquid impurities from precipitates by adding washing solution to the precipitates, mixing, then decanting, and repeating the operation as often as needed. **2.** The removal of soluble components from a mixture of solids by using the effect of differential solubility. { 'wäsh·iŋ }

washing soda [INORGANIC CHEMISTRY] *See* sal soda. { 'wäsh·iŋ ,sōd·ə }

water [CHEMISTRY] H_2O Clear, odorless, tasteless liquid that is essential for most animal and plant life and is an excellent solvent for many substances; melting point 0°C (32°F), boiling point 100°C (212°F); the chemical compound may be termed hydrogen oxide. { 'wȯd·ər }

water absorption tube [ANALYTICAL CHEMISTRY] A glass tube filled with a solid absorbent (calcium chloride or silica gel) to remove water from gaseous streams during or after chemical analyses. { 'wȯd·ər əb'sȯrp·shən ,tüb }

watercolor pigment [INORGANIC CHEMISTRY] A permanent pigment used in watercolor painting, for example, titanium oxide (white). { 'wȯd·ər,kəl·ər ,pig·mənt }

water of crystallization [CHEMISTRY] *See* water of hydration. { 'wȯd·ər əv ,krist·əl·ə'zā·shən }

water glass [INORGANIC CHEMISTRY] *See* sodium silicate. { 'wȯd·ər ,glas }

water of hydration [CHEMISTRY] Water present in a definite amount and attached to a compound to form a hydrate; can be removed, as by heating, without altering the composition of the compound. Also known as water of crystallization. { 'wȯd·ər əv hī'drā·shən }

water saturation [CHEMISTRY] **1.** A solid adsorbent that holds the maximum possible amount of water under specified conditions. **2.** A liquid solution in which additional water will cause the appearance of a second liquid phase. **3.** A gas that is at or just under its dew point because of its water content. { 'wȯd·ər ,sach·ə'rā·shən }

water softening [CHEMISTRY] Removal of scale-forming calcium and magnesium ions from hard water, or replacing them by the more soluble sodium ions; can be done by chemicals or ion exchange. { 'wȯd·ər ˌsȯf·ə·niŋ }

water-wettable [CHEMISTRY] Denoting the capability of a material to accept water, or of being hydrophilic or hydrophoric. { 'wȯd·ər 'wed·ə·bəl }

water white [CHEMISTRY] A grade of color for liquids that has the appearance of clear water; for petroleum products, a plus 21 in the scale of the Saybolt chromometer. { 'wȯd·ər 'wīt }

Watson equation [PHYSICAL CHEMISTRY] Calculation method to extend heat of vaporization data for organic compounds to within 10 or 15°C of the critical temperature; uses known latent heats of vaporization and reduced temperature data. { 'wät·sən iˌkwā·zhən }

wavelength standards [SPECTROSCOPY] Accurately measured lengths of waves emitted by specified light sources for the purpose of obtaining the wavelengths in other spectra by interpolating between the standards. { 'wāvˌleŋkth ˌstan·dərdz }

weak acid [CHEMISTRY] An acid that does not ionize greatly; for example, acetic acid or carbonic acid. { 'wēk 'as·əd }

wedge spectrograph [SPECTROSCOPY] A spectrograph in which the intensity of the radiation passing through the entrance slit is varied by moving an optical wedge. { 'wej 'spek·trəˌgraf }

weight titration [ANALYTICAL CHEMISTRY] A titration in which the amount of titrant required is determined in terms of the weight that must be added to reach the end point. { 'wāt tī'trā·shən }

Weisz ring oven [ANALYTICAL CHEMISTRY] A device for vaporization of solvent from filter paper, leaving the solute in a ring (circular) shape; used for qualitative analysis of very small samples. { 'vīs 'riŋ ˌəv·ən }

Werner band [SPECTROSCOPY] A band in the ultraviolet spectrum of molecular hydrogen extending from 116 to 125 nanometers. { 'ver·nər ˌband }

Werner complex [CHEMISTRY] *See* coordination compound. { 'ver·nər ˌkämˌpleks }

wet ashing [ORGANIC CHEMISTRY] The conversion of an organic compound into ash (decomposition) by treating the compound with nitric or sulfuric acid. { 'wet 'ash·iŋ }

wettability [CHEMISTRY] The ability of any solid surface to be wetted when in contact with a liquid; that is, the surface tension of the liquid is reduced so that the liquid spreads over the surface. { ˌwed·ə'bil·əd·ē }

wetted [CHEMISTRY] Pertaining to material that has accepted water or other liquid, either on its surface or within its pore structure. { 'wed·əd }

white copperas [INORGANIC CHEMISTRY] *See* zinc sulfate. { 'wīt 'käp·rəs }

white lead [INORGANIC CHEMISTRY] Basic lead carbonate of variable composition, the oldest and most important lead paint pigment; also used in putty and ceramics. { 'wīt 'led }

white phosphorus [CHEMISTRY] The element phosphorus in its allotropic form, a soft, waxy, poisonous solid melting at 44.5°C; soluble in carbon disulfide, insoluble in water and alcohol; self-igniting in air. Also known as yellow phosphorus. { 'wīt 'fä·sfə·rəs }

white vitriol [INORGANIC CHEMISTRY] *See* zinc sulfate. { 'wīt 'vi·trēˌȯl }

Wiedemann's additivity law [PHYSICAL CHEMISTRY] The law that the mass (or specific) magnetic susceptibility of a mixture or solution of components is the sum of the proportionate (by weight fraction) susceptibilities of each component in the mixture. { 'vēd·əˌmänz ˌad·ə'tiv·əd·ē ˌlȯ }

Wien effect [PHYSICAL CHEMISTRY] An increase in the conductance of an electrolyte at very high potential gradients. { 'vēn iˌfekt }

Wijs' iodine monochloride solution [ANALYTICAL CHEMISTRY] A solution in glacial acetic acid of iodine monochloride; used to determine iodine numbers. Also known as Wijs' special solution. { vīs 'ī·əˌdīn ¦män·ə'klȯrˌīd səˌlü·shən }

Wijs' special solution [ANALYTICAL CHEMISTRY] *See* Wijs' iodine monochloride solution. { vīs 'spesh·əl səˌlü·shən }

Williamson synthesis [ORGANIC CHEMISTRY] The synthesis of ethers utilizing an alkyl iodide and sodium alcoholate. { 'wil·yəm·sən 'sin·thə·səs }

Winkler titration [ANALYTICAL CHEMISTRY] A chemical method for estimating the dissolved oxygen in seawater: manganous hydroxide is added to the sample and reacts with oxygen to produce a manganese compound which in the presence of acid potassium iodide liberates an equivalent quantity of iodine that can be titrated with standard sodium thiosulfate. { 'viŋ·klər tī,trā·shən }

wintergreen oil [ORGANIC CHEMISTRY] *See* methyl salicylate. { 'win·tər,grēn ,oil }

Wittig ether rearrangement [ORGANIC CHEMISTRY] The rearrangement of benzyl and alkyl ethers when reacted with a methylating agent, producing secondary and tertiary alcohols. { 'vid·ik 'ē·thər ə,rānj·mənt }

Witt theory [CHEMISTRY] A theory of the mechanism of dyeing stating that all colored organic compounds (called chromogens) contain certain unsaturated chromophoric groups which are responsible for the color, and if these compounds also contain certain auxochromic groups, they possess dyeing properties. { 'wit ,thē·ə·rē }

Wolf-Kishner reduction [ORGANIC CHEMISTRY] Conversion of aldehydes and ketones to corresponding hydrocarbons by heating their semicarbazones, phenylhydrazones, and hydrazones with sodium ethoxide or by heating the carbonyl compound with excess sodium ethoxide and hydrazine sulfate. { 'wu̇lf 'kish·nər ri'dək·shən }

wolfram [CHEMISTRY] *See* tungsten; wolframite. { 'wu̇l·frəm }

wolframic acid [INORGANIC CHEMISTRY] *See* tungstic acid. { wu̇l'fram·ik 'as·əd }

wolfram white [INORGANIC CHEMISTRY] *See* barium tungstate. { 'wu̇l·frəm wīt }

wood alcohol [ORGANIC CHEMISTRY] *See* methyl alcohol. { 'wu̇d 'al·kə,hȯl }

wood ether [ORGANIC CHEMMISTRY] *See* dimethyl ether. { 'wu̇d 'ē·thər }

wood vinegar [ORGANIC CHEMISTRY] *See* pyroligneous acid. { 'wu̇d 'vin·ə·gər }

Woodward-Hoffmann rule [ORGANIC CHEMISTRY] A concept which can predict or explain the stereochemistry of certain types of reactions in organic chemistry; it is also described as the conservation of orbital symmetry. { 'wu̇d·wərd 'häf·mən ,rül }

Woodward's Reagent K [ORGANIC CHEMISTRY] *See* N-ethyl-5-phenylisoxazolium- 3′-sulfonate. { 'wu̇d·wərdz rē¦ā·jənt 'kā }

working electrode [PHYSICAL CHEMISTRY] The electrode used in corrosion testing by an electrochemical cell. { 'wərk·iŋ i'lek,trōd }

Wurtz-Fittig reaction [ORGANIC CHEMISTRY] A modified Wurtz reaction in which an aromatic halide reacts with an aklyl halide in the presence of sodium and an anhydrous solvent to form alkylated aromatic hydrocarbons. { 'wərtz 'fid·ig rē,ak·shən }

Wurtz reaction [ORGANIC CHEMISTRY] Synthesis of hydrocarbons by treating alkyl iodides in ethereal solution with sodium according to the reaction $2CH_3I + 2Na \rightarrow CH_3CH_3 + 2NaI$. { 'wərts rē,ak·shən }

X

X [SPECTROSCOPY] *See* siegbahn.

xanthan gum [ORGANIC CHEMISTRY] A high-molecular-weight (5-10 million) water-soluble natural gum; a heteropolysaccharide made up of building blocks of D-glucose, D-mannose, and D-glucuronic acid residues; produced by pure culture fermentation of glucose with *Xanthomonas campestris*. { 'zan·thən ˌgəm }

xanthate [ORGANIC CHEMISTRY] A water-soluble salt of xanthic acid, usually potassium or sodium; used as an ore-flotation collector. { 'zanˌthāt }

xanthene [ORGANIC CHEMISTRY] $CH_2(C_6H_4)_2O$ Yellowish crystals that are soluble in ether, slightly soluble in water and alcohol, melts at 100°C; used as a fungicide and chemical intermediate. Also known as tricyclic dibenzopyran. { 'zanˌthēn }

xanthene dye [ORGANIC CHEMISTRY] Any of a family of dyes related to the xanthenes; the chromophore groups are (C_6H_4). { 'zanˌthēn 'dī }

xanthine [ORGANIC CHEMISTRY] $C_5H_4N_4O_2$ A toxic yellow-white purine base that is found in blood and urine, and occasionally in plants; it is a powder, insoluble in water and acids, soluble in caustic soda; sublimes when heated; used in medicine and as a chemical intermediate. Also known as dioxopurine. { 'zanˌthēn }

xanthone [ORGANIC CHEMISTRY] $CO(C_6H_4)_2O$ White needle crystals that are found in some plant pigments; insoluble in water, soluble in alcohol, chloroform, and benzene; melts at 173°C, sublimes at 350°C; used as a larvicide, as a dye intermediate, and in perfumes and pharmaceuticals. { 'zanˌthōn }

Xe [CHEMISTRY] *See* xenon.

xenon [CHEMISTRY] An element, symbol Xe, member of the noble gas family, group 0, atomic number 54, atomic weight 131.30; colorless, boiling point −108°C (1 atmosphere, or 101,325 pascals), noncombustible, nontoxic, and nonreactive; used in photographic flash lamps, luminescent tubes, and lasers, and as an anesthetic. { 'zēˌnän }

xenyl [ORGANIC CHEMISTRY] The functional group $C_6H_5C_6H_4$—. { 'zen·əl }

xerogel [CHEMISTRY] **1.** A gel whose final form contains little or none of the dispersion medium used. **2.** An organic polymer capable of swelling in suitable solvents to yield particles possessing a three-dimensional network of polymer chains. { 'zer·əˌjel }

XPS [SPECTROSCOPY] *See* x-ray photoelectron spectroscopy.

x-ray crystal spectrometer [SPECTROSCOPY] An instrument designed to produce an x-ray spectrum and measure the wavelengths of its components, by diffracting x-rays from a crystal with known lattice spacing. { 'eksˌrā 'krist·əl spek'träm·əd·ər }

x-ray fluorescence analysis [ANALYTICAL CHEMISTRY] A nondestructive physical method used for chemical analyses of solids and liquids; the specimen is irradiated by an intense x-ray beam and the lines in the spectrum of the resulting x-ray fluorescence are diffracted at various angles by a crystal with known lattice spacing; the elements in the specimen are identified by the wavelengths of their spectral lines, and their concentrations are determined by the intensities of these lines. Also known as x-ray fluorimetry. { 'eksˌrā flu̇'res·əns əˌnal·ə·səs }

x-ray fluorescent emission spectrometer [SPECTROSCOPY] An x-ray crystal spectrometer used to measure wavelengths of x-ray fluorescence; in order to concentrate beams of low intensity, it has bent reflecting or transmitting crystals arranged so that the theoretical curvature required can be varied with the diffraction angle of a spectrum line. { 'eks‚rā flu̇'res·ənt i¦mish·ən spek'träm·əd·ər }

x-ray fluorimetry [ANALYTICAL CHEMISTRY] *See* x-ray fluorescence analysis. { 'eks‚rā flu̇'räm·ə·trē }

x-ray image spectrography [SPECTROSCOPY] A modification of x-ray fluorescence analysis in which x-rays irradiate a cylindrically bent crystal, and Bragg diffraction of the resulting emissions produces a slightly enlarged image with a resolution of about 50 micrometers. { 'eks‚rā ¦im·ij spek'träg·rə·fē }

x-ray microprobe [SPECTROSCOPY] *See* microprobe. { 'eks‚rā 'mī·krə‚prōb }

x-ray photoelectron spectroscopy [SPECTROSCOPY] A form of electron spectroscopy in which a sample is irradiated with a beam of monochromatic x-rays and the energies of the resulting photoelectrons are measured. Abbreviated XPS. Also known as electron spectroscopy for chemical analysis (ESCA). { 'eks‚rā ¦fōd·ō·i¦lek‚trän spek'träs·kə·pē }

x-ray spectrograph [SPECTROSCOPY] An x-ray spectrometer equipped with photographic or other recording apparatus; one application is fluorescence analysis. { 'eks‚rā 'spek·trə‚graf }

x-ray spectrometer [SPECTROSCOPY] An instrument for producing the x-ray spectrum of a material and measuring the wavelengths of the various components. { 'eks‚rā spek'träm·əd·ər }

x-ray spectrometry [SPECTROSCOPY] The measure of wavelengths of x-rays by observing their diffraction by crystals of known lattice spacing. Also known as roentgen spectrometry; x-ray spectroscopy. { 'eks‚rā spek'träm·ə·trē }

x-ray spectroscopy [SPECTROSCOPY] *See* x-ray spectrometry. { 'eks‚rā spek'träs·kə·pē }

x-ray spectrum [SPECTROSCOPY] A display or graph of the intensity of x-rays, produced when electrons strike a solid object, as a function of wavelengths or some related parameter; it consists of a continuous bremsstrahlung spectrum on which are superimposed groups of sharp lines characteristic of the elements in the target. { 'eks‚rā ‚spek·trəm }

XU [SPECTROSCOPY] *See* siegbahn.

X unit [SPECTROSCOPY] *See* siegbahn. { 'eks ‚yü·nət }

xylene [ORGANIC CHEMISTRY] $C_6H_4(CH_3)_2$ Any one of the family of isomeric, colorless aromatic hydrocarbon liquids, produced by the destructive distillation of coal or by the catalytic reforming of petroleum naphthenic fractions; used for high-octane and aviation gasolines, solvents, chemical intermediates, and the manufacture of polyester resins. Also known as dimethylbenzene; xylol. { 'zī‚lēn }

***meta*-xylene** [ORGANIC CHEMISTRY] 1,3-$C_6H_4(CH_3)_2$ A flammable, toxic liquid; insoluble in water, soluble in alcohol and ether; boils at 139°C; used as an intermediate for dyes, a chemical intermediate, and a solvent, and in insecticides and aviation fuel. { ¦med·ə 'zī‚lēn }

***ortho*-xylene** [ORGANIC CHEMISTRY] 1,2-$C_6H_4(CH_3)_2$ A flammable, moderately toxic liquid; insoluble in water, soluble in alcohol and ether; boils at 144°C; used to make phthalic anhydride, vitamins, pharmaceuticals, and dyes, and in insecticides and motor fuels. { ¦ȯr·thō 'zī‚lēn }

***para*-xylene** [ORGANIC CHEMISTRY] 1,4-$C_6H_4(CH_3)_2$ A toxic, combustible liquid; insoluble in water, soluble in alcohol and ether; boils at 139°C; used as a chemical intermediate, and to synthesize terephthalic acid, vitamins, and pharmaceuticals, and in insecticides. { ¦par·ə 'zī‚lēn }

xylenol [ORGANIC CHEMISTRY] $(CH_3)_2C_6H_3OH$ Highly toxic, combustible crystals; slightly soluble in water, soluble in most organic solvents; melts at 20-76°C; used as a chemical intermediate, disinfectant, solvent, and fungicide, and for pharmaceuticals and dyestuffs. { 'zī·lə‚nȯl }

xylidine [ORGANIC CHEMISTRY] $(CH_3)_2C_6H_3NH_2$ A toxic, combustible liquid; soluble in alcohol and ether, slightly soluble in water; boils about 220°C; used as a chemical intermediate and to make dyes and pharmaceuticals. { 'zī·lə,dēn }

xylite [ORGANIC CHEMISTRY] *See* xylitol. { 'zī,līt }

xylitol [ORGANIC CHEMISTRY] $CH_2OH(CHOH)_3CH_2OH$ Pentahydric alcohols derived from xylose. Also known as xylite. { 'zī·lə,tȯl }

xylol [ORGANIC CHEMISTRY] *See* xylene. { 'zī,lȯl }

Y

Y [CHEMISTRY] *See* yttrium.

yacca gum [ORGANIC CHEMISTRY] *See* acaroid resin. { 'yak·ə ˌgəm }

Yb [CHEMISTRY] *See* ytterbium.

yellow phosphorus [CHEMISTRY] *See* white phosphorus. { 'yel·ō 'fä·sfə·rəs }

yellow precipitate [INORGANIC CHEMISTRY] *See* mercuric oxide. { 'yel·ō pri'sip·əˌtāt }

yellow prussiate of potash [INORGANIC CHEMISTRY] *See* potassium ferrocyanide. { 'yel·ō 'prəs·ēˌāt əv 'pädˌash }

yellow prussiate of soda [INORGANIC CHEMISTRY] *See* sodium ferrocyanide. { 'yel·ō 'prəs·ēˌāt əv 'sōd·ə }

yellow pyoktanin [ORGANIC CHEMISTRY] *See* auramine hydrochloride. { 'yel·ō pī'äk·tə·nən }

yellow salt [INORGANIC CHEMISTRY] *See* uranyl nitrate. { 'yel·ō 'sölt }

ylide [ORGANIC CHEMISTRY] An organic compound which contains two adjacent atoms bearing formal positive and negative charges, and in which both atoms have full octets of electrons. { 'i·līd }

ylium ion [ORGANIC CHEMISTRY] *See* enium ion. { 'ī·lē·əm 'īˌän }

ytterbia [INORGANIC CHEMISTRY] *See* ytterbium oxide. { i'tər·bē·ə }

ytterbium [CHEMISTRY] A rare-earth metal of the yttrium subgroup, symbol Yb, atomic number 70, atomic weight 173.04; lustrous, malleable, soluble in dilute acids and liquid ammonia, reacts slowly with water; melts at 824°C, boils at 1427°C; used in chemical research, lasers, garnet doping, and x-ray tubes. { i'tər·bē·əm }

ytterbium oxide [INORGANIC CHEMISTRY] Yb_2O_3 A colorless compound, melts at 2346°C, dissolves in hot dilute acids; used to prepare alloys, ceramics, and special glasses. Also known as ytterbia. { i'tər·bē·əm 'äkˌsīd }

yttria [INORGANIC CHEMISTRY] *See* yttrium oxide. { 'i·trē·ə }

yttrium [CHEMISTRY] A rare-earth metal, symbol Y, atomic number 39, atomic weight 88.905; dark-gray, flammable (as powder), soluble in dilute acids and potassium hydroxide solution, and decomposes in water; melts at 1500°C, boils at 2927°C; used in alloys and nuclear technology and as a metal deoxidizer. { 'i·trē·əm }

yttrium acetate [ORGANIC CHEMISTRY] $Y(C_2H_3O_2)_3 \cdot 8H_2O$ Colorless, water-soluble crystals used as an analytical reagent. { 'i·trē·əm 'as·əˌtāt }

yttrium chloride [INORGANIC CHEMISTRY] $YCl_3 \cdot 6H_2O$ Reddish, transparent, water- and alcohol-soluble prisms; decomposes at 100°C; used as an analytical reagent. { 'i·trē·əm 'klörˌīd }

yttrium oxide [INORGANIC CHEMISTRY] Y_2O_3 A yellowish powder, insoluble in water, soluble in dilute acids; used as television tube phosphor and microwave filters. Also known as yttria. { 'i·trē·əm 'äkˌsīd }

yttrium sulfate [INORGANIC CHEMISTRY] $Y_2(SO_4)_3 \cdot 8H_2O$ Reddish crystals that are soluble in concentrated sulfuric acid, slightly soluble in water; decomposes at 700°C; used as an analytical reagent. { 'i·trē·əm 'səlˌfāt }

Z

ZAA spectrometry [SPECTROSCOPY] *See* Zeeman-effect atomic absorption spectrometry. { ¦zē¦ā¦ā spek'träm·ə·trē }

Zeeman displacement [SPECTROSCOPY] The separation, in wave numbers, of adjacent spectral lines in the normal Zeeman effect in a unit magnetic field, equal (in centimeter-gram-second Gaussian units) to $e/4\pi mc^2$, where e and m are the charge and mass of the electron, or to approximately 4.67×10^{-5} (centimeter)$^{-1}$(gauss)$^{-1}$. { 'zā·mən diˌsplās·mənt }

Zeeman effect [SPECTROSCOPY] A splitting of spectral lines in the radiation emitted by atoms or molecules in a static magnetic field. { 'zā·mən iˌfekt }

Zeeman-effect atomic absorption spectrometry [SPECTROSCOPY] A type of atomic absorption spectrometry in which either the light source or the sample is placed in a magnetic field, splitting the spectral lines under observation into polarized components, and a rotating polarizer is placed between the source and the sample, enabling the absorption caused by the element under analysis to be separated from background absorption. Abbreviated ZAA spectrometry. { ¦zē·mən iˌfekt ə¦täm·ik əp¦sȯrp·shən spek'träm·ə·trē }

zeolite catalyst [INORGANIC CHEMISTRY] Hydrated aluminum and calcium (or sodium) silicates (for example, $CaO \cdot 2Al_2O_3 \cdot 5SiO_2$ or $Na_2O \cdot 2Al_2O_3 \cdot 5SiO_2$) made with controlled porosity; used as a catalytic cracking catalyst in petroleum refineries, or loaded with catalyst for other chemical reactions. { 'zē·əˌlīt 'kad·əl·əst }

zeotrope [PHYSICAL CHEMISTRY] A nonazeotropic liquid mixture which may be separated by distillation, and in which the components are miscible in all proportions (homogeneous zeotrope or homozeotrope) or not miscible in all proportions (heterogeneous zeotrope or heterozeotrope). { 'zē·əˌtrōp }

Zerewitinoff reagent [ANALYTICAL CHEMISTRY] A light-colored methylmagnesium iodide-*n*-butyl ether solution that reacts rapidly with moisture and oxygen, used to determine water, alcohols, and amines in inert solvents. { ˌzir ə'wit·ənˌȯf rēˌā·jənt }

zero branch [SPECTROSCOPY] A spectral band whose Fortrat parabola lies between two other Fortrat parabolas, with its vertex almost on the wave number axis. { 'zir·ō 'branch }

zerogel [CHEMISTRY] A gel which has dried until apparently solid; sometimes it will swell or redisperse to form a sol when treated with a suitable solvent. { 'zir·ōˌjel }

zero-order reaction [PHYSICAL CHEMISTRY] A reaction for which reaction rate is independent of the concentrations of the reactants; for example, a photochemical reaction in which the rate is determined by the intensity of light. { 'zir·ō ¦ȯrd·ər rē'ak·shən }

Zimm plot [ANALYTICAL CHEMISTRY] A graphical determination of the root-square-mean end-to-end distances of coillike polymer molecules during scattered-light photometric analyses. { 'zim ˌplät }

zinc [CHEMISTRY] A metallic transition element, symbol Zn, atomic number 30, atomic weight 65.37; explosive as powder; soluble in acids and alkalies, insoluble in water; strongly electropositive; melts at 419°C; boils at 907°C. { ziŋk }

zinc acetate [ORGANIC CHEMISTRY] $Zn(C_2H_3O_2)_2 \cdot 2H_2O$ Pearly-white crystals with an astringent taste; soluble in water and alcohol; decomposes at 200°C; used to preserve wood in textile dyeing, and as an analytical reagent, a feed additive, and a polymer cross-linking agent. { 'ziŋk 'as·ə‚tāt }

zinc arsenite [INORGANIC CHEMISTRY] $Zn(AsO_2)_2$ A toxic white powder that is insoluble in water, soluble in alkalies; used as an insecticide and timber preservative. Also known as zinc meta-arsenite. { 'ziŋk 'ärs·ən‚īt }

zincate [INORGANIC CHEMISTRY] A reaction product of zinc with an alkali metal or with ammonia; for example, sodium zincate, Na_2ZnO_2. { 'ziŋ‚kāt }

zinc borate [INORGANIC CHEMISTRY] $3ZnO \cdot 2B_2O_3$ A white, amorphous powder that is soluble in dilute acids, slightly soluble in water; melts at 980°C; used in medicine, as a ceramics flux, as an inhibitor for mildew, and to fireproof textiles. { 'ziŋk 'bȯr‚āt }

zinc bromide [INORGANIC CHEMISTRY] $ZnBr_2$ Water- and alcohol-soluble, white crystals that melt at 294°C; used in medicine, manufacture of rayon, and photography, and in a radiation viewing screen. { 'ziŋk 'brō‚mīd }

zinc carbonate [INORGANIC CHEMISTRY] $ZnCO_3$ White crystals that are insoluble in water, soluble in alkalies and acids; used in ceramics and ointments, and as a fireproofing agent and feed additive. { 'ziŋk 'kär·bə‚nāt }

zinc chloride [INORGANIC CHEMISTRY] $ZnCl_2$ Water- and alcohol-soluble, white, fire-hazardous crystals that melt at 290°C, and are irritating to the skin; used as a catalyst and in electroplating, wood preservation, textile processing, petroleum refining, medicine, and feed additives. { 'ziŋk 'klȯr‚īd }

zinc chromate [INORGANIC CHEMISTRY] $ZnCrO_4$ A toxic, yellow powder that is insoluble in water, soluble in acids; used as a pigment in paints (artists', automotive, primer), varnishes, linoleum, and epoxy laminates. { 'ziŋk 'krō‚māt }

zinc cyanide [INORGANIC CHEMISTRY] $Zn(CN)_2$ A toxic, white powder that is insoluble in water and alcohol, soluble in alkalies and dilute acids; melts at 800°C; used as an analytical reagent and insecticide, and in medicine and metal plating. { 'ziŋk 'sī·ə‚nīd }

zinc fluoride [INORGANIC CHEMISTRY] ZnF_2 A toxic white powder that is slightly soluble in water and melts at 872°C; used in enamels, ceramic glazes, and galvanizing. { 'ziŋk 'flu̇r‚īd }

zinc formate [ORGANIC CHEMISTRY] $Zn(CHO_2)_2 \cdot 2H_2O$ Toxic, white crystals that are soluble in water, insoluble in alcohol; used as a catalyst, weatherproofing agent, and wood preservative. { 'ziŋk 'fȯr‚māt }

zinc halide [INORGANIC CHEMISTRY] A binary compound of zinc and a halogen; for example, $ZnBr_2$, $ZnCl_2$, ZnF_2, and ZnI_2. { 'ziŋk 'ha‚līd }

zinc hydroxide [INORGANIC CHEMISTRY] $Zn(OH)_2$ Colorless, water-soluble crystals that decompose at 125°C; used as a chemical intermediate and in rubber compounding and surgical dressings. { 'ziŋk hī'dräk‚sīd }

zinc meta-arsenite [INORGANIC CHEMISTRY] *See* zinc arsenite. { 'ziŋk ¦med·ə'ärs·ən‚āt }

zinc naphthenate [ORGANIC CHEMISTRY] $Zn(C_6H_5COO)_2$ A combustible, viscous, acetone-soluble solid; used in paints, varnishes, and resins, and as a drier and wetting agent, insecticide, fungicide, and mildewstat. { 'ziŋk 'naf·thə‚nāt }

zinc orthoarsenate [INORGANIC CHEMISTRY] $Zn_3(AsO_4)_2$ A toxic white powder that is insoluble in water, soluble in alkalies; used as an insecticide and wood preservative. { 'ziŋk ¦ȯr·thō'ärs·ən‚āt }

zinc orthophosphate [INORGANIC CHEMISTRY] *See* zinc phosphate. { 'ziŋk ¦ȯr·thō'fä‚sfāt }

zinc oxide [INORGANIC CHEMISTRY] ZnO A bitter-tasting, white to gray powder that is insoluble in water, soluble in alkalies and acids; melts at 1978°C; used as a pigment, mold-growth inhibitor, and dietary supplement, and in cosmetics, electronics, and color photography. { 'ziŋk 'äk‚sīd }

zinc phosphate [INORGANIC CHEMISTRY] $Zn_3(PO_4)_2$ A white powder that is insoluble in water, soluble in acids and ammonium hydroxide; melts at 900°C; used in coatings for steel, aluminum, and other metals, and in dental cements and phosphors. Also known as tribasic zinc phosphate; zinc orthophosphate. { 'ziŋk 'fä‚sfāt }

zinc phosphide [INORGANIC CHEMISTRY] Zn_3P_2 A toxic, alcohol-insoluble, gray gritty powder that reacts violently with oxidizing agents; melts at over 420°C, decomposes in water; used as a rat poison and in medicine. { 'ziŋk 'fä‚sfīd }

zinc selenide [INORGANIC CHEMISTRY] ZnSe A water-insoluble, moderately toxic, yellow to reddish solid that is a fire hazard when in contact with water and acids; melts above 1100°C; used as infrared optical windows. { 'ziŋk 'sel·ə‚nīd }

zinc sulfate [INORGANIC CHEMISTRY] $ZnSO_4 \cdot 7H_2O$ Efflorescent, water-soluble, colorless crystals with an astringent taste; used to preserve skins and wood and as a paper bleach, analytical reagent, feed additive, and fungicide. Also known as white copperas; white vitriol; zinc vitriol. { 'ziŋk 'səl‚fāt }

zinc sulfide [INORGANIC CHEMISTRY] ZnS A yellowish powder that is insoluble in water, soluble in acids; exists in two crystalline forms (alpha, or wurtzite, and beta, or sphalerite); beta becomes alpha at 1020°C, and sublimes at 1180°C; used as a pigment for paints and linoleum, in opaque glass, rubber, and plastics, for hydrosulfite dyeing process, as x-ray and television screen phosphor, and as a fungicide. { 'ziŋk 'səl‚fīd }

zinc telluride [INORGANIC CHEMISTRY] ZnTe Moderately toxic, reddish crystals that melt at 1238°C and decompose in water. { 'ziŋk 'tel·yə‚rīd }

zinc vitriol [INORGANIC CHEMISTRY] *See* zinc sulfate. { 'ziŋk 'vi·trē‚ōl }

zinc white [CHEMISTRY] *See* Chinese white. { 'ziŋk 'wīt }

zirconia [INORGANIC CHEMISTRY] *See* zirconium oxide. { ‚zər'kō·nē·ə }

zirconic anhydride [INORGANIC CHEMISTRY] *See* zirconium oxide. { ‚zər'kän·ik an'hī‚drīd }

zirconium [CHEMISTRY] A metallic transition element, symbol Zr, atomic number 40, atomic weight 91.22; occurs as crystals, flammable as powder; insoluble in water, soluble in hot, concentrated acids; melts at 1850°C, boils at 4377°C. { ‚zər'kō·nē·əm }

zirconium boride [INORGANIC CHEMISTRY] ZrB_2 A hard, toxic, gray powder that melts at 3000°C; used as an aerospace refractory, in cutting tools, and to protect thermocouple tubes. Also known as zirconium diboride. { ‚zər'kō·nē·əm 'bór‚īd }

zirconium carbide [INORGANIC CHEMISTRY] ZrC Hard, gray crystals that are soluble in water, soluble in acids; as powder, it ignites spontaneously in air; melts at 3400°C, boils at 5100°C; used as an abrasive, refractory, and metal cladding, and in cermets, incandescent filaments, and cutting tools. { ‚zər'kō·nē·əm 'kär‚bīd }

zirconium chloride [INORGANIC CHEMISTRY] *See* zirconium tetrachloride. { ‚zər'kō·nē·əm 'klór‚īd }

zirconium diboride [INORGANIC CHEMISTRY] *See* zirconium boride. { ‚zər'kō·nē·əm dī'bór‚īd }

zirconium dioxide [INORGANIC CHEMISTRY] *See* zirconium oxide. { ‚zər'kō·nē·əm dī'äk‚sīd }

zirconium halide [INORGANIC CHEMISTRY] A compound of zirconium with a halogen; for example, $ZrBr_2$, $ZrCl_2$, $ZrCl_3$, $ZrCl_4$, $ZrBr_2$, $ZrBr_3$, ZrF_4, and ZrI_4. { ‚zər'kō·nē·əm 'ha‚līd }

zirconium hydride [INORGANIC CHEMISTRY] ZrH_2 A flammable, gray-black powder; used in powder metallurgy and nuclear moderators, and as a reducing agent, vacuum-tube getter, and metal-foaming agent. { ‚zər'kō·nē·əm 'hī‚drīd }

zirconium hydroxide [INORGANIC CHEMISTRY] $Zr(OH)_4$ A toxic, amorphous white powder; insoluble in water, soluble in dilute mineral acids; decomposes at 550°C; used in pigments, glass, and dyes, and to make zirconium compounds. { ‚zər'kō·nē·əm hī'dräk‚sīd }

zirconium nitride [INORGANIC CHEMISTRY] ZrN A hard, brassy powder that is soluble in concentrated acids; melts at 2930°C; used in refractories, cermets, and laboratory crucibles. { ‚zər'kō·nē·əm 'nī‚trīd }

zirconium orthophosphate [INORGANIC CHEMISTRY] *See* zirconium phosphate. { ‚zər'kō·nē·əm ¦ór·thō'fä‚sfāt }

zirconium oxide [INORGANIC CHEMISTRY] ZrO_2 A toxic, heavy white powder that is insoluble in water, soluble in mineral acids; melts at 2700°C; used in ceramic glazes, special glasses, and medicine, and to make piezoelectric crystals. Also known as zirconia; zirconic anhydride; zirconium dioxide. { ˌzərʹkō·nē·əm ʹäkˌsīd }

zirconium oxychloride [INORGANIC CHEMISTRY] $ZrOCl_2 \cdot 8H_2O$ White crystals that are soluble in water, insoluble in organic solvents, and acidic in aqueous solution; used for textile dyeing and oil-field acidizing, in cosmetics and greases, and for antiperspirants and water repellents. Also known as zirconyl chloride. { ˌzərʹkō·nē·əm ¦äk·sēʹklȯrˌīd }

zirconium phosphate [INORGANIC CHEMISTRY] $ZrO(H_2PO_4)_2 \cdot 3H_2O$ A toxic, dense white powder that is insoluble in water, soluble in acids and organic solvents; decomposes on heating; used as an analytical reagent, coagulant, and radioactive-phosphor carrier. Also known as zirconium orthophosphate. { ˌzərʹkō·nē·əm ʹfäˌsfāt }

zirconium tetrachloride [INORGANIC CHEMISTRY] $ZrCl_4$ Toxic, alcohol- soluble, white lustrous crystals; sublimes above 300°C and decomposes in water; used to make pure zirconium and for water-repellent textiles and as an analytical reagent. Also known as zirconium chloride. { ˌzərʹkō·nē·əm ¦te·trəʹklȯrˌīd }

zirconyl chloride [INORGANIC CHEMISTRY] *See* zirconium oxychloride. { ʹzər·kən·əl ʹklȯrˌīd }

Zn [CHEMISTRY] *See* zinc.

zone [ANALYTICAL CHEMISTRY] *See* band. { zōn }

Zr [CHEMISTRY] *See* zirconium.

Zsigmondy gold number [CHEMISTRY] The number of milligrams of protective colloid necessary to prevent 10 milliliters of gold sol from coagulating when 0.5 milliliter of 10% sodium chloride solution is added. { ʹzig·mȯn·dē ʹgōld ˌnəm·bər }

zwitterion [CHEMISTRY] *See* dipolar ion. { ʹtsfid·ərˌīˌän }

Appendix

Appendix

Equivalents of commonly used units for the U.S. Customary System and the metric system

1 inch = 2.5 centimeters (25 millimeters)	1 centimeter = 0.4 inch	1 inch = 0.08 foot
1 foot = 0.3 meter (30 centimeters)	1 meter = 3.3 feet	1 foot = 0.3 yard (12 inches)
1 yard = 0.9 meter	1 meter = 1.1 yards	1 yard = 3 feet (36 inches)
1 mile = 1.6 kilometers	1 kilometer = 0.6 mile	1 mile = 5280 feet (1750 yards)
1 acre = 0.4 hectare	1 hectare = 2.47 acres	
1 acre = 4047 square meters	1 square meter = 0.0002 acre	
1 gallon = 3.8 liters	1 liter = 0.26 gallon	1 quart = 0.25 gallon (32 ounces; 2 pints)
1 fluid ounce = 29.6 milliliters	1 milliliter = 0.03 fluid ounce	1 pint = 0.125 gallon (16 ounces)
32 fluid ounces = 946.4 milliliters	1 liter = 1.1 quarts (0.3 gallon)	1 gallon = 4 quarts (8 pints)
1 quart = 0.9 liter	750 milliliters = 25.36 fluid ounces	
1 ounce = 28.4 grams	1 gram = 0.04 ounce	1 ounce = 0.6 pound
1 pound = 0.5 kilogram	1 kilogram = 2.2 pounds	1 pound = 16 ounces
1 ton = 907.18 kilograms	1 kilogram = 1.1×10^{-3} ton	1 ton = 2000 pounds
°F = (1.8 × °C) + 32	°C = (°F − 32) ÷ 1.8	

Appendix

Conversion factors for the U.S. Customary System, metric system, and International System

A. UNITS OF LENGTH

Units	cm	m	in	ft	yd	mi
1 cm	= 1	0.01*	0.39	0.033	0.01	6.21×10^{-6}
1 m	= 100.	1	39.37	3.28	1.09	6.21×10^{-4}
1 in	= 2.54	0.03	1	0.08...	0.03...	1.58×10^{-5}
1 ft	= 30.48	0.30	12.	1	0.33...	$1.89... \times 10^{-4}$
1 yd	= 91.44	0.91	36.	3.	1	$5.68... \times 10^{-4}$
1 mile	$= 1.61 \times 10^{5}$	1.61×10^{3}	6.34×10^{4}	5280.	1760.	1

B. UNITS OF AREA

Units	cm^2	m^2	in^2	ft^2	yd^2	mi^2
1 cm^2	= 1	10^{-4}	0.16	1.08×10^{-3}	1.20×10^{-4}	3.86×10^{-11}
1 m^2	$= 10^{4}$	1	1550.00	10.76	1.30	3.86×10^{-7}
1 in^2	= 6.45	6.45×10^{-4}	1	6.94×10^{-3}...	7.72×10^{-4}	2.49×10^{-10}
1 ft^2	= 929.03	0.09	1.44.	1	0.11...	3.59×10^{-8}
1 yd^2	= 8361.27	0.84	1296.	9.	1	3.23×10^{-7}
1 mi^2	$= 2.59 \times 10^{10}$	2.59×10^{6}	4.01×10^{9}	2.79×10^{7}	3.10×10^{6}	1

C. UNITS OF VOLUME

Units	m^3	cm^3	liter	in^3	ft^3	qt	gal
1 m^3	= 1	10^6	10^3	6.10×10^4	35.31	1.057×10^3	264.17
1 cm^3	= 10^{-6}	1	10^{-3}	0.061	3.53×10^{-5}	1.057×10^{-3}	2.64×10^{-4}
1 liter	= 10^{-3}	1000.	1	61.02374	0.03531467	1.056688	0.26
1 in^3	= 1.64×10^{-5}	16.39	0.02	1	5.79×10^{-4}	0.02	4.33×10^{-3}
1 ft^3	= 2.83×10^{-2}	28316.85	28.32	1728.		2.99	7.48
1 qt	= 9.46×10^{-4}	946.35	0.95	57.75	0.03	1	0.25
1 gal (U.S.)	= 3.79×10^{-3}	3785.41	3.79	231.	0.13	4.	1

D. UNITS OF MASS

Units	g	kg	oz	lb	metric ton	ton
1 g	= 1	10^{-3}	0.04	2.20×10^{-3}	10^{-6}	1.10×10^{-6}
1 kg	= 1000.	1	35.27	2.20	10^{-3}	1.10×10^{-3}
1 oz (avdp)	= 28.35	0.028	1	0.06	2.83×10^{-5}	$5. \times 10^{-4}$
1 lb (avdp)	= 453.59	0.45	16.	1	4.54×10^{-4}	0.0005
1 metric ton	= 10^6	1000.	35273.96	2204.62	1	1.10
1 ton	= 907184.7	907.18	32000.	2000.	0.91	1

Conversion factors for the U.S. Customary System, metric system, and International System (cont.)

E. UNITS OF DENSITY

Units	$g \cdot cm^{-3}$	$g \cdot L^{-1}$, $kg \cdot m^{-3}$	$oz \cdot in^{-3}$	$lb \cdot in^{-3}$	$lb \cdot ft^{-3}$	$lb \cdot gal^{-1}$
1 $g \cdot cm^{-3}$	= 1	1000.	0.58	0.036	62.43	8.35
1 $g \cdot L^{-1}$, $kg \cdot m^{-3}$	= 10^{-3}	1	5.78×10^{-4}	3.61×10^{-5}	0.06	8.35×10^{-3}
1 $oz \cdot in^{-3}$	= 1.729994	1730	1	0.06	108.	14.44
1 $lb \cdot in^{-3}$	= 27.68	27679.91	16.	1	1728.	231.
1 $lb \cdot ft^{-3}$	= 0.02	16.02	9.26×10^{-3}	5.79×10^{-4}	1	0.13
1 $lb \cdot gal^{-1}$	= 0.12	119.83	4.75×10^{-3}	4.33×10^{-3}	7.48	1

F. UNITS OF PRESSURE

Units	Pa, $N \cdot m^{-2}$	$dyn \cdot cm^{-2}$	bar	atm	kg (wt) $\cdot cm^{-2}$	mmHg (torr)	in Hg	lb (wt) $\cdot in^{-2}$
1 Pa, 1 $N \cdot m^{-2}$	= 1	10	10^{-5}	9.87×10^{-6}	1.02×10^{-5}	7.50×10^{-3}	2.95×10^{-4}	1.45×10^{-4}
1 $dyn \cdot cm^{-2}$	= 0.1	1	10^{-6}	9.87×10^{-7}	1.02×10^{-6}	7.50×10^{-4}	2.95×10^{-5}	1.45×10^{-5}
1 bar	= 10^{5}	10^{6}	1	0.99	1.02	750.06	29.53	14.50
1 atm	= 101325.0	1013250.	1.01	1	1.03	760.	29.92	14.70
1 kg (wt) $\cdot cm^{-2}$	= 98066.5	980665.	0.98	0.97	1	735.56	28.96	14.22
1 mmHg (torr)	= 133.32	1333.22	1.33×10^{-3}	1.32×10^{-3}	1.36×10^{-3}	1	0.04	0.02
1 in Hg	= 3386.39	33863.88	0.03	0.03	0.03	25.4	1	0.49
1 lb (wt) $\cdot in^{-2}$	= 6894.76	68947.57	0.07	0.07	0.07	51.71	2.04	1

G. UNITS OF ENERGY

Units	g mass		int J	cal	cal_{IT}	Btu_{IT}	kWh	hp h	ft-lb (wt)	cu ft-lb (wt) in^2	liter-atm
1 g mass	= 1	8.99 $\times 10^{13}$	8.99 $\times 10^{13}$	2.15 $\times 10^{13}$	2.15 $\times 10^{13}$	8.52 $\times 10^{10}$	2.50 $\times 10^{7}$	3.35 $\times 10^{7}$	6.63 $\times 10^{13}$	4.60 $\times 10^{11}$	8.87 $\times 10^{11}$
1 J	= 1.11 $\times 10^{-14}$	1	1.00	0.24	0.24	9.48 $\times 10^{-4}$	2.78 $\times 10^{-7}$	3.73	0.74	5.12 $\times 10^{-3}$	9.87 $\times 10^{-3}$
1 int J	= 1.11 $\times 10^{-14}$	1.00	1	0.24	0.24	9.48 $\times 10^{-4}$	2.78 $\times 10^{-7}$	3.73 $\times 10^{-7}$	0.74	5.12 $\times 10^{-3}$	9.87 $\times 10^{-3}$
1 cal	= 4.66 $\times 10^{-14}$	4.18	4.18	1	1.00	3.97 $\times 10^{-3}$	1.16... $\times 10^{-6}$	1.56 $\times 10^{-6}$	3.09	2.14 $\times 10^{-2}$	0.04
1 cal_{IT}	= 4.66 $\times 10^{-14}$	4.19	4.19	1.00	1	3.97 $\times 10^{-3}$	1.16 $\times 10^{-6}$	1.56 $\times 10^{-6}$	3.09	2.14 $\times 10^{-2}$	0.04
1 Btu_{IT}	= 1.17 $\times 10^{-11}$	1055.06	1054.88	252.16	252	1	2.93 $\times 10^{-4}$	3.93 $\times 10^{-4}$	778.17	5.40	10.41
1 kWh	= 4.01 $\times 10^{-8}$	3600000.	3599406.	860420.7	859845.2	3412.14	1	1.34	2655224.	18439.06	35529.24
1 hp h	= 2.99 $\times 10^{-8}$	2684519.	2684077.	641615.5	641186.5	2544.33	0.75	1	1980000.	13750.	26494.15
1 ft-lb (wt)	= 1.51 $\times 10^{-14}$	1.36	1.36	0.32	0.32	1.29 $\times 10^{-3}$	3.77 $\times 10^{-7}$	5.05... $\times 10^{-7}$	1	6.94... $\times 10^{-3}$	0.01
1 cu ft-lb (wt) in^2	= 2.17 $\times 10^{-12}$	195.24	195.21	46.66	46.63	0.19	5.42 $\times 10^{-5}$	7.27... $\times 10^{-5}$	144.	1	1.93
1 liter-atm	= 1.13 $\times 10^{-12}$	101.33	101.31	24.22	24.20	0.10	2.8[illegible] $\times 10^{-5}$	3.77 $\times 10^{-5}$	74.73	0.52	1

Appendix

Dimensional formulas of common quantities

Quantity	*Definition*	*Dimensional formula*
Mass	Fundamental	M
Length	Fundamental	L
Time	Fundamental	T
Velocity	Distance/time	LT^{-1}
Acceleration	Velocity/time	LT^{-2}
Force	Mass × acceleration	MLT^{-2}
Momentum	Mass × velocity	MLT^{-1}
Energy	Force × distance	ML^2T^{-2}
Angle	Arc/radius	0
Angular velocity	Angle/time	T^{-1}
Angular acceleration	Angular velocity/time	T^{-2}
Torque	Force × lever arm	ML^2T^{-2}
Angular momentum	Momentum × lever arm	ML^2T^{-1}
Moment of inertia	Mass × radius squared	ML^2
Area	Length squared	L^2
Volume	Length cubed	L^3
Density	Mass/volume	ML^{-3}
Pressure	Force/area	$ML^{-1}T^{-2}$
Action	Energy × time	ML^2T^{-1}
Viscosity	Force per unit area per unit velocity gradient	$ML^{-1}T^{-1}$

Internal energy and generalized work

Type of energy	*Intensive factor*	*Extensive factor*	*Element of work*
Mechanical			
Expansion	Pressure (P)	Volume (V)	$-PdV$
Stretching	Surface tension (γ)	Area (A)	γdA
Extension	Tensile stretch (F)	Length (l)	Fdl
Thermal	Temperature (T)	Entropy (S)	TdS
Chemical	Chemical potential (gm)	Moles (n)	μdn
Electrical	Electric potential (E)	Charge (Q)	EdQ
Gravitational	Gravitational field strength (mg)	Height (h)	$mgdh$
Polarization			
Electrostatic	Electric field strength (E)	Total electric polarization (P)	EdP
Magnetic	Magnetic field strength (H)	Total magnetic polarization (M)	HdM

Appendix

Defining fixed points of the International Temperature Scale of 1990 (ITS-90)

	Temperature		
Equilibrium state	K	°C	°F
Vapor pressure equation of helium	3 to 5	−270.15 to −268.15	−454.27 to 450.67
Triple point of equilibrium hydrogen	13.80	−259.35	−434.81
Vapor pressure point of equilibrium hydrogen	≈17	≈−256.15	−447.09
(or constant volume gas thermometer point of helium)	≈20.3	≈−252.85	−423.13
Triple point of neon	24.56	−248.59	−415.46
Triple point of oxygen	64.37	−218.79	361.82
Triple point of argon	83.81	−189.34	308.81
Triple point of mercury	234.32	−38.83	−37.89
Triple point of water	273.16	0.01	32.02
Melting point of gallium	302.91	29.78	85.60
Freezing point of indium	429.75	156.60	313.88
Freezing point of tin	505.08	232.93	449.47
Freezing point of zinc	692.68	419.53	787.15
Freezing point of aluminum	933.47	660.32	1220.58
Freezing point of silver	1234.93	961.78	1763.20
Freezing point of gold	1337.33	1064.18	1947.52
Freezing point of copper	1357.77	1084.62	1984.32

Appendix

Primary thermometry methods

Method	*Approximate useful range of T, K*	*Principal measured variables*	*Relation of measured variables to T*
Gas thermometry	1,3–1400	Pressure P and volume V	Ideal gas law plus corrections: $PV \propto k_B T$ plus corrections
Acoustic interferometry	1,5–20 300–17,000	Speed of sound W	$W^2 \propto k_B T$ plus corrections
Magnetic thermometry Electron paramagnetism Nuclear paramagnetism	 0.001–35 0.000001–1	Magnetic susceptibility χ	Curie's law plus corrections: $\chi \propto 1/k_B T$ plus corrections
Gamma-ray anisotropy or nuclear orientation thermometry	0.01–1	Spatial distribution of gamma-ray emission	Spatial distribution related to Boltzmann factor for nuclear spin states
Thermal electric noise thermometry Josephson junction point contact Conventional amplifier	 0.01–4 4–1000	Mean square voltage fluctuation $\bar{V}^2$	Nyquist's law: $\bar{V}^2 \propto k_B T$
Optical pyrometry (visual or photoelectric)	500–50,000	Spectral intensity J at wavelength λ	Planck's radiation law, related to Boltzmann factor for radiation quanta
Infrared spectroscopy	100–1500	Intensity I of rotational lines of light molecules	Boltzmann factor for rotational levels related to I
Ultraviolet and x-ray spectroscopy	5000–2,000,000	Emission spectra from ionized atoms—H, He, Fe, Ca, and so on	Boltzmann factor for electron states related to band structure and line density

Periodic table of the elements

Block	1	2	3	4	5	6	7	8	9	10	11	12	13	14	15	16	17	18
1s																	1 H Hydrogen	2 He Helium
s / d / p	3 Li Lithium	4 Be Beryllium											5 B Boron	6 C Carbon	7 N Nitrogen	8 O Oxygen	9 F Fluorine	10 Ne Neon
	11 Na Sodium	12 Mg Magnesium											13 Al Aluminum	14 Si Silicon	15 P Phosphorus	16 S Sulfur	17 Cl Chlorine	18 Ar Argon
	19 K Potassium	20 Ca Calcium	21 Sc Scandium	22 Ti Titanium	23 V Vanadium	24 Cr Chromium	25 Mn Manganese	26 Fe Iron	27 Co Cobalt	28 Ni Nickel	29 Cu Copper	30 Zn Zinc	31 Ga Gallium	32 Ge Germanium	33 As Arsenic	34 Se Selenium	35 Br Bromine	36 Kr Krypton
	37 Rb Rubidium	38 Sr Strontium	39 Y Yttrium	40 Zr Zirconium	41 Nb Niobium	42 Mo Molybdenum	43 Tc Technetium	44 Ru Ruthenium	45 Rh Rhodium	46 Pd Palladium	47 Ag Silver	48 Cd Cadmium	49 In Indium	50 Sn Tin	51 Sb Antimony	52 Te Tellurium	53 I Iodine	54 Xe Xenon
	55 Cs Cesium	56 Ba Barium	71 Lu Lutetium	72 Hf Hafnium	73 Ta Tantalum	74 W Tungsten	75 Re Rhenium	76 Os Osmium	77 Ir Iridium	78 Pt Platinum	79 Au Gold	80 Hg Mercury	81 Tl Thallium	82 Pb Lead	83 Bi Bismuth	84 Po Polonium	85 At Astatine	86 Rn Radon
	87 Fr Francium	88 Ra Radium	103 Lr Lawrencium	104	105	106	107	108	109									

f													
57 La Lanthanum	58 Ce Cerium	59 Pr Praseodymium	60 Nd Neodymium	61 Pm Promethium	62 Sm Samarium	63 Eu Europium	64 Gd Gadolinium	65 Tb Terbium	66 Dy Dysprosium	67 Ho Holmium	68 Er Erbium	69 Tm Thulium	70 Yb Ytterbium
89 Ac Actinium	90 Th Thorium	91 Pa Protactinium	92 U Uranium	93 Np Neptunium	94 Pu Plutonium	95 Am Americium	96 Cm Curium	97 Bk Berkelium	98 Cf Californium	99 Es Einsteinium	100 Fm Fermium	101 Md Mendelevium	102 No Nobelium

Appendix

Natural isotopic compositions of the elements

Atomic no.	*Element symbol*	*Mass no.*	*Isotopic abundance, %*
1	H	1	99.985
		2	0.015
2	H	3	0.000138
		4	99.999862
3	Li	6	7.5
		7	92.5
4	Be	9	100
5	B	10	19.9
		11	80.1
6	C	12	98.90
		13	1.10
7	N	14	99.634
		15	0.366
8	O	16	99.762
		17	0.038
		18	0.200
9	F	19	100
10	Ne	20	90.51
		21	0.27
		22	9.22
11	Na	23	100
12	Mg	24	78.99
		25	10.00
		26	11.01
13	Al	27	100
14	Si	28	92.23
		29	4.67
		30	3.10
15	P	31	100
16	S	32	95.02
		33	0.75
		34	4.21
		36	0.02
17	Cl	35	75.77
		37	24.23
18	Ar	36	0.337
		38	0.063
		40	99.600
19	K	39	93.2581
		40	0.0117
		41	6.7302
20	Ca	40	96.941
		42	0.647
		43	0.135
		44	2.086
		46	0.004
		48	0.187
21	Sc	45	100
22	Ti	46	8.0
		47	7.3
		48	73.8
		49	5.5
		50	5.4
23	V	50	0.250
		51	99.750
24	Cr	50	4.35
		52	83.79
		53	9.50
		54	2.36
25	Mn	55	100
26	Fe	54	5.8
		56	91.72
		57	2.2
		58	0.28
27	Co	59	100
28	Ni	58	68.27
		60	26.10
		61	1.13
		62	3.59
		64	0.91
29	Cu	63	69.17
		65	30.83
30	Zn	64	48.6
		66	27.9
		67	4.1
		68	18.8
		70	0.6
31	Ga	69	60.1
		71	39.9
32	Ge	70	20.5
		72	27.4
		73	7.8
		74	36.5
		76	7.8
33	As	75	100
34	Se	74	0.9
		76	9.0
		77	7.6
		78	23.5
		80	49.6
		82	9.4
35	Br	79	50.69
		81	49.31
36	Kr	78	0.35
		80	2.25
		82	11.6
		83	11.5
		84	57.0
		86	17.3
37	Rb	85	72.165
		87	27.835
38	Sr	84	0.56
		86	9.86
		87	7.00
		88	82.58
39	Y	89	100
40	Zr	90	51.45
		91	11.27
		92	17.17
		94	17.33
		96	2.78

Natural isotopic compositions of the elements (cont.)

Atomic no.	Element symbol	Mass no.	Isotopic abundance, %
41	Nb	93	100
42	Mo	92	14.84
		94	9.25
		95	15.92
		96	16.68
		97	9.55
		98	24.13
		100	9.63
44	Ru	96	5.52
		98	1.88
		99	12.7
		100	12.6
		101	17.0
		102	31.6
		104	18.7
45	Rh	103	100
46	Pd	102	1.02
		104	11.14
		105	22.33
		106	27.33
		108	24.46
		110	11.72
47	Ag	107	51.839
		109	48.161
48	Cd	106	1.25
		108	0.89
		110	12.49
		111	12.80
		112	24.13
		113	12.22
		114	28.73
		116	7.49
49	In	113	4.3
		115	95.7
50	Sn	112	1.0
		114	0.7
		115	0.4
		116	14.7
		117	7.7
		118	24.3
		119	8.6
		120	32.4
		122	4.6
		124	5.6
51	Sb	121	57.3
		123	42.7
52	Te	120	0.096
		122	2.60
		123	0.908
		124	4.816
		125	7.14
		126	18.95
		128	31.69
		130	33.80
53	I	127	100
54	Xe	124	0.10
		126	0.09
		128	1.91
		129	26.4
		130	4.1
		131	21.2
		132	26.9
		134	10.4
		136	8.9
55	Cs	133	100
56	Ba	130	0.106
		132	0.101
		134	2.417
		135	6.592
		136	7.854
		137	11.23
		138	71.70
57	La	138	0.09
		139	99.91
58	Ce	136	0.19
		138	0.25
		140	88.48
		142	11.08
59	Pr	141	100
60	Nd	142	27.13
		143	12.18
		144	23.80
		145	8.30
		146	17.19
		148	5.76
		150	5.64
62	Sm	144	3.1
		147	15.0
		148	11.3
		149	13.8
		150	7.4
		152	26.7
		154	22.7
63	Eu	151	47.8
		153	52.2
64	Gd	152	0.20
		154	2.18
		155	14.80
		156	20.47
		157	15.65
		158	24.84
		160	21.86
65	Tb	159	100
66	Dy	156	0.06
		158	0.10
		160	2.34
		161	18.9
		162	25.5
		163	24.9
		164	28.2

Natural isotopic compositions of the elements (cont.)

Atomic no.	*Element symbol*	*Mass no.*	*Isotopic abundance, %*	*Atomic no.*	*Element symbol*	*Mass no.*	*Isotopic abundance, %*
67	Ho	165	100	76	Os	184	0.02
68	Er	162	0.14			186	1.58
		164	1.61			187	1.6
		166	33.6			188	13.3
		167	22.95			189	16.1
		168	26.8			190	26.4
		170	14.9			192	41.0
69	Ta	169	100	77	Ir	191	37.3
70	Yb	168	0.13			193	62.7
		170	3.05	78	Pt	190	0.01
		171	14.3			192	0.79
		172	21.9			194	32.9
		173	16.12			195	33.8
		174	31.8			196	25.3
		176	12.7			198	7.2
71	Lu	175	97.40	79	Au	197	100
		176	2.60	80	Hg	196	0.15
72	Hf	174	0.16			198	10.1
		176	5.2			199	17.0
		177	18.6			200	23.1
		178	27.1			201	13.2
		179	13.74			202	29.65
		180	35.2			204	6.8
73	Ta	180	0.012	81	Tl	203	29.524
		181	99.988			205	70.467
74	W	180	0.13	82	Pb	204	1.4
		182	26.3			206	24.1
		183	14.3			207	22.1
		184	30.67			208	52.4
		186	28.6	83	Bi	209	100
75	Re	185	37.40	90	Th	232	100
		187	62.60	92	U	234	0.0055
						235	0.7200
						236	99.2745

Electrochemical series of the elements

Element	*Symbol*	*Element*	*Symbol*
Lithium	Li	Zinc	Zn
Potassium	K	Chromium	Cr
Rubidium	Rb	Gallium	Ga
Cesium	Cs	Iron	Fe
Radium	Ra	Cadmium	Cd
Barium	Ba	Indium	In
Strontium	Sr	Thallium	Tl
Calcium	Ca	Cobalt	Co
Sodium	Na	Nickel	Ni
Lanthanum	La	Molybdenum	Mo
Cerium	Ce	Tin	Sn
Magnesium	Mg	Lead	Pb
Scandium	Sc	Germanium	Ge
Plutonium	Pu	Tungsten	W
Thorium	Th	Hydrogen	H
Beryllium	Be		
Uranium	U	Copper	Cu
Hafnium	Hf	Mercury	Hg
Aluminum	Al	Silver	Ag
Titanium	Ti	Gold	Au
Zirconium	Zr	Rhodium	Rh
Manganese	Mn	Platinum	Pt
Vanadium	V	Palladium	Pd
Niobium	Nb	Bromine	Br
Boron	B	Chlorine	Cl
Silicon	Si	Oxygen	O
Tantalum	Ta	Fluorine	F

Average electronegativities from thermochemical data

Element	*Value*	*Element*	*Value*
H	2.20	Al	1.61
Li	0.98	Ga	1.81
Na	0.93	In	1.78
K	0.82	Tl	2.04
Rb	0.82	C	2.55
Cs	0.79	Si	1.90
Be	1.57	Ge	2.01
Mg	1.31	Sn	1.96
Ca	1.00	Pb	2.33
Sr	0.95	N	3.04
Ba	0.89	P	2.19
Sc	1.36	As	2.18
Ti	1.54	Sb	2.05
V	1.63	Bi	2.02
Cr	1.66	O	3.44
Mn	1.55	S	2.58
Fe	1.83	Se	2.55
Co	1.88	F	3.98
Ni	1.91	Cl	3.16
Cu	1.90	Br	2.96
Zn	1.65	I	2.66
B	2.04		

Appendix

Standard atomic weights

Name	Symbol	Atomic number	Atomic weight*
Actinium	Ac	89	227.0278
Aluminium	Al	13	26.98154
Americium	Am	95	(243)
Antimony (stibium)	Sb	51	121.75 ± 3
Argon	Ar	18	39.948
Arsenic	As	33	74.9216
Astatine	At	85	(210)
Barium	Ba	56	137.33
Berkelium	Bk	97	(247)
Beryllium	Be	4	9.01218
Bismuth	Bi	83	208.9804
Boron	B	5	10.81
Bromine	Br	35	79.904
Cadmium	Cd	48	112.41
Calcium	Ca	20	40.08
Californium	Cf	98	(251)
Carbon	C	6	12.011
Cerium	Ce	58	140.12
Cesium	Cs	55	132.9054
Chlorine	Cl	17	35.453
Chromium	Cr	24	51.996
Cobalt	Co	27	58.9332
Copper	Cu	29	63.546 ± 3
Curium	Cm	96	(247)
Dysprosium	Dy	66	162.50 ± 3
Einsteinium	Es	99	(252)
Erbium	Er	68	167.26 ± 3
Europium	Eu	63	151.96
Fermium	Fm	100	(257)
Fluorine	F	9	18.998403
Francium	Fr	87	(223)
Gadolinium	Gd	64	157.25 ± 3
Gallium	Ga	31	69.72
Germanium	Ge	32	72.59 ± 3
Gold	Au	79	196.9665
Hafnium	Hf	72	178.49 ± 3
Helium	He	2	4.00260
Holmium	Ho	67	164.9304
Hydrogen	H	1	1.00794 ± 7
Indium	In	49	114.82
Iodine	I	53	126.9045
Iridium	Ir	77	192.22 ± 3
Iron	Fe	26	55.847 ± 3
Krypton	Kr	36	83.80
Lanthanum	La	57	138.9055 ± 3
Lawrencium	Lr	103	(260)
Lead	Pb	82	207.2
Lithium	Li	3	6.941 ± 3
Lutetium	Lu	71	174.967
Magnesium	Mg	12	24.305
Manganese	Mn	25	54.9380
Mendelevium	Md	101	(258)
Mercury	Hg	80	200.59 ± 3
Molybdenum	Mo	42	95.94
Neodymium	Nd	60	144.24 ± 3
Neon	Ne	10	20.179

Appendix

Standard atomic weights (cont.)

Name	*Symbol*	*Atomic number*	*Atomic weight**
Neptunium	Np	93	237.0482
Nickel	Ni	28	58.69
Niobium	Nb	41	92.9064
Nitrogen	N	7	14.0067
Nobelium	No	102	(259)
Osmium	Os	76	190.2
Oxygen	O	8	15.9994 ± 3
Palladium	Pd	46	106.42
Phosphorus	P	15	30.97376
Platinum	Pt	78	195.08 ± 3
Plutonium	Pu	94	(244)
Polonium	Po	84	(209)
Potassium (kalium)	K	19	39.0983
Praseodymium	Pr	59	140.9077
Promethium	Pm	61	(145)
Protactinium	Pa	91	231.0359
Radium	Ra	88	226.0254
Radon	Rn	86	(222)
Rhenium	Re	75	186.207
Rhodium	Rh	45	102.9055
Rubidium	Rb	37	85.4678 ± 3
Ruthenium	Ru	44	101.07 ± 3
Samarium	Sm	62	150.36 ± 3
Scandium	Sc	21	44.9559
Selenium	Se	34	78.96 ± 3
Silicon	Si	14	28.0855 ± 3
Silver	Ag	47	107.8682 ± 3
Sodium (natrium)	Na	11	22.98977
Strontium	Sr	38	87.62
Sulfur	S	16	32.06
Tantalum	Ta	73	180.9479
Technetium	Tc	43	(98)
Tellurium	Te	52	127.60 ± 3
Terbium	Tb	65	158.9254
Thallium	Tl	81	204.383
Thorium	Th	90	232.0381
Thulium	Tm	69	168.9342
Tin	Sn	50	118.69 ± 3
Titanium	Ti	22	47.88 ± 3
Tungsten (wolfram)	W	74	183.85 ± 3
(Unnilhexium)	(Unh)	106	(263)
(Unnilpentium)	(Unp)	105	(262)
(Unnilquadium)	(Unq)	104	(261)
Uranium	U	92	238.0289
Vanadium	V	23	50.9415
Xenon	Xe	54	131.29 ± 3
Ytterbium	Yb	70	173.04 ± 3
Yttrium	Y	39	88.9059
Zinc	Zn	30	65.38
Zirconium	Zr	40	91.22

* Values in parentheses are used for radioactive elements whose atomic weights cannot be quoted precisely without knowledge of the origin of the elements; the value given is the atomic mass number of the isotope of that element of longest known half-life.

Appendix

Common acid-base indicators

Common name	pH range	Color change (acid to base)	pK	Chemical name	Structure
Methyl violet	0–2, 5–6	Yellow to blue violet to violet		Pentamethylbenzylpararosaniline hydrochloride	Base
Metacresol purple	1.2–2.8, 7.3–9.0	Red to yellow to purple	1.5	*m*-Cresolsulfonphthalein	Acid
Thymol blue	1.2–2.8, 8.0–9.6	Red to yellow to blue	1.7	Thymolsulfonphthalein	Acid
Tropeoline 00 (Orange IV)	1.4–3.0	Red to yellow		Sodium *p*-diphenylaminoazobenzenesulfonate	Base
Bromphenol blue	3.0–4.6	Yellow to blue	4.1	Tetrabromophenolsulfonphthalein	Acid
Methyl orange	2.8–4.0	Orange to yellow	3.4	Sodium *p*-dimethylaminoazobenzenesulfonate	Base
Bromcresol green	3.8–5.4	Yellow to blue	4.9	Tetrabromo-*m*-cresolsulfonphthalein	Acid
Methyl red	4.2–6.3	Red to yellow	5.0	Dimethylaminoazobenzene-*o*-carboxylic acid	Base
Chlorphenol red	5.0–6.8	Yellow to red	6.2	Dichlorophenolsulfonphthalein	Acid
Bromcresol purple	5.2–6.8	Yellow to purple	6.4	Dibromo-*o*-cresolsulfonphthalein	Acid
Bromthymol blue	6.0–7.6	Yellow to blue	7.3	Dibromothymolsulfonphthalein	Acid
Phenol red	6.8–8.4	Yellow to red	8.0	Phenolsulfonphthalein	Acid
Cresol red	2.0–3.0, 7.2–8.8	Orange to amber to red	8.3	*o*-Cresolsulfonphthalein	Acid
Orthocresolphthalein	8.2–9.8	Colorless to red		—	Acid
Phenolphthalein	8.4–10.0	Colorless to pink	9.7	—	Acid
Thymolphthalein	10.0–11.0	Colorless to red	9.9	—	Acid
Alizarin yellow GG	10.0–12.0	Yellow to lilac		Sodium *p*-nitrobenzeneazosalicylate	Acid
Malachite green	11.4–13.0	Green to colorless		*p,p'*-Benzylidenebis-N,N-dimethylaniline	Base

Appendix

The pH scale

Solution	pH
Battery acid	0
	1
Stomach acid	2
Lemon juice	3
	4
Black coffee	5
	6
Pure water	7
Seawater	8
Baking soda	9
Toilet soap	10
	11
	12
Household ammonia	13
Drain cleaner	14

Uses and effects of selected organic solvents

Solvent	Uses	Effects
Aliphatic hydrocarbons		
Pentane, hexanes, heptanes, octanes	Commercial products and solvents	Depression of central nervous system and liver pathology
Halogenated aliphatic hydrocarbons		
Methylene chloride	Paint remover, aerosol solvent	Depression of central nervous system, respiratory poison
Chloroform	Chemical intermediate, solvent	Liver pathology, carcinogen
Carbon tetrachloride	Industrial and laboratory solvent	Lipid peroxidation, liver and renal pathology
Aromatic hydrocarbons		
Benzene	Organic synthesis, solvent, printing	Hematopoietic toxicity, immunosuppression, leukemia
Toluene	Solvent, chemical intermediate, paints, rubber	Narcotic, central nervous system pathology
Xylene	Solvent, chemical intermediate, pesticides, adhesives	Central nervous system pathology
Alcohols		
Methanol	Commercial and laboratory solvent, paints, fuel additive	Toxic metabolites (formaldehyde), optic nerve damage
Isopropyl alcohol	Cosmetics, glass cleaning solutions	Central nervous system depressant
Glycols		
Ethylene glycol	Heat exchangers, antifreeze, hydraulic fluids	Toxic metabolites

Appendix

Safety properties of some organic solvents

Solvent	Flash point, K (°F)		Flammable limits, vol %		Minimum ignition, K (°F)	Threshold limit value	
	TCC*	TOC*	Lower	Upper		Vapor, g/m^3	Carcinogenic, mg/m†
Hexane	247 (−14.8)	—	1.18	7.43	533 (500)	360	—
Benzene	262 (−12.2)	—	1.4	7.1	835 (1044)	30†	32
Methanol	285 (53.6)	—	6.72	36.50	1140 (1593)	260†	—
Ethanol	286 (55.4)	—	3.28	18.95	712 (822)	1900	—
Phenol	352 (174)	—	—	—	988 (1319)	19†	—
Ethyl ether	228 (−49)	—	1.85	36.50	447 (345)	1200	—
p-Dioxane	285 (53.6)	—	1.97	22.25	539 (511)	180†	1015
Acetone	273 (32)	—	2.55	12.80	843 (1058)	2400	—
Chloroform	nf‡	—	nf	nf	nf	120	—
Carbon tetrachloride	nf	—	nf	nf	nf	65†	—
Trichloroethylene	nf	—	nf	nf	736 (865)	535	900
2-Nitropropane	312 (102)	—	2.5	—	—	90	—
Nitrobenzene	361 (190)	—	—	—	755 (900)	5†	—
Aniline	349 (169)	—	—	—	—	19†	—
Pyridine	296 (73.4)	—	1.81	12.40	755 (900)	15	—
N,N-Dimethylacetamide	—	350 (171)	1.70	18.5§	693 (788)	35†	—
Dimethyl sulfoxide	—	368 (203)	—	—	—	—	—
Methyl Cellosolve	315 (108)	—	—	—	—	80†	—
Epichlorohydrin	—	314 (106)	—	—	—	20†	5 ppm
Hexamethylphosphoric triamide	—	—	—	—	—	nf	>1

*Tag closed cup, tag open cup.
†Can be attributed to cutaneous absorption, including mucous membranes.
‡Nonflammable.
§At 373 K.

Physical properties of some organic solvents

Organic solvent	*Boiling point,* °C (°F)	*Freezing point,* °C (°F)	*Viscosity, cgs,* 25°C (77°F)	*Dielectric constant,* 25°C (77°F)
Benzene	80.100 (176.18)	5.533 (41.959)	0.6028	2.275
1,2-Dichloroethane	83.483 (182.269)	−35.66 (−32.19)	0.730, 30°C	10.36
Methanol	64.70 (148.46)	−97.68 (−143.82)	0.5445	32.70
1,2-Ethanediol	197.3 (387.14)	−13 (8.6)	13.55, 30°C	37.7
Acetic acid	117.90 (244.22)	16.66 (61.99)	1.040, 20°C	6.15, 20°C (68°F)
Phenol	181.839 (359.310)	40.90 (105.62)	4.076	9.78, 60°C (140°F)
Acetone	56.29 (133.32)	−94.7 (−138.46)	0.3040	20.70
2-Propanol	82.26 (180.07)	−88.0 (−126.4)	1.765, 30°C	19.92
Ethanol	78.29 (172.92)	−114.1 (−173.4)	1.078	24.55
1,2-Dimethoxybenzene	206.25 (403.25)	22.5 (72.5)	3.281	4.09
Fluorobenzene	84.734 (184.521)	−42.21 (−43.98)	0.517, 30°C	5.42
Pyridine	115.256 (239.47)	−41.55 (−42.79)	0.884	12.4
2-Ethoxyethanol	135.6 (276.1)	<−90 (−130)	1.85	29.6
N,N-Dimethylacetamide	166.1 (331.0)	−20 (−4)	0.838, 30°C	37.78
Dimethyl sulfoxide	189.0 (372.2)	18.54 (65.37)	1.996	46.68
2-Nitropropane	120.25 (248.45)	−91.32 (−132.38)	0.721	25.52

Principal organic functional groups

Compound class	Group	Structure
Alkene	Double bond	$>C=C<$
Alkyne	Triple bond	$-C\equiv C-$
Alcohol	Hydroxyl	$-OH$
Amine	Amino	$-NH_2(-NR_2)$*
Aldehyde	Carbonyl	$-\overset{O}{\overset{\|}{C}}H$
Ketone	Carbonyl	$-\overset{O}{\overset{\|}{C}}R$
Acid	Carboxyl	$-\overset{O}{\overset{\|}{C}}OH$

Compound class	Group	Structure
Ester	Alkoxy-carbonyl	$-\overset{O}{\overset{\|}{C}}OR$
Amide	Carbamoyl	$\overset{O}{\overset{\|}{C}}N<$
Nitrite	Cyano	$-C\equiv N$
Azide	Azido	$-N=N=N$
Nitro		$-NO_2$
Sulfide		$-S-$
Sulfoxide		$-\overset{O}{\overset{\|}{S}}-$
Sulfonic acid		$-SO_3H$

*R = any carbon group, for example, CH_3.

Compounds containing functional groups

Group*	Suffix	Prefix	Structure	Name
$-\overset{O}{\overset{\|}{C}}OH$	-oic acid	carboxy-	$CH_3CH_2\overset{O}{\overset{\|}{C}}OH$	Propanoic acid
$-\overset{O'}{\overset{\|}{C}}OR$	alkyl -oate	alkoxycarbonyl-	$CH_3CH_2\overset{O}{\overset{\|}{C}}OCH_3$	Methyl propanoate
			$CH_3O\overset{O}{\overset{\|}{C}}CH_2CH_2\overset{O}{\overset{\|}{C}}OH$	3-Methoxycarbonyl propanoic acid
$-C\equiv N$	-nitrile	cyano-	$CH_3CH_2C\equiv N$	Propanenitrile
			$CH_3CHCN\overset{O}{\overset{\|}{C}}OH$	2-Cyanopropanoic acid
$-\overset{O}{\overset{\|}{C}}H$	-al	formyl-	$CH_3CH_2\overset{O}{\overset{\|}{C}}H$	Propanal
$-\overset{O}{\overset{\|}{C}}R$	-one	oxo-	$CH_3CH_2\overset{O}{\overset{\|}{C}}CH_3$	Butanone
			$CH_3\overset{O}{\overset{\|}{C}}CH_2\overset{O}{\overset{\|}{C}}OH$	3-Oxobutanoic acid
$-OH$	-ol	hydroxy-	$CH_3CH_2CH_2OH$	1-Propanol
			$HOCH_2CH_2\overset{O}{\overset{\|}{C}}H$	3-Hydroxypropanal
$-SH$	-amine	amino-	$CH_3CH_2CH_2NH_2$	1-Propanamine
$-OR$	—	alkoxy-	$NH_2CH_2CH_2CH_2OCH_3$	3-Methoxy-1-propanamine

*R = any alkyl group.

Principal spectral regions and fields of spectroscopy

Spectral region	Approx. wavelength range	Typical source	Typical detector	Energy transitions studied in matter
Gamma	1–100 pm	Radioactive nuclei	Geiger counter; scintillation counter	Nuclear transitions and disintegrations
X-rays	6 pm–100 nm	X-ray tube (electron bombardment of metals)	Geiger counter	Ionization by inner electron removal
Vacuum ultraviolet	10–200 nm	High-voltage discharge; high-vacuum spark	Photomultiplier	Ionization by outer electron removal
Ultraviolet	200–400 nm	Hydrogen-discharge lamp	Photomultiplier	Excitation of valence electrons
Visible	400–800 nm	Tungsten lamp	Phototubes	Excitation of valence electrons
Near-infrared	0.8–2.5 μm	Tungsten lamp	Photocells	Excitation of valence electrons; molecular vibrational overtones
Infrared	2.5–50 μm	Nernst glower; Globar lamp	Thermocouple; bolometer	Molecular vibrations; stretching, bending, and rocking
Far-infrared	50–1000 μm	Mercury lamp (high-pressure)	Thermocouple bolometer	Molecular rotations
Microwave	0.1–30 cm	Klystrons; magnetrons	Silicon-tungsten crystal; bolometer	Molecular rotations; electron spin resonance
Radio-frequency	10^{-1}–10^{3} m	Radio transmitter	Radio receiver	Molecular rotations; nuclear magnetic resonance

Most commonly used radioisotopes and major applications

Isotope	Half-life	Average γ-ray energy, MeV	Main uses
Radium-266 (^{226}Ra)	1602 years	0.83	Medical (brachytherapy)
Gold-198 (^{198}Au)	2.697 days	0.412	Medical (brachytherapy-permanent implants)
Iridium-192 (^{192}Ir)	74.2 days	0.38	Medical (brachytherapy)
Cesium-137 (^{137}Cs)	30.3 years	0.662	Industrial, medical (brachytherapy), research irradiators
Cobalt-60 (^{60}Co)	5.261 years	1.25	Industrial, medical (teletherapy), research irradiators
Iodine-125 (^{125}I)	60 days	0.035	Medical (brachytherapy—permanent implants)
Iodine-131 (^{131}I)	8.065 days	0.187 (av. β-ray energy)	Medical (dispersal technique)

Some trace analysis methods and limits of detection

Method	Limits of detection	
	Absolute, g	*Concentration, ppm*
Chromatography		
Thin-layer	10^{-5} to 10^{-3}	
Gas-liquid		10 to 10^{6}
Liquid-liquid		10^{-3} to 10^{0}
Electrochemical		
Coulometry	10^{-9} to 10^{0}	
Ion selective electrode		10^{-2} to 10^{2}
Polarography		
Conventional		10^{0} to 10^{3}
Modern		10^{-3} to 10^{3}
Laser probe microanalysis		10^{2} to 10^{4}
Nuclear		
Neutron activation		10^{-3} to 10^{-1}
Electron probe		10^{2} to 10^{3}
Ion probe		10^{-1} to 10^{1}
Mass spectrometry		
Isotope dilution		10^{-5} to 10^{6}
Spark source		10^{-3} to 10^{1}
Electrical plasma*		10^{-5} to 10^{0}
Organic microanalysis	$>10^{-5}$	
Optical-absorption		
Atomic		
Flame		10^{-3} to 10^{1}
Nonflame	10^{-15} to 10^{-9}	
Molecular		
UV-visible		10^{-3} to 10^{2}
Infrared		10^{3} to 10^{6}
Microwave		10^{0} to 10^{3}
Optical-emission		
AC spark		10^{1} to 10^{3}
DC arc		10^{-2} to 10^{2}
Electrical plasma†		10^{-4} to 10^{2}
Optical-fluorescence		
Atomic		
Flame		10^{-3} to 10^{2}
Nonflame	10^{-15} to 10^{-9}	
Molecular		10^{-3} to 10^{-1}
Optical-phosphorescence		10^{-3} to 10^{2}
Optical-Raman		10^{0} to 10^{5}
Spectrometric-resonance		
Nuclear magnetic		10^{1} to 10^{5}
Electron spin	10^{-9} to 10^{-6}	
Thermal analysis	10^{-5} to 10^{-4}	
Wet chemistry		
Gravimetry	10^{-3} to 10^{-2}	
Titrimetry		10^{-2} to 10^{4}
X-ray spectrometry		
Auger		10^{3} to 10^{5}
ESCA		10^{3} to 10^{5}
Fluorescence		10^{-1} to 10^{2}
Mössbauer		10^{0} to 10^{3}
Photoelectron		10^{0} to 10^{3}
Immunoassay	10^{-2} to 10^{-9}	10^{-3} to 10^{0}

*Inductively coupled plasma or microwave-induced plasma.
†Inductively coupled plasma or direct-current plasma.